Ecology and Conservation

The Natural History of the Crustacea Series

SERIES EDITOR:
Martin Thiel

Functional Morphology and Diversity (Volume 1)
Edited by Les Watling and Martin Thiel

Lifestyles and Feeding Biology (Volume 2)
Edited by Martin Thiel and Les Watling

Nervous Systems and Control of Behavior (Volume 3)
Edited by Charles Derby and Martin Thiel

Physiology (Volume 4)
Edited by Ernest S. Chang and Martin Thiel

Life Histories (Volume 5)
Edited by Gary Wellborn and Martin Thiel

Reproductive Biology (Volume 6)
Edited by Rickey Cothran and Martin Thiel

Developmental Biology and Larval Ecology (Volume 7)
Edited by Klaus Anger, Steffen Harzsch, and Martin Thiel

Evolution and Biogeography (Volume 8)
Edited by Gary Poore and Martin Thiel

Fisheries and Aquaculture (Volume 9)
Edited by Gustavo Lovrich and Martin Thiel

Ecology and Conservation (Volume 10)
Edited by Lars Gutow and Martin Thiel

Ecology and Conservation

The Natural History of The Crustacea
Volume 10

EDITED BY LARS GUTOW
AND MARTIN THIEL

OXFORD
UNIVERSITY PRESS

OXFORD
UNIVERSITY PRESS

Oxford University Press is a department of the University of Oxford.
It furthers the University's objective of excellence in research, scholarship,
and education by publishing worldwide. Oxford is a registered trade mark of
Oxford University Press in the UK and in certain other countries.

Published in the United States of America by Oxford University Press
198 Madison Avenue, New York, NY 10016, United States of America.

Library of Congress Cataloging-in-Publication Data
Names: Thiel, Martin, 1962- editor. | Gutow, Lars, editor.
Title: Ecology and conservation / edited by Martin Thiel and Lars Gutow.
Description: New York, NY: Oxford University Press, [2025] | Series: Natural history of the crustacea series; Volume 10 | Includes bibliographical references and index.
Identifiers: LCCN 2024055744 (print) | LCCN 2024055745 (ebook) | ISBN 9780197768242 (hardback) | ISBN 9780197768266 (epub) | ISBN 9780197768273
Subjects: LCSH: Crustacea—Ecology. | Crustacea—Conservation.
Classification: LCC QL435 .E36 2025 (print) | LCC QL435 (ebook) | DDC 595.3—dc23/eng/20250403
LC record available at https://lccn.loc.gov/2024055744
LC ebook record available at https://lccn.loc.gov/2024055745

DOI: 10.1093/oso/9780197768242.001.0001

Printed by Integrated Books International, United States of America

MIX
Paper
FSC® C183721

CONTENTS

PREFACE

This tenth and final volume constitutes the end of a long and exciting journey through *The Natural History of the Crustacea*. Previous volumes along this journey addressed, among other topics, fundamental facets of the natural history of crustaceans, including their morphology and function, development, physiology, life histories, reproductive biology, developmental biology and larval ecology, evolution and biogeography, and fisheries and aquaculture. These ecological and evolutionary aspects justify the important role of the Crustacea in terrestrial and aquatic ecosystems. Furthermore, they are key to understanding how crustaceans perform in a changing environment and to developing efficient conservation strategies. This present volume, therefore, addresses the role and functions of crustaceans in ecosystems, how they are threatened by man-made degradation of the environment, and efforts that are being made to protect and restore endangered and declining crustacean populations and habitats.

The flow of energy and matter through trophic networks is a crucial characteristic of ecosystems. Crustaceans are present at various trophic levels of food webs as both consumers and prey, linking primary production with higher trophic levels. Rochelle Seitz, Romuald Lipcius, Jainita Patel, and Gina Ralph summarize the role of crustaceans as herbivores, predators, and general consumers, while the editors of this volume (Martin Thiel and Lars Gutow) address crustaceans as prey in aquatic and terrestrial food webs. Jonas Geburzi and Martin Zimmer describe how crustacean detritivores contribute to the trophic utilization of dead plant and animal material, thereby supporting important biogeochemical processes such as the oceanic carbon pump.

Apart from their contribution to predator–prey interactions, crustaceans are involved in a great variety of non-trophic ecological processes. Hazel Needham, Michael Townsend, and Andrew Lohrer demonstrate how bioturbation and bioirrigation by crustaceans promote sediment-bound biogeochemical processes and sculpt benthic habitats. Parasites require a specific sequence of hosts during their ontogenesis. Robert Poulin explains how certain groups of parasites essentially depend on crustacean hosts to complete their development. Competition for resources and habitat, and also for protection from predators, shapes the structure of species communities. Stuart Jenkins characterizes the various types of competition and provides examples of competitive interactions involving crustaceans. The functioning of ecosystems critically depends on the diversity of the species involved. Rebecca Best and Moritz Lürig evaluate the importance of functional and structural diversity of crustaceans for ecosystem integrity.

Worldwide, anthropogenic activities are changing the living conditions of terrestrial and aquatic species. Kevin Hovel summarizes examples of loss, fragmentation, and modification of crustacean habitats. Christian Pansch, Marian Hu, Dirk Weihrauch, and Christian Buschbaum demonstrate how crustaceans perform in a changing ocean characterized by elevated temperatures, acidification, oxygen depletion, and species invasions, while Dietmar Straile focuses on freshwater systems when summarizing anthropogenic effects on lake crustacean zooplankton. Different classes of contaminants and their sometimes catastrophic but often chronic sublethal effects on crustaceans are reviewed by Alex Ford and Thomas Knigge.

As a result of the many environmental threats, crustacean species and their habitats are critically endangered worldwide. Diverse conservation measures have been taken that aim at protecting biodiversity in general and that also protect specific crustacean species or populations. Sound conservation requires a comprehensive knowledge of species inventories. Keith Crandall demonstrates how modern molecular biological methods can help to comprehensively assess crustacean species inventories. Edwin Grosholz provides a global overview of vectors and consequences of crustacean invasions and introduces programs for the prevention, eradication, containment, and suppression of invasions. The persistence of species can be tightly bound to the integrity of their habitats. Thomas Iliffe, Alejandro Martínez, Fernando Alvarez, and William Humphreys describe the specific and often highly endemic crustacean fauna of anchialine caves, how this stygobiont biota is threatened by the destruction of their unique cave environments, and the various legislations that acknowledge the necessity to protect these vulnerable habitats. Promising examples of successful recovery of lobster populations in dedicated Marine Protected Areas are presented by Even Moland, Raquel Goñi, Benjamin Hanns, and Nick Shears, while Christopher Rogers provides an overview of the various national and international legislation relevant for the protection of crustacean populations. Social perception of nature may be decisive in humans' willingness to protect species. Christopher Boyko and Shane Ahyong provide exciting examples from the media and popular culture of a wide public awareness of crustaceans in various cultures.

At the end of this journey through *The Natural History of the Crustacea*, we acknowledge that countless aspects of this field had to remain unconsidered. Each chapter of this volume and of all previous volumes provides an outlook and perspectives for future research. We are ending this journey with great expectations for the future investigations to come on the most fascinating world of the Crustacea.

ACKNOWLEDGMENTS

We thank our contributors for graciously sharing their time, energy, knowledge, insights, confidence, and patience in order to make this volume possible—it has been both a pleasure and an honor to work with each of them. We also thank our editorial assistants from those present at the start of the book series, especially to Miguel Angel Penna-Díaz who has assisted us since the end of Volume 5, to those who worked on this concluding volume: all our editorial assistants were impeccably skilled and organized and always kept us moving forward. We thank our external reviewers for their valuable and generous feedback. Finally, we express our appreciation to our publisher, Oxford University Press, for its commitment to this project.

Editing of this book was generously supported by Universidad Católica del Norte, Chile.

CONTRIBUTORS

EDITORS

Lars Gutow
Alfred Wegener Institute
Helmholtz Centre for Polar and Marine Research
Bremerhaven, Germany

Martin Thiel
MarineGEO Program
Smithsonian Environmental Research Center
Edgewater, Maryland, USA
Departamento de Biología Marina
Facultad Ciencias del Mar
Universidad Católica del Norte
Coquimbo, Chile
Center of Ecology and Sustainable Management of Oceanic Island (ESMOI)
Coquimbo, Chile

AUTHORS

Shane T. Ahyong
Australian Museum Research Institute
The Australian Museum
Sydney, New South Wales, Australia
School of Biological, Earth and Environmental Sciences
University of New South Wales
Kensington, New South Wales, Australia

Fernando Alvarez
Instituto de Biología
Universidad Nacional de México
Ciudad Universitaria, Alcaldía Coyoacán
Ciudad de México, Mexico

Rebecca J. Best
School of Earth and Sustainability
Northern Arizona University
Flagstaff, Arizona, USA

Christopher B. Boyko
Department of Biology
Hofstra University
Hempstead, New York, USA
Division of Invertebrate Zoology
American Museum of Natural History
Manhattan, New York, USA

Christian Buschbaum
Alfred Wegener Institute
Helmholtz Centre for Polar and Marine Research
Wadden Sea Station
Sylt, List, Germany

Keith A. Crandall
Computational Biology Institute
Department of Biostatistics & Bioinformatics
George Washington University
Washington, D.C., Maryland, USA
Department of Zoology

National Museum of Natural History
Smithsonian Institution
Washington, D.C., Maryland, USA

Alex T. Ford
University of Portsmouth
Portsmouth, UK

Jonas C. Geburzi
Leibniz Centre for Tropical Marine Research
Bremen, Germany
Harvard University, Faculty of Arts and Sciences, Department of Organismic and Evolutionary Biology
Cambridge, Maryland, USA

Raquel Goñi
Centro Oceanográfico de Baleares
Instituto Español de Oceanografía
Palma de Mallorca, Spain

Edwin Grosholz
Department of Environmental Science and Policy
University of California
Davis, California, USA

Lars Gutow
Alfred Wegener Institute
Helmholtz Centre for Polar and Marine Research
Bremerhaven, Germany

Benjamin J. Hanns
Leigh Marine Laboratory
University of Auckland
Leigh, New Zealand

Kevin Hovel
Department of Biology & Coastal and Marine Institute
San Diego State University
San Diego, California, USA

Marian Hu
Institute of Physiology
Christian-Albrechts-University Kiel
Kiel, Germany

William F. Humphreys
Western Australian Museum
Collections and Research
Welshpool D.C., Western Australia, Australia
School of Animal Biology
The University of Western Australia
Crawley, Perth, Western Australia, Australia

Thomas M. Iliffe
Department of Invertebrate Zoology
Smithsonian Institution
National Museum of Natural History
Washington, D.C., Maryland, USA

Stuart R. Jenkins
School of Ocean Sciences
Bangor University
Menai Bridge, UK

Thomas Knigge
Stress Environnementaux et Biosurveillance des Milieux Aquatiques (SEBIO)
Université Le Havre Normandie
Le Havre, Normandie, France

Romuald N. Lipcius
Virginia Institute of Marine Science
William & Mary
Gloucester Point, Virginia, USA

Andrew M. Lohrer
National Institute of Water and Atmospheric Research
Hamilton, New Zealand

Moritz D. Lürig
Florida Museum of Natural History
University of Florida
Gainesville, Florida, USA
Department of Biology
Lund University
Lund, Sweden

Alejandro Martínez
Marine Biological Section
Department of Biology
University of Copenhagen
Copenhagen, Denmark
Institute for Ecosystems Study

Italian National Research Council
Verbania, Italy

Even Moland
Institute of Marine Research
His, Norway
Centre for Coastal Research
Department of Natural Sciences
University of Agder
Kristiansand, Norway

Hazel R. Needham
University of Waikato
Hamilton, New Zealand

Christian Pansch
Environmental and Marine Biology
Faculty of Science and Engineering
Åbo Akademi University
Turku, Finland

Jainita Patel
Virginia Institute of Marine Science
William & Mary
Gloucester Point, Virginia, USA

Robert Poulin
University of Otago
Dunedin, New Zealand

Gina M. Ralph
Virginia Institute of Marine Science
William & Mary
Gloucester Point, Virginia, USA

D. Christopher Rogers
Kansas Biological Survey and Center for Ecological Research
The University of Kansas
Lawrence, Kansas, USA

Rochelle D. Seitz
Virginia Institute of Marine Science
William & Mary
Gloucester Point, Virginia, USA

Nick T. Shears
Leigh Marine Laboratory
University of Auckland
Leigh, New Zealand

Dietmar Straile
Limnological Institute
Department of Biology
University of Konstanz
Konstanz, Germany

Martin Thiel
MarineGEO Program
Smithsonian Environmental Research Center
Edgewater, Maryland, USA
Departamento de Biología Marina
Facultad Ciencias del Mar
Universidad Católica del Norte
Coquimbo, Chile
Center of Ecology and Sustainable Management of Oceanic Island (ESMOI)
Coquimbo, Chile

Michael Townsend
Waikato Regional Council
Hamilton, New Zealand

Dirk Weihrauch
Biological Sciences
University of Manitoba
Winnipeg, Canada

Martin Zimmer
Leibniz Centre for Tropical Marine Research
Bremen, Germany
University of Bremen, Faculty 2 Biology/Chemistry
Bremen, Germany

1

CRUSTACEANS AS PREY IN AQUATIC AND TERRESTRIAL FOOD WEBS

Martin Thiel and Lars Gutow

Abstract

Crustaceans are essential components in food webs, linking benthic and pelagic primary production with consumers at higher trophic levels. This chapter focuses on the role of crustaceans as prey in terrestrial and aquatic food webs. Many small crustaceans are herbivores that convert plant and algal biomass into nutritionally valuable food for carnivorous consumers. Their biochemical body composition, with high lipid and protein content, as well as their wide range of body sizes, make crustaceans a valuable food source for a great variety of predators, including both invertebrates and vertebrates. Crustaceans occur in diverse habitats and ecosystems where they are involved in both regular predator–prey interactions and intraguild predation and cannibalism. They have evolved various mechanisms to minimize predation risk, including morphological, behavioral, and developmental defense strategies. Due to their central position in many food webs, crustacean prey can critically mediate the effects of environmental variability through trophic cascades. Crustacean populations can be efficiently controlled by their consumers but may increase markedly when released from top-down control, for example, after over-exploitation of predator stocks. Ongoing environmental changes, including global climate change, are expected to affect predator–prey interactions involving crustaceans, potentially resulting in severe perturbations of ecosystem functioning. Accordingly, a proper understanding of the significant trophic role of crustaceans will be key to predicting the responses of terrestrial and aquatic ecosystems to environmental changes.

Martin Thiel and Lars Gutow, *Crustaceans as Prey in Aquatic and Terrestrial Food Webs*. In: *Ecology and Conservation*.
Edited by: Lars Gutow and Martin Thiel, Oxford University Press. © Oxford University Press (2025).
DOI: 10.1093/oso/9780197768242.003.0001

INTRODUCTION

Crustaceans constitute important links in aquatic food webs, where they are typically found at lower and intermediate trophic levels. Many crustaceans feed on micro- and macroalgae in a world where higher-level consumers prey abundantly on crustaceans. Crustaceans are attractive prey for a very wide diversity of predators ranging from invertebrates to fishes, birds, and marine mammals (Thiel and Kruse 2001, Sabat et al. 2003, Kramer et al. 2015, Bashevkin and Morgan 2020, Torres et al. 2020). They owe this, on the one hand, to their biochemical composition, which is rich in proteins and lipids (Jeffs and O´Rorke 2020), and, on the other hand, to their mostly small sizes making them ideal food parcels for many predators, including fish larvae (Jepsen et al. 2020).

Since crustaceans are important consumers of primary production in many aquatic (but less in terrestrial) food webs, they represent major prey for secondary and tertiary consumers (Fig. 1.1; Potapov et al. 2019). On land, the trophic role of crustaceans is often less pronounced because of the great abundance and diversity of similarly sized consumers (primarily insects) dominating the lower trophic levels of terrestrial food webs. Crustacean scavengers are preyed upon by higher-level consumers that are attracted to marine carrion (Havermans and Smetacek 2018), and in some terrestrial and freshwater environments detritivorous crustaceans can be very common, likely representing an important food source for consumers (Lancaster and Robertson 1995).

Crustaceans are positioned at the interface between primary production and higher trophic levels. As primary consumers, crustaceans transfer primary production into valuable food for consumers from higher trophic levels (e.g., Kramer et al. 2013). They often convert carbon-rich material into stoichiometrically more valuable food for animal consumers. For example, krill contain high concentrations of energetically rich lipids (Färber-Lorda et al. 2009), making them attractive to a wide range of different predators (Bertolin and Casaux 2019). Similarly, crustaceans preyed upon by tuna are rich in lipids and fatty acids, providing high-value energy sources for these pelagic predators (Dhurmeea et al. 2020).

The important role of crustaceans in food webs is also due to their wide range of body sizes, providing appropriate food portions for a great range of consumers, small and large. Crustaceans occur individually or in large groups and are, thus, appropriate prey for consumers with different foraging strategies. Dense aggregations of benthic amphipods are food sources for whales (Coyle et al. 2007) or provide fuel for large numbers of migrating shorebirds or fishes (McCurdy et al. 2005, Hamilton et al. 2006). Similarly, concentrated swarms of pelagic krill are the target of whales, fish, and seabirds (Trathan and Hill 2016), while widely dispersed crustaceans are consumed by

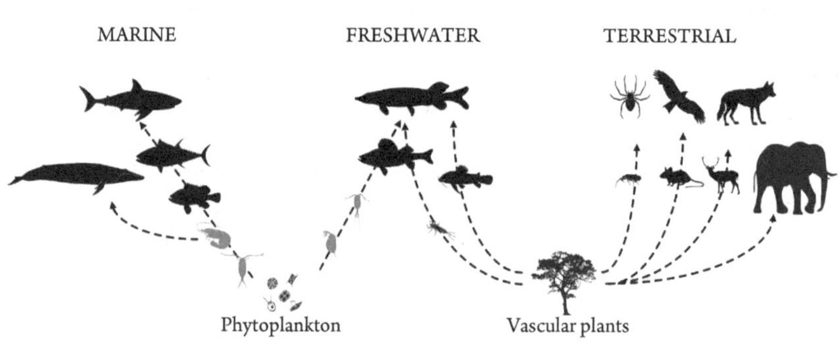

Fig. 1.1.
Crustaceans constitute crucial trophic links primarily in aquatic food webs, whereas in terrestrial food webs, the role of crustaceans as prey is often less pronounced. Source: data from Potapov et al. (2019).

active pursuit predators, such as sea otters, that may roam over extensive areas in search of their prey (Bodkin et al. 2004). Other predators wait in ambush for crustacean prey passing by before striking out (e.g., Coulson et al. 2015, Jurcak and Moore 2018). Planktivorous fishes capture individual crustacean prey in the water column (Bashevkin and Morgan 2020), whereas carnivorous gastropods forage on barnacles, which are immobile but possess protective calcareous shells (Lively 1986).

Through their presence and activities, predators select for the important life history traits of crustaceans (e.g., Glazier et al. 2020), thereby affecting the evolutionary ecology of crustaceans (Wellborn 2018). For example, predators can provoke vertical migration (Dawidowicz and Pijanowska 2018, Bandara et al. 2021), induce chemical and morphological variations (Weiss and Tollrian 2018), and shape crustacean life histories, for example, size or age at maturity (Wellborn 2018).

When predators feed intensively on specific crustacean species or taxonomic groups, they can regulate prey populations, resulting in top-down control. This becomes evident when crustacean populations increase in response to the decline of major predators (Worm and Myers 2003). However, the role of crustaceans as prey is not a simply passive one. Crustaceans have developed diverse strategies and defense mechanisms against predation that have a strong influence on the availability and accessibility of the prey to its predator, thereby potentially weakening the control of the predator over crustacean prey populations. These effects also mediate the role of crustacean prey in food webs.

In this contribution, we examine the position and importance of crustacean prey in aquatic and terrestrial food webs, focusing on their trophic role. We compare crustacean prey and their trophic level in different ecosystems in order to explore their functional role and susceptibility to ecosystem disturbance (habitat destruction, over-exploitation, and climate change). Crustaceans are prey to many different organisms, and thus changes affecting the populations of their predators will not only affect crustacean populations but also entire food webs, and thus the integrity of ecosystems.

TROPHIC RELATIONSHIPS INVOLVING CRUSTACEANS AS PREY

Predators Consuming Crustacean Prey

Typical predator–prey relationships involve the consumption of a live prey organism by a consumer from another species and from a higher trophic level. Different from specific cases of predator–prey interactions such as intraguild predation (see below), prey consumption in a typical predator–prey relationship serves, from the perspective of the predator, to acquire food but is not related to a competitive interaction between the consumer and the prey. The predator–prey interaction entails capturing, killing, processing, and assimilation of the prey organism and results in the flow of energy and matter from a lower to a higher trophic level.

Predators can be dietary specialists or generalists. Accordingly, crustaceans, or certain crustacean species, may constitute the prime food category for specialized predators, whereas for generalists, crustacean prey may be just a dietary supplement. The level of dietary specialization may be determined by the metabolic requirements of a consumer. Alternatively, prey availability can determine the diet composition of a consumer, which may simply be dominated by those prey items that are available at the lowest energetic cost for prey acquisition and handling. Consequently, a predator may be only temporarily specialized on a certain prey species when the availability of the prey species is fluctuating in time or space. For example, two species of reef fish (lionfish, *Pterois volitans/miles*, and schoolmaster snapper, *Lutjanus apodus*) in The Bahamas show different levels of specialization for crustacean prey depending on the site fidelity of the fish and prey distribution at different spatial scales (Layman and Allgeier 2012).

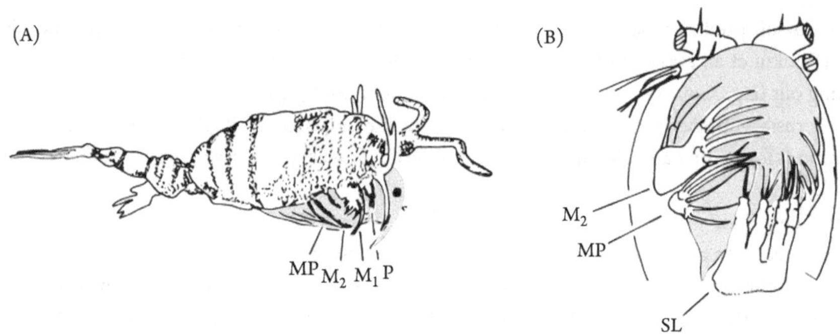

Fig. 1.2.
Handling of a relatively large prey, *Bosmina* sp., by two predatory copepods, (A) *Epischura* sp. and (B) *Cyclops* sp. For simplification, not all appendages of the predators are shown. MP = maxilliped, M_1 = first maxilla, M_2 = second maxilla, P = mandibular palp, SL = swimming leg. Adapted from Kerfoot (1978), with permission from © The American Society for Limnology and Oceanography.

When a consumer and its crustacean prey are of similar body size, the acquisition of single prey individuals likely requires some effort, but the consumer is rewarded by a relatively large food portion. For example, capturing and processing of cladocerans (*Bosmina*) by copepods (*Cyclops, Epischura*) in freshwater systems is an elaborate process but allows the predator to forage on a prey organism that is about half its own body size (Kerfoot 1978; Fig. 1.2). When the size difference between the predator and the prey is huge, a single prey individual provides only little energy but is often available in enormous quantities that require relatively little energy expenditure for acquisition. For example, feeding by baleen whales on dense aggregations of small benthic (e.g., ampeliscid amphipods) or pelagic (e.g., krill) crustaceans provides considerable food supply for large consumers from small prey organisms (Miller et al. 2019, Hildebrand et al. 2021).

In terrestrial systems, crustaceans are also preyed upon by numerous predators. For example, Gorvett (1956) summarizes from various previous studies the spectrum of predators of the cosmopolitan woodlouse *Porcellio scaber*, which comprises various vertebrate consumers, such as birds, mice, shrews, toads, frogs, hedgehogs, newts, and lizards, and also spiders (Carbines et al. 1992). An impressive terrestrial trophic interaction that involves crustaceans as both prey and predator is the predation of red crabs *Gecarcoidea natalis* by the opportunistic omnivorous giant robber crab *Birgus latro* in the forests of Christmas Island, eastern Indian Ocean (Fig. 1.3). Here, predation on red crabs involves two strategies, with robber crabs either killing and consuming whole prey individuals or provoking prey individuals to autotomize limbs, which are subsequently consumed by the predator (Krieger et al. 2016).

Intraguild Predation

According to Polis et al. (1989), intraguild predation is defined as "a combination of [competition and predation/parasitism], that is, the killing and eating of species that use similar, often limiting, resources and are thus potential competitors. Intraguild predation is distinguished from traditional concepts of competition by the immediate energetic gains for one participant (the predator). It differs from classical predation because the act reduces potential exploitation competition."

Intraguild predation can occur among primarily herbivorous species as a mechanism of strong interspecific interference competition. For example, in the North Sea, the herbivorous isopod *Idotea balthica* is rapidly displaced from its preferred habitat by the congener *I. emarginata* as a consequence of highly asymmetric predation on conspecific and heterospecific juveniles by both

Fig. 1.3.
A male robber crab, *Birgus latro*, handling and killing a freshly caught red crab, *Gecarcoidea natalis*, on Christmas Island. Reproduced from Krieger et al. (2016), with permission from © Academia Sinica; see a color version of this figure in the centerfold.

species (Franke and Janke 1998). Intraguild predation can also occur as a consequence of species invasions when competition emerges between native and invasive consumers that share similar food resources. For example, differential predation by adult males on congeneric molting females results in the replacement of the native *Gammarus duebeni celticus* by the invasive congener *Gammarus pulex* in Irish rivers (Dick 1992). Interestingly, the outcome of this mutual predation is mediated by infestation by the acanthocephalan parasite *Echinorhynchus truttae*, which is more prevalent in invasive *G. pulex* than in native *G. duebeni celticus*, and apparently attenuates the predatory behavior of the invader (MacNeil et al. 2003). Strong intraguild predation has also been observed between two predatory crabs along the US Atlantic coast where the abundances of blue crabs *Callinectes sapidus* and invasive green crabs *Carcinus maenas* were negatively correlated (DeRivera et al. 2005). Strong predation by *C. sapidus* on *C. maenas* suggests a strong natural biotic resistance to green crab invasion.

From the perspective of a shared prey, intraguild predation among two predatory species may be beneficial as it may reduce predation rates and predator redundancy in multiple predator assemblages. For example, when the two predatory crabs *C. maenas* and *Hemigrapsus sanguineus* foraged together on their shared amphipod prey, the specific predation rate of each predator was lower than expected from crabs foraging in isolation (Griffen and Byers 2006). Thus, intraguild predation among crustaceans can have important effects on the crustacean populations, possibly cascading through the food web to lower trophic levels.

Cannibalism

Cannibalism is particularly common in crustaceans because most crustacean species show some level of predatory behavior, even herbivores (see, e.g., MacNeil et al. 1997). Moreover, cannibalism is promoted in crustaceans because the growth of crustaceans involves molting, which makes individuals temporarily defenseless against predatory conspecifics. The high nutritional gain from foraging on conspecifics with similar elemental body composition and the high likelihood of prey encounter in a shared habitat makes cannibalism a particularly attractive type of alimentation. According to the prevalence of cannibalism in crustaceans, this topic has been extensively addressed in previous dedicated reviews (Bleakley 2018).

Not all individuals in a population are at equal risk of being cannibalized as predation pressure varies among different life history stages and sexes. Often, large adult individuals forage on small, less defensive (juvenile) conspecifics (Reaka 1987). Accordingly, cannibalism can be the major source of juvenile mortality in size-structured populations. However, early life history stages may be too small or too inaccessible to be profitable food for large adults (Méndez Casariego et al. 2009). Female crustaceans are also frequently attacked by conspecific adult males (Jormalainen and Shuster 1997). By targeting specific life history stages, cannibalism is likely to have strong demographic effects. Injured individuals are at particularly high risk of being cannibalized (Tran 2014).

Cannibalism in crustaceans is often density dependent (Moksnes 2004), indicating that, at high densities, individuals of a population compete for limited resources, potentially resulting in nutritional stress (Polis 1981). Similarly, cannibalism on females in precopulatory amplexus may indicate nutritional deficiency in the male, as mate guarding requires considerable energetic effort (Jormalainen 1998). Accordingly, the intensity of cannibalism in a population also varies with the availability of alternative food resources (Sparrevik and Leonardsson 1998).

Individuals may change their behavior to mitigate the risks of cannibalism. For example, individuals may molt in the dark (Lipcius and Herrnkind 1982), at a specific water depth (Tarling et al. 1999), or in hidden refuges (Marshall et al. 2005) to minimize encounters with conspecifics during vulnerable periods. The risk of being cannibalized may be reduced in heterogeneous environments with high availability of protective microhabitats (Dittel et al. 1995). Density-dependent cannibalism is particularly relevant in crustacean aquaculture, and there are multiple strategies (e.g., stocking densities, shelter provision, diet and feeding frequency) to reduce the risk of cannibalism (Romano and Zeng 2017).

FOOD QUALITY

From the perspective of the consumer, food quality depends on the handling/processing effort and the nutritional quality of the prey. Therefore, the food source with the highest nutritional quality is not necessarily of the highest food quality if the energy yield from the assimilated biomass is low compared to the energetic effort required for prey handling and processing. Accordingly, predators constantly have to face a trade-off between handling effort and nutritional quality to maximize the energetic yield from the prey. A prey organism is not determined to play a simply passive role in the interaction with a predator but has ample opportunities to minimize its attractiveness as food.

Accessibility

Crustaceans have developed defense strategies to reduce their accessibility to predators that increase the effort needed for detection, handling, and processing of the prey. Defense against predators can be either constitutive or inducible. Constitutive defenses provide protection against constant predation pressure but may be costly for the prey organisms when the development and maintenance of the defense requires permanent energy expenditure. In contrast, inducible defense is activated by the presence of the predator or chemical cues (kairomones) thereof. Consequently, the prey may save energy in the absence of the predator but may be temporarily unprotected in the presence of a predator until the induced defense is fully operative. Accordingly, defense mechanisms involve a trade-off between maximum protection and energy savings. A well-studied example of inducible defense in crustaceans is phenotypic plasticity in freshwater water fleas *Daphnia* spp.

and involves the formation of protective morphotypes in the presence of predators. A considerable number of vertebrate and invertebrate species prey on *Daphnia* spp., such as *Chaoborus* larvae (Insecta: Diptera), backswimmers *Notonecta* spp. (Insecta: Hemiptera), tadpole shrimp *Triops* spp., and planktivorous fish (Black 1993, Tollrian 1994, Rabus and Laforsch 2012). In response to predator kairomones, morphological changes are induced in *Daphnia* spp., such as expression of tail spines, "neckteeth," helmets, or a change in body size (Weiss and Tollrian 2018). These morphological changes complicate detection, capturing, and processing of the prey.

The presence of predators may also induce behavioral changes in crustaceans. In an experiment, the presence of a predator, the planktonic water flea *Bythotrephes longimanus* induced vertical migration in *Daphnia (Daphnia) mendotae* from Lake Michigan, US (Pangle and Peacor 2006). *Daphnia (Daphnia) mendotae* occupies the warmer top water layers in the absence of the predator but switched to the middle region of the water column during the night and to cold and low regions during daytime when *B. longimanus*, or just kairomones of that species, were introduced to the experiment. Predator avoidance by diel vertical migration is common in pelagic crustaceans not only in freshwater but also in the marine environment, where, for example, krill performs extensive periodic movements through the water column (Zhou and Dorland 2004). The result of diel vertical migration is spatiotemporal segregation of predators and prey (Fig. 1.4), which is maintained by predator presence, but rapidly disappears when predators are absent (Dawidowicz and Pijanowska 2018, Weiss and Tollrian 2018, Bandara et al. 2021).

Benthic crustaceans may evade areas with high densities of predators, resulting in spatiotemporal segregation of predators and prey. For example, gammarid amphipods emigrate from mussel clumps with high densities of nemertean predators, aggregating in clumps without these benthic predators (Thiel and Reise 1993). Similarly, in the presence of their fish predators, amphipods *Echinogammarus marinus* sought refuge in sheltering microhabitats and reduced their feeding activity (Beermann et al. 2018). Other epiphytic amphipods reduce the predation risk by hiding on morphologically complex algal hosts (Cacabelos et al. 2010). Additionally, small crustacean herbivores may associate with chemically rich seaweed species that efficiently deter herbivorous or omnivorous fish (Hay 1997).

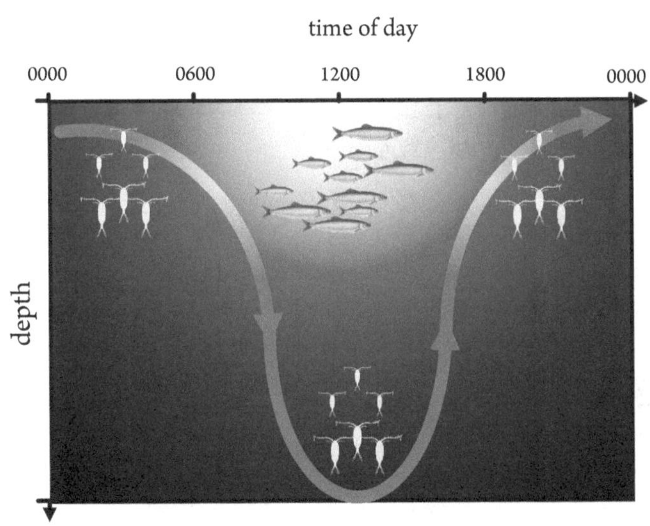

Fig. 1.4.
Predators (e.g., planktivorous fish) induce diel vertical migration in many pelagic crustaceans (e.g., copepods), both in freshwater and marine environments. Source: data from Bandara et al. (2021).

The amphipod *Pseudamphithoides incurvaria* constructs portable domiciles from the tissue of the chemically defended brown alga *Dictyota bartayresiana*, thereby achieving "behavioral sequestering" of predator deterrent seaweed chemicals (Hay et al. 1990). Construction of protective domiciles on seaweeds is common among ampithoid amphipods. For example, *Sunamphitoe stypotrupetes* burrow into the stipes of large kelp species and reside inside these burrows (Chess 1993). By constantly feeding on the stipe tissue inside the burrows, the amphipods enlarge their domiciles to provide safe living space for juveniles from successive reproductive cohorts. Other species construct protective nest-like domiciles by folding the blades of the kelp ("kelp curler"). Field experiments in Chilean kelp forests demonstrated that the domiciles provide efficient shelter from predation, especially when the domiciles are located in the upper sections of the kelp thallus outside the common foraging habitat of demersal predators (Gutow et al. 2012). However, occasional excursions of individual amphipods outside the protective domiciles are unavoidable, for example for mating or when the domicile is about to lose its structural integrity. To avoid visually hunting predators, migrations of small crustaceans within kelp forests occur primarily at night, while during the daytime, phytal crustaceans remain mostly hidden in safe microhabitats provided by structurally complex kelp thalli (Taylor 1998).

Adult individuals of large decapod crustaceans are protected from predation by their body size. Often, mechanical constraints of the predator (e.g., small mouth openings) prevent efficient handling and ingestion of the large prey. However, small individuals may be well within an appropriate size range for numerous predators so that, in particular, the smallest, freshly settled individuals are at high risk of being preyed upon. Accordingly, juvenile decapods often display a cryptic behavior and coloration that reduces the probability of being detected by predators. For example, adult American lobsters, *Homarus americanus*, hardly experience any predation whereas juveniles have to remain hidden in sheltering habitats such as cobble fields. The residence time of the lobsters inside shelters decreases with increasing body size. Simultaneously, the aggressiveness of the lobsters toward predatory fish increases with body size, indicating that growing lobsters progressively overcome their vulnerability (Wahle 1992). In response to the high vulnerability of the smallest individuals, rock crabs, *Cancer irroratus*, have evolved "developmental crypsis," which involves a cryptic coloration of freshly settled individuals (Palma and Steneck 2001). Juveniles of this species display a color polymorphism, which allows them to hide on heterogeneous, polychromatic habitats such as shell hash (Fig. 1.5), but is lost as the crabs grow.

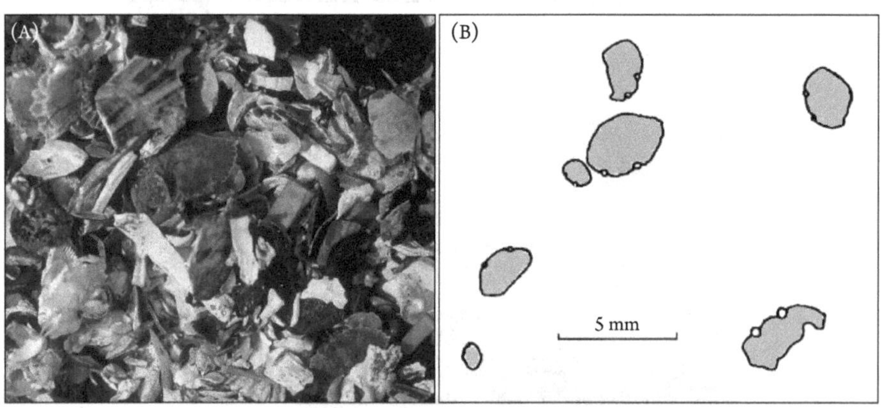

Fig. 1.5.
Juvenile rock crabs, *Cancer irroratus*, with cryptic coloration on shell hash. The cryptic individuals in (A) are outlined in (B). Adapted from Palma and Steneck (2001), with permission from © The American Society for Limnology and Oceanography; see a color version of this figure in the centerfold.

Cryptic coloration is also displayed by brown shrimp, *Crangon crangon*, which can rapidly adjust their coloration to match the background color of the underlying sediment (Siegenthaler et al. 2018). Additionally, individuals of *C. crangon* rapidly bury into sediments to reduce their detectability for visually hunting predators (Pinn and Ansell 1993). Morphologically undefended small crustaceans have to rely on other defense mechanisms. For example, pea crabs, *Zaops ostreus*, seem to be chemically defended against predation. In experiments, the great majority (>90%) of individuals were rejected by predatory mummichog, *Fundulus heteroclitus*, while individuals of the chemically unprotected blue crab, *C. sapidus*, were always consumed (Luckenbach and Orth 1990).

Nutritional Quality

Many crustaceans are consumers at the base of the food web, foraging directly on primary producers. Pelagic and benthic uni- and multicellular algae, as well as vascular plants, are usually of poor nutritional quality for consumers. Tissues of plants and algae have a relatively high carbon content and are often chemically enriched with grazer deterrent plant secondary metabolites (Hay and Fenical 1988). Vascular plant material, in particular, may contain high quantities of fibrous material, which is hard to digest (Linton and Greenaway 2007).

The nutritional requirements of a consumer are, to a large extent, determined by its elemental composition. Accordingly, the elemental composition of high-quality food ideally matches the composition of its consumer. Animals usually have a higher protein content than plants and, thus, commonly have lower tissue C:N ratios than carbon-rich plant tissues. C:N ratios of phytoplankton and aquatic macrophytes, including seaweeds and seagrasses, vary widely (Redfield et al. 1963, Atkinson and Smith 1983) but are mostly clearly above the C:N ratios of crustaceans (e.g., Nyssen et al. 2002, Evans-White et al. 2005), indicating that herbivorous crustaceans convert macrophyte tissue into animal tissue, which is a stoichiometrically more valuable food resource for consumers at higher trophic levels. Crustaceans also display wide interspecific variations in C:N ratios (Walve and Larsson 1999), demonstrating that the C:N ratio alone only partly determines the nutritional quality of food. However, it displays the stoichiometric mismatch between primary producers and consumers, highlighting the important role of herbivores, including crustaceans, in making the energy from primary production accessible for higher trophic levels. Additionally, herbivorous crustaceans supply predators with essential micronutrients from primary producers, which is indicated by, for example, limited growth performance of predatory fish under conditions of phosphorus limitation (Malzahn et al. 2007). Similarly, essential fatty acids, synthesized by planktonic algae, are transferred from pelagic crustaceans to upper trophic levels, particularly fish feeding on zooplankton (Kainz et al. 2004).

Various pelagic crustacean species are used as food for aquaculture organisms (Jepsen et al. 2020). Accordingly, the biochemical compositions of pelagic crustaceans have been analyzed to evaluate their suitability for use as a diet of cultured organisms. For example, mysids have high protein and lipid contents and a relatively low carbohydrate content (Mauchline 1980), suggesting that energy is stored by these crustaceans primarily as proteins and lipids but not as long-chained carbohydrates (Guevara et al. 2005). A high proportion of the lipids in *Metamysidopsis insularis* consists of phospholipids (Guevara et al. 2005), which are essential for the constitution and function of cellular membranes. Moreover, mysids contain a considerable amount of essential mono- and polyunsaturated fatty acids, which cannot be synthesized by consumers but have to be transferred over trophic levels from primary producers to consumers. A similar biochemical composition with considerable amounts of unsaturated fatty acids has been determined for Antarctic krill, *Euphausia superba*, from the Southern Ocean (Clarke 1980). The proportions of proteins, lipids, and carbohydrates vary somewhat among krill species from different oceanic regions but remain largely within a similar range (Mauchline 1980).

The biochemical composition can also vary within individual crustaceans or between the sexes of conspecific individuals. For example, the muscle of *Cancer pagurus* has a higher protein content than the gonads and the hepatopancreas, while females' ovaries have higher contents of proteins, fat, cholesterol, and amino acids than males' testes (Barrento et al. 2010). A comparative study of the biochemical compositions of woodlice, *Porcellio scaber*, and various insects, including cockroaches, beetles, fruit flies, and katydids, revealed no outstanding nutritional quality of the terrestrial isopods except for a high calcium content and a correspondingly high calcium:phosphorous ratio, which may provide a useful dietary source of calcium for insectivores (Oonincx and Dierenfeld 2011). The chitinous exoskeletons of crustaceans are rich in polysaccharides, which are, however, difficult to digest for many consumers. Certain consumers, including fish and seabirds, are equipped with a set of enzymes that allows them to digest and assimilate the chitinous structures of crustaceans (Lindsay and Gooday 1985, Jackson et al. 1992).

CRUSTACEAN PREY IN AQUATIC AND TERRESTRIAL FOOD WEBS

Crustaceans as Prey in Pelagic Food Webs

Copepods and Cladocerans

Pelagic crustaceans occur in marine and freshwater systems all over the world and are major contributors to the total zooplankton communities in terms of abundance and biomass (see Chapter 10 in this volume). For example, in the Southern Ocean pelagic copepods dominate the total biomass of the mesozooplankton community and are locally exceeded only by Antarctic krill (Atkinson et al. 2012). In other oceanic regions, too, pelagic copepods are extremely abundant, occasionally reaching densities of hundreds of thousands of individuals per cubic meter in surface waters (Turner 2004). Copepods are major consumers of pelagic primary production and contribute substantially to pelagic secondary production. They are consumed by a great variety of consumers, thereby establishing a trophic link that connects the base of the aquatic food web with higher trophic levels. Pelagic copepods can sustain huge populations of commercially important planktivorous fish species, such as herring, sardine, and mackerel (Patel et al. 2023).

Additionally, pelagic copepods are intensively preyed upon by gelatinous plankton, especially ctenophores, which can locally remove more than 50% of the crustacean prey stock (Alldredge 1984). By consuming planktonic copepods, jellyfish can induce cascading effects in pelagic food webs (Stibor et al. 2004). Significant reduction of copepod biomass and shifts in copepod feeding preferences resulted in two alternate states of marine phytoplankton communities, depending on nutrient availability (Fig. 1.6A, B). Thus, predation on planktonic crustaceans can have profound effects on pelagic food webs. In freshwater food webs copepods are often replaced by cladocerans, which are the main link between phytoplankton and higher trophic levels (Fig. 1.6C).

Krill

Species of krill are found in all regions of the oceans from the equator to polar waters, and are often associated with upwelling waters along continental margins (Mauchline and Fisher 1969). The most intensively studied euphausiid species is Antarctic krill, *E. superba*, which reaches an exceptionally high total biomass in the Southern Ocean (Siegel and Watkins 2016).

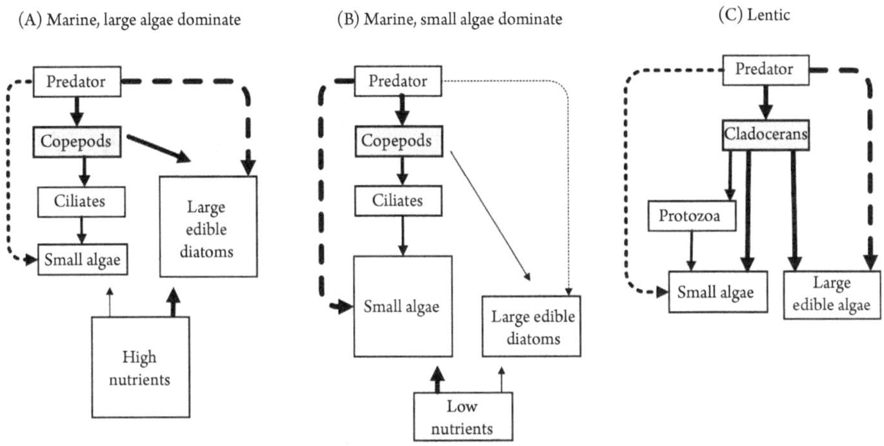

Fig. 1.6.
In marine systems with (A) high nutrient availability, copepods feed primarily on abundant large diatoms, leading to more direct transfer of nutrients to higher consumer levels, whereas in systems with (B) low nutrient availability, copepods feed on ciliates, which consume small algae leading to longer trophic chains to higher consumers. In freshwater systems with (C) abundant cladocerans, they feed nonselectively on all sizes of algae and ciliates, leading to shorter trophic links. Source: data from Stibor et al. (2004).

Euphausia superba primarily occupies the upper 200 m of the water column but may also migrate to greater depths (Tarling and Fielding 2016). Additionally, aggregations of foraging krill are also found under sea ice (O'Brien 1987) and on the seabed (Schmidt et al. 2011). Accordingly, *E. superba* is available for consumers in different habitats of the ocean, resulting in the species' key role in the food web of the Southern Ocean (Fig. 1.7).

Krill often occurs in large aggregations, providing huge amounts of local, easily accessible biomass for large-bodied consumers with high energetic demands such as baleen whales and seals, including crabeater seals (*Lobodon carcinophagus*), Antarctic fur seals (*Arctocephalus gazella*), and sea leopards (*Hydrurga leptonyx*) (Trathan and Hill 2016). Examinations of the stomach contents of several baleen whale species indicate that whales selectively target krill swarms of larger individuals (Mackintosh [1974], cited in Mauchline [1980]), suggesting that intensive predation by whales may have some effect on the structure of krill populations. Laws (1977) estimated that the industrial removal of whales has substantially increased the availability of krill to other consumers.

Additionally, seabirds (e.g., Macaroni penguins, *Eudyptes chrysolophus*), myctophid fish, and cephalopods are common consumers of Antarctic krill (Permitin 1970, Croxall et al. 1985, Boyd 2002, Trathan and Hill 2016). For many seabirds in the Southern Ocean, krill constitutes not the sole but a major fraction of the overall diet. In addition to krill, the diet of seabirds comprised other crustacean species, primarily gammarid and hyperiid amphipods, fish, and squid (Ridoux and Offredo 1989).

Similar to adult krill, larvae of *E. superba* constitute an important food resource for various pelagic consumers of the Southern Ocean, such as myctophid fish (Pakhomov et al. 1996), hyperiid amphipods (Pakhomov and Perissinotto 1996), and chaetognaths (Øresland 1990). Accordingly, Tarling et al. (2007) concluded that predation on larvae may have demographic consequences for the krill population.

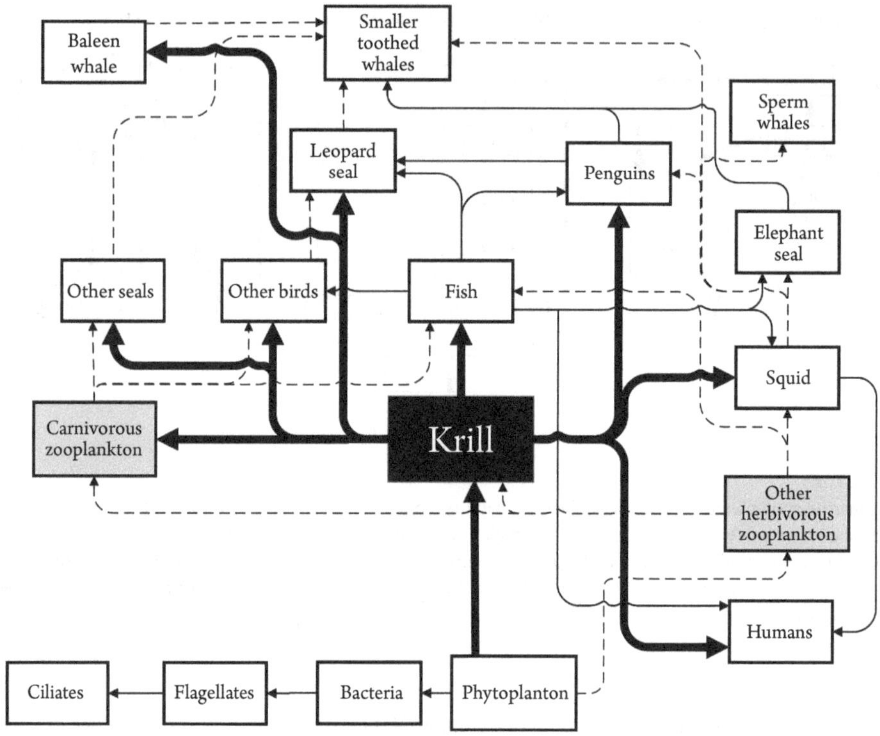

Fig. 1.7.
The central role of Antarctic krill, *Euphausia superba,* in the Antarctic food web as consumer of pelagic primary production and microbiota and as food for a great variety of vertebrate and invertebrate consumers. In addition to krill, pelagic amphipods (carnivorous plankton) and copepods (other herbivorous plankton) are important components of the Antarctic pelagic food web (gray boxes). Source: data from Trathan and Hill (2016).

Hyperiid amphipods, especially from the genus *Themisto,* can also be important prey items in pelagic ecosystems. In the Southern Ocean, they are preyed upon by squid, fishes, seabirds, and marine mammals, whereas in the northern oceans they are important food for commercially important fish species (Havermans et al. 2019).

Crustaceans as Prey in Benthic Food Webs

Small Crustacean Mesoherbivores in Macrophyte Beds

Kelp forests extend along temperate to polar rocky shores where they constitute highly productive and diverse ecosystems (Steneck et al. 2002) that are used by small herbivorous crustaceans (mainly amphipods and isopods) as both food and habitat (Arrontes 1999). Despite their small size, these herbivores can substantially damage large kelp species (Poore et al. 2014), but their effects are often less obvious because they are strongly controlled by predators. When released from predator control, small crustacean herbivores can be highly destructive by removing substantial amounts of kelp biomass (Tegner and Dayton 1987, Duffy and Hay 2000). Herbivorous crustaceans convert algae into a stoichiometrically more valuable food for heterotrophs (Norderhaug et al. 2003), linking kelp forest primary production with higher trophic levels of the food web (Koenigs et al. 2015).

In kelp forests and temperate reefs, crustaceans are primarily preyed upon by a diverse spectrum of fish, including carnivorous and omnivorous species, and even species that are considered herbivores. In a Norwegian kelp forest, crustacean prey constitutes up to 75% of the weight of the stomach content of Ballan wrasse *Labrus bergylta,* with the herbivorous peracarid crustaceans *Idotea balthica* (Isopoda) and *Ampithoe rubricata* (Amphipoda) alone contributing up to 35% of the stomach content (Norderhaug et al. 2005). In kelp forests near Santa Barbara, California, the herbivorous isopod *Pentidotea resecata* alone comprised 78% of the diet of kelp rockfish, *Sebastes atrovirens,* and, together with the kelp-rolling amphipod *Sunamphitoe humeralis,* more than 40% of the diet of kelp bass, *Paralabrax clathratus* (Koenigs et al. 2015). In a New England kelp forest, pollock, *Pollachius virens,* intensively preyed on crustaceans, with gammarid amphipods being the most preferred crustacean prey species (Ojeda and Dearborn 1991). In a kelp forest in northern Chile (SE Pacific), the maximum contribution of benthic crustaceans to the overall stomach content was about 5% in herbivorous fish, 55% in omnivorous fish, and 91% in carnivorous fish (Angel and Ojeda 2001).

The contribution of benthic crustaceans to the overall diet also depends on the differential availability of the various food resources in the kelp forest (Pérez-Matus et al. 2012). A comparative study on two subtidal reefs in northern Chile revealed that the proportion of benthic crustaceans in the diet of fish predators is much higher in an extensive and dense kelp forest (Fig. 1.8A) than in a sparse subtidal forest where kelp occurs only sporadically (Fig. 1.8B) (Angel and Ojeda 2001).

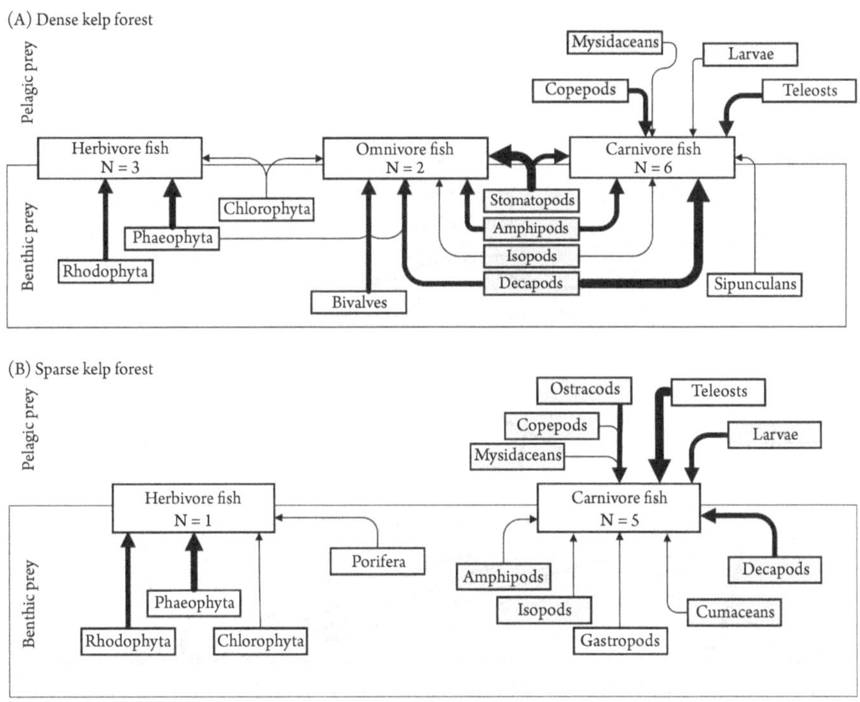

Fig. 1.8.
Food resources of herbivorous, omnivorous, and carnivorous fish in subtidal kelp forests in northern Chile. Caleta Errázuriz (A) is characterized by a dense kelp forest, whereas in Carrizal Bajo (B) kelp occurs only sporadically. The thickness of the arrows indicates the contributions (by weight) of different prey organisms to the total stomach content of the fish. Diverse groups of crustaceans (gray boxes) are important food resources for fish. Source: data from Angel and Ojeda (2001).

The contribution of benthic crustaceans to the overall diet of predatory fish may also vary within fish species depending on the trophic requirements, which may change with body size and ontogenetic stage of the fish. For example, in northeastern New Zealand, juveniles of the carnivorous Red Moki, *Chirodactylus spectabilis,* forage primarily on gammarid amphipods whereas adult individuals switch to alternative prey organisms, which consist primarily of ophiuroids (McCormick 1998).

In addition to fish, various other groups of mobile consumers, including decapod crustaceans, sea otters (Calkins 1978), and cephalopods (Barshaw et al. 2003) prey on benthic crustaceans in subtidal reefs and kelp forests. In a New England kelp forest, crabs, barnacles, and isopods are important food resources for the American lobster *Homarus americanus* and the congeneric crabs *Cancer irroratus* and *C. borealis* (Ojeda and Dearborn 1991). Only juvenile or small crabs were preyed upon by predatory decapods and fish, resulting in a "size escape" (sensu Connell 1975) of crustacean prey from predation. By removing only juveniles, the predators may have a critical effect on the age structure of the prey population. In choice feeding assays, lobsters and fish preferred small crabs over mussels, sea urchins, and brittle stars (Ojeda and Dearborn 1991). This preference matches with the high nutritional value of the crustaceans having a caloric content per unit tissue, which is about 2.5 times higher than in mussels and about five times higher than in sea urchins.

Seagrass beds are highly productive coastal marine ecosystems in tropical to boreal shallow coastal sediments (Hemminga and Duarte 2000) and host a great diversity of trophically interlinked species. Seagrass benefits from small herbivorous crustaceans that efficiently remove other autotrophic competitors or epiphytic microalgae covering the leaves (Hughes et al. 2004). Simultaneously, herbivorous crustaceans are prey for numerous consumers, primarily fish and decapod crustaceans, thereby constituting a central and diverse component of seagrass ecosystems (Fig. 1.9; Duffy 2006).

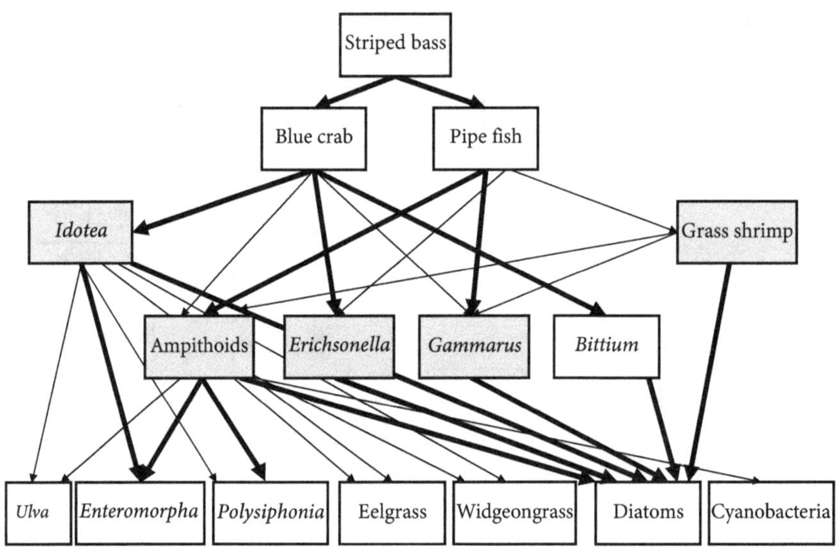

Fig. 1.9.
Schematic presentation of trophic structuring and biodiversity of a seagrass bed in Chesapeake Bay, USA, displaying the central role of small herbivorous crustaceans (gray boxes) as consumers of primary production and prey for predatory fish and decapods. Source: data from Duffy (2006).

In Australian seagrass beds, predatory fish remove an essential proportion of the total crustacean secondary production, with the amphipods *Paradexamine churinga* and *Tethygeneia* spp. being consumed by the majority of the fish species (Edgar and Shaw 1995). Similarly, crustaceans are the most important food source for numerous fish species in Mediterranean seagrass beds (Bell and Harmelin-Vivien 1983), suggesting that even mostly herbivorous fishes utilize crustacean prey to compensate for the relatively low nutritional quality of marine macrophytes (Klumpp et al. 1989). In a seagrass bed in North Carolina, USA, a considerable spectrum of fish species but also several decapod crustaceans, including various shrimp species and the blue crab *C. sapidus*, feed on epiphytic amphipods (Nelson 1979). In experiments, survival of amphipods was positively related to the shoot density of the seagrass bed, suggesting that habitat complexity is an important determinant of predation success (Nelson 1979).

In seagrass beds in Western Port, Australia, caridean shrimp are key components of the benthic food web as consumers of both plant and animal biomass and as prey for predatory fish and birds (Howard 1984). Feeding shrimp remove a considerable amount of the invertebrate secondary production of the seagrass bed. A substantial proportion of the shrimp diet consisted of epibenthic crustaceans, including peracarid crustaceans (amphipods and mysids), ostracods, and copepods. The shrimp themselves are preyed upon by predatory fish and wading birds. The high dominance of the shrimp *Palaemon intermedius* in the stomach contents of fish and wading birds matched the species' numerical dominance in the field and, thus, its availability to consumers.

Small Infaunal Crustaceans in Soft-Bottom Habitats

Crustaceans can be extremely abundant in sediments and are, thus, structurally and functionally essential components of soft-bottom communities. The omnipresence of crustaceans in and on aquatic soft bottoms, their wide range of body sizes, and their often-high biomass makes them a valuable food resource for a great variety of consumers. Nemerteans are particularly successful invertebrate predators of crustaceans, with hoplonemerteans feeding almost exclusively on amphipods but also isopods (reviewed by Thiel and Kruse 2001). Similarly, several predatory polychaetes forage on amphipods. For example, the "errant" species *Glycera alba* preys on ampeliscid amphipods (Ockelmann and Vahl 1970). Similar to nemerteans, the attacked organism is killed by a neurotoxic secretion produced by the paragnathal glands of the worm.

One of the most remarkable marine trophic interactions involving crustacean prey is the feeding of gray whales, *Eschrichtius robustus*, on benthic amphipods in the region of the northern Bering Sea and the southern Chukchi Sea. Among baleen whales, *E. robustus* is unique for its primarily benthic feeding. Gray whales preferentially feed on ampeliscid amphipods but also on other amphipod species that form dense tube mats on soft sediments. In common gray whale feeding grounds, amphipods dominate the total benthic biomass with densities of 10,000 to 15,000 individuals per square meter (Oliver et al. 1984, Dunham and Duffus 2002).

Single gray whales consume hundreds of kilograms of amphipods per day (Oliver et al. 1984). Amphipods are extracted from sediments by the whales via suction feeding (Fig. 1.10), a foraging technique that results in extensive feeding pits on the seafloor (Oliver and Kvitek 1984). In addition to benthic amphipods, *E. robustus* also exploits other food resources, including benthic, hyperbenthic, and pelagic crustaceans. For example, along the coast of British Columbia, gray whales feed preferentially on ampeliscid amphipods, but may switch to larvae of porcelain crabs, mysids, and ghost shrimp (Dunham and Duffus 2002). However, the extremely high densities and their easy availability likely make benthic amphipods the most valuable food source.

After feeding by gray whales, the seafloor excavations are rapidly colonized by voracious scavenging lysianassid amphipods, primarily *Anonyx* spp., that attack the injured and dislodged benthic fauna, especially smaller individuals of amphipods (Oliver and Slattery 1985). In the northern

Fig. 1.10.
Gray whale, *Eschrichtius robustus*, disturb sediments while foraging on benthic organisms. Tube-building ampeliscid amphipods are the major prey organisms of gray whales in the northern Bering Sea. Photo reproduced courtesy of Flip Nicklin, with permission from © Smithsonian Institution under Creative Commons Zero (CC0); see a color version of this figure in the centerfold.

Bering and southern Chukchi Sea around Alaska, individuals of the most common local seabird species have been observed to associate with the feeding plumes of foraging gray whales, probably for feeding on suspended benthic organisms (Harrison 1979). The author estimated that the association with feeding whales can support several hundred thousand seabirds from nearby colonies with a long-lasting and reliable food source. Accordingly, a diverse community of different consumer species can benefit from the disturbing benthic feeding habit of gray whales, with benthic amphipods being an essential prey item for most of the consumers.

The availability of crustacean prey may essentially determine the distribution of seabirds and their capacity to perform seasonal long-distance migrations. Hamilton et al. (2003) quantified shorebird activities on four intertidal mud flats in the Bay of Fundy, Canada, and identified a positive correlation between densities of shorebirds and the infaunal amphipod *Corophium volutator*, which is an important prey species of, for example, semipalmated sandpipers (*Calidris pusilla*). Feeding on this amphipod species in the Bay of Fundy allows the seabirds to double their body mass for long southward migrations (Hicklin 1987). The authors explained a long-term decline in abundance of *C. pusilla* in that region along with the collapse of the *C. volutator* population as a consequence of human activities, indicating a bottom-up control of the consumer population by crustacean prey availability.

Myers et al. (1980) investigated factors that determine the foraging success of seabirds feeding on crustaceans in sediments. In experiments, sanderlings (*Calidris alba*) from Bodega Bay, California, were allowed to prey on the infaunal isopods *Excirolana linguifrons* and *E. kincaidii* as well as on the sand crabs *Emerita analoga*, which are important local components of the sanderlings' diet.

The experiments revealed that the probing depth of sanderlings depends on the bill length, while sediment penetrability essentially determines the effort and handling time required to acquire the prey organisms. Moreover, larger prey was more easily detectable and, thus, more vulnerable to predation than smaller prey.

Large (Decapod) Crustaceans as Prey

Many large crustacean species, such as decapods and stomatopods, are often at an intermediate trophic position foraging on small invertebrates, including crustaceans, but also representing prey for larger predators (Ojeda and Dearborn 1991). While predation on decapod crustaceans is most intense during their pelagic larval phase (reviewed in Bashevkin and Morgan 2020), this section will focus on the consumption of large benthic individuals. Many large benthic crustaceans, including brachyuran and anomuran crabs as well as clawed and achelate lobsters, have narrow predator spectra because of their considerable body size and their feistiness. Additionally, the strong cuticula and the armored carapace of large decapods (e.g., spiny lobsters and king crabs) may be hard to process for most consumers. Accordingly, predators of large benthic crustaceans are mostly large-bodied species, primarily vertebrates such as fish and seabirds.

In a review, Boudreau and Worm (2012) emphasize the central position of benthic decapods in the network of biological interactions in coastal marine ecosystems (Fig. 1.11). Crabs and lobsters are voracious consumers of a wide spectrum of benthic invertebrates (i.e., primary consumers). At the same time, they are preyed upon by various top predators, primarily vertebrates, such as marine mammals, predatory fish, and seabirds, and also invertebrates, such as cephalopods

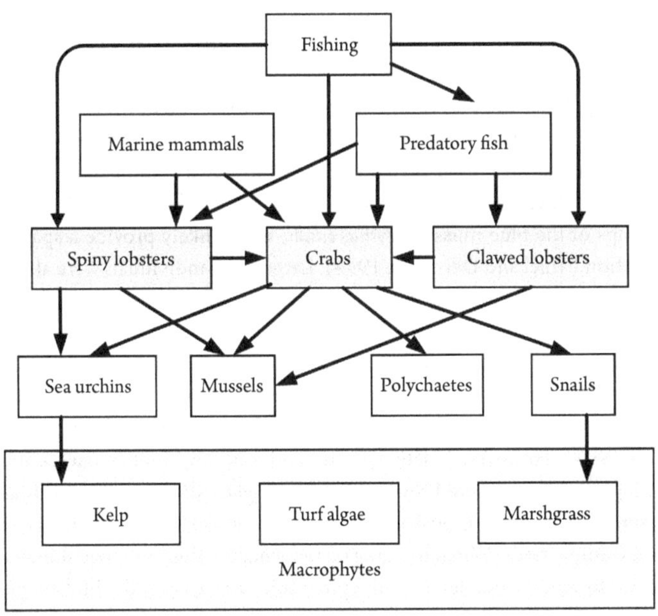

Fig. 1.11.
Conceptual model visualizing the central role of large benthic decapod species (gray boxes) in the interaction web of benthic marine ecosystems. Crabs and lobsters consume a large variety of benthic invertebrates and function as prey for primarily vertebrate predators, including mammals and fish. Additionally, benthic decapod crustaceans face intense intraguild predation. Source: data from Boudreau and Worm (2012).

(Butler and Lear 2009), thereby constituting a crucial trophic link between the lower levels of marine food webs and top predators, and also providing target species for commercial fisheries.

Stomach content analyses reveal that many demersal fish prey on large benthic crustaceans. American lobsters (*Homarus americanus*) have been found in the stomachs and intestines of several demersal fish species in the southern Gulf of St. Lawrence (Hanson and Lanteigne 2000), although lobsters were not the major food source but rather a food supplement. Also, based on stomach content analyses, Hanson et al. (2014) identified Atlantic rock crabs (*Cancer irroratus*) to be a trophically important crustacean species as both prey for various demersal fish species and as a predator of benthic invertebrate species while the subordinate role of American lobsters as food for demersal fish species in that region was confirmed.

Many bird species forage on crustaceans in marine and freshwater systems. While for some bird species, crustaceans may constitute (at least temporarily) the prime food source, they may be just a valuable dietary supplement in other bird species. Often, the diet composition and, thus, the contribution of crustacean prey varies temporally according to seasonal prey availability. Additionally, closely related bird species may prefer different crustacean prey species even when foraging at the same location. Miranda and Collazo (1997) investigated the diet composition of four ardeid wading birds in a mangrove swamp in Puerto Rico. While little blue herons (*Egretta caerulea*) foraged exclusively on crabs of the genus *Uca*, crustacean prey contributed differently to the diet of other species of herons and egrets.

In field experiments, Dumas and Witman (1993) showed strong predation by herring gulls (*Larus argentatus*) on two intertidal crab species, *C. maenas* and *Cancer irroratus*. *Cancer irroratus* was much more intensively preyed upon, potentially because *C. maenas* has a more cryptic coloration of the carapace in intertidal mussel beds. Crabs may not be the preferred food of herring gulls, but at the experimental sites the predation rate for the crabs increased when the predatory gulls were attracted by an alternative prey, the seastar *Asterias* spp. Similarly, the invasive green crab *C. maenas* is merely a food supplement of kelp gulls, *Larus dominicanus*, in Argentine Patagonia, SW Atlantic (Yorio et al. 2020), which preferentially preyed on mussels. However, depending on the sampling site, *C. maenas* was consumed by most individuals of the local gull population, as determined from the contents of regurgitated pellets. On intertidal flats of the Wadden Sea, North Sea, predation by seabirds and other predators shaped the distribution and behavior of juvenile intertidal green crabs. Juvenile *C. maenas* were predominantly found within or in close vicinity of scattered clumps of the blue mussel, *Mytilus edulis*, which likely provide a spatial shelter from epibenthic predation (Thiel and Dernedde 1994). Larger crab individuals were also observed outside the mussel clumps, indicating that growing crabs are progressively safe from predation by epibenthic predators, including seabirds.

Shorebirds may adjust their feeding strategy to the seasonal availability of crustacean prey. For example, the common feeding grounds of dunlin (*Calidris alpina*) in the Wadden Sea at the island of Sylt, Germany, are low intertidal mud flats where the birds forage primarily on polychaetes (Nehls and Tiedemann 1993). However, in late summer and autumn, dunlins have a high preference for feeding on high sandflats where brown shrimp, *Crangon crangon*, and small (carapace width: 5–10 mm) juvenile green crabs, *C. maenas*, abound. This seasonal diet change is also associated with a change in feeding mode. When foraging on polychaetes, dunlins move slowly and probe the sediment deeply. However, when feeding on epibenthic crustaceans, the birds walk quickly over the sandy sediment, displaying frequent shallow pecking for their primarily epibenthic crustacean prey.

Terrestrial Crustaceans as Prey

Diverse arthropod predators (centipedes, beetles, spiders) feed on woodlice, but preferentially on small juveniles (Sunderland and Sutton 1980). However, it has been estimated that this has only minor effects on the population dynamics of woodlice. Similarly, land crabs are prey

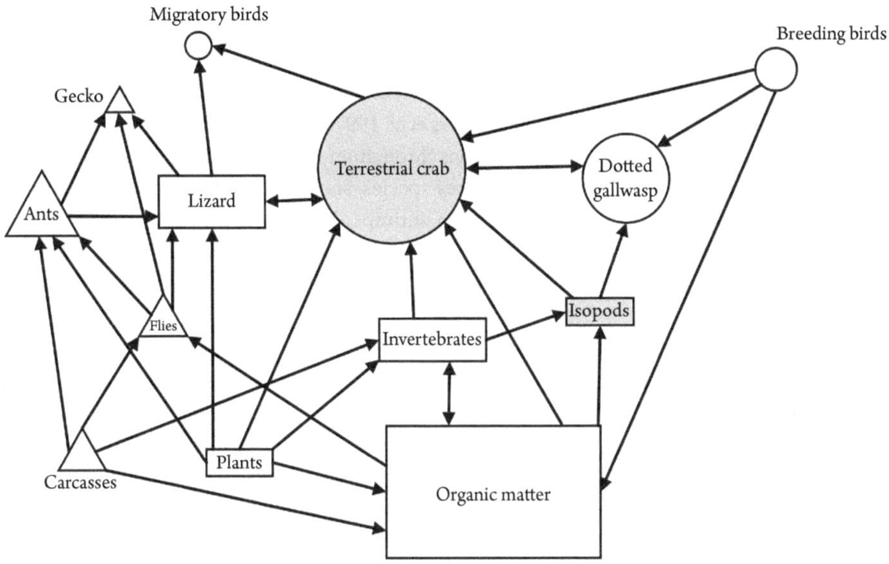

Fig. 1.12.
Simplified depiction of the terrestrial food web on Malpelo Island, an oceanic island off the coast of Colombia, with terrestrial crabs (gray circle) occupying a central position at a higher trophic level of the food web, and isopods (gray box) at an intermediate trophic position. Source: data from Estupiñán-Montaño et al. (2023).

of a wide range of predators, including lizards (López-Victoria et al. 2011). It is mostly small crabs that are susceptible to lizard predation, whereas larger crabs might actually prey on the lizards (López-Victoria et al. 2011). In semi-terrestrial crayfish *Procambarus clarkii,* predation by wading birds and mammals is also primarily on juveniles and subadults, likely because adults are spending more time in their deep burrows (Correia 2001). Cannibalism is very common among terrestrial crustaceans, and in woodlice it was shown to facilitate the transfer of bacterial parasites (Le Clec'h et al. 2013). Partial brood release is viewed as a possible strategy to reduce intra-brood cannibalism in woodlouse females (Telford and Dangerfield 1994). Predation has likely also resulted in the evolution of jumping behavior in terrestrial isopods (Hassall et al. 2006).

Terrestrial crabs, which feed on detritus and carrion, are consumed by migratory birds and are thus at higher trophic levels in terrestrial food webs (Fig. 1.12). Woodlice share an intermediate position with other arthropods and invertebrate decomposers, being preyed upon by lizards, wasps, and the crab *Johngarthia malpilensis*. Limited food quality also enhances the risk of cannibalism in land crabs on oceanic islands, which also primarily affects the survival of juvenile crabs (Wolcott and Wolcott 1984).

CRUSTACEANS AS PREY IN A CHANGING WORLD

Over-exploitation of Consumers and Crustacean Responses

Planktonic crustaceans often rely on pelagic primary production, and their populations are thus often strongly bottom-up controlled by phytoplankton availability and dynamics. Additionally, there is growing evidence for top-down control of both pelagic and benthic crustacean populations in aquatic food webs (Baum and Worm 2009). The controlling effect of a consumer can become evident from demographic responses of a prey when the predator population declines, for example, as a consequence of over-exploitation. Time series of stock developments indicate that predation by demersal fish can effectively control the populations of large benthic crustaceans. For example, landing statistics confirm that stocks of Atlantic cod (*Gadus morhua*) have declined dramatically after decades of intense commercial fishing (Myers et al. 1997). Cod is a voracious omnivorous demersal feeder, and the decline in cod populations throughout the North Atlantic was followed by an increase in numbers of major crustacean prey species, snow crabs, *Chionoecetes opilio*, American lobsters, *Homarus americanus*, and northern shrimps, *Pandalus* spp. (Fig. 1.13; Worm and Myers 2003, Frank et al. 2005). The negative correlation of the numbers of cod and benthic crustaceans suggests a release of the crustacean prey from top-down control after the breakdown of cod populations.

Correspondingly, growth or recovery of a predator population likely intensifies the control of a prey, resulting in strong population declines. For example, recovery of the Eastern North Pacific gray whale population from the 1980s onwards resulted in a drastic decline of the ampeliscid amphipod population, the major benthic prey of gray whales, in the Chirikov Basin of the northern Bering Sea (Coyle et al. 2007). The authors estimated that around the year 2000 only a minor fraction of the regional gray whale population was sufficient to remove a substantial fraction of the ampeliscid population. Recovery of the amphipod population from intense predation by gray whales was affected by slow population cycles in the cold boreal waters.

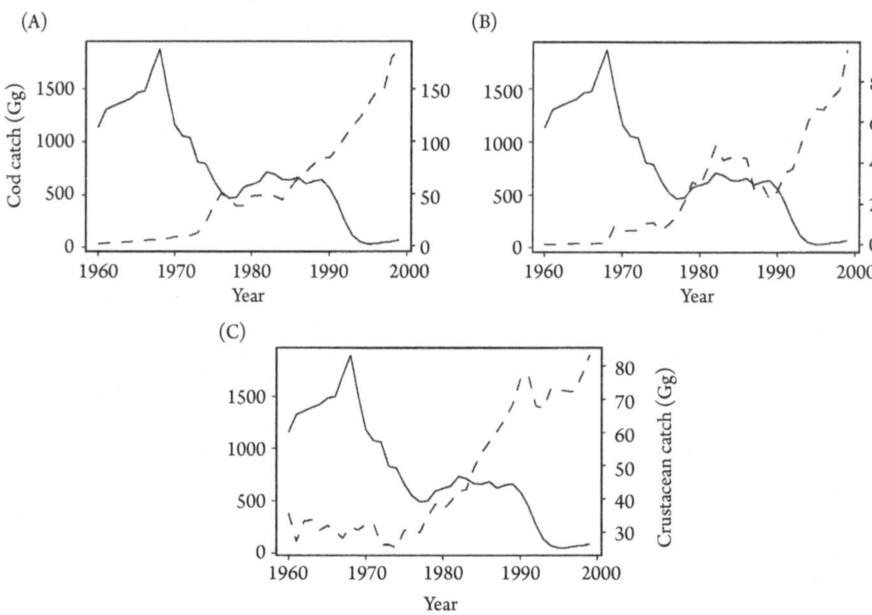

Fig. 1.13.
Opposing population developments of predatory cod (*Gadus morhua*; solid line in each panel) and its crustacean prey species (broken lines): (A) northern shrimp (*Pandalus* spp.), (B) snow crab (*Chionoecetes opilio*), and (C) American lobster (*Homarus americanus*) in the North Atlantic, showing prey population recovery in response to commercial consumer exploitation. Adapted from Worm and Myers (2003), with permission from © John Wiley & Sons.

The control of crustacean populations by consumers can be a direct effect of predator–prey interaction. For example, in management areas that are protected for benthic resources, demersal fish predators reach higher abundances and larger sizes (Gelcich et al. 2008), whereas the densities and sizes of their prey, the rock shrimp *Rhynchocinetes typus*, are lower than in open-access areas where fewer fish predators are found (Ory et al. 2012). Additionally, the effect can be indirect and mediated through trophic cascades involving several species from multiple trophic levels. Trophic cascades can be complex and involve potentially contrasting demographic responses of the consumer and prey species involved, and the controlling effects of predators may strongly depend on environmental conditions and the food availability at lower levels of the trophic cascade. For example, the exploitation of top predators, such as cod and haddock, in the North Sea resulted in changes in the planktonic copepod community, which were mediated by planktivorous fish and modulated by climate-induced phytoplankton dynamics (Lynam et al. 2017).

The over-exploitation of top predators can have long-lasting effects on the structure of benthic ecosystems with perennial vegetation. The removal of piscivorous fish by commercial fisheries enhances predation on crustacean mesograzers, such as amphipods and isopods, by mesopredatory fish species, which are released from predation by top predators. The removal of crustacean mesograzers allows fast-growing ephemeral epiphytes, promoted by eutrophication, to overgrow perennial seaweeds and seagrass resulting in a long-term degradation of benthic phytal habitats (Östman et al. 2016). Given their crucial position in food webs, crustaceans essentially shape the responses of aquatic systems to the removal or over-exploitation of consumers either through direct top-down control of crustacean populations by predators or in trophic cascades with crustaceans at an intermediate trophic position.

Crustacean Consumption and Pollutant Transfer

In their natural environments, crustaceans are exposed to a great variety of pollutants from both chronic exposure and accidental pulse pollution, and consequences for crustacean populations are discussed in detail by Ford and Knigge (see Chapter 11 in this volume). Exposed crustaceans may accumulate contaminants in their tissues (Pinheiro et al. 2012), potentially resulting in the trophic transfer of contaminants to their predators (Soto-Jiménez et al. 2011) and biomagnification along trophic pathways (Porte and Albaigés 1993). According to their crucial positions in food webs and their wide spectrum of predators, crustaceans likely play a key role in the transfer of contaminants through food webs with potential implications for apex predators, seafood safety, and humans (Marques et al. 2010). In many aquatic systems, small crustaceans are at the base of the food chain (e.g., copepods and krill in pelagic food webs), transferring pollutants to upper-level predators (Alava et al. 2017). For example, pelagic copepods and amphipods are central components of Arctic marine food webs with calanoid copepods, such as *Calanus hyperboreus*, functioning as consumers of pelagic primary production and as major prey for planktivorous amphipods, *Themisto libellula*. Within the Arctic marine food web, persistent organochlorine contaminants are biomagnified with crustacean prey as vector for the trophic transfer of contaminants to consumers of higher trophic levels, including seabirds, pinnipeds, and commercially exploited fish species, such as polar cod (Borgå et al. 2004). In terrestrial food webs, too, the trophic transfer of contaminants involves crustaceans at the base of the food web. For example, woodlice process detrital food and transfer assimilated pollutants along the food chain to their predators (Fig. 1.14).

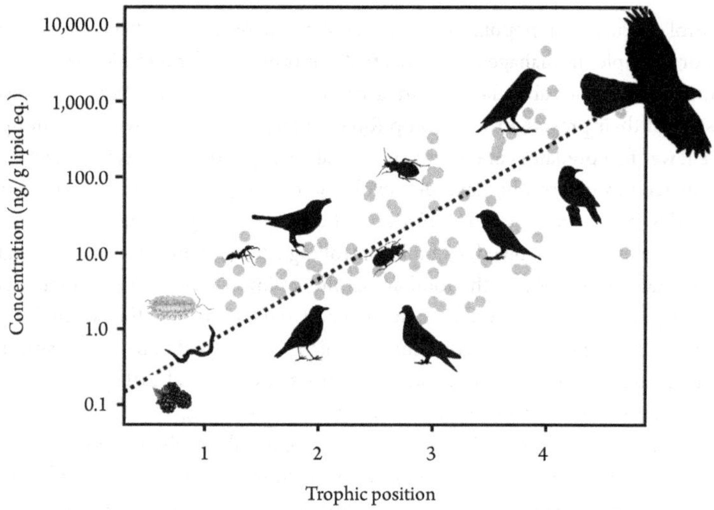

Fig. 1.14.
Biomagnification of pollutants across trophic levels in an urban terrestrial food web in British Columbia, Canada, involving detritivorous woodlice. Source: data from Fremlin et al. (2020).

Invasive Predators Consuming Crustaceans

In many aquatic and terrestrial food webs, invasive species prey extensively on crustaceans. In aquatic systems, invasive crustaceans not only prey on native species, but also compete with them for resources (for specific examples see Chapters 2, 9, and 13 in this volume). In terrestrial systems, other arthropods, or vertebrates, such as rats and cats, are typical invaders that prey on crustaceans (Doherty et al. 2015, Harper and Bunbury 2015). For example, invasive ants have been observed to massively attack land crabs on Christmas Island, causing a rapid demise of the crab populations (O'Dowd et al. 2003). Predation by invasive ants and feral cats on land crabs has also been observed on other islands (Ortega-Rubio et al. 1997, Lee et al. 2021). Following the removal of rats from oceanic islands, populations of land crabs have often rebounded (Samaniego-Herrera and Bedolla-Guzman 2012). On an island in the Gulf of California with two species of land crabs, the larger *Johngarthia planata* apparently did not suffer from rat predation, whereas populations of the smaller *Gecarcinus quadratus* significantly increased after rat extirpation (Samaniego-Herrera and Bedolla-Guzman 2012). Given the important trophic role of these crabs and other terrestrial crustaceans, their release from invasive predators likely has important effects on the food webs of oceanic islands.

Climate Change and Crustaceans as Prey

Consumption rates of carnivorous consumers are expected to increase with rising temperatures to meet elevated metabolic demands (Nagelkerken and Connell 2015), likely increasing predation pressure on crustacean prey. For example, juvenile lionfish *Pterois volitans* consumed significantly more shrimp *Palaemon varians* at elevated temperatures (South et al. 2017). However, there are also contrasting examples as, for example, a blenniid fish consumed fewer amphipods when temperatures exceeded the thermal optimum of the predator (South et al. 2018).

Temperature-induced shifts in predator–prey relationships are not controlled by the thermal response of the predator alone but may also change the behavior and, thus, the exposition of crustacean prey to its predator. For example, predation on *Daphnia pulex* by pelagic invertebrate predators increased with temperature as a consequence of enhanced activity of the ectotherms, which resulted in an elevated encounter rate between predators and prey (Twardochleb et al. 2020). Marangon et al. (2020), in an extensive mesocosm experiment, showed that elevated ocean temperature induces riskier behavior in two shrimp species, *Palaemon intermedius* and *P. serenus*. In warmer waters, the foraging activity and foraging duration of the shrimp increased in both the presence and absence of the predator, the South Australian cobbler *Gymnapistes marmoratus*. The body mass of the prey did not change over the experimental period, indicating that the shrimp intensified food uptake to meet the warming-induced elevated metabolic demands.

Warming waters also lead to range expansions, either via planktonic larvae transported into new areas by ocean currents, and/or because juveniles and adults can colonize new areas. Range expansions have been reported for decapod crustaceans, including species belonging to groups that are common prey of fish predators (Montagne and Cadien 2001). Increasing temperatures in Antarctic shallow waters also cause concern because king crabs might invade the continental shelf. The occurrence of remnants of lithodid crabs in the stomach contents of Antarctic predatory fish, as well as a considerable rate of limb damage, indicates that the emergence of the king crabs on the Antarctic shelf entails a considerable predation risk (Aronson et al. 2015).

The frequency and intensity of storms are predicted to increase in many regions of the world as a consequence of global climate change (Knutson et al. 2010). Storms are a major reason for the dislodgement of kelp in shallow coastal waters (Graham et al. 1997), which likely results in more frequent exposition of small kelp-associated crustaceans to predators (Miranda and Thiel 2008). However, elevated storm activity may also promote turbidity in shallow coastal waters through enhanced particle resuspension. High turbidity affects the detectability of prey for visually hunting fish (De Robertis et al. 2003) and may, thus, decrease the predation risk for crustaceans performing migrations outside their protective habitats.

Climate change also affects crustacean prey populations through shifts in the predator communities. In Chesapeake Bay on the NW Atlantic coast, changing climatic conditions induced a shift in the gelatinous plankton from Scyphomedusae to ctenophores. The more efficiently filtering ctenophores substantially reduced the local population of the copepod *Acartia tonsa* (Purcell and Decker 2005).

Climate-induced changes in metabolic demands, behavior, and community composition of both predators and prey can enhance the predation risk for crustacean prey. However, thermal responses are diverse, and it remains difficult to predict the outcome of predator–prey relationships under changing environmental conditions, especially when different environmental variables interactively affect trophic interactions (Rosenblatt and Schmitz 2014).

OUTLOOK AND CONCLUSIONS

Crustaceans as prey illustrate some of the most fascinating features of food webs on earth, namely a direct link between microscopic microalgae and the largest living organisms on this planet. Additionally, many crustaceans process dead organic matter and reconvert it back into living biomass, fueling extensive food webs. They owe this diversity of functional roles to their ability to process a wide variety of food items, extract nutritious compounds, and make them available in bite-sized food parcels, thereby converting organic material of limited nutritional value into valuable food for a wide range of consumers.

Due to their universal role in food webs from all major ecosystems on land and in water, changes in the taxonomic composition, population, or function of crustacean prey can have far-reaching consequences, and the implications of additions of crustacean species to established food webs, for example, due to species invasions or their loss due to extinctions, are as yet hard to predict and should, thus, be a priority of future research. In this context, it will be essential to understand the biotic and abiotic factors that influence crustacean prey populations to forecast the development of food resources and populations of consumers.

Naturally, humans rarely compete with other predators for crustacean prey. However, following over-exploitation of commercially used fish species, humans increasingly extract considerable amounts of large crustaceans that function as predators in global food webs. Accordingly, by fishing down the marine food webs (sensu Pauly et al. 1998), humans extract essential ecosystem components that crucially mediate bottom-up and top-down regulating effects through trophic cascades, thereby potentially altering the flow of energy and matter. This affects the functioning of ecosystems with implications across all trophic levels, including humans themselves as the ultimate consumers.

Global ecosystems are currently undergoing dramatic changes that also impact crustaceans. In many marine systems, due to the wide diversity of differently adapted crustaceans, there might be a high potential of redundancy, that is, if one species goes missing, others might be able to replace its function. However, in freshwater and terrestrial ecosystems, there are often few crustacean prey species, which furthermore might be highly specialized and endemic to small habitats, thereby reducing the potential for functional redundancy. A high degree of endemism in combination with multiple anthropogenic stressors results in a considerable risk of extinction of crustacean species, especially in freshwater systems, indicative of a developing biodiversity crisis (Cumberlidge et al. 2009) likely accompanied by the loss of crucial ecological functions. Accordingly, understanding the biology and ecology of crustaceans and their role as prey in both terrestrial and aquatic food webs is key for the development of appropriate conservation measures to avoid the irrecoverable loss of sensitive crucial components of various ecosystems.

ACKNOWLEDGMENTS

We are very grateful to Taizhu Zhou for excellent preparation of the figures for this contribution.

REFERENCES

Alava, J.J., W.W.L Cheung, P.S. Ross, and U.R. Sumaila. 2017. Climate change-contaminant interactions in marine food webs: Toward a conceptual framework. Global Change Biology 23:3984–4001.

Alldredge, A.L. 1984. The quantitative significance of gelatinous zooplankton as pelagic consumers. Pages 407–433 *in* M.J.R. Fasham, editor. Flows of energy and matter in marine ecosystems: Theory and practice. Plenum Press, New York, USA.

Angel, A., and F.P. Ojeda. 2001. Structure and trophic organization of subtidal fish assemblages on the northern Chilean coast: The effect of habitat complexity. Marine Ecology Progress Series 217:81–91.

Aronson, R.B., K.E. Smith, S.C. Vos, J.B. McClintock, M.O. Amsler, P.O. Moksnes, D.S. Ellis, J. Kaeli, H. Singh, J.W. Bailey, J.C. Schiferl, R. van Woesik, M.A. Martin, B.V. Steffel, M.E. Deal, S.M. Lazarus, J.N. Havenhand, R. Swalethorp, S. Kjellerup, and S. Thatje. 2015. No barrier to emergence of bathyal king crabs on the Antarctic shelf. Proceedings of the National Academy of Sciences 112:12997–13002.

Arrontes, J. 1999. On the evolution of interactions between marine mesoherbivores and algae. Botanica Marina 42:137–155.

Atkinson, A., P. Ward, B.P.V. Hunt, E.A. Pakhomov, and G.W. Hosie. 2012. An overview of Southern Ocean zooplankton data: Abundance, biomass and functional relationships. CCAMLR Science 19:171–218.

Atkinson, M.J., and S.V. Smith. 1983. C:N:P ratios of benthic marine plants. Limnology and Oceanography 28:568–574.

Bandara, K., Ø. Varpe, L. Wijewardene, V. Tverberg, and K. Eiane. 2021. Two hundred years of zooplankton vertical migration research. Biological Reviews 96:1547–1589.

Barrento, S., A. Marques, B. Teixeira, R. Mendes, N. Bandarra, P. Vaz-Pires, and M.L. Nunes. 2010. Chemical composition, cholesterol, fatty acid and amino acid in two populations of brown crab *Cancer pagurus*: Ecological and human health implications. Journal of Food Composition and Analysis 23:716–725.

Barshaw, D.E., K.L. Lavalli, and E. Spanier. 2003. Offense versus defense: Responses of three morphological types of lobsters to predation. Marine Ecology Progress Series 256:171–182.

Bashevkin, S.M., and S.G. Morgan. 2020. Predation and competition. Pages 360–382 *in* K. Anger, S. Harzsch, and M. Thiel, editors. The natural history of the Crustacea, Volume 7: Developmental biology and larval ecology. Oxford University Press, New York, USA.

Baum, J.K., and B. Worm. 2009. Cascading top-down effects of changing oceanic predator abundances. The Journal of Animal Ecology 78:699–714.

Beermann, J., K. Boos, L. Gutow, M. Boersma, and A.C. Peralta. 2018. Combined effects of predator cues and competition define habitat choice and food consumption of amphipod mesograzers. Oecologia 186:645–654.

Bell, J.D., and M.L. Harmelin-Vivien. 1983. Fish fauna of French Mediterranean *Posidonia oceanica* seagrass meadows. II: Feeding habits. Tethys 11:1–14.

Bertolin, M.L., and R. Casaux. 2019. Diet overlap among top predators at the South Orkney Islands, Antarctica. Polar Biology 42:371–383.

Black, A.R. 1993. Predator-induced phenotypic plasticity in *Daphnia pulex*: Life history and morphological responses to *Notonecta* and *Chaoborus*. Limnology and Oceanography 38:986–996.

Bleakley, B. 2018. Cannibalism in crustaceans. Pages 347–373 *in* G.A. Wellborn and M. Thiel, editors. The natural history of the Crustacea, Volume 5: Life histories. Oxford University Press, New York, USA.

Bodkin, J.L., G.G. Esslinger, and D.H. Monson. 2004. Foraging depths of sea otters and implications to coastal marine communities. Marine Mammal Science 20:305–321.

Borgå, K., A.T. Fisk, P.F. Hoekstra, and D.C.G. Muir. 2004. Biological and chemical factors of importance in the bioaccumulation and trophic transfer of persistent organochlorine contaminants in Arctic marine food webs. Environmental Toxicology and Chemistry 23:2367–2385.

Boudreau, S.A., and B. Worm. 2012. Ecological role of large benthic decapods in marine ecosystems: A review. Marine Ecology Progress Series 469:195–213.

Boyd, I.L. 2002. Estimating food consumption of marine predators: Antarctic fur seals and macaroni penguins. Journal of Applied Ecology 39:103–119.

Butler, M.J. IV, and J.A. Lear. 2009. Habitat-based intraguild predation by Carribean reef *Octopus briareus* on juvenile Carribean spiny lobster *Panulirus argus*. Marine Ecology Progress Series 386:115–122.

Cacabelos, E., C. Olabarria, M. Incera, and J.S. Troncoso. 2010. Effects of habitat structure and tidal height on epifaunal assemblages associated with macroalgae. Estuarine, Coastal and Shelf Science 89:43–52.

Calkins, D.G. 1978. Feeding behavior and major prey species of the sea otter, *Enhydra lutris*, in Montague Strait, Prince William Sound, Alaska. United States National Marine Fisheries Service Fishery Bulletin 76:125–131.

Carbines, G.D., R.M. Dennis, and R.R. Jackson. 1992. Increased turn alternation by woodlice (*Porcellio scaber*) in response to a predatory spider, *Dysdera crocata*. International Journal of Comparative Psychology 5:138–144.

Chess, J.R. 1993. Effects of the stipe-boring amphipod *Peramphithoe stypotrupetes* (Corophioidea: Ampithoidae) and grazing gastropods on the kelp *Laminaria setchellii*. Journal of Crustacean Biology 13:638–646.

Clarke, A. 1980. The biochemical composition of krill, *Euphausia superba* Dana, from South Georgia. Journal of Experimental Marine Biology and Ecology 43:221–236.

Connell, J.H. 1975. Some mechanisms producing structure in natural communities: A model and evidence from field experiments. Pages 460–367 *in* M.L. Cody and J.M. Diamond, editors. Ecology and evolution of communities. Belknap Press of Harvard University, Cambridge, MA, USA.

Correia, A.M. 2001. Seasonal and interspecific evaluation of predation by mammals and birds on the introduced red swamp crayfish *Procambarus clarkii* (Crustacea, Cambaridae) in a freshwater marsh (Portugal). Journal of Zoology 255:533–541.

Coulson, P.G., M.E. Platell, K.R. Clarke, and I.C. Potter. 2015. Dietary variations within a family of ambush predators (Platycephalidae) occupying different habitats and environments in the same geographical region. Journal of Fish Biology 86:1046–1077.

Coyle, K.O., B. Bluhm, B. Konar, A. Blanchard, and R.C. Highsmith. 2007. Amphipod prey of gray whales in the northern Bering Sea: Comparison of biomass and distribution between the 1980s and 2002–2003. Deep-Sea Research II 54:2906–2918.

Croxall, J.P., P.A. Prince, C. Ricketts. 1985. Relationships between prey life cycles and the extent, nature and timing of seal and seabird predation in the Scotia Sea. Pages 516–533 *in* W.R. Siegfried, P.R. Condy, and R.M. Laws, editors. Antarctic nutrient cycles and food webs. Springer, Berlin, Germany.

Cumberlidge, N., P.K.L. Ng, D.C.J. Yeo, C. Magalhães, M.R. Campos, F. Alvarez, T. Naruse, S.R. Daniels, L.J. Esser, F.Y.K. Attipoe, F.-L. Clotilde-Ba, W. Darwall, A. McIvor, J.E.M. Baillie, B. Collen, and M. Ram. 2009. Freshwater crabs and the biodiversity crisis: Importance, threats, status, and conservation challenges. Biological Conservation 142:1665–1673.

Dawidowicz, P., and J. Pijanowska. 2018. Diel vertical migration of aquatic crustaceans—Adaptive role, underlying mechanisms, and ecosystem consequences. Pages 232–257 G.A. Wellborn and M. Thiel, editors. The natural history of the Crustacea, Volume 5: Life histories. Oxford University Press, New York, USA.

DeRivera, C.E., G.M. Ruiz, A.H. Hines, and P. Jivoff. 2005. Biotic resistance to invasion: Native predator limits abundance and distribution of an introduced crab. Ecology 86:3364–3376.

Dhurmeea, Z., H. Pethybridge, E.V. Romanov, C. Appadoo, and N. Bodin. 2020. Diet of albacore *Thunnus alalunga* from the waters of Mauritius (western Indian Ocean) inferred from stomach contents and fatty acid analysis. African Journal of Marine Science 42:131–140.

De Robertis, A., C.H. Ryer, A. Veloza, and R.D. Brodeur. 2003. Differential effects of turbidity on prey consumption of piscivorous and planktivorous fish. Canadian Journal of Fisheries and Aquatic Sciences 60:1517–1526.

Dick, J.T.A. 1992. The nature and implications of differential predation between *Gammarus pulex* and *G. duebeni celticus*. Journal of Zoology 227:171–183.

Dittel, A.I., A.H. Hines, G.M. Ruiz, and K.K. Ruffin. 1995. Effects of shallow water refuge on behavior and density-dependent mortality of juvenile blue crabs in Chesapeake Bay. Bulletin of Marine Science 57:902–916.

Doherty, T.S., R.A. Davis, E.J. van Etten, D. Algar, N. Collier, C.R. Dickman, G. Edwards, P. Masters, R. Palmer, and S. Robinson. 2015. A continental-scale analysis of feral cat diet in Australia. Journal of Biogeography 42:964–975.

Duffy, J.E. 2006. Biodiversity and the functioning of seagrass ecosystems. Marine Ecology Progress Series 311:233–250.

Duffy, J.E., and M.E. Hay. 2000. Strong impacts of grazing amphipods on the organization of a benthic community. Ecological Monographs 70:237–263.

Dumas, J.V., and J.D. Witman. 1993. Predation by herring gulls (*Larus argentatus* Coues) on two rocky intertidal crab species [*Carcinus maenas* (L.) & *Cancer irroratus* Say]. Journal of Experimental Marine Biology and Ecology 169:89–101.

Dunham, J.S., and D.A. Duffus. 2002. Diet of gray whales (*Eschrichtius robustus*) in Clayoquot Sound, British Columbia, Canada. Marine Mammal Science 18:419–437.

Edgar, G.J., and C. Shaw. 1995. The production and trophic ecology of shallow-water fish assemblages in southern Australia III. General relationships between sediments, seagrasses, invertebrates and fishes. Journal of Experimental Marine Biology and Ecology 194:107–131.

Estupiñán-Montaño, C., F.R. Elorriaga-Verplancken, F. Galván-Magaña, A. Sánchez-González, C.J. Polo-Silva, and M.J. Zetina-Rejón. 2023. Terrestrial food web of the Malpelo Fauna and Flora Sanctuary, Colombia: An analysis from a topological approach. Ecological Informatics 75:102074.

Evans-White, M.A., R.S. Stelzer, G.A. Lamberti. 2005. Taxonomic and regional patterns in benthic macroinvertebrate elemental composition in streams. Freshwater Biology 50:1786–1799.

Färber-Lorda, J., R. Gaudy, and P. Mayzaud. 2009. Elemental composition, biochemical composition and caloric value of Antarctic krill: Implications in energetics and carbon balances. Journal of Marine Systems 78:518–524.

Frank, K.T., B. Petrie, J.S. Choi, and W.C. Leggett. 2005. Trophic cascades in a formerly cod-dominated ecosystem. Science 308:1621–1623.

Franke, H.-D., and M. Janke. 1998. Mechanisms and consequences of intra- and interspecific interference competition in *Idotea baltica* (Pallas) and *Idotea emarginata* (Fabricius) (Crustacea: Isopods): A laboratory study of possible proximate causes of habitat segregation. Journal of Experimental Marine Biology and Ecology 227:1–21.

Fremlin, K.M., J.E. Elliott, D.J. Green, K.G. Drouillard, T. Harner, A. Eng, and F.A. Gobas. 2020. Trophic magnification of legacy persistent organic pollutants in an urban terrestrial food web. Science of the Total Environment 714:136746.

Gelcich, S., N. Godoy, L. Prado, and J.C. Castilla. 2008. Add-on conservation benefits of marine territorial user rights fishery policies in central Chile. Ecological Applications 18:273–281.

Glazier, D.S., J.J. Borrelli, and C.L. Hoffman. 2020. Effects of fish predators on the mass-related energetics of a keystone freshwater crustacean. Biology 9:40.

Gorvett, H. 1956. Tegumental glands and terrestrial life in woodlice. Proceedings of the Zoological Society of London 126:291–314.

Graham, M.H., C. Harrold, S. Lisin, K. Light, J.M. Watanabe, and M.S. Foster. 1997. Population dynamics of giant kelp *Macrocystis pyrifera* along a wave exposure gradient. Marine Ecology Progress Series 148:269–279.

Griffen, B.D., and J.E. Byers. 2006. Intraguild predation reduces redundancy of predator species in multiple predator assemblages. Journal of Animal Ecology 75:959–966.

Guevara, M., C. Lodeiros, M.D. Donato, C. Graciani, and L. Freites. 2005. Nutritional quality of *Metamysidopsis insularis* Brattegard (Crustacea: Mysidacea). Aquaculture Nutrition 11: 315–319.

Gutow, L., J.D. Long, O. Cerda, I.A. Hinojosa, E. Rothäusler, F. Tala, and M. Thiel. 2012. Herbivorous amphipods inhabit protective microhabitats within thalli of giant kelp *Macrocystis pyrifera*. Marine Biology 159:141–149.

Hamilton, D.J., M.A. Barbeau, and A.W. Diamond. 2003. Shorebirds, mud snails, and *Corophium volutator* in the upper Bay of Fundy, Canada: Predicting bird activity on intertidal mud flats. Canadian Journal of Zoology 81:1358–1366.

Hamilton, D.J., A.W. Diamond, and P.G. Wells. 2006. Shorebirds, snails, and the amphipod (*Corophium volutator*) in the upper Bay of Fundy: Top-down vs. bottom-up factors, and the influence of compensatory interactions on mudflat ecology. Hydrobiologia 567:285–306.

Hanson, J.M., M. Comeau, and A. Rondeau. 2014. Atlantic rock crab, unlike American lobster, is important to ecosystem functioning in Northumberland Strait. Transactions of the American Fisheries Society 143:1266–1279.

Hanson, J.M., and M. Lanteigne. 2000. Evaluation of Atlantic cod predation on American lobster in the southern Gulf of St. Lawrence, with comments on other potential fish predators. Transactions of the American Fisheries Society 129:13–29.

Harper, G.A. and N. Bunbury. 2015. Invasive rats on tropical islands: Their population biology and impacts on native species. Global Ecology and Conservation 3:607–627.

Harrison, C.S. 1979. The association of marine birds and feeding gray whales. Condor 81:93–95.

Hassall, M., D.T. Jones, S. Taiti, Z. Latipi, S.L. Sutton, and M. Mohammed. 2006. Biodiversity and abundance of terrestrial isopods along a gradient of disturbance in Sabah, East Malaysia. European Journal of Soil Biology 42:S197–S207.

Havermans, C., H. Auel, W. Hagen, C. Held, N.S. Ensor, and G.A. Tarling. 2019. Predatory zooplankton on the move: *Themisto* amphipods in high-latitude marine pelagic food webs. Advances in Marine Biology 82:51–92.

Havermans, C., and V. Smetacek. 2018. Bottom-up and top-down triggers of diversification: A new look at the evolutionary ecology of scavenging amphipods in the deep sea. Progress in Oceanography 164:37–51.

Hay, M.E. 1997. The ecology and evolution of seaweed-herbivore interactions on coral reefs. Coral Reefs 16:S67–S76.

Hay, M.E., J.E. Duffy, and W. Fenical. 1990. Host-plant specialization decreases predation on a marine amphipod: An herbivore in plant's clothing. Ecology 71:733–743.

Hay, M.E., and W. Fenical. 1988. Marine plant-herbivore interactions: The ecology of chemical defense. Annual Reviews in Ecology and Systematics 19:111–145.

Hemminga, M.A., and C.M. Duarte. 2000. Seagrass ecology. Cambridge University Press, New York.

Hicklin, P.W. 1987. The migration of shorebirds in the Bay of Fundy. Wilson Bulletin 99:540–570.

Hildebrand, L., K.S. Bernard, and L.G. Torres. 2021. Do gray whales count calories? Comparing energetic values of gray whale prey across two different feeding grounds in the eastern North Pacific. Frontiers in Marine Science 8:683634

Howard, R.K. 1984. The trophic ecology of caridean shrimps in an eelgrass community. Aquatic Botany 18:155–174.

Hughes, A.R., K.J. Bando, L.F. Rodriguez, and S.L. Williams. 2004. Relative effects of grazers and nutrients on seagrasses: A meta-analysis approach. Marine Ecology Progress Series 282:87–99.

Jackson, S, A.R. Place, and L.J. Seiderer. 1992. Chitin digestion and assimilation by seabirds. The Auk 109:758–770.

Jeffs, A., and R. O'Rorke. 2020. Feeding and nutrition of crustacean larvae. Pages 309–331 *in* K. Anger, S. Harzsch, and M. Thiel, editors. The natural history of the Crustacea, Volume 7: Developmental biology and larval ecology. Oxford University Press, New York, USA.

Jepsen, P.M., K. Syberg, G. Drillet, and B.W. Hansen. 2020. Planktonic crustacean culture—Live planktonic crustaceans as live feed for finfish and shrimps in aquaculture. Pages 341–366 *in* G. Lovrich and M. Thiel, editors. The natural history of the Crustacea, Volume 9: Fisheries and aquaculture. Oxford University Press, New York, USA.

Jormalainen, V. 1998. Precopulatory mate guarding in crustaceans: Male competitive strategy and intersexual conflict. The Quarterly Review of Biology 73:275–304.

Jormalainen, V., and S.M. Shuster. 1997. Microhabitat segregation and cannibalism in an endangered freshwater isopod, *Thermosphaeroma thermophilum*. Oecologia 111:271–279.

Jurcak, A.M., and P.A. Moore. 2018. Sensory signals and the reaction space in predator–prey interactions. Hydrobiologia 816:137–152.

Kainz, M., M.T. Arts, and A. Mazumder. 2004. Essential fatty acids in the planktonic food web and their ecological role for higher trophic levels. Limnology and Oceanography 49:1784–1793.

Kerfoot, W.C. 1978. Combat between predatory copepods and their prey: *Cyclops*, *Epischura*, and *Bosmina*. Limnology and Oceanography 23:1089–1102.

Klumpp, D.W., R.T. Howard, and D.A. Pollard. 1989. Trophodynamics and nutritional ecology of seagrass communities. Pages 394–457 *in* A.W.D. Larkum, A.J. McComb, and S.A. Shepherd, editors. Biology of seagrasses: A treatise on the biology of seagrasses with special reference to the Australian region. Elsevier, Amsterdam, The Netherlands.

Knutson, T.R., J.L. McBride, J. Chan, K. Emanuel, G. Holland, C. Landsea, I. Held, J.P. Kossin, A.K. Srivastava, and M. Sugi. 2010. Tropical cyclones and climate change. Nature Geoscience 3: 157–163.

Koenigs, C., R.J. Miller, and H.M. Page. 2015. Top predators rely on carbon derived from giant kelp *Macrocystis pyrifera*. Marine Ecology Progress Series 537:1–8.

Kramer, M.J., O. Bellwood, and D.R. Bellwood. 2013. The trophic importance of algal turfs for coral reef fishes: The crustacean link. Coral Reefs 32:575–583.

Kramer, M.J., O. Bellwood, C.J. Fulton, and D.R. Bellwood. 2015. Refining the invertivore: Diversity and specialisation in fish predation on coral reef crustaceans. Marine Biology 162:1779–1786.

Krieger, J., M.M. Drew, B.S. Hansson, and S. Harzsch. 2016. Notes on the foraging strategies of the giant robber crab *Birgus latro* (Anomala) on Christmas Island: Evidence for active predation on red crabs *Gecarcoidea natalis* (Brachyura). Zoological Studies 55:6.

Lancaster, J., and A.L. Robertson. 1995. Microcrustacean prey and macroinvertebrate predators in a stream food web. Freshwater Biology 34:123–134.

Laws, R.M. 1977. Seals and whales of the Southern Ocean. Philosophical Transactions of the Royal Society Series B 279:81–96.

Layman, C.A., and J.E. Allgeier. 2012. Characterizing trophic ecology of generalist consumers: A case study of the invasive lionfish in The Bahamas. Marine Ecology Progress Series 448: 131–141.

Le Clec'h, W., F.D. Chevalier, L. Genty, J. Bertaux, D. Bouchon, and M. Sicard. 2013. Cannibalism and predation as paths for horizontal passage of *Wolbachia* between terrestrial isopods. PloS ONE 8:e60232.

Lee, C.C., M.C. Chiu, C.H. Shih, C.C.S. Yang, H.C. Liu, and C.C. Lin. 2021. The role of anthropogenic disturbance and invasion of yellow crazy ant in a recent decline of land crab population. Scientific Reports 11:12234.

Lindsay, G.J.H., and G.W. Gooday. 1985. Chitinolytic enzymes and the bacterial microflora in the digestive tract of cod, *Gadus morhua*. Journal of Fish Biology 26:255–265.

Linton, S.M., and P. Greenaway. 2007. A review of feeding and nutrition of herbivorous land crabs: Adaptations to low quality plant diets. Journal of Comparative Physiology B 177:269–286.

Lipcius, R., and W. Herrnkind. 1982. Molt cycle alterations in behavior, feeding and diel rhythms of a decapod crustacean, the spiny lobster *Panulirus argus*. Marine Biology 68:241–252.

Lively, C.M. 1986. Predator-induced shell dimorphism in the acorn barnacle *Chthamalus anisopoma*. Evolution 40:232–242.

López-Victoria, M., P.A. Herrón, and J.C. Botello. 2011. Notes on the ecology of the lizards from Malpelo Island, Colombia. Boletín de Investigaciones Marinas y Costeras-INVEMAR 40:79–89.

Luckenbach, M.W., and R.J. Orth. 1990. A chemical defense in Crustacea? Journal of Experimental Marine Biology and Ecology 137:79–87.

Lynam, C.P., M. Llope, C. Möllmann, P. Helaouët, G.A. Bayliss-Brown, and N.C. Stenseth. 2017. Interaction between top-down and bottom-up control in marine food webs. Proceedings of the National Academy of Science 114:1952–1957.

Mackintosh, N.A. 1974. Sizes of krill eaten by whales in the Antarctic. Discovery Reports 36:157–178.

MacNeil, C., J.T. Dick, and R.W. Elwood. 1997. The trophic ecology of freshwater *Gammarus* spp. (Crustacea: Amphipoda): Problems and perspectives concerning the functional feeding group concept. Biological Reviews of the Cambridge Philosophical Society 72:349–364.

MacNeil, C., N.J. Fielding, J.T.A. Dick, M. Briffa, J. Prenter, M.J. Hatcher, and A.M. Dunn. 2003. An acanthocephalan parasite mediates intraguild predation between invasive and native freshwater amphipods (Crustacea). Freshwater Biology 48:2085–2093.

Malzahn, A.M., N. Aberle, C. Clemmesen, and M. Boersma. 2007. Nutrient limitation of primary producers affects planktivorous fish condition. Limnology and Oceanography 52:2062–2071.

Marangon, E., S.U. Goldenberg, and I. Nagelkerken. 2020. Ocean warming increases availability of crustacean prey via riskier behavior. Behavioral Ecology 31:287–291.

Marques, A., M.L. Nunes, S.K. Moore, and M.S. Strom. 2010. Climate change and seafood safety: Human health implications. Food Research International 43:1766–1779.

Marshall, S., K. Warburton, B. Paterson, and D. Mann. 2005. Cannibalism in juvenile blue-swimmer crabs *Portunus pelagicus* (Linnaeus, 1766): Effects of body size, moult stage and refuge availability. Applied Animal Behaviour Science 90:65–82.

Mauchline, J. 1980. The biology of mysids and euphausiids. Advances in Marine Biology 18:1–677.

Mauchline, J., and L.R. Fisher. 1969. The biology of euphausiids. Advances in Marine Biology 7:1–454.

McCormick, M.I. 1998. Ontogeny of diet shifts by a microcarnivorous fish, *Cheilodactylus spectabilis*: Relationship between feeding mechanics, microhabitat selection and growth. Marine Biology 132:9–20.

McCurdy, D.G., M.R. Forbes, S.P. Logan, D. Lancaster, and S.I. Mautner, 2005. Foraging and impacts by benthic fish on the intertidal amphipod *Corophium volutator*. Journal of Crustacean Biology 25:558–564.

Méndez Casariego, A., J. Alberti, T. Luppi, and O. Iribarne. 2009. Stage-dependent interactions between intertidal crabs: From facilitation to predation. Journal of the Marine Biological Association of the United Kingdom 89:781–788.

Miller, E.J., J.M. Potts, M.J. Cox, B.S. Miller, S. Calderan, R. Leaper, P.A. Olson, R.L. O'Driscoll, and M.C. Double. 2019. The characteristics of krill swarms in relation to aggregating Antarctic blue whales. Scientific Reports 9:16487.

Miranda, L., and J.A. Collazo. 1997. Food habits of 4 species of wading birds (Ardeidae) in a tropical mangrove swamp. Colonial Waterbirds 20:413–418.

Miranda, L., and M. Thiel. 2008. Active and passive migration in boring isopods *Limnoria* spp. (Crustacea, Peracarida) from kelp holdfasts. Journal of Sea Research 60:176–183.

Moksnes, P.-O. 2004. Self-regulating mechanisms in cannibalistic populations of juvenile shore crabs *Carcinus maenas*. Ecology 85:1343–1354.

Montagne, D.E., and D.B. Cadien. 2001. Northern range extensions into the southern California Bight of ten decapod Crustacea related to the 1991/92 and 1997/98 El Niño events. Bulletin of the Southern California Academy of Sciences 100:199–212.

Myers, J.P., S.L. Williams, and F.A. Pitelka. 1980. An experimental analysis of prey availability for sanderlings (Aves: Scolopacidae) feeding on sandy beach crustaceans. Canadian Journal of Zoology 58:1564–1574.

Myers, R.A., J.A. Hutchings, and N.J. Barrowman. 1997. Why do fish stocks collapse? The example of cod in Atlantic Canada. Ecological Applications 7:91–106.

Nagelkerken, I. and S.D. Connell. 2015. Global alteration of ocean ecosystem functioning due to increasing human CO_2 emissions. Proceedings of the National Academy of Sciences 112:13272–13277.

Nehls, G., and R. Tiedemann. 1993. What determines the densities of feeding birds on tidal flats? A case study on dunlin, *Calidris alpina*, in the Wadden Sea. Netherlands Journal of Sea Research 31:375–384.

Nelson, W.G. 1979. Experimental studies of selective predation on amphipods: Consequences for amphipod distribution and abundance. Journal of Experimental Marine Biology and Ecology 38:225–245.

Norderhaug, K.M., S. Fredriksen, and K. Nygaard. 2003. Trophic importance of *Laminaria hyperborea* to kelp forest consumers and the importance of bacterial degradation to food quality. Marine Ecology Progress Series 255:135–144.

Norderhaug, K.M., H. Christie, J.H. Fosså, and S. Fredriksen. 2005. Fish-macrofauna interactions in a kelp (*Laminaria hyperborea*) forest. Journal of the Marine Biological Association of the United Kingdom 85:1279–1286.

Nyssen, F., T. Brey, G. Lepoint, J.-M. Bouquegneau, C. De Broyer, and P. Dauby. 2002. A stable isotope approach to the eastern Weddell Sea trophic web: Focus on benthic amphipods. Polar Biology 25:280–287.

O'Brien, D.P. 1987. Direct observations of the behavior of *Euphausia superba* and *Euphausia crystallorophias* (Crustacea: Euphausiacea) under pack ice during the Antarctic spring of 1985. Journal of Crustacean Biology 7:437–448.

Ockelmann, K.W., and O. Vahl. 1970. On the biology of the polychaete *Glycera alba*, especially its burrowing and feeding. Ophelia 8:275–294.

O'Dowd, D.J., P.T. Green, and P.S. Lake. 2003. Invasional 'meltdown' on an oceanic island. Ecology Letters 6:812–817.

Ojeda, P.F., and J.H. Dearborn. 1991. Feeding ecology of benthic mobile predators: Experimental analyses of their influence in rocky subtidal communities of the Gulf of Maine. Journal of Experimental Marine Biology and Ecology 149:13–44.

Oliver, J.S., and R.G. Kvitek. 1984. Side-scan sonar records and diver observations of the gray whale (*Eschrichtius robustus*) feeding grounds. Biological Bulletin 167:264–269.

Oliver, J.S., and P.N. Slattery. 1985. Destruction and opportunity on the sea floor: Effects of gray whale feeding. Ecology 66:1965–1975.

Oliver, J.S., P.N. Slattery, M.A. Silberstein, and E.F. O'Connor. 1984. Gray whale feeding on dense ampeliscid amphipod communities near Bamfield, British Columbia. Canadian Journal of Zoology 62:41–49.

Oonincx, D.G.A.B., and E.S. Dierenfeld. 2011. An investigation into the chemical composition of alternative invertebrate prey. Zoo Biology 29:1–15.

Øresland, V. 1990. Feeding and predation impact of the chaetognath *Eukrohnia hamata* in Gerlache Strait, Antarctic Peninsula. Marine Ecology Progress Series 63:201–209.

Ortega-Rubio, A., M.L. Jímenez, J. Llinas, and G. Arnaud. 1997. Some ecological aspects of the land crab, *Gecarcinus planatus* Stimpson, at Socorro Island, Colima, Mexico. Journal of the Arizona-Nevada Academy of Science 30:17–22.

Ory, N.C., D. Dudgeon, C.P. Dumont, L. Miranda, and M. Thiel. 2012. Effects of predation and habitat structure on the abundance and population structure of the rock shrimp *Rhynchocinetes typus* (Caridea) on temperate rocky reefs. Marine Biology 159:2075–2089.

Östman, Ö., J. Eklöf, B.K. Eriksson, J. Olsson, P.-O. Moksnes, and U. Bergström. 2016. Top-down control as important as nutrient enrichment for eutrophication effects in North Atlantic coastal ecosystems. Journal of Applied Ecology 53:1138–1147.

Pakhomov, E.A., and R. Perissinotto. 1996. Trophodynamics of the hyperiid amphipod *Themisto gaudichaudii* in the South Georgia region during late austral summer. Marine Ecology Progress Series 134:91–100.

Pakhomov, E.A., R. Perissinotto, and C.D. McQuaid. 1996. Prey composition and daily rations of myctophid fishes in the Southern Ocean. Marine Ecology Progress Series 134:1–14.

Palma, A.T., and R.S. Steneck. 2001. Does variable coloration in juvenile marine crabs reduce risk of visual predation? Ecology 82:2961–2967.

Pangle, K.L., and S.D. Peacor. 2006. Non-lethal effect of the invasive predator *Bythotrephes longimanus* on *Daphnia mendotae*. Freshwater Biology 51:1070–1078.

Patel, R., S. Roy, E. Capuzzo, and J. van der Kooij. 2023. Seasonality of diet overlap among small pelagic fish in the waters southwest of the UK. Marine Ecology Progress Series 708:101–123.

Pauly, D., V. Christensen, J. Dalsgaard, R. Froese, and F. Torres Jr. 1998. Fishing down marine food webs. Science 279:860–863.

Permitin, Y.E. 1970. The consumption of krill by Antarctic fishes. Pages 177–182 *in* M.W. Holdgate, editor. Antarctic ecology, Volume 1. Academic Press, London, UK, and New York, USA.

Pérez-Matus, A., S. Pledger, F.J. Díaz, L.A. Ferry, and J.A. Vásquez. 2012. Plasticity in feeding selectivity and trophic structure of kelp forest associated fishes from northern Chile. Revista Chilena de Historia Natural 85:29–48.

Pinheiro, M.A.A., P.D.G.E. Silva, L.F. de Almeida Duarte, A.A. Almeida, and F. Pinheiro Zanotto. 2012. Accumulation of six metals in the mangrove crab *Ucides cordatus* (Crustacea: Ucididae) and its food source, the red mangrove *Rhizophora mangle* (Angiosperma: Rhizophoracea). Ecotoxicology and Environmental Safety 81:114–121.

Pinn, E.H., and A.D. Ansell. 1993. The effect of particle size on the burying ability of the brown shrimp *Crangon crangon*. Journal of the Marine Biological Association of the United Kingdom 73:365–377.

Polis, G.A. 1981. The evolution and dynamics of intraspecific predation. Annual Review in Ecology and Systematics 12:225–251.

Polis, G.A., C.A. Myers, and R.D. Holt. 1989. The ecology and evolution of intraguild predation: Potential competitors that eat each other. Annual Review of Ecology and Systematics 20:297–330.

Poore, A.G.B., L. Gutow, J.F. Pantoja, F. Tala, D.J. Madariaga, and M. Thiel. 2014. Major consequences of minor damage: Impacts of small grazers on fast-growing kelps. Oecologia 174:789–801.

Porte, C., and J. Albaigés. 1993. Bioaccumulation patterns of hydrocarbons and polychlorinated biphenyls in bivalves, crustaceans, and fishes. Archives of Environmental Contamination and Toxicology 26:272–281.

Potapov, A.M., U. Brose, S. Scheu, and A.V. Tiunov. 2019. Trophic position of consumers and size structure of food webs across aquatic and terrestrial ecosystems. The American Naturalist 194:823–839.

Purcell, J.E., and M.B. Decker. 2005. Effects of climate on relative predation by scyphomedusae and ctenophores on copepods in Chesapeake Bay during 1987–2000. Limnology and Oceanography 50:376–387.

Rabus, M., and C. Laforsch. 2012. Growing large and bulky in the presence of the enemy: *Daphnia magna* gradually switches the mode of inducible morphological defences. Functional Ecology 25:1137–1143.

Reaka, M.L. 1987. Adult-juvenile interactions in benthic reef crustaceans. Bulletin of Marine Science 41:108–134.

Redfield, A.C., B.H. Ketchum, and F.A. Richards. 1963. The influence of organisms on the composition of seawater. Pages 26–77 *in* M.N. Hill, editor. The sea, Volume 2: The composition of sea-water; comparative and descriptive oceanography. Wiley, New York, USA.

Ridoux, V., and C. Offredo. 1989. The diets of five summer breeding seabirds in Adélie Land, Antarctica. Polar Biology 9:137–145.

Romano, N, and C. Zeng. 2017. Cannibalism of decapod crustaceans and implications for their aquaculture: A review of its prevalence, influencing factors, and mitigating methods. Reviews in Fisheries Science and Aquaculture 25:42–69.

Rosenblatt, A.E., and O.J. Schmitz. 2014. Interactive effects of multiple climate change variables on trophic interactions: A meta-analysis. Climate Change Responses 1:8.

Sabat, P., J.M. Fariña, and M. Soto-Gamboa. 2003. Terrestrial birds living on marine environments: Does dietary composition of *Cinclodes nigrofumosus* (Passeriformes: Furnariidae) predict their osmotic load? Revista Chilena de Historia Natural 76:335–343.

Samaniego-Herrera, A., and Y. Bedolla-Guzman. 2012. Land crabs (Decapoda, Brachyura, Gecarcinidae) on Isabel Island, Mexico, including a new record, and its relation to the removal of invasive rats. Crustaceana 85:1007–1011.

Schmidt, K., A. Atkinson, S. Steigenberger, S. Fielding, M.C.M. Lindsay, D.W. Pond, G.A. Tarling, T.A. Klevjer, C.S. Allen, S. Nicol, and E.P. Achterberg. 2011. Seabed foraging by Antarctic krill: Implications for stock assessment, bentho-pelagic coupling and the vertical transfer of iron. Limnology and Oceanography 56:1411–1428.

Siegel, V., and J.L. Watkins. 2016. Distribution, biomass and demography of Antarctic krill, *Euphausia superba*. Pages 21–100 *in* V. Siegel, editor. Biology and ecology of Antarctic krill. Springer, Cham, Switzerland.

Siegenthaler, A., A. Mastin, C. Dufaut, D. Mondal, and C. Benvenuto. 2018. Background matching in the brown shrimp *Crangon crangon*: Adaptive camouflage and behavioural plasticity. Scientific Reports 8:3292.

Soto-Jiménez, M.F., C. Arellano-Fiore, R. Rocha-Velarde, M.E. Jara-Marini, J. Ruelas-Inzunza, and F. Páez-Osuna. 2011. Trophic transfer of lead through a model marine four-level food-chain: *Tetraselmis suecica*, *Artemia franciscana*, *Litopenaeus vannamei*, and *Haemulon scudderi*. Archives of Environmental Contamination and Toxicology 61:280–291.

South, J., J.T.A. Dick, M. McCard, D. Barrios-O'Neill, and A. Anton. 2017. Predicting predatory impact of juvenile invasive lionfish (*Pterois volitans*) on a crustacean prey using functional response analysis: Effect of temperature, habitat complexity and light regimes. Environmental Biology of Fishes 100:1155–1165.

South, J., D. Welsh, A. Anton, J.D. Sigwart, and J.T.A. Dick. 2018. Increasing temperature decreases the predatory effect of the intertidal shanny *Lipophrys pholis* on an amphipod prey. Journal of Fish Biology 92:150–164.

Sparrevik, E., and K. Leonardsson. 1998. Recruitment in the predacious isopod *Saduria entomon* (L.): Alternative prey reduces cannibalism. Journal of Experimental Marine Biology and Ecology 221:117–130.

Steneck, R.S., M.H. Graham, B.J. Bourque, D. Corbett, J.M. Erlandson, J.A. Estes, and M.J. Tegner. 2002. Kelp forest ecosystems: Biodiversity, stability, resilience and future. Environmental Conservation 29:436–459.

Stibor, H., O. Vadstein, S. Diehl, A. Gelzleichter, T. Hansen, F. Hantzsche, A. Katechakis, B. Lippert, K. Løseth, C. Peters, W. Roederer, M. Sandow, L. Sundt-Hansen, and Y. Olsen. 2004. Copepods act as a switch between alternative trophic cascades in marine pelagic food webs. Ecology Letters 7:321–328.

Sunderland, K.D., and S.L. Sutton. 1980. A serological study of arthropod predation on woodlice in a dune grassland ecosystem. The Journal of Animal Ecology 49:987–1004.

Tarling, G.A., J. Cuzin-Roudy, and F. Buchholz. 1999. Vertical migration behaviour in the northern krill *Meganyctiphanes norvegica* is influenced by moult and reproductive processes. Marine Ecology Progress Series 190:253–262.

Tarling, G.A., J. Cuzin-Roudy, S.E. Thorpe, R.S. Shreeve, P. Ward, and E.J. Murphy. 2007. Recruitment of Antarctic krill *Euphausia superba* in the South Georgia region: Adult fecundity and the fate of larvae. Marine Ecology Progress Series 331:161–179.

Tarling, G.A., and S. Fielding. 2016. Swarming behaviour in Antarctic krill. Pages 279–319 *in* V. Siegel, editor. Biology and ecology of Antarctic krill. Advances in polar biology, Springer, Cham, Switzerland.

Taylor, R.B. 1998. Short-term dynamics of a seaweed epifaunal assemblage. Journal of Experimental Marine Biology and Ecology 227:67–82.
Tegner, M.J., and P.K. Dayton. 1987. El Niño effects on southern California kelp forest communities. Advances in Ecological Research 17:243–279.
Telford, S.R., and J.M. Dangerfield. 1994. Partial brood release in woodlice: A bethedging tactic? South African Journal of Zoology 29:223–224.
Thiel, M., and T. Dernedde. 1994. Recruitment of shore crabs *Carcinus maenas* on tidal flats: Mussel clumps as refuge for juveniles. Helgoländer Meeresuntersuchungen 48:321–332.
Thiel, M., and I. Kruse. 2001. Status of the Nemertea as predators in marine ecosystems. Hydrobiologia 456:21–32.
Thiel, M., and K. Reise. 1993. Interactions of nemerteans and their prey on tidal flats. Netherlands Journal of Sea Research 31:163–172.
Tollrian, R. 1994. Fish-kairomone induced morphological changes in *Daphnia lumholtzi* (Sars). Archiv für Hydrobiologie 130:69–75.
Torres, L.G., D.R. Barlow, T.E. Chandler, and J.D. Burnett. 2020. Insight into the kinematics of blue whale surface foraging through drone observations and prey data. PeerJ 8:e8906.
Tran, M.V. 2014. The scent of cannibalism: The olfactory basis of cannibalism in hermit crabs. Journal of Experimental Marine Biology and Ecology 457:8–14.
Trathan, P.N., and S.L. Hill. 2016. The importance of krill predation in the Southern Ocean. Pages 321–350 *in* V. Siegel, editor. Biology and ecology of Antarctic krill. Advances in polar biology, Springer, Cham, Switzerland.
Turner, J.T. 2004. The importance of small planktonic copepods and their roles in pelagic marine food webs. Zoological Studies 43:255–266.
Twardochleb, L.A., T.C. Treakle, and P.L. Zarnetske. 2020. Foraging strategy mediates ectotherm predator-prey responses to climate warming. Ecology 101:e03146.
Wahle, R.A. 1992. Body-size dependent anti-predator mechanisms of the American lobster. Oikos 65:52–60.
Walve, J., and U. Larsson. 1999. Carbon, nitrogen and phosphorus stoichiometry of crustacean zooplankton in the Baltic Sea: Implications for nutrient recycling. Journal of Plankton Research 21:2309–2321.
Weiss, L.C., and R. Tollrian. 2018. Predator-induced defenses in Crustacea. Pages 303–321 *in* G.A. Wellborn and M. Thiel, editors. The natural history of the Crustacea, Volume 5: Life histories. Oxford University Press, New York, USA.
Wellborn, G.A. 2018. Life history adaptation in prey. Pages 323–346 *in* G.A. Wellborn and M. Thiel, editors. The natural history of the Crustacea, Volume 5: Life histories. Oxford University Press, New York, USA.
Wolcott, D.L., and T.G. Wolcott. 1984. Food quality and cannibalism in the red land crab, *Gecarcinus lateralis*. Physiological Zoology 57:318–324.
Worm, B., and R.A. Myers. 2003. Meta-analysis of cod-shrimp interactions reveals top-down control in oceanic food webs. Ecology 84:162–173.
Yorio, P., N. Suárez, T. Kasinsky, M. Pollicelli, C. Ibarra, and A. Gatto. 2020. The introduced green crab (*Carcinus maenas*) as a novel food resource for the opportunistic kelp gull (*Larus dominicanus*) in Argentine Patagonia. Aquatic Invasions 15:140–159.
Zhou, M., and R.D. Dorland. 2004. Aggregation and vertical migration behavior of *Euphausia superba*. Deep-Sea Research II 51:2119–2137.

2

CRUSTACEANS AS HERBIVORES, PREDATORS, AND GENERAL CONSUMERS

Rochelle D. Seitz, Romuald N. Lipcius, Jainita Patel, and Gina M. Ralph

Abstract

This chapter reviews the ecological role of crustacean consumers. First, we discuss herbivores and their role in structuring phytal habitats and phytoplankton communities. Next, we consider predatory crustaceans and their effects on prey communities, including omnivory versus specialization, and cannibalism. Finally, we discuss the role of crustaceans in food webs, contrasting top-down and bottom-up effects, trophic cascades, behavioral (non-consumptive) effects, and the impact of invasive predatory crustaceans on ecosystems. Herbivorous crustaceans are dominated by smaller species (e.g., amphipods, isopods), which inhabit structured habitats and graze on the macrophytes themselves or on their epiphytes. Other herbivores include zooplankton, whose biomass correlates with their phytoplankton prey. Herbivores (e.g., mesograzers) often reduce the abundance of epiphytes and phytoplankton either directly or indirectly through trophic cascades, which can significantly alter community structure. Predatory crustaceans have dramatic effects on prey communities through top-down control and indirect interactions through behavioral (non-consumptive) effects. Cannibalistic crustaceans are common and can have strong control over population dynamics. Invasive crustacean predators can lower fishery productivity, decimate native species, and prevent ecosystem recovery by outcompeting or killing native crustaceans. In food webs, crustaceans are often biomass dominants. Crustaceans are thus prominent consumers that can structure their communities, play an important direct role as herbivores and omnivores, as well as general predators and cannibals, and produce indirect effects through trophic cascades and non-consumptive impacts. Through herbivory, predation, and general consumption, crustaceans are key members of diverse habitats and communities worldwide.

Rochelle D. Seitz et al., *Crustaceans as Herbivores, Predators, and General Consumers*. In: *Ecology and Conservation*. Edited by: Lars Gutow and Martin Thiel, Oxford University Press.
DOI: 10.1093/oso/9780197768242.003.0002

INTRODUCTION

Crustaceans are nothing short of remarkable in their many roles in ecosystems in which they optimize foraging and consumption. The ecological role of crustaceans as herbivores, predators, and general consumers is diverse and varied, yet extremely important. Predatory crustaceans have dramatic effects on prey communities, and these predators can specialize on certain prey, such as mollusks (Schweitzer and Feldmann 2010), can be omnivores (Vannini et al. 1989, Lipcius et al. 2007), or can even be cannibalistic (Mendonça and Vinagre 2018). They are often thought to dominate ecosystems through "top-down" control of abundance and diversity (Menge et al. 1994), though evidence for the importance of "bottom-up" control in many systems has been increasing (Menge et al. 2011). Complex interactions through "trophic cascades" (predators indirectly regulating ecosystems by controlling herbivores; Carpenter et al. 1985) can lead to crucial community dynamics. In addition, indirect interactions through the behavioral (or non-consumptive) effects of predators can be just as important to survival as consumptive effects (Wirsing et al. 2021).

The ecosystem impacts of crustaceans as predatory parasites (and viewed here in their role as consumers; Nowak et al. 2020), predatory invasive species, and cannibals (e.g., Bleakley 2018) are also topics that have been gaining attention. In a recent review, predatory parasites exert control on food webs (Dunne et al. 2013), and some parasitic crustaceans (e.g., isopods) can sterilize their hosts through consumption of hemolymph or ovarian fluids (Williams and Boyko 2012). Invasive predators can reduce fishery productivity (Veldhuizen and Stanish 1999) and harm ecosystems by either outcompeting native crustaceans or preventing ecosystem recovery (Gilbey et al. 2008). In addition to other predators (e.g., fish), crustaceans can be dominant (e.g., as parasitic consumers; Morton et al. 2021), and regulate ecosystem processes (Menge et al. 1994). Key examples of these ecological interactions will be discussed in this chapter, along with knowledge gaps and suggestions for future research.

CRUSTACEANS AS HERBIVORES

Crustaceans can influence food webs as herbivores; mesograzers (e.g., amphipods and isopods) are vital secondary producers that transfer primary production to higher trophic levels (Edgar and Shaw 1995). This is potentially because of their small size, high abundance, and short generation times (Fredette and Diaz 1986). Many mesograzers, especially within seagrass beds and coral reefs, feed on diverse algae from microalgae to macroalgae (Orth and van Montfrans 1984, Jernakoff et al. 1996). We explore the importance of these food web interactions below. Herein, we will elaborate on how herbivorous crustaceans shape ecosystems.

Amphipods and Isopods as Herbivores

Crustacean herbivores often choose macrophytes that can serve as both food and habitat. Their preference for one algal species over another may be determined either by (1) the ability of an alga to provide refuge from predators (Poore et al. 2000), or (2) the algal food quality. An example of the first aspect lies in co-occurring omnivorous fish and amphipods, such as *Ampithoe longimana*, which feed on similar algal species. In North Carolina, USA, *A. longimana* occur in higher abundances on algae species that are unpalatable to omnivorous fish (e.g., *Dictyota* sp. and *Sargassum* sp.) and provide protection from fish consumption. As examples of the second aspect, when Atlantic amphipod species were given a choice of four brown rockweed *Fucus* species, they preferred *F. vesiculosus* and other *Fucus* spp. that had lower anti-herbivory phlorotannins (chemical

Fig. 2.1.
Male (bottom) and female (top) *Idotea balthica* on a blade of the brown alga *Fucus vesiculosus* taken through a microscope. Photo reproduced courtesy of Lars Gutow; see a color version of this figure in the centerfold.

deterrents) than *F. distichus* (Denton and Chapman 1991). Additionally, *Ampithoe* sometimes chose plants based on protection and sometimes based on plant food quality (e.g., tender blades) (Nicotri 1980). Moreover, distinct amphipod assemblages occur on various host species (macroalgae and sponges) in Australia, where some amphipods (e.g., Icilidae and Leucothoidae) displayed host specificity for habitat. Herbivorous amphipods living on algae (e.g., *Ampithoe caddi, A. kava, A. ngana, Exampithoe kutti,* and *Hyale nigra*), however, displayed host specificity for feeding (Poore et al. 2000).

Like amphipods, isopods use algae as both habitat and nutrition and can display host preferences. Some isopods also prefer specific green and brown seaweeds (e.g., *Ulva* sp., and *Nereocystis luetkeana*), which may be based on factors outside of nutritional value, such as accessibility (Carefoot 1973). The isopod *Idotea balthica* (Fig. 2.1) prefers robust algae, such as *Codium* and *Sargassum*, that provide shelter and predator avoidance. For *Idotea* in the Baltic Sea, *F. vesiculosus* is preferred to *F. serratus* in high currents, as thalli of *F. vesiculosus* are smaller and easier to clench (Engkvist et al. 2004). *Idotea balthica* also utilizes *F. vesiculosus* as food. *Fucus vesiculosus* contains anti-herbivory chemicals, phlorotannins, which lower grazing and herbivory (Yun et al. 2007), as does tissue toughness (Tuomi et al. 1989). *Idotea balthica* persistence on *F. vesiculosus* suggests that feeding preferences and habitat choice evolved together, driven by predation avoidance and stability of host algae (Jormalainen et al. 2001a). Furthermore, male and female isopods graze on newly grown parts of algae, with males having stronger preferences and weight gain when feeding on apical rather than basal parts of the plant (Jormalainen et al. 2001b). This grazing preference could control algal biomass in ecosystems, and reduction of algal biomass could be stronger if grazers feed primarily on older, basal parts of algae (Gutow et al. 2020).

Zooplankton as Herbivores

Zooplankton crustaceans (e.g., krill and copepods) can substantially impact ecosystems, particularly through size selection. Large zooplankton, such as suspension-feeding krill *Euphausia superba,* or smaller zooplankton, such as the copepod *Acartia tonsa,* may structure phytoplankton assem-

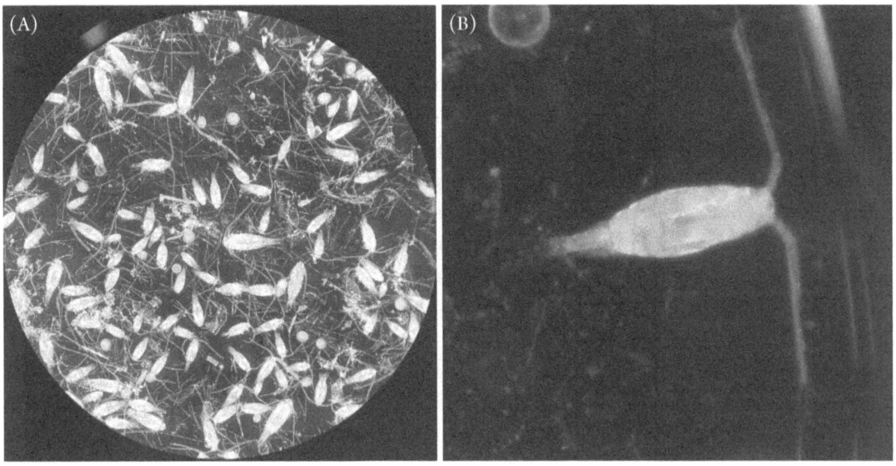

Fig. 2.2.
(A) Plankton tow sample under the microscope including a mixed copepod assemblage with many life stages, a long diatom chain in the upper middle, a single, round, centric diatom in the left center, and many pieces of microplastics; (B) Copepod *Acartia tonsa* on right side and centric diatom in upper left. Photos reproduced courtesy of Grace Saba; see a color version of this figure in the centerfold.

blages through selective feeding on centric diatoms (Fig. 2.2; Ryther and Sanders 1980, Haberman et al. 2003). Barnacles also are selective, suspension-feeding omnivores on zooplankton and suspended organic matter (Richoux et al. 2014). Additionally, copepod grazing is related to the size of zooplankton and phytoplankton. Off the Massachusetts coast, over 40% of > 10 μm phytoplankton are grazed by medium-sized copepods before the spring bloom (Bautista and Harris 1992). Grazing by copepods of all life stages can increase the proportion of smaller phytoplankton, shifting the phytoplankton community structure away from larger organisms like diatoms (Bergquist et al. 1985). Moreover, mesozooplankton may feed at the ocean surface during daylight but conduct diel vertical migration at night; they are important in the "biological pump," which transports primary production from the photic zone to deeper zones (e.g., Castro et al. 2022).

Competition among herbivores alters phytoplankton community structure in a subtle manner until predation arises. In a community of rotifers and macrozooplankton, such as *Daphnia longispina*, macrozooplankton depressed the rotifer population through competition and controlled occurrence of cyanobacterial colonies. When fish were introduced, a trophic cascade resulted where macrozooplankton (*D. longispina*) were consumed by the fish and cyanobacterial mats bloomed (Christoffersen et al. 1993). In larger systems, slight changes in zooplankton size can alter the phytoplankton community. Though phytoplankton decrease through zooplankton grazing regardless of zooplankton structure, phytoplankton can increase in response to nutrient enrichment at a faster rate than they can be consumed (Hunt and Matveev 2005) until zooplankton density subsequently increases.

Zooplankton biomass is positively correlated with the phytoplankton communities they feed upon, possibly because they lack specialization and feed in the relatively unstructured habitat of the open ocean (Irigoien et al. 2004). Benthic herbivores (mesograzers on macrophytes) generally exhibit a similar lack of specialization (Poore et al. 2008), with a few exceptions, such as when host-plant specialization helps the host survive (Hay et al. 1990). In natural systems, herbivores often alter the abundance and diversity of communities either directly for the next lower trophic level (primary producers) or indirectly through trophic cascades (Carpenter et al. 1985).

Crabs as Herbivores

Herbivorous crabs influence ecosystems based on which producer they consume (Table 2.1). In mangroves, crabs feeding on plants return ~79% of litter matter produced by the plants as fecal matter (Robertson and Daniel 1989), which enhances transfer of carbon from the coastal to the marine ecosystem. Some crabs, such as grapsids (*Metopograpsus latifrons, M. thukuhar*) and sesarmids (*Parasesarma messa, Leptarma moluccense, Neosarmatium smithi,* and *Aratus pisonii*), selectively consume mangrove seeds or leaves, and influence mangrove forest composition by preferentially eating seeds of specific species (Smith et al. 1989, Cannicci et al. 2008). Grapsids such as *Pachygrapsus crassipes* and *P. transversus* feed on macroalgae in the rocky intertidal of temperate zones and impact the community (Sousa 1979, Menge and Lubchenco 1981). In tropical communities, crabs, such as the majid (spider crabs) *Mithraculus sculptus* in the Caribbean, prevent algal overgrowth on coral reefs (Coen 1988). In coral reef crevices, small crabs (e.g., *P. transversus*) prevent algal colonization, which enables early colonizing non-algal species to find favorable conditions (Menge et al. 1983).

CRUSTACEANS AS PREDATORS

Predatory crustaceans can have dramatic effects on community structure and sometimes specialize on prey such as mollusks (Schweitzer and Feldmann 2010), or can be omnivores (Vannini et al. 1989, Lipcius et al. 2007), including those that are cannibalistic (Mendonça and Vinagre 2018). There are also environmental and ontogenetic effects on consumption.

Omnivory vs. Specialization

Crabs, lobsters, and other crustaceans are dominant consumers throughout marine and estuarine systems. Feeding modes are highly variable, ranging from generalist to specialist, with perhaps the most specialized species being parasitic crustaceans (see below). Generalist feeding is common, particularly within the decapods (crabs, lobsters, shrimp), and their diets can be diverse and dependent on prey availability (e.g., Vannini et al. 1989, Lipcius et al. 2007). Drawing broad conclusions across crustacean orders, families, and even species is challenging, however, as substantial intra- and interspecific variability exists in feeding strategies. For example, some families of Amphipoda can be assigned to a general feeding type (i.e., detritivore, carnivore, herbivore), while others include a range of feeding types (Guerra-García et al. 2014).

Evolution and Definitions of Feeding Strategies

Early arthropods were suspension (filter) feeders that collected small particles from suspension (Caron and Aria 2017). Many extant crustaceans, including shrimp and krill, are also suspension feeders and are important links in food webs. For example, Antarctic krill *E. superba* account for a large percentage of oceanic biomass, estimated at 10^7–10^8 tons (Siegel and Watkins 2016).

Most "predators" are not exclusively carnivorous, but predation satisfies only one facet of nutritional demands; most predatory crustaceans are omnivorous. The evolution of feeding strategies involves complex interactions between a species and its biotic and abiotic environment. Trade-offs are inherent in this evolution: As a species evolves to better exploit a given food resource, its ability to use other food resources diminishes (Chubaty et al. 2014). One operational definition of specialization is the following: Use of a relatively restricted subset of resources or habitats in the field by the focal species compared with other species (Colles et al. 2009). A simplified, binary approach comparing specialization to omnivory, however, is rarely accurate.

Table 2.1 Examples of plant material consumed by crabs. Source: Data from Wolcott and O'Connor (1992).

Plant material eaten	Crab	Family	Specific material consumed	References
Phytoplankton	*Gaetice depressus*	Grapsidae		Depledge 1989
	Petrolisthes cinctipes	Porcellanidae		Hartman and Hartman 1977
Benthic microalgae and detritus	*Uca* spp.	Ocypodidae		Crane 1975, Montague 1980
	Scopimera inflata	Ocypodidae		Zimmer-Faust 1987
	Grapsus tenuicrustatus	Grapsidae	Blue-green algae	Alexander 1979
Macroalgae	*Caphyra rotundifrons*	Portunidae		Hay et al. 1989
	Mithrax sculptus	Majidae		Coen 1988
	Microphrys bicornutus	Majidae		Kilar and Lou 1986
	Pugettia producta	Majidae	Kelp	Hines 1982
	Pachygrapsus crassipes	Grapsidae		Sousa 1979
Vascular plants Foliage	*Nectocarcinus integrifrons*	Portunidae	Seagrasses	Klumpp and Nichols 1983
	Sesarma reticulatum	Grapsidae	Marsh grasses	Seiple 1981
	Chasmagnathus convexus	Grapsidae	Marsh grasses	Nakasone et al. 1983
	Aratus pisonii	Grapsidae	Mangroves	Beever et al. 1979, Wilson 1989
Seeds and fruits	*Birgus latro*	Coenobitidae	Coconut	Vogel and Kent 1971, Alexander 1979
	Coenobita compressus	Coenobitidae		Kurta 1982
	Cardisoma carnifex	Gecarcinidae		Lee 1985
	Neosarmatium smithi	Grapsidae		Giddins et al. 1986, Smith 1987, Neilson et al. 1986

(Continued)

Table 2.1 Continued

Plant material eaten	Crab	Family	Specific material consumed	References
Leaf litter	*Gecarcoidea natalis*	Gecarcinidae	Rainforest	O'Dowd and Lake 1989
	Chiromanthes onychophorum	Grapsidae	Mangroves	Malley 1978, Lee 1989
	Sesarma erythrodactyla	Grapsidae	Mangroves	Camilleri 1989
	Cardisoma guanhumi	Gecarcinidae		Henning 1975, Wolcott and Wolcott 1987

Omnivory, a generalist feeding strategy defined by consumption at more than one trophic level (Pimm and Lawton 1978), was historically considered to be "rare and destabilizing" in food webs (Pimm 1980), but it was later deemed important and ubiquitous (e.g., Agrawal 2003). As food webs become complex, omnivory and related feedback loops can stabilize the system (Neutel et al. 2002). Omnivory is more common in marine food webs relative to lake, terrestrial, and stream systems. The underlying factors that drive high levels of omnivory are poorly understood (Thompson et al. 2007), but it may be a function of opportunistic feeding based on the size of the mouth opening (Mendonça and Vinagre 2018).

Prevalence of Specialization

Excluding parasites, few crustaceans have evolved specialized feeding behaviors. One example is the shallow-water, hydrothermal-vent crab *Xenograpsus testudinatus*, which exhibits a unique adaptation to its shallow-water, nutrient-poor, and toxic environment. During slack tide, thousands of individuals swarm the benthos, scavenging zooplankton (pelagic copepods) that were killed or narcotized by the vent plume (Jeng et al. 2004). Another specialization involves a powerful claw and blunt molar, ideal for crushing the intertidal snail *Littorina sitkana*, in the panopeid crab *Lophopanopeus bellus* and cancrid crabs *Glebocarcinus oregonensis* and *Cancer productus* (Yamada and Boulding 1998).

Prevalence of Omnivory

Many species exhibit some degree of diet generalization. For example, in Europe two opportunistic generalists (brown shrimp *Crangon crangon* and European green crab *Carcinus maenas*) exhibited high prey overlap, but clear niche separation: green crabs had a balanced diet of plants and animals, whereas brown shrimps preferentially consumed arthropods and annelid worms (Siegenthaler et al. 2022). Other species vary temporally in foraging patterns. For example, the pebble crab (*Eriphia smithi*) changes foraging patterns daily, consuming small algal fragments and epibionts during diurnal low tides and various invertebrate prey during nocturnal low tides (Vannini et al. 1989). Including plant material in crustaceans' diets allowed diversification of the group, as lineages with diets that include plant material have more species than closely related lineages with diets devoid of plant material (Poore et al. 2017).

Feeding rates of generalists are mediated by various factors, including habitat complexity and the presence of other predators or competitors. For instance, crustacean mesopredators (mud crabs) are directly affected by their predators, oyster toadfish (*Opsanus tau*), which also indirectly affect bivalve prey (e.g., hard clams *Mercenaria mercenaria* and eastern oysters *Crassostrea virginica*). On a low-complexity reef, oyster toadfish controlled the abundance of mud crabs (*Panopeus herbstii*), which indirectly enhanced oyster survival. However, although increased reef complexity reduced oyster toadfish predation on mud crabs, it concurrently reduced mud crab predation on oysters, such that total oyster survival was high (Grabowski 2004). In crushed-stone-aggregate trays, oyster toadfish reduced the mortality of juvenile hard clams by controlling the number, size, and feeding efficiencies of blue crabs (*Callinectes sapidus*) and mud crabs (*Dyspanopeus sayi*) preying on small hard clams (Bisker and Castagna 1989). Intraguild predation, whereby a predator species consumes both a prey species and another common predator, can further affect feeding behaviors, as occurs for mud crabs in the presence of blue crabs (intraguild predator) in oyster (prey) habitats (Hill and Weissburg 2013). Competition can also be important in omnivory. For example, *D. longispina* grazes on algae and ciliates. In one experiment, exploitative competition occurred between rotifers and *D. longispina*; *D. longispina* fed on small ciliates and algae, preventing rotifers from obtaining those phytoplankton resources (Fussmann 1996).

Environmental and Ontogenetic Effects on Consumption

Environmental conditions or degradation may also be reflected in diets of generalists. The St. Paul rock lobster (*Jasus paulensis*) typically consumes various faunal prey. At the Tristan da Cunha archipelago, however, its diet included a low diversity of animal prey and a large proportion of kelp, relative to other spiny lobsters (Blamey et al. 2019). This may reflect the depauperate marine fauna at these isolated islands and impoverished reef habitats (Blamey et al. 2019). The influence of environmental changes on diets of generalists, however, is likely mediated by species-specific traits. For example, the diet of only one of two spiny lobsters differed between healthy and degraded reefs: the spotted spiny lobster *Panulirus guttatus*, a smaller, reef-restricted species with a limited foraging range, had a smaller isotopic niche but higher trophic position than *P. argus* (Fig. 2.3), a larger, highly mobile species (Briones-Fourzán et al. 2019). While the ability to consume and assimilate a variety of food items may provide a buffer against habitat degradation (Briones-Fourzán et al. 2019), not all omnivores are flexible in diet and trophic position (Johnston et al. 2011).

Fig. 2.3.
Spiny lobster, *Panulirus argus*, foraging in a red algae (*Laurencia* sp.) bed. Photo reproduced courtesy of Romuald Lipcius; see a color version of this figure in the centerfold.

Although not well documented for most crustaceans, ontogenetic changes in diet and feeding preferences may contribute to the relatively high proportion of omnivorous species. For example, the opossum shrimp (*Mysis relicta*) exhibits "life-history omnivory," consuming algae but typically becoming more zooplanktivorous as they mature (Branstrator et al. 2000). Other species, such as the blue crab *C. sapidus*, opportunistically scavenge and prey upon diverse assemblages of food items that vary with the crab's ontogeny, as well as spatially and temporally; juvenile crabs predominantly consume benthic infaunal and epifaunal invertebrates, fish remains, and some plant and detrital matter, whereas adult diets primarily comprised bivalves (Hines 2007). Blue crab diets are likely based on availability of food items, which is habitat dependent and varies spatiotemporally (Lipcius et al. 2007). Similarly, large individuals of St. Paul rock lobster (*J. paulensis*) tended to consume more "large" prey items such as urchins, whelks, and other lobsters, while small lobsters were restricted to smaller prey, though the diversity of their diets did not differ significantly (Blamey et al. 2019).

Cannibalism in Marine Crustaceans

Cannibalism involves consumption of dead conspecifics or shed exoskeletons by scavenging and of live conspecifics by predation. It can significantly alter population dynamics through population regulation, the triggering of population cycles, and the alteration of stock-recruitment relationships (Claessen et al. 2004). Population regulation via density-dependent processes generally occurs when large conspecifics cannibalize smaller individuals of a cohort or younger year classes. Cannibalism can also arise independently of density when individuals are susceptible, such as when molting and not segregated from foraging conspecifics. Conspecifics, whether live individuals or molted exuvia, offer the highest nutritional quality of a food source because they match the elemental composition and nutritional requirements of the consumer (Polis 1988). Cannibalism is common, particularly when a potential prey is molting or small compared to the size of the cannibal; thus, the risk of injury or death for the cannibal is low (Moksnes et al. 1998, Fernández 1999). In addition, in times of scarcity of heterospecific prey, cannibalism becomes a productive alternative food source (Amaral et al. 2009).

Herein, we deal with the drivers and consequences of cannibalism in decapod crustaceans, with examples from some well-studied species (Fig. 2.4). Our focus is on cannibalism in the wild, not on cannibalism in aquaculture operations (recently reviewed by Romano and Zeng [2017]). For a broader overview and general review of cannibalism in crustaceans, see Bleakley (2018). Our review differs from that of Bleakley (2018) in that we focus on the main features, ecological drivers, and ecological effects of cannibalism in a few well-studied species.

Cannibalism in the wild is prevalent in crabs, lobsters, and shrimp (Table 2.2), but the frequency of cannibalism depends on the taxon and species. The best-studied cannibalistic species include nephropid lobsters (e.g., American lobster *Homarus americanus* and Norway lobster *Nephrops norvegicus*), portunid crabs (e.g., blue crab *C. sapidus* and green crab *Carcinus maenas*), lithodid crabs (e.g., southern king crab *Lithodes santolla*), cancrid crabs (e.g., Dungeness crab *Metacarcinus magister*), and majid crabs (e.g., snow crab *Chionoecetes opilio*; see Table 2.2); for further details see table 14.1 in Bleakley (2018).

The main ecological drivers of the intensity of cannibalism in marine decapods include predator–prey size ratio, availability of alternative prey, habitat type, and molting frequency. Intensity of cannibalism as a function of predator–prey size ratio is usually along a continuum such that intermediate size ratios are most common, because the trade-offs between energetic benefit and risk of injury are maximized, as is the case for most predator–prey interactions. In the case of small ratios (i.e., cannibal and prey are similar in size), the risk of injury or mortality for the cannibal

Fig. 2.4.
Adult blue crab (*Callinectes sapidus*) cannibalizing a young juvenile conspecific in a mesocosm with seagrass. Photo reproduced courtesy of Romuald Lipcius; see a color version of this figure in the centerfold.

is high (Moksnes et al. 1997, Fernández 1999). In addition, for small juveniles that are similar in size and do not cannibalize each other, they can profit from group defense, such as aggregations of spiny lobster (Mintz et al. 1994). For large ratios (i.e., cannibal is much larger than prey), the energetic benefit is too small due to low energetic value of the prey or because of long search times resulting from difficulty in detecting and capturing the small prey.

The extent of cannibalism depends on the availability of alternative prey, namely both conspecific and heterospecific density (Amaral et al. 2009). Cannibalism rates decrease when densities of heterospecific prey increase. For conspecific prey, total prey cannibalism consumption increases with conspecific prey density, whereas the per capita cannibalism rate decreases due to satiation (Moksnes et al. 1997, Sainte-Marie and Lafrance 2002) and mutual interference by cannibals (Mansour and Lipcius 1991).

Habitat type strongly influences cannibalism rates, such that cannibalism rates are lower in structurally complex habitats, and are also lowered by size-specific segregation among habitats (Fernández 1999, Hovel and Lipcius 2001, 2002, Sainte-Marie and Lafrance 2002, Longmire et al. 2021). For instance, young juvenile blue crabs inhabit seagrass beds, salt marshes, algal complexes, and very shallow waters, whereas older juveniles and adults, which cannibalize young juveniles, expand their habitat breadth to include unstructured soft bottoms, such as subtidal sand and mud flats (Johnston and Lipcius 2012, Bromilow and Lipcius 2017).

In most crustaceans, the molt cycle imposes an asymmetry in the likelihood of being cannibalized. Crustaceans in the process of molting or having recently molted are especially susceptible to cannibalism even by conspecifics of similar size (Lipcius and Herrnkind 1982). Consequently, both temporal and spatial behaviors have arisen to reduce the probability of being cannibalized. Caribbean spiny lobsters molt at night when conspecifics are out foraging in adjacent seagrass beds (Lipcius and Herrnkind 1982). Another temporal strategy that reduces cannibalism is synchronized molting, such as that in crabs and lobsters during late summer or early fall (Reaka 1976). Other crustaceans segregate themselves both spatially and temporally when molting to lower predation and cannibalism rates, such as northern krill *Meganyctiphanes norvegica*, which emigrate to greater depths at night (Tarling et al. 1999).

Table 2.2 Well-studied cannibalistic species of marine crustaceans, emphasizing crabs, lobsters, and shrimp.

Order			Common name	Species	Reference
Decapoda					
	Achelata				
		Palinuridae			
			Cape rock lobster	*Jasus lalandii*	Haley et al. 2011
			Spiny lobster	*Panulirus regius*	Lawal-Are and Adefule 2017
	Anomura				
		Lithodidae			
			Southern king crab	*Lithodes santolla*	Pardo et al. 2021
	Astacidea				
		Nephropidae			
			American lobster	*Homarus americanus*	Hanson 2009
			Norway lobster	*Nephrops norvegicus*	Cristo 1998
	Brachyura				
		Portunidae			
			Green crab	*Carcinus maenas*	Moksnes et al. 1998
			Blue crab	*Callinectes sapidus*	Hines and Ruiz 1995
			Paddle crab	*Ovalipes catharus*	Wear and Haddon 1987

Order			Common name	Species	Reference
		Majidae			
			Snow crab	*Chionoecetes opilio*	Lovrich and Sainte-Marie 1997
		Cancridae			
			Dungeness crab	*Metacarcinus magister*	Fernández 1999
		Ocypodidae			
			Ring-legged fiddler crab	*Uca (Austruca) annulipes*	Milner et al. 2010
			Tetragonal fiddler crab	*Uca (Gelasimus) tetragonon*	Koga et al. 1995
			Atlantic ghost crab	*Ocypode quadrata*	Wolcott and Wolcott 1988
		Sesarmidae			
			Semi-terrestrial grapsid crab	*Armases (= Sesarma) miersii*	Anger 1995
	Caridea				
		Crangonidae			
			Brown shrimp	*Crangon crangon*	del Norte-Campos and Temming 1994
		Hyppolytidae			
			Caridean shrimp	*Nauticaris marionis*	Vumazonke et al. 2003
Stomatopoda					
		Squillidae			
			Japanese mantis shrimp	*Oratosquilla oratoria*	Hamano and Matsuura 1986

Population-Level Consequences of Cannibalism

The population-level consequences of cannibalism in crustaceans are diverse, and can include regulation of population size, stabilization or destabilization of population dynamics, production of population cycles, and modification of population size structure (Claessen et al. 2004). Cannibalism has a dual advantage for the cannibal, in that intraspecific competition for resources such as heterospecific prey is minimized, and it receives an energetic gain from eating conspecifics (Polis 1988). The avenues through which cannibalism affect population dynamics include (i) mortality impact on prey, (ii) the size-specific feature of cannibal–prey interactions, (iii) energetic gains by cannibals, and (iv) intraspecific competition (Claessen et al. 2004). Regulation of population size can occur through density-dependent cannibalism, as demonstrated in the Dungeness crab, in which density-dependent cannibalism on young juveniles was associated with population cycles (Higgins et al. 1997). Density-dependent, size-structured cannibalism can regulate population density only if the cannibalism rate is sufficiently high (Diekmann et al. 1986), which may explain the diversity of cannibalism intensity in different crustacean species. A final consequence of cannibalism deals with the interaction between cannibalism and parasites. Van Allen et al. (2017) posits that "the interaction between cannibalism and parasites" can "reduce the prevalence of parasites and, thus, infection risk for cannibals. Cannibalism does this by both directly killing parasites in infected victims and by reducing the number of susceptible hosts, often enhanced by the stage-structured nature of cannibalism and infection." This is particularly effective in a few species where cannibalism of infected hosts does not transmit disease to the cannibal, as for blue crabs infected with *Hematodinium* sp. (Li et al. 2011), and in which parasitized individuals are more susceptible to cannibalism and predation (Butler et al. 2014).

TOP-DOWN VS. BOTTOM-UP CONTROL

Crustaceans often dominate ecosystems through "top-down" control (from higher trophic levels to lower ones) of abundance and diversity (Menge et al. 1994). Evidence for the importance of "bottom-up" control (from lower trophic levels to higher ones) in many systems (i.e., when crustaceans respond to resources from lower trophic levels; Eastman and Thiel 2015), however, has been increasing (Menge et al. 2011).

Herbivores in Top-Down and Bottom-Up Control

Herbivorous crustaceans (e.g., amphipods and isopods) can exert bottom-up or top-down control on an ecosystem, shifting the species composition of the habitat (Barros-Alves et al. 2018). Marine herbivores reduce producer abundance in many habitats, with the least impact in vascular-plant-dominated habitats and the strongest impacts in rocky intertidal habitats (Poore et al. 2012).

While herbivorous crustaceans reduce macroalgal biomass through either direct consumption or breaking of blades and stipes (Poore et al. 2018, Gutow et al. 2020), they can have positive effects on their environment by keeping their host algae or plants healthy and serving as prey for higher trophic levels. For example, grazers such as isopods (e.g., *Idotea*) are important parts of the Baltic Sea community because they provide positive top-down control by removing epiphytes (Jernakoff et al. 1996, Leidenberger et al. 2012). Herbivorous isopods like *I. balthica* also benefit higher trophic levels (bottom-up effects) by serving as prey for fish species (Leidenberger et al. 2012).

In some ecosystems, grazing preferences have ecosystem-wide effects. *Idotea balthica* grazes on algal species *F. radicans* and *F. vesiculosus* in the Baltic Sea. At high densities, *I. balthica* has top-down control and greatly reduces *F. vesiculosus* stands (Engkvist et al. 2000). *Idotea balthica* and *F. radicans* populations rarely overlap in Swedish coastal areas, suggesting that grazing pressure by *I. balthica* restricts the distribution of *F. radicans* (Gunnarsson and Berglund 2012). When both *Fucus* species occur, *I. balthica* is more abundant on *F. radicans* than on *F. vesiculosus*, suggesting a habitat preference, though further work is needed to definitively assess herbivore preferences (Forslund 2009). Moreover, biomass production of *F. radicans* was limited by both grazing of *I. balthica* and shading from taller *F. vesiculosus* in the species' southern distribution (Sommer 1997).

There are instances where grazing by crustaceans does not have a large-scale ecosystem-wide influence. Despite the presence of *I. balthica* and *Gammarus oceanicus* in the Baltic Sea, *Ulva intestinalis*, *Cladophora* spp., and *F. vesiculosus* remain uncontrolled in that region (Goecker and Kåll 2003). This lack of control on algae populations by mesoherbivores may be due to one of two effects. First, the Baltic Sea may be experiencing bottom-up nutrient enrichment. Mesoherbivores can help control algal biomass through grazing at moderate levels of eutrophication but have little effect in extreme eutrophication events because they cannot keep up with the rate of algal growth (Worm and Sommer 2000). Second, there may be top-down control of isopods by piscivorous fish, as densities of *I. balthica* may be strongly regulated by predation (Jormalainen et al. 2001b), reducing their abundance and thus their ability to control seaweeds. Reduced grazer pressure was responsible for the mass accumulation of the macroalga *U. lactuca* in a eutrophic Danish estuary (Geertz-Hansen et al. 1993), as occurs in other Scandinavian estuaries (Moksnes et al. 2008, Andersson et al. 2009, Östman et al. 2016).

Predators in Top-Down and Bottom-Up Control

Predators can be dominant forces in structuring the communities they inhabit, and many crustaceans exert control over benthic communities in coastal areas and the deep sea (Micheli et al. 2002, Boudreau and Worm 2012). The factors controlling community productivity is a subject of debate among marine scientists: whether marine communities are driven by top-down (predatory) or bottom-up (resource-driven) processes (Fraser et al. 2020). Many crustaceans are omnivorous, but decapod crustaceans (e.g., crabs and lobsters) often feed on mollusks or other crustaceans (Eastman and Thiel 2015). For crustacean predators, results are mixed, with many crustaceans exerting top-down pressures to control ecosystems, and other crustaceans controlled by resources. Initial experiments manipulating predators to examine top-down control used cages to exclude predatory sea stars in the rocky intertidal in the Pacific (Connell 1961, Paine 1966, Menge and Sutherland 1976). Later experiments using prey tethering to examine crab predation in the Pacific (Yamada and Boulding 1996, Musrri et al. 2019) show the importance of crustaceans in structuring hard-bottom systems. This top-down control by crustaceans is also common in marsh systems (Beheshti et al. 2021) and many other marine systems (Heck and Valentine 2007), and it can be important in diverse communities dominated by crustacean grazers (Duffy et al. 2015, Östman et al. 2016; also see the section on herbivores earlier in this chapter). Often, the "keystone predator" concept has been invoked, which occurs when the magnitude of a predator's effect on prey is disproportionately large relative to the number of predators (Paine 1969). The important roles of keystone predators in ecosystems can sometimes be filled by crustaceans. For example, in California rocky intertidal habitats, lobsters can drive community structure, as differences in mussel density and size frequency with shore elevation were attributed to spiny lobster, *Panulirus interruptus*, predation (Robles et al. 1990; Fig. 2.5). In addition, rock crabs, *Romaleon antennarium* and *C. productus*, in the northwest Pacific, despite being lower in abundance than sea stars, can play the role of a keystone predator in hard-bottom habitats by consuming 25 times more mussels per day than sea stars (Hull and Bourdeau 2017).

Soft-bottom marine communities differ substantially from other marine communities (Wilson 1991, Borja et al. 2003), and interspecific competition may not be as important as it is in

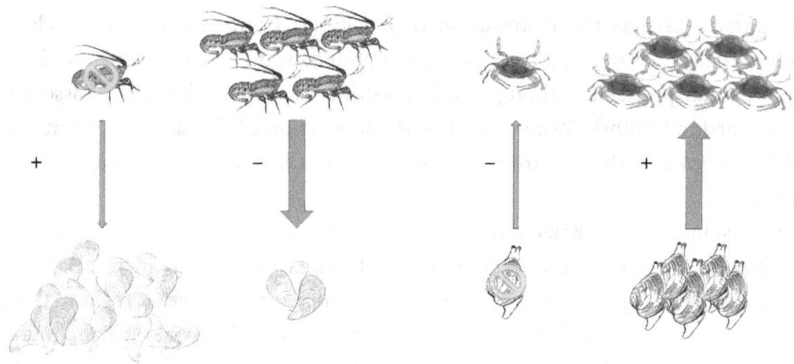

Fig. 2.5.
Diagram depicting top-down vs. bottom-up control in crustaceans. Example of top-down control by spiny lobsters (*Panulirus interruptus*) feeding on mussels (Robles et al. 1990), with left-most interaction showing few lobsters leading to many mussels (essentially a positive interaction), and second-from-left interaction showing many lobsters leading to few mussels (a negative interaction). Example of bottom-up control in blue crabs (*Callinectes sapidus*; Seitz et al. 2003), with second-from-right interaction showing few *Macoma* clams (*Macoma balthica*) leading to few crabs (essentially a negative interaction), and right-most interaction showing many clams leading to many crabs (a positive interaction). Figure reproduced courtesy of Taizhu Zhou; see a color version of this figure in the centerfold.

hard-bottom systems. Thus, even though predation occurs in these habitats, predators may or may not control soft-bottom community structure (Sih et al. 1985, Olsson et al. 2013). Species diversity either increases or remains unchanged when epibenthic predators, such as crabs, are excluded (Virnstein 1979, Hines et al. 1990, Wilson 1991), indicating the absence of both keystone predators and interspecific competition. Invasive crustaceans can impart substantial top-down control never seen before in some communities. For example, in a previously bottom-up control system with no similar predators, the Baltic Sea, the invasive mud crab *Rhithropanopeus harrisii* suppressed benthic invertebrate richness and biomass, which resulted in elevated nutrients and biomass of phytoplankton, degrading the ecosystem (Kotta et al. 2018).

Alternatively, in some cases bottom-up forces (Power 1992) may be more important in structuring benthic communities, and crustaceans may rely on abundant prey resources to increase their population size (Seitz 2011). In Chesapeake Bay, blue crabs *C. sapidus* and their bivalve prey, principally *Macoma balthica*, were positively correlated and abundant in upriver locations with high sedimentary carbon (Seitz et al. 2003; Fig. 2.5). The relative importance of top-down and bottom-up forces can depend on the scale of the study (Menge et al. 2011), with bottom-up forces being primal at large spatial scales and where nutrients and resources are abundant. In addition, productivity may dictate the relative importance of consumers or resources, with high-nutrient areas often controlled by bottom-up forces (Worm et al. 2002).

TROPHIC CASCADES

Trophic cascades can cause ecosystem-level effects when a top predator is removed or added, and subsequent changes in prey trickle down through the food chain at multiple levels. These interactions often result in indirect effects on the trophic level above or below the level where a prey or consumer has been removed or added.

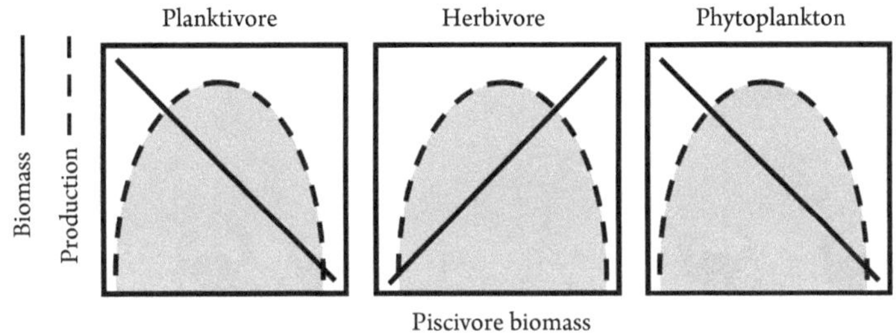

Fig. 2.6.
Example of a trophic cascade of piscivore (e.g., large fish) biomass in relation to biomass (solid line) and production (dashed line and shaded area) of vertebrate zooplanktivores (e.g., small fish), large herbivores (e.g., crustacean cladocerans), and phytoplankton (e.g., algae or cyanobacteria). As piscivore biomass increases, concurrently planktivore biomass decreases, herbivore biomass increases, and phytoplankton biomass decreases. Maximum production is at intermediate levels of the biomass of each group's predators. Source: data from Carpenter et al. (1985).

Crustacean Herbivores in Trophic Cascades

Top-down effects can be seen indirectly through trophic cascades (indirect effects) in communities with abundant crustacean grazers or mesopredators (Östman et al. 2016), and such effects can sometimes be more influential than direct consumptive effects (Preisser et al. 2005). When predators reduce the abundances of herbivorous crustaceans (amphipods and isopods), they can exert control indirectly by maintaining kelp forests, marsh habitats, and algal turfs through trophic cascades (Boudreau and Worm 2012). In freshwater systems, herbivorous crustaceans can also have important effects on algae through trophic cascades (Carpenter et al. 1985) when a top predator (zooplanktivorous fish) feeds on an herbivore (cladoceran), releasing the algae (phytoplankton) from herbivory and allowing them to proliferate in biomass and diversity (Fig. 2.6).

Decapods provide other examples of the influence of herbivorous crustacean trophic cascades on various ecological communities. In kelp forest ecosystems, herbivorous/omnivorous crustaceans (e.g., kelp crabs *Pugettia producta* and *P. gracilis*) can influence ecosystem structure by consuming snails in addition to kelp and indirectly facilitating the growth and survival of adult kelp species like *Nereocystis luetkeana* (Dobkowski 2017).

Crustacean Predators in Trophic Cascades

Crustaceans are often important mesopredators in trophic cascades. For example, increased lobster abundance reduced urchin grazing on kelp and allowed kelp beds to proliferate (Mann and Breen 1972). Additionally, turbidity reduced feeding rates of top predators (fish) on mesopredatory crabs, which increased their consumption of oysters and decreased oyster reef biodiversity (Reustle and Smee 2020). In the southeastern USA, predators such as blue crabs (*C. sapidus*) along with terrapin turtles (*Malaclemys terrapin*) consume periwinkle snails (*Littoraria irrorata*), releasing salt marsh cordgrass (*Spartina alterniflora*) from herbivory, and this promotes lush marsh habitats; disrupting this trophic cascade by overfishing blue crabs could lead to massive marsh die-offs (Silliman and Bertness 2002). In addition, blue crabs feed on mud crabs and reduce their predation on oysters (Fig. 2.7), but the effects of this cascade are dampened by omnivory (Johnson et al. 2014).

A review of four infraorders of large benthic marine decapods concluded that decapods (such as rock lobsters) can suppress other herbivores (such as urchins), which results in a trophic cascade that is beneficial to the maintenance of kelp forests (Babcock et al. 1999), salt marshes

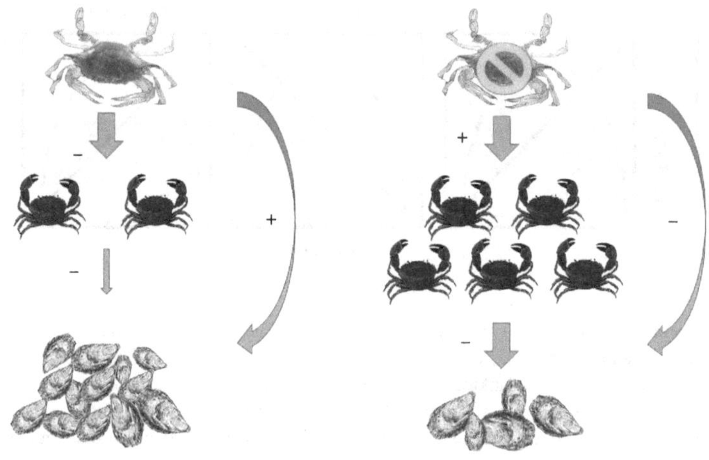

Fig. 2.7.
Trophic cascade involving blue crab (*Callinectes sapidus*), mud crab (*Panopeus herbstii*), and Eastern oyster (*Crassostrea virginica*) (Johnson et al. 2014). The left panel shows that when blue crabs are abundant, they greatly reduce mud crab densities (depicted with the negative sign next to the thick top arrow), which in turn exerts a minimal negative effect on oysters (depicted with the negative sign next to the thin bottom arrow), allowing oysters to proliferate in numerous, three-dimensional reefs (indirect positive effect depicted by the plus sign next to the curved arrow). In contrast, the right panel shows that in the absence of blue crabs, mud crabs occur in abundance (depicted by a positive sign next to the thick top arrow), and mud crabs exert a strong negative effect on oysters, resulting in fewer oysters (indirect negative curved arrow). Figure reproduced courtesy of Taizhu Zhou; see a color version of this figure in the centerfold.

(Silliman and Bertness 2002), and algal turf habitats (Robles and Robb 1993). This balance is impacted by the presence of predators, the level of competition, and habitat availability (Boudreau and Worm 2012). Furthermore, increased complexity of the food web can change the effects of trophic cascades. For example, in rocky intertidal habitats when the predatory crab *C. maenas* caused carnivorous snails *Nucella lapillus* to consume fewer barnacles, the elevated barnacle presence interrupted consumption of algae by herbivorous snails (*Littorina*). Accordingly, without *C. maenas*, there were additive effects of consumers when negative indirect effects of *N. lapillus* combined with direct negative effects of *Littorina*, and algal density was reduced by 81% compared with no-consumer controls (Trussell et al. 2017).

CONSUMPTIVE VS. NON-CONSUMPTIVE EFFECTS (NCES)

Direct consumptive effects of consumers have predictable responses, but behavioral responses (non-consumptive effects, NCEs) occur in both herbivores and carnivores. Such NCEs can be just as important to the foraging choices of consumers as consumptive effects (Wirsing et al. 2021).

NCEs and Herbivores

Predator cues combined with competition can affect habitat use and food choice of herbivorous crustaceans. In the presence of predator cues, the amphipod *Hyalella azteca* altered habitat use and occupied mats of a cyanobacterium that is toxic to predators (i.e., sunfish; Camacho and Thacker 2013). The predatory fish reduced amphipod abundance on highly palatable species and restricted these mesograzers to less-palatable species. Similarly, in the absence of predator cues, the Antarctic amphipod *Gondogeneia antarctica* preferred the non-chemically defended and palatable *Palmaria decipiens*, whereas with predator cues, the amphipods preferred the chemically defended and structurally complex *Desmarestia menziesii* (Zamzow et al. 2010). When the amphipod *Echinogammarus marinus* was in the presence of a fish mesopredator, most individuals reduced food consumption and chose a sheltered microhabitat. Furthermore, when an amphipod competitor was present, *E. marinus* reduced feeding rates and changed habitat preference (Beermann et al. 2018).

NCEs and Carnivores

Behavioral responses to avoid predators are also common in predatory crustaceans in ways other than through direct consumption (Preisser et al. 2005). Responses to the threat of a predator can take time and energy away from activities important for survival (e.g., consumption), and thus should be moderated in proportion to the threat imposed. These effects can cascade through not only the prey, but also to basal resources (Weissburg and Beauvais 2015), and anti-predator avoidance tactics of prey can be plastic (adaptable to varying conditions; Kishida et al. 2010). A recent thorough review of NCEs and their dependence on context concluded that their effects depend on prey characteristics, predator characteristics, and the environmental setting (Wirsing et al. 2021). Behavioral modifications are often driven by chemical cues (Weissburg et al. 2014), as when the predator's feeding status can alter prey activity. Mud crabs (*P. herbstii*), prey items of blue crabs and predators of oysters, detected blue crabs that recently consumed mud crabs from a farther distance than unfed blue crabs, and detection of its predator subsequently decreased the mud crabs' consumption of oysters through the trophic cascade (Weissburg and Beauvais 2015).

Direct consumptive effects may combine with indirect NCEs, where the presence of successful predators will result in strong modification of prey traits. For example, the amount of prey a blue crab consumed determined the strength and range of cascading NCEs in mud crabs. In that study, predators feeding on more prey biomass (mud crabs) decreased consumption of basal resources by mud crabs up to 1.5 m away, compared to 0.5 m when predators fed less (Weissburg and Beauvais 2015). Furthermore, invasive species can alter NCEs in native species. Invasive green crabs (*C. maenas*) use burrows of sesarmid crabs (*Sesarma*) and thereby reduce foraging by sesarmid crabs on marsh grass (Coverdale et al. 2013).

CLIMATE EFFECTS ON CRUSTACEAN CONSUMERS

Given the impact of crustaceans on trophic interactions and their role as consumers in coastal ecosystems, changes in crustacean abundance or composition as a result of climate stressors could cause shifts in community composition. Direct distributional shifts in crustaceans are likely to result from climate changes, such as warming. For example, higher temperatures lowered survival of the amphipod *Monoporeia affinis*, a plankton feeder, with subsequent effects on the food web (Wiklund et al. 2008). Climate change will likely also indirectly affect predator–prey relationships, as higher temperatures or differences in pH can alter feeding preferences. For instance, the combination of lower pH with high temperature decreased palatability of the algae on which *Sunamphitoe parmerong* grazed (Poore et al. 2013). Alternatively, long-term temperature stressors decreased algal defenses in *Fucus* spp. while grazing patterns of *I. balthica* remained unchanged, leaving *Fucus* more susceptible to isopod grazing. In the future, this could lead to losses of *Fucus*, and habitat shifts for the isopods that depend on them (Rothäusler et al. 2017).

Long-term temperature shifts also can disrupt trophic linkages between phytoplankton and zooplankton. For example, in Lake Washington, *Daphnia* populations have declined over the last 60 years because of a temporal mismatch with the spring phytoplankton bloom (Winder and Schindler 2004). As temperatures increase and coastal ecosystems undergo climate change, multiple stressors will also shift the roles of crustacean consumers in trophic cascades.

CRUSTACEANS AS PARASITES

Over 7,000 crustacean species are parasitic (Boxshall and Hayes 2019). The ecology of parasitic crustaceans is reviewed in a previous volume of this series (Nowak et al. 2020), so we focus on parasites as consumers and parasitism in ecosystems and food webs. Another review on parasites deals with their importance in controlling food webs (Dunne et al. 2013).

Complex trophic strategies are employed by parasites, but the analogy of parasites as predators can provide a useful framework (Raffel et al. 2008). Parasitic crustaceans directly consume host tissues: epicaridean isopods (Bopyroidea and Cryptoniscoidea) use other crustaceans as both intermediate and definitive hosts, consuming hemolymph and ovarian fluids (Williams and Boyko 2012). Diverse cymothoid isopods are fish ectoparasites, and most feed on the hosts as both juveniles and adults (Poore and Bruce 2012). These species may degrade commercial fisheries and aquaculture resources (Nowak et al. 2020). For example, *Olencira praegustator* parasitizes juvenile Atlantic menhaden (*Brevoortia tyrannus*) off Kiawah Island, South Carolina (Rose et al. 2020). Population-level impacts are uncertain, but the direct (tissue and fluid consumption) and indirect (secondary infections, increased mortality, reduced motility) costs of infection are substantial (Rose et al. 2020).

Parasites in Fish and Crabs

Some parasitic crustaceans (e.g., gnathiid isopods) have free-living, non-feeding adults and juveniles feeding on blood and lymph of fishes (Boxshall and Hayes 2019), which are food for adult cleaner wrasses on coral reefs (Lafferty and Kuris 2002) but might suppress recruitment of juvenile cleaner wrasses (Allan et al. 2020). Pinnotheridae (pea crabs) are often associated with invertebrate hosts (Williams and Bunkley-Williams 2019) in commensal or parasitic relationships. Whereas the African pea crab *Afropinnotheres monodi* is a commensal (González-Ortegón et al. 2021), the oyster pea crab *Zaops ostreum* is parasitic and may settle preferentially in juvenile oysters, with prevalence rates in juvenile oysters of 22–73% (Hanke et al. 2015). These high rates and reduced gonad size or condition index in infected oysters (O'Beirn and Walker 1999) may reduce abundance of oyster populations (Hanke et al. 2015).

Brood parasitism is common in crustaceans and can be coupled with egg predation. The Nicothoidae (cuckoo copepods) parasitize other marine crustaceans by mimicking eggs of the host, avoiding detection and removal, and feasting on the hosts' eggs (Williams and Bunkley-Williams 2019). Although *Choniosphaera indica* was the major source of egg mortality in the blue swimmer crab *Portunus pelagicus* in Moreton Bay, Australia, it had little impact on fecundity because of low infection intensities (Shields and Wood 1993).

Parasitic castration, or parasitic sterilization (Williams and Bunkley-Williams 2019), is a trophic strategy resulting in the elimination of reproductive ability in the host by a single parasitic individual. This strategy occurs in copepods, barnacles, brachyuran crabs, and isopods, and affects vertebrate and invertebrate hosts (Lafferty and Kuris 2009). The Rhizocephala, the most parasitic of crustaceans, comprises highly modified barnacles that parasitize crustacean hosts (Walker 2001, Nowak et al. 2020, Shields and Huchin-Mian 2020). With parasitic castration or sterilization, parasitized hosts can represent an extended phenotype of the parasite (Lafferty and Kuris 2009) and contribute substantial biomass to the ecosystem.

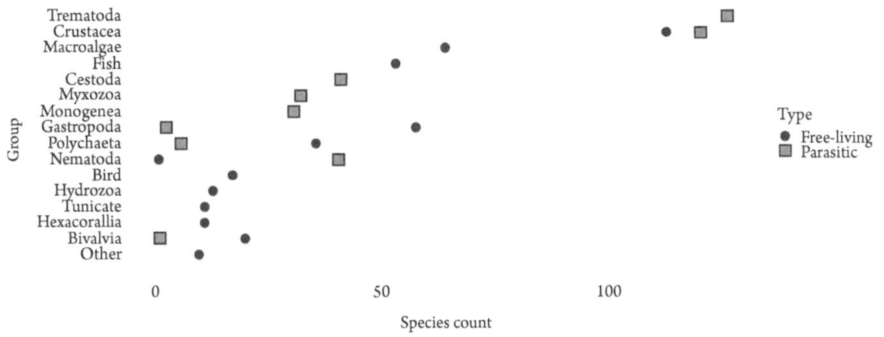

Fig. 2.8.
Species contributions to the food web of the Santa Barbara Channel, California, by organismal group (Top 15 groups contributing). Black circles indicate the number of free-living species in that group, and gray squares indicate the number of parasites. Source: data from Morton et al. (2021).

Importance of Parasitic Crustaceans in Food Webs

Parasites typically have been excluded from food web structure studies due to the small size, complex life history, and limited knowledge of their trophic position (Lafferty et al. 2008). However, parasites can contribute substantial biomass to ecosystems, even exceeding the biomass of top predators (e.g., California and Baja California estuaries; Kuris et al. 2008). Including parasites in food webs for kelp forests of the Santa Barbara Channel (Morton et al. 2021) nearly doubled the number of species relative to those that excluded parasites, with the number of free-living crustacean species at 112 and parasitic species at 120, totaling 232 species if parasitic Crustacea are included (Fig. 2.8). Within this food web, crustaceans (free-living and parasitic) were the most species-rich, with more parasitic crustaceans than free-living crustaceans. In coastal marine and estuarine food webs, including parasites increased link density, connectance, nestedness, and chain length (Lafferty et al. 2006, Dunne et al. 2013).

Inclusion of parasites in ecological studies can enhance understanding of ecosystem processes. Parasites may also be useful as bioindicators of environmental degradation or recovery (Lafferty et al. 2008, Vidal-Martínez et al. 2010, Navarro-Barranco et al. 2020). *Lamproglena* spp. (ectoparasitic copepods on the gills of several fishes) are potential bioindicators of heavy metal pollution (Pretorius and Avenant-Oldewage 2021).

Co-introductions and parasite spillover (when invasive parasites spill over to new hosts in invaded systems), such as in *Mytilicola orientalis* (Goedknegt et al. 2018), highlight potential indirect effects of species introductions on native species. Integrating aquatic parasites into ecological studies has provided new insights into ecosystem functioning, but cross-training and additional multidisciplinary collaborations are needed to continue developing synergies between these two fields (Poulin et al. 2016). Questions remain regarding the ecological role of parasitic crustaceans, and how parasites respond to environmental change.

CRUSTACEANS AS INVASIVE CONSUMERS

The impact of crustaceans as invasive species has occurred globally, and their importance as consumers can have profound effects on food webs. As predators with generalist diets, invasive crustaceans can transform ecosystems by outcompeting native predators and consuming prey from lower trophic levels. Comprehensive treatment of invasive crustaceans is reviewed by Grosholz (see Chapter 13 in this volume). Thus, we give a brief overview of invasive crustacean consumers, and review key examples of crustaceans as invasive predators.

Crustaceans possess characteristics that are advantageous for invasive success, such as a wide dietary niche (Romanuk et al. 2009), and diverse anatomical features adapted for consumption. For example, crabs have large pincers (e.g., *C. maenas*) and diverse mouthparts (mandibles, setae) adapted to feeding on multiple prey types (Mayer et al. 2008). Most invasive crustaceans are decapods and amphipods, but other examples include cladocerans, copepods, and mysids (Ricciardi 2011). Such opportunistic feeding can produce flexible diets depending on location, with extreme exploitation of dominant prey species in specific habitats (Gherardi and Barbaresi 2007). We illustrate the success of invasive crustacean consumers with a key example, the European green crab (*C. maenas*), though other key examples include the Asian shore crab (*Hemigrapsus sanguineus*), and the Chinese mitten crab (*Eriocheir sinensis*; see Chapter 13 in this volume).

Carcinus maenas has a particularly robust carapace (MacDonald et al. 2007) and a generalist diet. It is a voracious consumer that prefers juvenile clams, can consume prey at a faster rate than can native crabs (Walton et al. 2002), and exerts top-down control. For example, in the Gulf of Maine, juvenile rock crabs (*Cancer irroratus*) now inhabit deeper, less-suitable waters in response to high predation rates by *C. maenas* reducing their survival in the shallows (Griffen and Riley 2015). *Carcinus maenas* predation in New England caused soft-shell clams (*Mya arenaria*) to burrow deeper to avoid predation, which requires allocating more energy into siphon growth, rather than into body growth or reproduction (Whitlow 2010). *Carcinus maenas* may be a threat to blue crabs (*C. sapidus*) in temperate areas, as they are more aggressive competitors for prey, and they consume juvenile blue crabs (MacDonald et al. 2007).

Planktonic, invasive, predatory crustaceans can also be detrimental. In San Francisco Bay, *Tortanus dextrilobatus* is a large invasive copepod that favors preying on native copepods instead of invasive species such as *Oithona davisae*. It is responsible for shifting planktonic composition over 20 years, including facilitating the introduction of more invasive species (Bollens et al. 2002). Another community shift occurred in the Columbia-Snake River system (Washington state, USA) in 2005 (Cordell et al. 2008), where the first introduced calanoid copepod *Pseudodiaptomus inopinus* was later displaced by other invading calanoid copepods *P. forbesi* and *Sinocalanus doerrii*, and the cyclopoid copepod *Limnoithona tetraspina*, that compete with native copepods for diatoms and ciliates (Bowen et al. 2015).

ECOSYSTEM-LEVEL EFFECTS

Herbivorous and predatory crustacean consumers can impact ecosystems by preventing ecosystem recovery, outcompeting native crustacean species, and shifting community structure through top-down control or indirect effects. Herbivores have pronounced ecosystem-wide effects. In a review of over 600 marine herbivore exclusion experiments, producer abundance was reduced by herbivores by 68% globally, with small crustacean mesograzers responsible for a large percentage (Poore et al. 2012). The weakest effects were in vascular plant habitats (salt marshes and seagrass), and the strongest effects were in the rocky intertidal. The influence of latitude and temperature on herbivore effects were minimal, contrary to the prevailing hypothesis. Rather, the greatest effects were attributed to producer traits, with brown and green algae more affected by grazing than red algae (Poore et al. 2012). Moreover, some herbivorous crustaceans facilitated primary producers, such as by removing epiphytes from seagrass (Duffy et al. 2001).

Crustacean predators can dramatically affect ecosystems, with both direct top-down control of intermediate mesopredators, and indirect effects on basal species. For example, in a two-predator, two-prey experimental system, removal of the crab *C. maenas* and whelk *N. lapillus* predators alone decreased benthic primary producers when the assemblage (including prey) was intact; the loss of either of the prey species (mussels or grazing gastropods) led to different communities of basal primary producers (O'Connor et al. 2013). Crustacean predators can also have substantial NCEs on their prey species, including changes in habitat use and foraging behavior that can reduce survival rates and reproduction (Wirsing et al. 2021).

Parasitic crustaceans are among the least well-known of the crustaceans, including their role in structuring aquatic ecosystems. Crustacean parasites can be a dominant and species-rich component of the food web and can mediate how communities respond to human impacts such as fishery exploitation and global climate change (Kuris et al. 2008, Morton et al. 2021). Thus, a better understanding of these species with such complex life histories is needed.

Invasive crustaceans can have large, negative ecological and economic impacts, particularly in areas like San Francisco Bay, which are economically reliant on native species that are often susceptible to invasive crustaceans. Proposed solutions include controlling larval dispersal of invasive crustaceans, as population expansion is partly a function of larval dispersal (Epifanio et al. 1998). Methods have been devised to track larval dispersal using satellites (Natunewicz and Epifanio 2001) and with mathematical modeling (e.g., Tilburg et al. 2006). Additionally, for copepods and smaller crustaceans, extensive monitoring with a focus on microzooplankton and genetic studies to trace the vector of invasion for these species is recommended (Bollens et al. 2002). Many of these small crustaceans, particularly species with a calcified exoskeleton, are likely to be affected by global change, especially at the larval stage (Pansch et al. 2014).

FUTURE DIRECTIONS

Though current research has brought scientists a long way toward understanding the impacts of crustaceans as consumers, there remain areas where further research is warranted. Herbivorous decapods play critical roles in altering ecosystems, yet the grazing preferences of small crustaceans (e.g., amphipods and isopods) remain vague. Ecologists need to definitively assess herbivore host preferences and the mechanisms underlying them (Forslund 2009). An increased understanding of how biotic interactions of crustaceans change herbivore communities over the long term is also warranted (Poore et al. 2012). Debate is still ongoing regarding whether marine communities are driven by top-down or bottom-up forces (Fraser et al. 2020), and defining the various roles of crustacean consumers in these interactions is key. Additional studies on the long-term effects of invasive crustacean species (see Chapter 13 in this volume) are needed, including potentially beneficial cases where new fisheries have been founded, such as that of the blue crab's expansion northward along the northwestern Atlantic coast and eastward to the Mediterranean Sea (Marchessaux et al. 2023). Knowing the effects of depletion of crustaceans through anthropogenic effects, including fishing pressure and climate change, will be useful in understanding the roles of crustacean consumers. Further research on larval dispersion and population genetics, community-level impacts, and the translation of invasive ecology into economic impacts is warranted (Natunewicz and Epifanio 2001). Additional research is needed on the environmental effects on NCEs, including the collective effects of multiple drivers on NCEs (Wirsing et al. 2021). In parasitic crustaceans, host utilization and the use of multiple, alternate hosts may affect population dynamics, and additional research is needed, particularly for parasitic crabs such as pea crabs (Hanke et al. 2015). Additionally, further work is needed to describe parasitic crustacean diversity at all ecological scales, and their myriad life history strategies, to help inform broader ecosystem studies (Poulin et al. 2016). Given the advances in aquaculture, increased parasitism (e.g., in pen farming; Nowak et al. 2020) could have ramifications for the survival of farmed species, and this could be investigated further.

Finally, given predictions of future losses in species diversity with climate change (Bellard et al. 2012), losses to crustacean consumers could leave ecosystems without sufficient species redundancy to deal with novel conditions (Thompson and Shurin 2012). A further understanding of the impacts of and potential loss of crustacean consumers, therefore, is warranted.

CONCLUSIONS AND OUTLOOK

Crustaceans are adept at optimizing foraging and consumption within the ecosystems in which they live. They serve important roles as herbivores, as omnivores, or as specialized predators and cannibals, and can structure the faunal and floral communities in their habitats directly through consumption, and indirectly through trophic cascades and non-consumptive interactions. Smaller crustaceans, such as amphipods and isopods, are dominant herbivores (i.e., mesograzers) that inhabit structured habitats including seagrass and algal patches where they graze on the seagrass and algae or on epiphytes and reduce their abundance. Planktonic, herbivorous zooplankton are biomass dominants whose abundance is directly related to that of their main prey, phytoplankton. Herbivores can also reduce the abundance of epiphytes and phytoplankton indirectly through trophic cascades. Predators, mainly crabs and lobsters, can exert top-down control on benthic communities either directly through predation or indirectly through behavioral NCEs. Cannibalism is common in crustaceans and can produce complex population dynamics through inter-cohort, nonlinear mechanisms, which make management of cannibalistic crustaceans difficult due to the multifaceted processes. Invasive crustacean predators, particularly the green crab and blue crab, are continuing to expand their range globally, and may impact fishery productivity, deplete native species, and alter ecosystems. Parasitic crustaceans, particularly those on marine vertebrates, can reduce population growth rates of their hosts, while in other cases, they can comprise substantial biomass in the food webs. The ecological role of crustaceans as herbivores, predators, and general consumers has wide-ranging effects on communities in habitats worldwide.

Several issues regarding crustaceans as consumers and their effects on ecosystems remain to be addressed through further research. These issues include grazing and host preferences of small crustacean herbivores, top-down versus bottom-up control of communities by crustaceans, anthropogenic effects (e.g., habitat modifications and fishing pressure) on crustacean populations, population genetics, and dispersal. Increased focus on aquaculture of crustaceans could alleviate fishing pressures on some populations, thus leaving more crustacean consumers in communities, but it could also have negative effects stemming from high densities and increased parasitism. All of these new lines of research will become extremely important given the unknown species interactions likely to arise with human expansion and climate change. Fortunately, crustacean consumers appear to be robust to ecosystem fluctuations, in many cases can use alternative prey resources, and will likely be dominant members of ecosystems for decades to come.

REFERENCES

Agrawal, A.A. 2003. Why omnivory? Ecology 84:2521.

Alexander, H.G.L. 1979. A preliminary assessment of the role of the terrestrial decapod crustaceans in the Aldabran ecosystem. Philosophical Transactions of the Royal Society of London B 286:241–246.

Allan, B.J., B. Illing, E.P. Fakan, P. Narvaez, A.S. Grutter, P.C. Sikkel, E.C. McClure, J.L. Rummer, and M.I. McCormick. 2020. Parasite infection directly impacts escape response and stress levels in fish. Journal of Experimental Biology 223:jeb230904.

Amaral, V., J. Paula, S. Hawkins, and S. Jenkins. 2009. Cannibalistic interactions in two co-occurring decapod species: Effects of density, food, alternative prey and habitat. Journal of Experimental Marine Biology and Ecology 368:88–93.

Andersson, S., M. Persson, P.O. Moksnes, and S. Baden. 2009. The role of the amphipod *Gammarus locusta* as a grazer on macroalgae in Swedish seagrass meadows. Marine Biology 156:969–981.

Anger, K. 1995. Developmental biology of *Armases miersii* (Grapsidae), a crab breeding in supratidal rock pools. II. Food limitation in the nursery habitat and larval cannibalism. Marine Ecology Progress Series 117:83–89.

Babcock, R.C., S. Kelly, N.T. Shears, J.W. Walker, and T.J. Willis. 1999. Changes in community structure in temperate marine reserves. Marine Ecology Progress Series 189:125–134.

Barros-Alves, S.D.P., D.F.R. Alves, and V.J. Cobo. 2018. Brachyuran crab (Crustacea, Decapoda) assemblage associated with *Sargassum cymosum* in southeastern Brazil. Marine Biodiversity 48:2043–2055.

Bautista, B., and R.P. Harris. 1992. Copepod gut contents, ingestion rates and grazing impact on phytoplankton in relation to size structure of zooplankton and phytoplankton during a spring bloom. Marine Ecology Progress Series 82:41–50.

Beermann, J., K. Boos, L. Gutow, M. Boersma, and A.C. Peralta. 2018. Combined effects of predator cues and competition define habitat choice and food consumption of amphipod mesograzers. Oecologia 186:645–654.

Beever, J.W., III, D. Simberloff, and L.L. King. 1979. Herbivory and predation by the mangrove tree crab *Aratus pisonii*. Oecologia 43:317–328.

Beheshti, K.M., K. Wasson, C. Angelini, B.R. Silliman, and B.B. Hughes. 2021. Long-term study reveals top-down effect of crabs on a California salt marsh. Ecosphere 12:e03703.

Bellard, C., C. Bertelsmeier, P. Leadley, W. Thuiller, and F. Courchamp. 2012. Impacts of climate change on the future of biodiversity. Ecology Letters 15:365–377.

Bergquist, A.M., S.R. Carpenter, and J.C. Latino. 1985. Shifts in phytoplankton size structure and community composition during grazing by contrasting zooplankton assemblages. Limnology and Oceanography 30:1037–1045.

Bisker, R., and M. Castagna. 1989. Biological control of crab predation on hard clams *Mercenaria mercenaria* (Linnaeus, 1758) by the toadfish *Opsanus tau* (Linnaeus) in tray cultures. Journal of Shellfish Research 8:33–36.

Blamey, L.K., A.M. de Lecea, L.D.S. Jones, and G.M. Branch. 2019. Diet of the spiny lobster *Jasus paulensis* from the Tristan da Cunha archipelago: Comparisons between islands, depths and lobster sizes. Estuarine, Coastal and Shelf Science 219:262–272.

Bleakley, B. 2018. Cannibalism in crustaceans. Pages 347–373 *in* G. Wellborn and M. Thiel, editors. The natural history of the Crustacea, Volume 5: Life histories. Oxford University Press, New York, USA.

Bollens, S.M., J.R. Cordell, S. Avent, and R. Hooff. 2002. Zooplankton invasions: A brief review, plus two case studies from the northeast Pacific Ocean. Hydrobiologia 480:87–110.

Borja, A., I. Muxika, and J. Franco. 2003. The application of a marine biotic index to different impact sources affecting soft-bottom benthic communities along European coasts. Marine Pollution Bulletin 46:835–845.

Boudreau, S.A., and B. Worm. 2012. Ecological role of large benthic decapods in marine ecosystems: A review. Marine Ecology Progress Series 469:195–213.

Bowen, A., G. Rollwagen-Bollens, S.M. Bollens, and J. Zimmerman. 2015. Feeding of the invasive copepod *Pseudodiaptomus forbesi* on natural microplankton assemblages within the lower Columbia River. Journal of Plankton Research 37:1089–1094.

Boxshall, G., and P. Hayes. 2019. Biodiversity and taxonomy of the parasitic Crustacea. Pages 73–134 *in* N.J. Smit, N.L. Bruce, and K.A. Hadfield, editors. Parasitic Crustacea: State of knowledge and future trends. Springer, Cham, Switzerland.

Branstrator, D.K., G. Cabana, A. Mazumder, and J.B. Rasmussen. 2000. Measuring life-history omnivory in the opossum shrimp, *Mysis relicta*, with stable nitrogen isotopes. Limnology and Oceanography 45:463–467.

Briones-Fourzán, P., L. Álvarez-Filip, C. Barradas-Ortíz, P.S. Morillo-Velarde, F. Negrete-Soto, I. Segura-García, A. Sánchez-González, and E. Lozano-Álvarez. 2019. Coral reef degradation differentially alters feeding ecology of co-occurring congeneric spiny lobsters. Frontiers in Marine Science 5:516.

Bromilow, A.M., and R.N. Lipcius. 2017. Mechanisms governing ontogenetic habitat shifts: role of trade-offs, predation, and cannibalism for the blue crab. Marine Ecology Progress Series 584:145–159.

Butler, M.J., IV, J.M. Tiggelaar II, J.D. Shields. 2014. Effects of the parasitic dinoflagellate *Hematodinium perezi* on blue crab (*Callinectes sapidus*) behavior and predation. Journal of Experimental Marine Biology and Ecology 461:381–388.

Camacho, F.A., and R.W. Thacker. 2013. Predator cues alter habitat use by the amphipod *Hyalella azteca* (Saussure). Freshwater Science 32:1148–1154.

Camilleri, J. 1989. Leaf choice by crustaceans in a mangrove forest in Queensland. Marine Biology 102:453–459.

Cannicci, S., D. Burrows, S. Fratini, T.J. Smith III, J. Offenberg, and F. Dahdouh-Guebas. 2008. Faunal impact on vegetation structure and ecosystem function in mangrove forests: A review. Aquatic Botany 89:186–200.

Carefoot, T.H. 1973. Feeding, food preference, and the uptake of food energy by the supralittoral isopod *Ligia pallasii*. Marine Biology 18:228–236.

Caron, J.B., and C. Aria. 2017. Cambrian suspension-feeding lobopodians and the early radiation of panarthropods. BMC Evolutionary Biology 17:1–14.

Carpenter, S.R., J.F. Kitchell, and J.R. Hodgson. 1985. Cascading trophic interactions and lake productivity. BioScience 35:634–639.

Castro, L.R., J.L. Iriarte, S. Soto, and H. González. 2022. Role of vertically migrating and not-migrating large crustacean zooplankton as potential consumers of spring phytoplankton in an inshore Patagonian gulf. Estuaries and Coasts 46:463–476.

Christoffersen, K., B. Riemann, A. Klysner, and M. Søndergaard. 1993. Potential role of fish predation and natural populations of zooplankton in structuring a plankton community in eutrophic lake water. Limnology and Oceanography 38:561–573.

Chubaty, A.M., B.O. Ma, R.W. Stein, D.R. Gillespie, L.M. Henry, C. Phelan, E. Palsson, F.W. Simon, and B.D. Roitberg. 2014. On the evolution of omnivory in a community context. Ecology and Evolution 4:251–265.

Claessen, D., A.M. de Roos, and L. Persson. 2004. Population dynamic theory of size-dependent cannibalism. Proceedings of the Royal Society of London B: Biological Sciences 271:333–340.

Coen, L.D. 1988. Herbivory by crabs and the control of algal epibionts on Caribbean host corals. Oecologia 75:198–203.

Colles, A., L.H. Liow, and A. Prinzing. 2009. Are specialists at risk under environmental change? Neoecological, paleoecological and phylogenetic approaches. Ecology Letters 12:849–863.

Connell, J.H. 1961. The influence of interspecific competition and other factors on the distribution of the barnacle *Chthamalus stellatus*. Ecology 42:710–723.

Cordell, J.R., S.M. Bollens, R. Draheim, and M. Sytsma. 2008. Asian copepods on the move: Recent invasions in the Columbia–Snake River system, USA. ICES Journal of Marine Science 65:753–758.

Coverdale, T.C., E.E. Axelman, C.P. Brisson, E.W. Young, A.H. Altieri, and M.D. Bertness. 2013. New England salt marsh recovery: Opportunistic colonization of an invasive species and its non-consumptive effects. PLoS ONE 8:e73823.

Crane, J. 1975. Fiddler crabs of the world. Princeton University Press, Princeton, New Jersey, USA.

Cristo, M. 1998. Feeding ecology of *Nephrops norvegicus* (Decapoda: Nephropidae). Journal of Natural History 32:1493–1498.

del Norte-Campos, A.G.C., and A. Temming. 1994. Daily activity, feeding and rations in gobies and brown shrimp in the northern Wadden Sea. Marine Ecology Progress Series 115:41–53.

Denton, A.B., and A.R.O. Chapman. 1991. Feeding preferences of gammarid amphipods among four species of *Fucus*. Marine Biology 109:503–506.

Depledge, M.H. 1989. Observations on the feeding behaviour of *Gaetice depressus* (Grapsidae: Varuninae) with special reference to suspension feeding. Marine Biology 100:253–259.

Diekmann, O., R.M. Nisbet, W.S.C. Gurney, and F. Van den Bosch. 1986. Simple mathematical models for cannibalism: A critique and a new approach. Mathematical Biosciences 78:21–46.

Dobkowski, K.A. 2017. Biotic and abiotic influences on bull kelp (*Nereocystis luetkeana*) abundance and distribution in the Salish Sea. PhD dissertation, University of Washington, Seattle, USA.

Duffy, J.E., K.S. Macdonald, J.M. Rhode, and J.D. Parker. 2001. Grazer diversity, functional redundancy, and productivity in seagrass beds: An experimental test. Ecology 82:2417–2434.

Duffy, J.E., P.L. Reynolds, C. Boström, J.A. Coyer, M. Cusson, S. Donadi, J.G. Douglass, J.S Eklöf, A.H. Engelen, B.K. Eriksson, S. Fredriksen, L. Gamfeldt, C. Gustafsson, G. Hoarau, M. Hori, K. Hovel, K. Iken, J.S. Lefcheck, P. Moksnes, M. Nakaoka, M.I. O'Connor, J.L. Olsen, J.P. Richardson, J.L. Ruesink, E.E. Sotka, J. Thormar, M.A. Whalen, and J.J. Stachowicz. 2015. Biodiversity mediates top-down control in eelgrass ecosystems: A global comparative-experimental approach. Ecology Letters 18:696–705.

Dunne, J.A., K.D. Lafferty, A.P. Dobson, R.F. Hechinger, A.M. Kuris, N.D. Martinez, J.P. McLaughlin, K.N. Mouritsen, R. Poulin, K. Reise, D.B. Stouffer, D.W. Thieltges, R.J. Williams, and C.D. Zander. 2013. Parasites affect food web structure primarily through increased diversity and complexity. PLoS Biology 11:e1001579.

Eastman, L.B., and M. Thiel. 2015. Foraging behavior of crustacean predators and scavengers. Pages 535–556 *in* M. Thiel and L. Watling, editors. The natural history of the Crustacea, Volume 2: Lifestyles and feeding biology. Oxford University Press, New York, USA.

Edgar, G.J., and C. Shaw. 1995. The production and trophic ecology of shallow-water fish assemblages in southern Australia. III. General relationships between sediments, seagrasses, invertebrates and fishes. Journal of Experimental Marine Biology and Ecology 194:107–131.

Engkvist, R., T. Malm, and S. Tobiasson. 2000. Density dependent grazing effects of the isopod *Idotea baltica* Pallas on *Fucus vesiculosus* L in the Baltic Sea. Aquatic Ecology 34:253–260.

Engkvist, R., T. Malm, and J. Nilsson. 2004. Interaction between isopod grazing and wave action: A structuring force in macroalgal communities in the southern Baltic Sea. Aquatic Ecology 38:403–413.

Epifanio, C.E., A.I. Dittel, S. Park, S. Schwalm, and A. Fouts. 1998. Early life history of *Hemigrapsus sanguineus*, a non-indigenous crab in the Middle Atlantic Bight (USA). Marine Ecology Progress Series 170:231–238.

Fernández, M. 1999. Cannibalism in Dungeness crab *Cancer magister*: Effects of predator-prey size ratio, density, and habitat type. Marine Ecology Progress Series 182:221–230.

Forslund, H. 2009. Grazing and the geographical range of seaweeds: The introduced *Fucus evanescens* and the newly described *Fucus radicans*. PhD dissertation, Stockholm University, Stockholm, Sweden.

Fraser, K.M., J.S. Lefcheck, S.D. Ling, C. Mellin, R.D. Stuart-Smith, and G.J. Edgar. 2020. Production of mobile invertebrate communities on shallow reefs from temperate to tropical seas. Proceedings of the Royal Society B: Biological Sciences 287:20201798.

Fredette, T.J., and R.J. Diaz. 1986. Secondary production of *Gammarus mucronatus* (Amphipoda: Gammaridae) in warm temperate estuarine habitats, York River, Virginia. Journal of Crustacean Biology 6:729–741.

Fussmann, G. 1996. The importance of crustacean zooplankton in structuring rotifer and phytoplankton communities; an enclosure study. Journal of Plankton Research 18:1897–1915.

Geertz-Hansen, O., K.A.J. Sand-Jensen, D.F. Hansen, and A. Christiansen. 1993. Growth and grazing control of abundance of the marine macroalga, *Ulva lactuca* L. in a eutrophic Danish estuary. Aquatic Botany 46:101–109.

Gherardi, F., and S. Barbaresi. 2007. Feeding preferences of the invasive crayfish, *Procambarus clarkii*. Bulletin Français de la Pêche et de la Pisciculture 285:7–20.

Giddins, R.L., J.S. Lucas, M.J. Neilson, and G.N. Richards. 1986. Feeding ecology of the mangrove crab *Neosarmatium smithi* (Crustacea: Decapoda: Sesarmidae). Marine Ecology Progress Series 33:147–155.

Gilbey, V., M.J. Attrill, and R.A. Coleman. 2008. Juvenile Chinese mitten crabs (*Eriocheir sinensis*) in the Thames estuary: Distribution, movement and possible interactions with the native crab *Carcinus maenas*. Biological Invasions 10:67–77.

Goecker, M.E., and S.E. Kåll. 2003. Grazing preferences of marine isopods and amphipods on three prominent algal species of the Baltic Sea. Journal of Sea Research 50:309–314.

Goedknegt, M.A., S. Bedolfe, J. Drent, J. van der Meer, and D.W. Thieltges. 2018. Impact of the invasive parasitic copepod *Mytilicola orientalis* on native blue mussels *Mytilus edulis* in the western European Wadden Sea. Marine Biology Research 14:497–507.

González-Ortegón, E., M. Perez-Miguel, J.I. Navas, P. Drake, and J.A. Cuesta. 2021. Isotopic niche provides an insight into the ecology of a symbiont during its geographic expansion. Current Zoology 68:185–197.

Grabowski, J.H. 2004. Habitat complexity disrupts predator–prey interactions but not the trophic cascade on oyster reefs. Ecology 85:995–1004.

Griffen, B.D., and M.E. Riley. 2015. Potential impacts of invasive crabs on one life history strategy of native rock crabs in the Gulf of Maine. Biological Invasions 17:2533–2544.

Guerra-García, J.M., J.T. de Figueroa, C. Navarro-Barranco, M. Ros, J.E. Sánchez-Moyano, and J. Moreira. 2014. Dietary analysis of the marine Amphipoda (Crustacea: Peracarida) from the Iberian Peninsula. Journal of Sea Research 85:508–517.

Gunnarsson, K., and A. Berglund. 2012. The brown alga *Fucus radicans* suffers heavy grazing by the isopod *Idotea baltica*. Marine Biology Research 8:87–89.

Gutow, L., A.G.B. Poore, M.A. Díaz Poblete, V. Villalobos, and M. Thiel. 2020. Small burrowing amphipods cause major damage in a large kelp. Proceedings of the Royal Society B: Biological Sciences 287:20200330.

Haberman, K.L., R.M. Ross, and L.B. Quetin. 2003. Diet of the Antarctic krill (*Euphausia superba* Dana): II. Selective grazing in mixed phytoplankton assemblages. Journal of Experimental Marine Biology and Ecology 283:97–113.

Haley, C.N., L.K. Blamey, L.J. Atkinson, and G.M. Branch. 2011. Dietary change of the rock lobster *Jasus lalandii* after an 'invasive' geographic shift: Effects of size, density and food availability. Estuarine, Coastal and Shelf Science 93:160–170.

Hamano, T., and S. Matsuura. 1986. Food habits of the Japanese mantis shrimp in the benthic community of Hakata Bay. Nippon Suisan Gakkaishi 52:787–794.

Hanke, M.H., J.M. Hargrove, T.D. Alphin, and M.H. Posey. 2015. Oyster utilization and host variation of the oyster pea crab (*Zaops ostreum*). Journal of Shellfish Research 34:281–287.

Hanson, J.M. 2009. Predator-prey interactions of American lobster (*Homarus americanus*) in the southern Gulf of St. Lawrence, Canada. New Zealand Journal of Marine and Freshwater Research 43:69–88.

Hartman, B.H., and M.S. Hartman. 1977. The stimulation of filter feeding in the porcelain crab *Petrolisthes cinctipes* Randall by amino acids and sugars. Comparative Biochemistry and Physiology 56A:19–22.

Hay, M.E., J.E. Duffy, and W. Fenical. 1990. Host-plant specialization decreases predation on a marine amphipod: An herbivore in plant's clothing. Ecology 71:733–743.

Hay, M.E., J.R. Pawlik, J.E. Duffy, and W. Fenical. 1989. Seaweed-herbivore-predator interactions: Host-plant specialization reduces predation on small herbivores. Oecologia 81:418–427.

Heck, K.L., and J.F. Valentine. 2007. The primacy of top-down effects in shallow benthic ecosystems. Estuaries and Coasts 30:371–381.

Henning, H.G. 1975. Studies on the ecology, behavior, and sensory physiology of *Cardisoma guanhumi* (Decapoda, Brachyura), a terrestrial crab from north Colombia. Forma et Functio 8:253–304.

Higgins, K., A. Hastings, and L.W. Botsford. 1997. Density dependence and age structure: Nonlinear dynamics and population behavior. American Naturalist 149:247–269.

Hill, J.M., and M.J. Weissburg. 2013. Habitat complexity and predator size mediate interactions between intraguild blue crab predators and mud crab prey in oyster reefs. Marine Ecology Progress Series 488:209–219.

Hines, A.H. 1982. Coexistence in a kelp forest: Size, population dynamics, and resource partitioning in a guild of spider crabs (Brachyura, Majidae). Ecological Monographs 52:179–198.

Hines, A.H. 2007. Ecology of juvenile and adult blue crabs. Pages 564–573 *in* V.S. Kennedy and L.E. Cronin, editors. Biology and management of the blue crab. Maryland Sea Grant Press, College Park, Maryland, USA.

Hines, A.H., and G.M. Ruiz. 1995. Temporal variation in juvenile blue crab mortality: Nearshore shallows and cannibalism in Chesapeake Bay. Bulletin of Marine Science 57:884–901.

Hines, A.H., A.M. Haddon, and L.A. Wiechert. 1990. Guild structure and foraging impact of blue crabs and epibenthic fish in a subestuary of Chesapeake Bay. Marine Ecology Progress Series 67:105–126.

Hovel, K.A., and R.N. Lipcius. 2001. Habitat fragmentation in a seagrass landscape: Patch size and complexity control blue crab survival. Ecology 82:1814–1829.

Hovel, K.A., and R.N. Lipcius. 2002. Effects of seagrass habitat fragmentation on juvenile blue crab survival and abundance. Journal of Experimental Marine Biology and Ecology 271:75–98.

Hull, W.W., and P.E. Bourdeau. 2017. Can crabs kill like a keystone predator? A field-test of the effects of crab predation on mussel mortality on a northeast Pacific rocky shore. PloS ONE 12:e0183064.

Hunt, R.J., and V.F. Matveev. 2005. The effects of nutrients and zooplankton community structure on phytoplankton growth in a subtropical Australian reservoir: An enclosure study. Limnologica 35:90–101.

Irigoien, X., J. Huisman, and R.P. Harris. 2004. Global biodiversity patterns of marine phytoplankton and zooplankton. Nature 429:863–867.

Jeng, M.S., N.K. Ng, and P.K.L. Ng. 2004. Hydrothermal vent crabs feast on sea 'snow'. Nature 432:969.

Jernakoff, P., A. Brearley, and J. Nielsen. 1996. Factors affecting grazer–epiphyte interactions in temperate seagrass meadows. Oceanography and Marine Biology: An Annual Review 34:109–162.

Johnson, K.D., J.H. Grabowski, and D.L. Smee. 2014. Omnivory dampens trophic cascades in estuarine communities. Marine Ecology Progress Series 507:197–206.

Johnston, C.A., and R.N. Lipcius. 2012. Exotic macroalga *Gracilaria vermiculophylla* provides superior nursery habitat for native blue crab in Chesapeake Bay. Marine Ecology Progress Series 467:137–146.

Johnston, K., B.J. Robson, and P.G. Fairweather. 2011. Trophic positions of omnivores are not always flexible: Evidence from four species of freshwater crayfish. Austral Ecology 36:269–279.

Jormalainen, V., T. Honkanen, and N. Heikkilä. 2001a. Feeding preferences and performance of a marine isopod on seaweed hosts: Cost of habitat specialization. Marine Ecology Progress Series 220:219–230.

Jormalainen, V., T. Honkanen, A. Mäkinen, A. Hemmi, and O. Vesakoski. 2001b. Why does herbivore sex matter? Sexual differences in utilization of *Fucus vesiculosus* by the isopod *Idotea baltica*. Oikos 93:77–86.

Kilar, J.A., and R.M. Lou. 1986. The subtleties of camouflage and dietary preference of the decorator crab, *Microphrys bicornutus* Latreille (Decapoda: Brachyura). Journal of Experimental Marine Biology and Ecology 101:143–160.

Kishida, O., G.C. Trussell, A. Mougi, and K. Nishimura. 2010. Evolutionary ecology of inducible morphological plasticity in predator–prey interaction: Toward the practical links with population ecology. Population Ecology 52:37–46.

Klumpp, D.W., and P.D. Nichols. 1983. Utilization of the seagrass *Posidonia australis* as food by the rock crab *Nectocarcinus integrifons* (Latreille) (Crustacea: Decapoda: Portunidae). Marine Biology Letters 4:331–339.

Koga, T., S. Goshima, M. Murai, and S. Poovachiranon. 1995. Predation and cannibalism by the male fiddler crab *Uca tetragonon*. Journal of Ethology 13:181–183.

Kotta, J., T. Wernberg, H. Jänes, I. Kotta, K. Nurkse, M. Pärnoja, and H. Orav-Kotta. 2018. Novel crab predator causes marine ecosystem regime shift. Scientific Reports 8:1–7.

Kuris, A.M., R.F. Hechinger, J.C. Shaw, K.L. Whitney, L. Aguirre-Macedo, C.A. Boch, A.P. Dobson, E.J. Dunham, B.L. Fredensborg, T.C. Huspeni, J. Lorda, L. Mababa, F.T. Mancini, A.B. Mora, M. Pickerin, N.L. Talhouk, M.E. Torchin, and K.D. Lafferty. 2008. Ecosystem energetic implications of parasite and free-living biomass in three estuaries. Nature 454:515–518.

Kurta, A. 1982. Social facilitation of foraging behavior by the hermit crab, *Coenobita compressus*, in Costa Rica. Biotropica 14:132–136.

Lafferty, K.D., S. Allesina, M. Arim, C.J. Briggs, G. De Leo, A.P. Dobson, J.A. Dunne, P.T. Johnson, A.M. Kuris, D.J. Marcogliese, N.D. Martinez, J. Memmott, P.A. Marquet, J.P. McLaughlin, E.A. Mordecai, M. Pascual, R. Poulin, and D.W. Thieltges. 2008. Parasites in food webs: The ultimate missing links. Ecology Letters 11:533–546.

Lafferty, K.D., A.P. Dobson, and A.M. Kuris. 2006. Parasites dominate food web links. Proceedings of the National Academy of Sciences 103:11211–11216.

Lafferty, K.D., and A.M. Kuris. 2002. Trophic strategies, animal diversity and body size. Trends in Ecology and Evolution 17:507–513.

Lafferty, K.D., and A.M. Kuris. 2009. Parasitic castration: The evolution and ecology of body snatchers. Trends in Parasitology 25:564–572.

Lawal-Are, A.O., and A.O. Adefule. 2017. Growth pattern, fecundity, food and feeding habits of *Panulirus regius* (De Brito Capello) off the Coast of Nigeria. UNILAG Journal of Medicine, Science and Technology 5:62–73.

Lee, M.A.B. 1985. The dispersal of *Pandanus tectorius* by the land crab *Cardisoma carnifex*. Oikos 45:169–173.

Lee, S.Y. 1989. The importance of sesarminae crabs *Chiromanthes* spp. and inundation frequency on mangrove (*Kandelia candel* (L.) Druce) leaf litter turnover in a Hong Kong tidal shrimp pond. Journal of Experimental Marine Biology and Ecology 131:23–43.

Leidenberger, S., K. Harding, and P.R. Jonsson. 2012. Ecology and distribution of the isopod genus *Idotea* in the Baltic Sea: Key species in a changing environment. Journal of Crustacean Biology 32:359–389.

Li, C., K.N. Wheeler, and J.D. Shields. 2011. Lack of transmission of *Hematodinium* sp. in the blue crab *Callinectes sapidus* through cannibalism. Diseases of Aquatic Organisms 96:249–258.

Lipcius, R.N., and W.F. Herrnkind. 1982. Molt cycle alterations in behavior, feeding and diel rhythms of a decapod crustacean, the spiny lobster *Panulirus argus*. Marine Biology 68:241–252.

Lipcius, R.N., D.B. Eggleston, K.L. Heck, Jr., R.D. Seitz, and J. van Montfrans. 2007. Ecology of postlarvae and young juvenile blue crabs. Pages 535–565 *in* V.S. Kennedy and L.E. Cronin, editors. Biology and management of the blue crab. Maryland Sea Grant Press, College Park, Maryland, USA.

Longmire, K.S., R.D. Seitz, A. Smith, and R.N. Lipcius. 2021. Saved by the shell: Oyster reefs can shield juvenile blue crabs *Callinectes sapidus*. Marine Ecology Progress Series 672:163–173.

Lovrich, G.A., and B. Sainte-Marie. 1997. Cannibalism in the snow crab, *Chionoecetes opilio* (O. Fabricius) (Brachyura: Majidae), and its potential importance to recruitment. Journal of Experimental Marine Biology and Ecology 211:225–245.

MacDonald, J.A., R. Roudez, T. Glover, J.S. Weis. 2007. The invasive green crab and Japanese shore crab: Behavioral interactions with a native crab species, the blue crab. Biological Invasions 9:837–848.

Malley, D.F. 1978. Degradation of mangrove leaf litter by the tropical sesarmid crab *Chiromanthes onychophorum*. Marine Biology 49:377–386.

Mann, K.H., and P.A. Breen. 1972. The relation between lobster abundance, sea urchins, and kelp beds. Journal of the Fisheries Research Board of Canada 29:603–605.

Mansour, R.A., and R.N. Lipcius. 1991. Density-dependent foraging and mutual interference in blue crabs preying upon infaunal clams. Marine Ecology Progress Series 72:239–246.

Marchessaux, G., M.C. Mangano, S. Bizzarri, C. M'Rabet, E. Principato, N. Lago, D. Veyssiere, M. Garrido, S.B. Scyphers, and G. Sarà. 2023. Invasive blue crabs and small-scale fisheries in the Mediterranean Sea: Local ecological knowledge, impacts and future management. Marine Policy 148:105461.

Mayer, G., G. Maier, A. Maas, and D. Waloszek. 2008. Mouthparts of the ponto-caspian invader *Dikerogammarus villosus* (Amphipoda: Pontogammaridae). Journal of Crustacean Biology 28:1–15.

Mendonça, V., and C. Vinagre. 2018. Short food chains, high connectance and a high rate of cannibalism in food web networks of small intermittent estuaries. Marine Ecology Progress Series 587:17–30.

Menge, B.A., L.R. Ashkenas, and A. Matson. 1983. Use of artificial holes in studying community development in cryptic marine habitats in a tropical rocky intertidal region. Marine Biology 77:129–142.

Menge, B.A., E.L. Berlow, C.A. Blanchette, S.A. Navarrete, and S.B. Yamada. 1994. The keystone species concept: Variation in interaction strength in a rocky intertidal habitat. Ecological Monographs 64:249–286.

Menge, B.A., T.C. Gouhier, T. Freidenburg, and J. Lubchenco. 2011. Linking long-term, large-scale climatic and environmental variability to patterns of marine invertebrate recruitment: Toward explaining "unexplained" variation. Journal of Experimental Marine Biology and Ecology 400:236–249.

Menge, B.A., and J. Lubchenco. 1981. Community organization in temperate and tropical rocky inter-tidal habitats: Prey refuges in relation to consumer pressure gradients. Ecological Monographs 51:429–450.

Menge, B.A. and J.P. Sutherland. 1976. Species diversity gradients: Synthesis of the roles of predation, competition, and temporal heterogeneity. American Naturalist 110:351–369.

Micheli, F., C.H. Peterson, L.S Mullineaux, C.R. Fisher, S.W. Mills, G. Sancho, G.A. Johnson, and H.S. Lenihan. 2002. Predation structures communities at deep-sea hydrothermal vents. Ecological Monographs 72:365–382.

Milner, R.N.C., T. Detto, M.D. Jennions, and P.R.Y Backwell. 2010. Hunting and predation in a fiddler crab. Journal of Ethology 28:171–173.

Mintz, J.D., R.N. Lipcius, D.B. Eggleston, and M.S. Seebo. 1994. Survival of juvenile Caribbean spiny lobster: Effects of shelter size, geographic location and conspecific abundance. Marine Ecology Progress Series 112:255–266.

Moksnes, P.-O., M. Gullström, K. Tryman, and S. Baden. 2008. Trophic cascades in a temperate seagrass community. Oikos 117:763–777.

Moksnes, P.-O., R.N. Lipcius, L. Pihl, and J. van Montfrans. 1997. Cannibal–prey dynamics in young juveniles and postlarvae of the blue crab. Journal of Experimental Marine Biology and Ecology 215:157–187.

Moksnes, P.-O., L. Pihl, and J. van Montfrans. 1998. Predation on postlarvae and juveniles of the shore crab *Carcinus maenas*: Importance of shelter, size and cannibalism. Marine Ecology Progress Series 166:211–225.

Montague, C.L. 1980. A natural history of temperate western Atlantic fiddler crabs (genus *Uca*) with reference to their impact on the salt marsh. Contributions in Marine Science 23:25–55.

Morton, D.N., C.Y. Antonino, F.J. Broughton, L.N. Dykman, A.M. Kuris, and K.D. Lafferty. 2021. A food web including parasites for kelp forests of the Santa Barbara Channel, California. Scientific Data 8:1–14.

Musrri, C.A., A.G. Poore, I.A. Hinojosa, E.C. Macaya, A.S. Pacheco, A. Pérez-Matus, O. Pino-Olivares, N. Riquelme-Pérez, W.B. Stotz, N. Valdivia, V. Villalobos, and M. Thiel. 2019. Variation in consumer pressure along 2500 km in a major upwelling system: Crab predators are more important at higher latitudes. Marine Biology 166:1–17.

Nakasone, Y., Y. Ono, and S. Goshima. 1983. Daily activity and food consumption of the sesarmid crab *Chasmagnathus convexus* (Decapoda, Brachyura). Bulletin of the College of Education (University of the Ryukyus) 26:37–53.

Natunewicz, C.C., C.E. Epifanio. 2001. Spatial and temporal scales of zoeal patches in coastal waters. Marine Ecology Progress Series 212:217–222.

Navarro-Barranco, C., M. Ros, J. Tierno de Figueroa, and J. Guerra-García. 2020. Marine crustaceans as bioindicators: Amphipods as case study. Pages 435–462 *in* G. Lovrich and M. Thiel, editors. The natural history of the Crustacea, Volume 9: Fisheries and aquaculture. Oxford University Press, New York, USA.

Neilson, M.J., R.L. Giddins, and G.N. Richards. 1986. Effects of tannins on the palatability of mangrove leaves to the tropical sesarminid crab *Neosarmatium smithi*. Marine Ecology Progress Series 34:185–186.

Neutel, A.M., J.A. Heesterbeek, and P.C. De Ruiter. 2002. Stability in real food webs: Weak links in long loops. Science 296:1120–1123.

Nicotri, M.E. 1980. Factors involved in herbivore food preference. Journal of Experimental Marine Biology and Ecology 42:13–26.

Nowak, B.F., M.B. Martin, and S. Boltaña. 2020. Parasitic crustaceans. Pages 401–434 *in* G. Lovrich and M. Thiel, editors. The natural history of the Crustacea, Volume 9: Fisheries and aquaculture. Oxford University Press, New York, USA.

O'Beirn, F.X., and R.L. Walker. 1999. Pea crab, *Pinnotheres ostreum* Say, 1817, in the eastern oyster, *Crassostrea virginica* (Gmelin, 1791): Prevalence and apparent adverse effects on oyster gonad development. The Veliger 42:17–20.

O'Connor, N.E., M.C. Emmerson, T.P. Crowe, and I. Donohue. 2013. Distinguishing between direct and indirect effects of predators in complex ecosystems. The Journal of Animal Ecology 82:438–448.

O'Dowd, D.J., and P.S. Lake. 1989. Red crabs in rainforest, Christmas Island: Removal and relocation of leaf-litter. Journal of Tropical Ecology 5:337–348.

Olsson, J., L. Bergström, and A. Gårdmark. 2013. Top-down regulation, climate and multi-decadal changes in coastal zoobenthos communities in two Baltic Sea areas. PLoS ONE 8:e64767.

Orth, R.J., and J. van Montfrans. 1984. Epiphyte–seagrass relationships with an emphasis on the role of micrograzing: A review. Aquatic Botany 18:43–69.

Östman, Ö., J. Eklöf, B.K. Eriksson, J. Olsson, P.O. Moksnes, and U. Bergström. 2016. Top-down control as important as nutrient enrichment for eutrophication effects in North Atlantic coastal ecosystems. Journal of Applied Ecology 53:1138–1147.

Paine, R.T. 1966. Food web complexity and species diversity. American Naturalist 100:65–75.

Paine, R.T. 1969. A note on trophic complexity and community stability. American Naturalist 103:91–93.

Pansch, C., I. Schaub, J. Havenhand, and M. Wahl. 2014. Habitat traits and food availability determine the response of marine invertebrates to ocean acidification. Global Change Biology 20:765–777.

Pardo, L.M., C. Andrade, L. Zenteno-Devaud, B. Garrido, and C. Rivera. 2021. Trophic ecology of juvenile southern king crab associated with kelp forest: Evidence of cannibalism. Diversity 13:556.

Pimm, S.L. 1980. Properties of food webs. Ecology 61:219–225.

Pimm, S.L. and J.H. Lawton. 1978. On feeding on more than one trophic level. Nature 275:542–544.

Polis, G.A. 1988. Exploitation competition and the evolution of interference, cannibalism, and intraguild predation in age/size-structured populations. Pages 185–202 *in* B. Ebenman, editor. Size-structured populations. Springer, Berlin, Heidelberg, Germany.

Poore, A.G., A. Graba-Landry, M. Favret, H.S. Brennand, M. Byrne, and S.A. Dworjanyn. 2013. Direct and indirect effects of ocean acidification and warming on a marine plant-herbivore interaction. Oecologia 173:1113–1124.

Poore, A.G., A.H. Campbell, R.A. Coleman, G.J. Edgar, V. Jormalainen, P.L. Reynolds, E.E. Sotka, J.J. Stachowicz, R.B. Taylor, M.A. Vanderklift, and J.E. Duffy. 2012. Global patterns in the impact of marine herbivores on benthic primary producers. Ecology letters 15:912–922.

Poore, A.G., M.J. Watson, R. de Nys, J.K. Lowry, and P.D. Steinberg. 2000. Patterns of host use among alga- and sponge-associated amphipods. Marine Ecology Progress Series 208:183–196.

Poore, A.G., S.T. Ahyong, J.K. Lowry, and E.E. Sotka. 2017. Plant feeding promotes diversification in the Crustacea. Proceedings of the National Academy of Sciences, 114:8829–8834.

Poore, A.G., L. Gutow, A.N. Lörz, and M. Thiel. 2018. Nest building by a small mesograzer limits blade size of the giant kelp *Macrocystis pyrifera*. Marine Biology 165:1–9.

Poore, A.G., N.A. Hill, and E.E. Sotka. 2008. Phylogenetic and geographic variation in host breadth and composition by herbivorous amphipods in the family Ampithoidae. Evolution 62:21–38.

Poore, G.C., and N.L. Bruce. 2012. Global diversity of marine isopods (except Asellota and crustacean symbionts). PLoS ONE 7:e43529.

Poulin, R., I. Blasco-Costa, and H.S. Randhawa. 2016. Integrating parasitology and marine ecology: Seven challenges towards greater synergy. Journal of Sea Research 113:3–10.

Power, M.E. 1992. Top-down and bottom-up forces in food webs: Do plants have primacy? Ecology 73:733–746.

Preisser, E.L., D.I. Bolnick, and M.F. Benard. 2005. Scared to death? The effects of intimidation and consumption in predator-prey interactions. Ecology 86:501–509.

Pretorius, M., and A. Avenant-Oldewage. 2021. Parasites as biological indicators: The impact of environmental quality on the infections of *Lamproglena clariae* (Crustacea) on *Clarias gariepinus* along the Vaal River, South Africa. Biological Trace Element Research 200:2937–2947.

Raffel, T.R., L.B. Martin, and J.R. Rohr. 2008. Parasites as predators: Unifying natural enemy ecology. Trends in Ecology and Evolution 23:610–618.

Reaka, M.L. 1976. Lunar and tidal periodicity of molting and reproduction in stomatopod Crustacea: A selfish herd hypothesis. Biological Bulletin 150:468–490.

Reustle, J.W., and D.L. Smee. 2020. Turbidity and salinity influence trophic cascades on oyster reefs through modification of sensory performance and facilitation of different predator types. Marine Ecology Progress Series 639:127–136.

Ricciardi, A. 2011. Crustaceans (other). Pages 135–137 *in* D. Simberloff and M. Rejmánek, editors. Encyclopedia of biological invasions. University of California Press, Berkeley and Los Angeles, USA.

Robertson, A.I., and P.A. Daniel. 1989. The influence of crabs on litter processing in high intertidal mangrove forests in tropical Australia. Oecologia 78:191–198.

Robles, C., and J. Robb. 1993. Varied carnivore effects and the prevalence of intertidal algal turfs. Journal of Experimental Marine Biology and Ecology 166:65–91.

Robles, C., D. Sweetnam, and J. Eminike. 1990. Lobster predation on mussels: Shore-level differences in prey vulnerability and predator preference. Ecology 71:1564–1577.

Romano, N., and C. Zeng. 2017. Cannibalism of decapod crustaceans and implications for their aquaculture: A review of its prevalence, influencing factors, and mitigating methods. Reviews in Fisheries Science and Aquaculture 25:42–69.

Romanuk, T.N., Y. Zhou, U. Brose, E.L. Berlow, R.J. Williams, and N.D. Martinez. 2009. Predicting invasion success in complex ecological networks. Philosophical Transactions of the Royal Society B 364:743–1754.

Rose, D.P., D.M. Calhoun, and P.T. Johnson. 2020. Infection prevalence and pathology of the cymothoid parasite *Olencira praegustator* in Atlantic menhaden. Invertebrate Biology 139:e12300.

Rothäusler, E., F. Haavisto, and V. Jormalainen. 2017. Is the future as tasty as the present? Elevated temperature and hyposalinity affect the quality of *Fucus* (Phaeophyceae, Fucales) as food for the isopod *Idotea balthica*. Marine Biology 164:1–15.

Richoux, N.B., I. Vermeulen, and P.W. Froneman. 2014. Stable isotope ratios indicate differential omnivory among syntopic rocky shore suspension-feeders. Marine Biology 161:971–984.

Ryther, J.H., and J.G. Sanders. 1980. Experimental evidence of zooplankton control of the species composition and size distribution of marine phytoplankton. Marine Ecology Progress Series 3:279–284.

Sainte-Marie, B., and M. Lafrance. 2002. Growth and survival of recently settled snow crab *Chionoecetes opilio* in relation to intra- and intercohort competition and cannibalism: A laboratory study. Marine Ecology Progress Series 244:191–203.

Schweitzer, C.E., and R.M. Feldmann. 2010. The Decapoda (Crustacea) as predators on Mollusca through geologic time. Palaios 25:167–182.

Seiple, W.H. 1981. The ecological significance of the locomotor activity rhythms of *Sesarma cinereum* and *S. reticulatum*. Crustaceana 40:5–15.

Seitz, R.D. 2011. Gradient effects on structuring of soft-bottom benthic infauna: *Macoma balthica* and predation, recruitment, and food availability. Journal of Experimental Marine Biology and Ecology 409:114–122.

Seitz, R.D., R.N. Lipcius, W.T. Stockhausen, K.A. Delano, M.S. Seebo, and P.D. Gerdes. 2003. Potential bottom-up control of blue crab distribution at various spatial scales. Bulletin of Marine Science 72:471–490.

Shields, J.D., and J.P. Huchin-Mian. 2020. Ecological factors in the emergence of pathogens in commercially important crustaceans. Pages 367–399 *in* G. Lovrich and M. Thiel, editors. The natural history of the Crustacea, Volume 9: Fisheries and aquaculture. Oxford University Press, New York, USA.

Shields, J.D., and F.W. Wood. 1993. Impact of parasites on the reproduction and fecundity of the blue sand crab *Portunus pelagicus* from Moreton Bay, Australia. Marine Ecology Progress Series 92:159–170.

Siegel, V., and J.L. Watkins. 2016. Distribution, biomass and demography of Antarctic krill, *Euphausia superba*. Pages 21–100 *in* V. Siegel, editor. Biology and ecology of Antarctic krill. Springer, Cham, Switzerland.

Siegenthaler, A., O.S. Wangensteen, C. Benvenuto, R. Lollobrigidi, and S. Mariani. 2022. Niche separation between two dominant crustacean predators in European estuarine soft-bottom habitats. Ecological Indicators 138:108839.

Sih, A., P. Crowley, M. McPeek, J. Petranka, and K. Strohmeier. 1985. Predation, competition, and prey communities: A review of field experiments. Annual Review of Ecology and Systematics 16:269–311.

Silliman, B.R., and M.D. Bertness. 2002. A trophic cascade regulates salt marsh primary production. Proceedings of the National Academy of Sciences 99:10500–10505.

Smith, T.J., III. 1987. Seed predation in relation to tree dominance and distribution in mangrove forests. Ecology 68:266–273.

Smith, T.J., III, H.T. Chen, C.C. Mclvor, and M.B. Robblee. 1989. Comparisons of seed predation in tropical, tidal forests from three continents. Ecology 70:146–151.

Sommer, U. 1997. Selectivity of *Idothea chelipes* (Crustacea: Isopoda) grazing on benthic microalgae. Limnology and Oceanography 42:1622–1628.

Sousa, W.P. 1979. Experimental investigation of disturbance and ecological succession in a rocky intertidal algal community. Ecological Monographs 49:227–254.

Tarling, G.A., J. Cuzin-Roudy, and F. Buchholz. 1999. Vertical migration behaviour in the northern krill *Meganyctiphanes norvegica* is influenced by moult and reproductive processes. Marine Ecology Progress Series 190:253–262.

Thompson, P.L., and J.B. Shurin. 2012. Regional zooplankton biodiversity provides limited buffering of pond ecosystems against climate change. The Journal of Animal Ecology 81:251–259.

Thompson, R.M., M. Hemberg, B.M. Starzomski, and J.B. Shurin. 2007. Trophic levels and trophic tangles: The prevalence of omnivory in real food webs. Ecology 88:612–617.

Tilburg, C.E., L.T. Houser, C.N. Steppe, R.W. Garvine, and C.E. Epifanio. 2006. Effects of coastal transport on larval patches: Models and observations. Estuarine, Coastal and Shelf Science 67:145–160.

Trussell, G.C., C.M. Matassa, and P.J. Ewanchuk. 2017. Moving beyond linear food chains: Trait-mediated indirect interactions in a rocky intertidal food web. Proceedings of the Royal Society B: Biological Sciences 284:20162590.

Tuomi, J., H. Ilvessalo, P. Niemelä, S. Sirén, and V. Jormalainen. 1989. Within-plant variation in phenolic content and toughness of the brown alga *Fucus vesiculosus* L. Botanica Marina 32:505–510.

Van Allen, B.G., F.P. Dillemuth, A.J. Flick, M.J. Faldyn, D.R. Clark, V.H. Rudolf, and B.D. Elderd. 2017. Cannibalism and infectious disease: Friends or foes? The American Naturalist 190:299–312.

Vannini, M., G. Chelazzi, and F. Cherardi. 1989. Feeding habits of the pebble crab *Eriphia smithi* (Crustacea, Brachyura, Menippidae). Marine Biology 100:249–252.

Veldhuizen, T.C., and S. Stanish. 1999. Overview of the life history, distribution, abundance and impacts of the Chinese mitten crab, *Eriocheir sinensis*. California Department of Water Resources, Environmental Services Office, Sacramento, USA.

Vidal-Martínez, V.M., D. Pech, B. Sures, S.T. Purucker, and R. Poulin. 2010. Can parasites really reveal environmental impact? Trends in Parasitology 26:44–51.

Virnstein, R.W. 1979. Predation on estuarine infauna: Response patterns of component species. Estuaries 2:69–86.

Vogel, H.H., Jr., and J.R. Kent. 1971. A cunous case: The coconut crab. Fauna 2:3–11.

Vumazonke, L.U., E.A. Pakhomov, P.W. Froneman, and C.D. McQuaid. 2003. Diet and daily ration of male and female caridean shrimp *Nauticaris marionis* at the Prince Edward Archipelago. Polar Biology 26:420–422.

Walker, G. 2001. Introduction to the Rhizocephala (Crustacean: Cirripedia). Journal of Morphology 249:1–8.

Walton, W.C., C. MacKinnon, L.F. Rodriguez, C. Proctor, and G.M. Ruiz. 2002. Effect of an invasive crab upon a marine fishery: Green crab, *Carcinus maenas*, predation upon a venerid clam, *Katelysia scalarina*, in Tasmania (Australia). Journal of Experimental Marine Biology and Ecology 272:171–189.

Wear, R.G., and M. Haddon. 1987. Natural diet of the crab *Ovalipes catharus* (Crustacea, Portunidae) around central and northern New Zealand. Marine Ecology Progress Series 35:39–49.

Weissburg, M. and J. Beauvais. 2015. The smell of success: The amount of prey consumed by predators determines the strength and range of cascading non-consumptive effects. PeerJ 3:e1426.

Weissburg, M.J., D.L. Smee, and M.C. Ferner. 2014. The sensory ecology of non-consumptive effects. American Naturalist 182:141–157.

Whitlow, W.L. 2010. Changes in survivorship, behavior, and morphology in native soft-shell clams induced by invasive green crab predators. Marine Ecology 31:418–430.

Wiklund, A.-K.E., B. Sundelin, and R. Rosa. 2008. Population decline of amphipod *Monoporeia affinis* in Northern Europe: Consequence of food shortage and competition? Journal of Experimental Marine Biology and Ecology 367:81–90.

Williams, J.D., and C.B. Boyko. 2012. The global diversity of parasitic isopods associated with crustacean hosts (Isopoda: Bopyroidea and Cryptoniscoidea). PLoS ONE 7:p.e35350.

Williams, E.H., and L. Bunkley-Williams. 2019. Life cycle and life history strategies of parasitic Crustacea. Pages 179–266 *in* N.J. Smit, N.L. Bruce, and K.A. Hadfield, editors. Parasitic Crustacea state of knowledge and future trends. Zoological Monographs 3, Springer, Cham, Switzerland.

Wilson, K.A. 1989. Ecology of mangrove crabs: Predation, physical factors and refuges. Bulletin of Marine Science 44:263–273.

Wilson, W.H. 1991. Competition and predation in marine soft-sediment communities. Annual Review of Ecology and Systematics 21:221–241.

Winder, M., and D.E. Schindler. 2004. Climate change uncouples trophic interactions in an aquatic ecosystem. Ecology 85:2100–2106.

Wirsing, A.J., M.R. Heithaus, J.S. Brown, B.P. Kotler, and O.J. Schmitz. 2021. The context dependence of non-consumptive predator effects. Ecology Letters 24:113–129.

Wolcott, D.L., and N.J. O'Connor. 1992. Herbivory in crabs: Adaptations and ecological considerations. American Zoologist 32:370–381.

Wolcott, D.L., and T.G. Wolcott. 1987. Nitrogen limitation in the herbivorous land crab *Cardisoma guanhumi*. Physiological Zoology 60:262–268.

Wolcott, T.G., and D.L. Wolcott. 1988. When limiting factors aren't: Lessons from land crabs. Pages 135–151 *in* G. Chelazzi and M. Vannini, editors. Behavioral adaptation to intertidal life. Springer, Boston, Massachusetts, USA.

Worm, B., H.K. Lotze, H. Hillebrand, and U. Sommer. 2002. Consumer versus resource control of species diversity and ecosystem functioning. Nature 417:848–851.

Worm, B., and U. Sommer. 2000. Rapid direct and indirect effects of a single nutrient pulse in a seaweed-epiphyte-grazer system. Marine Ecology Progress Series 202:283–288.

Yamada, S.B., and E.G. Boulding. 1996. The role of highly mobile crab predators in the intertidal zonation of their gastropod prey. Journal of Experimental Marine Biology and Ecology 204:59–83.

Yamada, S.B., and E.G. Boulding. 1998. Claw morphology, prey size selection and foraging efficiency in generalist and specialist shell-breaking crabs. Journal of Experimental Marine Biology and Ecology 220:191–211.

Yun, H.Y., J. Cruz, M. Treitschke, M. Wahl, and M. Molis. 2007. Testing for the induction of anti-herbivory defenses in four Portuguese macroalgae by direct and water-borne cues of grazing amphipods. Helgoland Marine Research 61:203–209.

Zamzow, J.P., C.D. Amsler, J.B. McClintock, and B.J. Baker. 2010. Habitat choice and predator avoidance by Antarctic amphipods: The roles of algal chemistry and morphology. Marine Ecology Progress Series 400:155–163.

Zimmer-Faust, R.K. 1987. Substrate selection and use by a deposit-feeding crab. Ecology 68:955–970.

3

VULTURES OF THE SEA AND SHREDDERS IN THE WOODS: CRUSTACEANS AS DETRITIVORES AND SCAVENGERS

Jonas C. Geburzi and Martin Zimmer

Abstract

Detritus, that is, dead organic matter of any size and origin, forms an important energetic and nutritional basis for food webs in terrestrial, aquatic, and marine ecosystems. Across all ecosystems, crustaceans are strongly involved in the turnover of detritus, thus being important mediators of this fundamental ecosystem process. Crustaceans have evolved physiological, behavioral, and morphological adaptations toward a specialized detritivorous or scavenging lifestyle, yet omnivory is a unifying feature of their feeding ecology. Their diversity, nutritional flexibility, and individual motility assign detritivorous crustaceans a central position in many food webs, connecting them to particularly high numbers of other food web components, in some cases even across habitat boundaries. Large detritivorous crustaceans, in particular, act as shortcuts in detrital food web pathways by directly linking detritus-derived energy to the highest trophic levels. In some ecosystems, crustaceans dominate detritivore communities. In freshwater streams and on vegetated coasts, crabs, isopods, and amphipods shred and consume up to 100% of the plant litter input. In the pelagic zones of lakes and oceans, copepods, mysids, and euphausiids mediate the flux of particulate organic matter through the water column. In the deep sea, scavenging amphipods are the primary consumers of food falls, from jellyfish to whale carcasses, acting as spatial and temporal vectors of detritus-derived energy. Anthropogenic impacts, such as pollution, fisheries, species introductions, or climate change, already affect detritivorous crustacean communities and their contributions to detritus processing in various ways, but their long-term effects are still poorly understood.

Jonas C. Geburzi and Martin Zimmer, *Vultures of the Sea and Shredders in the Woods: Crustaceans as Detritivores and Scavengers*. In: *Ecology and Conservation*. Edited by: Lars Gutow and Martin Thiel, Oxford University Press.
 DOI: 10.1093/oso/9780197768242.003.0003

INTRODUCTION: WHAT ARE DETRITIVORES?

Dead organic matter is an important source of energy and nutrients in aquatic and terrestrial food webs. Ultimately, its decay provides the basis for new primary production, as it leads to the mineralization and release of nutrients (and CO_2) to the environment, making them available to macrophytes and phytoplankton (Zimmer 2019). While decay and mineralization are driven by physico-chemical processes (e.g., leaching and photo-/autoxidation) and microbes, animals feeding on dead biomass play a fundamental role in the decomposition pathway. They (pre)process, recycle, and redistribute detrital matter, thus mediating and modifying microbial decay and mineralization, and providing links between the dead and living resource pools across trophic levels, within and among food webs, and ecosystems (Moore et al. 2004, Benbow et al. 2019; compare Fig. 3.1).

The term *detritus*, while historically referring to dead plant material, is now commonly used for dead organic matter of any origin and size (reviewed, for example, in Moore et al. 2004). Autotrophically derived detritus (leaf and algal litter, dead wood, dead phytoplankton cells, etc.) and heterotrophically derived detritus (carrion, feces, animal tissues and excretions, dead microbes, etc.) differ in their chemical composition, energy and nutrient content, and recalcitrance (Moore et al. 2004, Benbow et al. 2019). In this chapter, we follow this holistic definition of detritus, and, accordingly, use the term *detritivore* for species that feed on any type of dead biomass, while we use *scavenger*, that is, feeding on carrion or parts of larger animals, when specifically referring to this particular type of detritivory.

Being one of the most diverse and widespread invertebrate groups, crustaceans represent a broad variety of types of detritivory across major marine, freshwater, and terrestrial ecosystems, from the polar regions to the tropics. Isopods (Oniscidea) process plant litter in terrestrial forests, as do amphipods (Gammaridea), crayfish (Astacidae, Cambaridae) and some anomurans (Aeglidae) in streams, and crabs (e.g., Sesarmidae) in salt marshes and mangrove forests (Hammann and Zimmer 2015). Pelagic copepods and krill (Euphausiidae) play a crucial role in the turnover

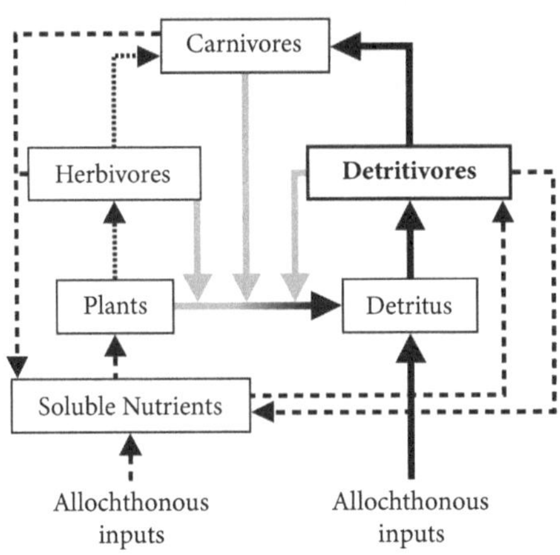

Fig. 3.1.
Generalized food web model illustrating the importance of dead organic matter (detritus) derived from autochthonous (gray arrows) or allochthonous sources, and of the decomposition (i.e., "brown") pathway (solid black arrows) for energy fluxes and food web integrity. Dotted arrows depict the grazer pathway originating from primary producers (plants, algae), dashed arrows depict mineralization of nutrients. Source: data from Moore et al. (2004).

and flux of particulate organic matter (POM) in the open ocean (Turner 2015), while amphipods (Lysianassoidae) are dominating benthic deep-sea scavengers (Havermans and Smetacek 2018). Various crustaceans are part of the unique successional detritivore communities that decompose whale and other megafaunal carcasses in the deep sea (Smith and Baco 2003).

Detritivory in crustaceans is generally not restricted to specific feeding habits. Many crustaceans that are commonly considered detritivores supplement their detrital diet with fresh food, while, for example, species generally characterized as omnivorous predators (e.g., large crabs, lobsters) opportunistically scavenge on carrion when it is available (see section "Detritivory vs. Omnivory"). Rather, crustacean detritivores may be characterized by their ecological role in contributing to, and mediating, the decomposition and turnover of dead biomass in their environment, which is based on their utilization of detrital food sources for at least parts of their life cycle.

In this chapter, we will briefly summarize how detritivorous crustaceans are adapted to their feeding habits, before exploring their ecological importance as drivers of decomposition processes across various ecosystems. We focus on the contribution of crustaceans to the assimilation and turnover of dead organic matter, their strategies to cope with scarcity and low nutritional quality of most forms of detritus, and highlight examples for competitive and synergistic interactions with other detritivores. We conclude this chapter by looking at how anthropogenic impacts might affect detritivorous crustaceans and their contribution to the essential ecosystem process of detritus turnover.

FEEDING ECOLOGY OF DETRITIVOROUS CRUSTACEANS

Detritivory vs. Omnivory

By narrow definition, a detritivore is an organism that exclusively feeds on dead organic matter. However, many crustacean detritivores at least occasionally include live prey, fresh plant material, etc. in their diet. Strictly speaking, they are therefore omnivores, that is, organisms that feed on several trophic levels, even leaving aside the long-standing discussion of whether the microbes colonizing dead biomass should be considered a part of detritus, or not (Zimmer 2019). For example, filter and deposit feeders will inevitably consume a mixture of organic particles and plankton or micro-/meiobenthos, respectively (Watling 2015). Detritivorous intertidal and semi-terrestrial crabs (mostly Grapsoidea), which mainly feed on algal and other plant litter, commonly supplement their diet by consuming amphipods and isopods inhabiting the litter patches. Such behavior probably compensates for the low nutritional value and high recalcitrance of their detrital food (e.g., Buck et al. 2003).

Scarcity and patchiness of detritus is another key factor leading to facultative omnivory in crustacean detritivores, particularly for scavengers in nutrient-poor environments. DNA metabarcoding of the gut contents of deep-sea amphipods provided evidence for predation (including cannibalism) and phytodetritivory in these supposedly highly specialized scavengers (Blankenship and Levin 2007). In fact, the question of whether an obligate crustacean scavenger even exists, and is energetically feasible, is still not fully resolved (compare Britton and Morton 1994, Ruxton and Houston 2004). At the same time, many generalist predators, particularly among decapods (e.g., crabs, lobsters, shrimp), are known to readily feed on carrion when it is available, a trait that is extensively used with baited traps in the fisheries of commercial species (Wahle et al. 2020). These opportunistic scavengers may contribute significantly to carrion removal in aquatic systems (Eastman and Thiel 2015).

Flexibility in the utilization of detrital and non-detrital food sources may thus be considered a unifying characteristic across crustacean detritivores. Through this, they affect decomposition processes in multiple ways. Directly, by consuming detritus themselves, and indirectly, by preying on other detritivores, thereby reducing the total number of organisms feeding on detritus, as well as their own detritus consumption, as nutritional requirements are partly met through their animal prey.

Finding Food

Spatiotemporal scarcity and patchiness of detrital food sources, as well as their rapid decrease in nutritional value, are common challenges for detritivores. The ability to find food quickly and effectively is, thus, pivotal to their foraging success. Crustacean detritivores have developed several physiological and behavioral adaptations that enhance their abilities to locate food (see e.g., Eastman and Thiel 2015, Hammann and Zimmer 2015 for extensive reviews).

Chemoreception (i.e., olfaction/"smelling") is the most important sensory mode that crustacean detritivores use to detect food, in both terrestrial and aquatic habitats. Multiple examples illustrate that advanced chemoreception enables crustaceans not only to effectively find detritus, but also to distinguish between different types and qualities of potential food sources. Terrestrial isopods are capable of olfactorily detecting areas of high microbial activity in a seemingly uniform matrix of leaf litter, which are indicative of detritus patches with lower recalcitrance and/or higher nutritional value (Zimmer et al. 2003). Similarly, olfaction enables sandhoppers (Amphipoda, Talitridae) to rapidly and selectively colonize freshly deposited patches of kelp-containing macroalgal detritus on sandy beaches (Michaud et al. 2019). For scavenging crustaceans, amino acids that exude from the tissues of dead and injured marine animals, or that are associated with decomposition processes, are important chemical foraging cues. Following concentration gradients of such cues, deep-sea scavenging amphipods are capable of localizing carrion over distances ranging from meters to kilometers (details in Eastman and Thiel 2015). Intertidal hermit crabs (family Diogenidae) can perceive subtle differences in the composition of cues exuded by different types of carrion. Their divergent preferences for these cues likely act as a mechanism to reduce resource competition among sympatric crustacean detritivores (Tran 2013).

Mechanoreception of hydroacoustic and hydrodynamic stimuli occurs as an alternative method of food detection (particularly of carrion) by crustacean detritivores across various habitats. Sound waves produced by carcasses hitting the seafloor, and the feeding noises of other scavengers likely alert deep-sea amphipods and shrimp of potential food sources over long distances (Klages et al. 2002, Premke et al. 2003). Calanoid copepods most commonly track down detrital particles with their ability to perceive the weak turbulences these particles cause when sinking through the water column (van der Jagt et al. 2020). "Eavesdropping" even allows the terrestrial hermit crab *Coenobita compressus* to discriminate between carrion patches of higher and lower resource values by sensing the level of commotion caused by already-feeding conspecifics (Laidre 2013a). Compared to the above-mentioned stimuli, visual detection of food appears to play only a minor role for terrestrial and coastal crustacean detritivores, and becomes insignificant for most aquatic and marine detritivores due to water turbidity and lack of light.

Coupled with their specific sensory abilities, many crustacean detritivores show behavioral adaptations that enhance their ability to locate food. This is particularly true for pelagic and deep-sea species, which need to orient themselves three-dimensionally in the open water. For example, suprabenthic lysianassoid amphipods engage in fast multidirectional swimming to scan larger areas for carrion. Upon detection of an odor plume, they switch to directional swimming toward the food source. The threshold concentration of foraging cues that triggers this behavioral switch corresponds to the motility of the species (Havermans and Smetacek 2018). Likewise, pelagic copepods change their swimming direction and speed (up to five times faster) when encountering the chemical trail of a sinking marine snow particle (Lombard et al. 2013).

A factor that warrants further exploration is the potential role of social cues and group formation behavior for the foraging success of crustacean scavengers, hinted at by observations of dense aggregations of the deep-sea amphipod *Eurythenes gryllus* approaching fish bait over larger distances (Premke et al. 2003), or the above-mentioned study on *Coenobita compressus*.

Acquiring Food

Relying on food sources that are nutrient-poor and usually restricted and unpredictable in space and time drives behavioral and morphological adaptations, optimizing food acquisition by crustacean detritivores. Such adaptations may either increase the amount of acquired food, to meet the detritivore's energy demands, or the quality of the ingested detritus.

Phytodetritivorous burrow-dwelling decapods, particularly mangrove crabs (mainly Sesarmidae), but also crayfish (family Parastacidae) pull leaf litter into their burrows and store it before consumption (e.g., Growns and Richardson 1988, Fratini et al. 2000). As microbial processing and decay increase the palatability and nutritional value of leaf litter over time, the burying behavior ensures access to the food, by preventing leaves from being removed by tides, currents, or competing detritivores (Forgeron et al. 2021).

In contrast to leaf litter, (large) carrion is a spatially and temporally highly unpredictable food source, and not easy to store away for scavenging crustaceans. Gluttonous feeding without satiation, as observed, for example, in squat lobsters and amphipods, maximizes energy uptake from a rare feeding occasion (Sainte-Marie 1992, Karas et al. 2007). Enlarged, shearing mandibles, an expansible fore- or midgut, and a poorly calcified exoskeleton are morphological adaptations found in scavenging lysianassoid amphipods that support their ability to quickly ingest large amounts of carrion (Fig. 3.2). In fact, some deep-sea species are capable of ingesting more than 100% of their body weight in a single meal (Dahl 1979, Eastman and Thiel 2015).

Specialization on small detrital particles as a major food source led to the development of convergent adaptations in mouthpart functional morphology across various benthic crustaceans.

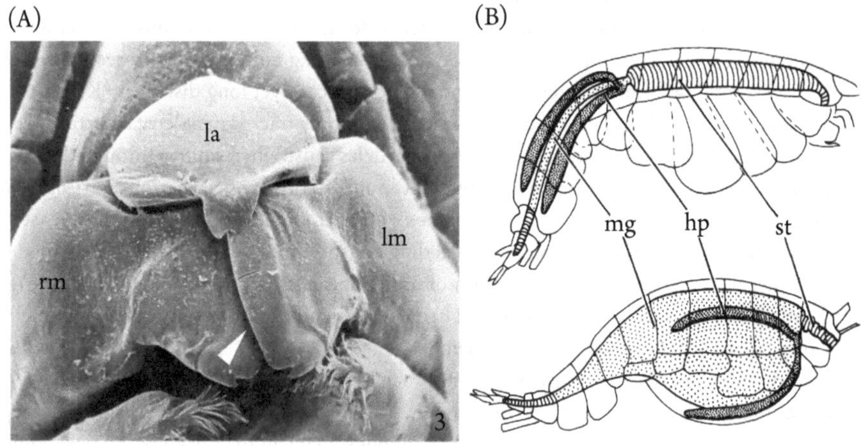

Fig. 3.2.
Morphological adaptations to scavenging in deep-sea Lysianassidae. (A) Shearing-type mandibles (rm, lm) of *Hirondellea gigas*, the elongated incisor edge of the left mandible is clearly visible (white arrow). (B) Schematic view of food storage by foregut expansion (e.g., *Orchomene*, top), and by midgut expansion (e.g., *Eurythenes, Hirondellea, Paralicella*, bottom). Adapted from Dahl (1979), with permission from © John Wiley & Sons.

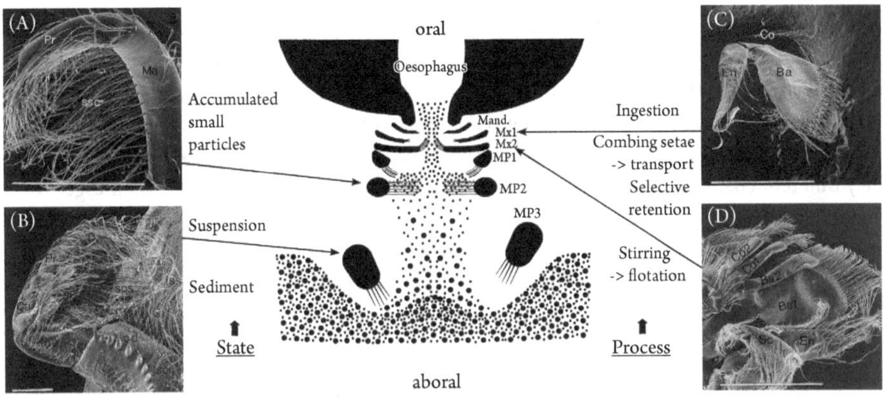

Fig. 3.3.
Functional morphology of the feeding appendages of the mud shrimp *Callianassa subterranea*. Center: schematic drawing of the particle sorting mechanism during sediment deposit feeding. Scanning electron microscopy images illustrate different setation patterns of the feeding appendages, depending on their position and function: (A) second maxilliped, (B) third maxilliped, (C) first maxilla, (D) second maxilla. Adapted from Stamhuis et al. (1998), with permission from © Springer Nature.

The third maxillipeds of various detritivorous Anomura, mud shrimps (Callianassidae), and the second gnathopods of Corophiidae (Amphipoda) alike are characterized by rows of long setae, used to effectively brush off deposited detrital particles from surfaces, or to filter suspended detritus from the water (Nicol 1932, Stamhuis et al. 1998, Watling 2015). The setation of the inner mouthparts in these species forms a mesh, which, together with coordinated mouthpart movements, leads to the selective ingestion of small- to medium-sized particles (Fig. 3.3). This is energetically beneficial as they have a larger detritus- or biofilm-covered surface relative to their weight and are thus of higher nutritional value. The ecological effects of this type of detritivory, particularly deposit feeding, can be substantial. By reaching either extremely high densities (Corophiidae) or high individual sediment turnover rates (Callianassidae), the feeding activity can control densities of sediment microbes, and even alter the sediment structure itself (Watling 2015).

Digesting Food

Upon fast leaching of soluble food compounds during the early stages of decay (Zimmer 2019), many remaining compounds of detritus are recalcitrant and difficult to digest. Such compounds can only be degraded by specialized digestive enzymes, such as "cellulases" (β-1,4-endoglucanase, cellobiohydrolase, and β-glucosidase) and phenol oxidases or peroxidases for lignin decay (Hammann and Zimmer 2015). Several studies provide evidence for the inducibility of digestive enzymes in mangrove and salt marsh crabs (e.g., Pavasovic et al. 2004, Hübner et al. 2015), leading to spatiotemporal variation in digestive enzyme activity; however, the potential effects of such variation on ecosystem processes remains essentially unknown.

While many of the enzymes required for the digestion of recalcitrant detrital matter are still considered to be of typically microbial origin, recent advances show that many detritivorous animals are also capable of producing such enzymes endogenously. Studies on various land crabs, freshwater decapods, and intertidal mangrove crabs strongly suggest that (at least) decapod crustaceans generally possess the genetic background for synthesizing endogenous cellulases (Linton et al. 2006, Bui and Lee 2015). The presence of cellulase enzymes in the digestive systems of deep-sea scavenging crustaceans such as *Chionoecetes* spp. (Brachyura) and *Hirondellea gigas* (Amphipoda) also indicates that "specialized" scavengers are, partly at least, capable of digesting sunken wood debris in adaptation to their highly oligotrophic habitats (Kobayashi et al. 2012, Adachi et al. 2016), and provides a striking example of adaptation to omnivory.

Relying on scarce and irregular feeding events, deep-sea scavenging crustaceans encounter entirely different challenges than phytodetritivores in less extreme environments. Their digestion is thought to be very effective, corresponding to the "batch reactor model" of assimilation. The energy gained from the digestive processes is mostly stored as lipids, allowing amphipods such as *Anonyx* spp. or *E. gryllus* to survive for five months (and potentially even longer) without feeding (Sainte-Marie 1992).

Interactions with, and Promotion of, Microorganisms

The low quality of plant detritus renders microbes important players in decaying the detritus itself, in supporting digestive processes of detritivores (see above), as indicators of high-quality food patches, or as additional food source to, or a high-quality alternative for, low-quality detritus (Zimmer et al. 2003). Toughness and toxic or repellent secondary compounds in the leaves of many plant species, particularly in the tropics, may even render microbial colonization a prerequisite for leaf litter consumption by crustaceans ("conditioning"; see Yang et al. 2020b for an example).

Microbial biofilms on detritus thus play an important role for crustacean detritivores and foster detritivory, whereas crustacean detritivores affect microbial abundance, community composition, and activity. Their effects are either direct, through (selective) feeding and digestion, or indirect, by changing their microenvironment physically and chemically (e.g., Marks 2019). Fenchel (1970) showed experimentally that the subtropical coastal amphipod *Parhyalella whelpleyi* only digests the microbes that colonize the seagrass detritus it ingests, but by breaking the detritus down into smaller particles during feeding, it increases the surface available for microbial colonization, resulting in an increase of microbial oxygen consumption that exceeds its own respiration. Examples from amphipods, isopods, and crabs in freshwater and semi-terrestrial habitats show that their effect on microbial respiration is species-specific, ranging from promotion to inhibition (Zimmer et al. 2002). Furthermore, crustacean detritivores mediate shifts in microbe communities, for example, from bacteria to fungi, through selective feeding and digestion (Moghadam and Zimmer 2016). The resulting changes in the composition and activity of the microbial community on detrital matter will have consequences for the mid- to long-term fate of the detritus.

THE ROLE OF CRUSTACEAN DETRITIVORES IN DIFFERENT ECOSYSTEMS

Terrestrial Ecosystems

In the terrestrial realm, crustaceans are mostly represented by isopods (Isopoda: Oniscidea), land crabs (Decapoda: Gecarcinidae and some Sesarmidae) and terrestrial hermit crabs (Decapoda: Coenobitidae). While the latter two live on tropical coasts and islands, the former colonize essentially all terrestrial habitats, from the intertidal zone through to montane and subalpine environment and deserts, except for permafrost soils and the polar environment. Terrestrial isopods are generally detritivores. As such, these crustaceans are embedded in a complex food web, linking their detrital and microbial food sources with invertebrate (e.g., carabid beetles, dysderid and other spiders, centipedes) and vertebrate (e.g., reptiles and amphibians, birds, shrews, and moles) predators.

Several studies have investigated the contribution of isopods to the processing of annual leaf litter fall in various forest ecosystems (see Richardson and Araujo 2015 for an overview). Quadros and Araujo (2008) estimated the oniscidean assemblage of a semi-deciduous forest in southern Brazil to process about 860 kg leaf litter ha^{-1} $year^{-1}$, the equivalent of about 16% of the annual leaf litter fall, which is among the highest estimates of processing rates (references in Zimmer 2002, Quadros and Araujo 2008). The ecological significance of terrestrial isopods lies not so much in the amount of consumed leaf litter, but its conversion into fecal material by the animals. First, oniscids redistribute detrital matter on the forest floor, as they move from feeding patches into shelter where they defecate, in order to avoid desiccation and possibly predation. These shelters represent microclimatic spaces that might foster microbial proliferation and detrital decay. Second, digestion by isopods changes the chemical composition of organic material during its conversion from ingestion (leaf litter) to egestion (feces) (Zimmer et al. 2004). While the effect of this conversion depends on the litter type and the detritivore species, it generally stimulates microbial activity, particularly when comparing microbial decay of fresh leaf litter and isopod feces derived from that litter (e.g., Hassall et al. 1987). Overall, the chemical conversion has far-reaching consequences for the terrestrial carbon cycle (Joly et al. 2020).

The decay of fecal pellets is further enhanced through coprophagy, that is, their consumption by the isopods themselves and by other detritivores. Coprophagy does not only promote digestive utilization of a low-quality food source ("external rumen"; Zimmer 2002) but also facilitates the turnover of organic matter toward mineralization and nutrient release, as is highlighted by an example of detritivore–herbivore interaction: the feeding preference of the isopods *Porcellio scaber* and *Oniscus asellus* for fecal pellets of a herbivorous caterpillar (*Operophtera fagata*) resulted in increased soil contents of calcium, magnesium, and potassium, and a decreased soil C:N ratio, compared to trials with isopods feeding only on beech leaf litter (Zimmer and Topp 2002).

Besides coprophagy, detritivorous terrestrial isopods are part of further interactions of ecosystemic relevance, within and across trophic levels. These encompass serving as a food source to predators, competition (for rare high-quality food sources in an excess of low-quality food), and also facilitation of other detritivores. Interactions of terrestrial isopods (*Porcellio scaber*) and earthworms result in synergistic effects on the decomposition of, and nutrient release from, high-quality leaf litter, whereas their interactive effects on the decomposition of low-quality detritus were only additive (Zimmer et al. 2005). These observations might hint at emergent effects of multiple detritivores, equivalent to multiple predator effects (MPE; Sih et al. 1998), but the consequences of an accelerated decomposition of high-quality detritus resulting from detritivore interactions remain poorly understood.

In contrast to the predominantly detritivorous terrestrial isopods, land crabs and hermit crabs are omnivorous, making their lives from detriti- and herbivory, scavenging, and, to a lesser extent, predation (e.g., Page and Willason 1983). Owing to their feeding ecology and often large size, crabs occupy relatively high trophic levels, thus exerting top-down effects on food webs and ecosystem processes. This is particularly prevalent in tropical island ecosystems with a reduced diversity of top consumers. There, terrestrial decapods can take key positions in detrital food web pathways.

For instance, the effectiveness of leaf litter removal by the emblematic red crabs (*Gecarcoidea natalis*) on Christmas Island (87% of the annual litter fall) significantly reduces the densities of other forest invertebrates (Green et al. 1999), and rapid carrion removal by terrestrial hermit crabs (*Coenobita* sp.) on the Marshall Islands even controls the populations of necrophagous flies (Page and Willason 1983). Similarly, *Johngarthia malpilensis* (Gecarcinidae), an endemic crab on Malpelo Island off Colombia, effectively removes almost 100% of Nacza booby chick carcasses, which occur in large numbers during the breeding season (López-Victoria and Werding 2008). The energy assimilated from the bird carrion is then transferred across the island by the crabs, and back to the sea, when females are preyed on by marine predators while spawning at the coast.

It may be due to the relatively low number of common or abundant terrestrial crustacean detritivores, that their contribution to, and effects on, ecosystem processes are mechanistically much less understood than those of other terrestrial detritivores. Yet, in some systems, crustaceans clearly take key positions in terrestrial detrital food web pathways, and even act as the dominant detritivores. It is, thus, important to better understand their overall role in terrestrial detritus fluxes.

Freshwater Ecosystems

Across freshwater ecosystems, the origin and characteristics of detritus inputs vary strongly, for example, between lakes, rivers, and wetlands, between small and large, or up- and downstream systems. With varying types, sources, and fluxes of detritus, the assemblages of crustaceans involved in detritus processing also vary across freshwater ecosystems.

Allochthonous (i.e., terrestrial) plant litter is the main form of energy input in low order streams and small ponds, as well as freshwater wetlands and the littoral zones of many larger rivers and lakes (Pieczyńska 1993, Marks 2019). These inputs of initially coarse particulate organic matter (CPOM) are primarily processed by detritivorous macroinvertebrates, including various crustaceans. They assimilate external carbon and nutrient inputs within the system, and, even more importantly, shred the CPOM into fine particulate organic matter (FPOM, Fig. 3.4A,B), which becomes available to detritivores relying on smaller food items, and is transferred downstream, or from the littoral into the pelagic zone of lakes (Bjelke et al. 2005, Marks 2019).

On a global scale, there is apparent biogeographic variation in the composition of crustacean detritivore assemblages in stream ecosystems. In temperate regions of the northern and southern hemisphere, leaf litter inputs are mainly processed by amphipods, isopods, and crayfish, together with the larvae of certain aquatic insects (e.g., Boyero et al. 2011, Encina et al. 2017). As these groups are less diverse or even absent in many tropical lakes and rivers, some authors conclude that macroinvertebrate shredders play only a minor role in leaf litter processing in (sub)tropical freshwater ecosystems (e.g., Boyero et al. 2011). A look beyond the well-investigated palaearctic taxa, however, indicates that diverse freshwater shredder communities do exist in some tropical regions (e.g., Masese et al. 2014). In particular, omnivorous freshwater shrimps and crabs, rather than specialist shredders, dominate the processing of plant litter in many tropical streams and wetlands (Moulton et al. 2010).

Estimates of direct leaf litter consumption by freshwater crustaceans in temperate and (sub)tropical stream systems range from about 1 to 10% of the respective total litter input (Table 3.1). While these estimates may appear low overall, crustaceans are regularly the most important macroinvertebrate consumers of leaf litter. The common European amphipod *Gammarus fossarum* was found to consume about 10 times more leaf litter per mg wet weight than the second most abundant insect shredder in a small stream in NE France (Dangles and Guérold 2001), and Moss (2005) experimentally showed that *Potamonautes* sp. (Brachyura) in East African highland streams reduced leaf area in litter bags by 78%, whereas leaf litter loss was negligible (< 0.5%) in litter bags inaccessible to crabs. Exclusion of crayfish (*Cambarus bartonii*) and freshwater crabs (Pseudothelphusidae) from mountain streams in the United States and Costa Rica, respectively, resulted in the respective decrease of overall leaf litter breakdown rates by 40% and 10% at exclusion sites (Yang et al. 2020b, Dudley et al. 2021; compare Fig. 3.4C–E).

Ecologically, non-consumptive litter processing (i.e., mechanical disruption by e.g., "sloppy feeding") by detritivorous crustaceans, particularly decapods, may be at least as important as direct litter consumption. These activities result in a large and diverse FPOM-pool (in addition to high fecal pellet production; see Oertli 1993), which facilitates detrital breakdown by

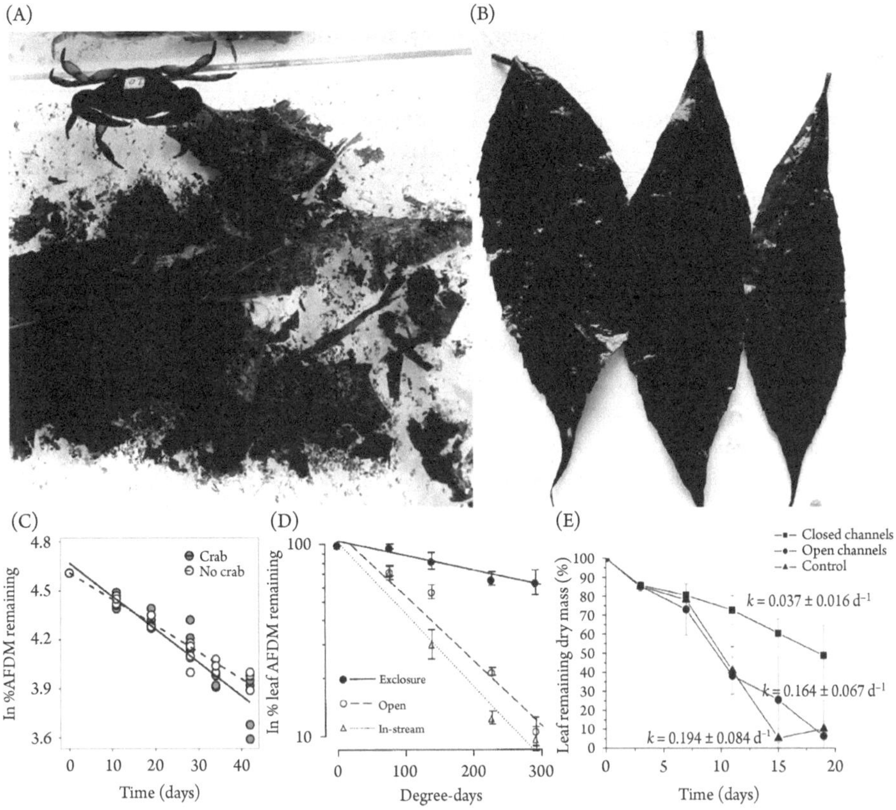

Fig. 3.4.
Effects of decapods on leaf litter breakdown in freshwater streams. (A) Leaves of *Koanophyllon pittieri* in aquarium tanks after nine days of incubation with *Ptychophallus tumimanus* (top left); (B) leaves after nine days of incubation in tanks without crabs. Remaining litter dry mass (in %) over time in exclusion/inclusion treatments with (C) *Ptychophallus tumimanus* (Brachyura, Pseudothelphusidae), (D) *Paranephrops zealandicus* (Astacidea, Parastacidae), and (E) *Aegla longirostri* (Anomura, Aeglidae). (A) and (B) reproduced from Yang et al. (2020a), with permission from © Sociedade Brasileira de Carcinologia under Creative Commons license (CC BY); (C) reproduced from Yang et al. (2020b), with permission from © John Wiley & Sons; (D) reproduced from Usio (2000), with permission from © Springer Nature; (E) reproduced from Cogo and Santos (2013), with permission from © Oxford University Press.

secondary consumers. The fact that crustaceans regularly dominate invertebrate communities in terms of biomass and abundance further increases their high impact on CPOM processing in stream ecosystems (e.g., Dangles and Guérold 2001, Masese et al. 2014). As a consequence, involvement of (crustacean) shredders in plant litter breakdown in freshwater ecosystems supports more complex food webs and higher transfer of detrital carbon to non-shredding invertebrates compared to systems with exclusively microbial/physical decay (Wissinger et al. 2021).

The overall impact of detritivorous freshwater crustaceans on plant litter-derived energy and nutrient fluxes in stream ecosystems strongly depends on detritivore identity and community structure. Macroconsumers like crabs and crayfish act as shortcut in freshwater food webs, as their detritus-derived biomass becomes exclusively available to top predators (fish, birds, and mammals, but also larger crustaceans), directly linking the lowest to the highest trophic levels (compare Usio 2000). Furthermore, decapods, and also amphipods and isopods, are not

Table 3.1 Estimates of leaf litter processing by freshwater crustaceans in stream ecosystems.

Region	Crab species	Higher taxonomy	Removal rate (% litter input)	Remarks	Reference
Temperate regions					
North Carolina (USA)	*Cambarus bartonii*	Decapoda, Astacoidea	4–6		Huryn and Wallace (1987)
Switzerland	*Gammarus pulex*	Amphipoda, Gammaridae	11	Combined with crane fly larvae	Oertli (1993)
Canada	*Gammarus pseudolimnaeus*	Amphipoda, Gammaridae	16		MacNeil et al. (1997)
New Zealand	*Paranephrops zealandicus*	Decapoda, Astacoidea	0.8	Daily removal from litter bags	Usio (2000)
Tropics and subtropics					
North Australia	*Cherax* sp.	Decapoda, Astacoidea	1		Coughlan et al. (2010)
SE Brazil	*Aegla longirostri*	Decapoda, Anomura	5		Cogo and Santos (2013)

specialized phytodetritivores, but omnivores that also act as herbivores, scavengers, or, most importantly in this context, as predators of other aquatic invertebrates (Kelly et al. 2002, Mancinelli et al. 2013). For example, freshwater shrimps of the circumtropical genus *Macrobrachium* primarily consume CPOM (e.g., Koussovi et al. 2020), but in some systems, they significantly reduce leaf litter breakdown rates as predators of specialist shredding shrimps (*Xhiphocaris* sp.) (March et al. 2001). This dual role of many crustaceans in detritus turnover might explain why they are still regularly neglected in studies on freshwater detritivore communities (e.g., Martins et al. 2015), despite their undeniable importance and complex role in the trophic structure and nutrient cycling of stream ecosystems.

The role of freshwater microcrustaceans (ostracods, copepods, and cladocerans) in stream ecosystems has been largely neglected so far, despite evidence for differential effects on litter breakdown arising from complex interactions with microbial decomposers and macroinvertebrate shredders (Chambord et al. 2017), and their high (functional) diversity in streams (Dole-Olivier et al. 2000). Microcrustaceans are, however, recognized as important agents of detritus processing in lake ecosystems. There, autochthonous inputs of FPOM become relatively more important, and its turnover by zooplankton in the water column and by benthic invertebrates at the lake bottom drives detrital food web pathways and dynamics. Freshwater copepods and cladocerans, which dominate lake zooplankton communities, are generally considered phytoplankton grazers or predators of other zooplankton (see Chapter 10 in this volume). Yet, many species, including the widespread *Daphnia* spp. and *Bosmina* spp. (Onchyura), include detritus in their diets, and utilize it as an important subsidy when higher-quality food is not available (Rautio et al. 2011).

Detritivory by cladocerans and copepods strongly mediates downward fluxes of carbon and nutrients in lakes through direct FPOM consumption and assimilation, through the conversion of well-defined detrital particles into amorphous, slowly sinking feces (Yoshimizu et al. 2001), as well as through considerable rates of coprophagy (Pilati et al. 2004). These activities may reduce gross sedimentation rates of detritus about three-fold, and deplete it of sugars and proteins, which might directly affect sediment mineralization (Danger et al. 2012). Moreover, Elgin and Jackson (2016) demonstrated in a mesocosm experiment that *Daphnia pulex* is able to decrease water turbidity by up to 75% by ingesting and repackaging detritus and other suspended particles, providing a water-clearing mechanism with strong implications for phytoplankton and hydrophyte growth.

Across temperate and boreal lakes, the turnover of detritus at lake bottoms is in large parts mediated by benthic amphipods (Pintoporeiidae and certain Gammaroidea), which are primarily detritivorous, and often dominate benthic invertebrate biomass (e.g., Johnson 1987, Scharold and Corry 2021). In turn, predatory fish rely heavily on benthic amphipods as food source (Berezina et al. 2018), making them a direct link between detrital carbon at the lake bottom and higher trophic levels of the lake food web (bentho-pelagic coupling). In deep lakes in northern North America and Europe, freshwater opossum shrimp (Mysidacea) promote bentho-pelagic coupling more directly as omnivorous consumers of zooplankton and considerable amounts of detritus (Griffin et al. 2020). By means of diel vertical migrations, the shrimps make carbon assimilated from detritus at the lake bottom available to consumers near the surface (Stockwell et al. 2020). Consequently, mysids and amphipods were identified as the most important link between detritus and pelagic consumers in a food web model of the North American Great Lakes (Stewart and Sprules 2011), a result that is likely transferable to other lake systems.

Studies addressing the role of crustaceans in processing carrion in freshwater ecosystems are generally scarce. Specialized scavengers do not seem to exist among freshwater crustaceans, with the sole exception of Lake Baikal, where two endemic scavenging amphipods (*Omatogammarus* spp.) evolved (Pekkoeva et al. 2021). Fenoglio et al. (2014) discuss the patchy and unpredictable occurrence of carrion, and its commonly rapid removal by currents and vertebrate scavengers (fish, mammals, birds) as important causes for the lack of specialized scavenging aquatic insects, and it seems plausible to assume similar reasons for crustaceans. Yet, there is evidence for opportunistic scavenging being widespread among omnivorous micro- and macrocrustaceans. Premke et al. (2010) found copepods, crustacean nauplii, and ostracods up to 10 times more abundant on fish carcasses than in adjacent sediments in the littoral zone of Lake Constance, Germany. Lotic amphipods feed on submerged mammal carcasses during all stages of decomposition (e.g., Barrios and Wolff 2011), and crayfish can be voracious scavengers. Being able to assimilate carrion-derived nutrients as efficiently as scavenging fish, crayfish act as an important short-term sink of carrion-derived nutrients by directly transferring them to high trophic levels (Boros et al. 2020). Furthermore, by breaking up fresh, intact carcasses, crayfish, and likely also crabs and large peracarids, may enhance accessibility for smaller scavenging, non-crustacean invertebrates that are unable to bite through scales, fur, and skin, a possible example of positive multiple detritivore effects. These few examples highlight the potential of crustaceans to contribute to carrion processing in freshwater ecosystems, but more systematic research is needed to better quantify their role in different systems.

Coastal Ecosystems

Salt marshes and mangrove forests, the dominating vegetated coastal ecosystems in temperate and (sub)tropical regions, respectively, are characterized by a constant supply and availability of high amounts of detritus resulting from autochthonous production of plant litter, as well as external inputs from the sea or the hinterland. Phytodetritus forms a basis for the high secondary production in these systems (Marinucci 1982).

Intertidal amphipods and semi-terrestrial isopods (in salt marshes), and sesarmid and ocypodid crabs (in mangrove forests and salt marshes), are dominating agents of the initial processing of plant litter. They mediate its fate and detrital fluxes in complex interactions with decomposing microorganisms and other detritivores. In fact, only the recognition of extensive leaf litter consumption and burial by Indo-Pacific mangrove crabs in the late 1970s and 1980s initiated the paradigm shift from mangrove forests being purely outwelling systems to acting as sinks with high internal carbon retention and turnover (reviewed in Cannicci et al. 2008).

Estimates of plant litter processed by crabs in mangrove forests vary by locality, season, and the dominating crab and tree species, but are usually substantial, reaching >80% in some cases (Table 3.2). For salt marshes, quantifications of detritus production and turnover by detritivores are generally scarce, likely being exacerbated by the vegetation structure. The available data suggest that crustaceans process only a small fraction of the available plant litter directly (Table 3.2), and that salt marsh gastropods are the more important detritivores in terms of absolute turnover rates (Graça et al. 2000, Zimmer et al. 2004). However, talitrid amphipods (*Uhlorchestia*/*Orchestia* spp.) and sesarmid crabs (*Armases cinereum*) are highly effective detritivores, ingesting up to 70% of their body mass as phytodetritus per day (Graça et al. 2000, Zimmer et al. 2004, Dias and Hassall 2005). Due to species-specific preferences for different types of plant litter, and for different salt marsh zones, which enhance the exploitation of detrital resources, the combined effect of the diverse assemblages of crustacean detritivores in salt marshes might be substantial (e.g., Zimmer et al. 2004, Dias and Hassall 2005).

Both salt marsh plant litter (mainly cordgrass *Spartina* spp.) and mangrove leaf litter are characterized by low nutritional value and palatability, with C:N ratios usually exceeding the range assumed energetically beneficial for metazoan nutrition, and high contents of deterrent substances. The immediate importance of plant litter for the nutrition of detritivorous crustaceans in salt marshes and mangrove forests is therefore a matter of ongoing debate. While there is evidence for the direct utilization of cordgrass litter by talitrids and crabs (Zimmer et al. 2004, Dias and Hassall 2005), as well as mangrove litter by crabs (Bui and Lee 2014), several studies question the importance of plant litter for crustacean nutrition, observing only little consumption and assimilation (Skov and Hartnoll 2002), or reduced survivorship when feeding on litter only (Parker et al. 2008).

Instead, decomposing microbes, particularly fungi, that colonize the plant litter might be a main food source for detritivorous salt marsh and mangrove crustaceans. For example, *Uhlorchestia spartinophila*, a common talitrid in NW Atlantic salt marshes, assimilates more energy from microbes associated with cordgrass litter and reduces their biomass much more than cordgrass biomass (Graça et al. 2000). At the same time, these amphipods, and equivalently litter-feeding crabs in mangrove forests, promote microbial growth and decay by shredding the plant litter into smaller pieces, much like macroinvertebrates in freshwater streams (Graça et al. 2000, Werry and Lee 2005). Additionally, ingestion and subsequent digestion by crustaceans chemically converts the plant litter, resulting in N-enriched and tannin-depleted feces, which further enhances microbial growth (Lee 1997, Graça et al. 2000). Feces of the litter-feeding mangrove crab *Parasesarma erythodactylum*, for instance, was colonized by microbes three times faster, and at densities 30 times higher, compared to whole-leaf samples (Werry and Lee 2005). In summary, litter processing by

Table 3.2 Plant litter removal, as a percentage of litter production, by crustaceans in salt marshes and mangrove forests.

Region	Species	Family	Plant species	Removal rate (% litter fall)	Reference
Salt marshes					
USA (Georgia)	*Uhlorchestia spartinophila*	Uhlorchestiidae	*Spartina alterniflora*	0.21–0.39[1]	Graça et al. (2000)
	Armases cinereum	Sesarmidae	*Spartina alterniflora*	0.72[1]	Zimmer et al. (2004)
	Venezillo parvus	Armadillidae	*Spartina alterniflora*	0.18[1]	
Mangrove forests					
Kenya	*Parasesarma guttatum, Neosarmatium smithi*	Sesarmidae	*Rhizophora mucoronata*	18.6	Slim et al. (1997)
China (Fujian)	*Parasesarma plicatum, Sesarma bidentata*	Sesarmidae	*Kandelia obovata*	31.9/16.7[4]	Chen et al. (2020)
India (Kerala)	*Neosarmatium malabricum*	Sesarmidae	natural mixture of eight species	80.2	Shanij et al. (2016)
Australia (Queensland)	*Neosarmatium fourmanoiri, Parasesarma messa*[3]	Sesarmidae	*Bruguiera exaristata*[2]	79	Robertson and Daniel (1989)
	Neosarmatium smithi, Parasesarma messa[3]	Sesarmidae	*Ceriops tagal*[2]	71	
	Paracleistostoma wardi, Neosarmatium fourmanoiri[3]	Ocypodidae, Sesarmidae	*Avicennia marina*[2]	33	
Brazil (Pará)	*Ucides cordatus*	Ocypodidae	*Rhizophora mangle*	81.3	Nordhaus et al. (2006)

[1] Removal rates based on estimates of *Spartina* production from previous studies; see respective references for details.
[2] This represents the dominating species per study site.
[3] Only the two dominating crab species are reported. See referenced publication for the entire community.
[4] In 48- and 24-year-old mangrove stands, respectively. Numbers represent yearly averages.

crustaceans in mangrove forests and salt marshes leads to a net increase of microbial biomass and activity, thus accelerating decomposition via a feedback loop.

Initial processing of salt marsh and mangrove litter by shredding peracarids and crabs not only promotes decomposing microorganisms but also a range of secondary metazoan detritivores, including deposit-feeding copepods and amphipods, as well as fiddler crabs and juvenile sesarmid and grapsid crabs. The combination of "sloppy feeding" and digestion results in a broad size range of POM, from undigested larger (cm to mm range) to much smaller (μm range) particles in feces, and dissolved organic matter (DOM) leachates, that may eventually re-aggregate into organic flakes. As a result, the diverse POM produced by the shredding and feeding activities of eight grapsid and sesarmid crabs in a NE Australian mangrove forest provides the nutritional basis for 38 invertebrate species (Camilleri 1992). Similarly, Lee (1997) suggested a coprophagous food chain of mangrove crustaceans, with amphipods and fiddler crabs consuming the feces of litter-feeding crabs. At high densities, fiddler crabs in salt marshes may even turn over the entire sedimentary pool of fine POM derived from initial plant litter breakdown over one tidal cycle, leaving patches of concentrated energy and nutrients in the form of fecal pellets that provide favorable conditions for further meio- and macrofaunal decomposition (Hoffman et al. 1984).

A distinct feature of leaf litter processing Indo-Pacific (e.g., *Neosarmatium* spp. and *Parasesarma* spp.) and Neotropical (*Ucides cordatus*) mangrove crabs is their behavior of pulling freshly fallen leaves into their burrows. Leaf burying may have evolved in response to competition among detritivores (other crabs, snails), or to prevent tidal removal of a valuable food source, given that leaching and microbial conditioning of leaves increase their nutritional quality and palatability over time (e.g., Fratini et al. 2000, Forgeron et al. 2021), and leads to the accumulation of carbon and nutrients and, subsequently, bacteria and microphytobenthos in the sediments around crab burrows. These enriched sediments provide a valuable food source for deposit feeders like fiddler crabs, and also for the burrowing crabs themselves (Skov and Hartnoll 2002). From an ecological perspective, leaf burying renders crabs the main agents of organic matter retention in mangrove forests.

In mangrove forests, direct consumption of plant litter inputs by detritivorous crabs may significantly contribute to detrital organic matter turnover (compare Table 3.2). Overall, however, the central role of crustaceans in detritus turnover in vegetated coastal ecosystems appears to involve mainly indirect effects, that is, by mediating the activity of decomposing microorganisms and the structure and (chemical) composition of the POM pool. Moreover, crustaceans link the detritus-processing microbial loop to higher trophic levels of salt marsh and mangrove forest food webs, and, by their mobility, connect detrital fluxes of salt marshes and mangrove forests to adjacent ecosystems (e.g., Guest and Connolly 2005, Parker et al. 2008). On an even broader scale, by mediating carbon fluxes and microbial decomposition, they also impact on these ecosystems' capacity to store and sequester CO_2 as "blue carbon," a currently highly active field of research.

In stark contrast to vegetated coastal ecosystems, sandy beaches have little to no primary production, but rely almost entirely on allochthonous inputs of organic matter. Deposits of detached macrophytes ("wrack") and carrion are conspicuous and nutritionally valuable organic inputs to beach ecosystems, but their occurrence is unpredictable, patchy, and ephemeral, leading to strong competition and resource partitioning among beach detritivores (for a summary see e.g., Hyndes et al. 2022).

Amphipods of the family Talitridae regularly dominate detritivore communities on sandy beaches in terms of abundance and biomass. Together with intertidal isopods, talitrids very rapidly and densely colonize fresh wrack patches on the lower beach (e.g., Colombini et al. 2009, Michaud et al. 2019), while they are replaced by other invertebrates (e.g., beetles) during later stages of decomposition (Colombini and Chelazzi 2003). Partitioning of wrack patches along the tidal gradient or between different decomposition stages may also reduce interspecific competition between co-occurring Talitridae, as illustrated by opposing density trends of beach fleas (*Traskorchestia traskiana*) and sand hoppers (*Megalorchestia california*) along the tidal gradient of northeast Pacific beaches (Pelletier et al. 2011).

Estimates of wrack consumption by sandy beach talitrids and isopods vary with location, detritus, and detritivore species. While *Talitrus saltator* (Amphipoda) and *Tylos europaeus* (Isopoda) in the Mediterranean use wrack of *Posidonia* seagrass only as shelter (Colombini et al. 2009), and *Transorchestia chilensis* consumption of *Ulva* wrack in the southeast Pacific appears insignificant (Catenazzi and Donnelly 2007), the talitrid community on Southern California beaches processes up to 81% of kelp wrack inputs (Lastra et al. 2008). The pronounced preferences of talitrids for certain macrophyte species (Fig. 3.5) strongly mediate the processing and composition of wrack deposits. Green algae (*Ulva* sp.) and kelp wrack (Laminariaceae) might undergo complete turnover in less than a day, because of their higher nutritional value and softer tissues, whereas the decomposition of other brown algae (*Fucus* sp.) and seagrass (*Phyllospadix* sp.) may take weeks to months (Mews et al. 2006), eventually allowing the latter to aggregate in sheltered shore sections.

Animal-derived detritus is a highly sought-after resource by beach crustaceans because of its high nutritional value, and wrack-colonizing peracarids readily consume carrion when it is available (Hyndes et al. 2022). Scavenging might be a principal feeding strategy for juvenile peracarids (e.g., *Tylos* sp. isopods, Kensley 1974), and animal-derived detritus essentially sustains the production of talitrids seasonally, for example, after near-shore fish spawning events (Fox et al. 2014). Owing to their larger size and high motility, decapods are better adapted to quickly exploiting fresh carrion inputs along longer stretches of coastline than the smaller peracarids. Large mole crabs (*Hippa pacifica*, Anomura) can mediate the supply of carrion to the beach by highly effective scavenging in the swash zone (Lastra et al. 2016). Intertidal land hermit crabs (*Coenobita* spp.) and ghost crabs (Ocypodidae), which are common generalist scavengers on (sub)tropical beaches, opportunistically utilize an extremely broad range of food sources, ranging from POM to various types of phytodetritus and carrion, as well as live prey, reflecting the unpredictable supply of (high-quality) food in their habitats (Laidre 2013b, Rae et al. 2019). While crustaceans are often among the first scavengers responding to fresh carrion inputs, and regionally dominate beach scavenger

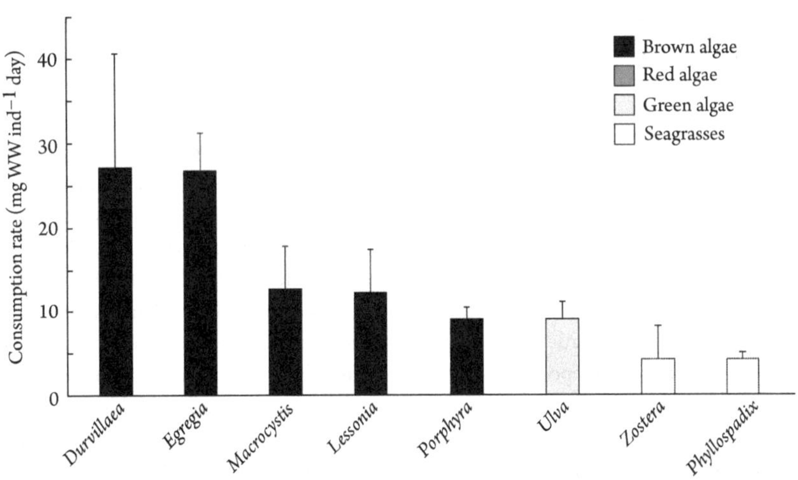

Fig. 3.5.
Feeding preferences of beach amphipods for different types of macrophyte wrack, expressed as mean daily individual consumption rate. Source: data from Hyndes et al. (2022).

assemblages (Laidre 2013b, Schlacher et al. 2013), they may be outcompeted by more specialized scavengers like whelks (Gastropoda) (Morton and Yuen 2000), or by large vertebrates like reptiles, birds, and mammals, which feed on both the carrion and scavenging crustaceans (Schlacher et al. 2013).

Although contributions of crustaceans to carrion turnover in beach ecosystems are likely lower compared to vertebrate scavengers or necrophagous insects (Colombini and Chelazzi 2003), detritivorous crustaceans on sandy beaches clearly provide a critical link between marine and terrestrial ecosystems by acting as vectors for the inland transfer of ocean-derived energy and nutrients of plant or animal origin. The crustacean-mediated detritus flux increases nutrient supply to beach pioneer plants (van Egmond et al. 2019) and is particularly vital to vertebrate consumers inhabiting dune fields or deserts adjacent to the sea (Catenazzi and Donnelly 2007, Schlacher et al. 2013), as well as to (migrating) shorebirds, which specifically seek wrack or carrion patches to forage on the highly abundant talitrids (Hyndes et al. 2022).

Pelagic Ecosystems

Zooplankton fecal pellets and marine snow (i.e., macroscopic aggregates of various detrital particles) are, with considerable seasonal and regional variation, the main components of detritus in the open ocean (Alldredge and Silver 1988, Turner 2015). The turnover of this large pool of POM is a key process in oceanic carbon cycling, and thus the global carbon budget. Epipelagic remineralization of nutrients promotes phytoplankton growth, and the oceanic capacity for carbon fixation and sequestration depends on the export flux of POM to deeper waters. Marine animals mediate a large part of pelagic POM turnover and fluxes, by processes summarized as the "biological carbon pump."

Pelagic crustaceans, particularly copepods and euphausiids, are the main components of this system (Fig. 3.6). They consume (and otherwise process) marine POM, they produce large amounts of fecal pellets, and their exudates, exuviae, and carcasses contribute to marine snow formation. Mass mortality events of crustacean zooplankton furthermore provide highly localized detritus pulses (reviewed, e.g., in Turner 2015). Estimates of crustacean contributions to POM attenuation and flux vary widely between regions, seasons, the studied species, and methodology, which might explain in parts the apparent paradigm shift in the perception of their role in the biological carbon pump over the last 30 to 40 years. Early studies saw the main role of crustacean zooplankton in enhancing downward flux of POM by packaging it into fecal pellets with accelerated sinking rates (reviewed in Turner and Ferrante 1979; examples in Table 3.3). More recent studies, however, recognized the major role of pelagic crustaceans, particularly copepods, in the attenuation of carbon export to depth, and its recycling and remineralization in the epipelagic zone.

Case studies in several oceanic regions found the respective dominating copepod species to attenuate up to 70% of the POM flux by feeding on marine snow aggregates (examples in Table 3.3). Congruently, Kiørboe (2000) noted concentration effects for crustacean nauplii and all major copepod groups on marine snow aggregates compared to background abundances in the water column, which were up to four orders of magnitude stronger than concentration effects for microbes. Particularly harpacticoid copepods, but also other copepods, as well as mysid and euphausiid zoea larvae, pelagic amphipods, and even adult euphausiids utilize aggregates for "benthic" foraging in a pelagic environment, which allows for much higher ingestion rates compared to suspension feeding (Lampitt et al. 1993, Kiørboe 2000). For the processing of fecal pellets, González and Smetacek (1994) postulated the existence of a "coprophagous filter." Due to their high consumption rates of fecal material in the uppermost ocean layers, cyclopoid copepods from the widespread genera *Oithona* and *Oncaea* supposedly retain most of the available POM in the euphotic zone, and effectively reduce and decelerate the export flux to deeper waters (e.g., Svensen and Nejstgaard 2003).

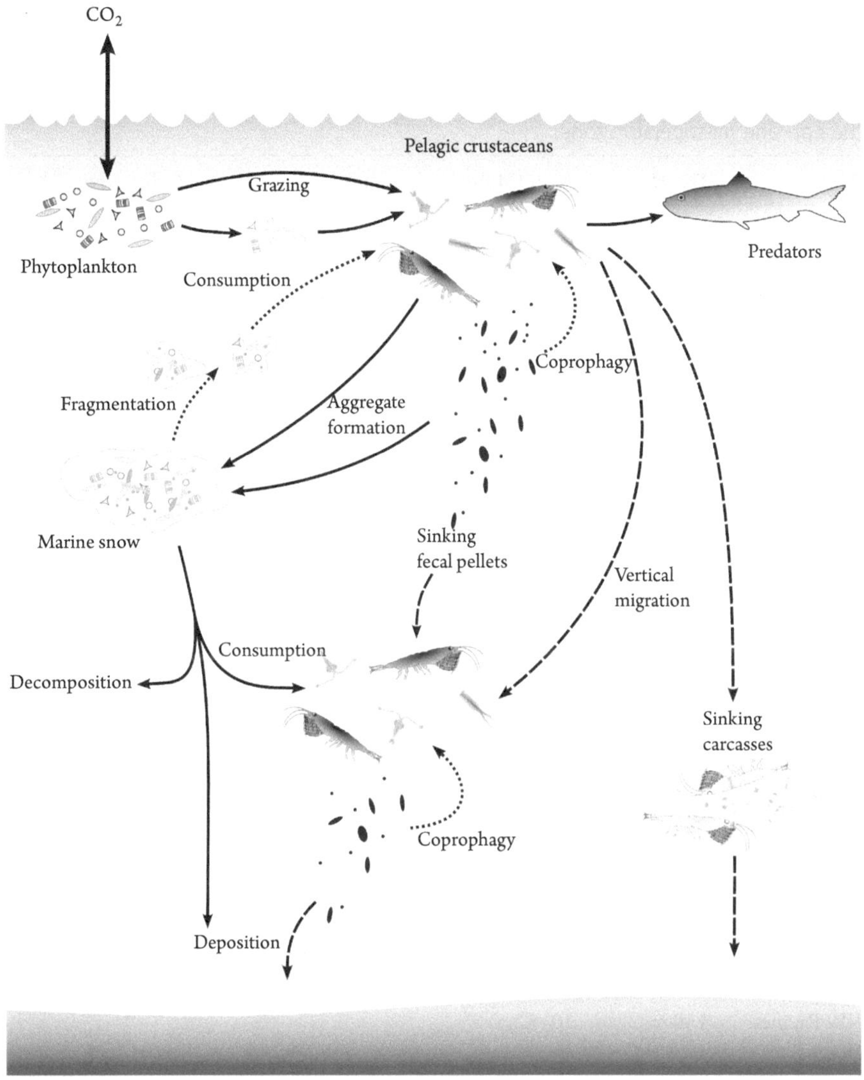

Fig. 3.6.
The role of pelagic crustaceans in oceanic carbon cycling (biological carbon pump). Crustaceans transfer carbon assimilated from atmospheric CO_2 to higher trophic levels by grazing on dead and live phytoplankton in the epipelagic zone. Production of fast-sinking fecal pellets by, and vertical migrations and sinking carcasses of copepods and euphausiids, enhance the flux of detrital organic matter to depth (dashed arrows). At the same time, fragmentation and consumption of marine snow aggregates and fecal pellets by pelagic crustaceans attenuate the flux of detrital organic matter to depth (dotted arrows). Source: data from Turner (2015), inspired by the descriptions of Steinberg and Landry (2017); see a color version of this figure in the centerfold.

By contrast, Reigstad et al. (2005) found no evidence of *Oithona* spp. feeding on calanoid feces in experiments mimicking natural conditions. Observations of enhanced coprophagy with increasing abundance of alternative food, as well as active rejection of fecal pellets during the feeding process, suggest that fecal pellet ingestion by pelagic copepods occurs rather unintentionally (Poulsen and Kiørboe 2005). Similarly, estimates of carbon remineralization based on measurements of in situ flux and respiration rate (instead of feeding rate) were considerably

Table 3.3 Contributions of detritivorous crustacean zooplankton to particulate organic matter fluxes in the ocean, given as % estimates of total POM exported to depth as fecal pellets or by vertical migration, and of total POM flux attenuation in the epi- and mesopelagic zone.

Region	Depth	% POM flux	Remarks	Reference
Export by fecal pellet production				
Subtropical E Pacific Ocean	120 m	41	Pellet production dominated by calanoids and cyclopoids	Small et al. (1983)
Baffin Bay	250 m	29	Measured after the phytoplankton spring bloom	Juul-Pedersen et al. (2006)
North Sea	50 m	3–39		Poulsen and Kiørboe (2005)
Southern Ocean	170 m	34/67	Winter/summer export flux	Gleiber et al. (2012)
Scotia Sea	30–180 m	up to 100		Cavan et al. (2015)
Export by vertical migration				
Sargasso Sea	0–175 m	3–18		Schnetzer and Steinberg (2002)
N Pacific Ocean	150 m	6–44		Kobari et al. (2008)
Subarctic Pacific Ocean	150 m	26–200		Steinberg et al. (2008)
Tropical Pacific Ocean	150 m	11–44		Kobari et al. (2013)
NW Pacific Ocean	150 m	10–52		
Attenuation by consumption/fragmentation				
North Sea	0–8 m	up to 99	Consumption of fecal pellets in mesocosms	Svensen and Nejstgaard (2003)
Beaufort Sea	210 m	70	Fragmentation by copepods, feeding by larvaceans	Sampei et al. (2009)
N Atlantic Ocean	50–1000 m	30		Giering et al. (2014)
Arctic Ocean	0–50 m	60–67		van der Jagt et al. (2020)

lower for zooplankton (max. 30%) compared to microorganisms (Giering et al. 2014). The dominant mechanism by which pelagic crustaceans attenuate POM flux is therefore still a matter of debate.

Rather than direct consumption, the main contribution of crustaceans to carbon retention in the upper pelagic zone could be by fragmenting fecal pellets and aggregates. Fragmentation reduces particle sizes and sinking rates and facilitates their consumption and degradation by microzooplankton and microbes (Giering et al. 2014, Turner 2015). Corroborating this, a recent modeling study indicated that particle fragmentation by pelagic copepods accounts for up to 80% of POM flux attenuation in the epipelagic zone (Mayor et al. 2020). Similarly, euphausiids may play an important role in the fragmentation of marine snow particles. Goldthwait et al. (2004) extrapolated from tethering experiments that the immense schools of large euphausiids (e.g., *Euphausia superba*) should be able to fragment up to 100% of organic aggregates within the first 15 m of the water column by the fluid shear forces created by their swimming movements.

Pelagic copepods and euphausiids furthermore contribute to the active export flux of carbon and nutrients from the surface ocean to deeper waters by performing diel and/or seasonal vertical migrations. Driven by food availability and predation pressure, these migrations can occur across several hundred meters in the water column and form an important component of the biological carbon pump, the "mesopelagic migrant pump" (e.g., Schnetzer and Steinberg 2002). Obviously, the efficiency of this biological export flux is highest when gut passage times (GPT) of food ingested in the epipelagic zone are longer than the time the organisms take to migrate downward. Strong vertical migrators like *Metridia* spp., *Pleuromamma* spp., and *Euphausia superba* have average GPTs of one hour or more, which is considered sufficiently long for direct transport of organic matter from the epi- to the mesopelagic zone (Schnetzer and Steinberg 2002 and references therein). Local contributions of zooplankton-mediated active flux of detritus to depth typically amounts to 3–50% of passive POM sedimentation rates (Table 3.3; see also Steinberg and Landry 2017). Estimates of active downward POM flux by copepods in winter being up to 200% of fecal pellet flux, (. . .), led Kobari et al. (2008, 2013) to suggest that, (. . .), at least seasonally, migrating crustacean detritivores might be more important for sustaining meso- and bathypelagic food webs than passively sinking detritus.

The ecological importance of vertically migrating crustacean detritivores is likely not (solely) defined by the amount of POM they provide, but also by its quality. Fresh fecal pellets that were produced at depth are likely of higher nutritional value to deep-water detritivores than detritus that had been subject to microbial and chemical decay while passively sinking the same distance (Steinberg and Landry 2017). Ochoa et al. (2013) even suggested that coprophagous copepods migrating between consecutive layers of the water column might act as a "bucket brigade" for rapid organic matter export to the deep sea. Further, high mortalities of copepods overwintering in the mesopelagic zone may provide important nutrient pulses to meso- and bathypelagic food webs in seasons with generally poor feeding conditions (Turner 2015).

Deep Sea

In the deep sea, detritivores act as foundation species for the entire ecosystem, as it entirely relies on detritus as allochthonous energy source due to the virtual absence of primary production (except for localized chemosynthetic hydrothermal vent and cold seep habitats). Detritus reaches the deep seafloor in the form of organic particles and aggregates (POM), as well as large animal (carcasses) and plant remains (e.g., wood, macroalgae) sinking down from the surface (Gooday and Turley 1990). While the large POM inputs are energetically and nutritionally depleted due to decomposition and remineralization in the water column (compare section "Pelagic Ecosystems"), food falls of larger organisms, particularly large vertebrate carrion, provide localized, but potentially long-lasting, pulses of high-quality organic matter (e.g., Scheer et al. 2022).

Only relatively few deep-sea squat lobsters (Munidopsidae), isopods (Ischnomesidae, Munnopsidae), and amphipods (Corophiidae, Gammaroidea) appear to feed primarily on benthic POM deposits (Sokolova 1997, Iken et al. 2001). Most crustacean detritivores in the deep sea act as scavengers on food falls, and thus take a key role in the processing and cycling of nutrients and energy from these high-quality sources. However, these "specialist necrophages" do also regularly consume particulate detritus to compensate for the scarcity of other food, particularly as juveniles when they require a more regular energy supply for successful growth (Blankenship and Levin 2007). Some squat lobsters, and notably the hadal amphipod *Hirondellea gigas*, are furthermore capable of digesting sunken wood, thereby transferring highly recalcitrant plant material into deep-sea energy flows (Hoyoux et al. 2009, Kobayashi et al. 2012). The crustacean scavenger guild comprises representatives of crabs, king crabs and squat lobsters, shrimp, isopods (family Cirolanidae), and, in particular, amphipods (mainly the superfamily Lysianassoidae). The latter are the most conspicuous, diverse, and abundant group of (crustacean) scavengers in many parts of the world's deep oceans, consequentially drawing much of the scientific attention (reviewed in Havermans and Smetacek 2018).

The occurrence and composition of deep-sea crustacean scavenger communities shows considerable (bio)geographic variation between, and within, ocean basins. Crustaceans dominate the scavenger/predator assemblages in the deep Weddell Sea, Antarctic Ocean (33 crustacean vs. 4 non-crustacean species; De Broyer et al. 2004). In several deep-sea trench systems, the scavenger community even appears to consist exclusively of amphipods, their abundances increasing with depth below ~5000 m (Jamieson and Weston 2023). Elsewhere, scavenging crustaceans and fish are equally diverse (e.g., in the NE Atlantic; Iken et al. 2001), and shifts from crustacean- to fish-dominated scavenger assemblages might occur between sites just about 500 km apart (e.g., Arabian Sea; Janßen et al. 2000). Linley et al. (2018) observed decapods, but no peracarids, on bait in the deep eastern Mediterranean Sea, and crustacean scavengers were absent from elasmobranch carcasses off the coast of Angola (Higgs et al. 2014). Where they are present, crustaceans usually dominate the scavenger community in terms of individual numbers, locally being up to two orders of magnitude more abundant than other taxa (e.g., De Broyer et al. 2004, Rohlfer et al. 2022).

Accurate estimates of crustacean contributions to the processing of natural deep-sea food falls are difficult to obtain due to limited observational and measuring capacities compared to the vastness of the environment. Studies using experimentally deployed carcasses strongly suggest that the crustacean scavenging guild is capable of processing up to 100% of the organic inputs from small to medium-sized food falls within very short time frames. Fish and squid carrion deployed in the Arabian Sea and the North Atlantic and Arctic Oceans at depths between 1200 and 2600 m was consumed at similar rates of 1300–5500 g d^{-1}. Their scavenger assemblages were strongly dominated by the widespread lysianassoids *E. gryllus* and *Orchomenella* spp., as well as decapods and isopods, while consumption rates and primary consumers on jellyfish falls varied between locations (Janßen et al. 2000, Premke et al. 2006, Rohlfer et al. 2022, Scheer et al. 2022). Complete removal of the baits' soft tissues in these studies was observed after about 5 to 23 hours (Fig. 3.7). Such rapid removal of small to medium-sized carrion pulses by macroconsumers is assumed to effectively prevent microbial colonization and sediment enrichment around and under the food fall (Gooday and Turley 1990).

By contrast, large megafaunal food falls (mostly whale carcasses) are processed over the course of weeks to years, leading to organic enrichment of the surrounding sediment and the development of a specialized detritivore community that undergoes distinct successional stages (Smith and Baco 2003). Crustaceans are components of these communities in all successional stages. Decapods and lysianassoid amphipods, together with sharks and fish, form the initial *mobile scavenger*

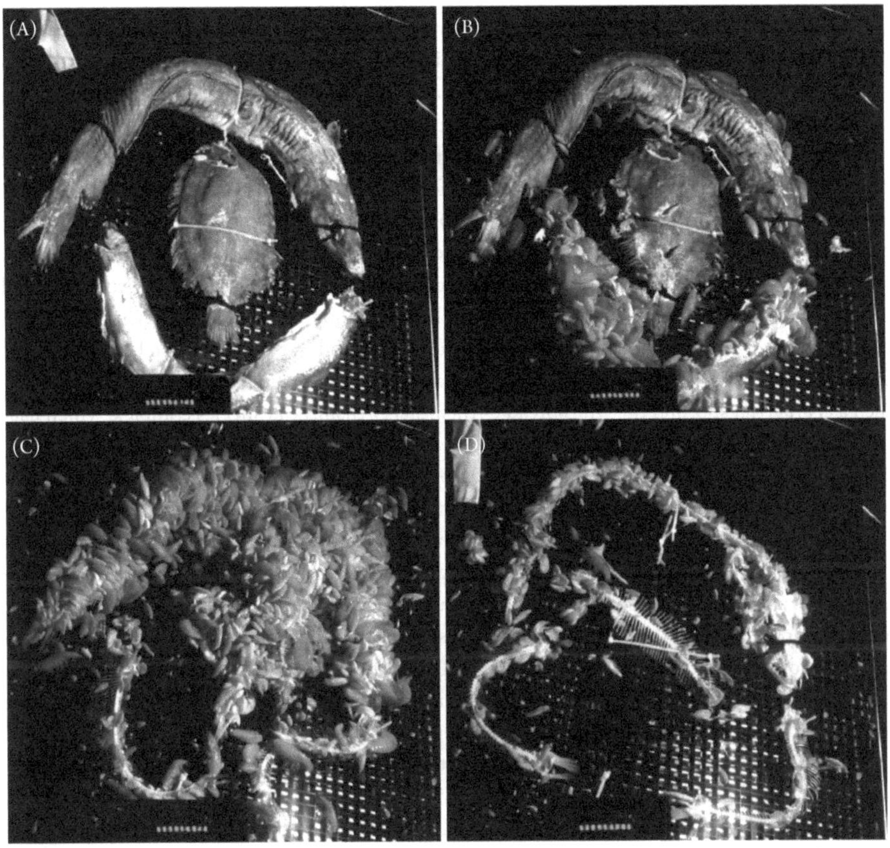

Fig. 3.7.
Processing of an experimental fish fall in the deep Arctic Ocean by aggregations of scavenging amphipods. The photographs show the state of the fish carcasses after (A) 0, (B) 6, (C) 10, and (D) 22.5 h of deployment, amphipods eventually cover entire carcasses (B, C). Reproduced from Premke et al. (2006), with permission from © Inter-Research Science Publisher; see a color version of this figure in the centerfold.

community, which feeds directly on the carcass. Lysianassoids, which occasionally form swarms of tens of thousands of individuals, along with brachyuran, galatheid, and stone crabs, typically dominate this phase after fish and sharks abandon the carcass and remove the bulk of its soft tissues (Smith and Baco 2003, Silva et al. 2021). These species are complemented, and partly replaced, by copepods, cumaceans, euphausiids, and generalist-detritivorous amphipods in the following *enrichment-opportunist* stage. Crustacean feeding activities gradually shift from the carcass itself to secondary detritus (tissue pieces scattered on the seafloor, detritus produced by microbes and invertebrates colonizing the bones and the organically enriched sediments) with ongoing skeletonization (e.g., Smith and Baco 2003, Smith et al. 2014). The shift to secondary consumption is completed in the final *sulphophilic* stage, with galatheids and skeleton-associated amphipods and isopods acting as microbivores, and potentially predators/scavengers, on the chemoautotrophic bone-colonizing fauna (Smith and Baco 2003, and references therein).

The significance of crustacean detritivores in deep-sea food webs is driven by the interplay of species-specific adaptations to the scavenging/detritivorous lifestyle (see section "Feeding Ecology of Detritivorous Crustaceans"), and their community composition, which in turn is shaped by complex interactions between food availability and other environmental factors (Ashford et al. 2019). Within scavenging deep-sea amphipods, Ingram and Hessler (1983) described two ecologically distinct groups, a bentho-pelagic (or scavenger-predator) group, and a demersal (or scavenger-detritivore) group. The bentho-pelagic group includes large species such as *E. gryllus*, *Hirondellea* spp., and the giant *Alicella gigantea*, which reach fresh food falls rapidly and over large distances, like vultures in terrestrial systems. Morphological adaptations enable them to initiate the breakdown of intact carcasses, and to rapidly engorge large amounts of food, while their highly efficient assimilation might turn them into short-term carbon sinks (Ingram and Hessler 1983, Sainte-Marie 1992). Vertical migrations of scavenger-predators, occasionally several hundred meters upward, furthermore promote bentho-pelagic coupling, making energy assimilated from detritus that had settled on the seafloor available to pelagic food webs (reviewed in Havermans and Smetacek 2018). The demersal group includes smaller, less motile, and more opportunistic species, such as *Paralicella* spp. and the widespread *Orchomene/Orchomenella* species complex, which usually arrive on food falls following initial breakdown by larger scavengers (e.g., Rohlfer et al. 2022). Their "sloppy feeding," combined with high fecal pellet production (resulting from rapid and rather ineffective digestion; Sainte-Marie 1992), and roosting behavior in the vicinity of a food fall, contribute to the short-term and short-range enrichment of energy around food falls. This facilitates the larger detritivore community and attracts predatory fishes and decapods that depend on these amphipods as primary food source (Havermans and Smetacek 2018, Jamieson and Weston 2023, and references therein). The ability of crustaceans to withstand extended periods of starvation by resting on or in the sediment at metabolic rates 30–70% lower than basal respiration rates (Christiansen and Diel-Christiansen 1993), may sustain the provision of detritus-derived energy to the food web for extended periods of time.

Taken together, amphipods, and also other detritivorous crustaceans, shape patterns and pathways of food fall processing, as well as the structure of predator communities in the deep sea, through their pivotal role in assimilating detrital energy in their biomass, thus keeping it available to deep-sea food webs over large spatial and/or temporal scales (Havermans and Smetacek 2018). However, the inherent challenges of deep-sea ecological research still limit our knowledge with regards to the natural feeding ecology of many deep-sea crustaceans and detrital fluxes on the ecosystem level. Much more data and novel approaches (e.g., "-omics"; see Jamieson and Weston 2023) will be required for a better understanding and quantification of crustacean contributions to the processing of detritus inputs to deep-sea food webs.

CONCLUSIONS AND FUTURE DIRECTIONS

Crustaceans act as detritivores in basically all major ecosystems, thus contributing to one of the most fundamental ecosystem processes as primary consumers on the "brown" food web pathways (compare Fig. 3.1). As highlighted by the examples provided throughout this chapter, turnover of major detrital inputs in several ecosystems is primarily mediated by representatives of three crustacean groups: the processing of leaf litter in tropical freshwater streams and mangrove forests by crabs, the processing of POM in the pelagic zones of lakes and oceans by copepods, and the processing of macro- and mega-carrion in the deep sea by amphipods. In adaptation to this role, remarkable behavioral, morphological, and physiological adaptations have evolved in many of the more specialized crustacean detritivores.

The ecological importance of crustacean detritivores and scavengers is defined not only by their absolute contributions to detritus consumption and turnover, but also the high prevalence of omnivory and the great diversity of feeding preferences found among sympatric crustaceans. The latter two traits, in particular, link crustacean detritivores to a disproportionately large number of other food web compartments. By their connections to the various sorts of detritus they feed on, to other detritivores they provide food for (in the form of feces and shredded particles), and to various predators that feed on them, crustaceans as a group appear to occupy key positions in numerous detrital food web pathways. Their ecological importance may be further enhanced by the high motility of many species, turning them into important vectors for spatial redistribution of detritus-derived energy within and between ecosystems.

Yet, despite decades of research, our knowledge about the quantitative and qualitative importance of crustaceans in detritus turnover remains fragmentary for many ecosystems and regions of the world, as it does with regards to regional and seasonal variation. Studies addressing crustacean feeding ecology usually focus on only a single or a few selected species, and regularly do not set their detritus turnover rates in relation to other detritivores and/or detritus inputs, while food web studies often treat crustacean, or even all detritivores, as a single food web component, making it difficult to quantify species- or group-specific contributions to detritus turnover. Joining both approaches might provide a step forward toward a more comprehensive understanding of the ecological role of crustaceans as detritivores and scavengers. Such joint approaches might particularly address the characterization of "multiple detritivore effects," that is, the emergent effects of multiple detritivores on detritus turnover (equivalent to MPE; Sih et al. 1998), which appear to be a common feature of crustacean detritivore communities (see examples above).

Recent studies provide mounting evidence that humans may strongly, and in various ways, affect crustacean detritivores. Chemical and light pollution potentially disrupt leaf litter processing by freshwater amphipods (Dangles et al. 2004, Czarnecka et al. 2021), and chemical pollutants as well as microplastics were found in high concentrations in the guts and tissues of various deep-sea scavengers and detritivores (Jamieson and Weston 2023). The impacts of invasive species on native crustacean detritivores may be complex, ranging from providing additional detritus (Haram et al. 2020) to the potential disruption of native detritivore food chains by large invasive omnivores (Jackson et al. 2016). The effects of fisheries are similarly complex. While industrial fishing and whaling is assumed to have significantly reduced the supply of macro- and mega-carrion falls to the seafloor (Havermans and Smetacek 2018), the concentrated release of fishing discards in heavily fished areas may promote benthic communities to shift toward scavenger dominance, namely of hermit crabs, crabs, and amphipods, and their increased reliance on discards (Catchpole et al. 2006). Finally, climate change is predicted to cause shifts in the amount and composition of detrital inputs, as well as the composition and activity of crustacean detritivore communities from stream to deep-sea ecosystems (Foucreau et al. 2013, Sweetman et al. 2017). Future lines of research on detritivorous crustaceans will have to explicitly account for the effects of anthropogenic impacts on detritivore communities and their contributions to detritus turnover.

REFERENCES

Adachi, K., K. Tanimura, T. Mitsui, T. Morita, I. Yosho, K. Ikejima, and K. Morioka. 2016. Cellulolytic activity in the hepatopancreas of *Chionoecetes opilio* and *Chionoecetes japonicus*: Enzymatic adaptations to deep sea environment. Fisheries Science 82:835–841.

Alldredge, A.L., and M.W. Silver. 1988. Characteristics, dynamics and significance of marine snow. Progress in Oceanography 20:41–82.

Ashford, O.S., A.J. Kenny, C.R.S. Barrio Froján, T. Horton, and A.D. Rogers. 2019. Investigating the environmental drivers of deep-seafloor biodiversity: A case study of peracarid crustacean assemblages in the Northwest Atlantic Ocean. Ecology and Evolution 9:14167–14204.

Barrios, M., and M. Wolff. 2011. Initial study of arthropods succession and pig carrion decomposition in two freshwater ecosystems in the Colombian Andes. Forensic Science International 212:164–172.

Benbow, M.E., P.S. Barton, M.D. Ulyshen, J.C. Beasley, T.L. DeVault, M.S. Strickland, J.K. Tomberlin, H.R. Jordan, and J.L. Pechal. 2019. Necrobiome framework for bridging decomposition ecology of autotrophically and heterotrophically derived organic matter. Ecological Monographs 89:e01331.

Berezina, N.A., A.P. Strelnikova, and A.A. Maximov. 2018. The benthos as the basis of vendace, *Coregonus albula*, and perch, *Perca fluviatilis*, diets in an oligotrophic sub-Arctic lake. Polar Biology 41:1789–1799.

Bjelke, U., I.M. Bohman, and J. Herrmann. 2005. Temporal niches of shredders in lake littorals with possible implications on ecosystem functioning. Aquatic Ecology 39:41–53.

Blankenship, L.E., and L.A. Levin. 2007. Extreme food webs: Foraging strategies and diets of scavenging amphipods from the ocean's deepest 5 kilometers. Limnology and Oceanography 52:1685–1697.

Boros, G., I. Czeglédi, T. Erős, and B. Preiszner. 2020. Scavenger-driven fish carcass decomposition and phosphorus recycling: Laboratory experiments with freshwater fish and crayfish. Freshwater Biology 65:1740–1751.

Boyero, L., R.G. Pearson, D. Dudgeon, M.A.S. Graça, M.O. Gessner, R.J. Albariño, V. Ferreira, C.M. Yule, A.J. Boulton, M. Arunachalam, M. Callisto, E. Chauvet, A. Ramírez, J. Chará, M.S. Moretti, J.F. Gonçalves Jr, J.E. Helson, A.M. Chará-Serna, A.C. Encalada, J.N. Davies, S. Lamothe, A. Cornejo, A.O.Y Li, L.M. Buria, V.D. Villanueva, M.C. Zúñiga, and C.M. Pringle. 2011. Global distribution of a key trophic guild contrasts with common latitudinal diversity patterns. Ecology 92:1839–1848.

Britton, J.C., and B. Morton. 1994. Marine carrion and scavengers. Oceanography and Marine Biology: An Annual Review 32:369–434.

Buck, T.L., G.A. Breed, S.C. Pennings, M.E. Chase, M. Zimmer, and T.H. Carefoot. 2003. Diet choice in an omnivorous salt-marsh crab: Different food types, body size, and habitat complexity. Journal of Experimental Marine Biology and Ecology 292:103–116.

Bui, T.H.H., and S.Y. Lee. 2014. Does 'You are what you eat' apply to mangrove grapsid crabs? PLoS ONE 9:e89074.

Bui, T.H.H., and S.Y. Lee. 2015. Endogenous cellulase production in the leaf litter foraging mangrove crab *Parasesarma erythodactylum*. Comparative Biochemistry and Physiology Part B: Biochemistry and Molecular Biology 179:27–36.

Camilleri, J.C. 1992. Leaf-litter processing by invertebrates in a mangrove forest in Queensland. Marine Biology 114:139–145.

Cannicci, S., D. Burrows, S. Fratini, T.J. Smith III, J. Offenberg, and F. Dahdouh-Guebas. 2008. Faunal impact on vegetation structure and ecosystem function in mangrove forests: A review. Aquatic Botany 89:186–200.

Catchpole, T., C. Frid, and T. Gray. 2006. Importance of discards from the English *Nephrops norvegicus* fishery in the North Sea to marine scavengers. Marine Ecology Progress Series 313:215–226.

Catenazzi, A., and M.A. Donnelly. 2007. Role of supratidal invertebrates in the decomposition of beach-cast green algae *Ulva* sp. Marine Ecology Progress Series 349:33–42.

Cavan, E.L., F.A.C. Le Moigne, A.J. Poulton, G.A. Tarling, P. Ward, C.J. Daniels, G.M. Fragoso, and R.J. Sanders. 2015. Attenuation of particulate organic carbon flux in the Scotia Sea, Southern Ocean, is controlled by zooplankton fecal pellets. Geophysical Research Letters 42:821–830.

Chambord, S., M. Tackx, E. Chauvet, G. Escolar, and F. Colas. 2017. Two microcrustaceans affect microbial and macroinvertebrate-driven litter breakdown. Freshwater Biology 62:530–543.

Chen, J., G. Chen, Y. Gu, H. Zhu, and Y. Ye. 2020. Fate of leaf litter in restored *Kandelia obovata* (S.L.) mangrove forests with different ages in Jiulong River Estuary, China. Restoration Ecology 28:369–377.

Christiansen, B., and S. Diel-Christiansen. 1993. Respiration of lysianassoid amphipods in a subarctic fjord and some implications on their feeding ecology. Sarsia 78:9–15.

Cogo, G.B., and S. Santos. 2013. The role of aeglids in shredding organic matter in neotropical streams. Journal of Crustacean Biology 33:519–526.

Colombini, I., and L. Chelazzi. 2003. Influence of marine allochthonous input on sandy beach communities. Oceanography and Marine Biology: An Annual Review 41:115–159.

Colombini, I., M.Á. Mateo, O. Serrano, M. Fallaci, E. Gagnarli, L. Serrano, and L. Chelazzi. 2009. On the role of *Posidonia oceanica* beach wrack for macroinvertebrates of a Tyrrhenian sandy shore. Acta Oecologica 35:32–44.

Coughlan, J.F., R.G. Pearson, and L. Boyero. 2010. Crayfish process leaf litter in tropical streams even when shredding insects are common. Marine and Freshwater Research 61:541.

Czarnecka, M., J. Kobak, M. Grubisic, and T. Kakareko. 2021. Disruptive effect of artificial light at night on leaf litter consumption, growth and activity of freshwater shredders. Science of the Total Environment 786:147407.

Dahl, E. 1979. Deep-sea carrion feeding amphipods: Evolutionary patterns in niche adaptation. Oikos 33:167–175.

Danger, M., B. Allard, M.B. Arnous, J.-F. Carrias, J. Mériguet, L. Ten-Hage, and G. Lacroix. 2012. Effects of food-web structure on the quantity and the elemental quality of sedimenting material in shallow lakes. Hydrobiologia 679:251–266.

Dangles, O., M.O. Gessner, F. Guerold, and E. Chauvet. 2004. Impacts of stream acidification on litter breakdown: Implications for assessing ecosystem functioning. Journal of Applied Ecology 41:365–378.

Dangles, O., and F. Guérold. 2001. Linking shredders and leaf litter processing: Insights from an acidic stream study. International Review of Hydrobiology 86:395–406.

De Broyer, C., F. Nyssen, and P. Dauby. 2004. The crustacean scavenger guild in Antarctic shelf, bathyal and abyssal communities. Deep-Sea Research Part II: Topical Studies in Oceanography 51:1733–1752.

Dias, N., and M. Hassall. 2005. Food, feeding and growth rates of peracarid macro-decomposers in a Ria Formosa salt marsh, southern Portugal. Journal of Experimental Marine Biology and Ecology 325:84–94.

Dole-Olivier, M.-J., D.M.P. Galassi, P. Marmonier, and M. Creuzé Des Châtelliers. 2000. The biology and ecology of lotic microcrustaceans. Freshwater Biology 44:63–91.

Dudley, M.P., K. Solomon, S. Wenger, C.R. Jackson, M. Freeman, K.J. Elliott, C.F. Miniat, and C.M. Pringle. 2021. Do crayfish affect stream ecosystem response to riparian vegetation removal? Freshwater Biology 66:1423–1435.

Eastman, L.B., and M. Thiel. 2015. Foraging behavior of crustacean predators and scavengers. Pages 535–556 *in* M. Thiel and L. Watling, editors. The natural history of the Crustacea, Volume 2: Lifestyles and feeding biology. Oxford University Press, New York, USA.

Elgin, E.L., and L.J. Jackson. 2016. Reduction of resuspended sediments by the pelagic herbivore, *Daphnia pulex*. Freshwater Science 35:589–596.

Encina, F., R. Vega, G. Lara, and P. De Los Ríos-Escalante. 2017. Ecological role of benthic crustaceans in Chilean north Patagonian lakes and rivers (Araucania region, 39°S). Crustaceana 90:437–447.

Fenchel, T. 1970. Studies on the decomposition of organic detritus derived from the turtle grass *Thalassia testudinum*. Limnology and Oceanography 15:14–20.

Fenoglio, S., R.W. Merritt, and K.W. Cummins. 2014. Why do no specialized necrophagous species exist among aquatic insects? Freshwater Science 33:711–715.

Forgeron, S.J., A.F. Quadros, and M. Zimmer. 2021. Crab-driven processing does not explain leaf litter-deposition in mangrove crab burrows. Ecology and Evolution 11:8856–8862.

Foucreau, N., C. Piscart, S. Puijalon, and F. Hervant. 2013. Effect of climate-related change in vegetation on leaf litter consumption and energy storage by *Gammarus pulex* from continental or Mediterranean populations. PLoS ONE 8:e77242.

Fox, C.H., R. El-Sabaawi, P.C. Paquet, and T.E. Reimchen. 2014. Pacific herring *Clupea pallasii* and wrack macrophytes subsidize semi-terrestrial detritivores. Marine Ecology Progress Series 495:49–64.

Fratini, S., S. Cannicci, and M. Vannini. 2000. Competition and interaction between *Neosarmatium smithi* (Crustacea: Grapsidae) and *Terebralia palustris* (Mollusca: Gastropoda) in a Kenyan mangrove. Marine Biology 137:309–316.

Giering, S.L.C., R. Sanders, R.S. Lampitt, T.R. Anderson, C. Tamburini, M. Boutrif, M.V. Zubkov, C.M. Marsay, S.A. Henson, K. Saw, K. Cook, and D.J. Mayor. 2014. Reconciliation of the carbon budget in the ocean's twilight zone. Nature 507:480–483.

Gleiber, M.R., D.K. Steinberg, and H.W. Ducklow. 2012. Time series of vertical flux of zooplankton fecal pellets on the continental shelf of the western Antarctic Peninsula. Marine Ecology Progress Series 471:23–36.

Goldthwait, S., J. Yen, J. Brown, and A. Alldredge. 2004. Quantification of marine snow fragmentation by swimming euphausiids. Limnology and Oceanography 49:940–952.

González, H.E., and V. Smetacek. 1994. The possible role of the cyclopoid copepod *Oithona* in retarding vertical flux of zooplankton faecal material. Marine Ecology Progress Series 113:233–246.

Gooday, A.J., and C.M. Turley. 1990. Responses by benthic organisms to inputs of organic material to the ocean floor: A review. Philosophical Transactions of the Royal Society of London Series A, Mathematical and Physical Sciences 331:119–138.

Graça, M.A., S.Y. Newell, and R.T. Kneib. 2000. Grazing rates of organic matter and living fungal biomass of decaying *Spartina alterniflora* by three species of salt-marsh invertebrates. Marine Biology 136:281–289.

Green, P.T., P.S. Lake, and D.J. O'Dowd. 1999. Monopolization of litter processing by a dominant land crab on a tropical oceanic island. Oecologia 119:435–444.

Griffin, J.E., B.P. O'Malley, and J.D. Stockwell. 2020. The freshwater mysid *Mysis diluviana* (Audzijonyte & Väinölä, 2005) (Mysida: Mysidae) consumes detritus in the presence of *Daphnia* (Cladocera: Daphniidae). Journal of Crustacean Biology 40:520–525.

Growns, I.O., and A.M.M. Richardson. 1988. Diet and burrowing habits of the freshwater crayfish, *Parastacoides tasmanicus tasmanicus* Clark (Decapoda: Parastacidae). Marine and Freshwater Research 39:525.

Guest, M.A., and R.M. Connolly. 2005. Fine-scale movement and assimilation of carbon in saltmarsh and mangrove habitat by resident animals. Aquatic Ecology 38:599–609.

Hammann, S., and M. Zimmer. 2015. Lifestyles of detritus-feeding crustaceans. Pages 479–501 *in* M. Thiel and L. Watling, editors. The natural history of the Crustacea, Volume 2: Lifestyles and feeding biology. Oxford University Press, New York, USA.

Haram, L.E., E.E. Sotka, and J.E. Byers. 2020. Effects of novel, non-native detritus on decomposition and invertebrate community assemblage. Marine Ecology Progress Series 643:49–61.

Hassall, M., J.G. Turner, and M.R.W. Rands. 1987. Effects of terrestrial isopods on the decomposition of woodland leaf litter. Oecologia 72:597–604.

Havermans, C., and V. Smetacek. 2018. Bottom-up and top-down triggers of diversification: A new look at the evolutionary ecology of scavenging amphipods in the deep sea. Progress in Oceanography 164:37–51.

Higgs, N.D., A.R. Gates, and D.O.B. Jones. 2014. Fish food in the deep sea: Revisiting the role of large food-falls. PLoS ONE 9:e96016.

Hoffman, J.A., J. Katz, and M.D. Bertness. 1984. Fiddler crab deposit-feeding and meiofaunal abundance in salt marsh habitats. Journal of Experimental Marine Biology and Ecology 82:161–174.

Hoyoux, C., M. Zbinden, S. Samadi, F. Gaill, and P. Compère. 2009. Wood-based diet and gut microflora of a galatheid crab associated with Pacific deep-sea wood falls. Marine Biology 156:2421–2439.

Hübner, L., S.C. Pennings, and M. Zimmer. 2015. Sex- and habitat-specific movement of an omnivorous semi-terrestrial crab controls habitat connectivity and subsidies: A multi-parameter approach. Oecologia 178:999–1015.

Huryn, A.D., and J.B. Wallace. 1987. Production and litter processing by crayfish in an Appalachian mountain stream. Freshwater Biology 18:277–286.

Hyndes, G.A., E.L. Berdan, C. Duarte, J.E. Dugan, K.A. Emery, P.A. Hambäck, C.J. Henderson, D.M. Hubbard, M. Lastra, M.A. Mateo, A. Olds, and T.A. Schlacher. 2022. The role of inputs of marine wrack and carrion in sandy-beach ecosystems: A global review. Biological Reviews 97: 2127–2161.

Iken, K., T. Brey, U. Wand, J. Voigt, and P. Junghans. 2001. Food web structure of the benthic community at the Porcupine Abyssal Plain (NE Atlantic): A stable isotope analysis. Progress in Oceanography 50:383–405.

Ingram, C.L., and R.R. Hessler. 1983. Distribution and behavior of scavenging amphipods from the central North Pacific. Deep Sea Research Part A: Oceanographic Research Papers 30:683–706.

Jackson, M.C., J. Grey, K. Miller, J.R. Britton, and I. Donohue. 2016. Dietary niche constriction when invaders meet natives: Evidence from freshwater decapods. Journal of Animal Ecology 85:1098–1107.

Jamieson, A.J., and J.N.J. Weston. 2023. Amphipoda from depths exceeding 6,000 meters revisited 60 years on. Journal of Crustacean Biology 43:1–28.

Janßen, F., T. Treude, and U. Witte. 2000. Scavenger assemblages under differing trophic conditions: A case study in the deep Arabian Sea. Deep-Sea Research Part II: Topical Studies in Oceanography 47:2999–3026.

Johnson, R.K. 1987. The life history, production and food habits of *Pontoporeia affinis* Lindström (Crustacea: Amphipoda) in mesotrophic Lake Erken. Hydrobiologia 144:277–283.

Joly, F.-X., S. Coq, M. Coulis, J.-F. David, S. Hättenschwiler, C.W. Mueller, I. Prater, and J.-A. Subke. 2020. Detritivore conversion of litter into faeces accelerates organic matter turnover. Communications Biology 3:1–9.

Juul-Pedersen, T., T.G. Nielsen, C. Michel, E.F. Møller, P. Tiselius, P. Thor, M. Olesen, E. Selander, and S. Gooding. 2006. Sedimentation following the spring bloom in Disko Bay, West Greenland, with special emphasis on the role of copepods. Marine Ecology Progress Series 314:239–255.

Karas, P., M. Gorny, and R. Alarcón-Muñoz. 2007. Experimental studies on the feeding ecology of *Munida subrugosa* (White, 1847) (Decapoda: Anomura: Galatheidae) from the Magellan region, southern Chile. Scientia Marina 71:187–190.

Kelly, D.W., J.T.A. Dick, and W.I. Montgomery. 2002. The functional role of *Gammarus* (Crustacea, Amphipoda): Shredders, predators, or both? Hydrobiologia 485:199–203.

Kensley, B. 1974. Aspects of the biology and ecology of the genus *Tylos* Latreille. Annals of the South African Museum 65:401–471.

Kiørboe, T. 2000. Colonization of marine snow aggregates by invertebrate zooplankton: Abundance, scaling, and possible role. Limnology and Oceanography 45:479–484.

Klages, M., S. Muyakshin, T. Soltwedel, and W.E. Arntz. 2002. Mechanoreception, a possible mechanism for food fall detection in deep-sea scavengers. Deep-Sea Research Part I: Oceanographic Research Papers 49:143–155.

Kobari, T., M. Kitamura, M. Minowa, H. Isami, H. Akamatsu, H. Kawakami, K. Matsumoto, M. Wakita, and M.C. Honda. 2013. Impacts of the wintertime mesozooplankton community to downward carbon flux in the subarctic and subtropical Pacific Oceans. Deep Sea Research Part I: Oceanographic Research Papers 81:78–88.

Kobari, T., D.K. Steinberg, A. Ueda, A. Tsuda, M.W. Silver, and M. Kitamura. 2008. Impacts of ontogenetically migrating copepods on downward carbon flux in the western subarctic Pacific Ocean. Deep Sea Research Part II: Topical Studies in Oceanography 55:1648–1660.

Kobayashi, H., Y. Hatada, T. Tsubouchi, T. Nagahama, and H. Takami. 2012. The hadal amphipod *Hirondellea gigas* possessing a unique cellulase for digesting wooden debris buried in the deepest seafloor. PLoS ONE 7:e42727.

Koussovi, G., F. Niass, C.J. Allozounhoue, M. Dahouda, C.A. Bonou, and E. Montchowui. 2020. Diet, ontogenetic changes and rhythm of feeding activities in *Macrobrachium macrobrachion* (Herklots, 1851) (Decapoda, Palaemonidae) in the waterbodies and rivers of Benin, West Africa. Crustaceana 93:999–1021.

Laidre, M.E. 2013a. Eavesdropping foragers use level of collective commotion as public information to target high quality patches. Oikos 122:1505–1511.

Laidre, M.E. 2013b. Foraging across ecosystems: Diet diversity and social foraging spanning aquatic and terrestrial ecosystems by an invertebrate. Marine Ecology 34:80–89.

Lampitt, R.S., K.F. Wishner, C.M. Turley, and M.V. Angel. 1993. Marine snow studies in the Northeast Atlantic Ocean: Distribution, composition and role as a food source for migrating plankton. Marine Biology 116:689–702.

Lastra, M., J. López, J. Troncoso, D.M. Hubbard, and J.E. Dugan. 2016. Scavenger and burrowing features of *Hippa pacifica* (Dana 1852) on a range of tropical sandy beaches. Marine Biology 163:212.

Lastra, M., H.M. Page, J.E. Dugan, D.M. Hubbard, and I.F. Rodil. 2008. Processing of allochthonous macrophyte subsidies by sandy beach consumers: Estimates of feeding rates and impacts on food resources. Marine Biology 154:163–174.

Lee, S.Y. 1997. Potential trophic importance of the faecal material of the mangrove sesarmine crab *Sesarma messa*. Marine Ecology Progress Series 159:275–284.

Linley, T.D., J. Craig, A.J. Jamieson, and I.G. Priede. 2018. Bathyal and abyssal demersal bait-attending fauna of the Eastern Mediterranean Sea. Marine Biology 165:159.

Linton, S.M., P. Greenaway, and D.W. Towle. 2006. Endogenous production of endo-β-1,4-glucanase by decapod crustaceans. Journal of Comparative Physiology B 176:339–348.

Lombard, F., M. Koski, and T. Kiørboe. 2013. Copepods use chemical trails to find sinking marine snow aggregates. Limnology and Oceanography 58:185–192.

López-Victoria, M., and B. Werding. 2008. Ecology of the endemic land crab *Johngarthia malpilensis* (Decapoda: Brachyura: Gecarcinidae), a poorly known species from the tropical eastern Pacific. Pacific Science 62:483–493.

MacNeil, C., J.T.A. Dick, and R.W. Elwood. 1997. The trophic ecology of freshwater *Gammarus* spp. (Crustacea: Amphipoda): Problems and perspectives concerning the functional feeding group concept. Biological Reviews of the Cambridge Philosophical Society 72:349–364.

Mancinelli, G., F. Sangiorgio, and A. Scalzo. 2013. The effects of decapod crustacean macroconsumers on leaf detritus processing and colonization by invertebrates in stream habitats: A meta-analysis. International Review of Hydrobiology 98:206–216.

March, J.G., J.P. Benstead, C.M. Pringle, and M.W. Ruebel. 2001. Linking shrimp assemblages with rates of detrital processing along an elevational gradient in a tropical stream. Canadian Journal of Fisheries and Aquatic Sciences 58:470–478.

Marinucci, A.C. 1982. Trophic importance of *Spartina alterniflora* production and decomposition to the marsh-estuarine ecosystem. Biological Conservation 22:35–58.

Marks, J.C. 2019. Revisiting the fates of dead leaves that fall into streams. Annual Review of Ecology, Evolution, and Systematics 50:547–568.

Martins, R.T., A.S. Melo, J.F. Gonçalves, and N. Hamada. 2015. Leaf-litter breakdown in urban streams of Central Amazonia: Direct and indirect effects of physical, chemical, and biological factors. Freshwater Science 34:716–726.

Masese, F.O., N. Kitaka, J. Kipkemboi, G.M. Gettel, K. Irvine, and M.E. McClain. 2014. Macroinvertebrate functional feeding groups in Kenyan highland streams: Evidence for a diverse shredder guild. Freshwater Science 33:435–450.

Mayor, D.J., W.C. Gentleman, and T.R. Anderson. 2020. Ocean carbon sequestration: Particle fragmentation by copepods as a significant unrecognised factor? BioEssays 42:2000149.

Mews, M., M. Zimmer, and D.E. Jelinski. 2006. Species-specific decomposition rates of beach-cast wrack in Barkley Sound, British Columbia, Canada. Marine Ecology Progress Series 328:155–160.

Michaud, K.M., K.A. Emery, J.E. Dugan, D.M. Hubbard, and R.J. Miller. 2019. Wrack resource use by intertidal consumers on sandy beaches. Estuarine, Coastal and Shelf Science 221:66–71.

Moghadam, F.S., and M. Zimmer. 2016. Effects of warming, nutrient enrichment and detritivore presence on litter breakdown and associated microbial decomposers in a simulated temperate woodland creek. Hydrobiologia 770:243–256.

Moore, J.C., E.L. Berlow, D.C. Coleman, P.C. Ruiter, Q. Dong, A. Hastings, N.C. Johnson, K.S. McCann, K. Melville, P.J. Morin, K. Nadelhoffer, A.D. Rosemond, D.M. Post, J.L. Sabo, K.M. Scow, M.J. Vanni, and D.H. Wall. 2004. Detritus, trophic dynamics and biodiversity. Ecology Letters 7:584–600.

Morton, B., and W.Y. Yuen. 2000. The feeding behaviour and competition for carrion between two sympatric scavengers on a sandy shore in Hong Kong: The gastropod, *Nassarius festivus* (Powys) and the hermit crab, *Diogenes edwardsii* (De Haan). Journal of Experimental Marine Biology and Ecology 246:1–29.

Moss, B. 2005. Rapid shredding of leaves by crabs in a tropical African stream. SIL Proceedings, 1922–2010 29:147–150.

Moulton, T.P., S.A.P. Magalhães-Fraga, E.F. Brito, and F.A. Barbosa. 2010. Macroconsumers are more important than specialist macroinvertebrate shredders in leaf processing in urban forest streams of Rio de Janeiro, Brazil. Hydrobiologia 638:55–66.

Nicol, E.A.T. 1932. The feeding habits of the Galatheidea. Journal of the Marine Biological Association of the United Kingdom 18:87–106.

Nordhaus, I., M. Wolff, and K. Diele. 2006. Litter processing and population food intake of the mangrove crab *Ucides cordatus* in a high intertidal forest in northern Brazil. Estuarine, Coastal and Shelf Science 67:239–250.

Ochoa, J., H. Maske, J. Sheinbaumc, and J. Candela. 2013. Diel and lunar cycles of vertical migration extending to below 1000 m in the ocean and the vertical connectivity of depth-tiered populations. Limnology and Oceanography 58:1207–1214.

Oertli, B. 1993. Leaf litter processing and energy flow through macroinvertebrates in a woodland pond (Switzerland). Oecologia 96:466–477.

Page, H.M., and S.W. Willason. 1983. Feeding activity patterns and carrion removal by terrestrial hermit crabs at Enewetak Atoll, Marshall Islands. Pacific Science 37:151–155.

Parker, J.D., J.P. Montoya, and M.E. Hay. 2008. A specialist detritivore links *Spartina alterniflora* to salt marsh food webs. Marine Ecology Progress Series 364:87–95.

Pavasovic, M., N.A. Richardson, A.J. Anderson, D. Mann, and P.B. Mather. 2004. Effect of pH, temperature and diet on digestive enzyme profiles in the mud crab, *Scylla serrata*. Aquaculture 242:641–654.

Pekkoeva, S.N., V.P. Voronin, Z.M. Shatilina, E.V. Madyarova, D.V. Axenov-Gribanov, Y.A. Shirokova, M.A. Timofeyev, N.N. Nemova, and S.A. Murzina. 2021. Lipid and fatty acid composition of scavenging amphipods *Ommatogammarus* spp. from different depths of lake Baikal. Limnology 22:299–311.

Pelletier, A.J.D., D.E. Jelinski, M. Treplin, and M. Zimmer. 2011. Colonisation of beach-cast macrophyte wrack patches by talitrid amphipods: A primer. Estuaries and Coasts 34:863–871.

Pieczyńska, E. 1993. Detritus and nutrient dynamics in the shore zone of lakes: A review. Hydrobiologia 251:49–58.

Pilati, A., W.A. Wurtsbaugh, and N.R. Brindza. 2004. Evidence of coprophagy in freshwater zooplankton. Freshwater Biology 49:913–918.

Poulsen, L.K., and T. Kiørboe. 2005. Coprophagy and coprorhexy in the copepods *Acartia tonsa* and *Temora longicornis*: Clearance rates and feeding behaviour. Marine Ecology Progress Series 299:217–227.

Premke, K., P. Fischer, M. Hempel, and K.-O. Rothhaupt. 2010. Ecological studies on the decomposition rate of fish carcasses by benthic organisms in the littoral zone of Lake Constance, Germany. Annales de Limnologie—International Journal of Limnology 46:157–168.

Premke, K., M. Klages, and W.E. Arntz. 2006. Aggregations of Arctic deep-sea scavengers at large food falls: Temporal distribution, consumption rates and population structure. Marine Ecology Progress Series 325:121–135.

Premke, K., S. Muyakshin, M. Klages, and J. Wegner. 2003. Evidence for long-range chemoreceptive tracking of food odour in deep-sea scavengers by scanning sonar data. Journal of Experimental Marine Biology and Ecology 285–286:283–294.

Quadros, A.F., and P.B. Araujo. 2008. An assemblage of terrestrial isopods (Crustacea) in southern Brazil and its contribution to leaf litter processing. Revista Brasileira de Zoologia 25:58–66.

Rae, C., G.A. Hyndes, and T.A. Schlacher. 2019. Trophic ecology of ghost crabs with diverse tastes: Unwilling vegetarians. Estuarine, Coastal and Shelf Science 224:272–280.

Rautio, M., H. Mariash, and L. Forsström. 2011. Seasonal shifts between autochthonous and allochthonous carbon contributions to zooplankton diets in a subarctic lake. Limnology and Oceanography 56:1513–1524.

Reigstad, M., C.W. Riser, and C. Svensen. 2005. Fate of copepod faecal pellets and the role of *Oithona* spp. Marine Ecology Progress Series 304:265–270.

Richardson, A., and P.B. Araujo. 2015. Lifestyles of terrestrial crustaceans. Pages 299–336 *in* M. Thiel and L. Watling, editors. The natural history of the Crustacea, Volume 2: Lifestyles and feeding biology. Oxford University Press, New York, USA.

Robertson, A.I., and P.A. Daniel. 1989. The influence of crabs on litter processing in high intertidal mangrove forests in tropical Australia. Oecologia 78:191–198.

Rohlfer, E.K., S.L. Scheer, M. Bergmann, A.K. Sweetman, and H.J.T. Hoving. 2022. Contrasting residence time and scavenging communities of experimental invertebrate food falls in the Arctic deep sea. Deep Sea Research Part I: Oceanographic Research Papers 189:103832.

Ruxton, G.D., and D.C. Houston. 2004. Energetic feasibility of an obligate marine scavenger. Marine Ecology Progress Series 266:59–63.

Sainte-Marie, B. 1992. Foraging of scavenging deep-sea lysianassoid amphipods. Pages 105–124 *in* G.T. Rowe and V. Pariente, editors. Deep-sea food chains and the global carbon cycle. Kluwer Academic Publishers, Dordrecht, The Netherlands.

Sampei, M., A. Forest, H. Sasaki, H. Hattori, R. Makabe, M. Fukuchi, and L. Fortier. 2009. Attenuation of the vertical flux of copepod fecal pellets under Arctic sea ice: Evidence for an active detrital food web in winter. Polar Biology 32:225–232.

Scharold, J.V., and T.D. Corry. 2021. Status of the amphipod *Diporeia* spp. in Lake Superior, 2006–2016. Journal of Great Lakes research 47:1033–1039.

Scheer, S.L., A.K. Sweetman, U. Piatkowski, E.K. Rohlfer, and H.J.T. Hoving. 2022. Food fall-specific scavenging response to experimental medium-sized carcasses in the deep sea. Marine Ecology Progress Series 685:31–48.

Schlacher, T.A., S. Strydom, and R.M. Connolly. 2013. Multiple scavengers respond rapidly to pulsed carrion resources at the land–ocean interface. Acta Oecologica 48:7–12.

Schnetzer, A., and D.K. Steinberg. 2002. Active transport of particulate organic carbon and nitrogen by vertically migrating zooplankton in the Sargasso Sea. Marine Ecology Progress Series 234:71–84.

Shanij, K., V.P. Praveen, S. Suresh, M.M. Oommen, and T.S. Nayar. 2016. Leaf litter translocation and consumption in mangrove ecosystems: The key role played by the sesarmid crab *Neosarmatium malabaricum*. Current Science 110:1969–1976.

Sih, A., G. Englund, and D. Wooster. 1998. Emergent impacts of multiple predators on prey. Trends in Ecology and Evolution 13:350–355.

Silva, A.P., A. Colaço, A. Ravara, J. Jakobsen, K. Jakobsen, and D. Cuvelier. 2021. The first whale fall on the Mid-Atlantic Ridge: Monitoring a year of succession. Deep Sea Research Part I: Oceanographic Research Papers 178:103662.

Skov, M.W., and R.G. Hartnoll. 2002. Paradoxical selective feeding on a low-nutrient diet: Why do mangrove crabs eat leaves? Oecologia 131:1–7.

Slim, F.J., M.A. Hemminga, C. Ochieng, N.T. Jannink, E. Cocheret de la Morinière, and G. van der Velde. 1997. Leaf litter removal by the snail *Terebralia palustris* (Linnaeus) and sesarmid crabs in an East African mangrove forest (Gazi Bay, Kenya). Journal of Experimental Marine Biology and Ecology 215:35–48.

Small, L.F., S.W. Fowler, S.A. Moore, and J. LaRosa. 1983. Dissolved and fecal pellet carbon and nitrogen release by zooplankton in tropical waters. Deep Sea Research Part A: Oceanographic Research Papers 30:1199–1220.

Smith, C.R., and A.R. Baco. 2003. Ecology of whale falls at the deep-sea floor. Oceanography and Marine Biology: An Annual Review 41:311–354.

Smith, C.R., A.F. Bernardino, A. Baco, A. Hannides, and I. Altamira. 2014. Seven-year enrichment: Macrofaunal succession in deep-sea sediments around a 30 tonne whale fall in the Northeast Pacific. Marine Ecology Progress Series 515:133–149.

Sokolova, M.N. 1997. Trophic structure of abyssal macrobenthos. Advances in Marine Biology 32:427–525.

Stamhuis, E.J., B. Dauwe, and J.J. Videler. 1998. How to bite the dust: Morphology, motion pattern and function of the feeding appendages of the deposit-feeding thalassinid shrimp *Callianassa subterranea*. Marine Biology 132:43–58.

Steinberg, D.K., and M.R. Landry. 2017. Zooplankton and the ocean carbon cycle. Annual Review of Marine Science 9:413–444.

Steinberg, D.K., B.A.S. Van Mooy, K.O. Buesseler, P.W. Boyd, T. Kobari, and D.M. Karl. 2008. Bacterial vs. zooplankton control of sinking particle flux in the ocean's twilight zone. Limnology and Oceanography 53:1327–1338.

Stewart, T.J., and W.G. Sprules. 2011. Carbon-based balanced trophic structure and flows in the offshore Lake Ontario food web before (1987–1991) and after (2001–2005) invasion-induced ecosystem change. Ecological Modelling 222:692–708.

Stockwell, J.D., B.P. O'Malley, S. Hansson, R.J. Chapina, L.G. Rudstam, and B.C. Weidel. 2020. Benthic habitat is an integral part of freshwater *Mysis* ecology. Freshwater Biology 65:1997–2009.

Svensen, C., and J.C. Nejstgaard. 2003. Is sedimentation of copepod faecal pellets determined by cyclopoids? Evidence from enclosed ecosystems. Journal of Plankton Research 25:917–926.

Sweetman, A.K., A.R. Thurber, C.R. Smith, L.A. Levin, C. Mora, C.-L. Wei, A.J. Gooday, D.O.B. Jones, M. Rex, M. Yasuhara, J. Ingels, H.A. Ruhl, C.A. Frieder, R. Danovaro, L. Würzberg, A. Baco, B.M. Grupe,

A. Pasulka, K.S. Meyer, K.M. Dunlop, L.-A. Henry, and J.M. Roberts. 2017. Major impacts of climate change on deep-sea benthic ecosystems. Elementa: Science of the Anthropocene 5:4.

Tran, M.V. 2013. Divergent reactions to olfactory foraging cues between two ecologically similar, sympatric hermit crab species. Journal of Crustacean Biology 33:512–518.

Turner, J.T. 2015. Zooplankton fecal pellets, marine snow, phytodetritus and the ocean's biological pump. Progress in Oceanography 130:205–248.

Turner, J.T., and J.G. Ferrante. 1979. Zooplankton fecal pellets in aquatic ecosystems. BioScience 29:670–677.

Usio, N. 2000. Effects of crayfish on leaf processing and invertebrate colonisation of leaves in a headwater stream: Decoupling of a trophic cascade. Oecologia 124:608–614.

van der Jagt, H., I. Wiedmann, N. Hildebrandt, B. Niehoff, and M.H. Iversen. 2020. Aggregate feeding by the copepods *Calanus* and *Pseudocalanus* controls carbon flux attenuation in the Arctic shelf sea during the productive period. Frontiers in Marine Science 7:543124.

van Egmond, E.M., P.M. van Bodegom, J.R. van Hal, R.S.P. van Logtestijn, R.A. Broekman, M.P. Berg, and R. Aerts. 2019. Growth of pioneer beach plants is strongly driven by buried macroalgal wrack, whereas macroinvertebrates affect plant nutrient dynamics. Journal of Experimental Marine Biology and Ecology 514–515:87–94.

Wahle, R.A., A.J. Linnane, and A.M. Harrington. 2020. Lobster fisheries. Pages 55–90 *in* G. Lovrich and M. Thiel, editors. The natural history of the Crustacea, Volume 9: Fisheries and aquaculture. Oxford University Press, New York, USA.

Watling, L. 2015. Deposit feeding: Obtaining nutrition from sediment. Pages 464–478 *in* M. Thiel and L. Watling, editors. The natural history of the Crustacea, Volume 2: Lifestyles and feeding biology. Oxford University Press, New York, USA.

Werry, J., and S.Y. Lee. 2005. Grapsid crabs mediate link between mangrove litter production and estuarine planktonic food chains. Marine Ecology Progress Series 293:165–176.

Wissinger, S.A., A.J. Klemmer, A. Braccia, B.M. Bush, and D.P. Batzer. 2021. Relationships between macroinvertebrates and detritus in freshwater wetlands. Freshwater Science 40:681–698.

Yang, C., I.S. Wehrtmann, S.J. Wenger, and A.T. Rugenski. 2020a. Neotropical freshwater crabs (Decapoda: Pseudothelphusidae) shred leaves. Nauplius 28:e2020020.

Yang, C., S.J. Wenger, A.T. Rugenski, I.S. Wehrtmann, S. Connelly, and M.C. Freeman. 2020b. Freshwater crabs (Decapoda: Pseudothelphusidae) increase rates of leaf breakdown in a neotropical headwater stream. Freshwater Biology 65:1673–1684.

Yoshimizu, C., T. Yoshida, M. Nakanishi, and J. Urabe. 2001. Effects of zooplankton on the sinking flux of organic carbon in Lake Biwa. Limnology 2:37–43.

Zimmer, M. 2002. Nutrition in terrestrial isopods (Isopoda: Oniscidea): An evolutionary-ecological approach. Biological Reviews 77:455–493.

Zimmer, M. 2019. Detritus. Pages 292–301 *in* B.D. Fath, editor. Encyclopedia of ecology, 2nd edition, Volume 3. Elsevier, Amsterdam, the Netherlands.

Zimmer, M., G. Kautz, and W. Topp. 2003. Leaf litter-colonizing microbiota: Supplementary food source or indicator of food quality for *Porcellio scaber* (Isopoda: Oniscidea)? European Journal of Soil Biology 39:209–216.

Zimmer, M., G. Kautz, and W. Topp. 2005. Do woodlice and earthworms interact synergistically in leaf litter decomposition? Functional Ecology 19:7–16.

Zimmer, M., S.C. Pennings, T.L. Buck, and T.H. Carefoot. 2002. Species-specific patterns of litter processing by terrestrial isopods (Isopoda: Oniscidea) in high intertidal salt marshes and coastal forests. Functional Ecology 16:596–607.

Zimmer, M., S.C. Pennings, T.L. Buck, and T.H. Carefoot. 2004. Salt marsh litter and detritivores: A closer look at redundancy. Estuaries 27:753–769.

Zimmer, M., and W. Topp. 2002. The role of coprophagy in nutrient release from feces of phytophagous insects. Soil Biology and Biochemistry 34:1093–1099.

4

THE IMPORTANCE OF MARINE BENTHIC CRUSTACEANS AS BIOTURBATORS, HABITAT MODIFIERS, AND ECOSYSTEM FUNCTION REGULATORS

Hazel R. Needham, Michael Townsend, and Andrew M. Lohrer

Abstract

The activities of burrowing Crustacea have profound physical and geochemical effects on soft-sediment environments. The construction, maintenance, and ventilation of burrow structures translocate solutes and material between reaction zones and extends the sediment–water interface. Activities within the sediment matrix have multiple and interrelated consequences influencing processes, functions, and the functioning of marine seafloor habitats. In this chapter we illustrate how seemingly small-scale processes around individual burrow structures can scale up to have system-wide impacts on ecosystem functions from intertidal environments to the deep sea and from tropics to poles. Effects on the sediment environment include changes to geotechnical properties such as grain size distribution, sediment permeability, bulk density, and shear strength that result in changes to the stability of the sediment matrix. Carbon storage in coastal systems is increasingly viewed as an ecosystem service. Many burrowing crustaceans modify the rates of carbon sequestration, with the flux of carbon in salt marsh and mangrove habitats largely governed by bioturbating crabs through burial and consumption of organic matter. Nitrogen cycling is influenced by burrow depth, irrigation, and sharp oxic-anoxic boundaries in burrow walls, which encourages close coupling of nitrification-denitrification and creates a pathway for removal of dinitrogen gas. This is important in systems that are experiencing excessive nutrients as a result of human activities. Given that crustaceans are among the most successful and prevalent taxa affecting ecosystem functioning, they will factor into how coastal ecosystems respond to warmer temperatures, greater CO_2 concentrations, and sea level rises.

Hazel Needham et al., *The Importance of Marine Benthic Crustaceans as Bioturbators, Habitat Modifiers, and Ecosystem Function Regulators*. In: *Ecology and Conservation*. Edited by: Lars Gutow and Martin Thiel, Oxford University Press.
 DOI: 10.1093/oso/9780197768242.003.0004

INTRODUCTION

"Ecosystem engineering" is a term coined to describe the construction, modification, and maintenance of biologically generated habitat features by organisms (Jones et al. 1994). In marine soft-sediment environments, most examples of ecosystem engineering link directly or indirectly to bioturbation and bioirrigation. Bioturbation is the process by which burrowing organisms create and modify sedimentary environments by mixing and moving sediment particles, for example, when excavating burrows or deposit feeding. Bioirrigation refers to the flushing or ventilation of sediments with water for the purposes of filter feeding, removing waste, and accessing dissolved oxygen.

Both bioturbation and bioirrigation, and thus the activities of animals that move particles and fluids in taxon-specific ways, are fundamental to our understanding of bioengineering and habitat modification in soft-sediment marine habitats. These activities have multiple interrelated consequences for marine seafloor habitats, influencing processes such as sediment transport, carbon trapping, primary production, and nutrient recycling (Lohrer et al. 2004, Mermillod-Blondin and Rosenberg 2006, Thrush et al. 2006, Le Hir et al. 2007, Lohrer et al. 2010), ultimately altering the ways in which ecosystems function. The magnitude and direction of effects generally relate to the physical and behavioral attributes of the organisms involved (e.g., large body size, depth of burrowing or body position within the sediment, burrow complexity or frequency) and/or their population densities.

The modification of microtopography through the creation of pits and mounds influences the hydrodynamics of benthic boundary layer flows and creates spatial variation in hydraulic pressures that can move dissolved solutes (oxygen, nutrients, metabolic wastes) in and out of the sediment (Huettel et al. 2003). Burrows also create a heterogeneous three-dimensional matrix within the sediment column, extending the sediment–water interface. Burrow networks have been likened to veins and arteries in marine sediment systems as they act as conduits to transport oxygen, metabolites, particles, and nutrients between reaction zones (Reise 2002). They also provide extensive habitat for microbial communities that exert strong control over biogeochemical processes. Burrow ventilation, which can occur both passively and actively, is closely coupled to bioturbation, as the latter subducts labile material while the former speeds up reaction rates through increased oxygenation and exchange (D'Andrea et al. 2002, Pillay et al. 2011, Pillay 2019, and references therein). These processes alter the pathways and magnitude of solute fluxes and enhance benthic-pelagic coupling (Nielsen et al. 2003). The activities of bioturbators and bioirrigators can have spatially widespread influence (e.g., at the scale of entire coastal systems; Seike et al. 2020) and evolutionary-scale significance. The marked shifts in seafloor ecosystems at the Ediacaran–Cambrian boundary (635–541 mya) were thought to be due in large part to habitat modification by bioturbating and bioirrigating species (Dornbos et al. 2005, Herringshaw et al. 2017).

Several groups of Crustacea are well-known bioturbators and bioirrigators. These include many sesarmid and varunid crabs (Grapsoidea), most or all macrophthalmid and ocypodid crabs (Ocypodoidea), several lineages of burrowing shrimps (e.g., Gebiidea, Axiidea, Squilloidea, and Lysiosquilloidea), and burrowing/tube-building amphipods (e.g., *Corophium* spp., *Ampelisca* spp.). While key bioturbating crustaceans are generally more common in intertidal and shallow subtidal habitats, some create burrows in continental shelf sediments at depths of > 1500 m (e.g., *Nephrops* spp., stomatopods, Bathysquilloidea). Although not a primary focus of this chapter, terrestrial and semi-terrestrial burrowing crabs are also common in many parts of the world. These include massive coconut crabs (*Birgus latro*; tropical southwest Pacific and Indian Ocean Islands) that burrow up to 60 cm deep, the 40+ million red crabs (*Gecarcoidea natalis*) that migrate to the sea annually from their forest burrows on Christmas Island (Green 2004), and mountain crabs of Japan (*Chiromantes haematocheir*) that can live at high elevations and away from the coast. Some crustaceans are even able to engineer habitats in deserts and arid continental interiors (e.g., litter recycling isopods of the Negev Desert, whose burrows are oases of higher soil fertility; see Sagi et al. 2019). Nevertheless, although ecologists tend to herald the importance of bioengineers, burrowing can be a hazard to public infrastructure at times. For example, translocations of Chinese Mitten Crabs (*Eriocheir sinensis*) to Europe and North America have caused billions of dollars of damage to dykes and riverbanks (Dittel and Epifanio 2009, Bentley 2011).

INTERACTIONS BETWEEN CRUSTACEA AND THEIR SURROUNDING ENVIRONMENT: PROCESSES, FUNCTIONS, AND FUNCTIONING

The scope of this chapter is limited to marine benthic environments and the way that crustaceans affect these systems physically, geochemically, and ecologically. The discussion can be specific (e.g., effects of an individual shrimp on the egress of ammonium from its burrow) or much broader (e.g., coastal-scale influence of upogebiid shrimp populations on the abundance of primary producers in the water column (Seike et al. 2020); effects on the delivery of "ecosystem services" that support human well-being (Beaumont et al. 2007)). Clarifying terminology will help us discuss the roles of crustaceans at different spatial scales and across different systems. We follow Jax (2010) and refer to ecosystem processes as the interactions between two or more components of an ecosystem (components including organisms and abiotic elements) that result in state changes over time. Changes in surface sediment organic content associated with the downward translocation of reactive material, or elevated subsurface oxygen (O_2) concentrations associated with bioirrigation, are examples of ecosystem processes. In this sense, ecosystem processes are descriptive and pathway oriented.

Ecosystem functions encapsulate greater detail and are thus reflective of a larger system view, for example, moving from single to multiple processes and the interactions of different components within the ecosystem. In the scientific literature, process and function are used somewhat interchangeably as overlapping terms and it can be difficult to define the point where the use of one should switch to the other. Ecosystem functioning is a subtle shift again, whereby the perspective changes from part of the system to the performance of the whole system and the network of processes and interactions that this represents. In this sense, ecosystem functioning is movement away from describing Crustacea as "mere participants" toward assigning them specific roles within an ecosystem (Jax 2010). Commonly, we say that many crustaceans affect or even drive "nutrient recycling," knowing that this is the product of a vast and complex array of physical, chemical, and biological (including microbial) interactions but in which they are integral. Often this detail is incompletely or poorly understood, yet we can measure overall ecosystem change by analyzing alterations in sediment and water column properties, solute and particle fluxes, and communities of organisms.

Crustaceans have the capacity to alter their environment in myriad ways. In this chapter we discuss how crustaceans affect their surroundings, from intertidal environments to the deep sea and from tropics to poles, and how this affects marine ecosystem functions. In most cases we will give specific examples, intended to illustrate the effects of broader functional groups (classes of organisms with similar activities, behaviors, and habitat-modifying characteristics). Species that are distantly related in a taxonomic sense sometimes perform similar functional roles; that is, they may create similar types of sedimentary structures or move solutes and particles in a similar manner. We expect there to be relative similarity in the effects of these species on functions. However, this does not imply that the various taxa in a functional group affect functions at the same rates. It is also important to recognize that some species exhibit "functional plasticity" in that they change their behaviors and strategies depending on the prevailing physical conditions, which alters their functional roles in a context-dependent manner (Pearson 2001).

TRANSLOCATION OF MATERIAL THROUGH BIOTURBATION

A wide variety of crustaceans rework sediments using a suite of burrowing techniques (Atkinson and Eastman 2015) in order to create temporary or permanent burrow refuges of different sizes, shapes, and depths associated with their lifestyle traits (Fig. 4.1). Through the physical process of burrowing, sediment is mixed between different depths and reaction zones. This destabilization of the sediment matrix mixes both particles and porewaters and can have considerable effects on sediment transport, water column turbidity, nutrient dynamics, and benthic community composition, all of which can have system-level consequences for functioning. Burrow construction modifies the existing sediment matrix, altering some of the geotechnical properties of the sediment such as grain size distribution, sediment permeability, bulk density, and shear strength (Grabowski et al. 2011), ultimately influencing the geomorphic evolution of systems, particularly in shallow coastal zones (Crotty et al. 2020). Many burrowing crustaceans have been shown to increase sediment water content and penetrability due to the loosening of the sediment fabric while building or maintaining burrows. This in turn can stimulate microbial activity and organic matter processing (Watling 1998) and influence the rates and pathways of sediment diagenesis.

Due to the diverse behavior and lifestyles of burrowing crustaceans, the degree to which a species influences sediment translocation and sediment transport is widely variable, even within one family or genus. Generally speaking, thalassinidean shrimp (comprising the infraorders Gebiidea and Axiidea) are considered to have some of the greatest rates of sediment processing of all benthic organisms (Branch and Pringle 1987, Witbaard and Duineveld 1989, Rowden et al. 1998, Pillay 2019) and form some of the deepest and most complex burrow structures (Dworschak 1983,

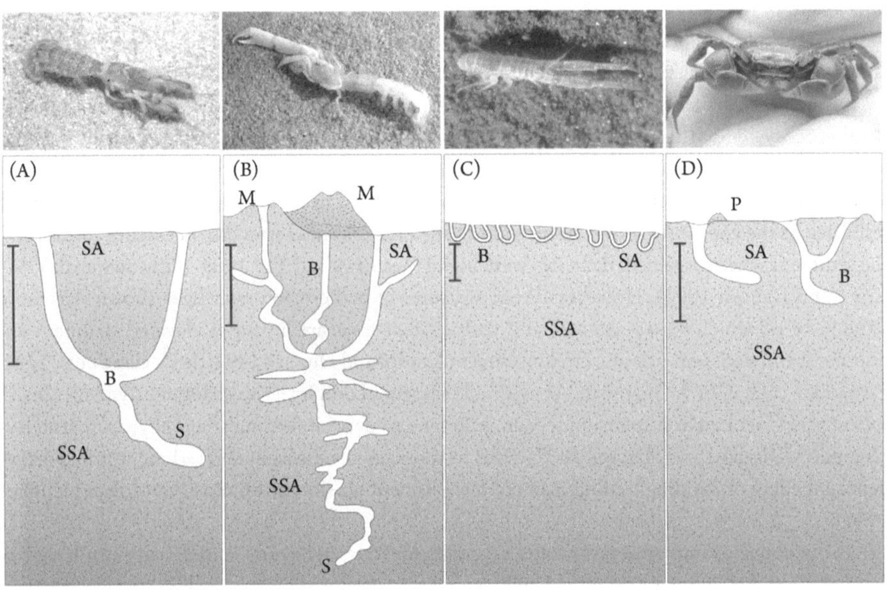

Fig. 4.1.
Photos of (A) *Upogebia africana*, (B) *Kraussillichirus kraussi*, (C) *Corophium volutator*, and (D) *Austrohelice crassa*. Below each photo is a two-dimensional generalized illustration of burrow structures typical in the genus. Burrow features include the lumen (B), surface mounds (M), feeding pellets (P), and sumps (S), as well as the ambient sediment surface (SA) and subsurface (SSA, shaded); scale bar = 10 cm. Source: data from Laverock et al. (2010). Photos reproduced courtesy of G.M. Branch (A, B); E. Gibcus (C), and H. Needham (D).

Atkinson and Taylor 1988). For example, the burrowing shrimp *Callianassa subterranea* creates deep gallery complexes (Ott et al. 1976, Witbaard and Duineveld 1989, Nickell and Atkinson 1995). These structures have both an inhalant and exhalant shaft visible at the surface. Burrows are continuously maintained by ejecting processed sediment and fecal deposits to the sediment surface. During periods of activity, the waste material is jettisoned to the surface where fine material may be laterally advected or forms a "volcano like" structure around the exhalant shaft (Fig. 4.2) (Stamhuis et al. 1996, Rowden et al. 1998).

In tropical lagoonal systems, *Callianassa*-induced sediment advection forms the primary mode of sediment transport outside of storm events (Roberts 1981), sculpting the landscape in which they live. Larger fragments of biogenic material such as coral or shell that is not returned to the surface forms characteristic lateral deposits at depth (Suchanek 1983) as well as deposits in deep, blind-ended burrow sumps (Atkinson and Nash 1990) altering local bulk sediment permeability (Ziebis et al. 1996a). Where mounds are dense, distinct bedding of fine to course layers of sediment are observed in cross-section. These features are used to identify callianassids in ichnological studies, demonstrating their clear effects on diagenetic processes (Curran and Martin 2003, Curran 2007). Excavation rates for *C. subterranea* have been estimated to be between 11 kg (dry weight) and 15.5 kg m^{-2} yr^{-1} (Stamhuis et al. 1996, Rowden et al. 1998). However, activity is highly seasonal with excavation and lateral advection rates peaking in the warmer months (Rowden et al. 1998). This change in activity may also induce alterations in burrow morphology. For example, the ghost shrimp *Filhollianassa filholi* has been shown to reduce lateral extent, shaft length, and overall burrow volume in association with a reduction in water temperatures and increased organic matter content (Berkenbusch and Rowden 2000).

Seabed roughness has been shown to increase substantially in the presence of *Callianassa* mounds (Fig. 4.2K). Rowden and Jones (1994) estimated that fecal deposits increased the roughness height of the seabed more than 1000-fold in the presence of high densities of *C. subterranea*, inducing turbulent benthic boundary layer flows. Porewater advection can also be altered through lateral velocity gradients created by the presence of *Callianassa* mounds (Ziebis et al. 1996b).

It is not only the large gallery-forming organisms that alter sediment structure and stability in benthic ecosystems. The burrowing amphipod *Corophium volutator* builds smaller, shallower, and less complex U-shaped burrows (Fig. 4.1C) with caps that sit above the sediment surface (Grant and Daborn 1994). Initially, observations of this species led to their categorization as a sediment stabilizer, as the caps were shown to increase the sediment's shear strength, in essence "armoring" the sediment where organism densities were high (Meadows and Tait 1989, Meadows et al. 1990, Grant and Daborn 1994). However, direct measures of bed erosion rates have shown that these same features, which vary seasonally and with current flow, are linked to elevated sediment entrainment and transport in some circumstances (Gerdol and Hughes 1994, de Deckere et al. 2001, Fernandez et al. 2006). Organism behavior adds further complexity to sediment dynamics, as *C. volutator* is a facultative suspension feeder, switching to deposit feeding when phytoplankton levels reduce (Riisgård and Schotge 2007). This in turn can affect suspended sediment transport, as deposit feeding on binding biofilms can reduce sediment erosion thresholds (Gerdol and Hughes 1994).

As the above examples demonstrate, organism lifestyle, behaviors, and density can have key consequences for sediment stability. Unlike many filter feeders, deposit feeding species (particularly mobile crab species) often modify the sediment surface by building burrows with a broad and deep aperture (Fig. 4.2). These high aspect ratio openings have been shown to serve as sediment traps, capturing fine particles and organic material at depth and altering the sediment granulometry over time (Yager et al. 1993, Le Minor et al. 2020). Subsurface burrow shape can also increase passive sediment retention, creating a highly effective and energy-efficient means of food capture and storage (Escapa et al. 2008). Material from the bottom of the burrow can be brought back to

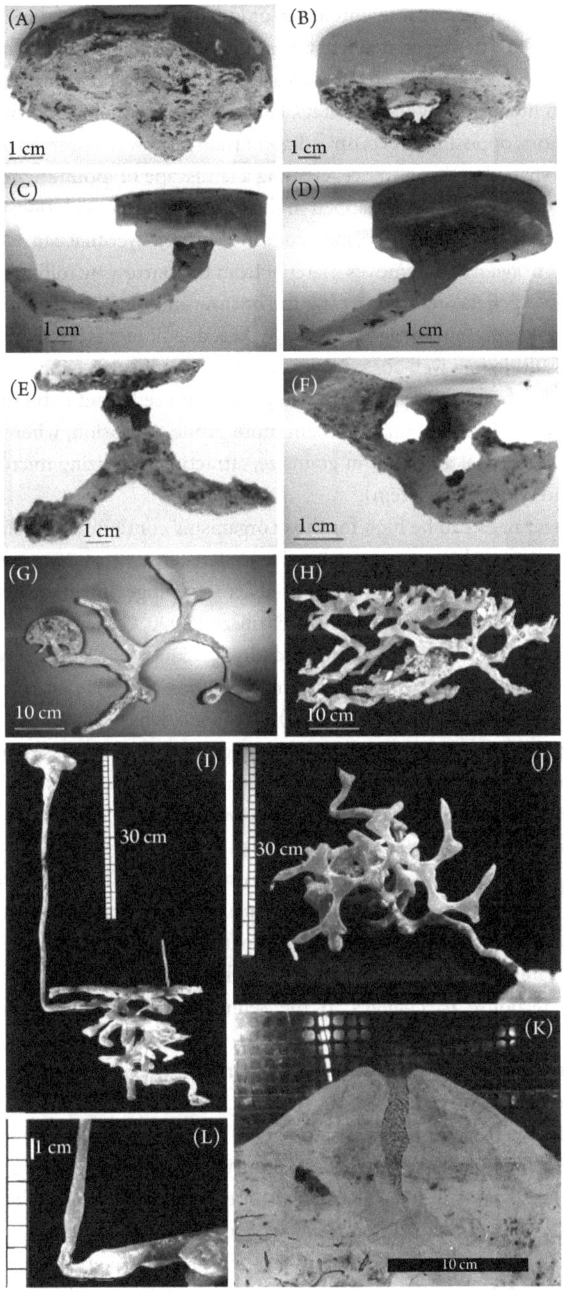

Fig. 4.2.
Resin casts of crustacean burrow structures. *Austrohelice crassa* burrow types from Needham et al. (2010) showing (A) cone, (B) "u" shape, (C) "j" shape, (D) "I" shape, (E) inverted "y" shape, (F) "y" shape, (G) branching, and (H) complex, joined with a second burrowing crustacean species (likely *Alpheus novaezealandiae*). Reproduced from Needham et al. (2010), with permission from © Inter-Research Science Publisher. Resin casts of typical *Callianassa subterranea* burrow structures. Photos show (I) lateral and (J) plan view, (K) the percolation mound, and (L) the base of the exhalent shaft. Reproduced from Nickell and Atkinson (1995) under a Creative Commons by Attribution License (CC BY).

the sediment surface during burrow maintenance or integrated into the sediment matrix at depth. Botto and Iribarne (2000) demonstrated that two seemingly similar intertidal crab species, *Leptuca uruguayensis* and *Neohelice granulata*, both increased sediment water content, penetrability, and organic matter content in their favored habitats. However, due to their differences in burrow morphology and behaviors, opposing effects on sediment transport were observed. *Neohelice granulata* burrows are funnel shaped at the entrance, creating a landscape of mounds and pits where densities are high. Fine cohesive sediment, trapped in the burrows during high tide, is pushed from the burrow entrance as the tide drops and forms a cohesive surface layer that can act to reduce erosion rates. The smaller *L. uruguayensis* removes material from the burrow by rolling it into pellets using its minor chela. These pellets are used to plug the burrow entrance when submerged to avoid an influx of material. However, this disaggregated, repackaged sediment is less cohesive and more easily eroded, often resulting in a well-sorted layer at the sediment surface. Similarly, pellets formed through feeding and defecation can have opposing effects on sediment entrainment; feeding pellets are stripped of organic matter making them more prone to erosion, whereas fecal pellets are often rich in organic material and finer in grainsize, attracting stabilizing microbial communities (Levinton 1972, and references therein).

Sediment turnover rates can be high for those organisms continuously maintaining their burrow structures (often referred to as "regenerators" sensu Kristensen et al. 2012) and rates associated with particular crab species have been documented across a range of habitats (Katz 1980, Takeda and Kurihara 1987, Iribarne et al. 1997, Botto and Iribarne 2000, Gutiérrez et al. 2006, Fanjul et al. 2007, Needham et al. 2010, Wang et al. 2010). Broadly speaking, differences in reworking rates are primarily linked to body size, organism density, and vegetation cover as well as root mass density in certain habitats (e.g., salt marsh; Wang et al. 2010). Organism density can also indirectly affect sediment stability through aerobic mineralization of below-ground plant biomass. For example, Wilson et al. (2012) measured a 50% reduction in sediment shear strength as a result of root degradation in salt marsh where *Sesarma reticulatum* bioturbation was high.

Sediment turnover rates can vary greatly between different species occupying similar habitats. For example, Takeda and Kurihara (1987), estimated mudflat turnover to a depth of 40 cm by *Helice tridens* to take 34.4 days whereas Wang et al. (2010) calculated mudflat turnover to be 1.8 years when examining a natural community of crabs including this species. Turnover rates within a species may be influenced by habitat type where organism behavior alters in response to changes to the sediment properties (Needham et al. 2010, 2013).

The ability of crustaceans to turn over the upper part of the sediment column has implications for disturbance and recovery dynamics. For example, crabs have been shown to be early colonizers of terrigenous sediment deposits on estuarine tidal flats in the immediate aftermath of storms (Norkko et al. 2002). Terrigenous sediment deposition is a major disturbance agent throughout the Pacific Rim where heavy rainfall, steep coastal catchments, and alterations of land have increased soil erosion rates. Terrigenous deposits on sandflats during the monsoon season in southern Japan can exceed 10–20 cm in thickness and can smother thousands of square meters of intertidal sandflat. Mud crabs, *Macrophthalmus (Mareotis) japonicus*, that can burrow through the deposits have been observed to survive these events, unlike many of the less capable burrowers (e.g., the bivalves *Ruditapes philippinarum*, *Mactra quadrangularis*, and *Meretrix lusoria*) (A.M. Lohrer, pers. obs.). Similarly, the common-to-abundant New Zealand grapsid crab *Austrohelice crassa*, which builds shallow, relatively uncomplicated burrows in the intertidal zone (Fig. 4.1D, Fig. 4.2; Needham et al. 2010), is an early colonizer of fresh terrigenous sediment deposits. Its burrowing tends to break up the surficial clay layers and enhance the erosion of the deposits, facilitating overall community recovery. Even in relatively thick deposits (up to 10 cm thickness), large specimens of *A. crassa* (2–3 cm carapace width) have been shown to penetrate the clay layer and mix the fine material into the sediment matrix at depth (Gibbs et al. 2001).

THE INFLUENCE OF CRUSTACEANS ON CARBON CYCLING

Coastal vegetated ecosystems are known repositories for carbon that can have greater capacity for storage than terrestrial forests (Donato et al. 2011). Therefore, carbon storage capacity in coastal systems is increasingly viewed as an ecosystem service in this era of high and rising atmospheric CO_2 concentrations. On average, burial rates in coastal vegetated environments range between 138 and 244g C m^{-2} y^{-1} (Martinetto et al. 2016), potentially providing carbon storage for millennia (Mateo et al. 1997). The degree to which organic carbon is sequestered or processed within a particular habitat will also have functional flow-on effects to connected ecosystems (Bulmer et al. 2020). Carbon cycling behavior differs by habitat type and condition (Lewis et al. 2018, Alongi 2020, Bulmer et al. 2020) so even small shifts in the proportions of different habitat types within an estuary could lead to large shifts in its overall carbon storage (Doughty et al. 2016, Kelleway et al. 2016).

Several studies have demonstrated the effects burrowing crustaceans can have on the rates of carbon sequestration in coastal habitats, and their ability to alter the cycling of organic carbon through marine systems can be substantial (Kristensen 2008, Kristensen et al. 2008a, and references therein). For example, Kinoshita et al. (2003) demonstrated that in muddy tidal flats dominated by the bioirrigating shrimp *Upogebia major*, their burrows acted as traps for labile material likely deposited through tidal action. They concluded that this deposition of phytodetritus supports heterotrophic bacterial communities within the burrow wall, reducing the overall outflow of organic carbon (and nitrogen) from the system (Kinoshita et al. 2003).

In salt marsh systems, the flux of organic carbon has been shown to be largely biologically controlled by burrowing crabs. For example, in the salt marshes of Mar Chiquita lagoon, Argentina, Gutiérrez et al. (2006) examined the effects of *Neohelice granulata* bioturbation on carbon storage and mobilization. They found that the quantity of material brought from depth to the surface was greater than that collected through passive trapping (resulting in net excavation) and that, relative to trapped material, excavate was lower in total and labile carbon content. In this habitat type, high crab densities may therefore promote a net decrease in surface sediment carbon stocks and hence a reduction in the transmission of carbon to neighboring habitats through tidal flow. *Neohelice granulata* populations therefore have the capacity to alter the functioning of this salt marsh in terms of carbon exchange.

Similar relationships have also been observed in another salt marsh crab species, *Minuca pugnax* (Montague 1982), as well as for multi-species assemblages (Wang et al. 2010). However, the latter study found higher total carbon in excavated material where dense vegetation existed. This actually reflected increased inorganic carbon content, likely driven by aerobic mineralization of organic matter. Inorganic carbon, dissolved organic carbon (and other solutes), can also be lost through lateral exchange of water along marsh channels where tidally driven hydraulic gradients create continuous porewater exchanges between burrows and the surrounding sediment (Xiao et al. 2021). Guimond et al. (2020) determined this to be governed by hydrobiological mechanisms that vary in association with crab activity (seasonal burrowing) and burrow density. Primarily, bioturbation increased sediment permeability and enhanced porewater exchange as a result.

Mangrove systems are also significant carbon reservoirs, and crabs burrowing within these habitats can serve to promote changes in intertidal carbon pathways (Nordhaus et al. 2006, Kristensen et al. 2008a). The integration of leaf litter into sediments is heavily associated with the passive and active burial of vegetation by sesarmid and ocypodid crabs from low to high shores. Adult sesarmids generally have a carapace width of around 3 cm and build complex burrows reaching as deep as 2 m. Here, leaf material, mangrove propagules, and other sources of organic matter are retained and integrated into the burrow structure or alternatively consumed by the occupant (Kristensen 2008). In terms of carbon storage, species from the genus *Sesarma* appear to be more important than members of the Ocypodidae (often from the genus *Uca*) that, although often present in high densities, are smaller in both size and burrow form and are more reliant on benthic microalgae as a food source (France 1998, Reinsel 2004). Sesarmids actively draw down a continuous fresh supply of leaf litter and organic material that, likely due to the presence of hypoxic water in the base of their deep burrows (even during periods of emersion; Andreetta et al. 2014), limits aerobic mineralization processes. Deposit feeders are therefore considered to be the "gatekeepers" of carbon cycling in these environments: these crabs hold strong control over organic matter subduction and vertical mixing, which fuels breakdown processes that link sediment systems to ocean and atmosphere.

Carbon dioxide (CO_2) release provides a pathway for carbon export from sediment systems and equates to the amount of organic matter mineralized through microbial respiration in aerobic conditions, once macrofaunal respiration has been accounted for. Several studies across a range of habitats have measured this exchange for crustacean beds. Generally speaking, when compared to neighboring non-bioturbated sediments, rates of CO_2 efflux are enhanced in the burrowed regions (Montague 1982, Kristensen et al. 2008b, Pülmanns et al. 2014, Ouyang et al. 2017), although the underpinning mechanisms driving these differences are rarely examined (Guimond et al. 2020). Enhanced CO_2 release from sediment is often attributed to increased rates of organic matter decomposition resulting from the burrow walls. For example, measurements of the influence of *Macrophthalmus (Mareotis) japonicus* on rates of CO_2 emission from the seabed showed significant influence on material cycling, as rates of organic carbon decomposition were 10% greater in burrowed mudflats than in surrounding non-bioturbated sediments (Otani et al. 2010). The efficiency of remineralization in burrowed sediments can therefore lead to rapid exportation of both CO_2 and dissolved organic carbon from soft-sediment systems (Fanjul et al. 2015). In salt marsh, Xiao et al. (2021) found that where (mixed assemblage) fiddler crab burrows were in densities greater than 300 per m^2, gas-phase concentrations of CO_2 were over six times greater than ambient air. This illustrates that, where burrows are dense enough, the processes occurring within the walls can contribute significantly to carbon cycling, in turn modifying coastal carbon budgets (Xiao et al. 2021).

CRUSTACEAN-DRIVEN REMINERALIZATION RATES AND OXYGEN FLUXES

The accumulation of organic matter in burrows is used as an energy source by heterotrophic microbes (Sasaki et al. 2014). Indeed, some species (most notably, shrimp) are thought to draw down fragments of vegetation into burrow sumps in order to augment microbial communities that can then be exploited as a food source; a process referred to as "gardening" (Griffis and Suchanek 1991, Bertics et al. 2010). Microbially driven breakdown processes that convert organic material to inorganic forms occur within burrow walls, where strong oxic-anoxic boundaries exist, promoting greater exchanges between the two chemical environments (Fig. 4.3). Studies on decapod crustaceans have reported increases in surface area of the sediment–water interface through burrow building, ranging from 20–772% (Katz 1980, Dworschak and Pervesler 1988, Griffis and Chavez 1988, Witbaard and Duineveld 1989, Coelho et al. 2000, Webb and Eyre 2004a). This results in significant acceleration of remineralization processes and fluxes of solutes between environments (Ziebis et al. 1996a). Where organisms are present in high densities, creating many burrow structures, these differences can have dramatic and marked effects on ecosystem function.

Bioirrigation, particularly the ventilation of burrows with oxygenated water, enhances the presence of high energy electron acceptors at depth in the sediment by regularly flushing surface water through the burrow structures. This is dependent on organism behavior and lifestyle mode as well as the complexity of the burrow form, all of which affect sediment geochemistry, creating a heterogeneous landscape of microbial assemblages throughout burrow matrices and across

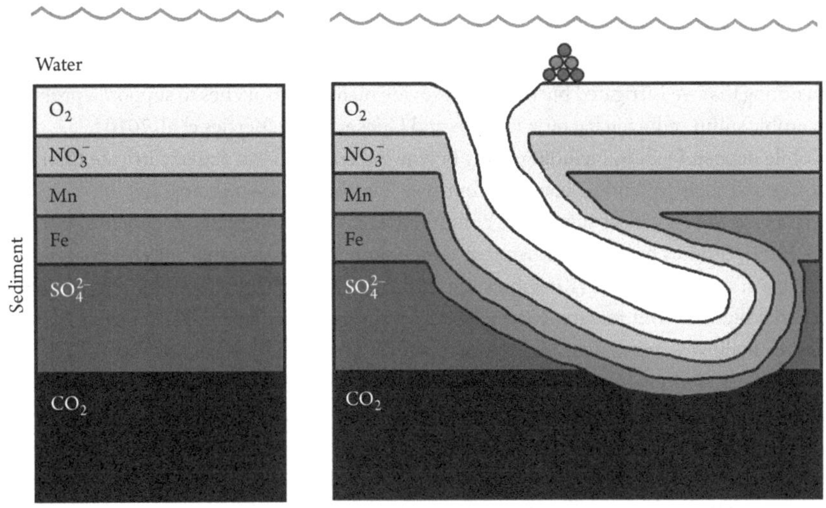

Fig. 4.3.
Left: Idealized, one-dimensional vertical succession of major oxidants association with heterotrophic metabolism and diagenetic reactions of organic matter in marine sediments. Source: data from Aller (1988). Right: Variation of diagenetic reaction geometry around a hypothetical crustacean burrow structure. Burrow construction, maintenance, and ventilation results in the translocation of material and variation in redox status.

burrow walls (Laverock et al. 2010). Certain crustaceans, such as the Australian yabby shrimp *Trypaea australiensis*, secrete mucous and line their burrows with it, which provides both a substrate and an energy source for enhanced microbial activity, relative to ambient sediment (Webb and Eyre 2004b). Other thalanassids, such as the upogebiids, induce a solid-phase iron precipitation (Fe III) in the outer burrow lumen to reduce porewater advection and increase water through-flow (Papaspyrou et al. 2005).

Although some studies have measured the irrigation rates of key engineering decapods in controlled laboratory experiments (Dworschak 1981, Gust and Harrison 1981, Koike and Mukai 1983, Allanson et al. 1992, Forster and Graf 1995), few to date have directly measured these rates in natural environments where behaviors and conditions may differ (Webb and Eyre 2004b, D'Andrea and DeWitt 2009). Species traits, such as feeding behavior, lifestyle mode, and irrigation rates interact to affect the degree of oxygenation occurring within burrows. Burrow shape and complexity have also been shown to alter the spatial distribution of oxic/anoxic zones within these structures (Astall et al. 1997). This in turn results in a complex patchwork of "hotspots" for nutrient cycling and remineralization associated with ecosystem-engineering species and their burrow densities (D'Andrea and DeWitt 2009). It stands to reason that an abundance of large burrowing shrimp will enhance sediment oxygen consumption. However, only a small proportion of this is through their own respiration requirements. Mostly this increased demand is due to enhanced microbial utilization (i.e., oxidation and respiration; see Webb and Eyre 2004b).

Several studies have demonstrated that, despite building deep and complex burrows, persistent irrigation by thalassinidean shrimp species such as *Neotrypaea californiensis*, *Upogebia deltaura*, and *C. subterranea* oxygenate their burrows to an extent that the microbial communities within them often reflect those found in the surface sediments (Bertics and Ziebis 2009, Laverock et al. 2010). The first few millimeters of the burrow wall have been shown to contain a far greater number of bacteria than the surrounding sediment, reaching a 100% increase in some cases (Branch and Pringle

1987). Increased taxonomic richness in bacteria has also been documented in association with thalassinid burrows when compared to uninhabited sediments (Laverock et al. 2010). However, it is worth noting that well-irrigated burrows still provide many micro-niches to support a proliferation of anaerobic sulfur reducing bacteria (Dobbs and Guckert 1988, Bertics et al. 2010).

Mobile deposit feeders, including many burrowing crabs, do not actively irrigate their burrow structures and their (often less complex) burrows have been shown to host microbial communities that more closely reflect subsurface communities in non-bioturbated sediments (Bertics and Ziebis 2009). Studies have also shown that temperature can have significant effects on microbial abundance of shrimp burrows (Kinoshita et al. 2003), suggesting that seasonal and latitudinal shifts in bacterial biomass and metabolism in burrow walls are also likely to occur, even for a single species. In shallow water systems, microphytobenthic communities (benthic diatoms, cyanobacteria, flagellates, and eukaryotic algae) are key primary producers that display seasonal patterns of abundance in both temperate and tropical environments (Uthicke and Klumpp 1998, Orvain et al. 2012). These tiny organisms can contribute significantly to the amount of fixed carbon in a system. For example, Underwood and Paterson (1993) estimated that one-third of estuarine fixed carbon came from this source. Top-down grazer control on microphyte biomass, resulting in lower benthic primary production, has been well documented in aquatic systems (Hillebrand 2002, and references therein). Nutrient dynamics can also cause bottom-up control of primary production, and these two pathways are often not independent of each other where engineering species occur. For instance, Needham et al. (2011) demonstrated that although an increase in the number of mud crabs (*A. crassa*) reduced the quantity of microphytobenthos in surface sediments through direct grazing pressure, the photosynthetic efficiency of the remaining microphytobenthos was seemingly enhanced by the increased availability of ammonium (NH_4^+, a preferentially used nitrogen source) associated with crab excretion.

Fiddler crabs (of which there are over 100 species) are nearly all deposit feeders that inhabit a wide range of habitats from unvegetated mudflats to mangroves and salt marshes. Their feeding is nonselective, and individuals can process several times their body weight of sediment each day (Lopez and Levinton 1987). Microphytobenthos provides a key nitrogen source for these organisms (Lopez and Levinton 1987, Rice 1982, Botto et al. 2005, Kanaya et al. 2008, Ribeiro et al. 2011) and the impact grazing has on primary production can be context- and scale-specific. Ribeiro et al. (2011) demonstrated that at small (mm) scales *Leptuca uruguayensis* was capable of stripping 60% of sediment chlorophyll content on days of high activity. However, this impact did not translate to system-level losses in productivity, as microphytes were replenished after each tidal cycle (partially on neap tides, completely on spring tides).

In tropical systems of Australasia, "armies" of soldier crabs are often observed in intertidal areas. Webb and Eyre (2004b) demonstrated that *Mictyris longicarpus* communities could reduce sediment chlorophyll by 77%, which in turn reduced benthic primary production by 71%. Other small but abundant burrowing crustaceans, such as the amphipod *C. volutator*, often consume considerable quantities of benthic diatoms each day (Smith et al. 1996). However, their regulation of primary production, at a system level, remains largely unexplored.

Deep marine systems are also reliant on supplies of particulate organic matter to drive benthic metabolism and, as with shallow water systems, burrow-building species are thought to enhance remineralization processes. Due to the logistical challenges associated with sampling these systems, few studies to date have been able to directly measure this. Leduc and Pilditch (2017) demonstrated the importance of the burrow-building shrimp *Eucalastacus* cf. *torbeni* (sampled at a depth of 489 m) on benthic metabolism rates. Only 1% of the measured daily sediment oxygen demand was attributable to shrimp respiration (based on allometric equations), with the rest ascribed to consumption and burial of microphytes during burrow maintenance and indirect stimulation of microbial communities (through enhanced oxidation/reduction pathways) in burrow walls. This indicates that burrowing crustaceans are likely to be of great importance to deep sea functioning; however, data on species density, behaviors, and community structure are still largely absent from the scientific literature.

THE INFLUENCE OF CRUSTACEANS ON NUTRIENT CYCLING

Nutrient exchange is a key ecosystem function often governed by engineering species, and the importance of burrowing crustaceans on nutrient cycling is now fairly well documented (Forster and Graf 1995, Nickell and Atkinson 1995, Webb and Eyre 2004a, Mermillod-Blondin and Rosenberg 2006, D'Andrea and DeWitt 2009, Dunn et al. 2009, Needham et al. 2011). As we've already touched upon, the temporal dynamics of solute exchanges within and around burrow structures is linked to labile carbon content, organism activity rates, and behavior, as motion alters porewater pressure gradients, particularly in advective sediments (Laverock et al. 2011, Volkenborn et al. 2012). At an ecosystem scale, active irrigators can pass significant volumes of the overlying water through their burrow structures, resulting in broad scale changes to nutrient dynamics (Fig. 4.4). Particulate organic matter captured in burrow structures is also transformed into inorganic nitrogen, a key process for fueling primary production.

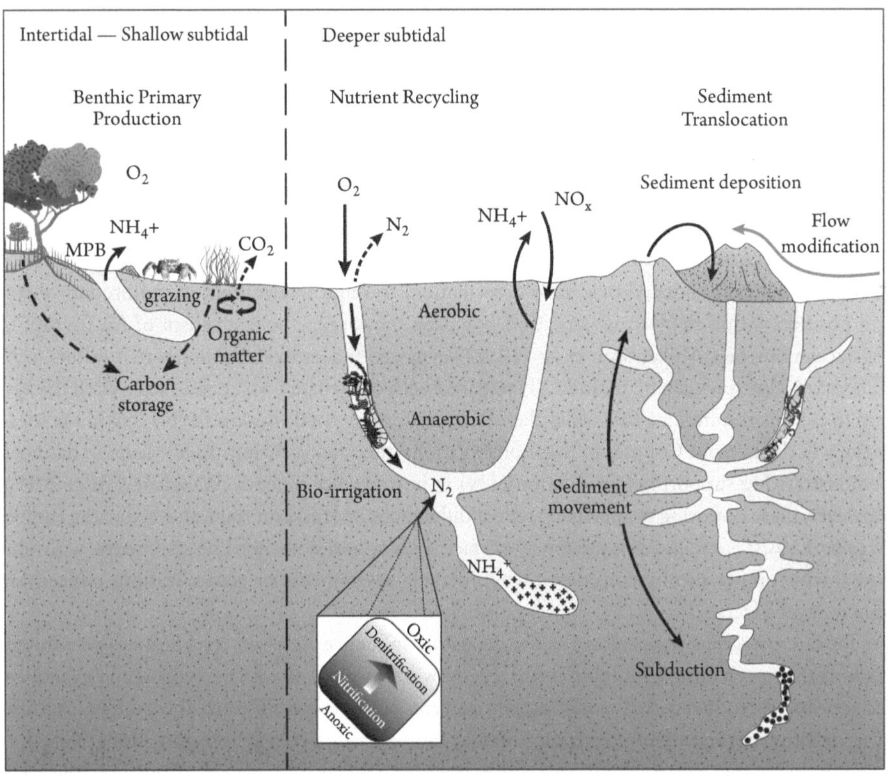

Fig. 4.4.
Illustration of the effects of Crustacea on the sediment environment affecting nutrient recycling processes and functions of intertidal and subtidal habitats. MPB = microphytobenthos.

As already discussed in this chapter, irrigators flush oxygen-rich waters to depths far greater than molecular diffusion alone (Revsbech et al. 1980, Glud et al. 1994, Ziebis et al. 1996b). For example, Dworschak (1981) calculated that the entire volume of a neap tide was passed through *Upogebia pusilla* burrows in an Adriatic tidal flat, clearly demonstrating how the presence of these shrimp increases the sediment–water interface and both the speed and pathway of geochemical exchange between these two environments. Strong exchange gradients can develop across the burrow wall and along the length of established and stable burrow structures (Aller 1988). This in turn can alter nitrogen pathways and speciation in marine systems. Nitrifying bacteria are stimulated by the co-occurrence of both oxygen (O_2) and ammonium (NH_4^+) (Aller 1982), often present in shallow or well-flushed structures, whereas reduced conditions in deeper anoxic portions of a burrow provides opportunity for this process to be reversed (nitrate reduction) (Fig. 4.4). Denitrifying bacteria also transform nitrate back into dinitrogen gas (N_2), facilitating a key pathway for nitrogen removal from marine environments. The balance between these processes can therefore alter the net gain or loss of nitrogen from a system, particularly in those dominated by a single engineering species (D'Andrea and DeWitt 2009).

Close coupling of nitrification/denitrification pathways often occurs in burrows due to the proximity of these differing microbial communities (Fig. 4.4). Webb and Eyre (2004a) demonstrated that the shrimp *Trypaea australiensis* is a key species in facilitating net loss of N_2 from subtropical estuarine environments. This was attributed to increased NH_4^+ (an excretion product) from the shrimp enhancing denitrification through quick transformation and utilization by microbes across the burrow walls. This close coupling has also been indirectly observed in non-irrigating species such as the estuarine crab *Neohelice granulata* (Botto et al. 2005). These relationships are likely dependent on burrow morphology and depth. However, the structure and complexity of burrows are often controlled by extrinsic factors, primarily sediment grain size (Waldbusser and Marinelli 2006). Bertics et al. (2010) proposed that for the shrimp *N. californiensis*, micro-niches of N_2 fixation by sulfate-reducing bacteria (as well as coupled zones) are present in burrows and suggested that N_2 fixation through this mechanism may be largely overlooked in coastal and continental shelf sediments.

Biohydraulically induced oxic-anoxic oscillation, through intermittent irrigation and organism position within the burrow structure, has been proposed as a mechanism of N_2 fixation in crustacean burrows. Volkenborn et al. (2012) suggested that anoxic, ammonia-rich waters in *N. californiensis* burrows can be periodically replaced with oxygenated water and thus switch the system to aerobic respiration. This could therefore create a temporal window for N_2 fixation in these regions of the burrow. Counter to these observations, the mantis shrimp *Squilla empusa* has been demonstrated to support net production of N_2 in burrowed sediments (Waugh and Aller 2017). However, these opposing results likely reflect differences in organism traits and behaviors (unlike *N. californiensis, S. empusa* is a predatory surface forager and does not line its burrow with mucous), which must also be considered when making assessments or generalizations concerning system-level effects.

Those crustaceans that are not active irrigators influence nitrogen transformations through several other mechanisms. Mobile deposit feeders will consume microphytobenthic and cyanobacterial primary producers that would otherwise promote nitrogen fixation. These organisms also strip detritus that fuels remineralization within the sediments (Kristensen and Alongi 2006). A key example of this comes from Webb and Eyre (2004b), who examined the tropical soldier crab *Mictyris longicarpus*. This organism makes transient igloo-shaped shelters that they take refuge in during high tide. These voracious mobile foragers were able to remove 94–96% of the organic matter and approximately 99% of the nitrogen transported to the sediment when compared to areas they were excluded from. By intercepting particulate organic matter and grazing on cyanobacteria

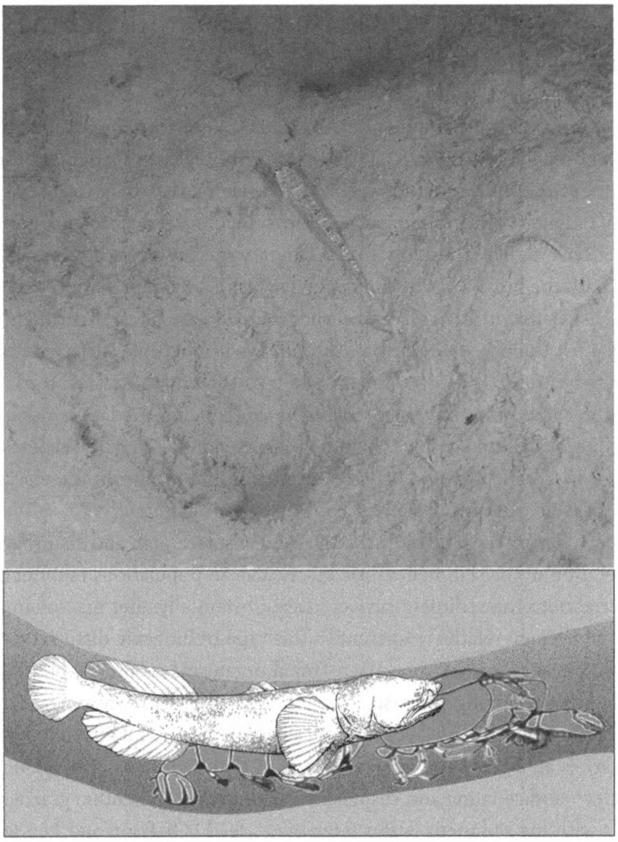

Fig. 4.5.
Top: An opalfish, *Hemerocoetes monopterygius*, using the refuge provided by the depression of a crustacean burrow opening. Many small fish, such as triple fins, and other taxa use crustacean burrows as temporary refugia. Bottom: A sketch of a blind goby, *Typhlogobius californiensis*, inside a ghost shrimp, *Neotrypaea biffari*, burrow.

as well as physically disturbing surface sediment, these soldier crabs significantly reduced sediment metabolism and the flux of solutes but promoted N_2 release (in the absence of microphytobenthic communities). Furthermore, *M. longicarpus* was shown to be under significant predation pressure from vertebrates (e.g., stingray), a mechanism by which carbon and nitrogen may be lost from intertidal environments.

Organism density also plays a significant part in the overall ecosystem level effect of nitrogen transformations and fluxes from and to the benthos. As an example, the small but often-abundant burrowing amphipod *C. volutator*, which continuously irrigates shallow U-shaped burrows, has been shown to enhance both nitrification and denitrification processes. Pelegri and Blackburn (1994) demonstrated a 1.5-fold increase in N_2 fluxes at high *C. volutator* densities (6000 individuals m^{-2}) relative to non-bioturbated microcosm sediments. Other corophiid species, such as *Monocorophium insidiosum*, have also demonstrated that their intensive bioirrigation has considerable effects on nitrification that sustains denitrification, resulting in overall N_2 production (Moraes et al. 2018). As the pervasive threat of coastal eutrophication grows globally, these tiny yet abundant creatures provide an important ecosystem service by reducing nitrogen loading through these pathways.

EFFECTS OF CRUSTACEAN BIOTURBATORS AND BIOIRRIGATORS ON COMMUNITY STRUCTURE

Habitat modification by large or abundant crustaceans is known to impact macrofaunal community composition (Brenchley 1981, 1982, Posey 1986, Posey et al. 1991, Berkenbusch and Rowden 2000, Widdicombe et al. 2000, Sumida et al. 2020). One element of this is the use of crustacean burrows by a range of sedentary and mobile species, from tubiculous polychaetes to gobiid fishes. These cohabitation relationships likely develop from secondary species taking advantage of existing crustacean burrows, with the burrows providing effective refuges from predation (e.g., from demersal fish). The gobiid fishes that utilize crab and shrimp burrows are often larger than the crustaceans themselves (Fig. 4.5). Gallery-style burrow structures, with multiple surface entrances/exits, are probably particularly useful as refuges for mobile taxa such as these (i.e., those hiding are not forced to leave the same hole they were chased into, reducing predation risk). The sedentary species that cohabit crustacean burrows, including suspension-feeding polychaetes and phoronids (Kinoshita 2002), may benefit from decreased predation but also from the elevated water flows generated by the crustaceans during feeding and burrow ventilation.

Another aspect of burrowing and bioturbation is the displacement and disturbance of sediment. The physical alteration of the sediment fabric by crustacean populations (subduction, reworking, ejection of material onto the sediment surface) can substantially alter macrofauna communities. Some macrofaunal taxa are relatively tolerant to this type of biogenic disturbance and can occur in highly bioturbated sediments. Other macrofaunal organisms require stable sediments and are negatively affected by the mixing and displacement of particles by bioturbators. This is the mechanism by which shifts in macrofaunal community structure have been linked to habitat-modifying crustaceans (Pillay et al. 2011). The general pattern is that high rates of biogenic sediment mixing negatively affect surface fauna and suspension-feeding and epibenthic grazing taxa that must maintain contact with the sediment–water interface to feed (Ólafsson and Persson 1986, Jensen and André 1993). In contrast, high biogenic disturbance can favor burrowing macrofaunal species such as subsurface predatory worms (Lohrer et al. 2008, 2013). Sediment disturbance can also have indirect effects on nektonic species. Pillay et al. (2012) found that, at high densities, bioturbation by the sand prawn *Kraussillichirus kraussi* (Fig. 4.1B) buried microphytobenthos and limited the food availability for grey mullet (*Chelon richardsonii*). This "bottom-up" control of energy exchange to pelagic environments may further serve to structure the ecosystem through trophic pathways.

Sediment disturbance through bioturbation can also weaken mutualistic relationships between species. For example, the relationship between filter-feeding mussels (*Geukensia demissa*) and cordgrass (*Spartina alterniflora*) in salt marshes of eastern North America is influenced by the presence of fiddler crabs (primarily *Minuca pugnax*). The mussels and cordgrass can each survive on their own, but both species benefit by growing in association with each other (cordgrass stems and shoots provide settlement structures and shade for mussels, while mussels stabilize the sediment and fertilize the cordgrass). Fiddler crabs have been shown to facilitate cordgrass growth at low to intermediate densities but appear to hamper cordgrass at high densities where rates of damage to roots outpace the ability of the cordgrass to regenerate. Derksen-Hooijberg et al. (2019) further demonstrated how high densities of fiddler crabs negatively affected both cordgrass and mussels, including the ability of cordgrass to re-sprout following disturbance.

Mutually beneficial relationships between crustaceans and vegetation have also been documented in mangrove ecosystems. The presence of fiddler crabs (*Gelasimus vocans*) has a positive effect on the growth of mangrove (*Avicennia marina*) saplings. Kristensen and Alongi (2006) suggested that due to the crabs' preference for burrowing near vegetative structures, oxidized conditions were enhanced for saplings, alleviating sulfide damage, and instead encouraging root growth and development. Crabs may also alleviate some of the competition for nutrients for saplings through consumption of microphytobethic populations that would otherwise utilize this resource. Crabs may also benefit from the presence of saplings as an increase in dissolved organic leachates around root structures may facilitate bacterial growth, providing an alternative food source for *G. vocans*. Density-dependent effects on the resilience of mutualistic foundational species, in critical and widespread ecosystem types, is another indicator of the fundamental importance of habitat-modifying crustaceans, particularly considering impending climate-related pressures.

In addition to the effects of habitat-modifying crustaceans themselves, there is emerging recognition of secondary ecosystem effects indirectly attributable to the crustaceans. One of the best-studied examples is the role of stingray foraging on tidal flats in southern Japan (Tamaki et al. 2020, Takeuchi and Tamaki 2014). Stingrays are known to feed on seafloor crustaceans (Fig. 4.6), and the bat ray *Hemitrygon akajei* specializes on ghost shrimps *Neotrypaea harmandi* in southern Japan. More than 18,000 new stingray feeding pits were observed each day on a Tomioka Bay sandflat, displacing 25–30% of the sediment's surface area per day across > 10 hectares. The distribution of feeding pits on the sandflat was correlated with ghost shrimp density, among other factors. The considerable alteration of seabed topography and the accumulation of detritus in the pits constitutes major habitat modification that has broad consequences for biogeochemistry, sediment transport, and faunal community composition (VanBlaricom 1982, Van Colen et al. 2015).

Gray whales (*Eschrichtius robustus*) are also known to disturb seafloor sediments while feeding on high-density populations of infaunal crustaceans. Individual feeding depressions each covering 3.7–5.6 m^2 have been observed in high-density ghost shrimp beds in Puget Sound (Weitkamp et al. 1992). Macrofaunal standing stock in the study area was dominated (55–85%) by ghost shrimps, and ghost shrimp biomass was five times less inside fresh pits relative to undisturbed adjacent sediment, suggesting that the whales were specifically targeting shrimp as prey. Gray whales have also been reported to target high-density *Ampelisca* amphipod mats in the Arctic, with excavation pits ranging from 2–20 m^2 (Oliver et al. 1984). Over time, these pits trap organic debris that further

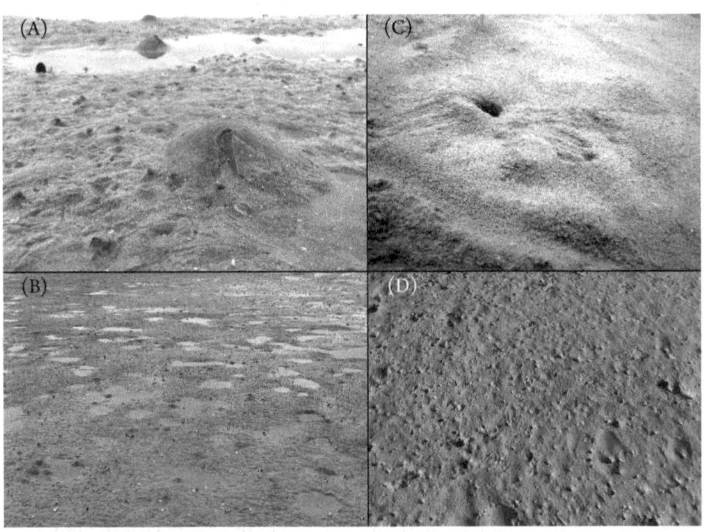

Fig. 4.6.
From burrow to habitat: (A) exhalent mounds likely from the ghost shrimp *Biffarius filholi*, covered with darker deposited sediment, (B) exhalent mounds covering the intertidal in addition to the feeding pits created by stingrays predating on infaunal crustaceans and mollusks, (C) an individual crab (*Hemiplax hirtipes*) burrow opening in very fine subtidal sand, and (D) high-density crab (*Austrohelice crassa*) habitat in intertidal muddy sand.

attracts amphipod recolonization and alters community succession. Crustaceans thus have the potential to alter the structure and function of seafloor soft-sediment habitats, directly and indirectly, across a large range of spatial and temporal scales (Nerini 1984, Oliver and Slattery 1985).

FUNCTIONAL PLASTICITY IN CRUSTACEA

Organism behavior can be directly affected by the sediment environment, in turn altering the degree to which species contribute to ecosystem functioning. This is particularly important when considering large-bodied "bioengineers" that live in high densities (Fig. 4.6) and across a wide habitat spectrum. Needham et al. (2010) demonstrated how the behavior of the burrowing mud crab *A. crassa* shifted as a consequence of sediment type, generating context-specific changes to ecosystem functioning. Burrows were observed to be over 18 times more stable in mud than sand, and crabs demonstrated greater sediment turnover and reworking rates in non-cohesive sediments. Muddy habitats also housed higher complexity burrow structures (Fig. 4.2) and produced greater NH_4^+ efflux attributed to the likelihood of stronger bacterial gradients along the burrow lumen due to the lack of regular disturbance (Needham et al. 2011). Therefore, *A. crassa* was shown to have distinctly different functional roles (i.e., a switch from burrow builders to "bulldozers/regenerators"; Kristensen et al. 2012) in different sediment environments.

Pillay (2019) speculated that much like *A. crassa*, thalassinid shrimp behavior is likely altered through changes in sediment properties. However, observing natural behavior in these organisms is inherently difficult due to their lifestyle traits. Likewise, it could be hypothesized that the facultative (and therefore context-specific) feeding behaviors of species such as the amphipod *C. volutator*, discussed earlier in this chapter, may also translate the changes in function under certain conditions, but empirical data are lacking. Indeed, few studies to date have examined context-specific behaviors of key burrow builders (Escapa et al. 2008, Sassa and Watabe 2008) or linked changes in behavior (of any key species) to ecosystem functioning (Biles et al. 2003, Cassidy et al. 2020). Hence, the ubiquity of this phenomenon remains largely unknown.

BEHAVIORAL SHIFTS IN A CHANGING CLIMATE

Mounting evidence suggests many key species are expanding their habitat ranges and their activities may be less seasonally constrained as a consequence of climate change. As global warming continues to accelerate, one of the most notable consequences in coastal ecosystems is sea level rise, which will likely have its greatest impacts on species inhabiting the upper part of the intertidal zone. Increased inundation of upper intertidal marshes is already causing cordgrass to alter its root versus rhizome allocations (Crotty et al. 2020). This seemingly makes the sediment softer and more readily burrowed by herbivorous marsh crabs such as *Sesarma reticulatum*. Increasingly this species is observed in high-density grazing fronts at the heads of tidal creeks in parts of North America. Crotty et al. (2020) estimate that crab grazing and burrowing has contributed to a tripling of the incision of creeks into some formerly vegetated areas, heavily modifying the landscape. In this way, facilitated by sea level rise, *Sesarma* is now functioning as a keystone species affecting species interactions and the geomorphic evolution of salt marshes in the southeastern United States. Shifts in the behavior and range of these key organisms will undoubtedly modify multiple ecosystem functions.

High densities of *Sesarma*, and denuded and burrow-pocked creek banks, also correspond with widespread declines in its predators. Many of the predators of *Sesarma* (e.g., striped bass, blue crabs) have been subjected to decades of high-intensity fishing pressure (Bertness et al. 2014). Without natural controls to keep *Sesarma* populations low, overgrazing and sediment erosion could outpace the landward migration of tidal marsh vegetation as sea levels rise. Effects may be magnified where marshes are directly adjacent to artificial coastal structures (Torio and Chmura 2013). This example serves to illustrate how anthropogenic activities and complex feedbacks may alter the roles of habitat-modifying crustaceans and potentially influence the responses of entire ecosystem types to change.

CONCLUSION

This chapter has outlined the roles of crustaceans as habitat modifiers and regulators of marine ecosystem functions. We have considered the overarching influences of crustaceans on the habitats in which they live and delved down to the scales of individual organisms, burrows, and processes. However, it is worth emphasizing that some key bioengineering crustaceans reach phenomenal densities (Fig. 4.6) that affect the development and distributions of entire ecosystem types, highlighting the global importance of key crustacean engineers. Processes that are governed by organism traits and densities in one system often have flow-on effects to other environments. In this way, key species, particularly some of the thalassinid shrimp, corophiid amphipods, and grapsid and ocypodid crabs can govern multiple functions interactively across habitats and scales and are therefore of global significance to ecosystem functioning and the delivery of goods and services to humankind.

The clearest illustrations of this functional dominance come from tidal flat, salt marsh, and mangrove ecosystems. We have highlighted that bioturbating crabs are key regulators of functioning in both unvegetated and vegetated intertidal flats. As key consumers of microphytobenthos and detritus, they regulate remineralization rates, solute exchanges, sediment stability, and primary production through their behaviors, some of which are modified by environmental conditions. Collectively, they regulate key functions such as carbon cycling, primary production, nutrient recycling, and sediment stability in these habitats. Generally speaking, these small crustaceans, in populations that can generate > 400 burrows per m^2 (Yoder et al. 2005), provide a suite of ecosystem services to humankind. In the case of fiddler crabs, these often-unseen effects are varied, for example, amelioration of high-salinity salt pan formation by reworking and increasing the drainage of the intertidal sediment (Yoder et al. 2005) and degradation of cordgrass wrack that would otherwise accumulate near the high tide mark in tidal marshes (Genoni and Montague 1995). This demonstrates how relatively small but widespread burrowing crustaceans can collectively have far-reaching and disproportionate ecosystem effects, many of which are dependent on the habitats in which they live and the densities in which they occur.

Crustaceans, particularly anomuran and brachyuran crabs, are among the most successful intertidal, supralittoral, and semi-terrestrial animal taxa. As such they will factor into how coastal ecosystems respond to warmer temperatures, greater CO_2 concentrations, and sea level rise. Yet, coastal environments are high-use, multi-use areas where large human populations impose a myriad pressures. Thus, crustaceans in the coastal zone suffer from multiple stressors and cumulative effects. Although understanding feedbacks between crustaceans and vegetation in complex coastal ecosystems will improve our ability to forecast change, holistic ecosystem-based conservation and management will be required to stem the tide of degradation and turn it toward recovery.

ACKNOWLEDGMENTS

We would like to thank the following people: Mr. M. Oulton for graphic design assistance with figures, Assoc. Prof. D. Pillay, Prof. G.M. Branch, Mr. E. Gibcus, Dr. J. van der Straaten, and the Saxifraga Foundation for assistance with or provision of photos. Many thanks to the editors of the volume for their invitation to contribute and their patience with us. Special thanks also to Prof. A. Tamaki for preliminary discussions on chapter content and for his decades of high-quality research on habitat-modifying crustaceans.

REFERENCES

Allanson, B.R., D. Skinner, and J. Imberger. 1992. Flow in prawn burrows. Estuarine, Coastal and Shelf Science 35:253–266.

Aller, R. 1988. Benthic fauna and biogeochemical processes in marine sediments: The role of burrow structures. Pages 301–338 *in* T.H. Blackburn and J. Sørensen, editors. Nitrogen cycling in coastal marine environments. John Wiley & Sons Ltd., New York, USA.

Aller, R.C. 1982. The effects of macrobenthos on chemical properties of marine sediment and overlying water. Pages 53–102 *in* P.L. McCall and M.J.S. Tevesz, editors. Animal-sediment relations. Springer, Boston, USA.

Alongi, D.M. 2020. Global significance of mangrove blue carbon in climate change mitigation. Sci 2:67.

Andreetta, A., M. Fusi, I. Cameldi, F. Cimò, S. Carnicelli, and S. Cannicci. 2014. Mangrove carbon sink. Do burrowing crabs contribute to sediment carbon storage? Evidence from a Kenyan mangrove system. Journal of Sea Research 85:524–533.

Astall, C.M., A.C. Taylor, and R.J.A. Atkinson. 1997. Behavioural and physiological implications of a burrow-dwelling lifestyle for two species of upogebiid mud-shrimp (Crustacea: Thalassinidea). Estuarine, Coastal and Shelf Science 44:155–168.

Atkinson, R.J.A., and L.B. Eastman. 2015. Burrow dwelling in Crustacea. Pages 100–140 in M. Thiel and L. Watling, editors. The natural history of the Crustacea, Volume 2: Life styles and feeding biology. Oxford University Press, New York, USA.

Atkinson, R.J.A., and R.D.M. Nash. 1990. Some preliminary observations on the burrows of *Callianassa subterranea* (Montagu) (Decapoda: Thalassinidea) from the west coast of Scotland. Journal of Natural History 24:403–413.

Atkinson, R.J.A., and A.C. Taylor. 1988. Physiological ecology of burrowing decapods. Symposia of the Zoological Society of London 59:201–226.

Beaumont, N.J., M.C. Austen, J.P. Atkins, D. Burdon, S. Degraer, T.P. Dentinho, S. Derous, P. Holm, T. Horton, E. Van Ierland, A.H. Marboe, D.J. Starkey, M. Townsend, and T. Zarzycki. 2007. Identification, definition and quantification of goods and services provided by marine biodiversity: Implications for the ecosystem approach. Marine Pollution Bulletin 54:253–265.

Bentley, M.G. 2011. The global spread of the Chinese mitten crab *Eriocheir sinensis*. Pages 107–127 *in* B.S. Galil, P.F. Clark, and J.T. Carlton, editors. In the wrong place—Alien marine crustaceans: Distribution, biology, and impacts. Springer, Dordrecht, The Netherlands.

Berkenbusch, K., and A.A. Rowden. 2000. Intraspecific burrow plasticity of an intertidal population of *Callianassa filholi* (Crustacea: Decapoda: Thalassinidea) in relation to environmental conditions. New Zealand Journal of Marine and Freshwater Research 34:397–408.

Bertics, V.J., J.A. Sohm, T. Treude, C.E.T. Chow, D.G. Capone, J.A. Fuhrman, and W. Ziebis. 2010. Burrowing deeper into benthic nitrogen cycling: The impact of bioturbation on nitrogen fixation coupled to sulfate reduction. Marine Ecology Progress Series 409:1–15.

Bertics, V.J., and W. Ziebis. 2009. Biodiversity of benthic microbial communities in bioturbated coastal sediments is controlled by geochemical microniches. The ISME Journal 3: 1269–1285.

Bertness, M.D., C.P. Brisson, T.C. Coverdale, M.C. Bevil, S.M. Crotty, and E.R. Suglia. 2014. Experimental predator removal causes rapid salt marsh die-off. Ecology Letters 17:830–835.

Biles, C.L., M. Solan, I. Isaksson, D.M. Paterson, C. Emes, and D.G. Raffaelli. 2003. Flow modifies the effect of biodiversity on ecosystem functioning: An in situ study of estuarine sediments. Journal of Experimental Marine Biology and Ecology 285:165–177.

Botto, F., and O. Iribarne. 2000. Contrasting effects of two burrowing crabs (*Chasmagnathus granulata* and *Uca uruguayensis*) on sediment composition and transport in estuarine environments. Estuarine, Coastal and Shelf Science 51:141–151.

Botto, F., I. Valiela, O. Iribarne, P. Martinetto, and J. Alberti. 2005. Impact of burrowing crabs on C and N sources, control, and transformations in sediments and food webs of SW Atlantic estuaries. Marine Ecology Progress Series 293:155–164.

Branch, G.M., and A. Pringle. 1987. The impact of the sand prawn *Callianassa kraussi* Stebbing on sediment turnover and on bacteria, meiofauna, and benthic microflora. Journal of Experimental Marine Biology and Ecology 107:219–235.

Brenchley, G. 1981. Disturbance and community structure: An experimental study of bioturbation in marine soft-bottom environments. Journal of Marine Research 39:767–790.

Brenchley, G. 1982. Mechanisms of spatial competition in marine soft-bottom communities. Journal of Experimental Marine Biology and Ecology 60:17–33.

Bulmer, R.H., F. Stephenson, H.F.E. Jones, M. Townsend, J.R. Hillman, L. Schwendenmann, and C.J. Lundquist. 2020. Blue carbon stocks and cross-habitat subsidies. Frontiers in Marine Science 7:380.

Cassidy, C., L.J. Grange, C. Garcia, S.G. Bolam, and J.A. Godbold. 2020. Species interactions and environmental context affect intraspecific behavioural trait variation and ecosystem function. Proceedings of the Royal Society B: Biological Sciences, 287:20192143.

Coelho, V.R., R.A. Cooper, and S.D. Rodrigues. 2000. Burrow morphology and behavior of the mud shrimp *Upogebia omissa* (Decapoda: Thalassinidea: Upogebiidae). Marine Ecology Progress Series 200:229–240.

Crotty, S.M., C. Ortals, T.M. Pettengill, L. Shi, M. Olabarrieta, M.A. Joyce, A.H. Altieri, E. Morrison, T.S. Bianchi, C. Craft, M.D. Bertness, and C. Angelini. 2020. Sea-level rise and the emergence of a keystone grazer alter the geomorphic evolution and ecology of southeast US salt marshes. Proceedings of the National Academy of Sciences 117:17891–17902.

Curran, H.A. 2007. Chapter 14. Ichnofacies, ichnocoenoses, and ichnofabrics of quaternary shallow-marine to dunal tropical carbonates: A model and implications. Pages 232–247 *in* W. Miller, editor. Trace fossils. Elsevier, Amsterdam, The Netherlands.

Curran, H.A., and A.J. Martin. 2003. Complex decapod burrows and ecological relationships in modern and Pleistocene intertidal carbonate environments, San Salvador Island, Bahamas. Palaeogeography, Palaeoclimatology, Palaeoecology 192:229–245.

D'Andrea, A.F., R.C. Aller, and G.R. Lopez. 2002. Organic matter flux and reactivity on a South Carolina sandflat: The impacts of porewater advection and macrobiological structures. Limnology and Oceanography 47:1056–1070.

D'Andrea, A.F., and T.H. DeWitt 2009. Geochemical ecosystem engineering by the mud shrimp *Upogebia pugettensis* (Crustacea: Thalassinidae) in Yaquina Bay, Oregon: Density-dependent effects on organic matter remineralization and nutrient cycling. Limnology and Oceanography 54:1911–1932.

De Deckere, E.M.G.T., T.J. Tolhurst, and J.F.C. De Brouwer. 2001. Destabilization of cohesive intertidal sediments by infauna. Estuarine, Coastal and Shelf Science 53:665–669.

Derksen-Hooijberg, M., T. van der Heide, L.P. Lamers, A. Borst, A.J. Smolders, L.L. Govers, J.R. Hoogveld, and C. Angelini. 2019. Burrowing crabs weaken mutualism between foundation species. Ecosystems 22:767–780.

Dittel, A.I., and C.E. Epifanio. 2009. Invasion biology of the Chinese mitten crab *Eriochier sinensis*: A brief review. Journal of Experimental Marine Biology and Ecology 374:79–92.

Dobbs, F.C., and J.B. Guckert. 1988. *Callianassa trilobata* (Crustacea: Thalassinidea) influences abundance of meiofauna and biomass, composition, and physiologic state of microbial communities within its burrow. Marine Ecology Progress Series 45:69–79.

Donato, D.C., J.B. Kauffman, D. Murdiyarso, S. Kurnianto, M. Stidham, and M. Kanninen. 2011. Mangroves among the most carbon-rich forests in the tropics. Nature Geoscience 4:293–297.

Dornbos, S.Q., D.J. Bottjer, and J.-Y. Chen. 2005. Paleoecology of benthic metazoans in the Early Cambrian Maotianshan Shale biota and the Middle Cambrian Burgess Shale biota: Evidence for the Cambrian substrate revolution. Palaeogeography, Palaeoclimatology, Palaeoecology 220:47–67.

Doughty, C.L., J.A. Langley, W.S. Walker, I.C. Feller, R. Schaub, and S.K. Chapman. 2016. Mangrove range expansion rapidly increases coastal wetland carbon storage. Estuaries and Coasts 39:385–396.

Dunn, R.J.K., D.T. Welsh, M.A. Jordan, P.R. Teasdale, and C.J. Lemckert. 2009. Influence of natural amphipod (*Victoriopisa australiensis*) (Chilton, 1923) population densities on benthic metabolism, nutrient fluxes, denitrification and DNRA in sub-tropical estuarine sediment. Hydrobiologia 628:95–109.

Dworschak, P.C. 1981. The pumping rates of the burrowing shrimp *Upogebia pusilla* (Petagna) (Decapoda: Thalassinidea). Journal of Experimental Marine Biology and Ecology 52:25–35.

Dworschak, P.C. 1983. The biology of *Upogebia pusilla* (Petagna) (Decapoda, Thalassinidea) I. The burrows. Marine Ecology 4:19–43.

Dworschak, P.C., and P. Pervesler. 1988. Burrows of *Callianassa bouvieri* Nobili 1904 from Safaga (Egypt, Red Sea) with some remarks on the biology of the species. Senckenbergiana Maritima 20:1–17.

Escapa, M., G.M.E. Perillo, and O. Iribarne. 2008. Sediment dynamics modulated by burrowing crab activities in contrasting SW Atlantic intertidal habitats. Estuarine, Coastal and Shelf Science 80:365–373.

Fanjul, E., M. Escapa, D. Montemayor, M. Addino, M.F. Alvarez, M.A. Grela, and O. Iribarne. 2015. Effect of crab bioturbation on organic matter processing in Southwest Atlantic intertidal sediments. Journal of Sea Research 95:206–216.

Fanjul, E., M.A. Grela, and O. Iribarne. 2007. Effects of the dominant SW Atlantic intertidal burrowing crab *Chasmagnathus granulatus* on sediment chemistry and nutrient distribution. Marine Ecology Progress Series 341:177–190.

Fernandez, R., J. Best, and F. López. 2006. Mean flow, turbulence structure, and bed form superimposition across the ripple-dune transition. Water Resources Research 42:W05406.

Forster, S., and G. Graf. 1995. Impact of irrigation on oxygen flux into the sediment: Intermittent pumping by *Callianassa subterranea* and piston-pumping by *Lanice conchilega*. Marine Biology 123:335–346.

France, R. 1998. Estimating the assimilation of mangrove detritus by fiddler crabs in Laguna Joyuda, Puerto Rico, using dual stable isotopes. Journal of Tropical Ecology 14:413–425.

Genoni, G.P., and C.L. Montague. 1995. Influence of the energy relationships of trophic levels and of elements on bioaccumulation. Ecotoxicology and Environmental Safety 30:203–218.

Gerdol, V., and R. Hughes. 1994. Effect of *Corophium volutator* on the abundance of benthic diatoms, bacteria and sediment stability in two estuaries in southeastern England. Marine Ecology Progress Series 144:109–115.

Gibbs, M., S. Thrush, and J. Ellis. 2001. Terrigenous clay deposition on estuarine sandflats: Using stable isotopes to determine the role of the mud crab, *Helice crassa* Dana, in the recovery process. Isotopes in Environmental and Health Studies 37:113–131.

Glud, R.N., J.K. Gundersen, B.B. Jørgensen, N.P. Revsbech, and H.D. Schulz. 1994. Diffusive and total oxygen uptake of deep-sea sediments in the eastern South Atlantic Ocean: In situ and laboratory measurements. Deep Sea Research Part I: Oceanographic Research Papers 41:1767–1788.

Grabowski, R.C., I.G. Droppo, and G. Wharton. 2011. Erodibility of cohesive sediment: The importance of sediment properties. Earth-Science Reviews 105:101–120.

Grant, J., and G. Daborn. 1994. The effects of bioturbation on sediment transport on an intertidal mudflat. Netherlands Journal of Sea Research 32:63–72.

Green, P.T. 2004. Burrow dynamics of the red land crab *Gecarcoidea natalis* (Brachyura, Gecarcinidae) in rain forest on Christmas Island (Indian Ocean). Journal of Crustacean Biology 24:340–349.

Griffis, R., and T. Suchanek. 1991. A model of burrow architecture and trophic modes in thalassinidean shrimp (Decapoda: Thalassinidea). Marine Ecology Progress Series 79:171–183.

Griffis, R.B., and F.L. Chavez. 1988. Effects of sediment type on burrows of *Callianassa californiensis* Dana and *C. gigas* Dana. Journal of Experimental Marine Biology and Ecology 117:239–253.

Guimond, J.A., A.L. Seyfferth, K.B. Moffett, and H.A. Michael. 2020. A physical-biogeochemical mechanism for negative feedback between marsh crabs and carbon storage. Environmental Research Letters, 15:034024.

Gust, G., and J.T. Harrison. 1981. Biological pumps at the sediment-water interface: Mechanistic evaluation of the alpheid shrimp *Alpheus mackayi* and its irrigation pattern. Marine Biology 64:71–78.

Gutiérrez, J.L., C.G. Jones, P.M. Groffman, S.E.G. Findlay, O.O. Iribarne, P.D. Ribeiro, and C.M. Bruschetti. 2006. The contribution of crab burrow excavation to carbon availability in surficial salt-marsh sediments. Ecosystems 9:647–658.

Herringshaw, L.G., R.H. Callow, and D. McIlroy. 2017. Engineering the Cambrian explosion: The earliest bioturbators as ecosystem engineers. Geological Society, London, Special Publications 448:369–382.

Hillebrand, H. 2002. Top-down versus bottom-up control of autotrophic biomass—A meta-analysis on experiments with periphyton. Journal of the North American Benthological Society 21:349–369.

Huettel, M., H. Røy, E. Precht, and S. Ehrenhauss. 2003. Hydrodynamical impact on biogeochemical processes in aquatic sediments. Springer, The Netherlands.

Iribarne, O., A. Bortolus, and F. Botto. 1997. Between-habitat differences in burrow characteristics and trophic modes in the southwestern Atlantic burrowing crab *Chasmagnathus granulata*. Marine Ecology Progress Series 155:137–145.

Jax, K. 2010. Ecosystem functioning. Cambridge University Press, Cambridge, UK.

Jensen, K.T., and C. André. 1993. Field and laboratory experiments on interactions among an infaunal polychaete, *Nereis diversicolor*, and two amphipods, *Corophium volutator* & *C. arenarium*: Effects on survival, recruitment and migration. Journal of Experimental Marine Biology and Ecology 168:259–278.

Jones, C.G., J.H. Lawton, and M. Shachak. 1994. Organisms as ecosystem engineers. Springer, New York, USA.

Kanaya, G., S. Takagi, and E. Kikuchi. 2008. Dietary contribution of the microphytobenthos to infaunal deposit feeders in an estuarine mudflat in Japan. Marine Biology 155:543–553.

Katz, L.C. 1980. Effects of burrowing by the fiddler crab, *Uca pugnax* (Smith). Estuarine and Coastal Marine Science 11:233–237.

Kelleway, J.J., N. Saintilan, P.I. Macreadie, C.G. Skilbeck, A. Zawadzki, and P.J. Ralph. 2016. Seventy years of continuous encroachment substantially increases 'blue carbon' capacity as mangroves replace intertidal salt marshes. Global Change Biology 22:1097–1109.

Kinoshita, K. 2002. Burrow structure of the mud shrimp *Upogebia major* (Decapoda: Thalassinidea: Upogebiidae). Journal of Crustacean Biology 22:474–480.

Kinoshita, K., T. Furota, and S. Nakayama. 2003. Life cycle characteristics of the deep-burrowing mud shrimp *Upogebia major* (Thalassinidea: Upogebiidae) on a tidal flat along the northern coast of Tokyo Bay. Journal of Crustacean Biology 23:318–327.

Koike, I., and H. Mukai. 1983. Oxygen and inorganic nitrogen contents and fluxes in burrows of the shrimps *Callianassa japonica* and *Upogebia major*. Marine Ecology Progress Series 12:185–190.

Kristensen, E. 2008. Mangrove crabs as ecosystem engineers; with emphasis on sediment processes. Journal of Sea Research 59:30–43.

Kristensen, E., and D.M. Alongi. 2006. Control by fiddler crabs (*Uca vocans*) and plant roots (*Avicennia marina*) on carbon, iron, and sulfur biogeochemistry in mangrove sediment. Limnology and Oceanography 51:1557–1571.

Kristensen, E., S. Bouillon, T. Dittmar, and C. Marchand. 2008a. Organic carbon dynamics in mangrove ecosystems: A review. Aquatic Botany 89:201–219.

Kristensen, E., Flindt, M., Ulomi, S., Borges, A., Abril, G., and S. Bouillon. 2008b. Emission of CO_2 and CH_4 to the atmosphere by sediments and open waters in two Tanzanian mangrove forests. Marine Ecology Progress Series 370:53–67.

Kristensen, E., G. Penha-Lopes, M. Delefosse, T. Valdemarsen, C. Quintana, and G. Banta. 2012. What is bioturbation? The need for a precise definition for fauna in aquatic sciences. Marine Ecology Progress Series 446:285–302.

Laverock, B., J.A. Gilbert, K. Tait, A.M. Osborn, and S. Widdicombe. 2011. Bioturbation: Impact on the marine nitrogen cycle. Biochemical Society Transactions 39:315–320.

Laverock, B., C.J. Smith, K. Tait, A.M. Osborn, S. Widdicombe, and J.A. Gilbert. 2010. Bioturbating shrimp alter the structure and diversity of bacterial communities in coastal marine sediments. The ISME Journal 4:1531–1544.

Le Hir, P., Y. Monbet, and F. Orvain. 2007. Sediment erodability in sediment transport modelling: Can we account for biota effects? Continental Shelf Research 27:1116–1142.

Le Minor, M., J.C. Mullarney, C.A. Pilditch, and K. Huhn. 2020. Crab burrow aspect ratio influences particle capture rates on intertidal sandflats. Geo-Marine Letters 40:197–216.

Leduc, D., and C.A. Pilditch. 2017. Estimating the effect of burrowing shrimp on deep-sea sediment community oxygen consumption. PeerJ 5:e3309.

Levinton, J. 1972. Stability and trophic structure in deposit-feeding and suspension-feeding communities. The American Naturalist 106:472–486.

Lewis, C.J.E., P.E. Carnell, J. Sanderman, J.A. Baldock, and P.I. Macreadie. 2018. Variability and vulnerability of coastal 'blue carbon' stocks: A case study from southeast Australia. Ecosystems 21:263–279.

Lohrer, A.M., L.D. Chiaroni, J.E. Hewitt, and S.F. Thrush. 2008. Biogenic disturbance determines invasion success in a subtidal soft-sediment system. Ecology 89:1299–1307.

Lohrer, A., N. Halliday, S. Thrush, J. Hewitt, and I. Rodil. 2010. Ecosystem functioning in a disturbance-recovery context: Contribution of macrofauna to primary production and nutrient release on intertidal sandflats. Journal of Experimental Marine Biology and Ecology 390:6–13.

Lohrer, A.M., I.F. Rodil, M. Townsend, L.D. Chiaroni, J.E. Hewitt, and S.F. Thrush. 2013. Biogenic habitat transitions influence facilitation in a marine soft-sediment ecosystem. Ecology 94:136–145.

Lohrer, A.M., S.F. Thrush, and M.M. Gibbs. 2004. Bioturbators enhance ecosystem function through complex biogeochemical interactions. Nature 431:1092–1095.

Lopez, G.R., and J.S. Levinton. 1987. Ecology of deposit-feeding animals in marine sediments. The Quarterly Review of Biology 62:235–260.

Martinetto, P., D.I. Montemayor, J. Alberti, C.S.B. Costa, and O. Iribarne. 2016. Crab bioturbation and herbivory may account for variability in carbon sequestration and stocks in South West Atlantic salt marshes. Frontiers in Marine Science 3:122.

Mateo, M.A., J. Romero, M. Pérez, M.M. Littler, and D.S. Littler. 1997. Dynamics of millenary organic deposits resulting from the growth of the Mediterranean seagrass *Posidonia oceanica*. Estuarine, Coastal and Shelf Science 44:103–110.

Meadows, P.S., and J. Tait. 1989. Modification of sediment permeability and shear strength by two burrowing invertebrates. Marine Biology 101:75–82.

Meadows, P.S., J. Tait, and S.A. Hussain. 1990. Effects of estuarine infauna on sediment stability and particle sedimentation. Hydrobiologia 190:263–266.

Mermillod-Blondin, F., and R. Rosenberg. 2006. Ecosystem engineering: The impact of bioturbation on biogeochemical processes in marine and freshwater benthic habitats. Aquatic Sciences 68:434–442.

Montague, C.L. 1982. The influence of fiddler crab burrows and burrowing on metabolic processes in salt marsh sediments. Pages 283–301 *in* V.S. Kennedy, editor. Estuarine comparisons. Proceedings of the sixth biennial international estuarine research conference, November 1–6, 1981, 1982. Gleneden Beach, Oregon, USA. Academic Press, New York, USA.

Moraes, P.C., M. Zilius, S. Benelli, and M. Bartoli. 2018. Nitrification and denitrification in estuarine sediments with tube-dwelling benthic animals. Hydrobiologia 819:217–230.

Needham, H.R, C.A. Pilditch, A.M. Lohrer, and S.F. Thrush. 2010. Habitat dependence in the functional traits of *Austrohelice crassa*, a key bioturbating species. Marine Ecology Progress Series 414:179–193.

Needham, H.R., C.A. Pilditch, A.M. Lohrer, and S.F. Thrush. 2011. Context-specific bioturbation mediates changes to ecosystem functioning. Ecosystems 14:1096–1109.

Needham, H.R., C.A. Pilditch, A.M. Lohrer, and S.F. Thrush. 2013. Density and habitat dependent effects of crab burrows on sediment erodibility. Journal of Sea Research 76:94–104.

Nerini, M. 1984. A review of gray whale feeding ecology. Pages 451–463 *in* M.L. Jones, S.L. Swartz, and S., Leatherwood, editors. The gray whale, *Eschrichtius robustus*. Academic Press Inc., Orlando, USA.

Nickell, L.A., and R.J.A. Atkinson. 1995. Functional morphology of burrows and trophic modes of three thalassinidean shrimp species, and a new approach to the classification of thalassinidean burrow morphology. Marine Ecology Progress Series 128:181–197.

Nielsen, O.I., E. Kristensen, and D.J. Macintosh. 2003. Impact of fiddler crabs (*Uca* spp.) on rates and pathways of benthic mineralization in deposited mangrove shrimp pond waste. Journal of Experimental Marine Biology and Ecology 289:59–81.

Nordhaus, I., M. Wolff, and K. Diele. 2006. Litter processing and population food intake of the mangrove crab *Ucides cordatus* in a high intertidal forest in northern Brazil. Estuarine, Coastal and Shelf Science 67:239–250.

Norkko, A., S. Thrush, J. Hewitt, V. Cummings, J. Norkko, J. Ellis, G. Funnell, D. Schultz, and I. Macdonald. 2002. Smothering of estuarine sandflats by terrigenous clay: The role of wind-wave disturbance and bioturbation in site-dependent macrofaunal recovery. Marine Ecology Progress Series 234:23–42.

Ólafsson, E.B., and L.-E. Persson. 1986. The interaction between *Nereis diversicolor* O.F. Müller and *Corophium volutator* Pallas as a structuring force in a shallow brackish sediment. Journal of Experimental Marine Biology and Ecology 103:103–117.

Oliver, J.S., and P.N. Slattery. 1985. Destruction and opportunity on the sea floor: Effects of gray whale feeding. Ecology 66:1965–1975.

Oliver, J.S., P.N. Slattery, M.A. Silberstein, and E.F. O'Connor. 1984. Gray whale feeding on dense ampeliscid amphipod communities near Bamfield, British Columbia. Canadian Journal of Zoology 62:41–49.

Orvain, F., S. Lefebvre, J. Montepini, M. Sébire, A. Gangnery, and B. Sylvand. 2012. Spatial and temporal interaction between sediment and microphytobenthos in a temperate estuarine macro-intertidal bay. Marine Ecology Progress Series 458:53–68.

Otani, S., Y. Kozuki, R. Yamanaka, H. Sasaoka, T. Ishiyama, Y. Okitsu, H. Sakai, and Y. Fujiki. 2010. The role of crabs (*Macrophthalmus japonicus*) burrows on organic carbon cycle in estuarine tidal flat, Japan. Estuarine, Coastal and Shelf Science 86:434–440.

Ott, J.A., B. Fuchs, R. Fuchs, and A. Malasek. 1976. Observations on the biology of *Callianassa stebbingi* Borrodaile and *Upogebia litoralis* Risso and their effect upon the sediment. Senckenbergiana Maritima 8:61–79.

Ouyang, X., S.Y. Lee, and R.M. Connolly. 2017. Structural equation modelling reveals factors regulating surface sediment organic carbon content and CO_2 efflux in a subtropical mangrove. Science of The Total Environment 578:513–522.

Papaspyrou, S., T. Gregersen, R. Cox, M. Thessalou-Legaki, and E. Kristensen. 2005. Sediment properties and bacterial community in burrows of the ghost shrimp *Pestarella tyrrhena* (Decapoda: Thalassinidea). Aquatic Microbial Ecology 38:181–190.

Pearson, T.H. 2001. Functional group ecology in soft-sediment marine benthos: The role of bioturbation. Oceanography and Marine Biology: An Annual Review 39:233–267.

Pelegri, S.P., and T.H. Blackburn. 1994. Bioturbation effects of the amphipod *Corophium volutator* on microbial nitrogen transformations in marine sediments. Marine Biology 121:253–258.

Pillay, D. 2019. Ecosystem engineering by thalassinidean crustaceans: Response variability, contextual dependencies and perspectives on future research. Diversity 11:64.

Pillay, D., G.M. Branch, J. Dawson, and D. Henry. 2011. Contrasting effects of ecosystem engineering by the cordgrass *Spartina maritima* and the sandprawn *Callianassa kraussi* in a marine-dominated lagoon. Estuarine, Coastal and Shelf Science 91:169–176.

Pillay, D., C. Williams, and A. Whitfield. 2012. Indirect effects of bioturbation by the burrowing sandprawn *Callichirus kraussi* on a benthic foraging fish, *Liza richardsonii*. Marine Ecology Progress Series 453:151–158.

Posey, M.H. 1986. Changes in a benthic community associated with dense beds of a burrowing deposit feeder, *Callianassa californiensis*. Marine Ecology Progress Series 31:15–22.

Posey, M.H., B.R. Dumbauld, and D.A. Armstrong. 1991. Effects of a burrowing mud shrimp, *Upogebia pugettensis* (Dana), on abundances of macro-infauna. Journal of Experimental Marine Biology and Ecology 148:283–294.

Pülmanns, N., K. Diele, U. Mehlig, and I. Nordhaus. 2014. Burrows of the semi-terrestrial crab *Ucides cordatus* enhance CO_2 release in a North Brazilian mangrove forest. PLoS ONE 9(10):e109532.

Reinsel, K.A. 2004. Impact of fiddler crab foraging and tidal inundation on an intertidal sandflat: Season-dependent effects in one tidal cycle. Journal of Experimental Marine Biology and Ecology 313:1–17.

Reise, K. 2002. Sediment mediated species interactions in coastal waters. Journal of Sea Research 48:127–141.

Revsbech, N.P., J. Sorensen, T.H. Blackburn, and J.P. Lomholt. 1980. Distribution of oxygen in marine sediments measured with microelectrodes. Limnology and Oceanography 25:403–411.

Ribeiro, S., T. Berge, N. Lundholm, T.J. Andersen, F. Abrantes, and M. Ellegaard. 2011. Phytoplankton growth after a century of dormancy illuminates past resilience to catastrophic darkness. Nature Communications 2:1–7.

Rice, D.L. 1982. The detritus nitrogen problem: New observations and perspectives from organic geochemistry. Marine Ecology Progress Series 9:153–162.

Riisgård, H.U., and P. Schotge. 2007. Surface deposit feeding versus filter feeding in the amphipod *Corophium volutator*. Marine Biology Research 3:421–427.

Roberts, H. 1981. Lagoon sediment transport: The significant effect of *Callianassa* bioturbation. Pages 459–465 *in* Proceedings of the fourth international coral reef symposium, Manila, Volume 1. Manila, Philippines.

Rowden, A.A., and M.B. Jones. 1994. A contribution to the biology of the burrowing mud shrimp, *Callianassa subterranea* (Decapoda: Thalassinidea). Journal of the Marine Biological Association of the United Kingdom 74:623–635.

Rowden, A.A., M.B. Jones, and A.W. Morris. 1998. The role of *Callianassa subterranea* (Montagu) (Thalassinidea) in sediment resuspension in the North Sea. Continental Shelf Research 18:1365–1380.

Sagi, N., J.M. Grünzweig, and D. Hawlena. 2019. Burrowing detritivores regulate nutrient cycling in a desert ecosystem. Proceedings of the Royal Society B: Biological Sciences 286:20191647.

Sasaki, A., H. Nakao, S. Yoshitake, and T. Nakatsubo. 2014. Effects of the burrowing mud shrimp, *Upogebia yokoyai*, on carbon flow and microbial activity on a tidal flat. Ecological Research 29:493–499.

Sassa, S., and Y. Watabe. 2008. Threshold, optimum and critical geoenvironmental conditions for burrowing activity of sand bubbler crab, *Scopimera globosa*. Marine Ecology Progress Series 354:191–199.

Seike, K., M. Banno, K. Watanabe, T. Kuwae, M. Arai, and H. Sato. 2020. Benthic filtering reduces the abundance of primary producers in the bottom water of an open sandy beach system (Kashimanada Coast, Japan). Geophysical Research Letters 47:e2019GL085338.

Smith, D., R. Hughes, and E. Cox. 1996. Predation of epipelic diatoms by the amphipod *Corophium volutator* and the polychaete *Nereis diversicolor*. Marine Ecology Progress Series 145:53–61.

Stamhuis, E.J., T. Reede-Dekker, Y. Van Etten, J.J. De Wiljes, and J.J. Videler. 1996. Behaviour and time allocation of the burrowing shrimp *Callianassa subterranea* (Decapoda, Thalassinidea). Journal of Experimental Marine Biology and Ecology 204:225–239.

Suchanek, T.H. 1983. Control of seagrass communities and sediment distribution by *Callianassa* (Crustacea, Thalassinidea) bioturbation. Journal of Marine Research 41:281–298.

Sumida, P.Y.G., A.Z. Güth, C.O. Quintana, and A.M.S. Pires-Vanin. 2020. Distribution and sediment selection by the mud shrimp *Upogebia noronhensis* (Crustacea: Thalassinidea) and the potential effects on the associated macroinfaunal community. Journal of Marine Science and Engineering 8:1032.

Takeda, S., and Y. Kurihara. 1987. The effects of burrowing of *Helice tridens* (De Haan) on the soil of a salt-marsh habitat. Journal of Experimental Marine Biology and Ecology 113:79–89.

Takeuchi, S., and A. Tamaki. 2014. Assessment of benthic disturbance associated with stingray foraging for ghost shrimp by aerial survey over an intertidal sandflat. Continental Shelf Research 84:139–157.

Tamaki, A., K. Harada, Y. Sogawa, and S. Takeuchi. 2020. Effect of stingray (*Hemitrygon akajei*) foraging on a ghost shrimp population (*Nihonotrypaea harmandi*) on an intertidal sandflat, western Kyushu, Japan. Marine and Freshwater Research 71:1128–1148.

Thrush, S.F., J.E. Hewitt, M. Gibbs, C. Lundquist, and A. Norkko. 2006. Functional role of large organisms in intertidal communities: Community effects and ecosystem function. Ecosystems 9:1029–1040.

Torio, D.D., and G.L. Chmura. 2013. Assessing coastal squeeze of tidal wetlands. Journal of Coastal Research 29:1049–1061.

Underwood, G.J., and D.M. Paterson. 1993. Seasonal changes in diatom biomass, sediment stability and biogenic stabilization in the Severn Estuary. Journal of the Marine Biological Association of the United Kingdom 73:871–887.

Uthicke, S., and D. Klumpp. 1998. Microphytobenthos community production at a near-shore coral reef: Seasonal variation and response to ammonium recycled by holothurians. Marine Ecology Progress Series 169:1–11.

VanBlaricom, G.R. 1982. Experimental analyses of structural regulation in a marine sand community exposed to oceanic swell. Ecological Monographs 52:283–305.

Van Colen, C., S.F. Thrush, S. Parkes, R. Harris, S.A. Woodin, D.S. Wethey, C.A. Pilditch, J.E. Hewitt, A.M. Lohrer, and M. Vincx. 2015. Bottom-up and top-down mechanisms indirectly mediate interactions between benthic biotic ecosystem components. Journal of Sea Research 98:42–48.

Volkenborn, N., L. Polerecky, D. Wethey, T. DeWitt, and S. Woodin. 2012. Hydraulic activities by ghost shrimp *Neotrypaea californiensis* induce oxic–anoxic oscillations in sediments. Marine Ecology Progress Series 455:141–156.

Waldbusser, G., and R. Marinelli. 2006. Macrofaunal modification of porewater advection: Role of species function, species interaction, and kinetics. Marine Ecology Progress Series 311:217–231.

Wang, J.Q., X.D. Zhang, L.F. Jiang, M.D. Bertness, C.M. Fang, J.K. Chen, T. Hara, and B. Li. 2010. Bioturbation of burrowing crabs promotes sediment turnover and carbon and nitrogen movements in an estuarine salt marsh. Ecosystems 13:586–599.

Waugh, S., and R.C. Aller. 2017. N_2 production and fixation in deep-tier burrows of *Squilla empusa* in muddy sediments of Great Peconic Bay. Journal of Sea Research 129:36–41.

Watling, L. 1998. Small-scale features of marine sediments and their importance to the study of deposit-feeding. Marine Ecology Progress Series 47:135.

Webb, A.P., and B.D. Eyre. 2004a. Effect of natural populations of burrowing thalassinidean shrimp on sediment irrigation, benthic metabolism, nutrient fluxes and denitrification. Marine Ecology Progress Series 268:205–220.

Webb, A.P., and B.D. Eyre. 2004b. The effect of natural populations of the burrowing and grazing soldier crab (*Mictyris longicarpus*) on sediment irrigation, benthic metabolism and nitrogen fluxes. Journal of Experimental Marine Biology and Ecology 309:1–19.

Weitkamp, L.A., R.C. Wissmar, C.A. Simenstad, K.L. Fresh, and J.G. Odell. 1992. Gray whale foraging on ghost shrimp (*Callianassa californiensis*) in littoral sand flats of Puget Sound, U.S.A. Canadian Journal of Zoology 70:2275–2280.

Widdicombe, S., M.C. Austen, M.A. Kendall, R.M. Warwick, and M.B. Jones. 2000. Bioturbation as a mechanism for setting and maintaining levels of diversity in subtidal macrobenthic communities. Hydrobiologia 440:369–377.

Wilson, C.A., Z.J. Hughes, and D.M. Fitzgerald. 2012. The effects of crab bioturbation on Mid-Atlantic saltmarsh tidal creek extension: Geotechnical and geochemical changes. Estuarine, Coastal and Shelf Science 106:33–44.

Witbaard, R., and G.C.A. Duineveld. 1989. Some aspects of the biology and ecology of the burrowing shrimp *Callianassa subterranea* (Montagu) (Thalassinidea) from the southern North Sea. Sarsia 74:209–219.

Xiao, K., A.M. Wilson, H. Li, I.R. Santos, J. Tamborski, E. Smith, S.Q. Lang, C. Zheng, X. Luo, M. Lu, and R.E. Correa. 2021. Large CO_2 release and tidal flushing in salt marsh crab burrows reduce the potential for blue carbon sequestration. Limnology and Oceanography 66:14–29.

Yager, P.L., A.R.M. Nowell, and P.A. Jumars. 1993. Enhanced deposition to pits: A local food source for benthos. Journal of Marine Research 51:209–236.

Yoder, J.A., K.A. Reinsel, J.M. Welch, D.M. Clifford, and E.J. Rellinger. 2005. Herding limits water loss in the sand fiddler crab, *Uca pugilator*. Journal of Crustacean Biology 25:141–145.

Ziebis, W., S. Forster, M. Huettel, and B.B. Jørgensen. 1996a. Complex burrows of the mud shrimp *Callianassa truncata* and their geochemical impact in the seabed. Nature 382:619–622.

Ziebis, W., M. Huettel, and S. Forster. 1996b. Impact of biogenic sediment topography on oxygen fluxes in permeable seabeds. Marine Ecology Progress Series 140:227–237.

5

CRUSTACEANS AS INTERMEDIATE HOSTS IN THE LIFE CYCLES OF PARASITES

Robert Poulin

Abstract

Many parasitic worms, or helminths, have evolved complex life cycles in which juvenile stages infecting an intermediate host must be transferred trophically to a definitive vertebrate host. The generally small-to-medium body sizes of crustaceans favor their role as primary or secondary consumers in food webs, making them ideal intermediate hosts for helminths and perfect vehicles for their prey-to-predator transmission. This chapter reviews the life cycles and biology of the four major taxa of helminths using crustaceans as intermediate hosts: acanthocephalans, nematodes, trematodes, and cestodes. The zoonotic helminth taxa that parasitize crustaceans before infecting humans are highlighted, followed by a summary of the most frequently studied helminth-crustacean associations. The many impacts of helminth parasites on crustacean biology range from effects on individual behavior and fitness, to potentially major influences on population dynamics and community structure and composition, especially when acting in combination with other stressors such as rising temperatures. There is currently little information available on biogeographic patterns in helminth infections of crustaceans, and there are glaring gaps in our knowledge of, and research on, helminth parasitism of crustaceans; brief suggestions for promising research directions to address these issues are given. The take-home message of this chapter is that parasites matter; they can affect multiple aspects of the physiology, behavior, survival, population biology, and community ecology of their crustacean hosts. All crustacean biologists should be strongly encouraged to take them into account in their future research.

Robert Poulin, *Crustaceans as Intermediate Hosts in the Life Cycles of Parasites*. In: *Ecology and Conservation*. Edited by: Lars Gutow and Martin Thiel, Oxford University Press.
DOI: 10.1093/oso/9780197768242.003.0005

INTRODUCTION

Most crustaceans have relatively small body sizes, from planktonic copepods measuring barely 1 mm to the vast majority of other crustaceans that fall between one and several cm in maximum linear dimensions (Klompmaker et al. 2015). This body size distribution makes them likely to act as either primary or secondary consumers in food webs and serve as food for higher trophic levels (Dunne et al. 2004, Kortsch et al. 2019). Crustaceans are therefore well-suited to play roles as intermediate hosts in the complex life cycles of parasitic organisms transmitted through food chains. Apart from very few microbes (viruses, bacteria, or protozoans), those that use crustaceans for trophic transmission are almost exclusively parasitic worms (helminths). These parasites have evolved multi-host life cycles in which they must pass through several host species in order to complete one generation, with larval or juvenile stages infecting intermediate hosts from which they are trophically transmitted to a predatory definitive host, which is always a vertebrate. Over evolutionary time, helminth life cycles have become embedded in food webs, and most helminths exploit well-established food chains for their transmission. The wide-ranging ecological implications of crustacean–helminth associations are slowly but increasingly being acknowledged by ecologists at large (e.g., Giari et al. 2020). At a time when all kinds of marine diseases appear to be on the rise (Harvell 2019), greater attention paid to crustacean parasites would be very timely.

The dependence of many helminths on crustaceans as intermediate hosts means that wherever suitable crustaceans live, their parasites may also exist and exploit them. Parasite-free crustacean populations do exist (e.g., Bryan-Walker et al. 2007), although extensive sampling might reveal that they are not. Helminths that use crustaceans as part of their life cycles show some degree of specificity: they may be able to use alternative crustacean species if their preferred target host is unavailable, but they *must* infect a suitable crustacean to survive and continue their development. The diversity and abundance of crustaceans in natural communities therefore determines which parasite species can occur locally, and what infection risk they present for wild vertebrates as well as humans. Indeed, humans too are part of food webs, and feed on species from multiple trophic levels. For some helminths, humans may serve as suitable definitive hosts in which survival and reproduction are possible; for others, humans are unsuitable dead-end hosts. Indeed, many zoonotic diseases are caused by food-borne helminth species that accidentally end up inside people who eat undercooked or raw crustacean meat, or fish that prey on crustaceans.

In this chapter, I review the helminth taxa commonly associated with crustaceans, their life cycles, and biology. Although parasites of all kinds negatively impact crustaceans in fisheries and aquaculture (Shields and Huchin-Mian 2020), the focus of the present chapter is on helminth parasitism in natural ecosystems. I present data on the most commonly reported crustacean–helminth associations and their frequency in natural communities. I then summarize the potential impact of helminth parasites on the biology of crustaceans at individual, population, and community levels, before presenting an overview of what little is known of the biogeography of helminth parasitism in crustaceans. After suggesting potential avenues for future research, I end with a plea, hopefully justified by then, for all researchers studying crustaceans to consider their parasites as potential sources of variation in whatever trait or variable is under investigation.

THE HELMINTH PARASITES OF CRUSTACEANS

Four large taxonomic groups of helminths may use crustaceans as intermediate hosts in their life cycle: acanthocephalans, nematodes, trematodes, and cestodes (Fig. 5.1). Depending on the species, they may be encysted or not, and found either lodged within tissue or free in the hemocoel. However, their location and whether or not they are encysted are not necessarily reliable predictors of

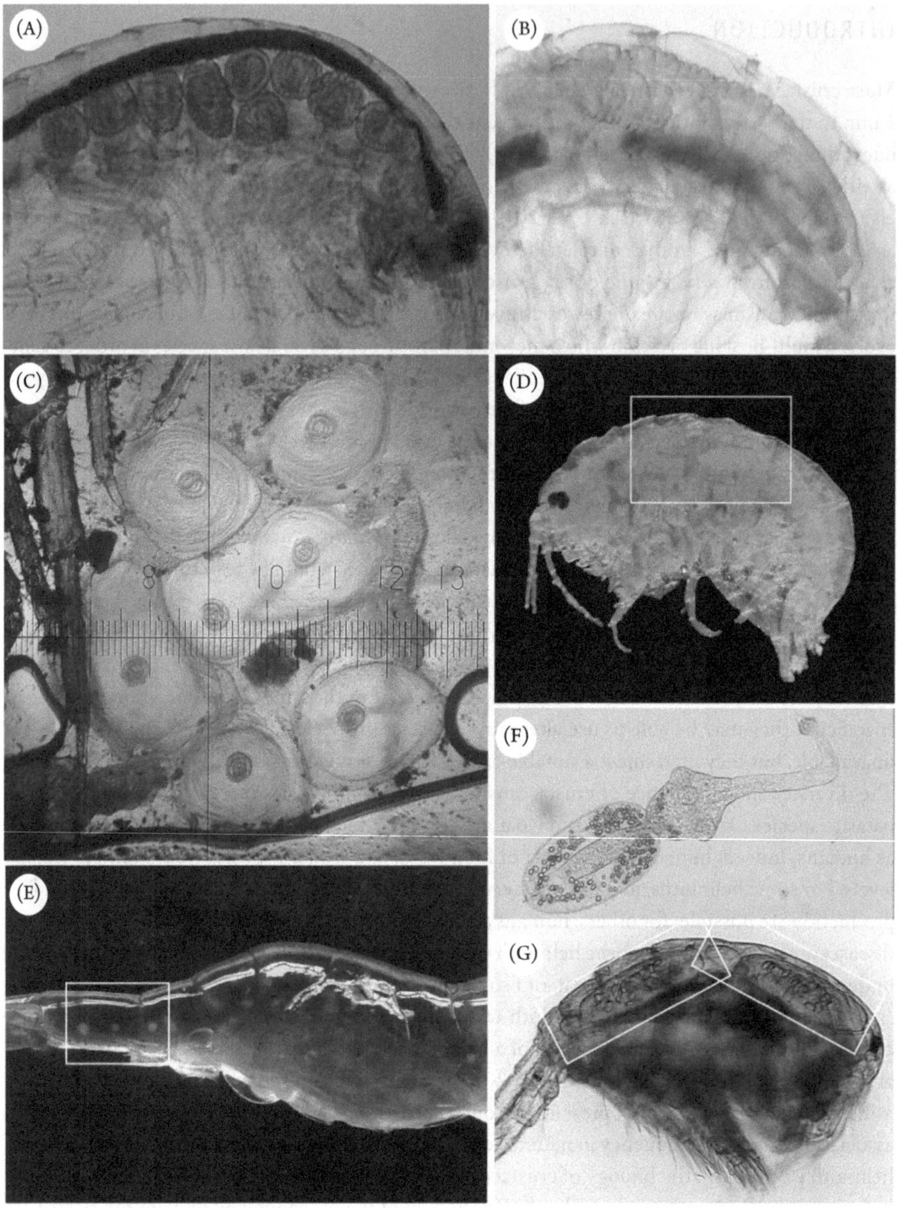

Fig. 5.1.
Various helminths in situ within their crustacean intermediate host. (A) Nine cystacanths of the acanthocephalan *Pomphorhynchus laevis* in the amphipod *Echinogammarus stammeri*, from Italy. Photo reproduced courtesy of B. Dezfuli. (B) Large cystacanth of the acanthocephalan *Acanthocephalus rhinensis* in the amphipod *Echinogammarus tibaldi*, from Italy. Photo reproduced courtesy of B. Dezfuli. (C) Encysted larvae of the nematode *Antechiniella septentrionalis* in the body cavity of the amphipod *Traskorchestia ditmari*, from Russia. Photo reproduced courtesy of N. Dokuchaev and E. Ivanova. (D) Juvenile nematode *Hedruris dratini* in the body cavity of the amphipod *Hyalella bonariensis*, from Argentina. Photo reproduced courtesy of E. Palumbo. (E) Metacercariae of the trematode *Microphallus szidati* (seen as white dots) in the body cavity of the freshwater shrimp *Palaemonetes argentinus*, from Argentina. Photo reproduced courtesy of M. Merlo. (F) Excysted cysticercoid of the cestode *Eurycestus avoceti* taken from the tissues of the brine shrimp *Artemia parthenogenetica*, from Spain. Photo reproduced courtesy of B. Georgiev and M. Sánchez. (G) Two procercoids of the cestode *Schistocephalus solidus* in the body cavity of the copepod *Macrocyclops albidus*, from Germany. Photo reproduced courtesy of D. Benesh. See a color version of this figure in the centerfold.

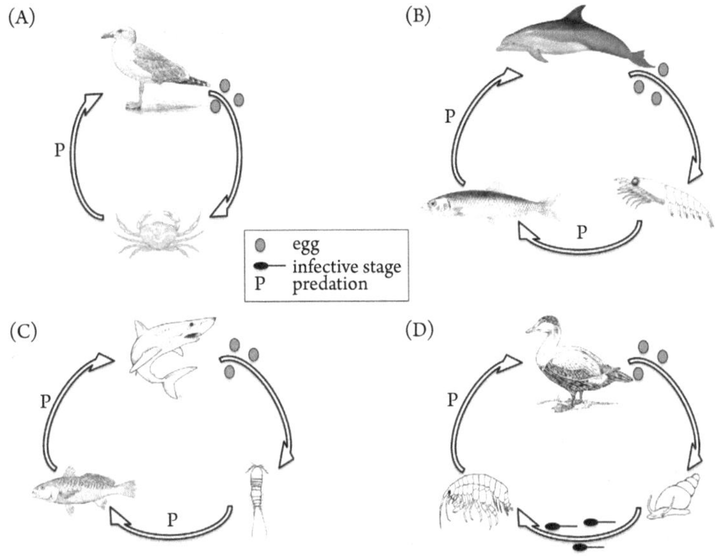

Fig. 5.2.
Representative life cycles of parasitic helminths using crustaceans as intermediate hosts. Adult worms live inside vertebrate definitive hosts (top of each figure), where they reproduce and lay eggs in host feces. Larvae from these eggs infect at least one intermediate host and return to the definitive host through one or two trophic transmission steps. (A) Acanthocephalan, illustrated by *Profilicollis* spp. which often infect crabs. (B) Nematodes, illustrated by *Anisakis* spp. which commonly infect euphausiids. (C) Cestodes, illustrated by members of the trypanorhynch genus *Lacistorhynchus* which have been shown to infect harpacticoid copepods. (D) Trematodes, illustrated by *Maritrema* spp. which frequently infect amphipods.

their physiological or behavioral impacts. And depending on the relative sizes of host and parasite and on local transmission patterns, the numbers of juvenile helminths per individual crustacean host can range from one to hundreds.

Acanthocephalans

The phylum Acanthocephala includes nearly 1500 described species and probably many more undescribed ones. The generalized acanthocephalan life cycle comprises only two hosts, an arthropod intermediate host and a vertebrate definitive host (Fig. 5.2A). However, in marine ecosystems, a facultative paratenic host is sometimes used to facilitate trophic transmission from the intermediate to the definitive host when these are separated by more than one trophic level. Paratenic hosts provide "transport" only, serving as trophic links between intermediate and definitive hosts; the parasites do not develop in them, and generally induce no significant pathology. For example, species in the genus *Corynosoma* use amphipods as intermediate hosts and pinnipeds as definitive hosts; small-to-medium sized teleost fish serve as paratenic hosts, consuming infected amphipods and later delivering the parasite to its definitive host when they are eaten by a seal or sea lion.

Crustaceans become infected when they accidentally ingest acanthocephalan eggs released in the feces of parasitized definitive hosts. Once in the crustacean gut, an acanthella larva emerges from the egg and burrows its way into the hemocoel, causing some damage to the gut wall. There, it will develop into a cystacanth, a resting and often encysted stage that is infective to the definitive host when the crustacean is eaten by a suitable vertebrate (Kennedy 2006).

Although rare, there are a few reports of acanthocephalans accidentally infecting humans, including species that use crustaceans early in their life cycle. For example, species in the genus *Corynosoma* can persist and cause abdominal pain in people that have eaten undercooked fish, though humans are unsuitable hosts, and the parasites fail to reproduce (Fujita et al. 2016).

Nematodes

Life cycle patterns vary greatly among the several thousand known species of parasitic nematodes. During their development, nematodes go through several larval stages separated by molts, but the number of host species required varies among species. Some taxa have simple, one-host life cycles whereas others use multiple hosts, with or without paratenic hosts (Fig. 5.2B). Consider members of the genus *Pseudoterranova*, which parasitize pinnipeds as definitive hosts, reaching them after up to three trophic transmission events (Anderson 2000). Eggs released in pinniped feces hatch into larvae that are then eaten by benthic harpacticoid or cyclopoid copepods, where they grow in the hemocoel until their copepod host is ingested by a larger benthic invertebrate, such as an amphipod or isopod, or a polychaete. In this new host, the larvae molt and grow further, until their host is eaten by small-to-medium sized demersal fish. Fish such as sculpins or plaice can accumulate hundreds of larvae (Jensen and Andersen 1992), which lodge themselves into, and achieve further growth, in the body cavity, mesenteries, and musculature of the fish. From that fish host, the larvae can be transmitted by predation directly to a pinniped definitive host, where they develop into adult worms and thus complete their life cycle. Alternatively, they may end up in large piscivorous fish; as seals and other pinnipeds feed mainly on smaller fish, larvae reaching large piscivorous fish may never complete their life cycle.

Most parasitic nematodes using crustaceans as intermediate hosts tend to have simpler life cycles involving only two or three hosts. In crustaceans, they almost invariably occupy the hemocoel, which they reach by penetrating the gut wall after their ingestion. Typically, they are enveloped in a protective sheath when ingested by the crustacean host, and conditions in the gut cause them to exsheath before burrowing out of the gut and into the hemocoel (Anderson 2000). When the crustacean host is eaten by another animal, the change in ambient conditions causes the nematode to either develop further (if ingested by an appropriate next host) or die and be consumed (if ingested by an unsuitable host).

Several zoonotic nematodes use crustaceans as intermediate hosts. The ones that most commonly infect humans are members of the genus *Anisakis*, which use a fish or squid after first developing inside a euphausiid or other small crustacean (Fig. 5.2B). These nematodes are acquired when people eat raw or undercooked seafood; the worms then penetrate the stomach or intestinal wall, where they cause moderate to severe pathology.

Cestodes

The complex life cycles of cestodes often involve crustaceans as intermediate hosts (Khalil et al. 1994). Examples include a well-studied cestode parasite (*Schistocephalus solidus*) of stickleback fish (which uses cyclopoid copepods as first intermediate hosts), the diverse cestode species infecting brine shrimp before completing their life cycle in flamingos and other waders, and the many species of tetraphyllidean (sensu lato) and trypanorhynch cestodes that can use small crustaceans in their transmission chain before maturing as adult worms in chondrichthyan fishes (Fig. 5.2C).

Within the crustacean host, larval cestodes may or may not be encysted, they can pass through a few developmental stages, and they can achieve considerable growth. They are generally located in the hemocoel, and usually cause little measurable pathology. Very few cestodes using crustaceans in their life cycle cause infections in humans, the notable exception being *Diphyllobothrium latum*, which has a copepod first intermediate host, a freshwater fish as second intermediate host, and a carnivore as definitive host. Here, again people ingesting undercooked fish incur a risk of infection, with ensuing symptoms including gastrointestinal complaints, fatigue, and weight loss.

Trematodes

There are around 20,000 described species of trematodes, only a fraction of which infect crustaceans. Those that do usually have a three-host life cycle (Galaktionov and Dobrovolskij 2003). Larvae issued from eggs infect a first intermediate host, usually a gastropod, in which they go through multiple rounds of asexual multiplication. This produces large numbers of free-swimming and short-lived (typically < 24 hours) infective stages, which leave the gastropod in order to seek and penetrate a crustacean second intermediate host. The external cues that trigger their exit from the snail host, their swimming behavior, and the cues they respond to in order to locate suitable crustacean targets vary among species (Combes et al. 1994). Typically, the cues used by trematode infective stages to get in close proximity to their target hosts include light intensity, hydrostatic pressure, water movement, or some combination of these factors. After they make contact with a crustacean, these infective stages generally enter through the gills or parts of the exoskeleton where penetration is easier, for example, at the joint membranes. Once inside the crustacean host, the parasites will migrate to their final site (hemocoel or other tissue), and grow into metacercariae, the encysted resting stages that await predation by the vertebrate definitive host (Fig. 5.2D). Members of the family Microphallidae (e.g., genera *Microphallus, Maritrema, Levinseniella*), which mature in shore bird definitive hosts after infecting crustaceans, are perfect examples of this common life cycle.

Trematodes of the genus *Paragonimus* are the causative agent of paragonimiasis, a food-borne infection affecting millions of people worldwide, mostly in eastern Asia. People acquire the worms by eating raw or inadequately cooked crayfish or freshwater crabs, in particular crabs from the family Potamidae. Pathology results from the migration and tissue invasion of the worms within the host, with symptoms ranging from abdominal pain to excess sputum production in the lungs, depending on the tissue entered by the parasites.

HOST–PARASITE ASSOCIATIONS

From an evolutionary perspective, helminths have evolved tighter associations with certain crustacean lineages than with others, resulting in the current distribution of host–parasite associations across the crustacean phylogenetic tree. Clearly, not all crustaceans are equally likely to be used as intermediate hosts by helminth parasites. A priori, we might expect smaller-bodied taxa to be more frequently used, since they can be consumed by a wider spectrum of predators and may thus be better conduits to predatory definitive hosts. However, current knowledge of which crustaceans serve as intermediate hosts is probably heavily biased by unequal study effort across taxa and geographical locations. Thus, the only question we may hope to answer is instead: which crustacean taxa are most often reported as intermediate hosts of parasitic helminths?

Intertidal and coastal crabs have generally been well-studied with respect to their parasites, and several taxa have been regularly found to harbor juvenile helminths (Fig. 5.3). Some helminth taxa appear to make frequent use of crab as intermediate hosts, notably the microphallid trematodes *Maritrema* spp. and *Microphallus* spp., the acanthocephalans *Polymorphus* spp. and *Profilicollis* spp., and the nematodes *Ascarophis* spp. Interestingly, cestodes are rarely reported from crabs.

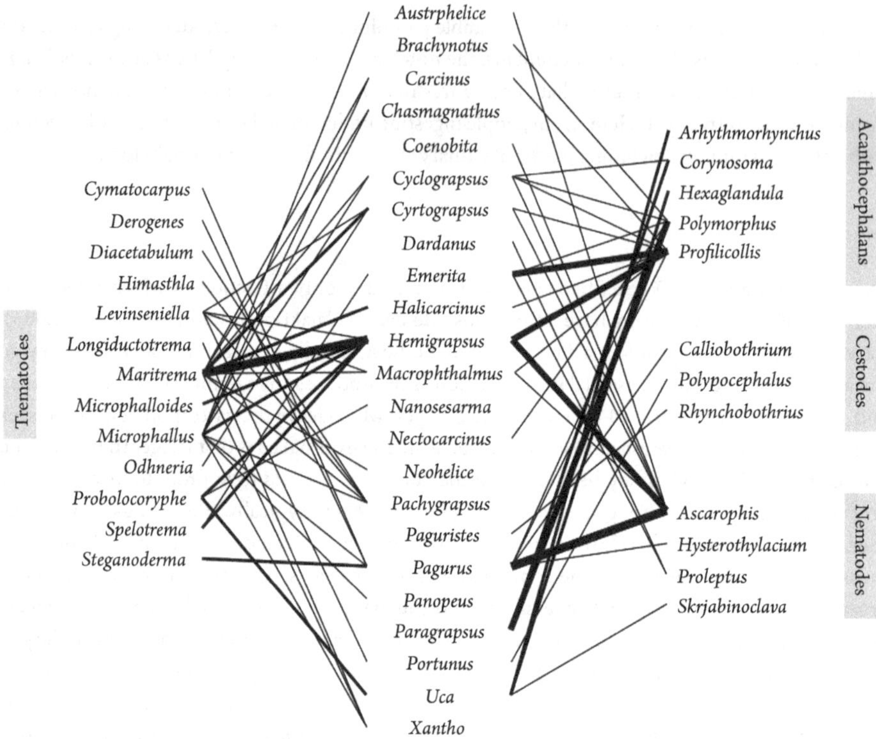

Fig. 5.3.
Association network for the most frequently studied genera of intertidal, estuarine, and coastal crabs (including mole crabs and hermit crabs), and the helminth genera occurring in them. Line thickness is proportional to the number of crab species (range 1 to 4) per genus serving as hosts for parasites of a given helminth genus. Data from a search of the Web of Science; lists of taxa are not comprehensive and exclude zoonotic helminths.

There are also numerous reports of helminths in amphipods and isopods (Fig. 5.4). Gammarid and corophiid amphipods, and idoteid isopods, have been particularly well-studied. Some of the same parasites frequently reported from crabs are also commonly found in amphipods and isopods. In particular, the acanthocephalan genera *Pomphorhynchus, Echinorhynchus,* and *Polymorphus* in amphipods, and *Acanthocephalus* in isopods; the microphallid trematodes *Maritrema* spp. and *Microphallus* spp., and the nematode *Hysterothylacium aduncum,* a globally distributed fish parasite. As with crabs, amphipods and isopods are rarely found to harbor cestodes. Figure 5.4 only shows aquatic taxa, although terrestrial taxa also serve as intermediate hosts for helminths; for example, acanthocephalans (Moore 1983) and nematodes (Moore and Lasswell 1986) can use oniscid isopods (woodlice or pill bugs) as part of their life cycle.

In addition to crabs, amphipods, and isopods, many other crustacean taxa also serve as intermediate hosts of helminths. These include copepods (Pulkkinen et al. 2000), ostracods (Valkounova 1985, Dezfuli 1996), brine shrimp (Georgiev et al. 2005), crayfish (Turner 2004, McAllister et al. 2011), and barnacles (Irwin and Irwin 1980, Sari and Malek 2000). It is unclear whether these crustaceans are less frequently reported as hosts of helminths because fewer parasites have adopted them as intermediate hosts over evolutionary time, or because they are the subject of much fewer parasitological investigations. There is no doubt that crustacean taxa like copepods and ostracods, which have generally very small body sizes, offer very limited space or nutrients to juvenile

Fig. 5.4.
Association network for the most frequently studied genera of aquatic amphipods and isopods, and the helminth genera occurring in them. Parasites are arranged by higher taxon (boxed, from top to bottom), in the following order: acanthocephalans, trematodes, nematodes, cestodes. Line thickness is proportional to the number of crustacean species (range 1 to 3) per genus serving as hosts for parasites of a given helminth genus. Data from a search of the Web of Science; lists of taxa are not comprehensive.

helminths; this may constrain their use as intermediate hosts to only a few helminth taxa capable of survival on such limited resources. Certainly, minimum resource requirements are needed for helminth development, as evidenced by the effect of intra-host competition. For example, even in larger hosts, like amphipods, the presence of just a few individual acanthocephalans of the same or different species results in significant reductions in average parasite size (Barger and Nickol 1999, Dezfuli et al. 2001). The hemocoel of a typical copepod could not even accommodate a single juvenile acanthocephalan.

Instead of focusing on associations between hosts and parasites from a taxonomic perspective, another way of looking at crustacean–helminth associations is to quantify the proportion of crustacean species in a local ecosystem that are used by helminths, or the average number of helminth species per crustacean species. Quantitative information on crustacean use by parasites is available for seven highly resolved estuarine or coastal marine food webs, from the Pacific coast of North America, northern Europe, and New Zealand (see Dunne et al. 2013). In each food web, all metazoans longer than a few millimeters have been sampled, identified to the lowest possible taxonomic level, and examined for parasite infection. Across all seven food webs combined, there were 13 crab populations, all of which were used by a minimum of three different species of trophically transmitted helminths. Twenty-six (65%) of the 40 amphipod populations were used as intermediate hosts by 1–5 helminth species, and 6 (50%) of the 12 isopod populations were used by 1–3 helminth species. One can therefore safely assume that in any marine (and probably any freshwater) ecosystem, the majority of crab, amphipod, and isopod species likely harbor one to several helminth species. As discussed in the following section, the ecological implications of this omnipresence of parasites in crustacean populations are far from negligible.

IMPACT OF HELMINTHS ON CRUSTACEAN HOSTS

Helminths do more than just pass through crustaceans on their way to a vertebrate host: they exploit their crustacean hosts and extract resources from them. What do juvenile helminths actually do to their host? The answer depends on the numbers of parasites per host, and can be addressed at the individual, population, or community level.

Host Individuals

Crustaceans possess a defense system against helminth parasites, a non-self-recognizing system catalyzed by the enzyme phenoloxidase that leads to the melanization and encapsulation, and thereafter the death, of invading helminth larvae (Cerenius and Söderhäll 2004). This defense system may be costly to maintain. In the absence of infections, individuals that invest more in defense may do so at the expense of long-term survival (Cornet et al. 2009). This trade-off can explain why the ability of crustaceans to melanize and kill invading helminths varies among populations (Cornet et al. 2009) and is greatly reduced in populations experiencing no parasitism (Bryan-Walker et al. 2007). In any event, no immune system is perfect, and parasites have evolved means to counter host defenses; therefore, crustaceans regularly encountering helminth infective stages almost inevitably become infected by one or many parasites, of the same or different species.

Individual crustaceans can incur loss of fitness as a direct outcome of the pathology resulting from tissue damage and exploitation of their resources by helminths. There are many correlational or experimental studies providing empirical evidence for a reduction in one or more fitness components in infected crustaceans. For example, infection by juvenile trematodes or acanthocephalans has been shown to result in elevated metabolic rates (Haye and Ojeda 1998), a lower number of offspring (Thomas et al. 1995a, Dezfuli et al. 1999), and smaller offspring (Dezfuli et al. 1999). Infection by juvenile helminths can also lower survival, either as a direct result of physiological stress resulting from infection (Fredensborg et al. 2004), or indirectly via parasite-induced increases in host susceptibility to predation (see below). However, these effects are often proportional to the intensity of infection, only becoming significant when a host is heavily infected, and they can differ between host sexes (Benesh and Valtonen 2007). Fitness impacts caused by helminth infections are therefore far from universal among helminth–crustacean associations.

In addition to the above pathological impacts of infection, many juvenile helminth parasites are capable of adaptive manipulation of the behavior or appearance of their crustacean intermediate host, modifying host phenotypic traits in ways that enhance trophic transmission to the definitive host (see reviews in Moore [2002], Poulin [2010]). These range from aberrant coloration in amphipods induced by acanthocephalans (Hindsbo 1972, Lagrue et al. 2016), to the altered swimming movements of cestode-infected copepods (Pulkkinen et al. 2000) and the reversed photophobia in trematode-infected amphipods (Johnson 2022). Acanthocephalans are particularly notorious for their ability to manipulate the behavior of their crustacean intermediate hosts (Fayard et al. 2020). The mechanisms used by helminths to induce phenotypic modifications in their crustacean hosts are still poorly understood. However, altered levels of the neurotransmitter serotonin, and/or altered serotonergic activity in certain parts of the host brain, have been linked to changes in the behavior of several crustacean species infected by a range of helminths (Helluy and Thomas 2003, Tain et al. 2006).

Whatever the underlying mechanisms involved, parasite-induced changes in crustacean behavior often leads to an increased risk of predation. For example, intertidal amphipods *Corophium volutator* infected by larval stages of the nematode *Skrjabinoclava morrisoni* spend more time actively moving outside their burrows during the day than their uninfected conspecifics (McCurdy et al. 1999). This altered behavior increases predation rates on infected amphipods by the nematode's definitive host, the sandpiper *Calidris pusilla*, and is only displayed by amphipods harboring nematode larvae that have developed to a stage where they are infective to birds. Since the changes in host behavior induced by juvenile helminths expose their crustacean hosts to greater danger from predation, the parasites may compromise the host's ability to show fear in the face of threats. Indeed, the acanthocephalan *Pomphorhynchus tereticollis* decreases the expression of anxiety-related responses in the amphipod *Gammarus fossarum* (Cozzarolo and Perrot-Minnot 2022). Infected amphipods showed reduced aversive responses to cues associated with danger. Parasite-induced alterations in crustacean behavior may also be context-dependent. For example, the personality of the crayfish *Faxonius rusticus* is shifted along the shy–bold continuum by trematode infection, but in different ways depending on whether the crayfish is in a novel environment or under threat (MacKay and Moore 2021). Manipulation of host behavior often benefits the parasite by enhancing its transmission success, but from the crustacean's perspective, this is just one more fitness cost of infection.

Host Populations

Inferring population-level impacts of helminths on crustaceans in nature often relies on indirect evidence. For instance, if hosts accumulate parasites over time, as they grow in size, and if the parasites induce no measurable increase in host mortality (whether directly due to pathology or indirectly due to increased risk of predation), then we would expect either a positive linear relationship between the average number of parasites per host and host body size, or a relationship leading to a plateau if the number of parasites reaches the maximum limit imposed by the space available within the host. In contrast, if the relationship is hump-shaped, then one likely explanation is that heavily infected individuals of larger size classes are taken out of the population through mortality, lowering the average number of parasites per host in these size classes below what would be expected in the absence of parasite-induced mortality (Thomas et al. 1995b, Latham and Poulin 2002). In nature, both patterns may be observed, depending on the host–parasite species combination or on the locality (Fig. 5.5). Therefore, in some cases, helminth infections can determine not only which individuals die, but also the overall local dynamics of crustacean populations.

In more extreme scenarios, helminth infections may even push crustacean populations to extinction. Higher than normal temperatures may provide the main condition necessary for helminths to cause crustacean die-offs. As part of their life cycle, trematodes multiply within-snail first intermediate hosts before infective larvae exit the snail to seek and infect crustaceans. The rates at which trematodes proliferate within snails, and the overall output of infective stages, are highly sensitive to temperatures: a rise in water temperature of just a few degrees can lead to many-fold increases in trematode multiplication rates (Poulin 2006). As a consequence of this response to temperature, populations of the amphipod *C. volutator* in the Wadden Sea (North Sea) can become locally extinct during summer heat waves, when they experience high infections by microphallid trematodes (Jensen and Mouritsen 1992). Modelling based on conservative parameter estimates

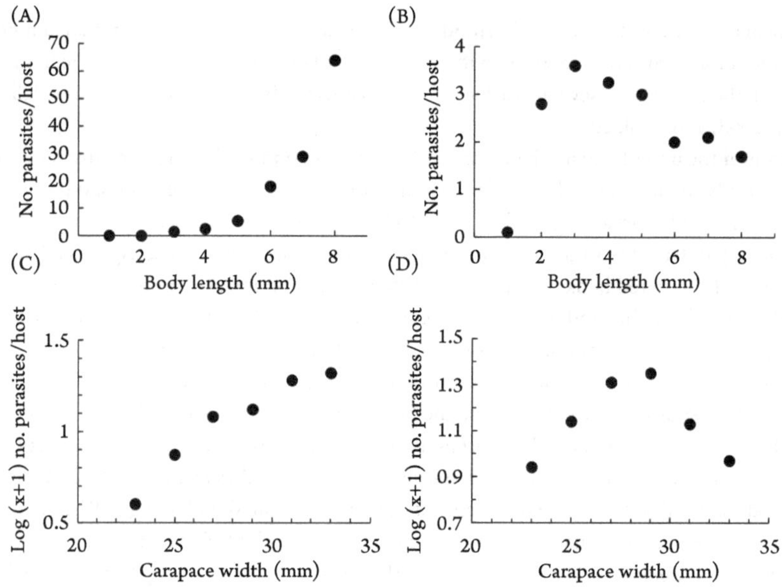

Fig. 5.5.
Mean number of parasites per individual host as a function of host size class. Trematode *Microphallus papillorobustus* in its amphipod hosts (A) *Gammarus aequicauda* and (B) *Gammarus insensibilis* from a spring 1994 sample collected in a French Mediterranean lagoon. Source: data from Thomas et al. (1995b). Acanthocephalan *Profilicollis* spp. in the crab *Macrophthalmus hirtipes* from winter 2001 samples collected at (C) Papanui Inlet and (D) Waipuna Bay, South Island, New Zealand. Source: data from Latham and Poulin (2002).

indicates that parasite-induced collapses of *C. volutator* populations may become frequent in the coming decades under the temperatures projected by climate warming scenarios (Mouritsen et al. 2005). Based on both field and modeling data, other coastal amphipod populations hosting trematode parasites are also expected to undergo frequent collapses and possible extinction during heat waves and under global warming (Bates et al. 2010, Studer et al. 2013). Several other abiotic factors, including water salinity and pH, can act additively or synergistically with helminth parasitism to regulate crustacean populations (Studer and Poulin 2012, Harland et al. 2015). If not on their own, parasites can thus act in combination with other external stressors to drive crustacean population dynamics.

Host Communities

If parasites impact the local population of a host species, then there are likely to be consequences for other species in the community. At the very least, an immediate prey or competitor of the host may be affected, either directly through a decrease in abundance of the host species, or in more subtle ways if the host's behavior is modified (Friesen et al. 2018). For example, laboratory experiments have demonstrated that predation on the invasive North American amphipod *Crangonyx pseudogracilis* by native European *Gammarus* spp. amphipods is greatly reduced when the latter are infected by the acanthocephalan *Polymorphus minutus* (MacNeil and Dick 2011). This may explain why these amphipods are more likely to co-occur in localities on the Isle of Man where the parasite *P. minutus* occurs than where it does not, since otherwise intense predation by *Gammarus* can lead to the local exclusion of *C. pseudogracilis* (MacNeil and Dick 2011).

If the parasite is shared by multiple sympatric crustacean species, that is, if the parasite can use any of these species as intermediate hosts, then the fitness and population-level impacts of infection may not be the same for all crustacean species. Some may be more tolerant of infection than others, and the presence of the parasite in the ecosystem can mediate interactions among those species and modify their relative abundances in the community (Hudson and Greenman 1998). For example, Friesen et al. (2020) have used a mesocosm experiment to test the impact of a shared trematode parasite on the relative abundances of four sympatric species of freshwater crustaceans, two amphipods and two isopods. Although all replicate mesocosms started with identical communities of the four species corresponding to their relative abundances in nature, those exposed to parasitism gradually developed different communities over time than those in the control group. In particular, the abundance of one amphipod species, which was originally the numerically dominant species, dropped dramatically in the treatment exposed to parasites. This was due to this species' greater susceptibility and lower tolerance to infection by the trematode, leading to both direct mortality as well as reduced competitiveness, relative to the other three crustacean species (Friesen et al. 2020). Thus, in a parasite-free natural environment, the composition of this crustacean assemblage might be very different.

The influence of helminths on communities of crustacean intermediate hosts may be amplified when combined with other environmental stressors. In a fully factorial mesocosm experiment, Mouritsen et al. (2018) quantified the separate and combined impacts of temperature and parasitism by a microphallid trematode on six sympatric species of intertidal amphipods, all suitable hosts for the parasites but varying in their susceptibility and tolerance to infection. Amphipods were maintained with either uninfected or infected snails, the latter serving as a source of trematode infective stages. Although the presence of the parasite and elevated temperature on their own had only modest effects on the composition of the amphipod community, their combined effect drastically altered the relative abundances of the different species (Fig. 5.6). Bird predation (i.e., trophic transmission of the parasite) was not a factor: all mortality was due to the direct deleterious effects of infection, temperature, or both. The interactive effect of temperature and parasitism is, in part, the outcome of the sensitivity of the within-snail multiplication of the parasite to external temperature: a difference of a few degrees can greatly amplify the numbers of parasites seeking to infect amphipods. In addition, the different trematode species have different thermal tolerances and resistance to parasite infection (Mouritsen et al. 2018). In this system, a slight rise in temperature combined with the continued presence of the trematode is likely to cause a decline in amphipod diversity, with infaunal species becoming over-represented relative to epibenthic ones. Comparable results were obtained in a similar mesocosm study testing the separate and combined effects of trematode infection and elevated temperature on a freshwater community of amphipod and isopods, with parasitism exacerbating the negative impact of warming on one amphipod species to the benefit of other community members (Friesen et al. 2021).

The impact of helminth infections in crustaceans can of course extend to non-crustacean species and the rest of the community. The parasite-induced die-off of the amphipod *C. volutator* mentioned in the previous section caused a marked change in the erosion and sediment properties of the soft-shore ecosystem, since the tube-dwelling amphipod acts to stabilize the substratum (Mouritsen et al. 1998). This had consequences for the rest of the infaunal invertebrate community since their small-scale distribution depends on sediment characteristics. Similarly, parasites infecting crustaceans that act as bioturbators of sediments (Dairain et al. 2019) or shredders of detritus (Hernandez and Sukhdeo 2008) can also indirectly impact ecosystem-wide functions by debilitating their hosts. Therefore, the impacts of any given parasite species may often extend well beyond its effects on its particular host.

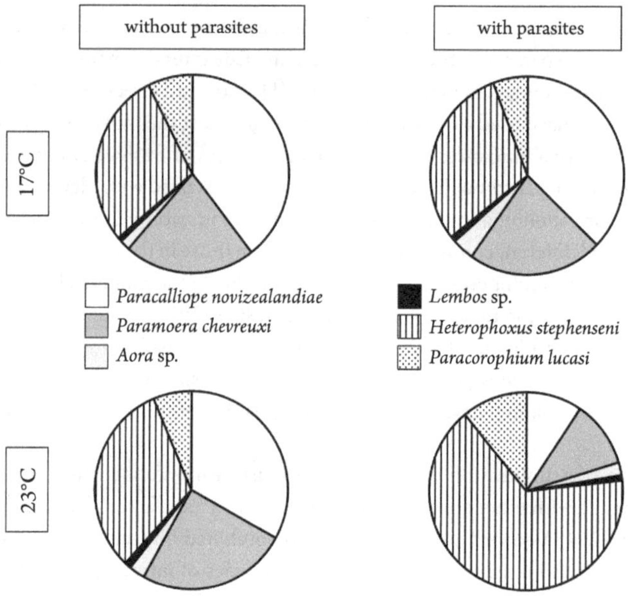

Fig. 5.6.
Average relative abundances of six intertidal amphipod species at the end of a long-term, replicated mesocosm experiment, in which the amphipod assemblages were maintained at normal or elevated temperature, and either exposed or not exposed to infection by the trematode *Maritrema novaezealandensis*. All replicates across all treatments had the same relative species compositions at the start of the experiment. Source: data from Mouritsen et al. (2018).

BIOGEOGRAPHY OF HELMINTH PARASITISM IN CRUSTACEANS

The importance of helminth parasites as agents of natural selection and determinants of fitness and population dynamics in crustaceans is not uniform across species and space. Crustacean species with certain traits may be more likely to be exploited either by many helminth species or at high intensities of infection. Similarly, crustaceans inhabiting certain geographical areas may generally experience a greater diversity of parasites and/or more intense infections. Despite a wealth of comparative studies on vertebrate hosts (e.g., Kamiya et al. 2014), there have been very few similar studies on crustacean hosts.

The best comparative data available for crustaceans comes from studies on trematodes compiled from 78 crustacean species (Thieltges et al. 2009a, b). Most crustacean populations that have been well-sampled for trematode parasites reveal that, typically, a crustacean population is exploited by one trematode species only, at low prevalence (< 20% individual hosts are infected) and low intensity (< 5 individual parasites per infected host) (Thieltges et al. 2009b). In rare cases, a crustacean population may harbor 5–6 trematode species, with a combined prevalence close to 100% and intensities exceeding 30 parasites per host. Across all crustacean species combined, average host body size covaried positively with intensity of infection, but not with trematode species richness or overall prevalence of infection (Thieltges et al. 2009b). Thus, individuals from a large crab species may harbor large numbers of parasites, but these are not necessarily from a greater number of trematode species. Values for intensity of infection also show a remarkably strong repeatability among populations of the same crustacean species: if infection intensity by a particular trematode species is high in a well-sampled host population, it is likely to be roughly equally high elsewhere in the same host species (Thieltges et al. 2009a). This suggests that host species traits must constrain tightly the

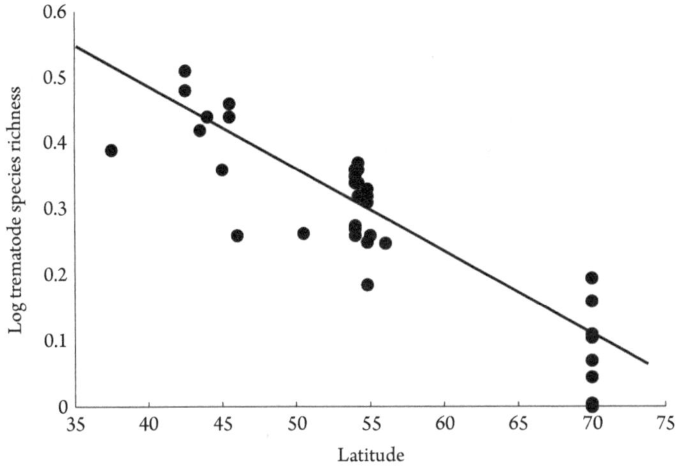

Fig. 5.7.
Latitudinal gradient in the species richness of trematode parasites infecting marine amphipods. The predicted values and best-fit line are from a general linear model that also accounts for sampling effort and host body size; the data are from 55 amphipod populations (representing 27 different species), with some points stacked on top of each other; no distinction is made between north and south latitudes. Source: data from Thieltges et al. (2009b).

burden that a host acquires, because one would expect that the local abundances of other hosts in the parasite's life cycle should also determine its intensity in crustaceans.

On a geographical scale, we may expect latitudinal gradients in helminth infections of crustaceans. The higher species diversity in the tropics than in temperate regions may lead to any focal crustacean population being exposed to a greater range of helminth species at low latitudes, although this is not always the pattern observed for other host taxa (Kamiya et al. 2014). Latitudinal gradients may exist for infection levels by particular species, and not just for parasite species richness. For instance, in the case of trematodes, the temperature-dependent multiplication rates of the parasites within their snail first intermediate host (Poulin 2006) may result in greater output of infective stages per snail in tropical climates, and possibly therefore in greater infections of crustacean second intermediate hosts. When comparing different amphipod populations and species sampled at different latitudes, and controlling for confounding variables, trematode species richness per host population is significantly higher at lower latitudes (Fig. 5.7). Similarly, prevalence and intensity of infection also increase toward lower latitudes (Thieltges et al. 2009b). Therefore, all else being equal, more individuals in tropical amphipod populations are infected, and they are infected at higher intensities, and by more trematode species, than their temperate counterparts. Trematodes may therefore exert stronger selective pressures on low-latitude crustacean populations, and play greater roles in their population dynamics, compared to those of higher latitudes. It remains to be determined whether these patterns also apply to other helminths, that is, cestodes, nematodes, and acanthocephalans.

Finally, in the context of comparisons among geographically distinct crustacean populations, it is important to consider the impact of species introductions on parasite infections. Introduced animal species often "escape" in their new range from the natural enemies that plagued them in their area of origin, which may be one reason why introduced species often attain high densities and become invasive (Colautti et al. 2004). The best-studied crustacean with respect to the enemy release hypothesis is the green shore crab, *Carcinus maenas*, native to Europe but now introduced to both the East and West coasts of North America, Japan, and the southern parts of Africa and

Australia. Comparisons between crab populations from native versus introduced regions indicate that both the prevalence and species richness of parasites were consistently much lower in introduced regions (Torchin et al. 2001, Blakeslee et al. 2009), although in some invaded localities *C. maenas* individuals can accumulate higher numbers of particular parasites than sympatric native crabs (Barnard et al. 2022). Generally speaking, the loss of parasites upon introduction to a new area can play a factor in facilitating the success and range expansion of invasive crustaceans.

FUTURE DIRECTIONS

Several glaring gaps exist in our knowledge of crustaceans as intermediate hosts of helminth parasites, all of which limit current understanding of their influence on crustacean behavior and health, on crustacean population dynamics, and on natural communities and food webs. The following list of promising future research directions is not meant to be exhaustive but provides some good starting points. First, despite many thousand helminth species from taxa presumed to use crustaceans as intermediate hosts having been discovered and formally described, knowledge of their life cycles is lagging far behind. In the vast majority of cases, helminth species are found as adult worms living in a vertebrate definitive host, and nothing is known of their juvenile stages or the intermediate hosts they use to reach the vertebrate (Blasco-Costa and Poulin 2017). For instance, well over one thousand cestode species using elasmobranch fishes as definitive hosts have been described, yet although many of them are assumed to use crustaceans as intermediate hosts, fewer than 10 species have had their life cycle elucidated and those intermediate hosts identified (Caira and Jensen 2014). Because juvenile and adult forms of the same helminth species are often very different morphologically, it is difficult to match them and confirm that they belong to the same species when they are found in different host species. However, the use of molecular markers is a very promising tool to achieve this and resolve life cycles (Blasco-Costa and Poulin 2017). As an example, similarity in sequences of 28S rDNA have been successfully used to match a larval tapeworm found in the crab *Nectocarcinus antarcticus* with the adult worm parasitizing skates (Bennett et al. 2019). Wider application of genetic tools would greatly improve our current knowledge of which crustaceans are used by which parasites.

Second, and related to the above, we need additional well-resolved food webs including parasites to determine exactly how the use of crustaceans at lower trophic levels by helminths (that are transmitted along predator–prey food chains) shapes the structure of the web and influences its stability (Marcogliese 2002, Sukhdeo 2012, Dunne et al. 2013). Does the presence of helminths change the energy flow through the food web by altering the vulnerability of infected crustacean prey and biasing their consumption by predators? What do the parasites themselves contribute to energy flow when they and their crustacean host are eaten by a predator that is not a suitable host for the predator (Thieltges et al. 2013)? How would the disappearance of key helminth species infecting crustaceans affect the structure of the food web? We currently only have partial answers to those questions, coming mostly from the study of a handful of estuarine and coastal marine food webs (Dunne et al. 2013). We need data from a broader range of ecosystems to achieve any general understanding of the roles of parasites in food webs.

Third, we still lack a solid understanding of the biochemical and physiological interactions at the interface of helminth–crustacean associations. This is particularly relevant to the widespread manipulation of crustacean behavior induced by parasitic helminths. In spite of recent advances in certain well-studied systems, such as the elucidation of the subtle role of altered serotonin levels in the *Gammarus pulex–Pomphorhynchus laevis* association (Tain et al. 2006, Perrot-Minnot et al. 2014), there is currently little evidence for the way in which the parasite initiates such changes. Modern -omics approaches offer many exciting possibilities to explore the role of gene regulation and expression in both host and parasite, before, during, and after the onset of behavioral manipulation.

Finally, as natural environments are under stress and changing rapidly in the face of threats ranging from habitat destruction and pollution to global warming and ocean acidification, there is an urgent need to better predict and mitigate negative impacts mediated through the effects of helminth parasites on key crustacean species. Parasites and diseases are often excluded from multi-stressor studies (Orr et al. 2020), and yet their integration into such research is not only feasible but highly desirable if we are to develop more comprehensive approaches to biodiversity conservation and ecosystem management.

SUMMARY AND CONCLUSIONS

Helminth parasites (acanthocephalans, nematodes, trematodes, and cestodes) are ubiquitous but invisible consumers of crustaceans worldwide, using them as stepping stones toward their final vertebrate host. Their effects are far from negligible: they can alter the behavior of their host, reduce their fitness, cause crashes in their populations, and impact the broader community and food web. All these effects can be amplified by environmental conditions; in particular, warmer temperatures can increase the severity and consequences of trematode infections in all kinds of crustaceans. In spite of all these important effects on crustacean biology, a look suffices at the past several volumes of journals dedicated to research on crustaceans (e.g., *Crustaceana*, or *Journal of Crustacean Biology*) to realize that biologists studying crustaceans do not often take helminths into account as part of their investigations. Quantifying helminth infections requires some additional work: each host must be dissected, the parasites must be recovered and counted (this is the easy part), and the parasites must then be identified to the lowest possible taxonomic level (even with molecular tools, that's the hard part!), to allow some inferences about the probable source of infection and the likely next host of the parasite. Given the demonstrated impacts of helminths on crustaceans at individual, population, and community levels, the extra work seems justified and may be rewarded with greater explanatory power. For example, I can easily imagine a scenario in which, as part of a study of some biological aspect of field-collected crustaceans, adding numbers of parasites per host to other standard predictors, like sex, age, or body size, might explain a greater percentage of variance in whatever trait or variable is under study. If the reader takes home one message from this chapter, let it be never to ignore the possible influence of parasites.

ACKNOWLEDGMENTS

I would like to thank the editors, Martin Thiel and Lars Gutow, for inviting me to contribute this chapter; Dan Benesh, Bahram Dezfuli, Elena Ivanova, Matias Merlo, Ezequiel Palumbo, and Marta Sánchez for allowing me to use their photos in Fig. 5.1; and Jerusha Bennett for constructive comments on an earlier version.

REFERENCES

Anderson, R.C. 2000. Nematode parasites of vertebrates: Their development and transmission. CABI Publishing, Wallingford, UK.

Barger, M.A., and B.B. Nickol. 1999. Effects of coinfection with *Pomphorhynchus bulbocolli* on development of *Leptorhynchoides thecatus* (Acanthocephala) in amphipods (*Hyalella azteca*). Journal of Parasitology 85:60–63.

Barnard, R.B., C.S. Moore, C.L. Keogh, and A.M.H. Blakeslee. 2022. If you encyst: Evidence of parasite escape and host-switching among three co-occurring crabs. Marine Ecology Progress Series 697:67–80.

Bates, A.E., R. Poulin, and M.D. Lamare. 2010. Spatial variation in parasite-induced mortality in an amphipod: Shore height versus exposure history. Oecologia 163:651–659.

Benesh, D.P., and E.T. Valtonen. 2007. Effects of *Acanthocephalus lucii* (Acanthocephala) on intermediate host survival and growth: Implication for exploitation strategies. Journal of Parasitology 93:735–741.

Bennett, J., F. Jorge, R. Poulin, and H. Randhawa. 2019. Revealing trophic transmission pathways of marine tapeworms. Parasitology Research 118:1435–1444.

Blakeslee, A.M.H., C.L. Keogh, J.E. Byers, A.M. Kuris, K.D. Lafferty, and M.E. Torchin. 2009. Differential escape from parasites by two competing introduced crabs. Marine Ecology Progress Series 393:83–96.

Blasco-Costa, I., and R. Poulin. 2017. Parasite life-cycle studies: A plea to resurrect an old parasitological tradition. Journal of Helminthology 91:647–656.

Bryan-Walker, K., T.L.F. Leung, and R. Poulin. 2007. Local adaptation of immunity against a trematode parasite in marine amphipod populations. Marine Biology 152:687–695.

Caira, J.N., and K. Jensen. 2014. A digest of elasmobranch tapeworms. Journal of Parasitology 100:373–391.

Cerenius, L., and K. Söderhäll. 2004. The prophenoloxidase-activating system in invertebrates. Immunological Reviews 198:116–126.

Colautti, R.I., A. Ricciardi, I.A. Grigorovich, and H.J. MacIsaac. 2004. Is invasion success explained by the enemy release hypothesis? Ecology Letters 7:721–733.

Combes, C., A. Fournier, H. Moné, and A. Théron. 1994. Behaviours in trematode cercariae that enhance parasite transmission: Patterns and processes. Parasitology 109:S3–S13.

Cornet, S., C. Biard, and Y. Moret. 2009. Variation in immune defence among populations of *Gammarus pulex* (Crustacea: Amphipoda). Oecologia 159:257–269.

Cozzarolo, C.-S., and M.J. Perrot-Minnot. 2022. Infection with an acanthocephalan helminth reduces anxiety-like behaviour in a crustacean host. Scientific Reports 12:21649.

Dairain, A., A. Legeay, and X. de Montaudouin. 2019. Influence of parasitism on bioturbation: From host to ecosystem functioning. Marine Ecology Progress Series 619:201–214.

Dezfuli, B.S. 1996. *Cypria reptans* (Crustacea: Ostracoda) as an intermediate host of *Neoechinorhynchus rutili* (Acanthocephala: Eoacanthocephala) in Italy. Journal of Parasitology 82:503–505.

Dezfuli, B.S., L. Giari, and R. Poulin. 2001. Costs of intraspecific and interspecific host sharing in acanthocephalan parasites. Parasitology 122:483–489.

Dezfuli, B.S., E. Rossetti, C.M. Bellettato, and B.J. Maynard. 1999. *Pomphorhynchus laevis* in its intermediate host *Echinogammarus stammeri* in the River Brenta, Italy. Journal of Helminthology 73:95–102.

Dunne, J.A., R.J. Williams, and N.D. Martinez. 2004. Network structure and robustness of marine food webs. Marine Ecology Progress Series 273:291–302.

Dunne, J.A., K.D. Lafferty, A.P. Dobson, R.F. Hechinger, A.M. Kuris, N.D. Martinez, J.P. McLaughlin, K.N. Mouritsen, R. Poulin, K. Reise, D.B. Stouffer, D.W. Thieltges, R.J. Williams, and C.D. Zander. 2013. Parasites affect food web structure primarily through increased diversity and complexity. PLoS Biology 11:e1001579.

Fayard, M., F.-X. Dechaume-Moncharmont, R. Wattier, and M.-J. Perrot-Minnot. 2020. Magnitude and direction of parasite-induced phenotypic alterations: A meta-analysis in acanthocephalans. Biological Reviews 95:1233–1251.

Fredensborg, B.L., K.N. Mouritsen, and R. Poulin. 2004. Intensity-dependent mortality of *Paracalliope novizealandiae* (Amphipoda: Crustacea) infected by a trematode: Experimental infections and field observations. Journal of Experimental Marine Biology and Ecology 311:253–265.

Friesen, O.C., R. Poulin, and C. Lagrue. 2018. Parasite-mediated microhabitat segregation between congeneric hosts. Biology Letters 14:20170671.

Friesen, O.C., R. Poulin, and C. Lagrue. 2021. Temperature and multiple parasites combine to alter host community structure. Oikos 130:1500–1511.

Friesen, O.C., S. Goellner, R. Poulin, and C. Lagrue. 2020. Parasites shape community structure and dynamics in freshwater crustaceans. Parasitology 147:182–193.

Fujita, T., E. Waga, K. Kitaoka, T. Imagawa, Y. Komatsu, K. Takanashi, F. Anbo, T. Anbo, S. Katuki, S. Ichihara, S. Fujimori, H. Yamasaki, Y. Morishima, H. Sugiyama, and H. Katahira. 2016. Human infection by acanthocephalan parasites belonging to the genus *Corynosoma* found from small bowel endoscopy. Parasitology International 65:491–493.

Galaktionov, K.V., and A.A. Dobrovolskij. 2003. The biology and evolution of trematodes. Kluwer Academic Publishers, Dordrecht, The Netherlands.

Georgiev, B.B., M.I. Sanchez, A.J. Green, P.N. Nikolov, G.P. Vasileva, and R.S. Mavrodieva. 2005. Cestodes from *Artemia parthenogenetica* (Crustacea: Branchiopoda) in the Odiel Marshes, Spain: A systematic survey of cysticercoids. Acta Parasitologica 50:105–117.

Giari, L., E.A. Fano, G. Castaldelli, D. Grabner, and B. Sures. 2020. The ecological importance of amphipod–parasite associations for aquatic ecosystems. Water 12:2429.

Harland, H., C.D. MacLeod, and R. Poulin. 2015. Non-linear effects of ocean acidification on the transmission of a marine intertidal parasite. Marine Ecology Progress Series 536:55–64.

Harvell, D. 2019. Ocean outbreak: Confronting the rising tide of marine disease. University of California Press, Berkeley, USA.

Haye, P.A., and F.P. Ojeda. 1998. Metabolic and behavioral alterations in the crab *Hemigrapsus crenulatus* (Milne-Edwards 1837) induced by its acanthocephalan parasite *Profilicollis antarcticus* (Zdzitowiecki 1985). Journal of Experimental Marine Biology and Ecology 228:73–82.

Helluy, S., and F. Thomas. 2003. Effects of *Microphallus papillorobustus* (Platyhelminthes: Trematoda) on serotonergic immunoreactivity and neuronal architecture in the brain of *Gammarus insensibilis* (Crustacea: Amphipoda). Proceedings of the Royal Society B: Biological Sciences 270:563–568.

Hernandez, A.D., and M.V.K. Sukhdeo. 2008. Parasite effects on isopod feeding rates can alter the host's functional role in a natural stream ecosystem. International Journal for Parasitology 38:683–690.

Hindsbo, O. 1972. Effects of *Polymorphus* (Acanthocephala) on colour and behaviour of *Gammarus lacustris*. Nature 238:333.

Hudson, P.J., and J.V. Greenman. 1998. Competition mediated by parasites: Biological and theoretical progress. Trends in Ecology and Evolution 13:387–390.

Irwin, S.W.B, and B.C. Irwin. 1980. The distribution of metacercariae of *Maritrema arenaria* (Digenea: Microphallidae) in the barnacle *Balanus balanoides* at three sites on the east coast of Northern Ireland. Journal of the Marine Biological Association of the United Kingdom 60:959–962.

Jensen, K.Y., and K.N. Mouritsen. 1992. Mass mortality in two common soft-bottom invertebrates, *Hydrobia ulvae* and *Corophium volutator*: The possible role of trematodes. Helgoländer Meeresuntersuchungen 46:329–339.

Jensen, T., and K. Andersen. 1992. The importance of sculpin (*Myoxocephalus scorpius*) as intermediate host and transmitter of the sealworm *Pseudoterranova decipiens*. International Journal for Parasitology 22:665–668.

Johnson, D.S. 2022. Are amphipods *Orchestia grillus* (Bosc, 1802) (Amphipoda: Talitridae) infected with the trematode *Levinseniella byrdi* (Heard, 1968) drawn to the light? Journal of Crustacean Biology 42:ruac017.

Kamiya, T., K. O'Dwyer, S. Nakagawa, and R. Poulin. 2014. What determines species richness of parasitic organisms? A meta-analysis across animal, plant and fungal hosts. Biological Reviews 89:123–134.

Kennedy, C.R. 2006. Ecology of the Acanthocephala. Cambridge University Press, Cambridge, UK.

Khalil, L.F., A. Jones, and R.A. Bray. 1994. Keys to the cestode parasites of vertebrates. CABI Publishing, Wallingford, UK.

Klompmaker, A.A., C.E. Schweitzer, R.M. Feldmann, and M. Kowalewski. 2015. Environmental and scale-dependent evolutionary trends in the body size of crustaceans. Proceedings of the Royal Society B: Biological Sciences 282:2015044.

Kortsch, S., R. Primicerio, M. Aschan, S. Lind, A.V. Dolgov, and B. Planque. 2019. Food-web structure varies along environmental gradients in a high-latitude marine ecosystem. Ecography 42:295–308.

Lagrue, C., K. Heaphy, B. Presswell, and R. Poulin. 2016. Strong association between parasitism and phenotypic variation in a supralittoral amphipod. Marine Ecology Progress Series 553:111–123.

Latham, A.D.M., and R. Poulin. 2002. Field evidence of the impact of two acanthocephalan parasites on the mortality of three species of New Zealand shore crabs (Brachyura). Marine Biology 141:1131–1139.

MacKay, R.N., and P.A. Moore. 2021. Parasites differentially impact crayfish personality in different contexts. Behaviour 158:921–943.

MacNeil, C., and J.T.A. Dick. 2011. Parasite-mediated intraguild predation as one of the drivers of co-existence and exclusion among invasive and native amphipods (Crustacea). Hydrobiologia 665:247–256.

Marcogliese, D.J. 2002. Food webs and the transmission of parasites to marine fish. Parasitology 124:S83–S99.

McAllister, C.T., H.W. Robison, and W.F. Font. 2011. Metacercaria of *Alloglossidium corti* (Digenea: Macroderoididae) from three species of crayfish (Decapoda: Cambaridae) in Arkansas and Oklahoma, USA. Comparative Parasitology 78:382–386.

McCurdy, D.G., M.R. Forbes, and J.S. Boates. 1999. Evidence that the parasitic nematode *Skrjabinoclava* manipulates host *Corophium* behavior to increase transmission to the sandpiper, *Calidris pusilla*. Behavioral Ecology 10:351–357.

Moore, J. 1983. Responses of an avian predator and its isopod prey to an acanthocephalan parasite. Ecology 64:1000–1015.

Moore, J. 2002. Parasites and the behavior of animals. Oxford University Press, New York, USA.

Moore, J., and J. Lasswell. 1986. Altered behavior in isopods (*Armadillidium vulgare*) infected with the nematode *Dispharynx nasuta*. Journal of Parasitology 72:186–189.

Mouritsen, K.N., L.T. Mouritsen, and K.T. Jensen. 1998. Changes of topography and sediment characteristics on an intertidal mud-flat following mass mortality of the amphipod *Corophium volutator*. Journal of the Marine Biological Association of the United Kingdom 78:1167–1180.

Mouritsen, K.N., D.M. Tompkins, and R. Poulin. 2005. Climate warming may cause a parasite-induced collapse in coastal amphipod populations. Oecologia 146:476–483.

Mouritsen, K.N., M.M. Sørensen, R. Poulin, and B.L. Fredensborg. 2018. Coastal ecosystem on a tipping point: Global warming and parasitism combine to alter community structure and function. Global Change Biology 24:4340–4356.

Orr, J.A., R.D. Vinebrooke, M.C. Jackson, K.J. Kroeker, R.L. Kordas, C. Mantyka-Pringle, P.J. Van den Brink, F. De Laender, R. Stoks, M. Holmstrup, C.D. Matthaei, W.A. Monk, M.R. Penk, S. Leuzinger, R.B. Schäfer, and J.J. Piggott. 2020. Towards a unified study of multiple stressors: Divisions and common goals across research disciplines. Proceedings of the Royal Society B: Biological Sciences 287:20200421.

Perrot-Minnot, M.-J., K. Sanchez-Thirion, and F. Cézilly. 2014. Multidimensionality in host manipulation mimicked by serotonin injection. Proceedings of the Royal Society B 281:20141915.

Poulin, R. 2006. Global warming and temperature-mediated increases in cercarial emergence in trematode parasites. Parasitology 132:143–151.

Poulin, R. 2010. Parasite manipulation of host behavior: An update and frequently asked questions. Advances in the Study of Behavior 41:151–186.

Pulkkinen, K., A.F. Pasternak, T. Hasu, and E.T. Valtonen. 2000. Effect of *Triaenophorus crassus* (Cestoda) infection on behavior and susceptibility to predation of the first intermediate host *Cyclops strenuus* (Copepoda). Journal of Parasitology 86:664–670.

Sari, A., and M. Malek. 2000. Occurrence of *Maritrema arenaria* (Digenea: Microphallidae) in the acorn barnacle, *Balanus perforatus* (Cirripedia: Balanidae) from south-west Wales. Journal of the Marine Biological Association of the United Kingdom 80:371–372.

Shields, J.D., and J.P. Huchin-Mian. 2020. Ecological factors in the emergence of pathogens in commercially important crustaceans. Pages 367–399 *in* G. Lovrich and M. Thiel, editors. The natural history of the Crustacea, Volume 9: Fisheries and aquaculture. Oxford University Press, New York, USA.

Studer, A., and R. Poulin. 2012. Effects of salinity on an intertidal host-parasite system: Is the parasite more sensitive than its host? Journal of Experimental Marine Biology and Ecology 412:110–116.

Studer, A., R. Poulin, and D.M. Tompkins. 2013. Local effects of a global problem: Modelling the risk of parasite-induced mortality in an intertidal trematode-amphipod system. Oecologia 172:1213–1222.

Sukhdeo, M.V.K. 2012. Where are the parasites in food webs? Parasites and Vectors 5:239.

Tain, L., M.-J. Perrot-Minnot, and F. Cézilly. 2006. Altered host behaviour and brain serotonergic activity caused by acanthocephalans: Evidence for specificity. Proceedings of the Royal Society B 273:3039–3045.

Thieltges, D.W., B.L. Fredensborg, and R. Poulin. 2009a. Geographical variation in metacercarial infection levels in marine invertebrate hosts: Parasite species character versus local factors. Marine Biology 156:983–990.

Thieltges, D.W., B.L. Fredensborg, A. Studer, and R. Poulin. 2009b. Large-scale patterns in trematode richness and infection levels in marine crustacean hosts. Marine Ecology Progress Series 389:139–147.

Thieltges, D.W., P.-A. Amundsen, R.F. Hechinger, P.T.J Johnson, K.D. Lafferty, K.N. Mouritsen, D.L. Preston, K. Reise, C.D. Zander, and R. Poulin. 2013. Parasites as prey in aquatic food webs: Implications for predator infection and parasite transmission. Oikos 122:1473–1482.

Thomas, F., A. Lambert, T. de Meeüs, F. Cézilly, and F. Renaud. 1995a. Influence of *Microphallus hoffmanni* (Trematoda, Microphallidae) on the survival, sexual selection, and fecundity of *Gammarus aequicauda* (Amphipoda). Canadian Journal of Zoology 73:1634–1639.

Thomas, F., F. Renaud, R. Rousset, F. Cézilly, and T. de Meeüs. 1995b. Differential mortality of two closely related host species induced by one parasite. Proceedings of the Royal Society B 260:349–352.

Torchin, M.E., K.D. Lafferty, and A.M. Kuris. 2001. Release from parasites as natural enemies: Increased performance of a globally introduced marine crab. Biological Invasions 3:333–345.

Turner, H.M. 2004. Seven-month survey of *Allocorrigia filiformis* (Trematoda: Dicrocoeliidae) infection in the crayfish *Procambarus clarkii*. Southwestern Naturalist 49:256–257.

Valkounova, J. 1985. Morphology and histochemistry of cysticercoids of four cestode species of the family Hymenolepididae Fuhrmann, 1907 from planktonic crustaceans (Copepoda, Ostracoda). Folia Parasitologica 32:333–340.

6

COMPETITION IN CRUSTACEANS: PATTERNS, MECHANISMS, AND CONSEQUENCES

Stuart R. Jenkins

Abstract

The study of competition in crustaceans spans many decades and has led to numerous advances in understanding how animals interact to secure limited resources, be they food, space, refugia, or access to a mate or host. This chapter considers different approaches to studying competition in crustaceans, from controlled experiments in mesocosms to models based on long-term, large-scale data sets, and introduces various model taxa, including barnacles, hermit crabs, and amphipods. Intraspecific and interspecific interactions are described that lead to changes in demography, behavior, morphology, and distribution, as well as consequences for evolutionary processes. In doing so, reference to both interference and exploitation competition is made. Crustacean research also provides insights into changing patterns of competition, principally due to global redistribution of taxa through human vectors. Notable global examples include the case of the shore crab *Carcinus maenas*, which has formed new populations around the globe since at least the 19th century. While true competition for a shared resource is the main focus here, apparent competition involving crustacean taxa is also described, whereby negative indirect interactions between species arise because they share a common enemy. This chapter brings together a diverse range of ecological work to demonstrate the pervasive influence of competition across crustacean taxa, from individuals through populations to communities and ecosystems. In addition, it shows that crustacean-focused research, and the adoption of particular model crustacean taxa, has had a significant impact in enhancing our understanding of competition in structuring populations, communities, and influencing the pathway of evolutionary processes.

Stuart R. Jenkins, *Competition in Crustaceans: Patterns, Mechanisms, and Consequences*. In: *Ecology and Conservation*. Edited by: Lars Gutow and Martin Thiel, Oxford University Press. © Oxford University Press (2025). DOI: 10.1093/oso/9780197768242.003.0006

INTRODUCTION

Competition for resources is a ubiquitous feature of life across all kingdoms. Crustaceans cover a wide range of body types and modes of life and occupy many different marine habitats. However, all require resources in order to grow and reproduce. Begon and Townsend (2020) define competition as "an interaction between individuals brought about by a shared requirement for a resource, and leading to a reduction in the survivorship, growth, and/or reproduction of at least some of the competing individuals concerned." Crustaceans compete for numerous resources including food (e.g., Pan et al. 2018), space (e.g., Jenkins et al. 2008a), a home or refuge (e.g., Jensen et al. 2002), and access to a mate (e.g., Jennions and Blackwell 1996) or host on, or in, which to live (e.g., De Bruyn et al. 2016). Winners in competition may be evidenced by a relative increase in growth or reproductive rate, by enhanced survival, or through such outcomes as may be inferred through indirect evidence such as retention of prized territory giving access to mates. In considering competition in crustaceans, this chapter takes a broad view across taxa and habitats. Its focus is predominantly marine, but also uses examples from freshwater and terrestrial environments to illustrate general points.

Characterization of Competition

Competition, whether between or within a species, can be usefully classified as either exploitation or interference. In exploitation competition, individuals interact indirectly through the exploitation of a limiting resource. Each individual is then affected by the amount of resource that remains after it has been exploited by others. For example, exploitation competition for phytoplanktonic food resources in the pelagic realm has been demonstrated among crustaceans both through experimental mesocosm-based manipulations (e.g., cladocerans; Pan et al. 2018) and in the wild (e.g., krill; Ryabov et al. 2017). In contrast to this indirect interaction, interference competition involves direct interactions: individuals, through interfering with others, prevent or limit access to resources. Such resources may hold real value and be in short supply, for example, barnacles competing for space through overgrowing or undercutting others. However, interference competition may also occur for resources that are essentially surrogates (e.g., a territory), which are valuable because they allow access to a real limiting resource. For example, fiddler crabs (genus *Uca*) typically live in crowded conditions where males defend burrows, thus securing shelter, which acts as a breeding resource (e.g., Christy 1982).

Many examples of competition in crustaceans include elements of both exploitation and interference and, in fact, they are likely in many cases to be inextricably linked. For example, hermit crabs fighting over shells in a highly ritualized behavior can be characterized as classic interference competition, but since use of shells makes this limiting resource unavailable for others, there is an important element of exploitation competition (e.g., Childress 1972, Bach et al. 1976). Subtleties of experimental design may also confuse both types of competition. Laboratory experiments examining interactions among individuals competing for prey (e.g., individuals of the shore crab *Carcinus maenas* competing for access to the blue mussel *Mytilus edulis*), which do not maintain prey at constant densities, may be demonstrating exploitation competition, when in fact reductions in feeding rate are more likely due to interference competition (Smallegange et al. 2006).

Other forms of competition have been described that do not fit within this basic categorization. Exploitation and interference competition can be considered "real" competition in the sense that competitors share a common resource. Apparent competition (see more details for this topic the respective section in this chapter), a term coined by Holt (1977), denotes negative indirect interactions between species that arise because they share a common enemy, be that a predator, parasite, or pathogen. For example, changes in the abundance of a prey species will have knock-on consequences for predator numbers or performance, which leads to changes in the abundance of other prey that share the same predator. Depending on the particular scenario, this may lead to symmetrical apparent competition, whereby each prey species has a negative effect on the other, or asymmetrical apparent competition where the negative effect between prey species is predominantly one-sided.

Approaches to Studying Competition in Crustaceans

A huge variety of approaches to examining competitive interactions in crustaceans have been adopted, from assessment of staged encounters in laboratory mesocosms between hermit crabs with manipulated shells (Briffa et al. 1998) and binocular-based observation of intertidal fiddler crabs on mud flats (Jennions and Blackwell 1996) to analysis of long-term data of krill dynamics in the Southern Ocean (Ryabov et al. 2017). Tightly controlled experiments in the laboratory (e.g., Steiner et al. 2007) or in the field (e.g., Roberts and Laidre 2019) can give unambiguous evidence of competition and insight into the mechanistic basis of competition, including identification of the limiting resource. Observational studies, for example, of the dynamics of potentially competing species (e.g., Ólafsson and Persson 1986), or of behavioral interactions over the short term (e.g., Elwood and Arnott 2007) can provide insight and inferential evidence that can be enhanced through modeling approaches (e.g., Marks et al. 2020).

Use of laboratory mesocosms allows manipulation of limiting resources and detailed observation of both short-term (e.g., behavior: Abrams 1987; physiology: Thorpe et al. 1995) and longer term (e.g., growth: Pan et al. 2018; mortality: Amaral et al. 2009) responses. Laboratory approaches also allow experimental assessment of competition in taxa from habitats, such as the open water column, where experimental manipulation in the field is challenging (e.g., interspecific competition among cladocerans; Dzialowski and O'Brien 2004, Pan et al. 2018). Implementation of the experimental approach in the field provides more realistic physical and biotic context, and there is a rich history of the use of crustaceans in tractable intertidal environments to address the presence, strength, and direction of competition. Intertidal rocky shores feature heavily in the scientific literature, with both sessile (e.g., barnacles; Connell 1961, Bertness et al. 1998, Jenkins et al. 2008a) and mobile crustaceans (e.g., crabs; Iribarne et al. 1994, Moksnes 2004a,b) being used to advance fundamental understanding of the role of competition in establishing population and community structure. In intertidal soft-sediment systems, investigation of competitive interactions between the polychaete worm *Hediste* (previously *Nereis*) *diversicolor* and the tube-forming amphipod *Corophium volutator* have made important contributions to the understanding of soft-sediment ecology (e.g., Ólafsson and Persson 1986). Terrestrial crustaceans such as terrestrial hermit crabs have also been the subject of study (e.g., *Coenobita rugosus* and *C. perlatus*; Steibl and Laforsch 2020). Subtidal experimental work is less common owing to obvious logistical challenges. Examples include the work of Steneck (2006) and Hovel and Wahle (2010), who manipulated the availability of refugia to examine intraspecific shelter competition in the American lobster *Homarus americanus*. In the pelagic realm, studies typically rely on observations and modeling. This approach enables work over large temporal and spatial scales. For example, Ryabov et al. (2017) used a combination of observations and modeling over five degrees of latitude and a 20-year time scale to examine intraspecific competition for food in Antarctic krill (*Euphausia superba*).

A number of crustacean taxa feature prominently in the ecological literature as model organisms in assessments of competition in the marine realm. These model taxa have developed as a consequence of their abundance and ubiquitous distribution (e.g., barnacles, amphipods, crabs), the ease with which they can be manipulated in the field (e.g., barnacles) or kept under laboratory conditions (e.g., hermit crabs, amphipods), and at times because of their commercial status (e.g., crabs, lobsters). Barnacles, in particular, are extremely well studied. Their abundance both as adults and larvae, their sessile nature, and the intertidal distribution of many species means they have been a focal taxon in thousands of studies addressing fundamental ecological processes in marine benthic invertebrates. One of the most prominent studies is that of Connell (1961), which features frequently in ecological textbooks as a field example of interspecific competition (expanded on below). Connell (1961) makes the point that intertidal barnacles are ideal model organisms with which to study survival (and by implication competition). Their sessile nature and intertidal habit mean they can be easily observed and mapped during low water periods, they occur in dense aggregations, and they can be easily transplanted. Barnacles have also developed as a model examining the consequences of variation in larval supply to the dynamics of adult populations (Jenkins et al. 2008a). Here, the presence and intensity of intraspecific competition is key to understanding whether the spatial and temporal variability in settlement from the plankton has a bearing on adult distribution and abundance.

Hermit crabs are another excellent example of a common crustacean that has become a highly effective model with which to investigate competition for limited resources, both within and among species. In nature, organisms are rarely limited by a single limiting resource. Thus, examination of competition for resources needs to take account of a multitude of abiotic and biotic variables, which may limit the fitness of a species and, hence, ultimately its population abundance. However, evidence suggests that hermit crabs in many habitats are only limited by a single resource, the availability of suitable shells in which to live (Vance 1972, Abrams 1980). As crabs grow, they need to change shells; individuals in shells that are too small show increased risk of mortality, reduced growth rate, and, in females, reduced fecundity. A shell that is too large brings the energetic costs of carrying it. Because shells are generally a limiting resource, and therefore unoccupied, attractive shells are rare or absent, hermit crabs have evolved complex ritualized shell fighting behavior to compete for already occupied shells (Hazlett 1966). Staged encounters between crabs in which the motivation of competitors is artificially manipulated are relatively straightforward to implement (e.g., Briffa et al. 1998) and allow a variety of indicators of competition, including short-term physiological status (Briffa and Elwood 2004), as well as longer term indicators of survivorship and/or fecundity (Bach et al. 1976), to be measured.

Amphipods provide numerous examples of crustacean competition, both among amphipod species (e.g., van Riel et al. 2007, Shucksmith et al. 2009, Beermann et al. 2018) and with unrelated taxa (e.g., with a polychaete worm; Ólafsson and Persson 1986, with a pulmonated snail; Wellborn 2002). They provide a tractable taxon for experimental assessment of competition in laboratory mesocosms (e.g., Dennert 1974, van Riel et al. 2007) and occur in sufficient abundance in many coastal habitats to allow direct observations to inform conclusions regarding the role of interspecific competition in coexistence (e.g., Lancellotti and Trucco 1993). In addition, particularly prominent examples of invasive amphipods such as the "killer shrimp" *Dikerogammarus villosus* and the Japanese skeleton shrimp *Caprella mutica* have provided the opportunity to understand the role of competition (e.g., via direct agonistic interactions) in successful invasion (Shucksmith et al. 2009, Rewicz et al. 2014).

INTRASPECIFIC COMPETITION

Individuals of the same species have the same resource requirements. Hence, if these resources are in short supply, then competition is a natural outcome, with consequences for the individual and for populations. Intraspecific competition in crustacean taxa has been examined from numerous perspectives, providing insight into density-dependent processes regulating population size, as well as behavioral and morphological responses.

Competition as a Determinant of Demography

Crustaceans, particularly barnacles, have proven an effective model organism to examine the role of intraspecific competition in determining population size in marine benthic invertebrates in general (e.g., Menge 2000, Jenkins et al. 2008a). The majority of marine benthic invertebrates have a complex life cycle that includes separate planktonic and bottom-dwelling phases linked by a settlement event (Thorson 1950). Larvae released into the water column by benthic adults are subject to a range of physical transport processes (see Pineda and Reyns [2018] for review), probably high levels of predation (Young and Chia 1987), and potentially competition for limiting resources (Olson and Olson 1989). Subsequent settlement to the adult habitat results in variation in recruitment over a number of spatial and temporal scales across crustacean taxa, including barnacles (Kendall et al. 1982, Jenkins et al. 2000), crabs (Botsford 2001), and lobsters (Katz et al. 1994).

Recognition of the huge temporal and spatial differences in the level of recruitment has led to questions regarding the consequences of such variability to local population dynamics and hence interest in the circumstances where density-dependent processes (predominantly competition) modify population patterns established at settlement. Pioneering work on rocky shore communities (e.g., Connell 1961, Southward 1964, Paine 1974) described populations as saturated and stable, controlled by density-dependent post-settlement processes such as competition and predation. An alternative view came to the fore as variability in recruitment was recognized. This view states that larval supply is insufficient for the total population size to reach a carrying capacity determined by resource levels (see Doherty and Williams 1988). In this situation, abundance and age structure should reflect variation in input of new recruits and be relatively independent of post-recruitment processes.

Several studies have used observations of barnacle settlement and subsequent survival to address these questions (e.g., Sutherland 1990, Menge 1991, Jeffery 2003). However, manipulative experiments provide greater insights. For example, Menge (2000) used transplant experiments to manipulate both the density of recently recruited barnacles and the environmental conditions. He found that tidal height (a proxy for physical conditions) mediated the impact of competition on adult abundance. Jenkins et al. (2008a) used a novel approach that recognized the potential confounding of numerous environmental variables when observing differences in the settler-survivor relationship across multiple sites, a phenomenon termed "cryptic density dependence" by Shima and Osenberg (2003). They identified an area in North Wales, UK, with predictably high recruitment of the barnacle *Semibalanus balanoides*. This species is a "pulse recruiter"; it releases larvae in the springtime to coincide with the spring phytoplankton bloom and, as a consequence, there is a short 6–8-week recruitment window (= pulse) when larvae settle to the shore. Jenkins et al. (2008a) cleared multiple areas at two shores on the west coast of the Isle of Anglesey, allowed settlement to occur, and then manipulated the density of recruits from very low to very high across 52 quadrats at each of the two shores. They then followed these patches over a two-year period using digital photography to determine growth and survival. They found that in this intertidal barnacle, recruit density was key to determining the importance of competition and hence the role of post-settlement processes in determining adult abundance. At low densities (up to approximately 10 settlers cm^{-2}) limited competition meant that adult abundance was a linear function of settlement density. Settler densities between approximately 10 and 25 individuals cm^{-2} resulted in strong competitive interactions such that adult abundance showed no increase with increasing settler number. At very high densities, adult abundance actually declined with increasing settlement (Fig. 6.1). In that work it was not possible to determine definitively the resource competed for. There is likely an element of exploitation competition as filter-feeding organisms compete for food in the boundary layer (Bertness et al. 1998), but at high densities, particularly in the early months after recruitment, interference competition for space is likely the dominant process.

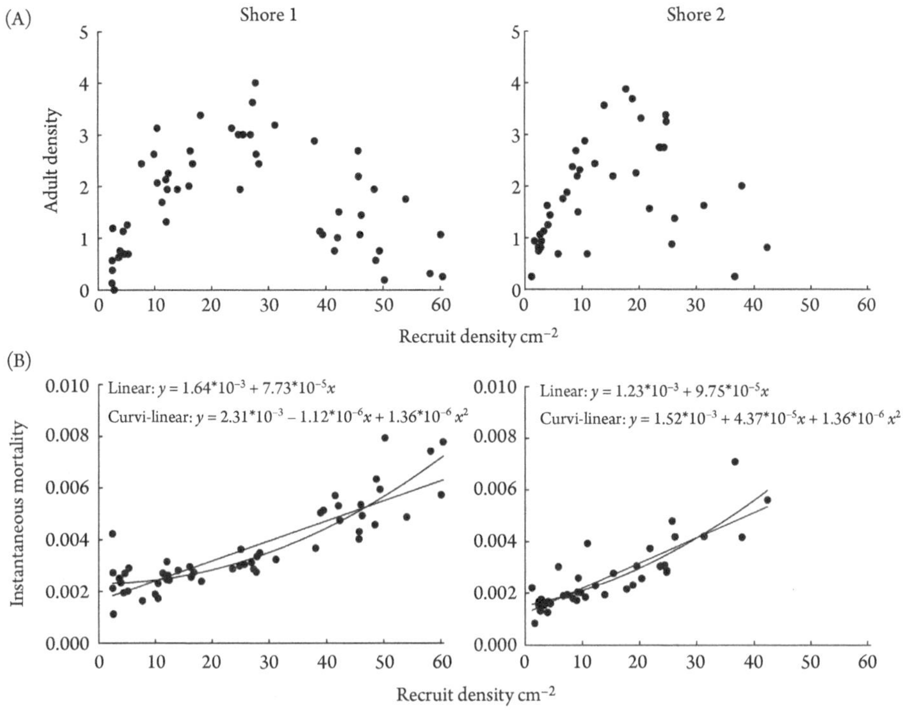

Fig. 6.1.
The relationship between the density of manipulated recruits of the barnacle *Semibalanus balanoides* in June 2002 on two shores of Anglesey, North Wales, UK, and (A) the density of surviving adults two years later; (B) the instantaneous mortality rate over the two-year period. On both shores, intense intraspecific competition for space led to strong density-dependent mortality and, consequently, an obvious hump-shaped relationship between recruitment and adult density. Reproduced from Jenkins et al. (2008a) with permission from © The British Ecological Society.

Assessing density-dependent growth and mortality in the field as a consequence of intraspecific competition for limiting resources in sessile crustaceans such as barnacles, particularly in the intertidal zone, can be relatively straightforward, but in mobile crustaceans, whether benthic or pelagic, it is more challenging. Assessment of age in many invertebrate taxa with hard body structures can be achieved using growth lines, which are formed on an annual or daily basis. Thus, growth can be described by fitting a model directly to length-at-age data (e.g., Oróstica et al. 2021). However, crustaceans do not retain any hard body structures during molting (Hartnoll 2001), and aging is reliant on direct observation of captive individuals, tagging of individuals, or modal progression (or length cohort) analysis through the fitting of growth models to modes in length-distribution data. Marks et al. (2020) used the latter approach to explore density-dependent growth effects in the commercially fished blue swimmer crab *Portunus armatus*. They exploited an 18-year data set from Western Australia showing a change from the second-largest commercial fishery for *P. armatus* in the late 1990s to recent periods of extremely low abundance and fishery closure. Substantial variation in mean size at age over those 18 years was observed along with crab densities that showed up to 60-fold and 270-fold variation in the 0+ and 1+ cohorts, respectively. Models indicated that the mean size of adults at 1.5 years was negatively related to their density as juveniles (0.5 years age). This, combined with positive relationships between juvenile abundance and mean summer chlorophyll concentration in the same year, led to the conclusion that high recruitment as a consequence of water column food availability leads to strong density-dependent competition among juveniles and a subsequent reduction in growth. This is consistent with the outcome of investigations into the blue crab *Callinectes sapidus* (Pile et al. 1996) and the shore crab *Carcinus maenas* (Moksnes 2004a).

High densities of successful crustacean species can lead to extreme levels of density-dependent competition. Successful recruitment in *C. maenas,* an epibenthic omnivore widely distributed across both soft and hard intertidal and shallow subtidal habitats, can lead to extremely high densities of juveniles, particularly in structurally complex habitats. For example, Moksnes (2002) reported densities of up to 500 crabs m^{-2} in mussel beds in Sweden. Investigations in the laboratory and field have demonstrated that such elevated densities can lead to intense interference competition for space, manifested through cannibalistic behavior, both among juveniles (e.g., Moksnes 2004b, Marshall et al. 2005) and size classes (Amaral et al. 2009). Moksnes (2004b) argued that such extreme manifestations of intraspecific competition can create a demographic bottleneck in many marine crustaceans (e.g., Wahle and Steneck 1991, Beck 1997) and can play an important role in the regulation of population sizes not only in *C. maenas* but also in many brachyuran crab fisheries including the blue crab *Callinectes sapidus* (Hines and Ruiz 1995), the stone crab *Homalaspis plana* (Fernández and Castilla 2000), the Dungeness crab *Cancer magister* (Botsford and Hobbs 1995), and the snow crab *Chionoecetes opilio* (Sainte-Marie and Lafrance 2002). For example, Fernández and Castilla (2000) considered cannibalism among juvenile stone crabs in high-quality habitat to be a contributory factor in the failure of Marine Protected Areas in Chile to enhance the abundance of stone crabs in comparison to other exploited benthic species.

In the pelagic environment certain crustacean taxa dominate, and understanding the dynamics of their populations can be critical to understanding ecosystem level patterns and changes in response to global change. This is certainly the case for Antarctic krill, *Euphausia superba,* which occupies a key position in Southern Ocean food webs, where it acts as a direct link between primary producers and apex predators. Attention has focused over a number of decades on the mechanistic basis for large (over an order of magnitude) interannual variation in its abundance over a variety of scales. Until recently, this variation was considered a consequence of periodic climatological factors such as the El Niño-Southern Oscillation index (Ross et al. 2014), or anomalies in primary production (Steinberg et al. 2015) or sea ice duration (Atkinson et al. 2004). Ryabov et al. (2017) pointed out logical inconsistencies in cycles of abundance solely being driven by the environment. For example, climatic factors cannot explain the observed negative effect of krill biomass on reproduction. In order to examine the underlying causes, Ryabov et al. (2017) constructed and tested a bioenergetic model of Antarctic krill for the West Antarctic Peninsula, which makes the critical assumption that all krill stages compete for their phytoplankton food resource from spring to autumn. Model outputs are closely aligned with observations and lead to the conclusion of strong intraspecific exploitation competition for limiting food resources, predominantly in the autumn, driving the patterns in krill abundance through time (Fig. 6.2). These conclusions have important implications for future changes in the entire Antarctic ecosystem, with predicted outcomes differing depending on, among other things, the level of krill mortality through either human exploitation or natural predation from recovering whale populations.

Behavioral and Morphological Responses to Intraspecific Competition

In mobile crustaceans, migration may be a more common response to intraspecific competition than changes in growth, reproductive output, or mortality (Ambrose 1986, Iribarne et al. 1994). Iribarne et al. (1994) used field experiments to determine the consequences of high recruitment to intertidal estuarine shell habitats in the Dungeness crab *C. magister*. They manipulated the density

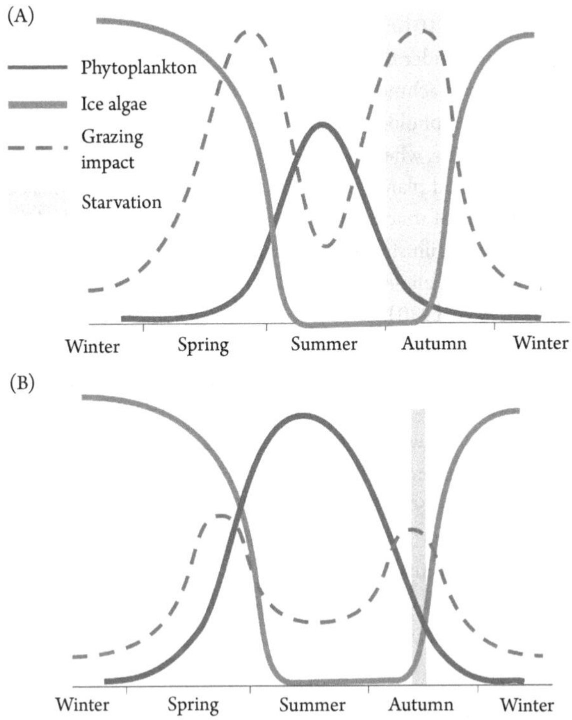

Fig. 6.2.
Graphical representation of seasonal changes in phytoplankton (dark gray), ice algae (light gray), and krill grazing impact (dashed line) during periods of (A) large and (B) small krill biomass. When the total krill biomass is large, autumn becomes the main bottleneck for population survival (indicated by the starvation period (gray bars)). Reproduced from Ryabov et al. (2017) with permission from © Springer Nature; see a color version of this figure in the centerfold.

of juvenile conspecifics while also controlling the level of food availability and conspecific physical contact. They found that crabs emigrated from high-density treatments irrespective of food availability. Observations and previous work of Jacoby (1983) had suggested that although *C. magister* juveniles do not hold a fixed territory, they defend their own individual space, which serves to regulate, through interference competition, overall density in this particular habitat. Such agonistic interactions, if not alleviated through emigration, can lead to cannibalism (Karpov 1983, Fernandez et al. 1993). The results of Iribarne et al. (1994) also suggested that emigration may be a response to avoid competition for food; at high densities (100 crabs m^{-2}) the per capita intake of clams was much reduced compared to low densities (25 crabs m^{-2}). Competition-induced emigration may lead to the redistribution of individuals into suboptimal habitats not normally occupied. For example, Ambrose (1986) speculated that density-dependent emigration of the amphipod *Rhepoxynius abronius* from the sediment into the water column may explain their occurrence in this environment.

Intense competition for space among crowded sessile crustaceans such as barnacles can lead to extreme forms of morphological modification. Intense settlement of barnacles can lead to a phenomenon known as hummocking (Barnes and Powell 1950), whereby vertical growth predominates since lateral expansion is limited. Bertness et al. (1998) described individuals of *Semibalanus balanoides* in hummocks to be 4–10 times taller than they are wide (in contrast to solitary barnacles, which are generally 2–4 times wider than tall). Hummocks of barnacles consist of elongated individuals with very little area of attachment, tightly crowded together. Bertness et al. (1998) made a direct analogy between the morphological modification in barnacles as a consequence of competition for space and that in plants, where crowded seedlings become tall and etiolated in response to low light levels. Like etiolated plants, such barnacles are structurally weak and dependent on neighbors for support. The lack of attachment strength can lead to wholesale detachment and hence mortality following disturbance, illustrating the consequence of intense intraspecific competition. However, there may be positive benefits resulting from such morphological modification (Wethey 1984, Bertness et al. 1998, Wu 1980). For example, changes in morphology can actually lead to enhanced feeding success. Bertness et al. (1998) showed that barnacles at the center of hummocks had enhanced particle capture success, presumably because these individuals are elevated above the surface and exposed to higher fluid velocities and hence particle fluxes (Fig. 6.3). Also, the structural support provided by crowded individuals means that less investment in shell structure is required and, consequently, hummocked individuals have been shown to invest greater resources in tissue growth (Bertness et al. 1998). Barnacle crowding can also increase survivorship by buffering individuals from thermal stress (Bertness 1989) and lead to increased reproductive output. For example, Wethey (1984) showed that in *S. balanoides* and *Chthamalus dalli*, columnar individuals produced larger clutches of eggs than conical uncrowded individuals of comparable somatic tissue weight.

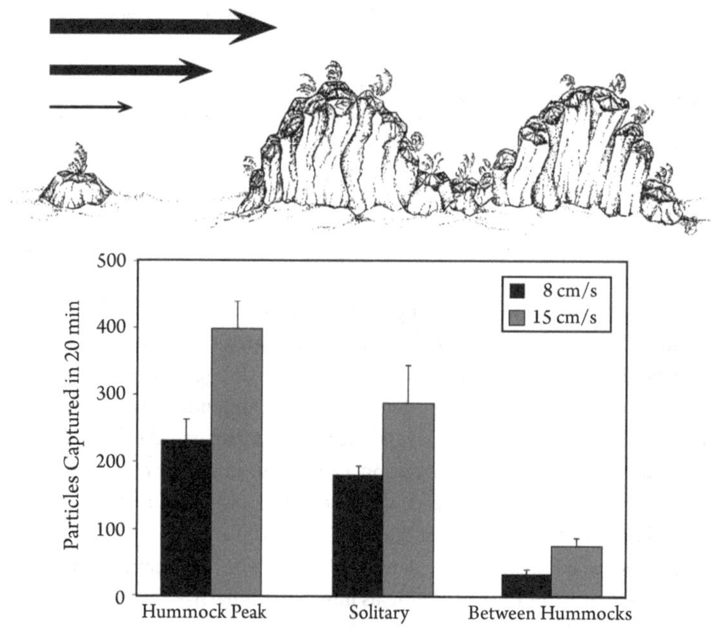

Fig. 6.3.
Feeding rates of solitary and hummocked barnacles *Semibalanus balanoides* at different flow rates in a recirculating flow tank. The effect of crowding on particle capture depends on position within the hummock, with a potentially positive effect of crowding for barnacles located centrally within a hummock. Adapted from Bertness et al. (1998) with permission from © John Wiley & Sons.

INTERSPECIFIC COMPETITION

Competition among crustacean species, or between crustaceans and other taxa, has received considerable attention in both pure and applied contexts. Some of the earliest experimental work demonstrating interspecific competition in the natural environment and its role in determining species distribution focused on Crustacea, and work on a range of crustacean taxa has allowed considerable insight into the evolutionary consequences of competition. The human-mediated transport of crustaceans and consequent prevalence of crustacean invasive species has led to an important body of work determining the mechanisms by which invasive species successfully establish in a novel environment.

Impacts on Species Distribution

Crustaceans have provided some of the most iconic examples of interspecific competition in the marine realm, most notably that of Joe Connell's work on intertidal barnacles on the Isle of Cumbrae, UK. Connell (1961) pointed out that, while at that time evidence existed for interspecific competition, it was drawn predominantly from laboratory populations, and hence its applicability in the natural environment was unknown. Thus, many ecologists viewed competition as occupying a minor role in determining the structure and composition of communities. Connell's work began with observations that two species of barnacles occupied separate zones on the rocky intertidal shore with only limited overlap. *Chthamalus stellatus* adults were limited to the upper part of the shore while *Balanus* (now *Semibalanus*) *balanoides* dominated lower down. Critically, *C. stellatus* settled on the low shore but did not survive there. Connell implemented a series of experiments to test the hypothesis "that the absence in the lower zone of adults of *Chthamalus* was due to interspecific competition with *Balanus* for space." Competition was assessed by following individuals of *C. stellatus* through time, in the presence and manipulated absence of *B. balanoides*. This was achieved at a range of tidal levels by transplanting rocks. Detailed observations of overgrowth, undercutting, and crushing by *B. balanoides* on *C. stellatus*, combined with very clear differences in its survival in the presence and absence of *B. balanoides*, provided evidence of interspecific competition for space at mid to low shore levels. Above this zone on the shore, where adults of both species naturally occur, no influence of *B. balanoides* on *C. stellatus* was observed. Here, the lower resistance of *B. balanoides* to desiccation (Foster 1971) meant its competitive advantage was nullified. Connell's work indicated that the lower limit of the *C. stellatus* distribution was determined by interspecific competition for space with *B. balanoides*: rapid growth of *B. balanoides* effectively eliminated *C. stellatus* by crowding.

Influential work from the 1970s onwards in soft-sediment habitats also demonstrated the role of interspecific competition in crustaceans in determining species distributions. In soft-sediment environments, the mechanisms outlined by Connell through which the dominant competitor exerted its effect, namely physical crowding, are unlikely to play such an important role. The three-dimensional nature of soft sediments, where space can be partitioned both vertically and horizontally, provides more potential to avoid direct competitive interactions. In fact, Wilson (1991) used this argument to suggest that in soft-sediment habitats, indirect competition through exploitation of limiting resources may be more likely on average than interference competition.

Probably the best-studied competitive interaction in infaunal biota is that between the amphipod *C. volutator* and the polychaete worm *H. diversicolor*. These are widely distributed intertidal infauna found in high abundance in estuarine and coastal mudflats where densities greater than 100,000 m^{-2} (Gerdol and Hughes 1993) and 50,000 m^{-2} (Möller 1985), respectively, have been reported. *Corophium volutator* are predominantly unselective deposit feeders (Gerdol and Hughes 1994a,b), although suspension feeding and epipsammic browsing may also occur. They occupy U-shaped burrows up to 5 cm deep (see Chapter 4 in this volume; Fig.4.2B) and form a major prey item for coastal seabirds and inshore fish (Wilson and Parker 1996). *Hediste diversicolor* has multiple modes of feeding, acting as a suspension feeder, surface deposit feeder (Fauchald and Jumars 1979, Reise 1979, Riisgård 1991, Jensen and André 1993), and scavenger/predator. It occupies semipermanent "J"-shaped or "U"-shaped burrows. *Corophium volutator* and *H. diversicolor* are sympatric species, but observations of local distributions indicate some form of negative relationship, with some suggestions that *Hediste* prey on *C. volutator* (e.g., Rönn et al. 1988). However, most evidence points to some form of interference (Ólafsson and Persson 1986) and potentially exploitation (Smith et al. 1996) competition. For example, Smith et al. (1996) demonstrated that interspecific competition for diatoms could be important under some natural conditions. Ólafsson and Persson (1986) made long-term observations over a period of three years of distributional patterns of *Hediste* and *Corophium*. They showed that the long-term local spatial separation of these two species (i.e., distinct *Hediste*-dominated and *Corophium*-dominated patches) was modified following a reduction in *Corophium* densities. This reduction led to an increase of *Hediste* in the area previously occupied by high levels of *Corophium*, suggesting that previous assumptions of a predatory relationship between the two species (e.g., Commito 1982) may be an oversimplification. Both Ólafsson and Persson (1986) and Jensen and André (1993) found that *Hediste* was an ineffective predator of *Corophium* and, thus, the local spatial separation of these sympatric organisms is unlikely a consequence solely of predation. Laboratory investigations prompted by these observations showed that *Hediste* burrowing activity frequently destroyed *Corophium* tubes. Although a negative effect of *Corophium* on *Hediste* was not revealed experimentally, Ólafsson and Persson (1986) consider that adult *Corophium* are probably able to limit *Hediste* adult densities through negative effects on recruitment.

Interactions between already established adults and newly settling larvae may have an important impact on the structure of invertebrate communities of soft sediments (Woodin 1976, Peterson 1979). Young settlers across a range of phyla are vulnerable not only to predation and accidental ingestion by deposit feeders, but also to interference and possibly exploitation competition from established macrofauna and meiofauna. However, it can be difficult to distinguish between predatory relationships and interference competition (e.g., the interaction between the common bivalve *Macoma balthica*, and the amphipod *Pontoporeia affinis*; Elmgren et al. 1986). Amphipods are clearly effective in reducing infaunal densities of other species through both burial and ingestion of larvae and juveniles (Oliver et al. 1982, Elmgren et al. 1986). Both Ambrose (1984) and Commito (1982) found a negative correlation between high densities of *C. volutator* and other infaunal species, suggesting that *Corophium* in some way inhibit their recruitment.

Evolutionary Consequences of Interspecific Competition

Debate over the role of interspecific competition in driving divergence in resource usage and hence niche differentiation has a long history dating back to Darwin's "The Origin of Species." Such a process leads naturally to a reduction in the strength of competition over time and facilitates the coexistence of species. The absence of contemporary evidence for competitive processes among species could then be a consequence of historical competition, which has waned through time as species have diverged. Weak or absent competitive interactions could then be considered "ghosts of competition past" (Connell 1980). It is worth noting that in defining this iconic phrase, Connell was arguing that there was a lack of evidence for competition driving coevolutionary divergence. Since his paper, however, there has been ample theoretical evidence to suggest competition as a coevolutionary selective process (e.g., Taper and Case 1985, Abrams 1987).

Providing direct empirical evidence for the decline in competitive interactions over time between coevolving species is challenging. Steiner et al. (2007), however, used a novel approach to examine this question, and in doing so provide an excellent example of the use of crustacean taxa to advance ecological/evolutionary understanding of competition. They take advantage of the fact that many zooplankton taxa produce diapausing eggs, a stage during which development is suspended, often during unfavorable environmental conditions. Such eggs can remain dormant for decades. They used diapausing eggs of three species of *Daphnia,* extracted from sediment cores of Sportsman's Lake, Illinois, to resurrect clones from different periods of community assembly since the lake's formation in 1953. Mesocosm manipulations of species identity and diversity were made to assess competitive interactions. Their results matched theoretical predictions: the clones of *Daphnia ambigua* isolated from early in the lake's assembly experienced strong competition with the current dominant species (*Daphnia dentifera*) and were driven to extinction over the period of the mesocosm experiment. In contrast, clones of *D. ambigua* retrieved from later in the lake's history fared well. Thus, the strength of competitive interactions declined through time, therefore supporting the role of competition in facilitating coexistence among different species (Fig. 6.4).

More indirect approaches to demonstrating the role of past competition include comparisons of competitive processes between sympatric and allopatric populations (reviewed in Schluter 2000, Dayan and Simberloff 2005), and inferences based on resource partitioning and the balance

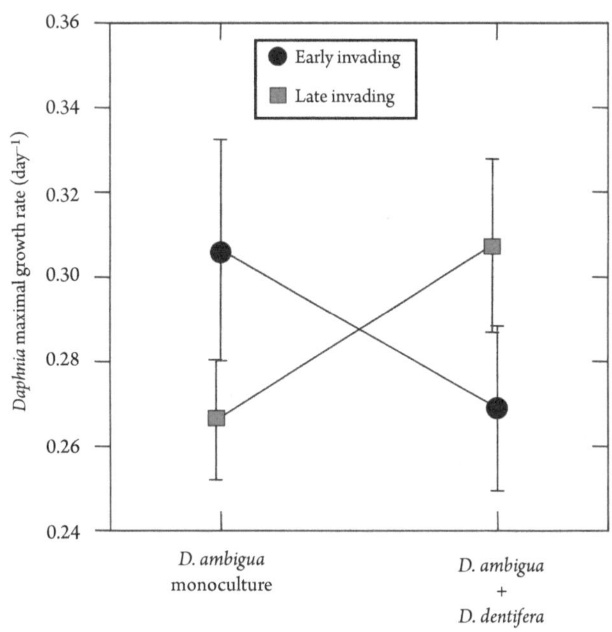

Fig. 6.4.
Outcome of competition experiments among and between different clones and species of *Daphnia* using diapausing eggs from different periods of community assembly in a freshwater lake in Illinois, USA. Population growth rates of early invading clones of *D. ambigua* were high in monoculture but they performed much worse in competition with *D. dentifera* (current lake dominants) than later invading forms, supporting the conclusion that natural selection favors reduced interaction strength among co-occurring species, facilitating coexistence and population persistence. Source: data from Steiner et al. (2007).

between intraspecific and interspecific competition. Inferences based on resource partitioning can be problematic because most taxa are typically limited by a multitude of biotic and abiotic factors. In this sense hermit crabs provide a useful model. Evidence suggests that many hermit crabs that co-occur are purely limited by the availability of a shell in which to live and are not limited by food or habitat conditions (e.g., Fotheringham 1976, Hazlett 1981). That shells are limiting is evidenced by observations in a number of different hermit crab habitats suggesting there are very few available empty shells (Provenzano 1960, Hazlett 1970, Ball 1972). Steibl and Laforsch (2020) investigated the preferences of two terrestrial hermit crab species, *Coenobita rugosus* and *C. perlatus*, which co-occur in the same beach habitat. Rather than focusing on preference for different species of gastropod shells, which can confound preference with availability, they assessed preference for particular shell parameters using shell morphometric data across multiple hermit crab and gastropod populations in the field and compared this with laboratory shell preference experiments. They found that the two species showed intrinsic preferences toward two distinct shell shapes (a short and globose shell morphology in *C. rugosus*, and an elongated shell morphology with narrow aperture in *C. perlatus*). This led to the conclusion that they were not in competition for limited shell resources but have evolved divergent preferences, which allows co-occurrence. Similar conclusions were reached by Abrams' (1987) work using a combination of field and laboratory observations on the relative magnitude of intra- and interspecific competition among three sympatric hermit crabs, *Pagurus hirsutiusculus*, *P. granosimanus*, and *P. beringanus*. Abrams found that intraspecific competition was approximately an order of magnitude more intense for all species than interspecific competition. These patterns were a consequence of a combination of species-specific shell shape/size preferences and differences in habitat use. Abrams used models of evolutionary responses based on a subtidal hermit crab assemblage (Abrams et al. 1986), combined with observations and inference, to conclude that the extremely low levels of observed competition among these sympatric species are a consequence of evolutionary responses to previous interspecific competition.

Changing Patterns of Interspecific Competition in Crustaceans

The distribution of crustaceans is changing as a consequence both of global climate change and transport through human vectors. With warming temperatures comes a gradual shift of taxa poleward; differences in responses among species can mean differential rates of movement and, consequently, the potential for interactions between novel competitors. Although this process is likely significant in changing the identity of competitors, it is the introduction of nonnative species through vectors such as hull fouling, ballast water, and movement of aquaculture products, that has produced the most profound changes in competitive interactions involving crustacean taxa. The lack of coevolutionary history between introduced and resident species means competitive interactions may be particularly strong. Invasive nonnative species, that is, those that colonize and then reach high population size, may achieve success through their ability to outcompete native species for resources (Parker et al. 2013).

Invasive crabs in intertidal and shallow subtidal habitats provide numerous examples of new competitive relationships. One of the most well-known examples involves global transport dating back to at least the 19th Century. The shore crab *C. maenas* is native to Atlantic Europe and has been widely introduced to North America and Australia, and more recently to South Africa, South America, East Asia, and elsewhere (Carlton and Cohen 1995). It was introduced to the East Coast of North America, probably around the start of the 19th century, but introductions to new areas have continued and it only arrived on the Pacific coast of North America in the latter part of the 20th century, being found in San Francisco Bay, California, in 1989. Measurable impacts on intertidal community structure on the East Coast over the last 200 years (see Jenkins et al. 2008b), and recent evidence for competitive dominance over other crustaceans, including species of commercial importance (e.g., over juvenile and subadult lobsters; Rossong et al. 2006, Williams et al. 2006), led to understandable concern among ecologists, resource managers, and the aquaculture industry following its arrival and expansion on the West Coast (Jensen et al. 2002). Here, experimental work suggests interactions with the native Dungeness crab *C. magister*, a species of commercial importance, have the potential to impact the fishery (McDonald et al. 2001). Laboratory experiments and infrared video observations of juvenile interactions revealed displacement of *C. magister* from refugia (Fig. 6.5) and domination by *C. maenas* in competition over food.

Despite its obvious success, *C. maenas* may not always dominate. Jensen et al. (2002) examined competitive interactions between the recently introduced *C. maenas* and the native grapsid crab *Hemigrapsus oregonensis* in Bodega Bay, California. Observations of *C. maenas* distribution in the presence and absence of grapsids, and use of lab trials, suggest, perhaps surprisingly for such a dominant invader, that *C. maenas* was outcompeted for limiting rock/shell shelter by *H. oregonensis*. Although lab trials examining competition for food showed *C. maenas* to dominate the native *H. oregonensis*, these mixed results suggest that the global success of *C. maenas* as an invader is not always simply a consequence of its competitive status, and that competition with native crabs, including *Cancer* spp. will limit their impacts and spread (Jensen et al. 2007).

Interestingly, in its natural range *C. maenas* may be under competitive pressure from invading species of *Hemigrapsus*. Both *H. takanoi* and *H. sanguineus* have expanded their ranges rapidly within Europe since arriving in the 1990s (Geburzi et al. 2018), with their success, potentially at

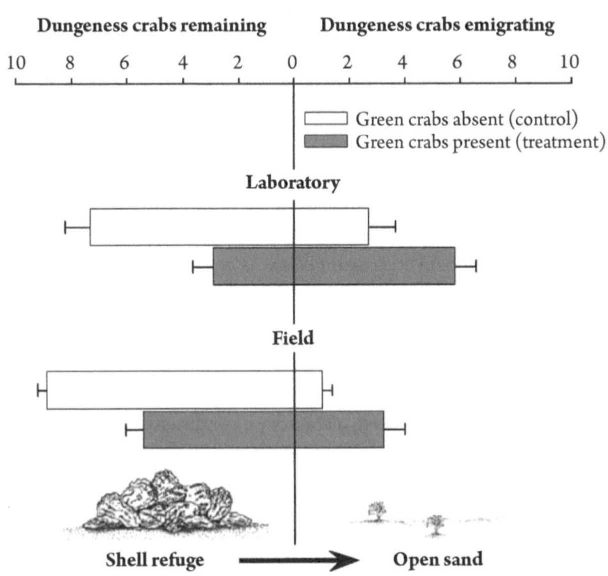

Fig. 6.5.
Impact of the nonnative European green crab *Carcinus maenas* on refugia occupancy by early benthic phase individuals of the native Dungeness crab *Cancer magister* from coastal estuaries of Washington State, USA. In both laboratory and field experiments, juvenile green crab displaced Dungeness crab of equal size from shelters during one-on-one competition. Loss of access to intertidal epibenthic shell refugia are likely to lead to enhanced mortality in more open sand habitat from fish and birds. Source: data from McDonald et al. (2001).

least, being a partial consequence of positive outcomes in competitive interactions with *C. maenas*. For example, van den Brink and Hutting (2017) showed *H. takanoi* to be a more aggressive competitor for food than *C. maenas* in oyster reef habitats in the Netherlands, while Gothland et al. (2014) speculated that biological traits of *H. takanoi*, including a short life cycle, rapid growth, and early sexual maturity, place this species at a competitive advantage over *C. maenas*. Cornelius et al. (2021) examined the potential overlap in prey preferences between *H. takanoi* and *C. maenas* in mixed oyster and mussel reefs and recorded the changes in density of the two species over an 11-year period in the Wadden Sea. They showed a more than 10-fold increase in *H. takanoi*, whereas densities of *C. maenas* declined. Laboratory experiments demonstrated a strong overlap in prey, and showed that at observed field densities, the impact of *H. takanoi* on prey populations was likely much higher than *C. maenas*.

The invasive barnacle *Austrominius* (previously *Elminius*) *modestus* colonized shores of Britain in the 1940s (Bishop 1947) and has since extended its distribution throughout Europe (Crisp 1958), where it now dominates in many locations (see e.g., Witte et al. 2010). In the years following its arrival in Britain, Crisp (1958) documented its rapid range expansion and speculated on the potential for exclusion of native species through competition from this southern species. Crisp (1958) observed rapid space domination of *Austrominius* as a consequence of its long breeding season and high intensity of settlement, but also remarked upon the ability of larger *Semibalanus balanoides* individuals to effectively compete for food through being able to "fish a layer of water beyond the reach of the cirri of *Elminius*." More recently, Gallagher et al. (2016) examined the co-existence of native and non-indigenous species of intertidal barnacles in Ireland and the potential exclusion of native species through competition for space. They demonstrated domination of space by *Austrominius* during clearance experiments, but the lack of total space saturation, plus use of different shore levels by different species, ensured coexistence. It is worth noting that competitive interactions may change with a warming climate. For example, Witte et al. (2010) documented the recent domination by *Austrominius* of shores on the island of Sylt in the southeastern North Sea. Despite arriving there in 1955, it has only very recently begun to dominate barnacle assemblages, potentially as a consequence of climate warming.

In many examples of colonization by non-indigenous crustaceans the mechanisms by which successful invasion occurs may be complex and the balance between predation and competition unclear. The "killer" shrimp *Dikerogammarus villosus* is a Ponto-Caspian amphipod that has invaded European freshwater systems with negative impacts on many resident amphipod species (see Rewicz et al. [2014] for a review). Dick and Platvoet (2000) consider that negative interactions within a guild of species are often wrongly assumed to be a consequence of interspecific competition. Their work on the localized elimination of *Gammarus duebeni* and *G. tigrinus* in the Netherlands by the invading *D. villosus* used laboratory experiments to demonstrate that intraguild predation rather than interspecific competition was the major process. However, stable isotope (Hellmann et al. 2015), diet analysis (Koester et al. 2016), observations of mouthpart morphology (Mayer et al. 2008), and field estimates of leaf litter consumption (Worischka et al. 2018) suggest that at least in many circumstances, interspecific competition for food resources plays an important role in domination over other amphipod species. Interspecific competition, through aggressive interactions over space usage, was described by Jermacz et al. (2015) in their investigations of the interaction between *D. villosus* and another Ponto-Caspian invasive gammarid, *Pontogammarus robustoides*. They showed experimentally that *P. robustoides* actively avoided *D. villosus*, even at low densities, and moved to a less favored habitat. Interestingly this interspecific aggression was reduced in the presence of the fish predator, the racer goby *Babka gymnotrachelus* (Kessler) (also a Ponto-Caspian invader), leading to coexistence of the two amphipod invaders in the same favored habitat.

Apparent Competition

Apparent competition, the process whereby species in similar trophic positions negatively interact, not through competing for a shared resource, but through enhancement of a shared predator, pathogen, or parasite is surprisingly common. For example, Menge's (1995) review of manipulative experiments in the intertidal zone found that approximately 25% of identified indirect effects demonstrated apparent competition, exceeding by far exploitative competition in terms of measurable effects. Numerous examples of apparent competition among crustacean groups exist, including between shrimp groups on marine temperate reefs, mediated by predatory rockfish (Frid and Marliave 2010), and between amphipod mesograzers in experimental mesocosms with a shared predator, the sea scorpion *Taurulus bubalis* (Beermann et al. 2018). Apparent competition among shrimp groups was described by Frid and Marliave (2010) through the use of structural equation modeling to describe the relationship between rockfish and two shrimp groups, *Pandalus* sp. and an aggregate of three smaller-bodied genera, on which it preys. Analyses indicated that the two shrimp groups were negatively correlated. However, rather than this negative relationship being a consequence of direct competition, apparent competition was invoked, caused by the positive influence of shrimp on their predator, the rockfish. This relationship was asymmetric: the indirect effect of the small-bodied shrimps on *Pandalus* was seven times stronger than the reverse, based on the asymmetry of positive effects on the predatory rockfish.

Distinguishing between apparent competition and "true" competition can be challenging and requires fundamental understanding of community dynamics and the direction and magnitude of species interactions. It is also likely that in many food webs, species may interact in both ways, potentially mediated by environmental conditions (e.g., Schoen et al. 2015). A freshwater example illustrates how distinguishing the form of competition is fundamental to decisions on management strategies. The opossum shrimp *Mysis diluviana* was introduced into many North American and European lakes as a food source for planktivorous fishes (Lasenby et al. 1986), but instead of a positive effect, has had a detrimental effect on some species. For example, the kokanee (lacustrine sockeye salmon *Oncorhynchus nerka*) has disappeared from many lake systems (Lasenby et al. 1986). Because mysids and planktivorous fish share the same prey and the same predators (cladocerans and the lake trout *Salvelinus namaycush*, respectively) they may interact through both exploitation competition (e.g., Spencer et al. 1991) and apparent competition (Ellis et al. 2011). Schoen et al. (2015), using observations of production and consumption rates in zooplankton and fish over a two-year period, concluded that the balance between apparent and exploitation competition is mediated by the environment, specifically temperature and lake depth. This has implications for management. If the environment dictates that mysids limit planktivorous fish through resource competition, then enhancement of fish production may be achieved through bottom-up interventions such as lake fertilization. However, if apparent competition dominates through mysid enhancement of predatory lake trout, then top-down strategies such as limiting predator density will be favored.

Crustacean–fish interactions have also been used in the marine pelagic realm to illustrate the importance of apparent competition. In the Barents Sea, a key pelagic fish species, capelin, predates on zooplankton, predominantly copepods and krill. Stige et al. (2018) use state space statistical modeling of time series of capelin, krill, and copepod dynamics over 30+ years where both top-down and bottom-up dynamics had previously been observed. Their models show that a temporary increase in krill biomass, for example, caused by good krill recruitment, enhances capelin survival. The subsequent increase in capelin biomass leads to increased predation and reduced biomass of copepods in the following years, resulting in apparent competition between krill and copepods (Fig. 6.6). Whether this interaction is asymmetric (i.e., negative effects of krill on copepods but

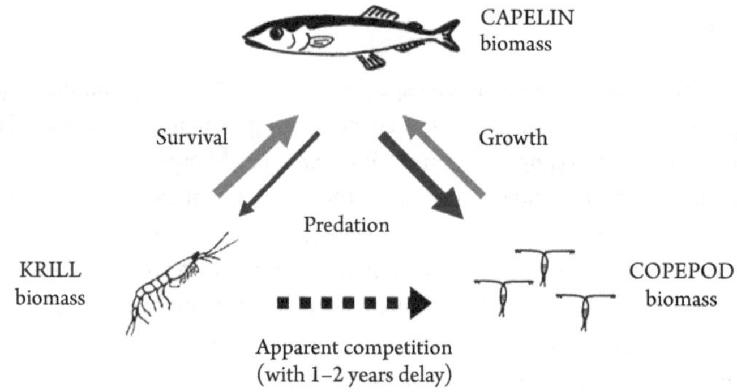

Fig. 6.6.
Schematic representation of apparent competition between krill and copepods in the Barents Sea mediated by their sharing of a predator, the capelin, based on inferences gained from state space statistical modeling of a 30+ year time series. Bottom-up (light gray) relationships between prey species and the capelin led to top-down control (dark gray). Arrow widths are approximately proportional to estimated effect sizes. Reproduced from Stige et al. (2018), with permission from © John Wiley & Sons; see a color version of this figure in the centerfold.

not vice versa) is unclear. Stige et al. (2018) argue that understanding of such indirect effects allows prediction of future ecosystem scale responses to both physical (e.g., climate change) and biological (fishing) drivers.

FUTURE DIRECTIONS

Consideration of the future direction of research into a topic as wide ranging as "competition" necessarily takes a broad view. First and foremost, despite considerable ecological research into the biology of crustaceans over decades, it is still true to say that advancing basic knowledge of biological processes across not only common/dominant taxa but also the rich diversity of crustaceans is at the heart of enhancing understanding of competitive relationships. Increasingly it is likely that such fundamental knowledge will be gained through investigations placed into the context of anthropogenic impacts such as climate change, eutrophication, fishing, and ocean sprawl. Such work will obviously also demonstrate the way in which competition will change in degraded systems or those undergoing rapid environmental change.

Predictions about the role of competition in changing ecosystems will also undoubtedly be influenced by the increasing prevalence of invasive nonnative species. As outlined above, global transport networks are modifying the distribution of taxa such as crustaceans at an alarming rate, providing numerous opportunities for novel interactions among previously separated taxa and hence changes in competitive relationships. Certain crustacean taxa, including the shore crab *C. maenas* and the shrimp *D. villosus*, have received considerable attention owing to their invasive success. Given the trajectory of global species movements, it seems likely that more invasive crustacean taxa will appear in the future. Understanding the nature and outcome of competitive interactions will be important in predicting their success and hence the potential impacts on native biota.

In many habitats, for example the open ocean, experimental determination of competitive relationships is challenging if not impossible. Mesocosm approaches may be possible for small taxa with short generation times, but in many cases investigators rely on observational and modeling techniques to determine the role of competition and other ecological processes. The scope of such approaches is limited by the length and quality of observational time series, the basic understanding of biological processes in key taxa, and the application of appropriate statistical and modeling techniques. Ecologists are keenly aware of the value of time series data, and it is to be hoped that current time series continue to be supported and new ones instigated.

CONCLUSIONS AND SUMMARY

Addressing the prevalence and importance of a fundamental ecological process like competition across a whole phylum is a challenging task. However, in doing so this chapter has demonstrated that competitive interactions involving crustacean taxa are an important driver of both population and community structure and of evolutionary processes in a variety of habitats, from the benthos to the pelagic realm. It has also clearly shown that study of such relationships in crustaceans has been important in advancing fundamental understanding in a number of areas, including population/community ecology, evolutionary biology, and invasion ecology.

Certain model taxa feature disproportionately in the literature, generally as a consequence of the ease with which they can be studied. Barnacles on hard substrata and amphipods in soft-sediment systems have proven useful taxa to consider the prevalence and importance of both intra- and interspecific competition in natural systems. Barnacles have provided one of the most iconic examples of interspecific competition in the animal kingdom owing to the groundbreaking work of Connell in the 1950s, and have also contributed to advancing marine population ecology through studies of larval supply and subsequent density-dependent competition following variable levels of settlement. The particular nature of the limiting resource for hermit crabs (i.e., a suitable shell in which to live) combined with the ease in which this resource can be manipulated have ensured this taxon has had an important role in understanding the drivers and consequence of individual competition.

Examples of both interference and exploitation competition are common across crustacean taxa, although demonstration of interference is much more common. In this regard, the agonistic behavior among brachyuran crabs in habitats that support high densities is striking and can lead to high levels of cannibalism. Competition through exploitation of limiting resources is generally logistically harder to demonstrate experimentally, although it is inferred in many cases. This is particularly the case in observational and modeling studies of pelagic crustaceans, including economically important taxa such as krill. Apparent competition, though not competition in the true sense, was considered in this chapter owing to its obvious prevalence across crustacean taxa.

A wide range of consequences of competitive interactions has been demonstrated in crustaceans from changes in demography to behavior, individual morphology, and evolutionary processes. Consequences at the individual level start with non-lethal effects on physiology and growth, perhaps best demonstrated in hermit crabs. Looking more broadly at population responses, there are numerous examples of small-scale experimental studies conducted in mesocosms or in controlled settings in the field. However, broad scale, often long-term monitoring of crustacean taxa demonstrates how competitive interactions can have significant large-scale consequences at the population level. Very general responses to competitive interactions in crustaceans are also demonstrated in the divergence in resource usage and hence niche differentiation, which leads naturally to a reduction in the strength of competition over time and facilitates the coexistence of species.

Finally, the nature of crustacean competitive interactions is changing globally as taxa are redistributed through multiple anthropogenic vectors. High profile examples exist, both of very recent species' range expansions in crustaceans leading to changes in competitive relationships, for example, the shrimp *Dikerogammarus villosus* and the brachyuran crab *Hemigrapsus sanguineus*, and those that have a much longer history, for example the shore crab *Carcinus maenas* and the barnacle *Austrominius modestus*.

Overall, it is clear from this review that competition plays a central role in the biology and ecology of crustacean taxa. Across many decades and considering numerous crustacean taxa, the scientific literature is full of fascinating examples of the way in which Crustacea from Antarctic krill to tropical hermit crabs compete for resources. The consequences of competition among the Crustacea are far-reaching, affecting individual physiology, behavior, and survival, and through doing so impact the distribution and hence biogeography of crustacean taxa, as well as fundamental evolutionary processes. In considering the patterns, mechanisms, and consequences of competition in the Crustacea, this review has effectively demonstrated the important role of crustacean research in advancing ecological and evolutionary understanding.

REFERENCES

Abrams, P.A. 1980. Resource partitioning and interspecific competition in a tropical hermit crab community. Oecologia 46:365–379.

Abrams, P.A. 1987. Alternative models of character displacement and niche shift. I. Adaptive shifts in resource use when there is competition for nutritionally non-substitutable resources. Evolution 41:651–661.

Abrams, P.A., C.F. Nyblade, and S. Sheldon. 1986. Resource partitioning and competition in a subtidal hermit crab assemblage. Oecologia 69:429–445.

Amaral, V., J. Paula, S. Hawkins, and S. Jenkins. 2009. Cannibalistic interactions in two co-occurring decapod species: Effects of density, food, alternative prey and habitat. Journal of Experimental Marine Biology and Ecology 368:88–93.

Ambrose, W.G. 1984. Influences of predatory polychaetes and epibenthic predators on the structure of a soft bottom community in a Maine estuary. Journal of Experimental Marine Biology and Ecology 81:115–145.

Ambrose, W.G. Jr. 1986. Experimental analysis of density dependent emigration of the amphipod *Rhepoxynius abronius*. Marine Behaviour and Physiology 12:209–216.

Atkinson, A., V. Siegel, E. Pakhomov, and P. Rothery. 2004. Long-term decline in krill stock and increase in salps within the Southern Ocean. Nature 432:100–103.

Bach, C, B.A. Hazlett, and D. Rittschof. 1976 Effects of interspecific competition on fitness of the hermit crab *Clibanarius tricolor*. Ecology 57:579–586.

Ball, E.E. 1972. Observations on the biology of the hermit crab, *Coenobita compressus* H. Milne Edwards (Decapoda; Anomura) on the west coast of the Americas. Revista de Biologia Tropical 20:265–273.

Barnes, H., and H.T. Powell. 1950 The development, general morphology and subsequent elimination of barnacle populations, *Balanus crenatus* and *B. balanoides*, after a heavy initial settlement. The Journal of Animal Ecology 19:175–179.

Beck, M.W. 1997. A test of the generality of the effects of shelter bottlenecks in four stone crab populations. Ecology 78:2487–2503.

Beermann, J., K. Boos, L. Gutow, M. Boersma, and A.C. Peralta. 2018 Combined effects of predator cues and competition define habitat choice and food consumption of amphipod mesograzers. Oecologia 186:645–654.

Begon, M., and C.R. Townsend. 2020. Ecology: From individuals to ecosystems, 15th edition. John Wiley and Sons, Oxford, UK.

Bertness, M.D. 1989. Positive and negative density-dependent mortality and the population structure of *Semibalanus balanoides* in a sheltered bay habitat. Ecology 70:257–268.

Bertness, M.D., S.D. Gaines, and S.M. Yeh. 1998. Making mountains out of barnacles: The dynamics of acorn barnacle hummocking. Ecology 79:1382–1394.

Bishop, M.W.H. 1947 Establishment of an immigrant barnacle in British coastal waters. Nature 159:501–502.

Botsford, L.W. 2001. Physical influences on recruitment to California Current invertebrate populations on multiple scales. ICES Journal of Marine Science 58:1081–1091.
Botsford, L.W., and R.C. Hobbs. 1995. Recent advances in the understanding of cyclic behavior of Dungeness crab *(Cancer magister)* populations. ICES Marine Science Symposium 199:157–166.
Briffa, M., and R.W. Elwood. 2004. Use of energy reserves in fighting hermit crabs. Proceedings of the Royal Society B: Biological Sciences 271:373–379.
Briffa, M, R.W. Elwood, and J.T.A. Dick. 1998. Analysis of repeated signals during shell fights in the hermit crab *Pagurus bernhardus*. Proceedings of the Royal Society B: Biological Sciences 265:1467–1474.
Carlton, J.T., and A.N. Cohen. 1995. Episodic global dispersal in shallow water marine organisms: The case history of the European shore crabs *Carcinus maenas* and *C. aestuarii*. Journal of Biogeography 30:1809–1820.
Childress, J.R. 1972. Behavioral ecology and fitness theory in a tropical hermit crab. Ecology 53:960–964.
Christy, J.H. 1982. Burrow structure and use in the sand fiddler crab, *Uca pugilator* (Bosc). Animal Behaviour 30:687–694.
Commito, J.A. 1982. Importance of predation by infaunal polychaetes in controlling the structure of a soft-bottom community in Maine, USA. Marine Biology 68:77–81.
Connell, J.H. 1961. The influence of interspecific competition and other factors on the distribution of the barnacle *Chthamalus stellatus*. Ecology 42:710–723.
Connell, J.H. 1980. Diversity and the coevolution of competitors, or the ghost of competition past. Oikos 35:131–138.
Cornelius, A., K. Wagner, and C. Buschbaum, 2021. Prey preferences, consumption rates and predation effects of Asian shore crabs (*Hemigrapsus takanoi*) in comparison to native shore crabs (*Carcinus maenas*) in northwestern Europe. Marine Biodiversity 51:1–17.
Crisp, D.J. 1958. The spread of *Elminius modestus* Darwin in north-west Europe. Journal of the Marine Biological Association. U.K. 37:483–520.
Dayan, T., and D. Simberloff. 2005. Ecological and community-wide character displacement: The next generation. Ecology Letters 8:875–894.
De Bruyn, C., B. David, S. Motreuil, G. Caulier, Q. Jossart, T. Rigaud, and C. De Ridder. 2016. Should I stay or should I go? Causes and dynamics of host desertion by a parasitic crab living on echinoids. Marine Ecology Progress Series 546:163–171.
Dennert, H.G. 1974. Tolerance differences and interspecific competition in three members of the amphipod genus *Gammarus*. Hydrobiological Bulletin 8:83–99.
Dick, J.T.A., and D. Platvoet. 2000. Invading predatory crustacean *Dikerogammarus villosus* eliminates both native and exotic species. Proceedings of the Royal Society B: Biological Sciences 267:977–983.
Doherty, P.J., and D.M. Williams. 1988 The replenishment of coral reef populations. Oceanography and Marine Biology—An Annual Review 26:487–551.
Dzialowski, A.R., and W.J. O'Brien. 2004. Is competition important to arctic zooplankton community structure? Freshwater Biology 49:1103–1111.
Ellis, B., J. Stanford, D. Goodman, C. Stafford, D. Gustafson, D. Beauchamp, D. Chess, J. Craft, M. Deleray, and B. Hansen. 2011. Long-term effects of a trophic cascade in a large lake ecosystem. Proceedings of the National Academy of Sciences USA 108:1070–1075.
Elmgren, R., S. Ankar, B. Marteleur, and G. Ejdung. 1986. Adult interference with postlarvae in soft sediments: The *Pontoporeia-Macoma* example. Ecology 67:827–836.
Elwood, R.W., and G. Arnott. 2007. Fighting for shells: How private information about resource value changes hermit crab pre-fight displays and escalated fight behaviour. Proceedings of the Royal Society B: Biological Sciences 271:3011–3017.
Fauchald, K., and P.A. Jumars. 1979. The diet of worms: A study of polychaete feeding guides. Oceanography and Marine Biology Annual Review 17:193–284.
Fernández, M., and J.C. Castilla. 2000. Recruitment of *Homalaspis plana* in intertidal habitats of central Chile and implications for the current use of management and marine protected areas. Marine Ecology Progress Series 208:157–170.

Fernandez, M., D. Armstrong, and O. Iribarne. 1993. First cohort of dungeness crab *Cancer magister* reduces abundance of subsequent cohorts in intertidal shell habitat. Canadian Journal of Fisheries and Aquatic Science 50:2100–2105.

Foster, B.A. 1971. Desiccation as a factor in the intertidal zonation of barnacles. Marine Biology 8:12–29.

Fotheringham, N. 1976. Hermit crab shells as a limiting resource (Decapoda, Paguridea). Crustaceana 31:193–199.

Frid, A., and J. Marliave. 2010. Predatory fishes affect trophic cascades and apparent competition in temperate reefs. Biology Letters 6:533–536.

Gallagher, M.C., S. Culloty, R. McAllen, and R. O'Riordan. 2016. Room for one more? Coexistence of native and non-indigenous barnacle species. Biological Invasions 18:3033–3046.

Geburzi, J.C., D. Brandis, and C. Buschbaum, 2018. Recruitment patterns, low cannibalism and reduced interspecific predation contribute to high invasion success of two Pacific crabs in northwestern Europe. Estuarine, Coastal and Shelf Science 200:460–472.

Gerdol, V., and R.G. Hughes. 1993. Effect of the amphipod *Corophium volutator* on the colonisation of mud by the halophyte *Salicornia europaea*. Marine Ecology Progress Series 97:61–69.

Gerdol, V., and R.G. Hughes. 1994a. Feeding behaviour and diet of *Corophium volutator* in an estuary in southeastern England. Marine Ecology Progress Series 114:103–108.

Gerdol, V., and R.G. Hughes. 1994b. Effect of *Corophium volutator* on the abundance of benthic diatoms, bacteria and sediment stability in two estuaries of southeastern England. Marine Ecology Progress Series 114:109–115.

Gothland, M., J.C. Dauvin, L. Denis, F. Dufossé, S. Jobert, J. Ovaert, J.P. Pezy, A.T. Rius, and N. Spilmont. 2014. Biological traits explain the distribution and colonisation ability of the invasive shore crab *Hemigrapsus takanoi*. Estuarine, Coastal and Shelf Science 142:41–49.

Hartnoll, R.G. 2001. Growth in Crustacea—Twenty years on. Hydrobiologia 449:111–122.

Hazlett, B.A. 1966. Social behavior of the Paguridae and Diogenidae of Curacao. Studies on the Fauna of Curaçao and other Caribbean Islands 23:1–143.

Hazlett, B.A. 1970. Interspecific shell fighting in three sympatric species of hermit crabs in Hawaii. Pacific Science 24:472–482.

Hazlett, B.A. 1981. The behavioral ecology of hermit crabs. Annual Review of Ecology, Evolution, and Systematics 12:1–22.

Hellmann, C., S. Worischka, E. Mehler, J. Becker, R. Gergs, and C. Winkelmann. 2015. The trophic function of *Dikerogammarus villosus* (Sowinsky, 1894) in invaded rivers: A case study in the Elbe and Rhine. Aquatic Invasions 10:385–397.

Hines, A.H., and G.M. Ruiz. 1995. Temporal variation in juvenile blue crab mortality: Nearshore shallows and cannibalism in Chesapeake Bay. Bulletin of Marine Science 57:884–901.

Holt, R.D. 1977. Predation, apparent competition, and the structure of prey communities. Theoretical Population Biology 12:197–229.

Hovel, K.A., and R.A. Wahle. 2010. Effects of habitat patchiness on American lobster movement across a gradient of predation risk and shelter competition. Ecology 91:1993–2002.

Iribarne, O., M. Fernandez, and D.A. Armstrong. 1994. Does space competition regulate density of juvenile Dungeness crab *Cancer magister* Dana in sheltered habitats? Journal of Experimental Marine Biology and Ecology 183:259–271.

Jacoby, C.A. 1983. Ontogeny of behavior in the crab instars of the Dungeness crab, *Cancer magister* Dana 1852. Zeitschrift für Tierpsychologie 63:1–16.

Jeffery, C.J. 2003. Determination of abundance and distribution of an intertidal barnacle: Settlement or post settlement mortality? Marine Ecology Progress Series 246:291–305.

Jenkins, S.R., P. Åberg, G. Cervin, R.A. Coleman, J. Delaney, P. Della Santina, S.J. Hawkins, E. LaCroix, A.A. Myers, M. Lindergarth, A.-M. Power, and R.G. Hartnoll. 2000. Spatial and temporal variation in the settlement and recruitment of the intertidal barnacle *Semibalanus balanoides* (L.) (Crustacea: Cirripedia) over a European scale. Journal of Experimental Marine Biology and Ecology 243:209–225.

Jenkins, S.R., J. Murua, and M.T. Burrows. 2008a Temporal changes in the strength of density-dependent mortality and growth in intertidal barnacles. Journal of Animal Ecology 77:573–584.

Jenkins, S.R., P. Moore, M.T. Burrows, D.J. Garbary, S.J. Hawkins, A. Ingólfsson, K.P. Sebens, P.V.R. Snelgrove, D.S. Wethey, and S.A. Woodin. 2008b. Comparative ecology of North Atlantic shores: Do differences in players matter for process? Ecology 89:S3–S23.

Jennions, M.D., and P.R.Y. Backwell. 1996. Residency and size affect fight duration and outcome in the fiddler crab *Uca annulipes*. Biological Journal of the Linnean Society 57:293–306.

Jensen, K.T., and C. André. 1993. Field and laboratory experiments on interactions among an infaunal polychaete, *Nereis diversicolor*, and two amphipods, *Corophium volutator* & *C. arenarium*: Effects on survival, recruitment and migration. Journal of Experimental Marine Biology and Ecology 168:259–278.

Jensen, G.C., S.P. McDonald, and D.A. Armstrong. 2002. East meets west: Competitive interactions between green crab *Carcinus maenas*, and native and introduced shore crab *Hemigrapsus* spp. Marine Ecology Progress Series 225:251–262.

Jensen, G.C., S.P. McDonald, and D.A. Armstrong. 2007. Biotic resistance to green crab, *Carcinus maenas*, in California bays. Marine Biology 151:2231–2243.

Jermacz, Ł., A. Dzierzyńska, T. Kakareko, M. Poznańska, and J. Kobak. 2015. The art of choice: Predation risk changes interspecific competition between freshwater amphipods. Behavioral Ecology 26:656–664.

Karpov, K. 1983. Effect of substrata type on survival and growth in high density communal cultures of juvenile Dungeness crab, *Cancer magister.* Pages 311–318 *in* P.W. Wild and R.N. Tasto, editors. Fish bulletin 172. Life history environment and mariculture studies of' the Dungeness crab, *Cancer magister*, with emphasis on the central Californian fishery resources. Department of Fish and Game, California, USA.

Katz, C.H., J.S. Cobb, and M. Spaulding. 1994. Larval behavior, hydrodynamic transport, and potential offshore-to-inshore recruitment in the American lobster *Homarus americanus*. Marine Ecology Progress Series 103:265–273.

Kendall, M.A., R.S. Bowman, P. Williamson, and J.R. Lewis. 1982. Settlement patterns, density and stability in the barnacle *Balanus balanoides*. Netherlands Journal of Sea Research 16:119–126.

Koester, M., B. Bayer, and R. Gergs. 2016. Is *Dikerogammarus villosus* (Crustacea, Gammaridae) a 'killer shrimp' in the River Rhine system? Hydrobiologia 768:299–313.

Lancellotti, D.A., and R.G. Trucco. 1993. Distribution patterns and coexistence of six species of the amphipod genus *Hyale*. Marine Ecology Progress Series 93:131–131.

Lasenby, D.C., T.G. Northcote, and M. Furst. 1986. Theory, practice, and effects of *Mysis relicta* introductions to North American and Scandinavian lakes. Canadian Journal of Fisheries and Aquatic Sciences 43:1277–1284.

Marks, R, S.A. Hesp, D. Johnston, A. Denham, and N. Loneragan. 2020. Temporal changes in the growth of a crustacean species, *Portunus armatus*, in a temperate marine embayment: Evidence of density dependence. ICES Journal of Marine Science 77:773–790.

Marshall, S., K. Warburton, B. Paterson, and D. Mann. 2005. Cannibalism in juvenile blue-swimmer crabs *Portunus pelagicus* (Linnaeus, 1766): Effects of body size, moult stage and refuge availability. Applied Animal Behaviour Science 90:65–82.

Mayer, G., G. Maier, A. Maas, and D. Waloszek. 2008. Mouthparts of the ponto-caspian invader *Dikerogammarus villosus* (Amphipoda: Pontogammaridae). Journal of Crustacean Biology 28:1–15.

McDonald, S.P., G.C. Jensen, and D.A. Armstrong. 2001. The competitive and predatory impacts of the nonindigenous crab *Carcinus maenas* (L.) on early benthic phase Dungeness crab *Cancer magister* Dana. Journal of Experimental Biology and Ecology 258:39–54.

Menge, B.A. 1991. Relative importance of recruitment and other causes of variation in rocky intertidal community structure. Journal of Experimental Marine Biology and Ecology, 146:69–100.

Menge, B.A. 1995. Indirect effects in marine rocky intertidal interaction webs: Patterns and importance. Ecological Monographs 65:21–74.

Menge, B.A. 2000. Recruitment vs. postrecruitment processes as determinants of barnacle population abundance. Ecological Monographs 70:265–288.

Moksnes, P.-O. 2002. The relative importance of habitat-specific settlement, predation and juvenile dispersal for distribution and abundance of young juvenile shore crabs *Carcinus maenas* L. Journal of Experimental Marine Biology and Ecology 271:41–73.

Moksnes, P.-O. 2004a Interference competition for space in nursery habitats: Density-dependent effects on growth and dispersal in juvenile shore crabs *Carcinus maenas*. Marine Ecology Progress Series 281:181–191.

Moksnes, P.-O. 2004b. Self-regulating mechanisms in cannibalistic populations of juvenile shore crabs *Carcinus maenas*. Ecology 85:1343–1354.

Möller, P. 1985. Production and abundance of juvenile *Nereis diversicolor* and oogenic cycles of adults in shallow waters of western Sweden. Journal of the Marine Biological Association of the United Kingdom 65:603–616.

Ólafsson, E.B., and L.-E. Persson. 1986. The interaction between *Nereis diversicolor* O.F. Müller and *Corophium volutator* Pallas as a structuring force in a shallow brackish sediment. Journal of Experimental Marine Biology and Ecology 103:103–117.

Oliver, J.S., J.M. Oakden, and P.N. Slattery. 1982. Phoxocephalid amphipod crustaceans as predators on larvae and juveniles in marine soft-bottom communities. Marine Ecology Progress Series 7:179–184.

Olson, R.R., and M.H. Olson. 1989. Food limitation of planktotrophic marine invertebrate larvae: Does it control recruitment success? Annual Review of Ecology and Systematics 20:225–247.

Oróstica, M.H., C.A. Richardson, J. Estrella-Martinez, S.R. Jenkins, and S.J. Hawkins. 2021. Shell growth and age determined from annual lines in the southern warm-water limpet *Patella depressa* at its poleward geographical boundaries. Journal of the Marine Biological Association 101:707–716.

Paine, R.T. 1974 Intertidal community structure: Experimental studies on the relationship between a dominant competitor and its principal predator. Oecologia 15:93–120.

Pan, Y., Y. Zhang, and S. Shucun. 2018 Habitat orientation alters the outcome of interspecific competition: A microcosm study with zooplankton grazers. Ecology and Evolution 8:3254–3269.

Parker, J.D., M.E. Torchin, R.A. Hufbauer, N.P. Lemoine, C. Alba, D.M. Blumenthal, O. Bossdorf, J.E. Byers, A.M. Dunn, R.W. Heckman, and M. Hejda. 2013. Do invasive species perform better in their new ranges? Ecology 94:985–994.

Peterson, C.H. 1979. Predation, competitive exclusion and diversity in soft-bottom benthic communities of estuaries and lagoons. Pages 233–264 *in* R.J. Livingstone, editor. Ecological processes in coastal and marine systems. Plenum, New York, USA.

Pile, A.J., R.N. Lipcius, J. van Montfrans, and R.J. Orth. 1996. Density-dependent settler-recruit-juvenile relationships in blue crabs. Ecological Monographs 66:277–300.

Pineda, J., and N. Reyns. 2018. Larval transport in the coastal zone: Biological and physical processes. Pages 147–163 *in*. T.J. Carrier, A.M. Reitzel, and A. Heyland, editors. Evolutionary ecology of marine invertebrate larvae. Oxford University Press, New York, USA.

Provenzano, A.J. Jr. 1960. Notes on Bermuda hermit crabs (Crustacea; Anomura). Bulletin of Marine Science 10:117–124.

Reise, K. 1979. Moderate predation on meiofauna by the macrobenthos of the Wadden Sea. Helgol Meeresunters 32:453–465.

Rewicz, T., M. Grabowski, C. MacNeil, and K. Bącela-Spychalska. 2014. The profile of a 'perfect' invader—The case of killer shrimp, *Dikerogammarus villosus*. Aquatic Invasions 9:267–288.

Riisgård, H.U. 1991. Suspension feeding in the polychaete *Nereis diversicolor*. Marine Ecology Progress Series 70:29–37.

Roberts, L., and M.E. Laidre. 2019. Get off my back: Vibrational assessment of homeowner strength. Biology Letters 15:20180819.

Rönn, C., E. Bonsdorff, and W.G. Nelson. 1988. Predation as a mechanism of interference within infauna in shallow brackish water soft bottoms; experiments with an infauna predator, *Nereis diversicolor* O.F. Müller. Journal of Experimental Marine Biology and Ecology 116:143–157.

Ross, R.M., L.B. Quetin, T. Newberger, and C.T. Shaw. 2014. Trends, cycles, interannual variability for three pelagic species west of the Antarctic Peninsula 1993–2008. Marine Ecology Progress Series 515:11–32.

Rossong, M.A., P.J. Williams, M. Comeau, S.C. Mitchell, and J. Apaloo. 2006. Agonistic interactions between the invasive green crab, *Carcinus maenas* (Linnaeus) and juvenile American lobster, *Homarus americanus* (Milne Edwards). Journal of Experimental Marine Biology and Ecology 329:281–288.

Ryabov, A.B., A.M de Roos, B. Meyer, S. Kawaguchi, and B. Blasius. 2017. Competition-induced starvation drives large-scale population cycles in Antarctic krill. Nature Ecology and Evolution 1:0177.

Sainte-Marie, B., and M. Lafrance. 2002. Growth and survival of recently settled snow crab *Chionoecetes opilio* in relation to intra- and intercohort competition and cannibalism: A laboratory study. Marine Ecology Progress Series 244:191–203.

Schluter, D. 2000. Ecological character displacement in adaptive radiation. The American Naturalist 156:S4–S16.

Schoen, E.R., D.A. Beauchamp, A.R. Buettner, and N.C. Overman. 2015. Temperature and depth mediate resource competition and apparent competition between *Mysis diluviana* and kokanee. Ecological Applications 25:1962–1975.

Shima, J.S., and C.W. Osenberg. 2003. Cryptic density dependence: Effects of covariation between density and site quality in reef fish. Ecology 84:46–52.

Shucksmith, R., E.J. Cook, D.J. Hughes, and M.T. Burrows. 2009. Competition between the non-native amphipod *Caprella mutica* and two native species of caprellids *Pseudoprotella phasma* and *Caprella linearis*. Journal of the Marine Biological Association of the United Kingdom 89:1125–1132.

Smallegange, I.M., J. van der Meer, and R.H.J.M. Kurvers. 2006. Disentangling interference competition from exploitative competition in a crab-bivalve system using a novel experimental approach. Oikos 113:57–167.

Smith, D., R.G. Hughes, and E.J. Cox. 1996. Predation of epipelic diatoms by the amphipod *Corophium volutator* and the polychaete *Nereis diversicolor*. Marine Ecology Progress Series 145:53–61.

Southward, A.J. 1964. Limpet grazing and the control of vegetation on rocky shores. Pages 265–273 *in* D.J. Crisp, editor. Grazing in terrestrial and marine environments, Blackwell Scientific Publications, Oxford, UK.

Spencer, C.N., B.R. McClelland, and J.A. Stanford. 1991. Shrimp stocking, salmon collapse, and eagle displacement. BioScience 41:14–21.

Steinberg, D.K., K.E. Ruck, M.R. Gleiber, L.M. Garzio, J.S. Cope, K.S. Bernard, S.E. Stammerjohn, O.M.E. Schofield, L.B. Quetin, and R.M. Ross. 2015. Long-term (1993–2013) changes in macrozooplankton off the Western Antarctic Peninsula. Deep-Sea Research Part I: Oceanographic Research Papers 101:54–70.

Steiner, C.G., C.E. Cáceres, and S.D.P. Smith. 2007. Resurrecting the ghost of competition past with dormant zooplankton eggs. The American Naturalist 169:416–412.

Steibl, S., and C. Laforsch. 2020. Shell resource partitioning as a mechanism of coexistence in two co-occurring terrestrial hermit crab species. BMC Ecology 20:1.

Steneck, R.S. 2006. Possible demographic consequences of intraspecific shelter competition among American lobsters. Journal of Crustacean Biology 26:628–638.

Stige, L.C., K.O. Kvile, B. Bogstad, and O. Langangen. 2018 Predator-prey interactions cause apparent competition between marine zooplankton groups. Ecology 99:632–641.

Sutherland, J.P. 1990. Recruitment regulates demographic variation in a tropical intertidal barnacle. Ecology 71:955–972.

Taper, M.L., and T.J. Case. 1985. Quantitative genetic models for the coevolution of character displacement. Ecology 66:355–371.

Thorpe, K.E., A.C. Taylor, and F.A. Huntingford. 1995. How costly is fighting? Physiological effects of sustained exercise and fighting in swimming crabs, *Necora puber* (L.) (Brachyura, Portunidae). Animal Behaviour 50:1657–1666.

Thorson, G. 1950. Reproductive and larval ecology of marine bottom invertebrates. Biological Review 25:1–45.

Vance, R.R. 1972. Competition and mechanism of coexistence in three sympatric species of intertidal hermit crabs. Ecology 53:1062–1074.

van den Brink, A., and S. Hutting. 2017. Clash of the crabs: Interspecific, inter-cohort competition between the native European green crab, *Carcinus maenas* and the exotic brush clawed crab *Hemigrapsus takanoi* on artificial oyster reefs. Journal of Sea Research 128:41–51.

van Riel, M.C., E.P. Healy, G. van der Velde, and A. bij de Vaate. 2007. Interference competition among native and invader amphipods. Acta Oecologica 31:282–289.

Wahle, R.A., and R.S. Steneck. 1991. Recruitment habitats and nursery grounds of the American lobster *Homarus americanus*: A demographic bottleneck? Marine Ecology Progress Series 69:231–243.

Wellborn, G.A. 2002. Trade-off between competitive ability and antipredator adaptation in a freshwater amphipod species complex. Ecology 83:129–136.

Wethey, D.S. 1984. Effects of crowding on fecundity in barnacles: *Semibalanus balanoides, Balanus balanoides, Balanus glandula,* and *Chthamalus dalli.* Canadian Journal of Zoology 62:1788–1795.

Williams, P.J., T.A. Floyd, and M.A. Rossong. 2006. Agonistic interactions between invasive green crabs, *Carcinus maenas* (Linnaeus) and sub-adult American lobsters, *Homarus americanus* (Milne Edwards). Journal of Experimental Marine Biology and Ecology 329:66–74.

Wilson, W.H. 1991. Competition and predation in marine soft-sediment communities. Annual Review of Ecology and Systematics 21:221–241.

Wilson, W.H. Jr., and K. Parker. 1996. The life history of the amphipod, *Corophium volutator*: The effects of temperature and shorebird predation. Journal of Experimental Marine Biology and Ecology 196:239–250.

Witte, S., C. Buschbaum, J.E.E. van Beusekom, and K. Reise. 2010. Does climatic warming explain why an introduced barnacle finally takes over after a lag of more than 50 years? Biological Invasions 12:3579–3589.

Woodin, S.A. 1976. Adult-larval interactions in dense infaunal assemblages: Patterns of abundance. Journal of Marine Research 34:25–41.

Worischka, S., L. Richter, A. Hänig, C. Hellmann, J. Becker, P. Kratina, and C. Winkelmann. 2018. Food consumption of the invasive amphipod *Dikerogammarus villosus* in field mesocosms and its effects on leaf decomposition and periphyton. Aquatic Invasions 13:261–275.

Wu, R.S.S. 1980. Effects of crowding on the energetics of the barnacle *Balanus glandula* Darwin. Canadian Journal of Zoology 58:559–566.

Young, C.M., and F.S. Chia. 1987. Abundance and distribution of pelagic larvae, as influenced by predation, behavior, and hydrographic factors. Pages 385–463 *in* A.C. Giese, J.S. Pearse, and V.B. Pearse, editors. Reproduction of marine invertebrates, Volume 9: General aspects: Seeking unity in diversity. Blackwell/Boxwood, Palo Alto, California, USA.

7

THE ECOLOGICAL IMPORTANCE OF CRUSTACEAN DIVERSITY

Rebecca J. Best and Moritz D. Lürig

Abstract

Crustaceans inhabit a wide variety of habitats, from the open ocean to ephemeral pools. Within these habitats, crustaceans play key roles as herbivores, detritivores, and predators, and interact with competitors ranging from conspecifics to other crustaceans to diverse assemblages of invertebrates. As with most communities and ecosystems, efforts to understand the impacts of crustacean community diversity on ecosystem functioning began with studying species richness and relative abundance and have progressed to quantifying multiple continuous functional traits. This shift has resulted in more direct tests of how environmental conditions and changing anthropogenic stressors shape community composition, and how functional diversity allows species to make complementary contributions to control of algae, nutrient cycling, and provision of food to higher trophic levels. Greater focus on the role of functional traits in communities has also led ecologists to think about the macroevolutionary origins of functional differences within and among crustacean lineages and the ways that evolutionary histories might predict contemporary community functioning. There is also increasing evidence that rapid genetic and phenotypic divergence within species can affect ecological function at a finer scale of biodiversity that will be important for adaptive responses to global change. In this chapter, we review evidence for the effects of genetic, taxonomic, and functional or phenotypic diversity within and among crustacean taxa on the food web structure and functioning of aquatic communities. We end by outlining emerging methods for measuring these integrated facets of biodiversity and emerging questions about ecological and evolutionary components of resilience in rapidly changing ecosystems.

Rebecca J. Best and Moritz D. Lürig, *The Ecological Importance of Crustacean Diversity*. In: *Ecology and Conservation*. Edited by: Lars Gutow and Martin Thiel, Oxford University Press.
DOI: 10.1093/oso/9780197768242.003.0007

INTRODUCTION

More diverse communities are predicted to have higher levels of ecosystem functioning, either because multiple species make complementary contributions to functioning, or because diverse communities are more likely to contain a species with a large individual effect (Stachowicz et al. 2007). Similarly, studies in ecosystems dominated by only a few species have shown that greater genetic diversity within those species can also increase ecosystem functioning, including resistance to and recovery from disturbance, and ecosystem productivity (Hughes et al. 2008). Although this prediction has been supported over a wide range of ecological systems, including those involving crustaceans, there are also frequently large effects as a result of particular species or combinations of species that matter more than species number (O'Connor and Crowe 2005). This has led to a focus on models that identify particularly important species or species interactions (Hillebrand et al. 2008, Connolly et al. 2013), or that replace taxonomic diversity with metrics of functional diversity (Crowe and Russell 2009). Whereas taxonomic diversity includes only the number of species (species richness) or their relative abundance in a community (diversity and evenness), functional diversity incorporates each species' traits to more accurately represent the range of ecological niches occupied in a community (Crowe and Russell 2009). Within species, taxonomic vs. functional diversity are paralleled by genetic vs. phenotypic diversity, where the former is the number and frequency of genotypes, and the latter is variation in traits that may be jointly determined by genetic and environmental factors.

Here, we review evidence for the ecological effects of diversity in communities of crustaceans, including both taxonomic and functional measures and both inter- and intraspecific variation. Because crustaceans occupy a range of trophic levels, including herbivores, detritivores, and predators, we focus on the effects of diversity on ecosystem function within the context of marine and freshwater food webs. Although research on genetic, taxonomic, and phenotypic/functional facets of biodiversity across orders of crustaceans is generally in proportion to the number of species in those orders, research focused specifically on the relationship between biodiversity and ecosystem function (BEF) has involved a smaller number of groups (Fig. 7.1). Amphipods, cyclopoid and harpacticoid copepods, and ostracods have been particularly well studied (Fig. 7.1A), and to a lesser extent cladocerans (Diplostraca), isopods, decapods, and barnacles (Sessilia). These lineages are found across both marine and freshwater habitats, and in communities with very few crustacean species (e.g., crayfish, amphipods, or isopods in freshwater benthic habitats) as well as communities with diverse representation of multiple orders (e.g., decapods, amphipods, and isopods in marine benthic habitats or copepods, cladocerans, and ostracods in pelagic habitats).

In this chapter, we review BEF research across these lineages and habitats. We begin with the assembly of diversity into communities and follow this by examining the consequences of that diversity for multiple ecosystem functions in the context of food web interactions. Looking forward, we include a review of emerging methods that make it possible to quantify and study both taxonomic and functional diversity at higher temporal and spatial resolutions using high-throughput genetic and phenotypic sampling. We end with exciting questions at the confluence of ecological and evolutionary approaches to changing biodiversity and its consequences for crustacean communities.

WHAT IS KNOWN?

Assembly of Crustacean Diversity Across Very Different Ecosystems

Assembly and Maintenance of Species and Genetic Diversity

The ecological functioning of a community depends on the assembly of species that can coexist. This assembly is first constrained by evolutionary processes that supply diversity within and across

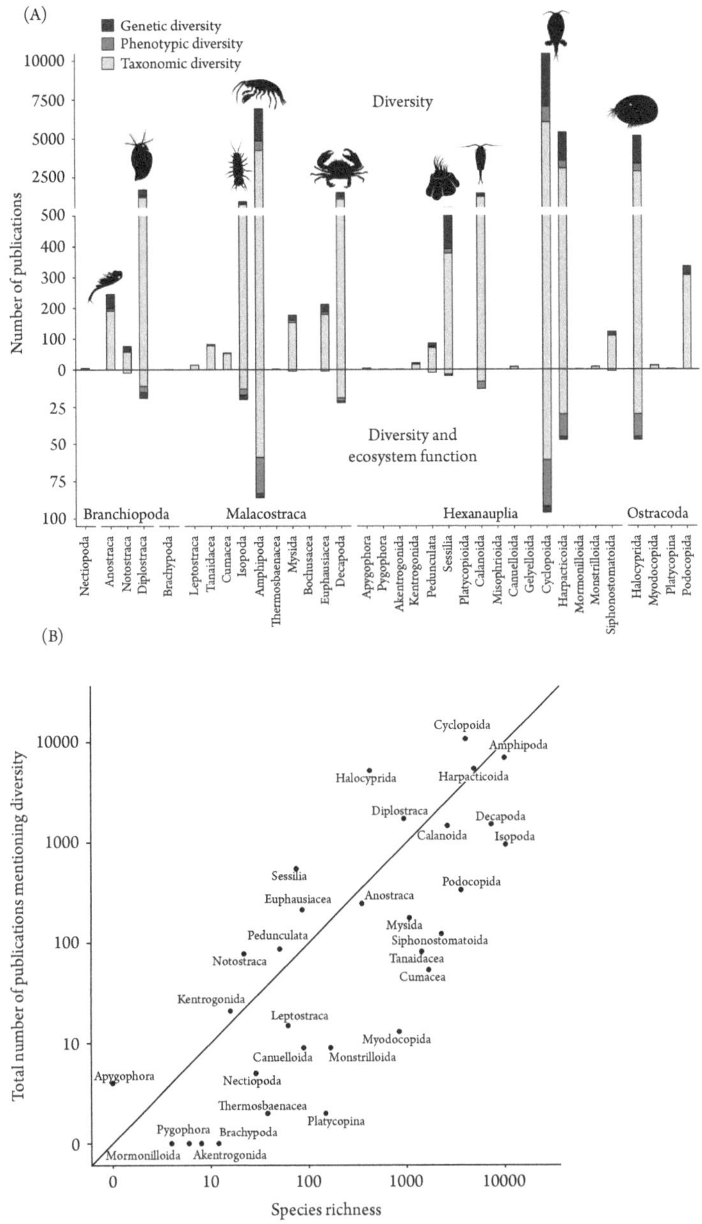

Fig. 7.1.
(B) Distribution of research examining different types of diversity (top) and their ecological effects (bottom) across crustacean orders. A list of search terms was created by combining names of all 52,800 crustacean species listed in the Catalogue of Life database (Bánki et al. 2021), with each of the following diversity-related search terms (exact search term given in parentheses): genetic diversity ("genetic diversity" OR "genetic richness" OR "genomic diversity" OR "genomic richness" OR "molecular diversity" OR "molecular richness"), phenotypic diversity ("functional diversity" OR "functional richness" OR "phenotypic diversity" OR "phenotypic richness" OR "trait diversity" OR "trait richness"), and taxonomic diversity ("species diversity" OR "species richness" OR "taxonomic diversity" OR "taxonomic richness"). In addition, all diversity terms were combined with ecosystem function search terms ("BDEF" OR "BEF" OR "ecosystem function*"). The results for each of the terms and term combinations were then aggregated at the order level. (B) Number of papers referencing diversity of any kind as a function of species richness in that order of crustaceans.

lineages, and these processes vary substantially across the variety of ecosystems that crustaceans inhabit. In the ocean, crustacean diversity tends to be higher at lower latitudes, but this pattern varies across orders and ocean basins (Rivadeneira and Poore 2020). In freshwater and terrestrial systems, however, crustacean diversity is limited to radiations within the subset of lineages that have colonized these environments and evolved the necessary physiological tolerances (Ahyong and Huang 2020). After these evolutionary processes set the regional diversity available for local communities (also known as gamma diversity), assembling these species into habitats then depends on dispersal, which is extremely variable among crustacean lineages and their habitats. At one end of this spectrum are pelagic marine zooplankton communities, with high mobility at all life stages and high habitat connectivity. This high connectivity can lead to low differentiation among communities (low beta diversity) but high local diversity (alpha diversity) within communities. At the other end are isolated freshwater springs, ponds, or cave systems inhabited by lineages with much lower dispersal abilities. These include direct-developing peracarids (amphipods and isopods) that are likely limited to low-frequency dispersal by waterfowl (Swanson 1984), and desiccation-adapted branchiopods (fairy shrimp, clam shrimp, tadpole shrimp) and ostracods, where local species richness (alpha diversity) depends on pond size and hydroperiod, and eggs are dispersed by wind (Altermatt and Ebert 2010).

Importantly, the ways that habitat connectivity and dispersal abilities shape crustacean community diversity are also changing due to anthropogenic influence. Climate change is altering habitat connectivity through space and time, as water temperatures warm beyond species' limits, or as freshwater habitats shift from perennial to ephemeral hydroperiods. In addition, anthropogenic species introductions can remove dispersal limitations, leading to increased homogeneity (decreased beta diversity) among habitats and potentially large effects of novel species interactions. These effects are likely to be especially strong in previously low-diversity and isolated habitats with limited resistance to invasion (Stachowicz et al. 2002, France and Duffy 2006) due to the availability of empty niche space (Shurin et al. 2000).

Just as the number of species can vary across regions and communities, genetic identity within species also varies. Although many crustacean taxa are fast growing (Jeffs and O'Rorke 2020) and good dispersers (Morgan 2020), leading to high genetic diversity (Wares 2011), there are also many examples where changing habitat connectivity and/or dispersal limitation have enabled genetic differentiation within taxa. At the coarsest level there is cryptic species diversity, where taxa previously thought to be a single species due to morphological similarity are discovered to be a complex of genetically differentiated species. Cryptic diversity in several freshwater crustacean taxa (e.g., *Hyalella* and *Gammarus* amphipods and *Moina* cladocerans) is well documented across a wide range of ecological, geographic, and taxonomic gradients (Wellborn and Cothran 2004, Wattier et al. 2020). Cryptic diversity has been studied mostly from an evolutionary perspective, with a focus on the mechanisms of phenotypic and molecular divergence. However, there is increasing interest in how intraspecific cryptic variation is ecologically relevant for community assembly or ecosystem dynamics. For instance, morphologically similar diaptomid copepod species can have contrasting effects on algal biomass, ammonium concentrations, and sedimentation rates (Matthews et al. 2011a). Cryptic species can also differ in their predation risk, such as the complex of freshwater amphipod species of the genus *Hyalella* (Wellborn and Cothran 2004), which adds complexity to our understanding of population dynamics and food web stability. Fortunately, identifying this diversity is increasingly feasible via high-throughput sequencing and open-source sequence-collections such as GenBank, opening the door to increased study of the causes and ecological consequences of such differentiation.

Even within taxa recognized as a single species, genetic diversity varies substantially across scales. For example, a widespread freshwater amphipod such as *Gammarus fossarum*, which plays an important role as a detritivore, shows very different levels of genetic diversity in individual populations across local, regional, and continental scales (Wattier et al. 2020). The patterns of diversity and divergence observed in gammarid amphipods and the isopod *Asellus aquaticus* (Sworobowicz et al. 2015) suggest that postglacial colonization history is a principal determinant of contemporary genetic structure on the continental scale (Taberlet et al. 1998). As for species diversity, the effects of regional and local geography, habitat connectivity and size, and hydroperiod may play important roles in governing the ecological consequences of genetic diversity, especially for marine and freshwater crustaceans that are extremely effective dispersers during the larval phase (Roman 2006). Environmental variation in connected water bodies may serve as a filter that can introduce mosaic patterns and stochasticity into local and regional population structure. This can lead to founder effects, that is, the loss of variation due to the establishment of populations by just a few individuals. This is common among crustaceans, for example, in *Daphnia* (Haileselasie et al. 2017), crabs (Roman 2006), or amphipods (Yampolsky et al. 1994). If such populations remain isolated or otherwise maintain very low levels of genetic diversity, they can suffer from inbreeding depression, and a range of other negative impacts while also making less diverse contributions to the ecological functions these species perform.

Assembly and Loss of Functional Diversity

Although most research on crustacean diversity so far has focused on taxonomic or genetic diversity (Fig. 7.1), understanding both the distribution and consequences of ecological diversity requires that we move beyond counting the number of species in a community or genotypes within a population and instead quantify functional diversity. Assessing the diversity in organismal traits allows for much clearer tests of mechanisms that assemble species into communities and govern their contribution to specific individual functions or composite measures of multifunctionality (Lefcheck et al. 2015). Understanding not only the diversity in individual traits but also the evolutionary links between multiple traits is critical to predicting both current ecological functioning and the consequences of environmental change. This is because evolution has produced a nonrandom set of trait combinations to supply ecosystems, and because these trait combinations constrain both community assembly and the functioning of assembled communities (Bannar-Martin et al. 2018, Fontana et al. 2021).

Ecologists have long been interested in the concept of functional redundancy, where a diverse local community or regional species pool will hold equivalent species that can substitute for each other if one is extirpated. To substitute for a missing species, however, the replacement will need to match both the environmental tolerance traits that allow a species to survive in a habitat (defined as response traits) and the traits that determine a species' contribution to ecosystem functions (defined as effect traits) (Altermatt et al. 2008, Thompson and Shurin 2012). Diversity in response traits among crustaceans (Connell 1961) can lead to a substantial shift in the functional diversity remaining after a disturbance. For example, crustaceans have been found to be more sensitive to pH but less sensitive to organic pollution than some insect larvae, leading to a shift in the functional composition of freshwater detritivore communities and altered decomposition rates (Elmqvist et al. 2003). Gradients in environmental stressors like temperature can also establish diversity gradients as fewer species are able to tolerate more stressful conditions. This can lead to a larger scope for BEF relationships involving other traits and other functions at the lower-stress end of the gradient. In contrast, ecosystem functioning under stronger filters of environmental stress will depend entirely on the availability of the stress-tolerant species. For example, the presence of herbivorous amphipod species with high temperature tolerance has a large effect on the control of algae in warmer waters (Fig. 7.2). Although early work in BEF research used comparisons of randomly assembled communities with few vs. many species to study the potential consequences of biodiversity loss, in reality species will not be randomly lost from their communities (Walker and Thompson 2010).

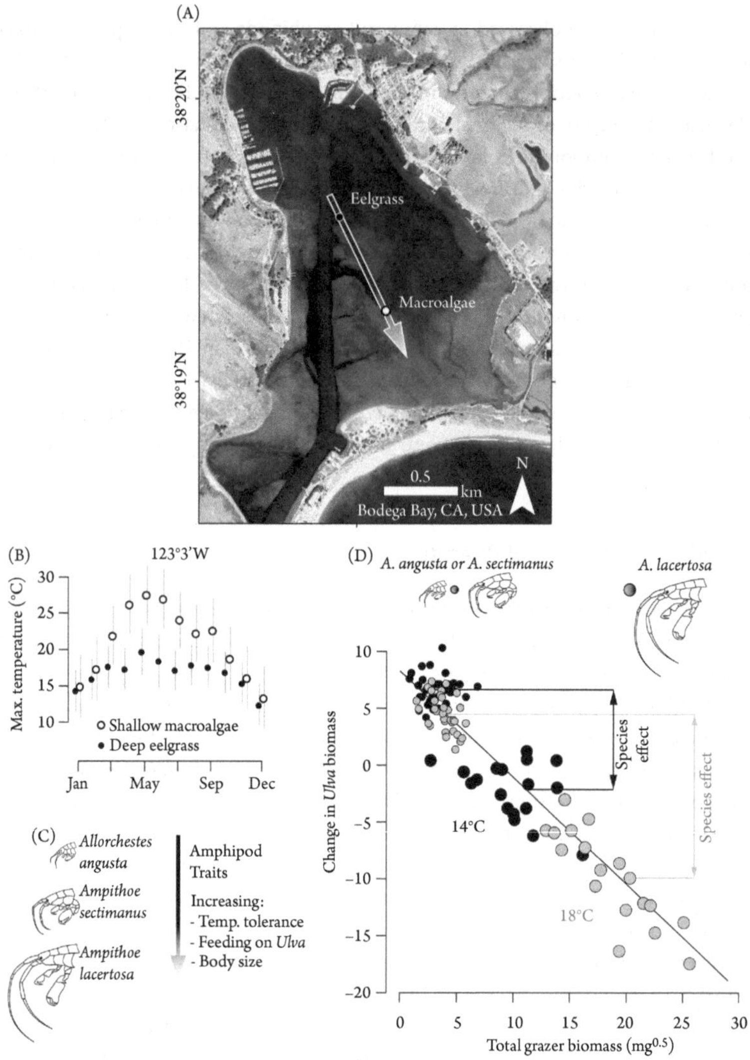

Fig. 7.2.

The community composition and ecological impact of herbivorous amphipods in Bodega Bay, California, is driven by variation in both temperature tolerance and feeding traits. (A) Environmental filtering occurs along a depth gradient from deep eelgrass beds to shallow mudflats covered with *Ulva* spp. macroalgae. Map reproduced courtesy of © Google Earth. (B) Especially during summer low tides, water temperature is substantially higher in shallow macroalgae habitats than deeper eelgrass beds. Source: data from Best and Stachowicz (2014). (C) This environmental variation interacts with trait variation among amphipod species, with *Ampithoe lacertosa* having substantially higher temperature tolerance and ability to consume *Ulva* than smaller species. (D) Over the course of a mesocosm experiment, communities containing *A. lacertosa* had a larger negative impact on macroalgal biomass than communities containing other herbivorous amphipods. This effect of species identity is shown for each temperature (black for 14° C and gray for 18° C) as a contrast between the mean *Ulva* change in communities with vs. without *A. lacertosa*. The effect was amplified at higher temperatures due to the joint effects of temperature tolerance and feeding traits, with *A. lacertosa* increasing its feeding rate on *Ulva* even more when raised at higher temperatures. Source: data from Best et al. (2015).

Within species, genetic diversity underpinning response traits can also affect the distribution of diversity among and within populations. For example, colonization patterns in the barnacle *Balanus improvisus* (Gamfeldt et al. 2005), persistence of the freshwater amphipod *Gammarus duebeni* when exposed to environmental stress (Gamfeldt and Källström 2007), or the response to toxins in the marine amphipod *Peramphithoe parmerong* (Pease et al. 2010) all affect genetic diversity. In the freshwater amphipod *Hyalella azteca,* which is often considered an indicator species due to sensitivity to pollution, some populations of isolated agricultural ditches have evolved much higher tolerance of pyrethroid pesticides, enabled by multiple known mutations (Weston et al. 2013). This evolutionary response has both expanded the fundamental niche of the species and increased genetic differentiation among populations in more vs. less impacted habitats. Genetic variation can also translate into fitness variation in the context of past and ongoing global climate change (Reusch et al. 2005). For example, genetic variation of thermal tolerance in the tidepool copepod *Tigriopus californicus* may be a good predictor for how well populations can cope with rising ambient temperatures (Kelly et al. 2012). This is important to consider because it means the adaptive capacity of a species will be governed by genotype frequencies within component populations. Especially in intertidal and supralittoral species like *T. californicus,* which are controlled by strong predation pressure and have limited dispersal ability, gene flow between populations is low and selection on standing genetic variation may govern the evolutionary response to global climate change. To predict both evolutionary and ecological responses to global change, it is important to increase efforts to quantify intraspecific levels of genetic variation and its ecological consequences.

Effects of Crustacean Diversity on Food Web Functioning

Effects of Species Diversity in Benthic Food Webs

In marine environments, crustaceans play important ecological roles both as predators and herbivores. For example, the number of amphipod and isopod mesograzers in seagrass communities can increase control of epiphytic algae (Fig. 7.3), increase control of multiple sessile algal and invertebrate colonizers at once (Duffy et al. 2007), and increase secondary production (Douglass et al. 2008). Amphipod and isopod diversity can also dampen the predicted impact of trophic cascades, which could otherwise translate a loss of predators into increased blooms of macroalgae. This is because amphipod species vary in the strength of their interactions with both algae and predators such as fish (Best and Stachowicz 2012), due partly to variation in microhabitat use (Lürig et al. 2016). This makes it possible for predators to increase the diversity and evenness (similarity in abundance) of crustacean herbivores through differential consumption rather than removing the entire grazer trophic level (Amundrud et al. 2015).

At the predator level, increased diversity of crustaceans (e.g., multiple species of *Cancer* crabs) can lead to decreased activity of herbivores (urchins, a kelp crab, and snails) and higher biomass of kelp (Byrnes et al. 2006). This type of interaction is called a trait-mediated indirect interaction because diversity affects the feeding rate (a trait) of consumers rather than or as well as their absolute abundance. In addition to horizontal diversity (within a trophic level), the vertical diversity (number of trophic levels) can also shape ecosystem functions like total secondary production or control of algae blooms (Duffy 2002), and diversity within and across trophic levels can interact. For example, predator survival and impact can depend on both the diversity of competing predators and the diversity of their herbivorous prey (Douglass et al. 2008).

The ultimate outcomes of diverse interactions across trophic levels can be hard to predict, especially in more complex communities with omnivory or dispersal (Bruno and O'Connor 2005). One promising avenue for increasing predictive power for multi-trophic diversity may be to quantify

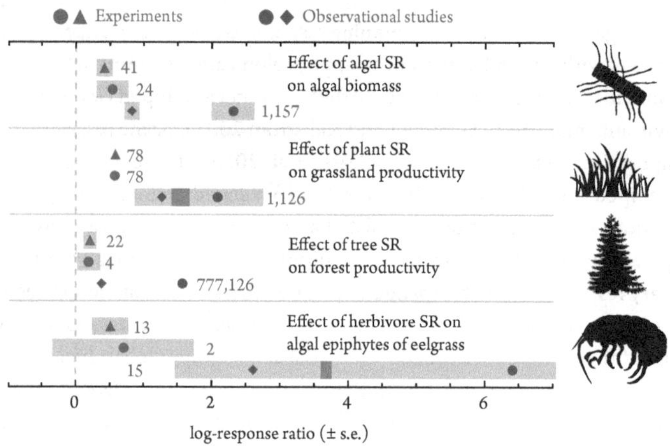

Fig. 7.3.
Species richness effects can be detected using experimental or observational approaches. The size of the species richness effect is given on the x-axis, and numbers indicate the number of experiments or sites studied. Although there are fewer studies of richness effects for crustacean herbivores than for producers, results have generally shown increased control of epiphytic algae with multiple herbivores, especially in the wild. Different symbols within the experiments and observational studies categories indicate different ways of calculating effect sizes; horizontal bands are standard errors. Reproduced from Duffy et al. (2017), with permission from © Nature Publishing Group.

network connections using a common metric of energy flux among food web members (Barnes et al. 2018). Another possible approach is to quantify species interactions across trophic levels in uniform ways using species-specific functional traits. Although the concept of using functional groups to identify species with different methods of obtaining food has a long history in marine and freshwater invertebrates (Cummins and Klug 1979), more recent approaches have sought to quantify the dietary roles of species in a community using continuous traits to construct functional diversity metrics. This can be done by measuring each species' feeding rates on multiple possible foods using lab assays (Best and Stachowicz 2014), or via gut contents or stable isotope ratios from wild-collected individuals (Jeong et al. 2012), or via standardized measurements of mouthparts (Watling 1993). Alternatively, we can construct and test models of physiological trade-offs that allow body size to predict quantitative variation in diet or rates of nutrient cycling (Tessier et al. 2000, Hébert et al. 2015). A continuous trait approach allows for greater resolution of variation in niche breadth among species and enables stronger tests of the hypothesis that increased functional diversity in competitors facilitates coexistence (Best et al. 2013).

As well as shaping coexistence via niche partitioning, feeding traits are important examples of effect traits because they determine the impact of species assemblages on ecosystem function, shaping community control of lower trophic levels or total secondary production. For example, across a wide range of marine habitats and latitudes, the feeding traits of herbivores account for much of the variation in consumption of producers (Poore et al. 2012). Variation in feeding niche also exists in groundwater amphipod species (Hutchins et al. 2014) and can impact water quality via variable control of microbial communities (Griebler et al. [2014] and references reviewed therein). In river ecosystems, peracarid crustaceans (isopods and amphipods) play an important role in the breakdown of organic material such as terrestrial leaf litter, thereby affecting the cycling of nutrients and carbon (Graça et al. 1994). Depending on stream conditions, dominance by a single *Gammarus* species can lead to higher rates of detritivory than communities with an even mixture of crustaceans and insects (Dangles and Malmqvist 2004). Where availability of organic material is high, crustacean detritivores can reach very high abundances (1000–10,000s per square meter) (Graça et al. 1994), representing a substantial portion of benthic biomass production, which is utilized by higher trophic levels.

Effects of Intraspecific Genetic and Phenotypic Diversity in Benthic Food Webs

In addition to variation among species, there is growing awareness that intraspecific diversity can have a wide array of effects that play out across all ecological scales: populations (Agashe 2009), communities (Bolnick et al. 2011), and ecosystems (Raffard et al. 2019). Parallel to taxonomic diversity at the species level, intraspecific variation can be measured as genetic diversity (e.g., allelic diversity, genotypic richness, or genetic variance; for an overview see Hughes et al. [2008]), or as phenotypic diversity, which (if ecologically relevant) is essentially functional diversity at the organismal level rather than averaged at the species level. The study of intraspecific phenotypic diversity has a long history in the field of population genetics, where it is considered to fuel natural selection and evolutionary change (Endler 1986). More recently, there has been increased interest in the effects of intraspecific trait variation on the assembly of and coexistence within communities (Bolnick et al. 2011) and the functioning of ecosystems (Matthews et al. 2011b).

There are multiple sources of phenotypic variation that can influence the impacts of key crustacean species on ecosystem functioning. For instance, the ecological role of the freshwater isopod *A. aquaticus* depends on predation pressure. Being an important detritivore that can be present in high densities, this species is frequently exposed to high predation rates (Hargeby et al. 2004, Lürig et al. 2019) as it can contribute up to 25% of the entire diet of fish (Rask and Hiisivuori 1985). Therefore, *A. aquaticus* needs to balance its microhabitat choice between availability of resources and the value as shelter from predation, a phenomenon that is also known from marine peracarid mesograzers (Boström and Mattila 1999, Lürig et al. 2016). Such behavioral trade-offs may affect ecosystem processes like the cycling of nutrients or dissolved organic carbon in ecosystems with strong predation pressure. Indeed, leaf processing and shredding rates in both *A. aquaticus* and *Gammarus pulex*, another keystone freshwater detritivore, are significantly decreased when predator cues are present (Rask and Hiisivuori 1985, Åbjörnsson et al. 2000), which can affect the transfer of phosphorus through the food web (Calizza et al. 2013).

Important phenotypic variation with ecological consequences can also arise in response to environmental cues during development. Crustaceans are more diverse than any other arthropod group with respect to their developmental biology, including cleavage and gastrulation patterns, segmentation processes, and larval types and developmental patterns (see Volume 7 of this book series). Developmental plasticity, the genotype-specific expression of alternative developmental trajectories in response to the environment, is likely an important source of intraspecific phenotypic variation (West-Eberhard 2003). Either through specific events or prolonged exposure, the environment experienced by juveniles can cue developmental switches, which regulate development at later life history stages. The complex, biphasic life cycles of most crustaceans include phases where developing the phenotype is especially responsive to environmental cues, such as larval or different molting stages (Mileikovsky 1971). Environmental conditions during such phases can trigger developmental switches that have a cascading effect on adult phenotypes.

One important type of environmental cue impacting development is resource quality and quantity, which can have large effects on the development of morphological, physiological, and behavioral traits of individuals. Detritivores often face stoichiometric mismatches (Martinson et al. 2008), which may affect their development. For instance, small-scale variation in detritus quality shapes developmental trajectories of pigmentation in *A. aquaticus* (Marcus et al. 1978, Lürig et al. 2019, Lürig and Matthews 2021) and leads to variation in growth of *G. pulex* (Graça et al. 1993).

The effect of variable diet quality may depend on diet specificity. *Gammarus pulex*, for example, has a relatively narrow food preference and may therefore express less phenotypic variation related to diet quality and quantity (Friberg and Jacobsen 1994). In contrast, *A. aquaticus* expresses a much wider range of feeding preferences (Graça et al. 1993), which may explain the high levels of intrapopulation variation in pigmentation and body size (Marcus et al. 1978, Lürig et al. 2019, Lürig and Matthews 2021). Since these two crustaceans are also important prey, such intraspecific phenotypic variation can affect the growth and the size spectra of subsequent trophic levels (Rask and Hiisivuori 1985).

Effects of Species Diversity in Pelagic Food Webs

In marine and freshwater benthic communities, crustaceans occupy many different trophic positions but share these roles with diverse invertebrates from many other phyla (in the ocean) or from the insect component of the arthropod phylum (in freshwater). In contrast, pelagic zooplankton communities are often dominated by crustaceans such as copepods, cladocerans, and ostracods, which play a critical role in pelagic food webs by grazing phytoplankton and providing biomass for higher trophic levels. Copepods in the epilimnion of lakes (Boxshall and Hayes 2019) and the epipelagic zone in the open ocean are sometimes referred to as the "heart of pelagic food webs" (Fig. 7.4) due to their central interactions with other species, be it through grazing, predator–prey relationships, or as vectors for parasites. Much work has been dedicated to uncovering the role of copepod communities in marine and freshwater food webs (Sommer and Sommer 2006, Lima-Mendez et al. 2015), and diversity in feeding traits among species has important ecological consequences. For example, understanding variation in feeding methods and the formation of fecal pellets among copepods can help predict effects on the ocean carbon cycle by altering the horizontal and vertical movement of carbon (Steinberg and Landry 2017).

Compared to open-ocean food webs, where assessing community diversity and interactions is more difficult, easily accessible and replicated lake ecosystems have allowed for both observational and experimental approaches to investigate the role of planktonic crustaceans in food webs. Calanoid copepods and cladocerans both have a long history in freshwater ecology research, including studies of top-down control and trophic integration (Sommer and Sommer 2006), stoichiometric specificity (Bergström et al. 2018), and ecosystem function (Walsh et al. 2016). Strong effects of complementary foraging and nutrient fluxes among species make small planktonic crustaceans the most important group for the control of planktonic algae. For example, calanoid copepods and cladocerans both control phytoplankton abundances, but from different ends of the size spectrum and through different mechanisms: cladocerans graze small phytoplankton and sequester more limiting nutrients, whereas copepods feed on larger phytoplankton and small protists, which releases small phytoplankton from predation (Sommer and Sommer 2006). Among multiple species of copepods there can also be substantial trait variation, including age and size at metamorphosis and of early postlarval stages (Twombly 1995). Little is known about how such variation may affect top-down control and nutrient cycling in aquatic ecosystems, but there is some evidence that even subtle phenotypic differences in copepods can have contrasting ecosystem effects (Fig. 7.4).

Effects of Intraspecific Genetic and Phenotypic Diversity in Pelagic Food Webs

Especially in cases where a few planktonic species dominate a food web, genetic and phenotypic variation within that species can have large effects on the strength and stability of trophic interactions and the food web as a whole (Moya-Laraño 2011). For instance, Antarctic krill can have tremendous bottom-up effects on intermediate predators and, via trophic cascades, the entire pelagic food web of the Scotia Sea ecosystem (Murphy et al. 2007). It is becoming increasingly clear that the effects of krill on ocean food webs are not only governed by changes in krill abundance (an estimated decrease by 50–80% over the last 30 years; Atkinson et al. 2004), but also

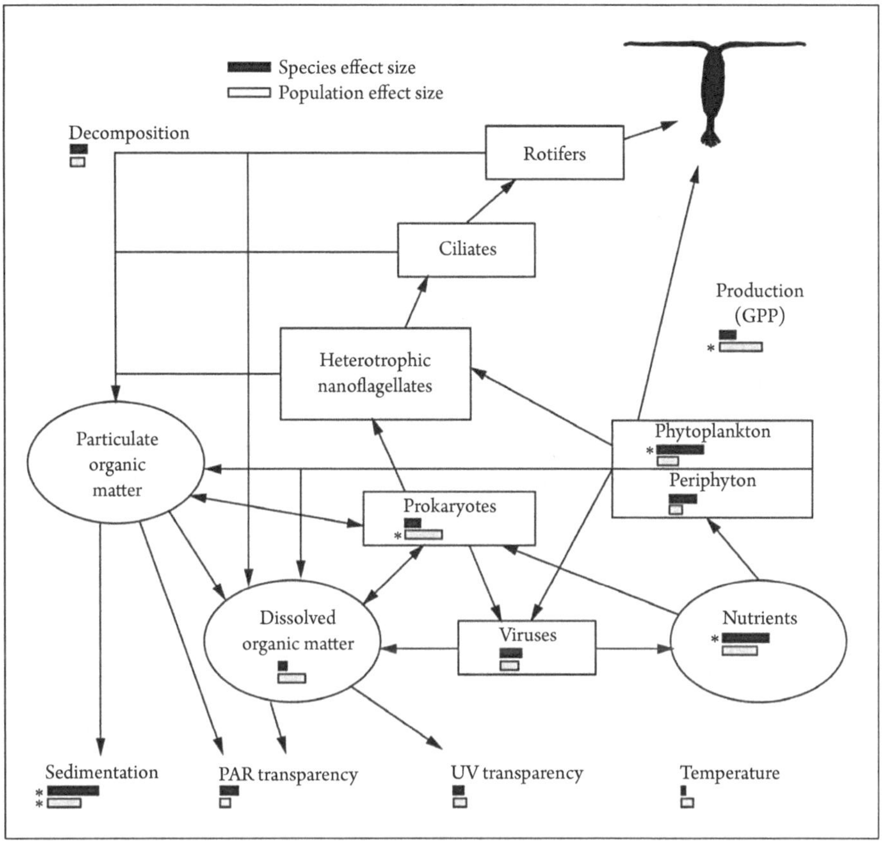

Fig. 7.4.
Zooplankton communities are sometimes referred to as "the heart of pelagic food webs" (Russo et al. 2016) and are relevant to a suite of ecosystem properties and processes. Trait variation both among species and among populations of the same species can impact a suite of biological, chemical, and physical ecosystem properties. Depicted here, within a network of ecosystem interactions, are average effect sizes (Cohen's d) for species contrasts (black bars) and population contrasts (gray bars) of three morphologically similar copepod species (*Leptodiaptomus ashlandi, Hesperodiaptomus franciscanus, Skistodiaptomus oregonensis*). Stars indicate if the ecosystem metric had at least one significant contrast (at the level of $p < 0.05$). Reproduced from Matthews et al. (2011a), under a Creative Commons by Attribution License (CC BY).

by within- and among-year variation in body size spectra, as observed in *Euphausia superba* (Shelton et al. 2013). This is important because the distribution of body sizes and other phenological traits within a species can affect the strength and timing of predator–prey interactions (Moya-Laraño 2011). Unfortunately, little is known about the drivers of variation in growth and other life history traits of open-ocean plankton, which are notoriously difficult to quantify (Shelton et al. 2013).

Just as the ecological impact of key benthic consumers such as freshwater isopods and amphipods is known to be shaped by developmental plasticity, pelagic cladocerans also provide a well-studied example of this source of functional diversity (Walsh et al. 2018). Several *Daphnia* species display an impressive diversity of morphological changes to protect themselves against gape-limited predators when they are exposed to their kairomones during early life stages (Boersma et al. 1998, Weiss and Tollrian 2018). This variation has obvious ecological consequences because it reduces the availability of food to predators. In addition, defense against predation is linked to other ecosystem effects via costs of plasticity and trade-offs among traits, such as increased resource requirement during development or decreased reproductive output (Weber et al. 2003). Costs can also manifest as maladaptive plasticity, which may move phenotypes away from their optima. For example, producing defended phenotypes often carries a cost of increased resource requirements during development or decreased reproductive output (Weber et al. 2003). As for trade-offs, *Daphnia hyalina* and *D. pulex* express more transparent bodies in the presence of predator kairomones, making them less conspicuous to visual predators but more vulnerable to harmful UV radiation, which may affect longevity in clear water with high predation pressure (Tollrian and Heibl 2004). Finally, plasticity in the cladoceran diet, and especially the ability to consume cyanobacteria, impacts the control of algae blooms. Plastic responses to diet can be detected in gene expression data (Walsh et al. 2018) and can evolve in response to the changing producer communities resulting from eutrophication (Hairston et al. 2001). Developmental plasticity and associated trade-off scenarios may be especially important contributors to the maintenance of ecosystem function in environments with high spatiotemporal variability, such as ponds that transition between clear and turbid water states (Hargeby et al. 2007), and in scenarios where standing genetic variation is low (Morris and Rogers 2013).

Another important role of intraspecific variation in food webs comes from the impact of genetic diversity on gut microbes. Little is known about the links between intraspecific variation in crustaceans and the specificity and functional diversity of their gut microbiome (GMB). However, there is mounting evidence for high specificity between *Daphnia* genotypes and their GMB, which affects susceptibility to pathogens (Sison-Mangus et al. 2018) and to toxic cyanobacteria (Macke et al. 2017). Such interdependencies tied to genetic variation of the host populations may affect whole ecosystem responses. For example, GMBs are crucial for the ability of hosts to digest leaf litter and break down cellulose, which has likely contributed to evolutionary diversification of marine isopod lineages that colonized freshwater and terrestrial habitats (Bouchon et al. 2016). Variation in host-genotype-GMB relationships can thus affect breakdown of organic matter (Vonk et al. 2016) and nutrient cycling in terrestrial and freshwater ecosystems (Zimmer et al. 2002).

Finally, high genetic diversity is also beneficial to parasite resistance (King and Lively 2012), which is also documented for cladocerans (Fig. 7.5). Since crustaceans are an important yet underexplored vector for parasites and viruses (Lima-Mendez et al. 2015), it seems likely that variation in parasite resistance may have consequences for community assembly. Moreover, genetic diversity can also affect resistance of local *Daphnia* communities to invasive *Daphnia* populations; more diverse populations have higher resistance against invaders (De Meester et al. 2007).

Emerging Methods for Quantifying Crustacean Diversity

To better assess how diversity is changing in response to climate change and other environmental stressors, and the consequences of this change for ecological function, we need methods that accurately quantify diversity in more habitats over time. Fortunately, emerging methods using genetic approaches to identify species, quantify genetic variation within species, and rapidly assess phenotypic or functional variation are increasingly available.

Quantifying Species Diversity Using Genetic Approaches

Genetic methods have long been important for identifying cryptic diversity in crustacean species complexes that could not be detected with morphological approaches (Witt et al. 2006, Knox et al. 2012). Increased efficiency and decreased cost of sequencing across a wide range of taxa, even without species-specific markers, now make it possible to quantify diversity across a wide range of taxonomic scales, from genetic diversity within and among populations of a single species, to

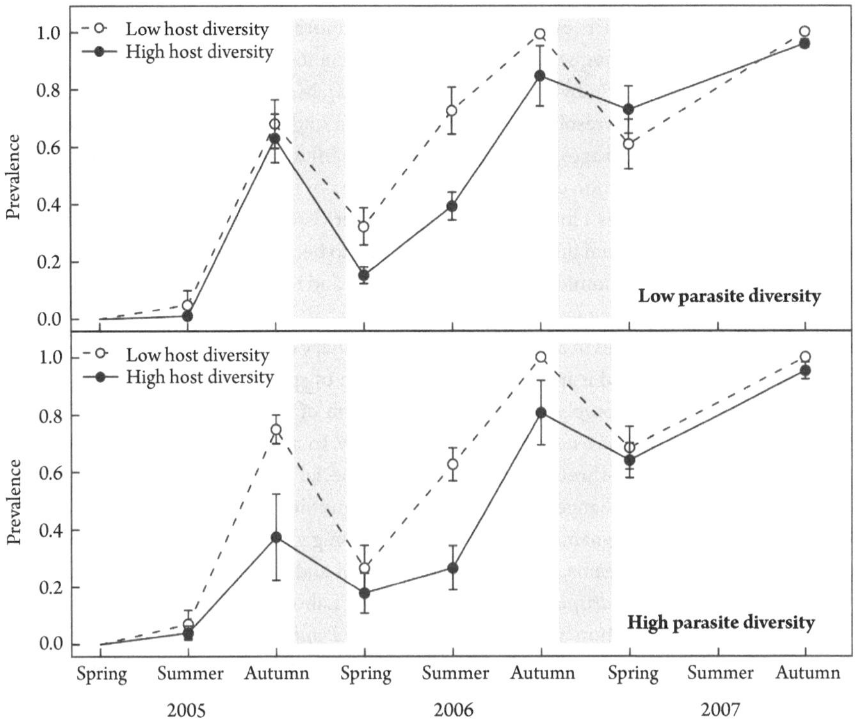

Fig. 7.5.
Variation in intraspecific diversity can have significant fitness consequences for crustacean hosts. For instance, *Daphnia magna* populations with low genetic diversity (open dots) have a higher susceptibility to parasites than populations with high genetic diversity (filled dots). This effect was evident both in parasite treatments with low (top panel) and high (bottom panel) genetic diversity. In the winter diapause, indicated by the gray area, the hosts only survived in sexually produced resting stages. Reproduced from Altermatt and Ebert (2008), with permission from © John Wiley & Sons.

differentiation across a species complex (Rahman et al. 2018). Using universal COI (cytochrome oxidase I) barcodes, it is also possible to use genetic methods to quantify whole-community taxonomic diversity rapidly and efficiently across distantly related species of crustaceans and their co-inhabitants (Lindeque et al. 2013). Two major options include eDNA, which is easy to collect from a water sample but difficult to connect to a precise origin in space and time, especially in large or well-connected water bodies like rivers and oceans (Bista et al. 2017), and direct sequencing of invertebrate community samples (Radulovici et al. 2009).

Quantifying Intraspecific Phenotypic Variation Using Computer Vision

Our capacity to quantify genotypic variation has greatly outpaced our capacity to quantify phenotypic variation. However, without highly resolved phenotypic data many of the standing questions regarding intraspecific variation will be impossible to answer—not just in crustaceans, but any taxonomic group. Therefore, the study of an organisms' phenome, that is, the phenotype as a whole, has been suggested as a way to reduce the current mismatch between the rates at which molecular and phenotypic data become available (Bilder et al. 2009, Houle et al. 2010). In practice, the assembly of phenomic data sets entails the collection of more traits (increased trait dimensionality), higher measurement frequencies (sampling the same individuals more often), and higher sample size (measuring more individuals from a population). Many biologists are taking digital images, which, due to their high resolution and pixel-depth (a single pixel can have over 16.7 million colors in a standard 24-bit image) provide an excellent medium to capture phenotypic information. However, manual extraction of this information is a major bottleneck, as it is time consuming, not reproducible, and requires a lot of expertise. Computer vision (CV), the automatic extraction of meaningful information from images, has the potential to become a key method to collect high-resolution, reproducible phenomic data on a massive scale and to relieve a long-standing analytical bottleneck in ecological and evolutionary research (Lürig et al. 2021).

When working with images in an ecological or evolutionary context, there are three basic types of tasks that can be addressed using CV: (i) the detection of species or specimens, (ii) the identification of species or phenotypes, and (iii) the extraction of phenotypic information. There is a growing body of work on crustaceans that has used CV to answer a broad array of biological questions. For example, high-throughput imaging using special scanning devices and custom software was used to quantify changes in zooplankton community structure (Gorsky et al. 2010). Community or population dynamics can also be traced using video cameras installed in the field. From the resulting video streams, individuals are counted and their phenotypes extracted, as in the squat lobster *Munida quadrispina* (Aguzzi et al. 2011). Laboratory video assays can also reveal intraspecific behavioral variation at the individual level in *Daphnia* (Huang et al. 2017). One advantage of images that are produced in the laboratory is their potentially high signal-to-noise ratio. The objects that are contained in images (specimens and their traits) typically contrast strongly with the background so that automatic image analysis with simple signal processing approaches become feasible (Lürig et al. 2021). For example, intra-population variation in pigmentation and body size in the freshwater isopod *A. aquaticus* was captured using a modified flatbed scanner and thresholding algorithms from an open-source CV package for Python (Fig. 7.6) (Lürig et al. 2019, Lürig and Matthews 2021, Lürig 2022). Artificial-intelligence-based CV algorithms provide a robust approach to capture individual-level variation in noisy images taken in field environments, and even have the potential to discriminate between individuals from the same populations, as in the southern rock lobster *Jasus edwardsii* (Vo et al. 2020).

Quantifying Interspecific Functional Diversity Using Phylogenetic Approaches

At a greater taxonomic scale, there has also been substantial interest in using evolutionary divergence as a proxy for ecological distances, under the assumption that closer evolutionary relatives will be more ecologically similar. The ecological differences that filter species into habitats and shape their contributions to ecosystem function can be difficult to measure, but the relative magnitude of these differences might be predictable from the amount of evolutionary time that species have had to diverge from each other. If it is possible to construct a well-resolved molecular phylogeny with accurate branch lengths connecting all community members, the branch distances among species can be summed or averaged to obtain a range of metrics capturing community phylogenetic diversity or dispersion (Tucker et al. 2017). We can then test whether ecosystem function is maximized by high phylogenetic diversity (as a proxy for the high functional diversity often found in groups of distantly related species). Community phylogenetics approaches also include testing whether coexisting species are more closely or distantly related than expected from random community assembly, as a possible indicator of whether species are coexisting due to high functional similarity (via environmental filtering) or high functional dispersion (via competitive exclusion of similar species). This interpretation depends critically on whether important traits show phylogenetic signal, meaning that trait differences are indeed correlated with evolutionary distances, an assumption easily violated by convergent evolution (Best and Stachowicz 2013). In addition,

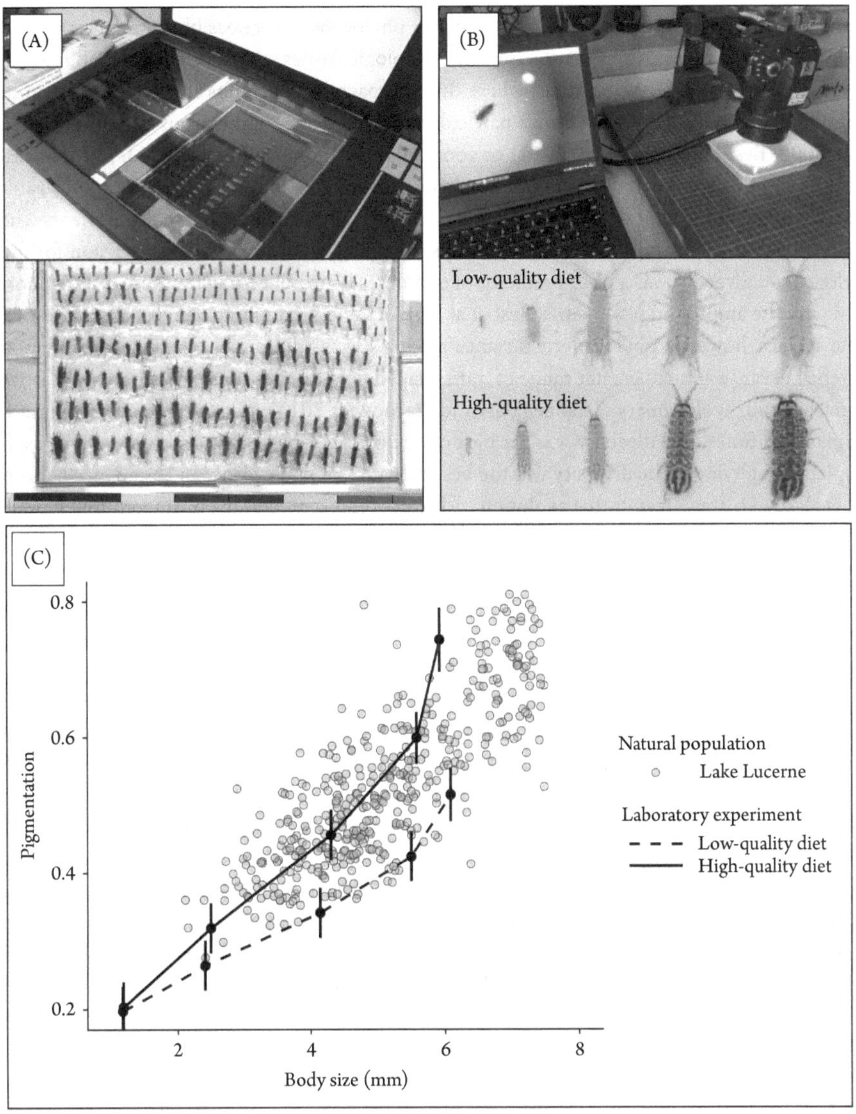

Fig. 7.6.
Computer vision can be used for high-throughput phenotyping in natural and experimental contexts to alleviate image analysis as an analytical bottleneck. Here, two methodological approaches were combined to investigate the contribution of phenotypic plasticity to variation of pigmentation and body size in natural populations of the freshwater isopod *Asellus aquaticus*. (A) Several hundred isopod specimens collected from Lake Lucerne (Kastanienbaum, Switzerland) were phenotyped simultaneously by placing them upside down on the wetted surface of a modified flatbed scanner. The positive relationship between body size and pigmentation can be explained with increasing buildup of pigmentation as isopods grow (Lürig et al. 2019). (B) To investigate the role of phenotypic plasticity in variation of pigmentation for a given body size, a subsequent laboratory experiment was conducted: juvenile siblings from the same family were reared on different diets (low vs. high protein-carbohydrate ratios, i.e., low vs. high diet quality) and photographed on a camera stand every three weeks. Source: data from Lürig and Matthews (2021). (C) Throughout the entire duration of the experiment, isopod siblings reared under the two diet qualities express considerable variation in pigmentation for a given body size, which spans almost the entire range of variation found in natural populations. Source: data from Lürig and Matthews (2021). In both studies, close to 10,000 isopods from almost 5000 pictures were analyzed using simple thresholding algorithms (Lürig 2022), an effort that would have been impossible to replicate by manual labor. It is likely that computer vision will become a basic component of the biologist's toolkit.

connecting all community members on an accurate phylogeny is not possible for many marine and freshwater communities where crustaceans share ecological roles with species from multiple other invertebrate phyla. The few examples where this approach has been possible have been communities dominated by crustaceans from a particular lineage, such as herbivorous amphipods (Poore et al. 2008, Best and Stachowicz 2014) or zooplankton (Thompson et al. 2015). In these cases, the utility of this approach depends on the phylogenetic signal in the particular traits relevant to the scale of the community processes under study. For example, in mesocosm experiments where realized community composition of amphipods was likely determined by competition for food, phylogenetic diversity was a poor predictor of outcomes because feeding traits are not well correlated with the amphipod phylogeny (Best et al. 2013). When examining field communities of the same species, however, phylogenetic distance among species did predict their probability of occurrence because a much greater range of traits shaped field distributions and some of these were correlated with evolutionary distance (Best and Stachowicz 2014). In natural pond communities, zooplankton functional diversity was the best predictor of zooplankton community biomass, but zooplankton phylogenetic diversity was the best predictor of phytoplankton abundance (Thompson et al. 2015). This suggests that evolutionary history represented by the phylogeny might predict variation in unmeasured traits important for control of algae better than available measures of feeding group and habitat-use traits.

COMPARISON WITH OTHER TAXA

The extent to which crustaceans make unique contributions to ecosystem functioning compared to other taxa will depend on their relative overlap in both response and effect traits. First, differences in response traits such as tolerance of temperature, salinity, pH, pollutants, desiccation, depth, and water flow will limit community membership in ways that may favor or exclude freshwater crustaceans relative to aquatic insects (Elmqvist et al. 2003). For example, egg diapause in many freshwater crustaceans plays a unique role in the persistence of genetic and community diversity over time in habitats subject to drying. Drought-resistant diapause stages are certainly present in other taxa of benthic macroinvertebrates, but to a lesser extent (Alekseev et al. 2007). On the other hand, the difference between direct development and planktonic larval stages also has a major impact on colonization and persistence in communities, and this life history trait varies as much within crustacean lineages as it does across phyla.

Once communities are assembled, their functioning is also shaped by effect traits that vary widely across both crustacean and non-crustacean community members. Benthic crustacean grazers share their feeding niche with gastropods across both marine and freshwater habitats, as well as additional mollusks and other phyla in marine systems. Filter feeders in benthic habitats include crustaceans (barnacles, benthic ostracods, and cladocerans), bivalves, bryozoans, cnidarians, sponges, and tunicates. In planktonic environments, crustacean filter feeders share their niche space with rotifers and/or with pelagic cnidarians. Thus, the extent to which food web functioning can be understood in these systems by focusing only on crustaceans depends on the extent to which crustaceans are numerically dominant.

FUTURE DIRECTIONS

Functional diversity across and within crustacean species is a critical tool for predicting both the changing distribution of aquatic biodiversity and its consequences for ecosystem functioning. Importantly, functional diversity depends on phenotypes that are modified by evolutionary change and by phenotypic plasticity, causing them to vary across space and over multiple timescales (Matthews et al. 2011b). Within species, evolutionary change can occur rapidly enough for feedbacks to emerge between the evolution of traits that shape ecosystems and the ecological conditions that shape selection. Eco-evolutionary feedback is a fascinating area of current and future study because it offers insight into the complexity of ecological interactions that may have shaped past micro- to macroevolution while also revealing evolutionary dimensions to the way whole communities and ecosystems might respond to new stressors. The best examples of this kind of feedback in crustaceans so far are probably in cladocerans such as *Daphnia*, where the evolution of dietary niche, stoichiometric ratios, and defense against predators can all evolve and diverge rapidly and interact with major ecosystem characteristics like lake eutrophication and fish biomass (Miner et al. 2012, Walsh et al. 2018). Progress on understanding such feedback between ecology and evolution should be greatly enhanced by the continued development of sequencing technologies and high-throughput phenotyping. In combination, these two advances should increase capacity to identify specific genes coding for evolving traits within species, and to track the distribution and changes in those genes in response to stressors such as increasing temperature or pollution (Berry et al. 2019). It is also important, however, to move beyond this single-species framework to ask questions about the simultaneous evolution of interacting community members (e.g., both *Daphnia* and their algal food source; Tseng et al. 2019). For example, how does the potential rate of adaptation to change (adaptive capacity) within a species depend on interactions with other crustacean or non-crustacean species structuring niche space? Conversely, how does the effect of interspecific diversity on ecosystem functioning depend on evolutionary or plastic changes in phenotypic values within certain keystone species? If the main objective is conserving or restoring ecosystem functioning, what is the relative importance of maintaining species diversity across closely and distantly related lineages vs. using natural or assisted evolution of key traits within species? These questions integrating different scales of functional diversity from population genetics to community ecology are likely to see a great deal of attention from both basic and applied perspectives.

SUMMARY AND CONCLUSIONS

Crustaceans contribute a fascinating breadth of functional trait diversity across their habitats and lineages, interacting with each other and with a range of other invertebrates to shape ecosystem functioning. As with biodiversity and ecosystem functioning (BEF) research more broadly, a great deal can be learned about the historical, present, and future functioning of crustacean communities by quantifying functional traits among and within species. To maximize this potential, it should be particularly valuable to prioritize (i) studies that link crustacean traits to their colonization and persistence in the face of global change, (ii) studies that investigate how the evolution of multivariate trait combinations constrains joint processes of community assembly and ecosystem functioning, and (iii) studies that consider functional trait variation across multiple spatial and temporal scales to compare the ecological contributions of changing species- and population-level trait distributions.

ACKNOWLEDGMENTS

We thank Jay Stachowicz for inspiring our interest in marine crustaceans as exciting examples of a wide range of species interactions linking community diversity and ecosystem function, and Blake Matthews for being excited about both freshwater copepods and genetic diversity. For enabling the translation of inspiration into data, we thank undergraduate, graduate, staff, and faculty colleagues at the Bodega Marine Laboratory and the Swiss Federal Institute for Aquatic Sciences and Technology in Kastanienbaum, two very different places that both allowed us to collaborate on multiple projects involving at least one crustacean species.

REFERENCES

Åbjörnsson, K., J. Dahl, P. Nyström, and C. Brönmark. 2000. Influence of predator and dietary chemical cues on the behaviour and shredding efficiency of *Gammarus pulex*. Aquatic Ecology 34:379–387.

Agashe, D. 2009. The stabilizing effect of intraspecific genetic variation on population dynamics in novel and ancestral habitats. The American Naturalist 174:255–267.

Aguzzi, J., C. Costa, K. Robert, M. Matabos, F. Antonucci, S.K. Juniper, and P. Menesatti. 2011. Automated image analysis for the detection of benthic crustaceans and bacterial mat coverage using the VENUS undersea cabled network. Sensors 11:10534–10556.

Ahyong, S.T. and C. Huang. 2020. Colonization, adaptation, radiation, and diversity in fresh water. Pages 303 -330 *in* G.C.B. Poore and M. Thiel, editors. The natural history of the Crustacea, Volume 8: Evolution and Biogeography. Oxford University Press, New York, USA

Alekseev, V.R., B.T. de Stasio, and J.J. Gilbert. 2007. Diapause in aquatic invertebrates—Theory and human use. Springer, Dordrecht, The Netherlands.

Altermatt, F., and D. Ebert. 2008. Genetic diversity of *Daphnia magna* populations enhances resistance to parasites. Ecology Letters 11:918–928.

Altermatt, F., and D. Ebert. 2010. Populations in small, ephemeral habitat patches may drive dynamics in a *Daphnia magna* metapopulation. Ecology 91:2975–2982.

Altermatt, F., V.I. Pajunen, and D. Ebert. 2008. Climate change affects colonization dynamics in a metacommunity of three *Daphnia* species. Global Change Biology 14:1209–1220.

Amundrud, S.L., D.S. Srivastava, and M.I. O'Connor. 2015. Indirect effects of predators control herbivore richness and abundance in a benthic eelgrass (*Zostera marina*) mesograzer community. The Journal of Animal Ecology 84:1092–1102.

Atkinson, A., V. Siegel, E. Pakhomov, and P. Rothery. 2004. Long-term decline in krill stock and increase in salps within the Southern Ocean. Nature 432:100–103.

Bánki, O., Y. Roskov, L. Vandepitt, R.E. DeWalt, D. Remsen, P. Schalk, T. Orrell, M. Keping, J. Miller, R. Aalbu, R. Adlard, E. Adriaenssens, C. Aedo, E. Aescht, N. Akkari, M.A. Alondo-Zarazaga, B. Alverez, F. Alverez, G. Anderson, et al. 2021. Catalogue of life checklist (annual checklist 2021). Catalogue of Life, Leiden, The Netherlands. Accessed November 16, 2024. [https://www.catalogueoflife.org/col].

Bannar-Martin, K.H., C.T. Kremer, S.K.M. Ernest, M.A. Leibold, H. Auge, J. Chase, S.A.J. Declerck, N. Eisenhauer, S. Harpole, H. Hillebrand, F. Isbell, T. Koffel, S. Larsen, A. Narwani, J.S. Petermann, C. Roscher, J.S. Cabral, and S.R. Supp. 2018. Integrating community assembly and biodiversity to better understand ecosystem function: The Community Assembly and the Functioning of Ecosystems (CAFE) approach. Ecology Letters 21:167–180.

Barnes, A.D., M. Jochum, J.S. Lefcheck, N. Eisenhauer, C. Scherber, M.I. O'Connor, P. de Ruiter, and U. Brose. 2018. Energy Flux: The link between multitrophic biodiversity and ecosystem functioning. Trends in Ecology and Evolution 33:186–197.

Bergström, A.-K., J. Karlsson, D. Karlsson, and T. Vrede. 2018. Contrasting plankton stoichiometry and nutrient regeneration in northern arctic and boreal lakes. Aquatic Sciences 80:24.

Berry, T.E., B.J. Saunders, M.L. Coghlan, M. Stat, S. Jarman, A.J. Richardson, C.H. Davies, O. Berry, E.S. Harvey, and M. Bunce. 2019. Marine environmental DNA biomonitoring reveals seasonal patterns in biodiversity and identifies ecosystem responses to anomalous climatic events. PLoS Genetics 15:e1007943.

Best, R.J., N.C. Caulk, and J.J. Stachowicz. 2013. Trait vs. phylogenetic diversity as predictors of competition and community composition in herbivorous marine amphipods. Ecology Letters 16:72–80.

Best, R.J., and J.J. Stachowicz. 2012. Trophic cascades in seagrass meadows depend on mesograzer variation in feeding rates, predation susceptibility, and abundance. Marine Ecology Progress Series 456:29–42.

Best, R.J., and J.J. Stachowicz. 2013. Phylogeny as a proxy for ecology in seagrass amphipods: Which traits are most conserved? PloS ONE 8:e57550.

Best, R.J., and J.J. Stachowicz. 2014. Phenotypic and phylogenetic evidence for the role of food and habitat in the assembly of communities of marine amphipods. Ecology 95:775–786.

Best, R.J., M.N. Stone, and J.J. Stachowicz. 2015. Predicting consequences of climate change for ecosystem functioning: Variation across trophic levels, species and individuals. Diversity and Distributions 21:1364–1374.

Bilder, R.M., F.W. Sabb, T.D. Cannon, E.D. London, J.D. Jentsch, D.S. Parker, R.A. Poldrack, C. Evans, and N.B. Freimer. 2009. Phenomics: The systematic study of phenotypes on a genome-wide scale. Neuroscience 164:30–42.

Bista, I., G.R. Carvalho, K. Walsh, M. Seymour, M. Hajibabaei, D. Lallias, M. Christmas, and S. Creer. 2017. Annual time-series analysis of aqueous eDNA reveals ecologically relevant dynamics of lake ecosystem biodiversity. Nature Communications 8:14087.

Boersma, M., P. Spaak, and L. De Meester. 1998. Predator-mediated plasticity in morphology, life history, and behavior of *Daphnia*: The uncoupling of responses. The American Naturalist 152:237–248.

Bolnick, D.I., P. Amarasekare, M.S. Araújo, R. Bürger, J.M. Levine, M. Novak, V.H.W. Rudolf, S.J. Schreiber, M.C. Urban, and D.A. Vasseur. 2011. Why intraspecific trait variation matters in community ecology. Trends in Ecology and Evolution 26:183–192.

Boström, C., and J. Mattila. 1999. The relative importance of food and shelter for seagrass-associated invertebrates: A latitudinal comparison of habitat choice by isopod grazers. Oecologia 120:162–170.

Bouchon, D., M. Zimmer, and J. Dittmer. 2016. The terrestrial isopod microbiome: An all-in-one toolbox for animal-microbe interactions of ecological relevance. Frontiers in Microbiology 7:1472.

Boxshall, G., and P. Hayes. 2019. Biodiversity and taxonomy of the parasitic Crustacea. Pages 73–134 *in* N.J. Smit, N.L. Bruce, and K.A. Hadfield, editors. Parasitic Crustacea: State of knowledge and future trends. Springer, Cham, Switzerland.

Bruno, J.F., and M.I. O'Connor. 2005. Cascading effects of predator diversity and omnivory in a marine food web. Ecology Letters 8:1048–1056.

Byrnes, J., J.J. Stachowicz, K.M. Hultgren, A. Randall Hughes, S.V. Olyarnik, and C.S. Thornber. 2006. Predator diversity strengthens trophic cascades in kelp forests by modifying herbivore behaviour. Ecology Letters 9:61–71.

Calizza, E., L. Rossi, and M.L. Costantini. 2013. Predators and resources influence phosphorus transfer along an invertebrate food web through changes in prey behaviour. PloS ONE 8:e65186.

Connell, J.H. 1961. The influence of interspecific competition and other factors on the distribution of the barnacle *Chthamalus stellatus*. Ecology 42:710–723.

Connolly, J., T. Bell, T. Bolger, C. Brophy, T. Carnus, J.A. Finn, L. Kirwan, F. Isbell, J. Levine, A. Lüscher, V. Picasso, C. Roscher, M.T. Sebastia, M. Suter, and A. Weigelt. 2013. An improved model to predict the effects of changing biodiversity levels on ecosystem function. The Journal of Ecology 101:344–355.

Crowe, T.P., and R. Russell. 2009. Functional and taxonomic perspectives of marine biodiversity. Pages 375–390 *in* M. Wahl, editor. Marine hard bottom communities: Patterns, dynamics, diversity, and change. Springer, Berlin, Heidelberg, Germany.

Cummins, K.W., and M.J. Klug. 1979. Feeding ecology of stream invertebrates. Annual Review of Ecology and Systematics 10:147–172.

Dangles, O., and B. Malmqvist. 2004. Species richness-decomposition relationships depend on species dominance. Ecology Letters 7:395–402.

De Meester, L., G. Louette, C. Duvivier, C. Van Damme, and E. Michels. 2007. Genetic composition of resident populations influences establishment success of immigrant species. Oecologia 153:431–440.

Douglass, J.G., J.E. Duffy, and J.F. Bruno. 2008. Herbivore and predator diversity interactively affect ecosystem properties in an experimental marine community. Ecology Letters 11:598–608.

Duffy, J.E. 2002. Biodiversity and ecosystem function: The consumer connection. Oikos 99:201–219.

Duffy, J.E., B.J. Cardinale, K.E. France, P.B. McIntyre, E. Thébault, and M. Loreau. 2007. The functional role of biodiversity in ecosystems: Incorporating trophic complexity. Ecology Letters 10:522–538.

Duffy, J.E., C.M. Godwin, and B.J. Cardinale. 2017. Biodiversity effects in the wild are common and as strong as key drivers of productivity. Nature 549:261–264.

Elmqvist, T., C. Folke, M. Nyström, G. Peterson, J. Bengtsson, B. Walker, and J. Norberg. 2003. Response diversity, ecosystem change, and resilience. Frontiers in Ecology and the Environment 1:488–494.

Endler, J.A. 1986. Natural selection in the wild. Princeton University Press, New Jersey, USA.

Fontana, S., S. Rasmann, F. de Bello, F. Pomati, and M. Moretti. 2021. Reconciling trait-based perspectives along a trait integration continuum. Ecology 102:e03472.

France, K., and J.E. Duffy. 2006. Consumer diversity mediates invasion dynamics at multiple trophic levels. Oikos 113:515–529.

Friberg, N., and D. Jacobsen. 1994. Feeding plasticity of two detritivore-shredders. Freshwater Biology 32:133–142.

Gamfeldt, L., and B. Källström. 2007. Increasing intraspecific diversity increases predictability in population survival in the face of perturbations. Oikos 116:700–705.

Gamfeldt, L., J. Wallén, P.R. Jonsson, K.M. Berntsson, and J.N. Havenhand. 2005. Increasing intraspecific diversity enhances settling success in a marine invertebrate. Ecology 86:3219–3224.

Gorsky, G., M.D. Ohman, M. Picheral, S. Gasparini, L. Stemmann, J.-B. Romagnan, A. Cawood, S. Pesant, C. García-Comas, and F. Prejger. 2010. Digital zooplankton image analysis using the ZooScan integrated system. Journal of Plankton Research 32:285–303.

Graça, M.A.S., L. Maltby, and P. Calow. 1993. Importance of fungi in the diet of *Gammarus pulex* and *Asellus aquaticus*. II. Effects on growth, reproduction and physiology. Oecologia 96:304–309.

Graça, M.A.S., L. Maltby, and P. Calow. 1994. Comparative ecology of *Gammarus pulex* (L.) and *Asellus aquaticus* (L.) I: Population dynamics and microdistribution. Hydrobiologia 281:155–162.

Griebler, C., F. Malard, and T. Lefébure. 2014. Current developments in groundwater ecology—from biodiversity to ecosystem function and services. Current Opinion in Biotechnology 27:159–167.

Haileselasie, T.H., J. Mergeay, J. Vanoverbeke, L. Orsini, and L. De Meester. 2017. Founder effects determine the genetic structure of the water flea *Daphnia* in Ethiopian reservoirs. Limnology and Oceanography 63:915–926.

Hairston, N.G., Jr, C.L. Holtmeier, W. Lampert, L.J. Weider, D.M. Post, J.M. Fischer, C.E. Cáceres, J.A. Fox, and U. Gaedke. 2001. Natural selection for grazer resistance to toxic cyanobacteria: Evolution of phenotypic plasticity? Evolution 55:2203–2214.

Hargeby, A., I. Blindow, and G. Andersson. 2007. Long-term patterns of shifts between clear and turbid states in Lake Krankesjön and Lake Tåkern. Ecosystems 10:29–36.

Hargeby, A., J. Johansson, and J. Ahnesjö. 2004. Habitat-specific pigmentation in a freshwater isopod: Adaptive evolution over a small spatiotemporal scale. Evolution 58:81–94.

Hébert, M.-P., B.E. Beisner, and R. Maranger. 2015. A meta-analysis of zooplankton functional traits influencing ecosystem function. Ecology 97:1069–1080.

Hillebrand, H., D.M. Bennett, and M.W. Cadotte. 2008. Consequences of dominance: A review of evenness effects on local and regional ecosystem processes. Ecology 89:1510–1520.

Houle, D., D.R. Govindaraju, and S. Omholt. 2010. Phenomics: The next challenge. Nature Reviews Genetics 11:855–866.

Huang, Y., O. Campana, and D. Wlodkowic. 2017. A millifluidic system for analysis of *Daphnia magna* locomotory responses to water-born toxicants. Scientific Reports 7:17603.

Hughes, A.R., B.D. Inouye, M.T.J. Johnson, N. Underwood, and M. Vellend. 2008. Ecological consequences of genetic diversity. Ecology Letters 11:609–623.

Hutchins, B.T., B.F. Schwartz, and W.H. Nowlin. 2014. Morphological and trophic specialization in a subterranean amphipod assemblage. Freshwater Biology 59:2447–2461.

Jeffs, A., and R. O'Rorke. 2020. Feeding and nutrition of crustacean larvae. Pages 309–331 *in* K. Anger, S. Harzsch, and M. Thiel, editors. The natural history of the Crustacea, Volume 7: Developmental biology and larval ecology. Oxford University Press, New York, USA.

Jeong, S.J., H.L. Suh, and C.K. Kang. 2012. Trophic diversity in amphipods within a temperate eelgrass ecosystem as determined by gut contents and C and N isotope analysis. Marine Biology 159:1943–1954.

Kelly, M.W., E. Sanford, and R.K. Grosberg. 2012. Limited potential for adaptation to climate change in a broadly distributed marine crustacean. Proceedings of the Royal Society B: Biological Sciences 279:349–356.

King, K.C., and C.M. Lively. 2012. Does genetic diversity limit disease spread in natural host populations? Heredity 109:199–203.

Knox, M.A., I.D. Hogg, C.A. Pilditch, A.-N. Lörz, P.D.N. Hebert, and D. Steinke. 2012. Mitochondrial DNA (COI) analyses reveal that amphipod diversity is associated with environmental heterogeneity in deep-sea habitats. Molecular Ecology 21:4885–4897.

Lefcheck, J.S., J.E.K. Byrnes, F. Isbell, L. Gamfeldt, J.N. Griffin, N. Eisenhauer, M.J.S. Hensel, A. Hector, B.J. Cardinale, and J.E. Duffy. 2015. Biodiversity enhances ecosystem multifunctionality across trophic levels and habitats. Nature Communications 6:6936.

Lima-Mendez, G., K. Faust, N. Henry, J. Decelle, S. Colin, F. Carcillo, S. Chaffron, J.C. Ignacio-Espinosa, S. Roux, F. Vincent, L. Bittner, Y. Darzi, J. Wang, S. Audic, L. Berline, G. Bontempi, A.M. Cabello, L. Coppola, F.M. Cornejo-Castillo, F. d'Ovidio, L. De Meester, I. Ferrera, M.-J. Garet-Delmas, L. Guidi, E. Lara, S. Pesant, M. Royo-Llonch, G. Salazar, P. Sánchez, M. Sebastian, C. Souffreau, C. Dimier, M. Picheral, S. Searson, S. Kandels-Lewis, Tara Oceans coordinators, G. Gorsky, F. Not, H. Ogata, S. Speich, L. Stemmann, J. Weissenbach, P. Wincker, S.G. Acinas, S. Sunagawa, P. Bork, M.B. Sullivan, E. Karsenti, C. Bowler, C. de Vargas, and J. Raes. 2015. Ocean plankton. Determinants of community structure in the global plankton interactome. Science 348:1262073.

Lindeque, P.K., H.E. Parry, R.A. Harmer, P.J. Somerfield, and A. Atkinson. 2013. Next generation sequencing reveals the hidden diversity of zooplankton assemblages. PloS ONE 8:e81327.

Lürig, M.D. 2022. Phenopype: A Phenotyping Pipeline for Python. Methods in Ecology and Evolution 13: 569–576.

Lürig, M.D., R.J. Best, and J.J. Stachowicz. 2016. Microhabitat partitioning in seagrass mesograzers is driven by consistent species choices across multiple predator and competitor contexts. Oikos 125:1324–1333.

Lürig, M.D., R.J. Best, M. Svitok, J. Jokela, and B. Matthews. 2019. The role of plasticity in the evolution of cryptic pigmentation in a freshwater isopod. The Journal of Animal Ecology 88:612–623.

Lürig, M.D., S. Donoughe, E.I. Svensson, A. Porto, and M. Tsuboi. 2021. Computer vision, machine learning, and the promise of phenomics in ecology and evolutionary biology. Frontiers in Ecology and Evolution 9:148.

Lürig, M.D., and B. Matthews. 2021. Dietary-based developmental plasticity affects juvenile survival in an aquatic detritivore. Proceedings of the Royal Society B: Biological Sciences 288:20203136.

Macke, E., M. Callens, L. De Meester, and E. Decaestecker. 2017. Host-genotype dependent gut microbiota drives zooplankton tolerance to toxic cyanobacteria. Nature Communications 8:1608.

Marcus, J.H., D.W. Sutcliffe, and L.G. Willoughby. 1978. Feeding and growth of *Asellus aquaticus* (Isopoda) on food items from the littoral of Windermere, including green leaves of *Elodea canadensis*. Freshwater Biology 8:505–519.

Martinson, H.M., K. Schneider, J. Gilbert, J.E. Hines, P.A. Hambäck, and W.F. Fagan. 2008. Detritivory: Stoichiometry of a neglected trophic level. Ecological Research 23:487–491.

Matthews, B., S. Hausch, C. Winter, C.A. Suttle, and J.B. Shurin. 2011a. Contrasting ecosystem-effects of morphologically similar copepods. PloS ONE 6:e26700.

Matthews, B., A. Narwani, S. Hausch, E. Nonaka, H. Peter, M. Yamamichi, K.E. Sullam, K.C. Bird, M.K. Thomas, T.C. Hanley, and C.B. Turner. 2011b. Toward an integration of evolutionary biology and ecosystem science. Ecology Letters 14:690–701.

Mileikovsky, S.A. 1971. Types of larval development in marine bottom invertebrates, their distribution and ecological significance: A re-evaluation. Marine Biology 10:193–213.

Miner, B.E., L. de Meester, M.E. Pfrender, W. Lampert, and N.G. Hairston Jr. 2012. Linking genes to communities and ecosystems: *Daphnia* as an ecogenomic model. Proceedings of the Royal Society B: Biological Sciences 279:1873–1882.

Morgan, S.G. 2020. Dispersal. Pages 383–407 *in* K. Anger, S. Harzsch, and M. Thiel, editors. The natural history of the Crustacea, Volume 7: Developmental biology and larval ecology. Oxford University Press, New York, USA.

Morris, M., and S.M. Rogers. 2013. Overcoming maladaptive plasticity through plastic compensation. Current Zoology 59:526–536.

Moya-Laraño, J. 2011. Genetic variation, predator-prey interactions and food web structure. Philosophical Transactions of the Royal Society Series B: Biological Sciences 366:1425–1437.

Murphy, E.J., J.L. Watkins, P.N. Trathan, K. Reid, M.P. Meredith, S.E. Thorpe, N.M. Johnston, A. Clarke, G.A. Tarling, M.A. Collins, J. Forcada, R.S. Shreeve, A. Atkinson, R. Korb, M.J. Whitehouse, P. Ward, P.G. Rodhouse, P. Enderlein, A.G. Hirst, A.R. Martin, S.L. Hill, I.J. Staniland, D.W. Pond, D.R. Briggs, N.J. Cunningham, and A.H. Fleming. 2007. Spatial and temporal operation of the Scotia Sea ecosystem: A review of large-scale links in a krill centred food web. Philosophical Transactions of the Royal Society Series B: Biological Sciences 362:113–148.

O'Connor, N.E., and T.P. Crowe. 2005. Biodiversity loss and ecosystem functioning: Distinguishing between number and identity of species. Ecology 86:1783–1796.

Pease, C.J., E.L. Johnston, and A.G.B. Poore. 2010. Genetic variability in tolerance to copper contamination in a herbivorous marine invertebrate. Aquatic Toxicology 99:10–16.

Poore, A.G.B., A.H. Campbell, R.A. Coleman, G.J. Edgar, V. Jormalainen, P.L. Reynolds, E.E. Sotka, J.J. Stachowicz, R.B. Taylor, M.A. Vanderklift, and J.E. Duffy. 2012. Global patterns in the impact of marine herbivores on benthic primary producers. Ecology Letters 15:912–922.

Poore, A.G.B., N.A. Hill, and E.E. Sotka. 2008. Phylogenetic and geographic variation in host breadth and composition by herbivorous amphipods in the family Ampithoidae. Evolution 62:21–38.

Radulovici, A.E., B. Sainte-Marie, and F. Dufresne. 2009. DNA barcoding of marine crustaceans from the Estuary and Gulf of St Lawrence: A regional-scale approach. Molecular Ecology Resources 9:181–187.

Raffard, A., F. Santoul, J. Cucherousset, and S. Blanchet. 2019. The community and ecosystem consequences of intraspecific diversity: A meta-analysis. Biological Reviews 94:648–661.

Rahman, S., D. Schmidt, and J. Hughes. 2018. De novo SNP discovery and strong genetic structuring between upstream and downstream populations of *Paratya australiensis* Kemp, 1917 (Decapoda: Caridea: Atyidae). Journal of Crustacean Biology 38:166–172.

Rask, M., and C. Hiisivuori. 1985. The predation on *Asellus aquaticus* (L.) by perch, *Perca fluviatilis* (L.), in a small forest lake. Hydrobiologia 121:27–33.

Reusch, T.B.H., A. Ehlers, A. Hämmerli, and B. Worm. 2005. Ecosystem recovery after climatic extremes enhanced by genotypic diversity. Proceedings of the National Academy of Sciences of the United States of America 102:2826–2831.

Rivadeneira, M.M., and G.C.B. Poore. 2020. Latitudinal gradient of diversity of marine crustaceans: Towards a synthesis. Pages 389–412 *in* G.C.B. Poore and M. Thiel editors. The natural history of the Crustacea, Volume 8: Evolution and biogeography. Oxford University Press, New York, USA.

Roman, J. 2006. Diluting the founder effect: Cryptic invasions expand a marine invader's range. Proceedings of the Royal Society B: Biological Sciences 273:2453–2459.

Russo, E., K. Franke, H. Hager, B. Espinasse, H. Stibor, and S. Schultes. 2016. Modifying the functional diversity in the zooplankton assemblage of an oligotrophic lake differentially affects pelagic community structure and biomass. Food Webs 8:23–31.

Shelton, A.O., D. Kinzey, C. Reiss, S. Munch, and M. Mangel. 2013. Among-year variation in growth of Antarctic krill *Euphausia superba* based on length-frequency data. Marine Ecology Progress Series 481:53–67.

Shurin, J.B., J.E. Havel, M.A. Leibold, and B. Pinel-Alloul. 2000. Local and regional zooplankton species richness: A scale-independent test for saturation. Ecology 81:3062–3073.

Sison-Mangus, M.P., C.M.J.A. Metzger, and D. Ebert. 2018. Host genotype-specific microbiota do not influence the susceptibility of *D. magna* to a bacterial pathogen. Scientific Reports 8:9407.

Sommer, U., and F. Sommer. 2006. Cladocerans versus copepods: The cause of contrasting top-down controls on freshwater and marine phytoplankton. Oecologia 147:183–194.

Stachowicz, J.J., J.F. Bruno, and J.E. Duffy. 2007. Understanding the effects of marine biodiversity on communities and ecosystems. Annual Review of Ecology, Evolution, and Systematics 38:739–766.

Stachowicz, J.J., H. Fried, R.W. Osman, and R.B. Whitlatch. 2002. Biodiversity, invasion resistance, and marine ecosystem function: Reconciling pattern and process. Ecology. 83:2575–2590.

Steinberg, D.K., and M.R. Landry. 2017. Zooplankton and the ocean carbon cycle. Annual Review of Marine Science 9:413–444.

Swanson, G. A. 1984. Dissemination of amphipods by waterfowl. The Journal of Wildlife Management 48:988–991.

Sworobowicz, L., M. Grabowski, T. Mamos, A. Burzyński, A. Kilikowska, J. Sell, and A. Wysocka. 2015. Revisiting the phylogeography of *Asellus aquaticus* in Europe: Insights into cryptic diversity and spatiotemporal diversification. Freshwater Biology 60:1824–1840.

Taberlet, P., L. Fumagalli, A.G. Wust-Saucy, and J.F. Cosson. 1998. Comparative phylogeography and postglacial colonization routes in Europe. Molecular Ecology 7:453–464.

Tessier, A.J., M.A. Leibold, and J. Tsao. 2000. A fundamental trade-off in resource exploitation by *Daphnia* and consequences to plankton communities. Ecology 81:826–841.

Thompson, P.L., T.J. Davies, and A. Gonzalez. 2015. Ecosystem functions across trophic levels are linked to functional and phylogenetic diversity. PloS ONE 10:e0117595.

Thompson, P.L., and J.B. Shurin. 2012. Regional zooplankton biodiversity provides limited buffering of pond ecosystems against climate change. The Journal of Animal Ecology 81:251–259.

Tollrian, R., and C. Heibl. 2004. Phenotypic plasticity in pigmentation in *Daphnia* induced by UV radiation and fish kairomones. Functional Ecology 18:497–502.

Tseng, M., J.R. Bernhardt, and A.E. Chila. 2019. Species interactions mediate thermal evolution. Evolutionary Applications 12:1463–1474.

Tucker, C.M., M.W. Cadotte, S.B. Carvalho, T.J. Davies, S. Ferrier, S.A. Fritz, R. Grenyer, M.R. Helmus, L.S. Jin, A.O. Mooers, S. Pavoine, O. Purschke, D.W. Redding, D.F. Rosauer, M. Winter, and F. Mazel. 2017. A guide to phylogenetic metrics for conservation, community ecology and macroecology. Biological Reviews of the Cambridge Philosophical Society 92:698–715.

Twombly, S. 1995. Phenotypic variation in metamorphosis in four species of freshwater copepods. Freshwater Biology 34:29–38.

Vonk, J.A., B.F. van Kuijk, M. van Beusekom, E.R. Hunting, and M.H.S. Kraak. 2016. The significance of linoleic acid in food sources for detritivorous benthic invertebrates. Scientific Reports 6:35785.

Vo, S.A., J. Scanlan, P. Turner, and R. Ollington. 2020. Convolutional neural networks for individual identification in the southern rock lobster supply chain. Food Control 118:107419.

Walker, M.K., and R.M. Thompson. 2010. Consequences of realistic patterns of biodiversity loss: An experimental test from the intertidal zone. Marine and Freshwater Research 61:1015–1022.

Walsh, J.R., S.R. Carpenter, and M.J. Vander Zanden. 2016. Invasive species triggers a massive loss of ecosystem services through a trophic cascade. Proceedings of the National Academy of Sciences of the United States of America 113:4081–4085.

Walsh, M.R., M. Packer, S. Beston, C. Funkhouser, M. Gillis, J. Holmes, and J. Goos. 2018. *Daphnia* as a model for eco-evolutionary dynamics. Pages 403–424 *in* G. Wellborn and M. Thiel, editors. The natural history of the Crustacea, Volume 5: Life histories. Oxford University Press, New York, USA.

Wares, J.P. 2011. Insights from population genetics: Are all crustaceans created equal? Journal of Crustacean Biology 31:339–344.

Watling, L. 1993. Functional morphology of the amphipod mandible. Journal of Natural History 27:837–849.

Wattier, R., T. Mamos, D. Copilaş-Ciocianu, M. Jelić, A. Ollivier, A. Chaumot, M. Danger, V. Felten, C. Piscart, K. Žganec, T. Rewicz, A. Wysocka, T. Rigaud, and M. Grabowski. 2020. Continental-scale patterns of hyper-cryptic diversity within the freshwater model taxon *Gammarus fossarum* (Crustacea, Amphipoda). Scientific Reports 10:16536.

Weber, A., S. Vesela, and S. Repka. 2003. The supposed lack of trade-off among *Daphnia galeata* life history traits is explained by increased adult mortality in *Chaoborus* conditioned treatments. Hydrobiologia 491:273–287.

Weiss, L.C., and R. Tollrian. 2018. Predator-induced defenses in Crustacea. Pages 303–321 *in* G. Wellborn and M. Thiel, editors. The natural history of the Crustacea, Volume 5: Life histories. Oxford University Press, New York, USA.

Wellborn, G.A., and R.D. Cothran. 2004. Phenotypic similarity and differentiation among sympatric cryptic species in a freshwater amphipod species complex. Freshwater Biology 49:1–13.

West-Eberhard, M.J. 2003. Developmental plasticity and evolution. Oxford University Press, New York, USA.

Weston, D.P., H.C. Poynton, G.A. Wellborn, M.J. Lydy, B.J. Blalock, M.S. Sepulveda, and J.K. Colbourne. 2013. Multiple origins of pyrethroid insecticide resistance across the species complex of a nontarget aquatic crustacean, *Hyalella azteca*. Proceedings of the National Academy of Sciences of the United States of America 110:16532–16537.

Witt, J.D.S., D.L. Threloff, and P.D.N. Hebert. 2006. DNA barcoding reveals extraordinary cryptic diversity in an amphipod genus: Implications for desert spring conservation. Molecular Ecology 15:3073–3082.

Yampolsky, L.Y., R.M. Kamaltynov, D. Ebert, D.A. Filatov, and V.I. Chernykh. 1994. Variation of allozyme loci in endemic gammarids of Lake Baikal. Biological Journal of the Linnean Society 53:309–323.

Zimmer, M., J. Danko, S. Pennings, A. Danford, T. Carefoot, A. Ziegler, and R. Uglow. 2002. Cellulose digestion and phenol oxidation in coastal isopods (Crustacea: Isopoda). Marine Biology 140:1207–1213.

8

CRUSTACEAN RESPONSES TO HABITAT LOSS, FRAGMENTATION, AND MODIFICATION

Kevin Hovel

Abstract

This chapter reviews the response of crustaceans to disturbances that affect the amount, configuration, and quality of their habitats. Anthropogenic and naturally occurring forces modify marine, freshwater, and terrestrial habitats by causing habitat loss, fragmentation, and degradation over many spatial scales. In this chapter, the focus is on widespread but commonly disturbed and degraded habitats essential for supporting crustacean abundance and diversity: freshwater wetlands, salt marshes, seagrasses, coral reefs, and sandy beaches. Sources of disturbance to these habitats include human development and agriculture, dredging and draining, destructive fishing practices, sedimentation, eutrophication, climate change, and trampling, which coincide with naturally occurring abiotic and biotic processes. Disturbances remove habitat; reduce habitat quality, complexity, and refuge value; and cause patchiness and edge proliferation, all of which influence crustacean population dynamics and biodiversity. Though many crustacean species exhibit resilience in the face of moderate habitat fragmentation and loss, extensive loss and degradation of habitats reduces the survival, abundance, and biodiversity of a wide variety of crustaceans across ecosystems. These effects are likely to accelerate as elements of climate change increasingly stress crustaceans and the habitat-forming species on which they depend.

INTRODUCTION

Crustaceans occupy a vast array of habitat types in marine, freshwater, and terrestrial ecosystems, where they often constitute some of the most abundant and diverse taxa and perform key ecosystem functions and services. Crustaceans act as herbivores, detritivores, scavengers, filter feeders, and

Kevin Hovel, *Crustacean Responses to Habitat Loss, Fragmentation, and Modification*. In: *Ecology and Conservation*.
Edited by: Lars Gutow and Martin Thiel, Oxford University Press. © Oxford University Press (2025).
DOI: 10.1093/oso/9780197768242.003.0008

predators, thus forming key linkages within food webs in a variety of ways. Many crustaceans, including marine amphipods and isopods, are primary consumers that link primary producers to predators in marine and freshwater habitats. Predatory crustaceans, notably crabs and lobsters, may provide top-down control of prey populations, including alien species and habitat-altering species (Shears and Babcock 2002, Silliman and Bertness 2002). Some ecological functions performed by crustaceans directly affect the health and persistence of the habitats they occupy: for instance, in seagrass habitats and on coral reefs, crustaceans act as mutualists with foundation species by consuming competing algae (Duffy et al. 2001, Spadaro and Butler 2021). These species thus form one component of a key feedback loop in which complex habitats support dense and diverse crustacean populations that, in turn, act to maintain the complex structure of the habitats themselves (Reynolds et al. 2014). In addition to the valuable ecosystem functions they provide, many crustacean taxa (particularly crabs, lobsters, and shrimps) are the target of fisheries and aquaculture and form an important source of protein and capital for humans.

Across all ecosystems, the habitats that crustaceans occupy are increasingly degraded, fragmented, or even removed entirely. In particular, marine, freshwater, and terrestrial habitats occupied by crustaceans and other fauna are increasingly being altered and eliminated by human activities, a process that has been viewed as the most long-standing and significant threat to biodiversity on the planet. My aim in this chapter is to describe the ways in which anthropogenic and naturally occurring habitat alterations affect marine (subtidal and intertidal) and freshwater crustaceans. I will focus mainly on large-scale habitat alterations involving habitat fragmentation and loss and will provide examples of their effects on different stages of the crustacean life cycle. Many crustacean species are of great value to humans; therefore, in addition to describing effects of habitat alteration on the behavior, survival, abundance, and diversity of crustaceans, I will touch on the ways in which the loss, fragmentation, and degradation of habitats affect ecosystem services and ecosystem functions performed by crustacean species.

COMPONENTS OF HABITAT ALTERATION

Habitat alteration involves changes to habitat cover (the amount of habitat) and to habitat configuration (the spatial arrangement of habitat) (Fig. 8.1). Reductions in habitat cover involve how much of the original habitat remains and is available to organisms, whereas changes to habitat configuration reflect increases in the patchiness of the habitat (also known as habitat fragmentation). Both of these components of alteration can strongly influence the survival, density, and diversity of organisms (Fahrig 2003, Roberts and Poore 2006). Habitat alteration can also involve changes to habitat structural complexity, which refers to the density, biomass, or surface area of structural elements, such as salt marsh stems, seagrass shoots, mangrove prop roots, or algal sporophytes. Changes to structural complexity strongly influence the refuge value of habitats for organisms; for instance, reductions in the density or biomass of seagrass shoots can increase the vulnerability of invertebrate and vertebrate fauna to predators (Heck and Crowder 1991).

Human-induced changes to habitats act on top of naturally occurring disturbances, and together these processes involve a variety of components (Fig. 8.1). At the most extreme level, *habitat loss* results in the entire removal of a foundation species or an extreme alteration of the ecosystem. What remains may be a less valuable alternative habitat: for instance, marshes and mudflats may be replaced with rocky structures to armor shorelines (Gittman et al. 2015, Prosser et al. 2018), seagrass habitat may become mudflat, and storms and herbivory may transform kelp forests to barren areas that are nearly devoid of habitat-forming algae (Foster 2006). In contrast, humans sometimes transform habitats to something entirely unsuitable for fauna, such as the widespread replacement of wetland habitat with cropland in the American Midwest (Jenkins et al. 2003).

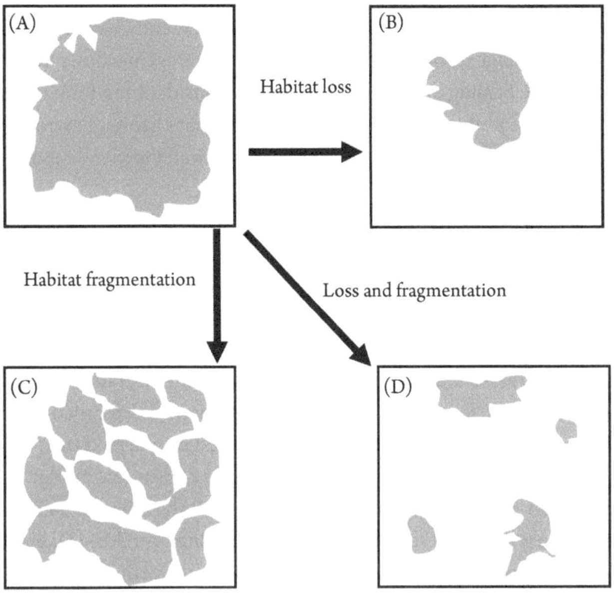

Fig. 8.1.
Intact habitats (A) can undergo habitat loss (B), which reduces their coverage but does not change their configuration (here, one intact habitat patch). When habitats are fragmented (C), they become patchy with little reduction in habitat coverage, and gain more edge. Loss and fragmentation typically occur together (D), reducing habitat coverage, reducing average patch size, and increasing patch isolation.

Less extreme alterations occur when habitats are degraded. *Habitat degradation* occurs when the original habitat remains in place but is compromised in some way, making it less valuable for fauna. Frequently, degradation occurs via reductions in structural complexity (Alvarez-Filip et al. 2009), but degradation may also involve pollution, compaction, or erosion, as often seen on sandy beaches (Schlacher et al. 2016). Degraded habitats offer fewer resources for fauna, such as refuge and food, and may contain contaminants or decomposing biological material, altering the chemical as well as the physical nature of the habitat.

Habitat loss is rarely instantaneous at all but the smallest spatial scales, and fragmented habitats are created when habitat loss is incremental or incomplete. *Habitat fragmentation* is a widespread phenomenon in which reductions in habitat cover are accompanied by changes in habitat configuration as habitats are broken apart into discrete patches (Fahrig 2003). Habitat fragmentation can have strong effects on faunal behavior, survival, abundance, and diversity, which may stem from reduced connectivity among patches, the proliferation of edge habitat, and reductions in average patch size. Habitat fragmentation can greatly alter predator–prey encounter rates, reducing the value of habitat for species vulnerable to predators. These effects can occur independently of reductions in habitat cover (Trzcinski et al. 1999), but the strongest effects of large-scale habitat alteration on fauna occur when changes to habitat cover and configuration co-occur and act synergistically (Andrén 1994, Hanski 2015, Yeager et al. 2016).

Scale is an important consideration when assessing the effects of habitat alteration on fauna. Natural and human-induced habitat alteration occur at many spatial scales in freshwater and marine (not to mention terrestrial) ecosystems. At small scales, individual vernal pools or ponds may dry up or be filled in, or gaps may develop within vegetated habitats. Small-scale disturbances to habitats often are common occurrences, and in some habitats may enhance recruitment and biodiversity

(Jerabek et al. 2017). At the scale of landscapes and seascapes, entire wetlands may be drained or infilled, coral reefs may change to algal reefs or rubble, and seagrass meadows may be lost to climate change and development. Scaling out further still, we can consider long-term trends of habitat loss worldwide. For example, about one-third of the world's seagrass habitat has been lost over the last several decades (Waycott et al. 2009), and vast tracts of freshwater wetlands and tidal marshes have been lost to agriculture, development, and climate change.

EFFECTS OF FRESHWATER WETLAND ALTERATION ON CRUSTACEANS

Freshwater wetland habitat varies greatly in cover and configuration. At one extreme, large, connected, semipermanent areas of marshland cover 100s to 1000s of hectares of flatlands. At the other extreme, wetlands can comprise small, isolated, ephemeral pools (also called vernal pools) on the order of a few square meters that form in depressions created by glacial retreat or by large animals such as buffalo that form wallows. Freshwater wetlands have been lost at high rates due to agriculture and urbanization. Over 50% of freshwater wetlands of the colonial United States were lost from the late 18th century to the present, with some areas losing over 90% (Dahl 1990). Landscape changes have strongly impacted ephemeral pool biota in central North America and Canada, where undulating topography and millions of buffalo created a huge swath of wetland habitat once spanning several Canadian provinces and much of the American Midwest. A massive effort to convert these freshwater wetlands to land suitable for agriculture resulted in the elimination of millions of vernal pools across the region. Vernal pools in areas with Mediterranean climates have been eliminated at alarming rates, and only a small fraction of their historical abundance remains: in Central California, current vernal pool abundance is likely less than 10% of pre-anthropogenic levels (Keeley and Zedler 1998).

Vernal pool crustacean fauna are dominated by shrimp (e.g., Branchiopoda, fairy shrimp; Conchostraca, clam shrimp; and Notostraca, tadpole shrimp), Ostracoda, Copepoda, and Cladocera (Fryer 1985, Keeley and Zedler 1998). Though few historical data on crustacean diversity in vernal pools exist, it is highly likely that extensive loss of vernal pools has strongly affected crustacean populations and communities. Severe effects of habitat loss are most likely to occur in habitats that exhibit high faunal endemism, are naturally isolated or exhibit poor connectivity among patches, and are vulnerable to human-induced alteration. Many freshwater wetland habitats exhibit these characteristics and provide examples of the strong effects of habitat alteration at the population, species, and community levels. For instance, Jenkins et al. (2003) sampled remaining Illinois ephemeral wetlands for crustacean taxa and used simulation modeling to estimate crustacean biodiversity loss due to the conversion of 2.8 million hectares of wetlands to agricultural fields. They estimated that eight to nine crustacean species went extinct due to wetland drainage, leaving approximately 75 crustacean species currently inhabiting the wetlands of Illinois. Importantly, their simulations suggest that present-day crustacean populations no longer function as connected metapopulations. Metapopulations are groups of spatially separated subpopulations, each inhabiting a habitat patch (or island) within a patchy landscape. Within metapopulations, connectivity among subpopulations in the form of dispersal and gene flow help prevent local extinctions. However, high rates of habitat loss may select against highly dispersing genotypes because the probability of being dispersed to small, isolated remaining patches is low. Modeling a generic freshwater crustacean, Jenkins et al. (2003) found that dispersal rate is the main determinant of metapopulation size at high rates of habitat loss, with highly dispersing genotypes going extinct (Fig. 8.2A). In contrast, the proportion of habitat patches in the landscape occupied by the species depended strongly on habitat loss and was largely invariant to dispersal rates at all the but the highest rates of habitat loss (Fig. 8.2B). Based on these simulations and estimates of crustacean diversity

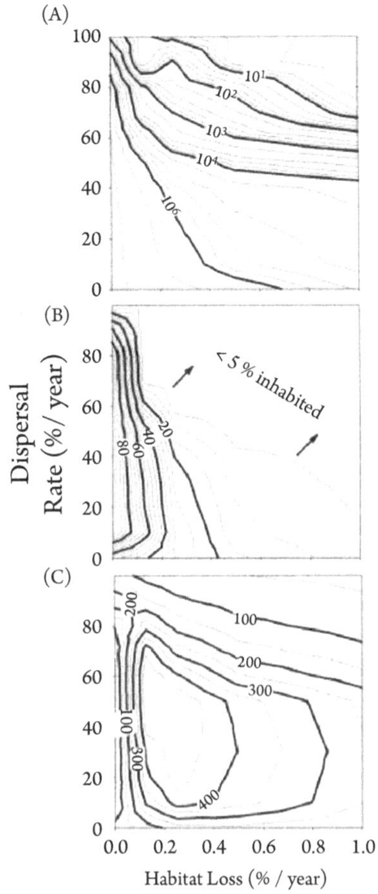

Fig. 8.2.
Results of the simulation modeling quantifying the effects of freshwater wetland loss on a generic crustacean metapopulation with a range of dispersal genotypes. Lines in each panel depict average state at the steady-state for each parameter. (A) Total metapopulation size depended largely on dispersal rate at high levels of habitat loss, because high-dispersing genotypes were selected against in a very patchy landscape. (B) Selection against these genotypes also resulted in few patches being occupied by the species at intermediate to high rates of habitat loss. (C) Time (years) to reach the steady-state asymptote for population size was longest at intermediate rates of habitat loss and dispersal: the metapopulation was stable at low rates of habitat loss and was rapidly extinguished at high rates of habitat loss. Reproduced from Jenkins et al. (2003), with permission from © John Wiley & Sons.

and populations size, Jenkins et al. (2003) predicted that isolated populations caused by habitat loss will be more vulnerable to future habitat loss, and that the greatest extinction rates have yet to occur for these populations.

Though some vernal pool species are widely distributed globally, due to the occurrence of vernal pool habitat on many continents, many vernal pool crustacean species are rare within their region, further enhancing the risk of extinction. A survey of 58 pools within California's Central Valley found that over a quarter of the species had only a single pool occurrence, and 40% of the species had only a single site occurrence (King 1998). Moreover, there has been a loss of 15% to 33% of the original crustacean biodiversity in this habitat in Central California, where most vernal pools have been replaced with agricultural fields and urbanized landscapes.

Two recent studies illustrate that threats to freshwater wetland crustaceans can involve habitat degradation as well as habitat loss. Simovich et al. (2013) studied the effects of landscape homogenization on the most prominent endangered species in Southern California's vernal pools, the San Diego fairy shrimp, *Branchinecta sandiegonensis*. This species is not only threatened by habitat loss, but by hybridization with a generalist congener, the versatile fairy shrimp *B. lindahli*. Habitat degradation plays a central role in this process: frequent vehicle incursions into coastal vernal pool habitats convert pools into turbid, less heterogeneous habitats that are conducive to generalist species such as *B. lindahli*, which now also occurs in a variety of human-created and disturbed pools. This research suggests that high rates of habitat loss, coupled with high rates of disturbance to remaining habitat, has led to increasing genetic and taxonomic homogenization of the assemblages inhabiting vernal pools in Southern California. Mabidi et al. (2018) studied the effects of salinization on the hatching success of crustacean resting eggs in depression wetlands of the Cape Karoo region of South Africa, where habitat degradation via increased salinity is accelerating due to aridity and the production of brines from hydraulic fracturing. Resting eggs are dormant life stages capable of surviving long periods of drought, and the emergence of juveniles from resting eggs is cued by changes in temperature, salinity, and light. High salinities (2.5 g L^{-1} and above) resulted in a decrease in the richness and abundance of emerged taxa as well as a change in community composition. In particular, branchiopod crustaceans exhibited reduced emergence from resting eggs in high salinity water, which was dominated by copepods, cladocerans, and ostracods. Only Anostraca, Notostraca, and Spinicaudata hatchlings were abundant at salinities of 0.5 g L^{-1} and below. Mabidi et al. (2018) suggested that ephemeral wetlands are particularly susceptible to degradation via pollution and climate change, which may greatly limit their contributions to invertebrate diversity and ecosystem function.

EFFECTS OF MARINE COASTAL WETLAND ALTERATION ON CRUSTACEANS

Coastal wetland habitats, also known as tidal marshes or salt marshes, are low-lying, intertidal habitats that are regularly flooded and drained by the tides. Cordgrass (*Spartina* spp., particularly *Spartina alterniflora*) typically forms the primary vegetation in salt marshes, particularly in the low intertidal areas that are regularly submerged and that house many crustaceans. Salt marshes provide a variety of key ecosystem functions and services, including water filtration, shoreline protection and stabilization, waste removal, carbon storage, secondary production, and provision of habitat for many invertebrate and vertebrate species. Salt marshes are primary habitat for commercially important crustaceans such as the blue crab (*Callinectes sapidus*), brown shrimp (*Farfantepenaeus aztecus*), white shrimp (*Litopenaeus setiferus*), and pink shrimp (*Farfantepenaeus duorarum*) (Zimmerman et al. 2000, Rozas et al. 2007, Roth et al. 2008). Other common inhabitants include fiddler crabs (*Uca* spp.), crabs of the family Sesarmidae (e.g., *Sesarma reticulatum*), and shrimps of the family Palaemonidae (e.g., "grass shrimp" or "glass shrimp" in the genus *Palaemonetes*). Sampling surveys indicate that the majority of the individual organisms found in salt marshes (sometimes over 90%) are crustaceans (Rozas et al. 2007).

Wide swaths of salt marshes have been lost from shorelines worldwide (Kennish 2001, Schultz et al. 2016). Salt marsh loss is a result of a mix of abiotic and biotic factors. Marshes have been filled in and fragmented to make way for coastal development, and rising sea levels combined with subsidence (often precipitated by extraction of subsurface water, petroleum, or gas) have eroded and flooded many marshes, eliminating dominant vegetation such as *Spartina* spp. In coastal Louisiana alone, sea level rise, subsidence, saltwater intrusion, and reduced sediment inflow have eliminated 1000s of km^2 of salt marsh habitat (Day et al. 2011). Kennish (2001) estimated that at the beginning of the 21st century, more than 50% of US tidal marsh habitat had been lost due to these abiotic factors. Additionally, it is now recognized that biotic, top-down factors play a large role in marsh habitat loss. Specifically, a great deal of marsh habitat loss has been precipitated by changes in the abundance and behavior of consumers in coastal waters, which has resulted in increased herbivory on *Spartina* (Fig. 8.3). Direct herbivory on *Spartina* by the crab *Sesarma reticulatum* in the northeastern United States has resulted in the loss of 1000s of hectares of salt marsh in that region. Increased herbivory by *Sesarma* is caused by overfishing of their fish and crab predators in shallow coastal waters, which leads to greatly increased *Sesarma* densities (Altieri et al. 2012). There also is evidence that increased herbivory on marsh plants may be caused by rising sea levels that provide increasing access to the marsh, and that synergistic effects of herbivory and sea level rise greatly accelerate marsh loss (Crotty et al. 2017, Raposa et al. 2018). Similarly, in southeastern US marshes overfishing for predatory blue crabs has led to increased density of their primary prey, the native snail *Littoraria irrorata*. *Littoraria* damages plants by using their radula to produce scars in which they cultivate fungi for consumption. Combined with increasing drought stress, this has resulted in massive die-offs of salt marsh on the Atlantic and Gulf Coasts of the US (Bertness and Silliman 2008). Thus, crustaceans are integral components of a major mechanism that results in the loss of tidal wetlands.

Salt marsh habitat loss is often strongly correlated with reduced crustacean populations. For instance, shrimp and crab populations declined substantially between 1982 and 1995 in Galveston Bay, Texas, which coincided with the loss of over 60% of the region's salt marsh habitat and

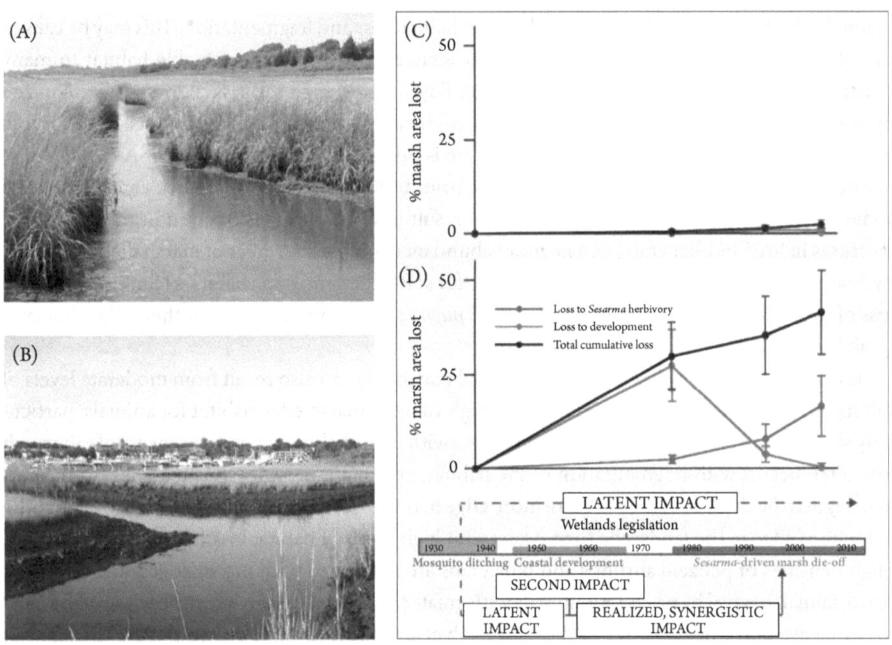

Fig. 8.3.
(A) Healthy *Spartina alterniflora* salt marsh, showing beneficial edge habitat formed by a tidal creek. (B) Marsh bank degradation due to herbivory by the crab *Sesarma reticulatum*. Both photographs from Cape Cod, Massachusetts, USA, reproduced from Altieri et al. (2012), with permission from © John Wiley & Sons. Also shown are time series of salt marsh habitat loss in (C) non-developed marshes, and (D) developed marshes in Cape Cod. Development was the chief agent of marsh loss until the mid-1970s, after which *Sesarma* herbivory accelerated due to overfishing of large predators. Reproduced from Coverdale et al. (2013), with permission from © John Wiley & Sons; see a color version of this figure in the centerfold.

its replacement with open water (Rozas et al. 2007). Subsequent restoration of some of the lost marsh habitat resulted in large increases in shrimp and crab populations. However, gauging the effects of salt marsh loss on crustaceans is complex for several reasons. First, in some cases, as noted above, it is *increased* abundance of crustaceans (particularly crabs such as *Sesarma* spp.) that help cause salt marsh degradation. Second, lost marsh habitat is sometimes replaced by different vegetated habitats that are as valuable or more valuable to fauna, including seagrasses or other types of submerged aquatic vegetation (SAV) (Jerabek et al. 2017). Third, some crustaceans, such as commercially important brown shrimp, use salt marsh habitat for only part of their lives, recruiting to salt marshes as postlarvae but eventually emigrating from marshes (Roth et al. 2008, Bauer 2018). Temporary use of marshes, though critical for robust shrimp populations, makes it challenging to directly tie changes in marsh cover and configuration to shrimp population dynamics and fishery landings (Haas et al. 2004). Fourth, the effects of marsh fragmentation and loss on crustacean populations may involve complex indirect mechanisms such as food web disruption (Hyndes et al. 2014).

Additionally, and counterintuitively, many studies have shown that seascape-level changes to marshes associated with the eventual loss of marsh habitat may enhance the density and survival of some crustacean species (Chesney et al. 2000). This can occur as saltwater inundation associated with sea level rise and marsh subsidence make marshes temporarily more accessible to fauna, though unchecked inundation will eventually result in marsh habitat loss (Minello et al. 1994, Roth et al. 2008). As stated above, this can also occur if newly opened benthic habitat is colonized by SAV. For instance, in coastal Louisiana, populations of commercially important fauna have remained robust in the face of extensive salt marsh habitat loss and fragmentation. This may be caused by colonization of open subtidal bottom by seagrasses, which provide valuable habitat to many crustaceans (Jerabek et al. 2017). In Galveston Bay, areas of high salt marsh loss were colonized by seagrasses (*Halodule wrightii* and *Halophila engelmanni*), which partially compensated for loss of marsh habitat (Rozas et al. 2007). It therefore is possible that the presence of SAV could ameliorate marsh fragmentation effects on nekton, prior to total marsh loss (Jerabek et al. 2017). For some crustacean species, tidal marsh loss may result in habitat expansion even in the absence of increases in SAV. Fiddler crab (*Uca pugnax*) abundance is highest in areas of marsh dieback caused by *Sesarma* overgrazing, due to their preference for soft-sediment, open habitats (Smith 2015), and loss of marsh habitat has actually resulted in *U. pugnax* range expansion in Southern New England (Luk and Zajac 2013).

Enhanced faunal densities in the face of tidal marsh loss can also result from moderate levels of salt marsh habitat fragmentation, due to the high value of marsh edge habitat for animals, particularly shrimps and crabs (Chesney et al. 2000). As with many habitat types, the process of salt marsh loss often begins with fragmentation of the habitat, creating more edges. As marshes break apart and begin to be degraded, they acquire more edge habitat before losing much habitat cover, but the amount of edge in the landscape then drops with high levels of habitat loss (Browder et al. 1985). High densities of penaeid shrimps and blue crabs are associated with marsh edges, and edges enhance faunal survival by providing animals with greater access to the marsh surface without the need for extensive movement (Rozas et al. 2007, Roth et al. 2008). Juvenile brown shrimp that use salt marshes as nursery habitat remain close to the interface between marsh vegetation and open water, where they find suitable foraging opportunities but also remain close to refuge. Using field sampling and an individual-based simulation model (IBM), Haas et al. (2004) found that brown shrimp survival depended more on habitat configuration than on habitat cover: marshes with higher amounts of edge permitted juvenile brown shrimp to remain in high-quality edge habitat, reduced their need to move through the marsh, and reduced their local densities, all of which increased shrimp survival and density-dependent growth. Roth et al. (2008) used an expanded version of this IBM to test how marsh inundation (seawater flooding) interacts with fragmentation and loss to dictate brown shrimp survival, growth, and export from the marsh (Fig. 8.4). Overall, inundation had stronger effects on shrimp survival than marsh habitat fragmentation and loss: inundation increased shrimp access to protective marsh habitat, reducing shrimp mortality and increasing shrimp export from salt marshes. These effects were strongest in highly fragmented marsh landscapes that contained high amounts of valuable marsh edge habitat.

To date, the overall degree to which tidal marsh loss has affected crustacean populations and biodiversity is difficult to judge, for the reasons listed above. However, there are reasons to be concerned that accelerating tidal marsh loss will harm a variety of crustacean species (Rozas et al. 2007, Baker et al. 2020). For instance, blue crab abundance and survival were lowest in bare sediment compared to tidal marsh and SAV habitat, suggesting strong negative impacts of marsh loss on blue crabs if lost marsh is not replaced by SAV (Shakeri et al. 2020). Additionally, accelerating climate change may interact with marsh habitat loss to alter crustacean populations. Though inundation may increase access to marsh habitat and, consequently, survival for fauna such as brown shrimp (Roth et al. 2008), the high rates of inundation expected with sea level rise can accelerate erosion and drown salt marsh grasses, resulting in complete habitat loss. Blue crabs and other crustaceans

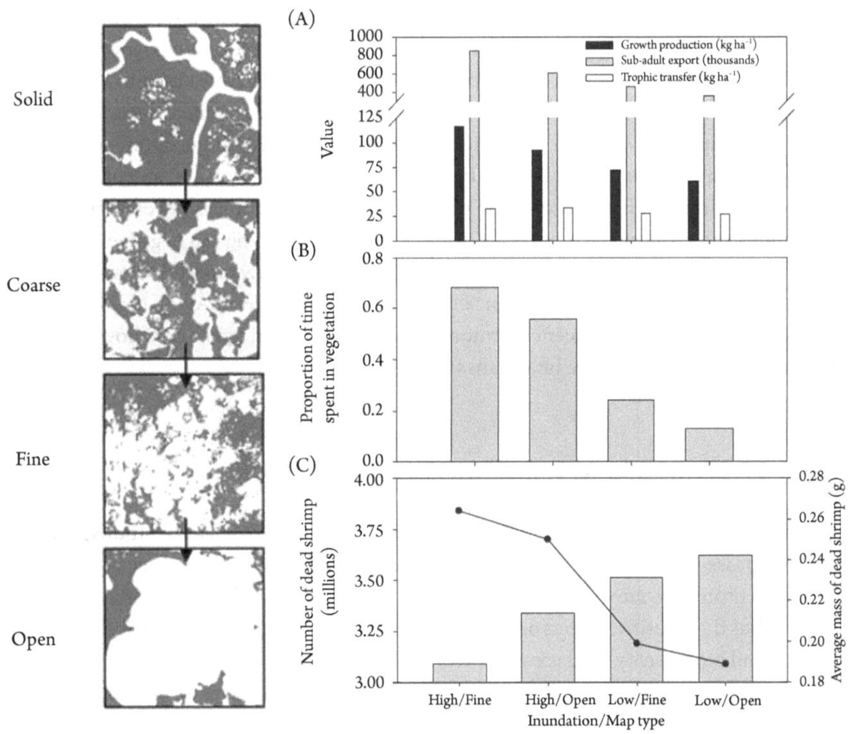

Fig. 8.4.
Left: Stages of salt marsh fragmentation and loss, progressing from solid (undisturbed marsh, about 75% *Spartina* cover) to open (about 20% of *Spartina* cover remaining). Pictures were generated using GIS analysis of orthophotos taken in coastal Louisiana. Right: Effects of simulated seawater inundation and marsh habitat fragmentation on brown shrimp *Farfantepenaeus aztecus*: (A) growth production, export, and trophic transfer; (B) proportion of time spent in marsh habitat; and (C) shrimp mortality (bars) and biomass of dead shrimp (line). High = high inundation, low = low inundation. "Fine" and "open" refer to marsh configurations shown in maps on left. Note break in y-axis in (A). Reproduced from Roth et al. (2008), with permission from © Inter-Research Science Publisher.

may also be susceptible to an alternative, growing threat to tidal marshes in the Gulf of Mexico and the Atlantic coast of Florida: *Spartina* replacement by red (*Rhizophora racemosa*) and black (*Avicennia germinans*) mangroves. Tropical mangroves are expanding poleward in response to climate change, colonizing temperate systems dominated by *Spartina* (Scheffel et al. 2018). Along the Atlantic coast of Florida, blue crabs prefer *Spartina* habitat over mangrove habitat, and their survival is higher in *Spartina* than in mangrove habitat (Johnston and Caretti 2017). In the Gulf of Mexico, Scheffel et al. (2018) found that black mangrove expansion into tidal marshes altered community composition for epifauna, infauna, and nekton. Effects on crustaceans were mixed: though amphipods readily used new mangrove habitat and were more abundant there than in marsh habitat, tanaids exhibited the opposite trend (Scheffel et al. 2018).

Shoreline hardening, also known as shoreline armoring, also poses an increasing threat to tidal marshes and the crustaceans that use marsh habitat. Shoreline hardening occurs when naturally occurring coastal habitat is replaced by concrete or stone structures designed to reduce erosion and wave activity. An extensive survey of continental US shorelines revealed that as of 2015, approximately 14% of the continental US shoreline has been hardened (Gittman et al. 2015). A relatively small fraction of tidal marshes has been hardened so far (ca. 1%), but this is likely an underestimate given the uncertainty about causes of tidal marsh loss in the past. Gittman et al. (2015) warn that shoreline hardening and the accompanying loss of tidal wetlands will accelerate, based on projections for human population growth, increases in storm frequency and severity, and lax environmental regulations. This is problematic because structures used to harden shorelines are generally less valuable for fauna than the naturally occurring habitats that they replace. In Chesapeake Bay and Delaware Bay, the majority of species were less abundant in benthic habitat adjacent to hardened shorelines vs. benthic habitat adjacent to tidal wetlands, and blue crab abundance rapidly declined with increases in the proportion of shoreline consisting of combinations of hard structures and cropland (Kornis et al. 2017). Shoreline hardening can negatively affect fauna via direct mechanisms (the direct replacement of habitat) and indirect mechanisms. For instance, areas adjacent to hardened shorelines typically contain less protective SAV (Landry and Golden 2018), and sediment erosion adjacent to armored structures results in reduced shallow-water refuge habitat for organisms like juvenile blue crabs (Prosser et al. 2018).

EFFECTS OF SEAGRASS ALTERATION ON CRUSTACEANS

The consequences of habitat alteration for marine fauna have been extensively studied in seagrass habitats. Seagrasses are heavily researched due to their value as a foundation species and their extensive use as protective nursery habitat by juvenile invertebrates and fishes (Williams and Heck 2001), as well as their positive effects on ecosystem functions (e.g., enhancing herbivory, secondary production, and biodiversity) and ecosystem services (e.g., sequestering carbon, slowing shoreline erosion, and enhancing survival and abundance of fishery species). Seagrass habitat is also inherently patchy (Fig. 8.5), drawing comparisons with fragmented forests in which the foundations of landscape ecology are based.

Coastal marine seagrasses form extensive meadows in shallow water that support the recruitment, growth, and survival of many crustacean species. Indeed, many crustaceans spend most or all of their benthic lives in seagrass habitat (e.g., gammarid amphipods, isopods, and grass shrimp), or visit it frequently for foraging or refuge (e.g., predatory crabs and lobsters). Unfortunately, the close proximity of seagrasses to human populations has led to high rates of degradation and loss: approximately one-third of the world's seagrass habitat has been lost over the last several decades (Waycott et al. 2009). Damage to seagrass habitat arises from many sources, including eutrophication, sedimentation, fishing, propeller and anchor scarring, vessel groundings, disease, mariculture, and climate change (Orth et al. 2006, Boström et al. 2011).

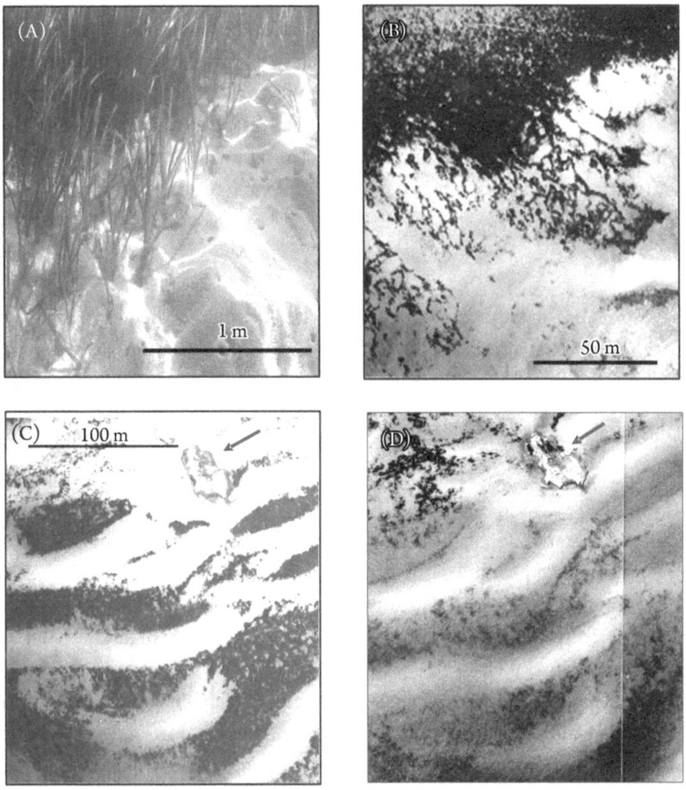

Fig. 8.5.
(A) Eelgrass (*Zostera marina*) shoots at a patch edge in sandy substratum at a depth of approximately 1.5 m in Mission Bay, California, USA. Small holes in sand are burrows of ghost shrimp *Neotrypaea californiensis*, which contributes to eelgrass patchiness via bioturbation. (B–D) Aerial photos of submerged eelgrass habitat in Chesapeake Bay, Virginia, USA, demonstrating different seascape configurations: (B) fragmented eelgrass along the edge of a continuous meadow; (C) large, isolated patches of eelgrass formed by current scouring; (D) the same seascape as shown in (C) after degradation (reductions in shoot density and biomass). Arrows point to a small island for orientation. Photos reproduced courtesy of K. Hovel; see a color version of this figure in the centerfold.

As for freshwater wetlands and tidal marshes, alteration of seagrass habitat can take on many forms. First, complete loss of seagrass habitat occurs at global scales and at local scales (e.g., within individual estuaries). The loss of a seagrass meadow may be permanent, which may occur when geomorphological changes (e.g., dredging, shoreline alteration, and infilling) alter sediment suspension or distribution, bathymetry, and patterns of shoreline erosion. Eutrophication promotes the growth of epiphytic algae and macroalgae, which can smother seagrasses (Reynolds et al. 2014). Climate change also can create inhospitable conditions for seagrasses, due to the relatively narrow thermal tolerances of major habitat-forming species like eelgrass (*Zostera marina*). Seagrass loss may also be temporary, due to factors such as heat waves and bioturbation.

Research on the effects of seagrass alteration has involved manipulative experimentation, such as simulating propeller scarring to study its effects on fauna (Kenworthy et al. 2002), restoring seagrass habitat into particular configurations that are then sampled for fauna (Lefcheck et al. 2016), and allowing fauna to colonize artificial seagrass patches that are formed into different shapes and sizes (Eggleston et al. 1998, Johnson and Heck 2006). However, researchers studying the effects of habitat loss, fragmentation, and degradation on fauna can only rarely extensively manipulate habitats, due to (i) the logistical difficulties of creating or simulating seagrass patches at large scales; and (ii) the ethical (and sometimes legal) issues associated with altering critical foundation species. Therefore, the vast majority of studies assessing the effects of seagrass habitat alteration on fauna rely on comparisons made across preexisting patches or seascapes that vary in size, structural complexity, habitat cover, or habitat configuration, with the assumption that such comparisons are a good proxy for the process of habitat fragmentation or degradation.

Seagrass Scarring

Scarring is the removal of discrete areas of a seagrass meadow due to boat propellers, anchors, and fishing gear. Scars can consist of small ($< 1\ m^2$) "holes" within a continuous area of seagrass, or long, linear gouges formed by propellers churning up sediment and seagrass rhizomes. Reed and Hovel (2006) quantified the effects of small-scale experimental eelgrass scarring (e.g., discrete, small-scale disturbances like those caused by anchoring) on epifauna in San Diego Bay, California. Experimental scarring reduced species richness and epifaunal density when 90% of the habitat was removed, but not at 10% or 50% removal (Reed and Hovel 2006). An important finding from this study was that a threshold of seagrass habitat loss was necessary before changes in epifaunal communities were observed. This finding is in line with studies from forests (see Radford et al. 2005) and seagrass habitat (see Yeager et al. 2016) also suggesting that changes in faunal population dynamics and community structure typically are not observed until relatively high rates of loss occur (the "threshold effect"; Andrén 1994).

The effects of seagrass scarring on crustaceans also have been evaluated by comparing the identity and density of fauna in scarred areas vs. unaltered reference areas. Propeller scarring of seagrasses is particularly problematic in shallow, heavily trafficked areas such as coastal Florida, where fragmentation of *Halodule wrightii* and *Thalassia testudinum* is extensive (Bell et al. 2002). In Tampa Bay and Charlotte Harbor, Florida, densities of shrimp (e.g., *Hippolyte zostericola, Gnathophylloides mineri*, and others) were not affected even by the highest levels of scarring (close to 50% of the habitat) (Bell et al. 2002). The authors suggested that higher levels of scarring than those found in the bays might be necessary before effects on nekton are observed, lending credence to the threshold hypothesis. In contrast, Uhrin and Holmquist (2003) found that propeller scarring resulted in lower densities of shrimp within scars compared to intact seagrass, and lower densities of crabs up to 5 m away from individual scars in Puerto Rico. Shrimp species inhabiting scars were different from those inhabiting intact seagrass.

Seagrass Patch Size and Patch Edge

The effects of seagrass patch size on faunal abundance and survival have received much attention, at least in part due to the wide range of patch sizes typically found within a seagrass seascape. In the field of landscape ecology, patch size effects on fauna are inherently tied to *edge effects*, because smaller patches, like those created in the process of habitat fragmentation, contain proportionally more edge habitat than do larger patches. Patch edges represent an ecotone, or transition habitat, in which processes may differ substantially from the patch interior and from the habitat surrounding patches, often referred to as matrix habitat. Some of the earliest landscape ecology studies involved quantifying how strongly patch edges influence predation rates on fauna, and it was common to find that prey survival rates were lower along patch edges than in patch interiors. Predators use the "matrix habitat" (agricultural land in terrestrial systems, muddy or sandy unvegetated habitat in marine systems) surrounding habitat patches as corridors to move among patches in search of prey. In doing so, they often attack prey along patch edges, making edges more risky than patch interiors, and small patches riskier than large patches, to the point where they may become "ecological traps"

(Gates and Gysel 1978). However, modern syntheses of edge effects have suggested that edge effects are highly variable among species, sites, and habitat types, and that edges can have either negative or positive effects on fauna (Lahti 2001, Ries et al. 2017).

A novel study by Lefcheck et al. (2016), conducted in Chesapeake Bay, involved transplanting eelgrass into a range of patch sizes (4–400 m^2) that also were fragmented in a controlled manner to standardize configuration. Their major finding was that species abundances and diversity were generally invariant to patch size and degree of seagrass fragmentation. However, one abundant crustacean taxon, *Palaemonetes* spp. (grass shrimp), was consistently densest in the smallest patches. The authors proposed that small patches may intercept shrimp larvae at higher rates than large patches due to higher perimeter-to-area ratios of small patches (Lefcheck et al. 2016). The same rationale was proposed to explain higher grass shrimp densities in small vs. large artificial seagrass patches in North Carolina (Eggleston et al. 1998). These findings correspond to an early prediction of seascape ecology: seascapes composed of many small habitat patches, like that found after a fragmentation event, would increase the probability of larval encounter with habitat due to the high perimeter-to-area ratio of small patches. Similarly, elongated or irregularly shaped patches with high perimeter-to-area ratios also might intercept settlers or immigrants at disproportionate rates (Tanner 2003).

Subsequent studies testing these predictions revealed that the effects of habitat configuration on settlement and recruitment are more complex and heavily depend on organism type or stage. Organisms that rely on currents for dispersal (e.g., amphipods that occupy seagrass habitat) are strongly affected by the amount of edge and the orientation of patches relative to prevailing currents, whereas organisms that are strong swimmers and disperse more actively are not (Tanner 2003). Working on the island of Hokkaido in northern Japan, Shinomiya et al. (2017) found that adult Hokkai shrimp (*Pandalus latirostris*) often occupied the smallest eelgrass patches in the meadow due to high rates of patch interception as they moved about the seascape to forage. The same patterns were not evident for juvenile *P. latirostris*, which were more resident in eelgrass. Additionally, though small patches may intercept more mobile individuals, they also contain less interior ("core") habitat. Patch interiors are often areas of reduced predator–prey encounter rates and enhanced foraging opportunities. For instance, small (0.25 m^2) patches of seagrass and oyster shell, representing fragmented settlement habitat, had higher rates of grass shrimp (*Palaemonetes* spp.) recruitment than did larger (4 m^2) patches (Eggleston et al. 1998). However, within the same experimental system, the density of newly recruited blue crabs was higher in large vs. small patches, which may have been caused by high grass shrimp predation on blue crab juveniles in small patches. A similar mechanism may have been responsible for the reduced abundance of newly settled blue crabs along the edge of seagrass patches in Chesapeake Bay: seagrass patch edges housed many shrimps, which are key predators of blue crab megalopae and early instars (Hovel and Lipcius 2002).

Some crustacean taxa can be considered "edge species" because they are commonly found in higher densities along seagrass patch edges. This includes predatory shrimp (e.g., grass shrimp *Palaemonetes* spp.) and herbivorous gammarid amphipods (Bologna and Heck 2002, Moore and Hovel 2010). Bologna and Heck (2002) found significantly higher densities and secondary production for amphipods and decapod crustaceans along *T. testudinum* patch edges in St. Joseph Bay, Florida. They theorized that edges actively accumulate mobile fauna that move throughout the seascape, and that seagrass patch edges may be areas of high trophic transfer if predation rates are high along edges. This idea is corroborated from a study using artificial seagrass patches in Southeast Australia: higher densities of predatory pipefish (*Stigmatopora argus*) along patch edges were explained by higher densities of their crustacean prey (copepods, amphipods, isopods, mysids, and others) along edges (Macreadie et al. 2010). In San Diego Bay eelgrass habitat, gammarid and caprellid amphipods were more abundant along patch edges than in patch interiors, but their abundance and distribution was not influenced by seagrass structural complexity (Moore and Hovel 2010). Even though some taxa are frequently the most abundant along seagrass patch edges, there is high variability in the effects of edge proximity on crustacean abundance, and little consistency among studies. For instance, Mahoney et al. (2018) sampled eelgrass patch edges and interiors in Back Sound, North Carolina, and found no effects of edge proximity on blue crab catch per unit effort (but see below for edge effects on blue crab survival). The authors also reviewed other studies on seagrass edge effects and found little consistency among studies for edge effects on crustacean densities, in addition to densities of bivalves and fishes (Mahoney et al. 2018).

The idea that patch edges are areas of high risk of predator-induced mortality has pervaded the field of landscape ecology for decades. Despite this common view, as mentioned above, reviews of the literature have suggested that effects of patch edges on predation risk are highly variable among species, habitats, and studies. This holds true for crustaceans in seagrass habitat. Peterson et al. (2001) found that tethered shrimp and crabs were found and consumed by predators much more quickly along *T. testudinum* patch edges vs. patch interiors in St. Joseph Bay, Florida. In the same study, however, no differences in survival time were seen for shrimps and crabs between eelgrass patch edges and interiors in the Damariscotta River, Maine. Though results from Hovel and Lipcius (2002) suggested that juvenile blue crabs experience reduced survival along eelgrass patch edges compared to patch interiors, Mahoney et al. (2018) found the opposite: mean time to consumption of tethered juvenile blue crabs was approximately 2.5 times longer along seagrass patch edges than in patch interiors (Fig. 8.6A). However, differences in time to consumption for blue crabs were not evident when seagrass structural complexity was experimentally reduced, suggesting that reductions in structural complexity altered the abundance and type of predators foraging within experimental patches (Mahoney et al. 2018).

Variability in edge effects on predation risk, and the potential influence of within-patch attributes like structural complexity, were assessed in a global-scale study assessing the effects of habitat degradation (shoot thinning) on the strength of edge effects on crustacean predation risk in eelgrass habitat. Hovel et al. (2021) quantified how edge proximity and shoot thinning interactively affect predation risk for shrimps and crabs using a comparative-experimental study spanning 18 eelgrass sites across North America, Europe, and Asia (Fig. 8.6B). Tethering was used to expose locally collected shrimp and crab species at each site to predators along eelgrass patch edges and within patch interiors. Results showed that location within the patch had variable but consistently stronger effects than habitat degradation on crustacean predation risk, with predation risk being higher within patch interiors at 10 sites, higher along patch edges at three sites, and invariant to edge proximity at the remaining sites. Though habitat degradation played a smaller role in crustacean predation risk than did proximity to patch edges, sites with high predation risk along patch edges were those with lower structural complexity along patch edges, suggesting a strong role of structural complexity in dictating edge effects (Hovel et al. 2021).

Seagrass Cover and Configuration

Seagrasses form the primary settlement habitat for blue crabs in many estuaries and survival of settling blue crab megalopae, as well as early instar juveniles, is higher in seagrass than in alternative structured habitat and in unvegetated sediment (van Montfrans et al. 1995). Hovel and Lipcius (2001) tested the effects of eelgrass cover and configuration on the survival and density of juvenile blue crabs in Chesapeake Bay. Aerial photography was used to identify distinct patch types within the seascape, which differed in their relative size and isolation from one another (see examples in Fig. 8.5). Juvenile blue crabs were exposed to predators in all patch types using tethering, which revealed the importance of patch size and connectivity in driving crab relative survival, but in an unexpected way: juvenile blue crab survival was highest in the smallest, most isolated patches, and lowest in the large, continuous areas of seagrass ("reverse fragmentation effects"). Sampling of the seascape using trawls revealed that predators of juvenile blue crabs (primarily

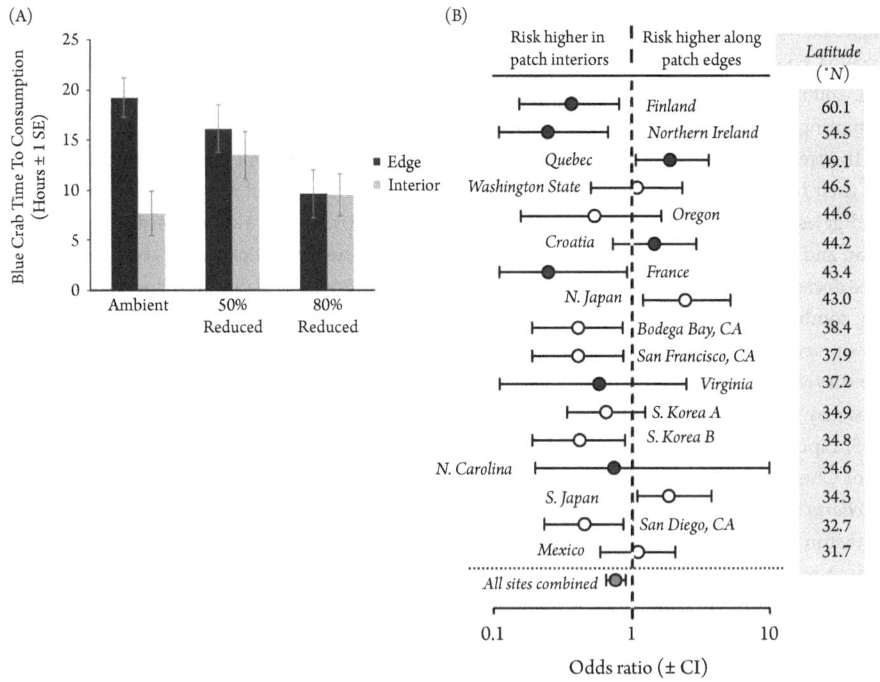

Fig. 8.6.
(A) Time to consumption of tethered blue crabs as a function of location within a seagrass meadow (edge or interior) and seagrass shoot density (ambient and reduced by 50% or 80%). Reproduced from Mahoney et al. (2018), with permission from © Elsevier. (B) Results of a global-scale study on the effects of proximity to seagrass patch edges on crustacean predation risk. Sites are arranged from north (top) to south (bottom). Odds ratios (±95% confidence intervals) represent the odds of being consumed by predators at the patch edge compared to the patch interior. Confidence intervals that do not overlap with the vertical dotted line indicate significant effects of edges on predation risk at alpha = 0.05. Overall odds ratio for all sites combined is shown at bottom beneath the horizontal dotted line. Solid circles = Atlantic sites; open circles = Pacific sites. Source: data from Hovel et al. (2021).

larger conspecifics) were least abundant in small, isolated patches, perhaps due to their vulnerability to higher-order predators like birds and large fishes that might attack crabs crossing expanses of unvegetated sediment between patches.

In temperate eelgrass habitat, predictable seasonal changes in eelgrass above-ground habitat structure can allow tests of how seagrass degradation and loss throughout an entire seagrass seascape may affect crustacean abundance and survival. For instance, in midsummer in the lower Chesapeake Bay, defoliation (senescence of seagrass blades) due to warm water temperatures, as well as bioturbation by schools of cownose rays (*Rhinoptera bonasus*) commonly reduce eelgrass structural complexity and increase eelgrass patchiness over a period of several weeks (see Fig. 8.5). When juvenile blue crabs were exposed to predators within a seagrass seascape subjected to these habitat alterations, none of the strong effects of seascape structure on predation risk remained (Hovel and Lipcius 2002).

Estuarine-Scale Studies in Seagrass Habitat

In contrast to the plethora of studies on the effects of seagrass patch size, seagrass edges, and seagrass fragmentation on crustaceans and other fauna, relatively few studies have addressed the effects of seagrass habitat alteration on crustaceans at the scale of entire estuaries. However, changes in seagrass cover and configuration have been extensive in areas like Chesapeake Bay, coastal North Carolina, coastal New England, and along other coastlines worldwide (e.g., in the Baltic Sea) (Orth et al. 2006, Pihl et al. 2006). Many of these areas have historically high abundance fishery species, prompting research on the effects of seagrass loss on fauna.

The tremendous economic value of blue crabs fuels a great deal of research on this species, and their heavy dependence on complex benthic habitat for successful recruitment (Etherington and Eggleston 2000, Moksnes and Heck 2006) has prompted speculation that seagrass fragmentation and loss have played a large role in the declines in blue crab stocks observed over the past three decades. Despite the value of seagrass habitat for blue crabs, field sampling and experimentation, combined with hydrodynamic modeling, have indicated that the effects of large-scale seagrass loss on settlers and recruits may be mitigated by other factors. For instance, though it is clear that survival of newly settled blue crabs is higher in structured habitat vs. unstructured habitat, crabs may still find refuge from predation in very shallow unvegetated sediments (Hines and Ruiz 1995, Lipcius et al. 2005). Juvenile blue crabs can also find refuge in novel habitats: in many areas of Chesapeake Bay, eelgrass nursery habitat has largely been replaced by the exotic macroalga *Gracilaria vermiculophylla,* which provides higher refuge value to juveniles than does native habitat (Johnston and Lipcius 2012). Though seagrass loss may result in higher densities of juvenile blue crabs in remaining seagrass, juveniles may use the prevailing currents to disperse from the initial seagrass settlement habitat to other habitats to avoid density-dependent mortality due to predation and competition (Reyns and Eggleston 2004). Megalopae may even bypass seagrass habitat and be advected to alternative habitats upriver, where survival can be high (e.g., low salinity areas like those found in the back of many estuaries) (Lipcius et al. 2005). Similarly, settling green crabs (*Carcinus maenas*) commonly move among potential settlement habitats before metamorphosing to the juvenile form, which may alleviate cannibalism and reductions in growth due to competition for limited habitat (Moksnes 2004).

Stockhausen and Lipcius (2003) investigated effects of the disappearance of upriver seagrass habitat on blue crab recruitment in the York River, a major tributary of Chesapeake Bay. Their model indicated smaller declines in blue crab recruitment than expected based on extensive seagrass habitat loss in this area. A major reason for this was high blue crab settlement in seagrass habitat near the river mouth, forming a "settlement shadow" in which settlement primarily occurs in the first suitable habitats encountered by larvae or postlarvae. A key message to arise from this research is the importance of considering spatial patterns of habitat (rather than simply the total amount of available habitat), as well as variation in larval and postlarval transport processes (prevailing currents) when considering both the effects of habitat loss and habitat restoration on crab recruitment and subsequent survival. Though there is little doubt that seagrass habitat loss has altered blue crab settlement dynamics, the relative role of habitat degradation and loss in the decline of blue crab stocks, in comparison to other factors such as overfishing, remains an open question.

Other studies at the estuarine scale have focused on the effects of seagrass spatial patterning on faunal assemblages. Santos et al. (2018) used multispectral satellite images of seagrasses (*T. testudinum* and *Halodule wrightii*) and a geographic information system to classify areas of Biscayne Bay, Florida, as either fragmented or continuous seagrass seascapes. They also grouped seascapes into two salinity zones: a zone of high and stable salinity, contrasted by a zone heavily influenced by pulsed freshwater inflow from canals, creating low and variable salinity. Seagrass fragmentation resulted in higher abundance and diversity of most taxa: in particular, pink shrimp (*Farfantepenaeus duorarum*), blue crabs, and caridean shrimp were more abundant in fragmented seagrass than in continuous seagrass. The authors suggested that prey for these species were abundant along the many edges found in fragmented seagrass seascapes, and that fragmentation of seagrass resulted in higher habitat heterogeneity that supported generalist predators and higher faunal diversity. Notably, the difference in abundance and diversity between fragmented and continuous seagrass seascapes were strongest where salinity was low and variable, demonstrating that seascape context plays a role in determining responses of fauna to seascape patterning (Santos et al. 2018).

The Caribbean spiny lobster (*Panulirus argus*) is an abundant resident of shallow hard bottom, reef, and seagrass habitats in the Caribbean, and they support Florida's most valuable fishery. Butler et al. (2001, 2005) investigated the effects of habitat loss on Caribbean spiny lobsters in the Florida Keys. In this region, blooms of cyanobacteria in 1991–1992 caused a large-scale die-off of sponges and the loss of thousands of hectares of seagrass, both of which provide habitat for juvenile spiny lobsters. Working across a stretch of the Florida Keys region about 200 km long, they combined repeated, long-term sampling with spatially explicit IBMs to evaluate the effects of habitat loss on lobster settlement, recruitment, and population dynamics. Surprisingly, field sampling for lobster recruits, juveniles, and adults suggested that the population of spiny lobsters in Florida increased during the period of habitat loss and degradation (Butler et al. 2005). Simulation modeling revealed that the resilience of lobsters to this major change to the seascape arose from high postlarval supply that occurred in the years before the blooms, but also from increased movement of lobsters out of heavily degraded areas and their subsequent use of alternative habitat types.

EFFECTS OF CORAL REEF ALTERATION ON CRUSTACEANS

Coral reefs famously house an astounding number of species from a wide variety of phyla, and their contributions to marine biodiversity are unparalleled in shallow coastal seas. A great deal of research on the effects of coral reef degradation has been performed on the most conspicuous residents of coral reefs: fishes, and the corals themselves. However, the diversity of motile invertebrates on coral reefs far surpasses that of vertebrate fauna (Kramer et al. 2014), and crustaceans are one of the most speciose groups on coral reefs. Stella et al. (2011) estimate that 20% of the ~168,000 species of motile invertebrates inhabiting the world's coral reefs are crustaceans. The vast majority of crustacean species are small (i.e., body lengths of millimeters) and cryptic, and include the Amphipoda, Cumacea, Copepoda, Isopoda, and Tanaidacea, which may inhabit living coral, dead coral, algae, sessile invertebrates (e.g., sponges), and coral rubble (Stella et al. 2011, Kramer et al. 2014). More conspicuous crustacean species are decapods and include crabs, shrimps, and spiny lobsters, which rely heavily on living host corals for food, habitat, and reproduction (Tsuchiya et al. 1992, Stella et al. 2011) and therefore may be more sensitive to coral reef degradation than the cryptofauna.

Coral reefs are in decline due to a variety of anthropogenic factors, including sedimentation, overfishing, destructive fishing practices, pollution, eutrophication, ocean warming, and acidification (Stella et al. 2011, Hughes et al. 2017). Reef-building corals are particularly vulnerable to anthropogenic disturbances due to the high sensitivity of the coral animals to perturbations and climatic extremes. Bleaching (the loss of symbiotic zooxanthellae) occurs in response to ocean warming and heatwaves, pollution, and disease. Corals are increasingly being overgrown by macroalgae, which in many areas are proliferating due to the loss of invertebrate and vertebrate herbivores and alterations to trophic structure. Episodic and chronic disturbances to coral reefs result in the loss of structural complexity, and "flattening" of coral reefs is a prominent aspect of their degradation (Alvarez-Filip et al. 2009).

Little research has been done on how coral reef degradation affects coral-associated invertebrates, despite their numerical dominance and overwhelming contributions to faunal diversity on reefs. However, the strength of effects of coral reef degradation on coral-associated invertebrates seems to depend on their level of dependence on a coral host, as well as their degree of specialization

(Stella et al. 2011). To this end, small, generalist, cryptic crustacean fauna (Copepoda, Amphipoda, Isopoda, etc.) are far less susceptible to coral reef degradation than the larger, commensal or predatory shrimps, crabs, and lobsters that depend on live coral. On Lizard Island (a mid-shelf reef on the Great Barrier Reef in Australia), abundance of cryptic crustaceans was two to three orders of magnitude higher on dead coral than on live coral and on algal matrix, and dead coral and coral rubble were deemed the most valuable microhabitats for crustaceans, including smaller decapods (Kramer et al. 2014; Fig. 8.7). Crustacean species richness is also higher on dead coral and coral rubble, suggesting that the loss of living coral on reefs may not cause catastrophic losses in faunal biodiversity, at least for cryptic motile epifauna (Enochs and Manzello 2012). On coral reefs in Culebra Bay, Costa Rica, species richness of malacostracan crustaceans increased by 50% after

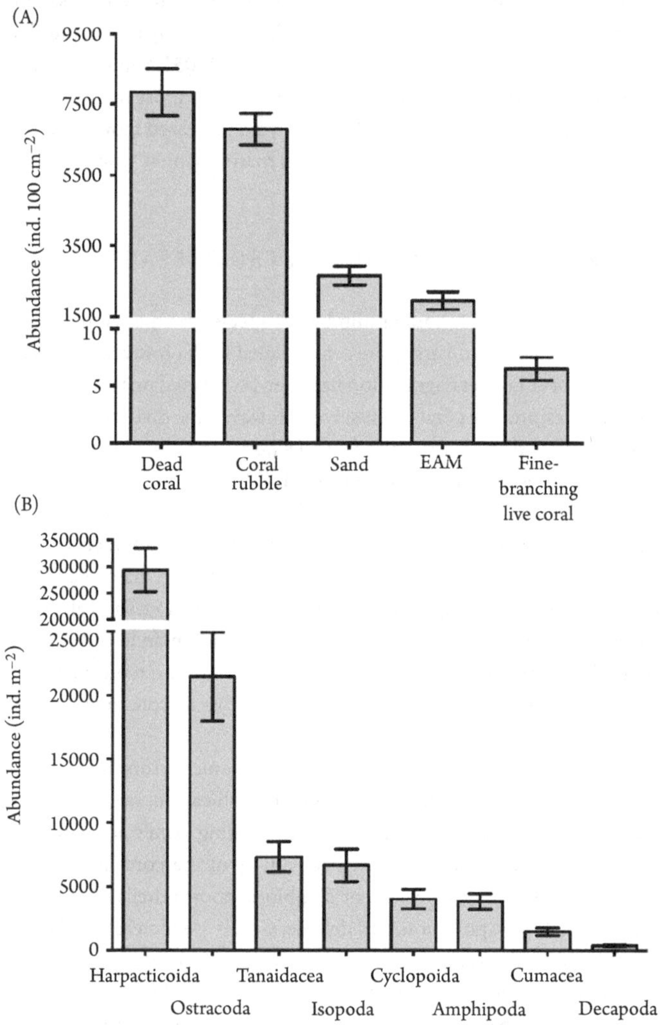

Fig. 8.7.
(A) Abundance of crustacean taxa on different coral reef substrata from Lizard Island on the Great Barrier Reef, Australia. EAM = epilithic algal matrix. (B) Relative abundance of crustacean taxa found on the surveys. Note axis breaks on y-axes. Adapted from Kramer et al. (2014), with permission from © Inter-Research Science Publisher under a Creative Commons by Attribution License (CC BY).

widespread reef degradation caused by El Niño events and harmful phytoplankton blooms (Salas-Moya et al. 2021). However, crustacean persistence relies on the replenishment of coral reef rubble that is derived from living coral, as erosional processes eventually will eliminate coral rubble habitat and flatten reefs (Kramer et al. 2014). Additionally, trophic linkages among corals, crustaceans, and predators are strongest on live coral substratum. Coral reefs composed of a variety of habitat types are necessary for the maintenance of crustacean populations and the ecosystem functions they support.

In contrast to cryptic motile epifaunal crustaceans, larger crustaceans on coral reefs may be highly susceptible to reef degradation due to their close association with corals and habitat specialization (Stella et al. 2011). Crustacean species that are commensal with corals, including crabs and shrimps, are particularly reliant on living coral, and often abandon corals undergoing degradation. One of the most comprehensive studies on the effects of coral habitat alteration on commensal crustaceans was performed by Caley et al. (2001) using the widely distributed coral *Stylophora pistillata* (Fig. 8.8). This study is notable for experimentally separating degradation effects from fragmentation effects at ecologically relevant scales. They tested for separate effects of coral degradation (death of living coral polyps) and fragmentation (the splitting apart of coral heads, which is caused by storms and destructive fishing practices) on the community of commensal macrofauna inhabiting *S. pistillata*. The study particularly focused on two prominent commensal species: the crab *Trapezia cymodoce*, which is sedentary within colonies and territorial, and the shrimp *Palaemonella* sp., which is more vagile. A primary finding of this study was that coral degradation had far stronger effects on species richness and commensal abundance than did coral head fragmentation. Species richness and commensal abundance rapidly declined after coral degradation, with commensal crabs and shrimps abandoning degraded corals within one month. In contrast, fragmentation had no effect on *Palaemonella* sp., and resulted in increased abundances of *T. cymodoce*. Fragmentation of corals resulted in more surface area for occupation by territorial crabs.

Other studies have documented the rapid loss of crustacean mutualists on degraded corals. Species of crabs such as those in the family Trapeziidae, as well as alpheid shrimps, form protection mutualisms with corals. Crustacean mutualists deter predators of corals, including the crown-of-thorns starfish *Acanthaster planci*, and clean corals of excess sediment in turbid conditions, and in return are provided with food (mucus which supplies lipids, and organic particles retained in mucus), refuge, and mating sites by corals (Stella et al. 2011). Very high sea surface temperatures, such as those seen in the 1983 El Niño event in the Eastern Pacific, resulted in widespread coral bleaching and death along the Panamanian coastline and elsewhere (Glynn 1984). In experiments in which the coral *Pocillopora damicornis* and crustacean mutualists were exposed to extreme temperatures like those observed during El Niño (> 30°C), corals died within five weeks and mutualistic xanthid crabs and alpheid shrimps died within nine weeks (Glynn and D'croz 1990). These results were similar to those of Glynn et al. (1985), who observed strong negative effects of ocean warming on *P. damicornis* lipid levels, and concomitantly lower lipid levels in the crustacean mutualists *Trapezia corallina* and *T. ferruginea*. This resulted in a decrease in ovigerous female crabs, high crab emigration, and declines in aggressive responses of crustacean mutualists to attacks by *A. planci*. *Pocillopora* is one of the most susceptible coral taxa to stress-induced bleaching and mortality, but is also the coral taxon used by the highest number of commensal and mutualistic taxa, suggesting that continued rise in coral bleaching may have disproportionate effects on some reef crustaceans (Stella et al. 2011). Despite the strong negative effects of coral degradation on crustacean mutualists, mutualistic crabs and shrimps have been observed to recolonize *Pocillopora* and other coral taxa recovering from bleaching (i.e., that have partially or completely regained zooxanthellae) (Tsuchiya et al. 1992).

Coral habitat degradation also influences the feeding ecology and habitat use of spiny lobsters, one of the most conspicuous predators on coral reefs. Spiny lobsters consume a wide variety of

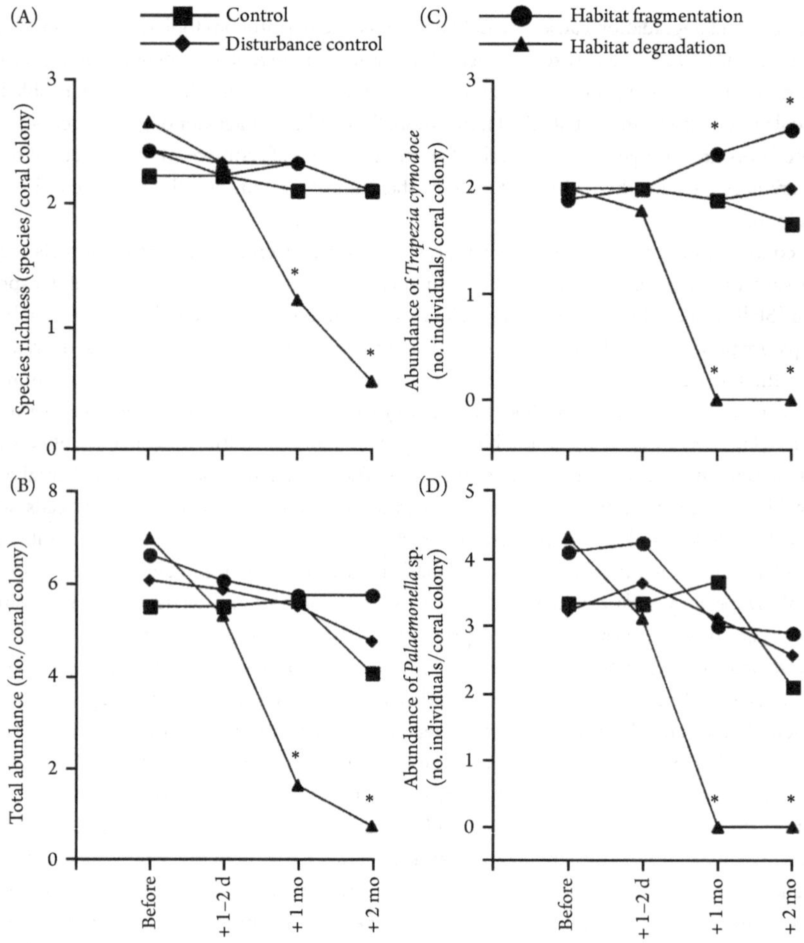

Fig. 8.8.
Effects of experimental coral habitat fragmentation and degradation on (A) crustacean species richness, (B) total crustacean abundance, (C) abundance of the commensal crab *Trapezia cymodoce*, and (D) abundance of the commensal shrimp *Palaemonella* sp. Disturbance control corals were handled similarly to degraded corals but were not killed. Habitat degradation (death of coral polyps), but not habitat fragmentation, caused the loss of crustaceans. Reproduced from Caley et al. (2001), with permission from © The Ecological Society of America.

benthic invertebrate prey, thereby playing a key role in the transfer of energy and biomass through reef food webs (Blamey et al. 2019, Sabino et al. 2021). Their generalist feeding and use of a wide variety of habitats suggests that coral reef degradation may not reduce lobster population size until degradation becomes extensive (Lozano-Álvarez et al. 2017). However, degradation has been shown to influence spiny lobster feeding ecology and their interactions with congeneric species. Degradation of back-reef habitat in the Puerto Morelos Reef National Park in Mexico resulted in smaller trophic niches and higher trophic position for the spiny lobster *Panulirus guttatus*, reflecting a narrower diet for lobsters on reefs transitioning from high coral cover to high algal cover (Briones-Fourzán et al. 2019). This may be a response to lower prey availability and reduced options for prey species. However, reef degradation did not influence the feeding ecology of the more generalist and wider-ranging congener *P. argus* on these reefs, perhaps due to the species' ability to forage in multiple habitat types. In the Seychelles of the Western Indian Ocean, coral reef degradation (primarily in the form of bleaching) increased the potential for competition between co-occurring spiny lobster species. Sabino et al. (2021) found that isotopic niche overlap between *Panulirus penicillatus* and *P. longipes* increased following a severe coral bleaching event in the Seychelles that reduced reef structural complexity and food availability. Spiny lobsters in this region are the basis of commercial fisheries that have been deemed unstable despite fishery management, leading the authors to express concern that continued reef degradation could add further instability to these populations (Sabino et al. 2021).

EFFECTS OF SANDY BEACH ALTERATION ON CRUSTACEANS

Sandy beaches are home to a diverse and specialized biota, including many crustaceans. The popularity of sandy beaches and their close proximity to human population centers has resulted in widespread alteration of these habitats, but the extent of this alteration, and its consequences for fauna, are underappreciated and understudied (Dugan et al. 2010). Beach alteration occurs due to coastal development, introduction of solid waste, trampling, altered morphodynamics, vehicle traffic, mechanical cleaning, dune removal, and a host of other disturbances (Schooler et al. 2017).

Sandy beaches are home to a variety of cryptic crustacean infauna, notably amphipods (particularly sandhoppers of the family Talitridae) and isopods. Ghost crabs (family Ocypodidae, subfamily Ocypodinae) and mole crabs (genus *Emerita*) are widely distributed, conspicuous crustaceans occupying the intertidal zone of sandy beaches. Ghost crabs form deep burrows in the high intertidal and emerge at night to scavenge and hunt for prey, whereas mole crabs rapidly burrow into the sand in the swash zone and use their antennae to filter feed within the incoming waves. All of these taxa are considered to be bioindicators of human impact (Cardoso et al. 2016, Suciu et al. 2018) and, additionally, they form key linkages in intertidal food webs.

Sandy beach crustaceans are strongly influenced by beach degradation in a variety of forms. Cardoso et al. (2016) tested whether abundances of the sandhopper *Atlantorchestoidea brasiliensis* and the mole crab *Emerita brasiliensis* are negatively affected by urbanization and recreation on sandy beaches in Brazil. Both species showed clear decreases in abundance with increasing levels of urbanization and human beach use. These species also exhibited greater degrees of fluctuating asymmetry, an indicator of unstable organismal development, on urbanized beaches compared to less impacted beaches (Frota et al. 2019). Working in the same geographic area, Veloso et al. (2006) documented the absence of the amphipod *Atlantorchestoidea brasiliensis* on heavily trafficked Rio de Janeiro beaches and the abundant populations of this amphipod on protected beaches. The sensitivity of amphipods to human trampling and vehicle traffic in Brazil and in other locations (e.g., the Iberian Peninsula: Reyes-Martínez et al. 2015; Okinawa, Japan: Mottaghi 2019; the Tuscany coast: Ugolini et al. 2008) suggest that amphipods are particularly useful bioindicators of human degradation of sandy beaches.

Ghost crabs have been the subject of many studies on the effects of beach degradation on crustacean fauna due to their widespread distribution and sensitivity to anthropogenic impacts. Degradation can not only reduce ghost crab densities, but also influence their distribution and behavior. On sandy beaches in South Carolina, USA, ghost crab burrow density (a proxy for crab abundance) decreased as human beach use increased (Gül and Griffen 2018). However, heavy human presence also caused crabs to move to the edges of heavily impacted areas, and to create deeper, steeper, and smaller burrows in response to sediment compaction caused by the weight of vehicles and pedestrians. Ghost crabs have also been useful in studies attempting to evaluate the relative strengths of different sources of beach degradation on fauna. This is important to do because different components of degradation and disturbance may differentially affect crustaceans but are often strongly correlated. However, restoration and mitigation efforts rely on an understanding of the relative effects of various aspects of human alteration on fauna. To this end, Suciu et al. (2018) used generalized linear mixed models and structural equation modeling to determine the relative effects of human visitation and amount of solid waste on the ghost crab *Ocypode quadrata*, as well as on amphipods and isopods, on Brazilian beaches. Their models suggested that reduced ghost crab abundance was directly caused by high amounts of solid waste on beaches, which in turn was correlated with beach visitation. They therefore cautioned against assessing beach disturbance using only human visitation and trampling as proxies (Suciu et al. 2018). Schlacher et al. (2016) conducted a meta-analysis on human threats to sandy beaches using ghost crabs as an indicator of beach degradation (Fig. 8.9). Their meta-analysis encompassed studies on five continents and included over two dozen types of human pressures on ghost crab abundance and size of crab burrow openings. Using log response ratios, the authors reported on strong and consistent negative effects of human stressors on ghost crab abundance (35 out of 37 studies), but not on crab size. Degradation due to vehicles, beach nourishment, and urbanization had the strongest effects on ghost crab abundance.

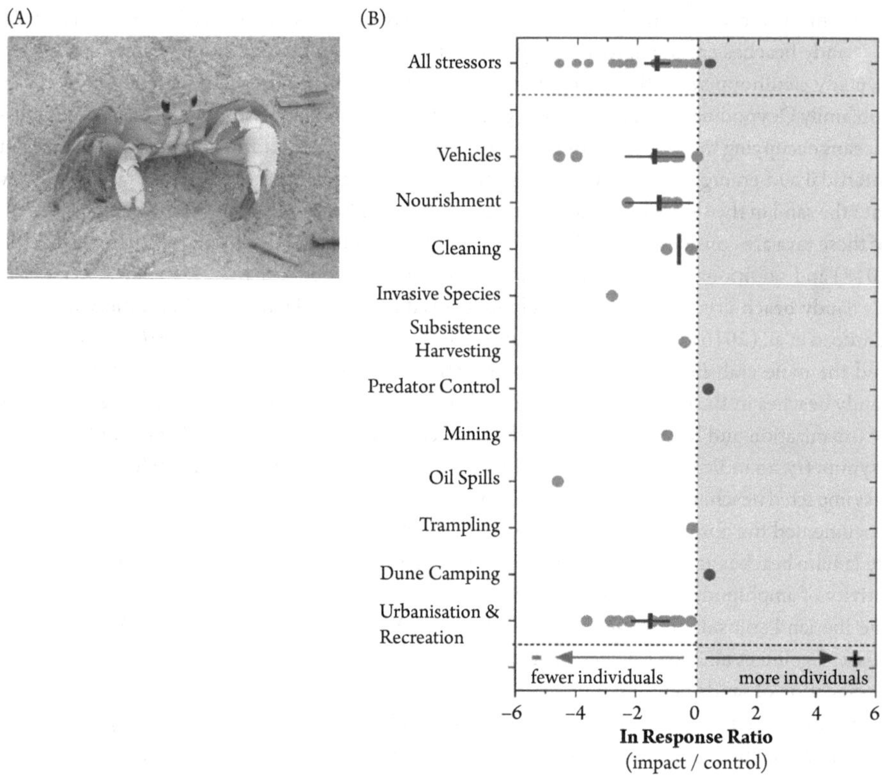

Fig. 8.9.
(A) The Atlantic ghost crab *Ocypode quadrata*. Reproduced courtesy of Antony-22 under Creative Commons 4.0 license. (B) Results of the meta-analysis on anthropogenic factors affecting ghost crab abundance on ocean beaches. Dots represent effect sizes as response ratios ($\ln R = \text{mean}_{\text{impact}}/\text{mean}_{\text{control}}$) from individual studies in the meta-analysis. Vertical lines are means and horizontal lines are 95% confidence intervals. Reproduced from Schlacher et al. (2016), with permission from © Elsevier.

UNIFYING PRINCIPLES

Though many types of alterations are specific to particular habitats and ecosystems, several unifying principles emerge when broadly considering how human and natural habitat alteration affects crustacean biology and ecology.

First, crustaceans are generally resilient to habitat alteration and loss. This is particularly true when suitable alternative habitats are available; for instance, if salt marshes are replaced by seagrass habitat (as seen in the Gulf of Mexico), or if seagrass is replaced by algal habitat (as seen in the Chesapeake Bay). Many crustaceans are generalists with a great deal of versatility in their habitat use and diet. This principle is not applicable, however, to crustacean taxa occupying isolated habitats such as vernal pools and sandy beaches, where specialization is high, human interference is extensive, and there are few alternative habitats available. The importance of degradation and destruction of specialized habitats, which have reached extreme levels for some habitats (e.g., freshwater wetlands), should not be downplayed.

Second, in many habitats, moderate habitat alteration or degradation does not result in reductions in crustacean populations or in reductions to their contributions to ecosystem function. Habitat fragmentation in habitats such as salt marshes, seagrass beds, and coral reefs does not often reduce crustacean populations. Indeed, crustaceans often thrive in fragmented habitats, and habitats containing many edges and high variability in patch size and isolation may be of high value to many species. This reflects the generalist nature of many crustacean species (e.g., the ability to forage or seek refuge in multiple habitat types) as well as their generally high levels of mobility. However, variability in the effects of habitat features (such as edges or patchiness) on abundance, survival, and diversity can be high across sites and through time.

Third, the effects of habitat alteration on crustaceans should not be assumed to be the same across body sizes and life history stages. Juveniles often have very different habitat requirements and nutritional requirements than adult conspecifics. Habitat edges may be beneficial to larger, predatory life history stages, but may be risky areas for juveniles. Degradation (e.g., via coral death) may result in higher abundances of small epifaunal crustaceans, but reductions in abundance and survival of larger crustacean fauna (which may be related more to degree of specialization than to body size). Studies will need to assess effects of habitat alteration on multiple life history stages to fully address anthropogenic effects on populations and communities.

Finally, crustaceans offer a plethora of opportunities to investigate how habitat alteration affects faunal abundance, diversity, survival, and behavior. The diversity of crustacean morphotypes and ethology can be extensive within a single habitat type (e.g., on coral reefs), allowing comparisons of how habitat alteration influences fauna using correlative and manipulative approaches. Though investigating the effects of habitat fragmentation and loss on fauna using manipulative experimentation can be challenging, creative approaches in the field and lab, and the use of simulation modeling (Hovel and Regan 2008, Roth et al. 2008), present a host of opportunities to advance our understanding of how anthropogenic habitat alteration impacts fauna.

REFERENCES

Altieri, A.H., M.D. Bertness, T.C. Coverdale, N.C. Herrmann, and C. Angelini. 2012. A trophic cascade triggers collapse of a salt-marsh ecosystem with intensive recreational fishing. Ecology 93:1402–1410.

Alvarez-Filip, L., N.K. Dulvy, J.A. Gill, I.M. Côté, and A.R. Watkinson. 2009. Flattening of Caribbean coral reefs: Region-wide declines in architectural complexity. Proceedings of the Royal Society B: Biological Sciences 276:3019–3025.

Andrén, H. 1994. Effects of habitat fragmentation on birds and mammals in landscapes with different proportions of suitable habitat: A review. Oikos 71:355–366.

Baker, R., M.D. Taylor, K.W. Able, M.W. Beck, J. Cebrian, D.D. Colombano, R.M. Connolly, C. Currin, L.A. Deegan, I.C. Feller, B.L. Gilby, M.E. Kimball, T.J. Minello, L.P. Rozas, C. Simenstad, R.E. Turner, N.J. Waltham, M.P. Weinstein, S.L. Ziegler, P.S.E. Zu Ermgassen, C. Alcott, S.B. Alford, M.A. Barbeau, S.C. Crosby, K. Dodds, A. Frank, J. Goeke, L.A.G. Gaines, F.E. Hardcastle, C.J. Henderson, W.R. James, M.D. Kenworthy, J. Lesser, D. Mallick, C.W. Martin, A.E. McDonald, C. McLuckie, B.H. Morrison, J.A. Nelson, G.S. Norris, J. Ollerhead, J.W. Pahl, S. Ramsden, J.S. Rehage, J.F. Reinhardt, R.J. Rezek, L.M. Risse, J.A.M. Smith, E.L. Sparks, and L.W. Staver. 2020. Fisheries rely on threatened salt marshes. Science 370:670–671.

Bauer, R.T. 2018. Life cycle and seasonal migrations. Pages 203–229 in G.A. Wellborn and M. Thiel, editors. The natural history of the Crustacea, Volume 5: Life histories. Oxford University Press, New York, USA.

Bell, S.S., M.O. Hall, S. Soffian, and K. Madley. 2002. Assessing the impact of boat propeller scars on fish and shrimp utilizing seagrass beds. Ecological Applications 12:206–217.

Bertness, M.D., and B.R. Silliman. 2008. Consumer control of salt marshes driven by human disturbance. Conservation Biology 22:618–623.

Blamey, L.K., A.M. de Lecea, L.D.S. Jones, and G.M. Branch. 2019. Diet of the spiny lobster *Jasus paulensis* from the Tristan da Cunha archipelago: Comparisons between islands, depths and lobster sizes. Estuarine, Coastal and Shelf Science 219:262–272.

Bologna, P.A.X., and K.L. Heck. 2002. Impact of habitat edges on density and secondary production of seagrass-associated fauna. Estuaries 25:1033–1044.

Boström, C., S.J. Pittman, C. Simenstad, and R.T. Kneib. 2011. Seascape ecology of coastal biogenic habitats: Advances, gaps, and challenges. Marine Ecology Progress Series 427:191–217.

Briones-Fourzán, P., L. Álvarez-Filip, C. Barradas-Ortíz, P.S. Morillo-Velarde, F. Negrete-Soto, I. Segura-García, A. Sánchez-González, and E. Lozano-Álvarez. 2019. Coral reef degradation differentially alters feeding ecology of co-occurring congeneric spiny lobsters. Frontiers in Marine Science 5:516.

Browder, J.A., H.A. Bartley, and K.S. Davis. 1985. A probabilistic model of the relationship between marshland-water interface and marsh disintegration. Ecological Modelling 29:245–260.

Butler IV, M.J., T. Dolan, W. Herrnkind, and J. Hunt. 2001. Modelling the effect of spatial variation in postlarval supply and habitat structure on recruitment of Caribbean spiny lobster. Marine and Freshwater Research 52:1243.

Butler, M.J., T.W. Dolan, J.H. Hunt, K.A. Rose, and W.F. Herrnkind. 2005. Recruitment in degraded marine habitats: A spatially explicit, individual-based model for spiny lobster. Ecological Applications 15:902–918.

Caley, M.J., K.A. Buckley, and G.P. Jones. 2001. Separating ecological effects of habitat fragmentation, degradation, and loss on coral commensals. Ecology 82:3435–3448.

Cardoso, R.S., C.A.M. Barboza, V.B. Skinner, and T.M.B. Cabrini. 2016. Crustaceans as ecological indicators of metropolitan sandy beaches health. Ecological Indicators 62: 154–162.

Chesney, E.J., D.M. Baltz, and R.G. Thomas. 2000. Louisiana estuarine and coastal fisheries and habitats: Perspectives from a fish's eye view. Ecological Applications 10:350–366.

Coverdale, T.C., M.D. Bertness, and A.H. Altieri. 2013. Regional ontogeny of New England salt marsh die-off. Conservation Biology 27:1041–1048.

Crotty, S.M., C. Angelini, and M. D. Bertness. 2017. Multiple stressors and the potential for synergistic loss of New England salt marshes. PLoS ONE 12:e0183058.

Dahl, T.E. 1990. Wetlands losses in the United States, 1780s to 1980s. U.S. Department of the Interior, Fish and Wildlife Service, Washington D.C. USA.

Day, J.W., G.P. Kemp, D.J. Reed, D.R. Cahoon, R.M. Boumans, J.M. Suhayda, and R. Gambrell. 2011. Vegetation death and rapid loss of surface elevation in two contrasting Mississippi delta salt marshes: The role of sedimentation, autocompaction and sea-level rise. Ecological Engineering 37:229–240.

Duffy, J.E., K.S. Macdonald, J.M. Rhode, and J.D. Parker. 2001. Gracer diversity, functional redundancy, and productivity in seagrass beds: An experimental test. Ecology 82:2417–2434.

Dugan, J.E., O. Defeo, E. Jaramillo, A.R. Jones, M. Lastra, R. Nel, C.H. Peterson, F. Scapini, T. Schlacher, and D.S. Schoeman. 2010. Give beach ecosystems their day in the sun. Science 329:1146–1146.

Eggleston, D.B., L.L. Etherington, and W.E. Elis. 1998. Organism response to habitat patchiness: Species and habitat-dependent recruitment of decapod crustaceans. Journal of Experimental Marine Biology and Ecology 223:111–132.

Enochs, I.C., and D.P. Manzello. 2012. Responses of cryptofaunal species richness and trophic potential to coral reef habitat degradation. Diversity 4:94–104.

Etherington, L.L., and D.B. Eggleston. 2000. Large-scale blue crab recruitment: Linking postlarval transport, post-settlement planktonic dispersal, and multiple nursery habitats. Marine Ecology Progress Series 204:179–198.

Fahrig, L. 2003. Effects of habitat fragmentation on biodiversity. Annual Review of Ecology, Evolution, and Systematics 34:487–515.

Foster, M.S. 2006. Top-down vs. bottom-up effects in kelp forests. Science 313:1737–1739.

Frota, G.P., T.M.B. Cabrini, and R.S. Cardoso. 2019. Fluctuating asymmetry of two crustacean species on fourteen sandy beaches of Rio de Janeiro State. Estuarine, Coastal and Shelf Science 223:138–146.

Fryer, G. 1985. Crustacean diversity in relation to the size of water bodies: Some facts and problems. Freshwater Biology 15:347–361.

Gates, J.E., and L.W. Gysel. 1978. Avian nest dispersion and fledging success in field-forest ecotones. Ecology 59:871–883.

Gittman, R.K., F.J. Fodrie, A.M. Popowich, D.A. Keller, J.F. Bruno, C.A. Currin, C.H. Peterson, and M.F. Piehler. 2015. Engineering away our natural defenses: An analysis of shoreline hardening in the US. Frontiers in Ecology and the Environment 13:301–307.

Glynn, P.W. 1984. Widespread coral mortality and the 1982–83 El Niño warming event. Environmental Conservation 11:133–146.

Glynn, P.W., and L. D'croz. 1990. Experimental evidence for high temperature stress as the cause of El Nino-coincident coral mortality. Coral reefs 8:181–191.

Glynn, P.W., M. Perez, and S.L. Gilchrist. 1985. Lipid decline in stressed corals and their crustacean symbionts. The Biological Bulletin 168:276–284.

Gül, M.R., and B.D. Griffen. 2018. Impacts of human disturbance on ghost crab burrow morphology and distribution on sandy shores. PLoS ONE 13:e0209977.

Haas, H.L., K.A. Rose, B. Fry, T.J. Minello, and L.P. Rozas. 2004. Brown shrimp on the edge: Linking habitat to survival using an individual-based simulation model. Ecological Applications 14:1232–1247.

Hanski, I. 2015. Habitat fragmentation and species richness. Journal of Biogeography 42:989–993.

Heck, K.L., and L.B. Crowder. 1991. Habitat structure and predator–prey interactions in vegetated aquatic systems. Pages 281–299 *in* S.S. Bell, E.D. McCoy, and H.R. Mushinsky, editors. Habitat structure: The physical arrangement of objects in space. Springer Netherlands, Dordrecht.

Hines, A.H., and G.M. Ruiz. 1995. Temporal variation in juvenile blue crab mortality: Nearshore shallows and cannibalism in Chesapeake Bay. Bulletin of Marine Science 57:884–901.

Hovel, K.A., J.E. Duffy, J.J. Stachowicz, P. Reynolds, C. Boström, K.E. Boyer, S. Cimon, M. Cusson, F.J. Fodrie, K. Gagnon, C.M. Hereu, M. Hori, P. Jorgensen, C. Kruschel, K.-S. Lee, M. Nakaoka, N.E. O'Connor, F. Rossi, J. Ruesink, F. Tomas, and S. Ziegler. 2021. Joint effects of patch edges and habitat degradation on faunal predation risk in a widespread marine foundation species. Ecology 102:e03316.

Hovel, K.A., and R.N. Lipcius. 2001. Habitat fragmentation in a seagrass landscape: Patch size and complexity control blue crab survival. Ecology 82: 1814–1829.

Hovel, K.A., and R.N. Lipcius. 2002. Effects of seagrass habitat fragmentation on juvenile blue crab survival and abundance. Journal of Experimental Marine Biology and Ecology 271:75–98.

Hovel, K.A., and H.M. Regan. 2008. Using an individual-based model to examine the roles of habitat fragmentation and behavior on predator–prey relationships in seagrass landscapes. Landscape Ecology 23:75–89.

Hughes, T.P., M.L. Barnes, D.R. Bellwood, J.E. Cinner, G.S. Cumming, J.B.C. Jackson, J. Kleypas, I.A. van de Leemput, J.M. Lough, T.H. Morrison, S.R. Palumbi, E.H. van Nes, and M. Scheffer. 2017. Coral reefs in the Anthropocene. Nature 546:82–90.

Hyndes, G.A., I. Nagelkerken, R.J. McLeod, R.M. Connolly, P.S. Lavery, and M.A. Vanderklift. 2014. Mechanisms and ecological role of carbon transfer within coastal seascapes. Biological Reviews 89:232–254.

Jenkins, D.G., S. Grissom, and K. Miller. 2003. Consequences of prairie wetland drainage for crustacean biodiversity and metapopulations. Conservation Biology 17:158–167.

Jerabek, A., K.M. Darnell, C. Pellerin, and T.J.B. Carruthers. 2017. Use of marsh edge and submerged aquatic vegetation as habitat by fish and crustaceans in degrading southern Louisiana coastal marshes. Southeastern Geographer 57:212–230.

Johnson, M.W., and K.L. Heck. 2006. Seagrass patch characteristics alter direct and indirect interactions in a tritrophic estuarine food web. Estuaries and Coasts 29:499–510.

Johnston, C.A., and O.N. Caretti. 2017. Mangrove expansion into temperate marshes alters habitat quality for recruiting *Callinectes* spp. Marine Ecology Progress Series 573:1–14.

Johnston, C.A., and R. Lipcius. 2012. Exotic macroalga *Gracilaria vermiculophylla* provides superior nursery habitat for native blue crab in Chesapeake Bay. Marine Ecology Progress Series 467:137–146.

Keeley, J.E., and P.H. Zedler. 1998. Characterization and global distribution of vernal pools. Pages 1–14 *in* C.W. Witham, E.T. Bauder, D. Belk, W.R. Ferren Jr., and R. Ornduff, editors. Ecology, conservation, and management of vernal pool ecosystems—Proceedings from a 1996 Conference. California Native Plant Society, Sacramento, CA, USA.

Kennish, M.J. 2001. Coastal salt marsh systems in the U.S.: A review of anthropogenic impacts. Journal of Coastal Research 17:731–748.

Kenworthy, W.J., M.S. Fonseca, P.E. Whitfield, and K.K. Hammerstrom. 2002. Analysis of seagrass recovery in experimental excavations and propeller-scar disturbances in the Florida Keys national marine sanctuary. Journal of Coastal Research 37:75–85.

King, J.L., 1998. Loss of diversity as a consequence of habitat destruction in California vernal pools. Pages 119–123 *in* C.W. Witham, editor. Ecology, conservation, and management of vernal pool ecosystems—Proceedings from a 1996 conference. California Native Plant Society, Sacramento, CA, USA.

Kornis, M.S., D. Breitburg, R. Balouskus, D.M. Bilkovic, L.A. Davias, S. Giordano, K. Heggie, A.H. Hines, J.M. Jacobs, T.E. Jordan, R.S. King, C.J. Patrick, R.D. Seitz, H. Soulen, T.E. Targett, D.E. Weller, D.F. Whigham, and J. Uphoff. 2017. Linking the abundance of estuarine fish and crustaceans in nearshore waters to shoreline hardening and land cover. Estuaries and Coasts 40:1464–1486.

Kramer, M.J., D.R. Bellwood, and O. Bellwood. 2014. Benthic Crustacea on coral reefs: A quantitative survey. Marine Ecology Progress Series 511:105–116.

Lahti, D.C. 2001. The "edge effect on nest predation" hypothesis after twenty years. Biological Conservation 99:365–374.

Landry, J.B., and R.R. Golden. 2018. In situ effects of shoreline type and watershed land use on submerged aquatic vegetation habitat quality in the Chesapeake and Mid-Atlantic coastal bays. Estuaries and Coasts 41:101–113.

Lefcheck, J.S., S.R. Marion, A.V. Lombana, and R.J. Orth. 2016. Faunal communities are invariant to fragmentation in experimental seagrass landscapes. PLoS ONE 11:e0156550.

Lipcius, R.N., R.D. Seitz, M.S. Seebo, and D. Colón-Carrión. 2005. Density, abundance and survival of the blue crab in seagrass and unstructured salt marsh nurseries of Chesapeake Bay. Journal of Experimental Marine Biology and Ecology 319:69–80.

Lozano-Álvarez, E., N. Luviano-Aparicio, F. Negrete-Soto, C. Barradas-Ortiz, S. Aguíñiga-García, P.S. Morillo-Velarde, L. Álvarez-Filip, and P. Briones-Fourzán. 2017. Does reef architectural complexity influence resource availability for a large reef-dwelling invertebrate? Journal of Sea Research 128:84–91.

Luk, Y.C., and R.N. Zajac. 2013. Spatial ecology of fiddler crabs, *Uca pugnax*, in Southern New England salt marsh landscapes: Potential habitat expansion in relation to salt marsh change. Northeastern Naturalist 20:255–274.

Mabidi, A., M.S. Bird, and R. Perissinotto. 2018. Increasing salinity drastically reduces hatching success of crustaceans from depression wetlands of the semi-arid Eastern Cape Karoo region, South Africa. Scientific Reports 8:5983.

Macreadie, P.I., J.S. Hindell, M.J. Keough, G.P. Jenkins, and R.M. Connolly. 2010. Resource distribution influences positive edge effects in a seagrass fish. Ecology 91:2013–2021.

Mahoney, R.D., M.D. Kenworthy, J.K. Geyer, K.A. Hovel, and F.J. Fodrie. 2018. Distribution and relative predation risk of nekton reveal complex edge effects within temperate seagrass habitat. Journal of Experimental Marine Biology and Ecology 503:52–59.

Minello, T.J., R.J. Zimmerman, and R. Medina. 1994. The importance of edge for natant macrofauna in a created salt marsh. Wetlands 14:184–198.

Moksnes, P.-O., and K.L. Heck Jr. 2006. Relative importance of habitat selection and predation for the distribution of blue crab megalopae and young juveniles. Marine Ecology Progress Series 308:165–181.

Moksnes, P.-O. 2004. Self-regulating mechanisms in cannibalistic populations of juvenile shore crabs *Carcinus maenas*. Ecology 85:1343–1354.

Moore, E.C., and K.A. Hovel. 2010. Relative influence of habitat complexity and proximity to patch edges on seagrass epifaunal communities. Oikos 119:1299–1311.

Mottaghi, A. 2019. Comparing the diversity of interstitial crustaceans at natural and artificial/degraded beaches in Okinawa. Regional Studies in Marine Science 29:100685.

Orth, R.J., T.J.B. Carruthers, W.C. Dennison, C.M. Duarte, J.W. Fourqurean, K.L. Heck, A.R. Hughes, G.A. Kendrick, W.J. Kenworthy, S. Olyarnik, F.T. Short, M. Waycott, and S.L. Williams. 2006. A global crisis for seagrass ecosystems. BioScience 56:987–996.

Peterson, B.J., K.R. Thompson, J.H. Cowan Jr., and K.L. Heck Jr. 2001. Comparison of predation pressure in temperate and subtropical seagrass habitats based on chronographic tethering. Marine Ecology Progress Series 224: 77–85.

Pihl, L., S. Baden, N. Kautsky, P. Rönnbäck, T. Söderqvist, M. Troell, and H. Wennhage. 2006. Shift in fish assemblage structure due to loss of seagrass *Zostera marina* habitats in Sweden. Estuarine, Coastal and Shelf Science 67:123–132.

Prosser, D.J., T.E. Jordan, J.L. Nagel, R.D. Seitz, D.E. Weller, and D.F. Whigham. 2018. Impacts of coastal land use and shoreline armoring on estuarine ecosystems: An introduction to a special issue. Estuaries and Coasts 41:2–18.

Radford, J.Q., A.F. Bennett, and G.J. Cheers. 2005. Landscape-level thresholds of habitat cover for woodland-dependent birds. Biological Conservation 124:317–337.

Raposa, K.B., R.A. McKinney, C. Wigand, J.W. Hollister, C. Lovall, K. Szura, J. John A. Gurak, J. McNamee, C. Raithel, and E.B. Watson. 2018. Top-down and bottom-up controls on southern New England salt marsh crab populations. PeerJ 6:e4876.

Reed, B., and K. Hovel. 2006. Seagrass habitat disturbance: How loss and fragmentation of eelgrass *Zostera marina* influences epifaunal abundance and diversity. Marine Ecology Progress Series 326:133–143.

Reyes-Martínez, M.J., M.C. Ruíz-Delgado, J.E. Sánchez-Moyano, and F.J. García-García. 2015. Response of intertidal sandy-beach macrofauna to human trampling: An urban vs. natural beach system approach. Marine Environmental Research 103:36–45.

Reynolds, P.L., J.P. Richardson, and J.E. Duffy. 2014. Field experimental evidence that grazers mediate transition between microalgal and seagrass dominance. Limnology and Oceanography 59:1053–1064.

Reyns, N.B., and D.B. Eggleston. 2004. Environmentally-controlled, density-dependent secondary dispersal in a local estuarine crab population. Oecologia 140:280–288.

Ries, L., S.M. Murphy, G.M. Wimp, and R.J. Fletcher. 2017. Closing persistent gaps in knowledge about edge ecology. Current Landscape Ecology Reports 2:30–41.

Roberts, D.A., and A.G.B. Poore. 2006. Habitat configuration affects colonization of epifauna in a marine algal bed. Biological Conservation 127:18–26.

Roth, B.M., K.A. Rose, L.P. Rozas, and T.J. Minello. 2008. Relative influence of habitat fragmentation and inundation on brown shrimp *Farfantepenaeus aztecus* production in northern Gulf of Mexico salt marshes. Marine Ecology Progress Series 359:185–202.

Rozas, L.P., T.J. Minello, R.J. Zimmerman, and P. Caldwell. 2007. Nekton populations, long-term wetland loss, and the effect of recent habitat restoration in Galveston Bay, Texas, USA. Marine Ecology Progress Series 344:119–130.

Sabino, M.A.,R. Govinden, H. Pethybridge, L. Blamey, F. Le Grand, F. Sardenne, M. Rose, P. Bustamante, and N. Bodin. 2021. Habitat degradation increases interspecific trophic competition between three spiny lobster species in Seychelles. Estuarine, Coastal and Shelf Science 256:107368.

Salas-Moya, C., S. Fabregat-Malé, R. Vargas-Castillo, J.M. Valverde, F. Vásquez-Fallas, J. Sibaja-Cordero, and J.J. Alvarado. 2021. *Pocillopora* cryptofauna and their response to host coral mortality. Symbiosis 84:91–103.

Santos, R.O., D. Lirman, S.J. Pittman, and J.E. Serafy. 2018. Spatial patterns of seagrasses and salinity regimes interact to structure marine faunal assemblages in a subtropical bay. Marine Ecology Progress Series 594:21–38.

Scheffel, W.A., K.L. Heck, and M.W. Johnson. 2018. Tropicalization of the northern Gulf of Mexico: Impacts of salt marsh transition to black mangrove dominance on faunal communities. Estuaries and Coasts 41:1193–1205.

Schlacher, T.A., S. Lucrezi, R.M. Connolly, C.H. Peterson, B.L. Gilby, B. Maslo, A.D. Olds, S.J. Walker, J.X. Leon, C.M. Huijbers, M.A. Weston, A. Turra, G.A. Hyndes, R.A. Holt, and D.S. Schoeman. 2016. Human threats to sandy beaches: A meta-analysis of ghost crabs illustrates global anthropogenic impacts. Estuarine, Coastal and Shelf Science 169:56–73.

Schooler, N.K., J.E. Dugan, D.M. Hubbard, and D. Straughan. 2017. Local scale processes drive long-term change in biodiversity of sandy beach ecosystems. Ecology and Evolution 7:4822–4834.

Schultz, R.A., S.C. Anisfeld, and T.D. Hill. 2016. Submergence and herbivory as divergent causes of marsh loss in long island sound. Estuaries and Coasts 39:1367–1375.

Shakeri, L.M., K.M. Darnell, T.J.B. Carruthers, and M.Z. Darnell. 2020. Blue crab abundance and survival in a fragmenting coastal marsh system. Estuaries and Coasts 43:1545–1555.

Shears, N.T., and R.C. Babcock. 2002. Marine reserves demonstrate top-down control of community structure on temperate reefs. Oecologia 132:131–142.

Shinomiya, Y., S. Chiba, M. Kanamori, S. Hashizume, K. Yoshino, and S. Goshima. 2017. Importance of patch size variation for the population persistence of a decapod crustacean in seagrass beds. Marine Ecology Progress Series 570:157–171.

Silliman, B.R., and M.D. Bertness. 2002. A trophic cascade regulates salt marsh primary production. Proceedings of the National Academy of Sciences 99:10500–10505.

Simovich, M.A., K.B. Davis, and A.J. Bohonak. 2013. Landscape homogenization threatens the genetic integrity of the endangered San Diego fairy shrimp *Branchinecta sandiegonensis* (Branchiopoda: Anostraca). Journal of Crustacean Biology 33:730–740.

Smith, S.M. 2015. Does loss of salt marsh vegetation caused by a native grapsid crab improve habitat suitability for the Atlantic mud fiddler (*Uca pugnax*)? Journal of Crustacean Biology 35:616–621.

Spadaro, A.J., and M.J. Butler. 2021. Herbivorous crabs reverse the seaweed dilemma on coral reefs. Current Biology 31:853–859.

Stella, J.S., M.S. Pratchett, P.A. Hutchings, and G.P. Jones. 2011. Coral-associated invertebrates: Diversity, ecological importance and vulnerability to disturbance. Pages 43–104 *in* R.N. Gibson, A.J. Atkinson, J.D.M. Gordon, I.P. Smith, and D.J. Hughes, editors. Oceanography and marine biology: An annual review. Taylor and Francis, Boca Raton, Florida, USA.

Stockhausen, W.T., and R.N. Lipcius. 2003. Simulated effects of seagrass loss and restoration on settlement and recruitment of blue crab postlarvae and juveniles in the York River, Chesapeake Bay. Bulletin of Marine Science 72:409–422.

Suciu, M.C., D.C. Tavares, and I.R. Zalmon. 2018. Comparative evaluation of crustaceans as bioindicators of human impact on Brazilian sandy beaches. Journal of Crustacean Biology 38:420–428.

Tanner, J.E. 2003. Patch shape and orientation influences on seagrass epifauna are mediated by dispersal abilities. Oikos 100:517–524.

Trzcinski, M.K., L. Fahrig, and G. Merriam. 1999. Independent effects of forest cover and fragmentation on the distribution of forest breeding birds. Ecological Applications 9:586–593.

Tsuchiya, M., Y. Yamauchi, F. Moretzsohn, and M. Tsukiji. 1992. Species composition and some population traits of obligate symbiotic xanthid crabs, *Trapezia* and *Tetralia*, associated with bleached corals. Proceedings of the Seventh International Coral Reef Symposium:56–63

Ugolini, A., G. Ungherese, S. Somigli, G. Galanti, D. Baroni, F. Borghini, N. Cipriani, M. Nebbiai, M. Passaponti, and S. Focardi. 2008. The amphipod *Talitrus saltator* as a bioindicator of human trampling on sandy beaches. Marine Environmental Research 65:349–357.

Uhrin, A., and J. Holmquist. 2003. Effects of propeller scarring on macrofaunal use of the seagrass *Thalassia testudinum*. Marine Ecology Progress Series 250:61–70.

van Montfrans, J., C.E. Epifanio, D.M. Knot, R.N. Lipcius, D.J. Mense, K.S. Metcalfem, E.J. OImi III, R.J. Orth, M.H. Posey, E.L. Wenner, and T.L. West. 1995. Settlement of blue crab postlarvae in western North Atlantic estuaries. Bulletin of Marine Science 57:834–854.

Veloso, V.G., E.S. Silva, C.H.S. Caetano, and R.S. Cardoso. 2006. Comparison between the macroinfauna of urbanized and protected beaches in Rio de Janeiro State, Brazil. Biological Conservation 127:510–515.

Waycott, M., C.M. Duarte, T.J.B. Carruthers, R.J. Orth, W.C. Dennison, S. Olyarnik, A. Calladine, J.W. Fourqurean, K.L. Heck, A.R. Hughes, G.A. Kendrick, W.J. Kenworthy, F.T. Short, and S.L. Williams. 2009. Accelerating loss of seagrasses across the globe threatens coastal ecosystems. Proceedings of the National Academy of Sciences 106:12377–12381.

Williams, S., and K. Heck. 2001. Seagrass community ecology. Pages 317–337 *in* M. Bertness, S. Gaines, and M. Hay, editors. Marine community ecology, 1st edition. Sinauer Associates, Sunderland, MA, USA.

Yeager, L.A., D.A. Keller, T.R. Burns, A.S. Pool, and F.J. Fodrie. 2016. Threshold effects of habitat fragmentation on fish diversity at landscapes scales. Ecology 97:2157–2166.

Zimmerman, R.J., T.J. Minello, and L.P. Rozas. 2000. Salt marsh linkages to productivity of penaeid shrimps and blue crabs in the northern Gulf of Mexico. Pages 293–314 *in* M.P. Weinstein and D.A. Kreeger, editors. Concepts and controversies in tidal marsh ecology. Springer, Dordrecht, The Netherlands.

9

IMPACTS OF OCEAN CHANGE ON CRUSTACEANS

Christian Pansch, Marian Hu, Dirk Weihrauch, and Christian Buschbaum

Abstract

This chapter addresses the direct and indirect effects of changing abiotic and biotic conditions on crustacean species in coastal ecosystems, and the ecological consequences from the individual to the ecosystem level. We focus on significant environmental drivers of ocean change, such as warming, acidification, and hypoxia, but also include the role of invasive species and their parasitic assemblages. Further, we review descriptive and experimental studies from coastal ecosystems worldwide to describe how these factors affect cell to organismic physiology, population and species performance, and species interactions. Three case studies highlight (i) the adaptation of brachyuran crabs to extreme, but natural, environmental acidification in a shallow-water hydrothermal vent system; (ii) the complexity of multiple environmental drivers acting upon multiple life-history stages of acorn barnacles, and the species' potential for acclimation and adaptation; as well as (iii) the competitive abilities and dynamics of two successful nonnative brachyuran crabs during a recent invasion process. Physiological pathways, tolerance, and capability for adaptation to changing ocean conditions determine population dynamics and changing distribution patterns of crustaceans. The successful establishment of crustaceans in new habitats entails shifts in species occurrence and interactions and, consequently, causes changes at the ecosystem level. To fill remaining research gaps, we propose experiments with long-term exposure of crustaceans to combinations of abiotic and biotic conditions. These studies should include multiple generations of a crustacean species to provide new insights into their ability to acclimate across ontogeny and to adapt evolutionarily over generations, in times of rapid environmental change.

Christian Pansch et al., *Impacts of Ocean Change on Crustaceans*. In: *Ecology and Conservation*. Edited by: Lars Gutow and Martin Thiel, Oxford University Press. © Oxford University Press (2025). DOI: 10.1093/oso/9780197768242.003.0009

INTRODUCTION

Global ocean change encompasses multiple abiotic alterations in coastal to offshore environments (Fox-Kemper et al. 2021), with implications for crustaceans in the world's oceans. Cellular and physiological processes determine the tolerance limits of individual crustaceans (Faria et al. 2020), defining distribution patterns of species (Stillman 2002) and population-specific sensitivities across ocean regions and latitudes (Faulkner et al. 2014). Crustaceans encompass key species in coastal, benthic, and pelagic ecosystems worldwide, and represent a central component in marine food webs. Therefore, shifts in performance traits, abundance, and distributions of crustaceans induced by ocean change will ultimately translate into ecosystem-level changes (Whiteley 2011).

Today, the oceans are warming and acidifying at the highest rates recorded over the past ~66 million years (Zeebe et al. 2016). Many crustacean species have high capacities to acclimate to changes in environmental conditions by altering their physiological properties within their life span and are therefore considered more tolerant of climate change. Through transgenerational plasticity, crustacean parents may prepare their offspring to cope with the surrounding stress regime of a changing ocean (Thor and Dupont 2015). In addition, crustaceans with short generation times may be able to rapidly adapt through evolutionary processes (Langer et al. 2019). Consequently, crustaceans can acclimate and/or adapt to future ocean changes, yet their capacity for adaptation may be exceeded in the Anthropocene (Klymasz-Swartz et al. 2019), with unknown consequences for marine ecosystems.

Substantial changes in environmental drivers (often referred to as stressors or factors; Boyd et al. 2018, Orr et al. 2020; Fig. 9.1) generally involve warming, acidification (a drop in pH and an increase in seawater pCO_2), and deoxygenation of the oceans, as well as multiple factors controlling primary production, which represents the basis of food webs and life in marine systems (Bopp et al. 2013). Additionally, coastal habitats are particularly impacted by multiple (more regional to local) environmental drivers such as eutrophication, shifts in salinity, regional upwelling of acidified and hypoxic seawater, and human activities such as construction work or tourism, all of which additively, synergistically, or antagonistically (Fig. 9.1) affect crustaceans across latitudes.

Environmental drivers may not solely occur as steady changes (trends) but may also appear as extreme events (Oliver et al. 2021), and many of these drivers may be present in the form of circadian (tidal to diurnal) or stochastic variability (Lima and Wethey 2012, Pansch and Hiebenthal 2019). Such fluctuations in environmental drivers may push crustaceans beyond their physiological limits, long before steady changes in the environment (climate trends) appear stressful. Whether the impacts of different drivers and different types of extreme events or variability patterns result in additive, synergistic, or antagonistic (Fig. 9.1) responses of marine crustaceans will depend on their intensity (magnitude x duration), their periodicity, and their timing (Gunderson et al. 2016, Wahl et al. 2021).

Abiotic environmental shifts interact with novel biotic driving forces (Fig. 9.1), of which the invasion of species and parasitism may be of particular importance for marine ecosystems. Crustaceans represent a conspicuous and successful taxonomic group of invaders during increasing seawater temperatures in coastal habitats. This facilitates the increased establishment of introduced crustacean species in new habitats, often originating from warmer regions (Witte et al. 2010). In addition, as vectors of species introductions, crustaceans can influence parasite–host dynamics and disease occurrence in native ecosystems, while nonnative crustaceans themselves can also be parasites, which have been co-introduced with species from other taxonomic groups (Shields 2019).

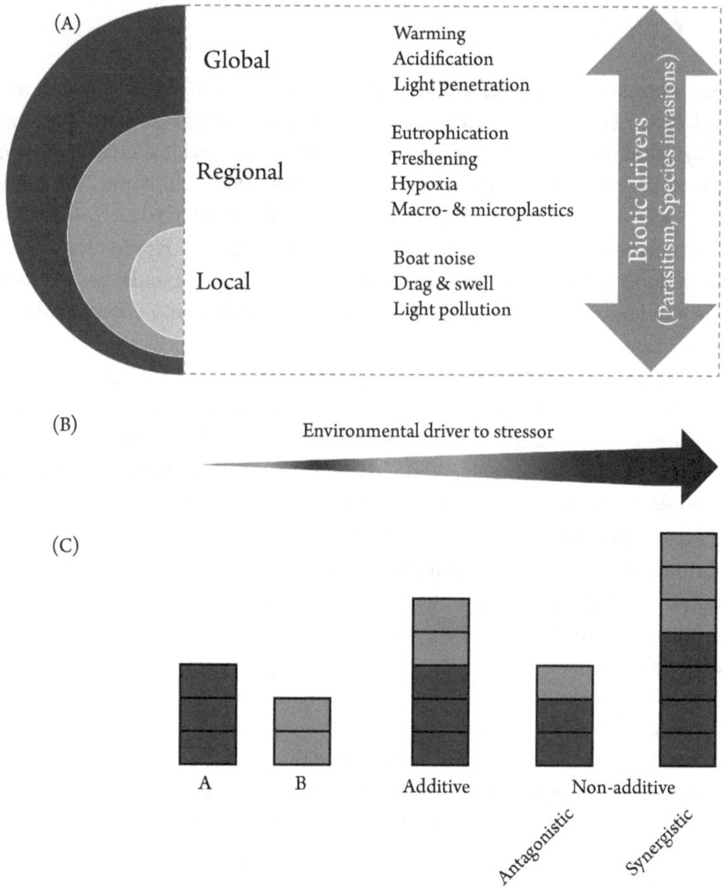

Fig. 9.1.
(A) Multiple abiotic drivers of ocean change that may be scaled from global to regional and local, and which may be influenced by the presence or absence of multiple biotic drivers. (B) Any environmental driver may turn into a stressor once tolerance limits of species and ecosystems are exceeded. (C) Abiotic and biotic drivers can act in an additive or non-additive (antagonistic, synergistic) way (exemplified as the interaction of one global environmental and one biotic driver). See a color version of this figure in the centerfold.

MULTIPLE DRIVERS ACT UPON MULTIPLE LEVELS OF ORGANIZATION

Ocean Warming—One of the Dominant Drivers of Species Responses and Ecosystem Changes

Crustaceans are ectothermic organisms, which means that their body temperature changes in response to their environment, and internal processes are directly linked to the water temperature (Dell et al. 2014). Thus, temperature represents one of the main environmental drivers (Brown et al. 2004) determining the stress tolerance of aquatic ectotherms (Stuart-Smith et al. 2017), and distributions and range shifts of species (Sunday et al. 2012).

The average global sea surface temperature currently increases at rates of about 0.1 °C per decade (Fox-Kemper et al. 2021). Importantly, warming trends are not homogeneous (Popova et al. 2016) but progress stronger in marginal seas and polar regions. Marginal seas such as the North Sea and

Baltic Sea can represent hotspots of ocean warming and may therefore represent a natural analog or a "time machine for the future coastal ocean" (Reusch et al. 2018). In the Baltic Sea, warming trends can be a magnitude higher (Reusch et al. 2018), with heatwaves lasting for days to weeks (Wolf et al. 2022). Yet such coastal habitats are among the most productive ecosystems worldwide, providing critical services (e.g., food production, energy transfer through food webs, biogeochemical cycling, water clearance, and provisioning of nursery grounds) for higher trophic levels (Griffiths et al. 2017 and references therein), with crustaceans playing a key role in many of these processes.

Commonly, the range of a species is determined by the temperature maxima and/or minima experienced in the habitat, yet physiological and/or behavioral properties may widen or contract the thermal niche of a species. To characterize tolerances of crustaceans to ocean warming, thermal performance curves (also known as TP-functions or response norms) are used in species distribution models from local to global spatial scales (Sinclair et al. 2016). Performance windows may differ with taxon or species but may also differ between performance traits in focus (e.g., growth vs. reproduction; Wahl et al. 2020), and across populations of crustacean species (Pansch et al. 2014). This performance strongly depends on the aerobic scope of a species (Pörtner and Farrell 2008). At both ends of the thermal range, performance drops to suboptimal levels, mainly because of a limited capacity for oxygen supply (Whiteley et al. 1997). Beyond this range, metabolic depression, anaerobic energy production, and stress protection mechanisms such as the action of heat shock proteins can achieve short- to medium-term tolerance of crustaceans to thermal extremes (Whiteley et al. 1997, Pörtner and Farrell 2008). Thus, temperature and future ocean warming will affect the many levels of biological organization, from energy uptake and conversion rates through population dynamics and species interactions to community functioning, and therefore the provision of services to humans (Brown et al. 2004, Schramski et al. 2015).

Crustaceans that evolved in environments of near-constant thermal conditions such as the deep sea or the tropics may have a limited tolerance to instant temperature changes. Intertidal Hawaiian ghost crabs, for example, can be found in thermally rather constant (26–28°C) habitats, and may thus suffer from small deviations from these temperatures (Florey and Hoyle 1976). At the other extreme, temperate shallow and subtidal environments exhibit seasonal thermal fluctuations of > 20°C and diurnal fluctuations of > 5°C (Pansch and Hiebenthal 2019). In the intertidal zone and rock pools, crustaceans may even experience extreme daily thermal gradients of more than 30°C (Wang et al. 2020). Thus, crustaceans have evolved physiological and behavioral mechanisms to cope with a wide range of thermal fluctuations (Whiteley et al. 1997, Lagerspetz and Vainio 2006). For example, porcelain crabs (genus *Petrolisthes*) living higher on the shore are thermally more tolerant than congeners from low intertidal habitats (Stillman 2002). Overall, tropical species may show the highest tolerable thermal maxima, yet their acclimation potential may be limited compared to temperate species (Stillman 2002). How much such existing traits manifest in the tolerance to future ocean warming remains to be investigated.

The frequency, duration, and intensity of marine heatwaves are increasing worldwide (Oliver et al. 2021), with particular implications for marginal seas. For two Indo-Pacific crustacean species, the shrimp *Rhynchocinetes durbanensis* and the hermit crab *Calcinus laevimanus*, physiological responses occurred mostly after seven days of exposure to a heatwave, while a change in the ultimate performance was detected in only one of the species (Madeira et al. 2018), indicating some potential for short-term acclimation (physiological adjustments). Yet the impact on the overall fitness may be determined by (i) the critical thermal limits of a species and/or (ii) the sum of thermal stress as a combination of the amplitude and duration of heat occurring outside the thermal optimum of a species. Both long-term (weeks to months) but also very short-term (days) marine heatwaves can have persistent consequences at the species-, the population-, and the ecosystem level (Pansch et al. 2018b, Smale et al. 2019).

Ocean Acidification—Impacts on Multiple Processes Across Life History

The phenomenon of ocean acidification is associated with multiple abiotic factors that can affect the physiology of seawater-breathing organisms including crustaceans. A rise in atmospheric CO_2 partial pressure increases the hydration of CO_2 in seawater, changing the carbonate chemistry of the ocean. This generates carbonic acid and bicarbonate and liberates protons, causing an increase in the concentration of free protons and a predicted decrease of the coastal surface seawater pH by 0.1 to 0.4 pH units by the year 2100 (Gattuso and Hansson 2011).

The decrease in seawater pH challenges the acid–base regulatory machinery of crustaceans, which is located in cells of specialized epithelia directly facing the seawater, such as gills or the integument (Melzner et al. 2009, Henry et al. 2012), and can, if not adequately countered, directly translate into systemic disturbances in acid–base homeostasis. Since intra- and extracellular pH homeostasis is essential for the proper functioning of enzymes, crustaceans must sustain pH homeostasis by increased proton transport and/or by buffering capacities to maintain intra- or extracellular pH within a tolerable range (Boron 2004). An increase in seawater pCO_2 (hypercapnia) directly translates into a higher pCO_2 in body fluids and the intracellular compartment to maintain a sufficient outward-directed diffusion gradient for CO_2 (Melzner et al. 2009). This increase in intra- and extracellular pCO_2 leads to enhanced hydration of CO_2-liberating protons in body fluids and the cytosol. Thus, a crustacean facing hypercapnia requires sufficient acid export or internal buffering to maintain pH homeostasis.

Decapod crustaceans can show remarkable tolerance to acidification. However, changes in the seawater carbonate system often affect crustaceans through a reallocation of energy budgets toward compensatory processes (i.e., homeostatic regulation and calcification) leaving less energy for growth and development. For example, larvae of the spider crab *Hyas araneus* reduce development and growth when exposed to increased pCO_2 of 710 μatm and 3000 μatm compared to 400 μatm pCO_2 (Walther et al. 2010). This reduction in somatic growth and developmental rate under conditions of ocean acidification is common in the early developmental stages of crustaceans (Walther et al. 2010, Pedersen et al. 2014, Schiffer et al. 2014).

In many marine invertebrates, the sensitivity to ocean acidification is highest in the early life-history stages (Fig. 9.2). Here, high energetic demands for developmental processes and metamorphosis may represent a critical factor (Walther et al. 2010, Whiteley 2011, Schiffer et al. 2014). In *H. araneus*, the megalopa stage, which is the transition from the planktonic to the benthic lifestyle, is the most vulnerable ontogenetic stage of this species, but this is probably also the case in the majority of the over 6700 other crab species (Ng et al. 2008, Fehsenfeld and Weihrauch 2017). This ontogenetic bottleneck in *H. araneus* is also reflected in the narrowest thermal tolerance in larval stages that is further reduced in combination with acidification scenarios (Fig. 9.2; Schiffer et al. 2014). In contrast, the juvenile (settled benthic) stage constitutes the bottleneck in acorn barnacles in response to low pH (Pansch et al. 2014, Pansch et al. 2018a). Given the general observation that early developmental stages are more sensitive to environmental hypercapnia than adults, future studies should focus on these ontogenetic stages to estimate the vulnerability of crustaceans to future ocean acidification.

Compared to other invertebrates, many crustaceans have a high degree of mobility associated with a large metabolic range, leading to naturally strong fluctuations in internal pCO_2 and hemolymph pH. In many crustaceans, this has led to the emergence of strong pH-regulatory capacities, partly explaining their relatively high tolerance to ocean acidification (Melzner et al. 2009). Specialized epithelia found in decapod gills and branchiostegites (Fehsenfeld and Weihrauch 2017) are responsible for ion exchange between the surrounding seawater and the body fluids. High tolerance of many decapod crustaceans including *Callinectes sapidus, Neohelice granulata, Carcinus maenas,* and *Xenograpsus testudinatus* to hypercapnic conditions (Fehsenfeld and Weihrauch 2013, Allen et al. 2020) is associated with their ability to accumulate large amounts

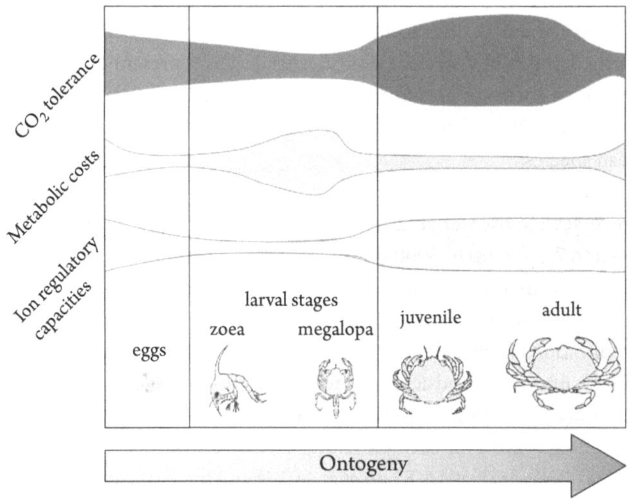

Fig. 9.2.
Ontogeny of metabolic costs (demands) and ion regulatory capacities associated with CO_2 tolerance from selected marine decapod crustaceans. CO_2 tolerance is summarized along different life-history stages in the spider crab *Hyas araneus* based on Walther et al. (2010) and Schiffer et al. (2013). Routine metabolic rates along the development of *H. araneus* (Jacobi and Anger 1985, Anger 1986, Schiffer et al. 2013) and *Carcinus maenas* (Dawirs 1987) reflect the energetic demands of the respective ontogenetic stages to support fundamental processes of maintenance, growth, and development. The ontogeny of osmoregulatory capacities is derived from selected decapod crustaceans including *C. maenas* (Cieluch et al. 2004), *Eriocheir sinensis* (Cieluch et al. 2007), and *Homarus* sp. (Charmantier et al. 2001) identifying larvae as the ontogenetic stage with the lowest ion regulatory capacities.

of HCO_3^- in their body fluids to buffer excess protons generated by the hydration of CO_2 (Henry et al. 2012). These physiological characteristics of adult stages will allow most crustaceans to maintain vital functions without extensively compromising energy for growth and reproduction under pH and pCO_2 changes predicted for the ocean of the coming century (Whiteley 2011).

Crustaceans have key functions in marine food webs as predators or as prey for many species of higher trophic levels such as fish, birds, and seals. Here, changes in energy turnover and defense traits (e.g., their exoskeleton) in response to ocean acidification may have important repercussions on their performance in an ecosystem context. Crustaceans show variable responses to ocean acidification with regard to calcification. For example, the Chinese mitten crab *Eriocheir sinensis* decreases its calcification rates when exposed to acidification (Quijada-Rodriguez et al. 2021). In contrast, several crustacean species were shown to be even more heavily calcified under such conditions (Ries et al. 2009). The change in density of the exoskeleton may have implications for the buoyancy and swimming ability of a crustacean, leading to changes in ecological performance. However, there is currently little information linking hyper-calcification in crustaceans, and other species, to ecologically relevant parameters.

The ecological consequences of an acidified ocean at the local to regional scale can already be seen today in acidification hotspots. As demonstrated in the example given below (on vent-endemic crabs *X. testudinatus*), volcanic activity may drive regional to local seawater carbonate chemistry (Aiuppa et al. 2020). Here, the extent of ocean acidification may vary at very small spatial scales, and seawater pH conditions predicted for the end of the 21st century (pH 7.7) and lower (pH 6.5) can be found locally (González-Delgado and Hernández 2018). Such systems have been widely used as natural analogs to demonstrate the impact of ocean acidification on entire communities in natural settings. In this context, crustaceans represent a ubiquitous taxon inhabiting naturally acidified environments from the deep sea to shallow-water vent habitats (Jeng et al. 2004, Martin and Haney 2005, Foo et al. 2018), highlighting the relatively high CO_2 tolerance of certain crustacean species.

Over larger spatial scales, coastal or marginal seas are hotspots for ongoing and future ocean acidification processes (Reusch et al. 2018). Here, upwelling is occasionally leading to pH and pCO_2 levels down to 7.3 and up to 3000 μatm, respectively (Thomsen et al. 2010), conditions expected to occur in the open ocean by the year 2100 or beyond. This may impact coastal species of the Baltic Sea such as the shore crab *C. maenas*, a benthic predator often confronted with unfavorable environmental conditions (Fehsenfeld et al. 2011). However, shore crabs and other crustaceans have evolved remarkable osmo- and acid-base regulatory abilities allowing them to occupy ecological niches with strong fluctuations in pH and pCO_2 (Henry et al. 2012). At an even larger spatial scale, coastal upwelling occurs, for example, along extensive stretches of the West Coast of North and South America (Feely et al. 2016, Köhn et al. 2017) and the East Australian Shelf (Schulz et al. 2019), with frequent shoaling of corrosive waters, and with impacts on species and ecosystems, as well as the aquaculture industry (Bednaršek et al. 2021).

Case Study I: Xenograpsus testudinatus—A Brachyuran Crab from One of the World's Most Acidic Marine Habitats

One example of a crustacean species' tolerance to acidified conditions is found in the shallow-water hydrothermal vent system off Kueishan Island (ROC Taiwan; Fig. 9.3). The volcanic vent system has been characterized as one of the world's most acidic (pH ~1.25) and sulfide-rich (up to 800 μM H_2S) environments with *X. testudinatus* (Fig. 9.3) being the only metazoan species endemic to this habitat. This highlights the extraordinary pH-regulatory and detoxification capacities of crustaceans within the animal kingdom. Due to the limitation of organic material in this hostile environment, *X. testudinatus* evolved a unique feeding behavior by scavenging on marine snow, consisting of dead zooplankton killed by the toxic discharges of the vents. In turn, this extreme environment protects the crabs from predators, leading to the presence of high densities (> 350 individuals per m^2) of this species in the vicinity of the vents (Jeng et al. 2004).

The ability of *X. testudinatus* to inhabit this extreme environment is associated with a range of physiological adaptations, including energy metabolism, sulfide detoxification, and acid–base regulation (Hu et al. 2012, 2016). This potential for physiological adaptations to extreme environmental conditions in a decapod crustacean reflects the high capacities of some crustacean species to tolerate increased environmental pCO_2 and acidity (see, e.g., Pansch et al. 2012, Fehsenfeld and Weihrauch 2013). When exposed to hypercapnic conditions of 24.6 kPa pCO_2 (pH 5.5), *X. testudinatus* elevates hemolymph HCO_3^- levels from 20 to 68 mM within one day and further to 88.5 mM within seven days, compared to control animals that maintain 20 mM HCO_3^- along the entire experimental period (Allen et al. 2020). There is, however, evidence that this exceptionally strong acid–base regulatory ability is associated with an upregulation of ATP-consuming ion pumps, including the Na^+/K^+-ATPase and the V-Type H^+-ATPase during hypercapnic conditions in gill tissues that are the predominant site of HCO_3^-/Cl^- exchange and proton secretion in crustaceans (Hu et al. 2016). Thus, under prolonged exposure, a higher fraction of energy must be allocated to acid–base regulatory conditions. Exposure to hypercapnic conditions led to an elevation in metabolic rate by approximately 1 mg O_2 g^{-1} h^{-1} under acute (0–1 days) and 0.5 mg O_2 g^{-1} h^{-1} under medium-term (14 days) compared to control animals that maintained oxygen consumption rates (MO_2) in the range of 0.7–1.2 mg O_2 g^{-1} h^{-1} (Allen et al. 2020). This observation suggests that the exceptional hypercapnia tolerance of this hydrothermal vent crab requires energetic strategies adapted to life in a food-limited environment.

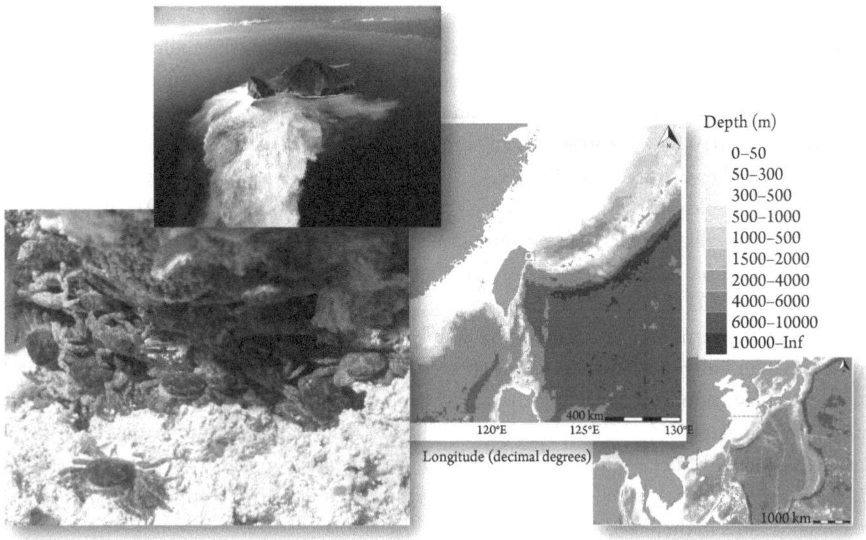

Fig. 9.3.
Specimen of the vent-endemic crab *Xenograpsus testudinatus* thriving under extreme environmental hypercapnia in a shallow-water hydrothermal vent system off Kueishan Island (ROC Taiwan). Maps were illustrated in R, using the packages "ggOceanMapsData" and "ggOceanMaps." Photos reproduced courtesy of Dr. Jeng M.-S.; see a color version of this figure in the centerfold.

In summary, this species has evolved behavioral and physiological adaptations allowing it to occupy an ecological niche in one of the world's most hostile marine ecosystems, which in turn results in low ecological interactions and release from interspecific competition and predation.

Responses of Crustaceans to Spreading Hypoxic Zones and Hypoxic Events

Ocean deoxygenation is mainly a consequence of stratification, a warming-induced reduction in the oxygen solubility in seawater, and a reduced exchange with the atmosphere. Ocean warming in combination with increasing eutrophication intensifies bottom hypoxia and promotes the expansion of hypoxic zones worldwide (Gilbert et al. 2010). Hypoxic zones may develop, particularly in marginal seas (Meier et al. 2017), and can favor coastal hypoxic upwelling (Bakun et al. 2015). In the lower York River (Chesapeake Bay, USA) for example, reduction in bottom seawater oxygen fosters migration of two crustacean species (mantis shrimp *Squilla empusa* and the blue crab *Callinectes sapidus*) from deeper to shallower waters (Pihl et al. 1991).

In the middle of the 20th century, many crustaceans were still assumed to be oxygen conformers, that is, not capable of modifying oxygen consumption independent of the surrounding environmental oxygen partial pressure (Waterman 1960). Today, the excellent regulatory capacities of crustaceans are well recognized (McMahon 2001, Falconer et al. 2019). In decapods, ventilation of the gills, oxygen uptake, transport effectiveness, and adjustments in the branchial hemolymph flow, as well as behavioral activities, may allow for a strong tolerance to environmental hypoxia (reviewed in McMahon 2001, Whiteley and Taylor 2015). Both, the hairy-handed crab *Hemigrapsus crenulatus* and the purple rock crab *H. sexdentatus* demonstrated tolerance to short-term hypoxia (Falconer et al. 2019). Yet, the number of hypoxic coastal sites is increasing worldwide (Vaquer-Sunyer and Duarte 2008), likely exceeding the threshold levels of hypoxia tolerance for many species in a future ocean.

Upwelling along the Humboldt Current System in the South Pacific Ocean shoals waters is characterized by low dissolved oxygen (Ulloa and Pantoja 2009). Along the Chilean coast, the commercially important red squat lobster *Pleuroncodes monodon* may, like other species, have adapted to low-oxygen conditions by, for example, brood-carrying during the more oxygenated season (Gallardo et al. 2019). At a smaller spatial scale, summer marine heatwaves may be interrupted by sporadic weather-driven events (Fig. 9.4), fostering regional coastal upwelling of bottom waters, and therefore leading to temporal deoxygenation (e.g., Seabra et al. 2019). In addition to hypoxia, such upwelling events may alter numerous environmental parameters that change sharply within hours to days (Fig. 9.4), with the characteristics of these changes differing between seasons (Wahl et al. 2021). This aspect has received little attention, and identifying the impacts on species and ecosystems remains challenging, mainly due to the complex changes in environmental drivers and their differential impacts on the species within a community. Crustaceans, unlike some fish species, may endure such events if their extent and duration remain within their physiological limits. Here, behavioral adaptation, such as accumulation at the water–air interface, may allow crustaceans to cope with such temporal events and potential ongoing ocean deoxygenation (Fig. 9.4; I. Bobsien, pers. observ.). Thus, ecological consequences may arise from reductions in crustacean predators or shifts in their feeding behavior.

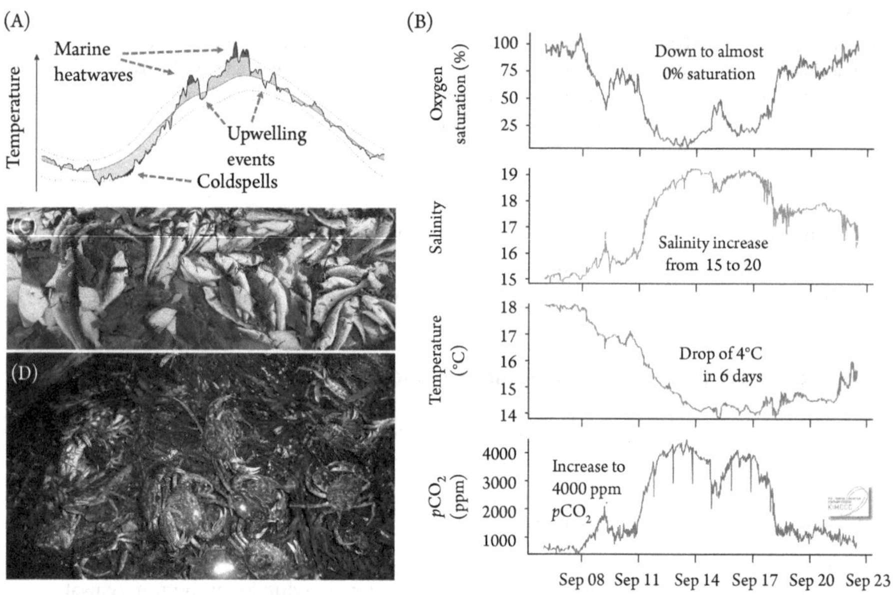

Fig. 9.4.
(A) Temperature course and marine heatwaves (dark shading above confidence interval) as recorded in the Western Baltic Sea Kiel Fjord in the year 2018 may be interrupted by (hypoxic) upwelling events, occasionally turning into cold spells (dark shading below confidence interval) of extremely low temperatures (F. Wolf, GEOMAR). (B) Seawater oxygen saturation, salinity, temperature, and pCO_2 were recorded at one meter depth in Kiel Fjord between September 6 and September 22, 2017, following an early-autumn upwelling event (plot in collaboration with M. Wall, GEOMAR, and C. Hiebenthal at KIMOCC GEOMAR; available at PANGAEA [https://doi.org/10.1594/PANGAEA.930979]). (C) Accumulations of dead fish washed ashore in September 2017 following the event. (D) *Carcinus maenas*, *Crangon crangon*, and fish aggregate on a stone in oxygenated surface water, inhabited by the bladderwrack *Fucus vesiculosus*, during a nocturnal upwelling event of hypoxic seawater in September 2020. Photos reproduced courtesy of Ivo Bobsien; see a color version of this figure in the centerfold.

Multiple Drivers Can Act Globally to Locally in Synergistic and Antagonistic Modes

Environmental drivers (stressors/factors; see Fig. 9.1) may be categorized regarding their spatial (and temporal) extent as global, regional, and local effects. While ocean warming, acidification, and deoxygenation, as well as sea-level rise, may represent a global-scale threat to marine crustaceans, coastal hypoxia, shifts in seawater salinity, or eutrophication may be drivers acting more regionally (Popova et al. 2016, Halpern et al. 2019). Temperature may act as an overriding driver of organism physiology due to the strong implications for enzymatic and cellular processes (see above). Seawater carbonate chemistry and salinity may disturb the osmo- and acid–base balance of organisms and rates of calcification, while oxygen availability in seawater may critically influence the supply of tissues with oxygen (see above). Eutrophication may act indirectly on the abundance and performance of crustaceans when nutrient availability leads to shifts in the performance of macrophytes (Wahl et al. 2020), which provide habitat and nutrition to many crustaceans such as amphipods and isopods. On a regional to local scale (yet, acting globally), stress from plastics (micro- and macroparticles), shipping, overfishing, chemical pollution, underwater noise, and artificial light impose additional threats to crustaceans by changing population survival, performance, and individual behaviors, and often amplifying global ocean stress.

Multiple drivers of ocean change will act in additive, antagonistic, or synergistic ways (Gunderson et al. 2016; Fig. 9.1), and the outcome for a species (or even the entire ecosystem) may depend on the combination of drivers as well as their intensity (Boyd et al. 2018). Meta-analyses and summary articles revealed that the interactive effects of ocean warming and acidification may be species-specific and become evident mainly at the stress extremes (Kroeker et al. 2013, Gravinese et al. 2018). For juvenile red king crabs (*Paralithodes camtschaticus*), ocean warming and acidification had antagonistic effects under a scenario of weak warming but a synergistic negative effect under a strong future warming scenario, representing extreme stress for this species (Swiney et al. 2017). While the combination of these two factors did not affect juveniles of the European lobster (*Homarus gammarus*; Small et al. 2016), adult American lobsters (*H. americanus*) were impacted (Klymasz-Swartz et al. 2019). The ability to compensate for CO_2-induced acid–base disturbances is tightly associated with the metabolic thermal window, involving shifts in the organismal energy budget (Metzger et al. 2007, Pörtner and Farrell 2008, Walther et al. 2010). However, more experimental data are needed to identify general patterns for combined stressors across crustacean early life-history stages. Examples show that early stages of crustaceans from deep-sea habitats and polar regions have a narrow aerobic scope and low acid–base regulatory capacities compared to coastal and temperate decapod crustaceans (Pane and Barry 2007, Walther et al. 2010), suggesting that polar and deep-sea species are generally more sensitive to CO_2-induced seawater acidification than those from temperate coastal habitats (Walther et al. 2010).

Particular drivers encompass multiple simultaneous changes such as ocean acidification, including changes in seawater pH (in form of H^+), pCO_2, total dissolved inorganic carbon, HCO_3^-, CO_3^{2-}, and saturation states of calcite and aragonite (see above; Hurd et al. 2020). Yet, pulse stress events may also appear sequentially. Marine heatwaves may be followed (interrupted) by upwelling events bringing colder, more saline, and often acidified and hypoxic water bodies into shallow-water ecosystems (Wahl et al. 2021; Fig. 9.4). As for the simultaneous occurrence of stress, such consecutive stressors can either have additive, antagonistic, or synergistic impacts, strongly depending on the timing of their occurrence (Gunderson et al. 2016). When temporally decoupled, lagged effects can be explained by an "ecological (stress) memory," a concept well studied in terrestrial plants (Walter et al. 2013) and some aquatic species (Hughes et al. 2019), but lacking empirical evidence in the marine realm (Jackson et al. 2021) and, to our knowledge, in marine crustaceans in particular. Thus, a multitude of environmental drivers may not act simultaneously but rather sporadically and at different seasons (sequential occurrence), leading to potentially synergistic and variable stress impacts, and making predictions on the implications for crustacean populations and ecosystems difficult.

In addition to environmental drivers acting at different spatial and temporal scales, crustacean species may be challenged by biological constraints in terms of species interactions such as competition and predation. The spectrum of biological interactions of a species may be amplified by the invasion of new species and their parasites into ecosystems (Goedknegt et al. 2016, Cornelius et al. 2021). Thus, abiotic environmental conditions and biotic interactions shape ecosystems today and in the future (Fig. 9.1), a field that had received strong attention in terrestrial systems (e.g., Soliveres et al. 2012) but for which little is known in the marine realm.

Case Study II: A Barnacle Story—Multiple Drivers Act Upon Multiple Life-History Stages

Intertidal barnacles are ecologically important ecosystem engineers (Gaonkar and Anil 2010). The acorn barnacle *Balanus (Amphibalanus) improvisus* is found worldwide in oceanic and brackish waters. In repeated experiments (Fig. 9.5), *B. improvisus* was exposed to ocean warming, acidification, and different salinities. Additionally, food availability, which determines energy availability and, therefore, the capacity to withstand persistent environmental perturbation, was manipulated. This was accomplished over multiple populations and across multiple life-history stages of this crustacean species, including the nauplius and cypris larval stages, as well as the calcified juveniles and adults. In addition, reproduction and the responses of F1 offspring were investigated.

\- not tested
/ not testable

		Kiel population				Tjärnö population			
		Temp	pH	Sal	Food	Temp	pH	Sal	Food
Nauplii	Survival	×	×	×	-	×	×	×	-
	Nauplius duration	×	×	×	-	×	×	×	-
	Metamorphosis to cyprids	×	×	×	-	×	×	×	-
	Development to settlement	×	×	×	-	×	×	×	-
Cyprids	Survival	×	×	×	/	×	×	×	/
	Settlement	×	×	×	/	×	×	×	/
Juvenile	Survival	×	×	×	×	×	×	#	×
to	Growth	×	×	×	×	×	×	#	×
Adults	Calcification	×	×	-	×	-	-	-	-
	Condition index	×	×	×	×	×	×	#	×
	Shell strength	×	×	-	×	-	-	#	-
	Respiration rates	-	-	-	-	-	×	-	-
	Gonad development	-	×	-	×	-	×	#	-
	Reproduction	-	×	-	×	-	×	-	-
	F1 offspring performance	-	×	-	-	-	-	-	-

Fig. 9.5.
Overview of the drivers applied and the response traits/variables tested on the model system, the acorn bay barnacle *Balanus (Amphibalanus) improvisus* across life-history stages (nauplii, cyprids, juveniles, and adults, to reproduction and the F1 generation) and populations (Kiel and Tjärnö). Information marked with # is given in Wrange et al. 2014. Photos of larval to adult stages reproduced courtesy of Jon Havenhand and Giannina Hattich.

Generally, warmer water temperature accelerates the development of *B. improvisus* nauplii and increases the survival and settlement success of the cyprids (Pansch et al. 2012; Nasrolahi et al. 2016). In addition, nauplius larvae appear robust to ocean acidification scenarios down to pH 7.3 (3280 µatm pCO_2) when reared at natural (benign) summer temperatures (Pansch et al. 2012, Pansch et al. 2013b). Yet, at more extreme experimental temperatures, representative of future marine heatwaves in the area (Pansch et al. 2018b) or during cold spells, acidification down to pH 7.4 (3110 µatm pCO_2) can negatively affect the development of the nauplii (Pansch et al. 2012). Cypris settlement success and survival are largely unaffected by similar levels of ocean acidification (Pansch et al. 2012). Yet many responses to perturbation appear to be population-specific. At a high-salinity site in the Skagerrak (Tjärnö Archipelago, salinity ~25), nauplius development and metamorphosis appear best at an intermediate salinity of 15. In contrast, a population from the Western Baltic Sea (Kiel Fjord, salinity ~15), on average, performs best in lower salinities of 5 to 15 (Nasrolahi et al. 2016). Thus, these crustacean larvae perform best at salinity levels lower than those found in their natural habitats (Nasrolahi et al. 2012, 2016, Sundell et al. 2019). While sensitivities to ocean acidification were generally weak and influenced by temperature, larvae of this species may be generally favored by near-future seawater warming and freshening.

Newly settled juvenile *B. improvisus* in Kiel Fjord generally benefited from warmer water temperatures (Nasrolahi et al. 2013, Pansch et al. 2013a, Johnson et al. 2021). At low spring seawater temperatures (12°C) the salinity tolerance of the barnacles was narrowed, demonstrating an interactive impact of these two drivers (Nasrolahi et al. 2013). Juvenile barnacles also responded weakly to ocean acidification in Kiel Fjord, which is characterized by significant natural pH and pCO_2 variability (Reusch et al. 2018). Even though molting frequency and shell corrosion increased with decreasing pH, survival, growth, reproduction, and shell integrity, all remained largely unaffected by acidification (pH 7.2, 2870 µatm pCO_2 for 8 weeks [Pansch et al. 2013a], pH 7.2, 2750 µatm pCO_2 for 20 weeks [Pansch et al. 2014]), across different nutrition levels (Pansch et al. 2014). Supporting earlier findings (Pansch et al. 2012, 2013b), larval development and juvenile growth of the F1 generation were also tolerant to ocean acidification, irrespective of the acidification level experienced by the parental generation (Pansch et al. 2014). In contrast, reduced pH had a strong negative impact on the survival of juvenile barnacles from the Tjärnö Archipelago, an area naturally much less impacted by upwelling of acidified waters (Pansch et al. 2014). Yet specimens from this population were able to withstand moderate levels of acidification (pH 7.8, 1142 µatm pCO_2 for 5 weeks) when food was plentiful but showed reduced growth under conditions of ocean acidification and food limitation (Pansch et al. 2014). Severe acidification (pH 7.3, 3256 µatm pCO_2 for 5 weeks) negatively impacted the growth of the barnacles at low and high nutritional levels (Pansch et al. 2014), providing evidence that energy availability can mediate the ability of organisms to withstand moderate stress.

During a one-year incubation, moderate ocean acidification (pH 7.5, 1600 µatm pCO_2) caused high mortality in the very early juvenile stage (as seen earlier; Pansch et al. 2014) and reduced growth of *B. improvisus* (Pansch et al. 2018a). However, some acclimation of the survivors was evident in their metabolic activity, measured as rates of respiration. Yet, the long-term impacts of ocean acidification were apparent and detrimental: adult *B. improvisus* developed mature gonads but failed to produce viable embryos and/or larvae (Pansch et al. 2018a), revealing a significant ontogenetic bottleneck of this barnacle population. Thus attempts to select genotypes capable of coping with scenarios of future ocean acidification failed, raising uncertainty about the fate of this barnacle population in an acidifying ocean.

This case study demonstrates population- and life-stage-specific sensitivities of a crustacean to ocean change with different drivers showing synergistic and antagonistic effects, depending on the combination and the strength of the drivers. *Balanus improvisus* may benefit from most but not all combinations of ocean warming and freshening predicted for the Baltic Sea as a consequence of climate change. Generally, populations from environments with naturally fluctuating pH (pCO_2) appear more tolerant to end-of-century acidification scenarios than populations from more stable pH (pCO_2) habitats (Pansch et al. 2014). Yet food availability may be more crucial for the persistence of *B. improvisus* in competition with other species than the species' sensitivity to ocean acidification (Thomsen et al. 2013), highlighting the multiple challenges crustaceans face in a future ocean.

EFFECTS OF OCEAN CHANGE ON THE ROLE OF CRUSTACEANS IN INVASION ECOLOGY

Background

In earth history, processes such as plate tectonics caused the geographical isolation of species, which could not be overcome by their limited dispersal capabilities, and these processes, in combination with environmental forces (see above), shaped the biogeography, evolution, and structure of today's marine ecosystems. Progressive globalization over the last decades, and primarily the transport of goods via trans-oceanic shipping, has caused a human-mediated breakdown of barriers of originally separate marine realms. Species are transported as hitchhikers in ballast waters or attached to the hulls of vessels across long distances, causing geographical range expansions and biological invasion of species into previously unoccupied areas (Galil et al. 2008, Miralles et al. 2021). Ocean warming accelerates the successful establishment of nonnative species in recipient systems, especially in temperate regions, because many invaders originating from warmer coasts of, for example, the Pacific Ocean, are transported to higher latitudes (Buschbaum et al. 2012). Additionally, introduced marine organisms, including crustaceans, are often tolerant toward environmental stress factors, giving them a competitive advantage over native species in the recipient marine ecosystem (Lenz et al. 2011). Thus, the interacting effects of global warming and transport initiate revolutionary processes worldwide because species invasions can change the structure and functioning of marine systems from the genome to the ecosystem level. This process is driven by, for example, changing species diversity, competition, and predation, and by causing introductions of diseases and parasites, which may finally result in altered trophic network structures (Reise et al. 2006).

Crustaceans comprise numerous successful invaders in marine systems, making them a taxonomic key group in marine invasion ecology (Bącela-Spychalska et al. 2020; see Chapter 13 in this volume). They possess biological and life-history traits that enable high survival and establishment potential in recipient areas. Additionally, crustaceans are often omnivorous and show high fecundity and phenotypic plasticity. Among the very diverse taxonomic group of crustaceans, the suborders Amphipoda and Decapoda in particular contain many successful invaders, yet examples from Cladocera, Copepoda, and Mysida are also known (Grabowski et al. 2007, Hänfling et al. 2011).

Large-sized nonnative crustaceans are comparatively easy to recognize in native habitats, whereas small-sized crustaceans and species with a cryptic lifestyle are much more difficult to detect in new environments. This is the case when they are transported as endosymbionts or endoparasites inside individuals of other taxonomic groups such as bivalves, which are common hosts for parasitic copepods (Goedknegt et al. 2016). Additionally, nonnative crustaceans may serve as hosts and vectors and, therefore, can promote the introduction of co-invading parasitic organisms into native ecosystems (Bojko 2020). Host–parasite relationships are an important but often neglected species interaction in coastal ecosystems and invasion biology. Introduced parasites can cause strong effects on the population dynamics of native organisms. This is the case when a parasite is co-introduced with a nonnative host and uses a native species as an alternative host, a process called spillover (Mouritsen and Poulin 2002, Goedknegt et al. 2016).

Thus the increasing exchange of crustaceans around the globe, as well as global changes in abiotic environmental conditions of the ocean, initiate a restructuring of marine communities. This may alter species diversity, with changes in abundance, distribution, and behavior of native species through resource competition and the introduction and spread of new parasites and pathogens (Stasolla et al. 2021). We give examples of how (i) ocean warming can facilitate warm-adapted introduced crustacean species; (ii) introduced crustaceans may affect native crustaceans; (iii) introduced crustaceans may affect native communities; and (iv) nonnative parasitic crustaceans may impact native host species.

Introduced species may have to await appropriate conditions before they are successful and develop from a rare to a common species with a high density of individuals. Such a retarded invasion happened with the barnacle *Austrominius modestus* in the North Sea (Witte et al. 2010). This species originates from southern Australia and New Zealand and has a thermal optimum above 20°C and a high sensitivity to cold winters. *Austrominius modestus* was introduced to the region in the mid-19th century but remained comparatively rare until the mid-1990s. In recent decades with mild winter temperatures, the population has increased exponentially and constitutes now the dominant barnacle species along the southeastern North Sea coast (Witte et al. 2010). Thus, warmer temperatures have transformed the barnacle from an "ecological sleeper" to a very common species that now interacts strongly with native barnacle species by, for example, competing for settlement substratum.

According to Darwin, similar species with similar food requirements and occurring at the same site should exclude each other unless habitat heterogeneity allows for niche portioning (Hutchinson 1978, Reise et al. 2017). Indeed, many nonnative species are known to be strong competitors having negative effects on native organisms when occupying the same niche as their native counterparts, and several studies describe direct predation by nonnative species on resident organisms within the same guild (Ruiz et al. 1999, Geburzi and McCarthy 2018). The strong competitiveness of successful nonnative species can be related to the fact that they often originate from regions with high biodiversity and are, therefore, well adapted to competitive pressure.

Case Study III: Three Direct Competitors in Oyster Reefs

Two nonnative brachyuran crabs have recently invaded the southeastern North Sea coast (Northeast Atlantic). The Asian brush-clawed shore crab *Hemigrapsus takanoi* and the Asian shore crab *H. sanguineus* (Fig. 9.6) were introduced to coastal areas of northern Europe in the 1990s, presumably by shipping. Both species originate from the northwestern Pacific Ocean. In the southeastern North Sea, these non-indigenous crabs were first detected in 2004 in the southern and 2009 in the northern part (Obert et al. 2007, Landschoff et al. 2013, Geburzi et al. 2018). Here, they were confronted with the native European shore crab *C. maenas*, which at that time was the most abundant and dominant crab species in the area. The interactions between the three crab species have been intensively studied in epibenthic oyster reefs. Oyster reefs are the preferred habitat of all three crab species and provide a model habitat for studying the interactions of introduced predatory crab species with a native counterpart with similar ecological requirements.

Direct interactions between introduced and native crustaceans may occur in different life-history stages of the species. Geburzi et al. (2018) investigated the recruitment patterns and interactions of juveniles with subadult individuals among both introduced Pacific crabs and *C. maenas* in the northern Wadden Sea. The native *C. maenas* and both *Hemigrapsus* species showed different seasonal recruitment patterns with temporally shifted recruitment peaks. Additionally,

Fig. 9.6.
Three ecologically similar predatory crabs in the same habitat: introduced Asian shore crabs *Hemigrapsus takanoi* (left), *H. sanguineus* (center; photo reproduced courtesy of Karsten Reise), and the native European shore crab *Carcinus maenas* (right, male and female) on oyster reefs in the northern Wadden Sea.

recruitment of juvenile *C. maenas* was reduced by predation of larger *Hemigrapsus* species, while newly recruited *Hemigrapsus* spp. had a much lower risk of being preyed upon by subadult conspecifics and native *C. maenas*. The high survival probability of juvenile Asian shore crabs may have led to the high constant densities of about 200 individuals per square meter reached by *H. takanoi* on oyster reefs 10 years after its introduction, while densities of *C. maenas* have declined over the same period (Cornelius et al. 2021). Considering adult *H. takanoi* and *H. sanguineus* and their interactions with adult native *C. maenas*, it was shown worldwide that Asian shore crabs prefer the same species and sizes of prey as similar-sized native *C. maenas* (Griffen et al. 2008, Nour et al. 2020, Cornelius et al. 2021).

Despite the same food and habitat preferences, all three species still co-occur on oyster reefs in the northern as well as the southern Wadden Sea (van den Brink and Hutting 2017, Cornelius et al. 2021) and the question is why? Cornelius et al. (2021) showed that *H. takanoi* and *C. maenas* use the same epibenthic food sources but that the native shore crab additionally consumed endobenthic organisms such as polychaetes. This may reduce interspecific competition between the native and the introduced crabs, potentially allowing co-occurrence on oyster reefs. The same development was also observed in other coastal systems, where *C. maenas* shifted its diet and preferred prey following the introduction of *Hemigrapsus* spp. (Griffen et al. 2008). These findings illustrate that the capacity for adaptive responses to changing living conditions enables native crabs to cope with the introduction of intraguild competitors without being displaced (O'Connor 2018).

In summary, density changes of native crabs may occur, which may cause effects at the ecosystem level due to their trophic importance. There is also evidence that an introduced crab may replace another nonnative crab (Lohrer and Whitlatch 2002, Griffen et al. 2011). However, to our knowledge, there is no case worldwide yet known where a native crab species has been displaced by an introduced crab. Thus, an adaptation process between introduced and native decapod crustaceans seems to be the common pattern rather than mutual exclusion of species, at least in benthic coastal habitats (Baillie and Grabowski 2018). However, this situation may change with global ocean warming, (i) challenging native species across temperate and polar regions; and (ii) increasing competitive advantage for introduced warm-adapted crustaceans.

Effects of Introduced Crustaceans on Native Communities

The consequences of species introductions are complex and range from the molecular level (e.g., by hybridization) to effects on individual native species (e.g., by competition, facilitation, evolutionary processes) and entire communities (e.g., by changing food webs). In this respect, nonnative crustaceans play a crucial role. In European brackish and coastal habitats, for example, crustaceans represent the second most common group of introduced species after mollusks (Hänfling et al. 2011). The number of introduced crustaceans in the Mediterranean Sea increased from 37 species in the 1950s to 159 in 2012 (Zenetos et al. 2012). On a global scale, marine crabs (infraorder Brachyura), in particular, have been introduced to native marine systems with an exponential increase of introduction events over the last 150 years (Howard et al. 2017). Introduced crabs can cause significant direct and indirect effects on native populations and communities. They can feed directly on native prey organisms and thus influence population dynamics, a pattern well described for native bivalves (Grosholz et al. 2000, Campbell et al. 2019). Therefore, crabs as nonnative consumers receive special attention when considering impacts on resident communities.

Introduced crabs can cause indirect effects by, for example, preying on native intermediate consumers, resulting in reduced predation pressure on native organisms (Trussell et al. 2002, Griffen and Byers 2009) or by consuming native ecosystem engineers and, thereby, affecting associated species. Most studies on introduced crustaceans support the general notion that nonnative crabs cause detrimental effects on native prey populations. A comparison of the effects of native predatory crabs is often lacking but is mandatory to distinguish the effects caused by introduced crabs from those of native crabs. Howard et al. (2017) conducted a meta-analysis of 834 crab-foraging experiments and found that direct consumption by nonnative crabs did not generally reduce prey abundance more than predation by native crabs. Despite this overarching result, a recent direct comparison of food preferences and predation rates of native and introduced crabs showed that consumption effects may depend on a variety of factors that should be additionally considered, such as prey preferences of introduced and native crabs (Cornelius et al. 2021). Prey preferences may change when native and introduced crab species interact and may also be sex-dependent within one crab species. Females of the Pacific shore crab *H. takanoi*, for example, prefer consuming amphipods, while male conspecifics consume more blue mussels (Cornelius et al. 2021). Thus, to assess the net consumption effects of an introduced predatory crustacean, the prey preference after its successful establishment and the sex ratio of this species must be known, which may differ in recipient ecosystems (Nour et al. 2020). Besides the consumption pattern of an introduced crab species in comparison to the native one, population development is also of high importance. Cornelius et al. (2021) showed that the consumption rates of a single native *C. maenas* are generally higher than those of the introduced *H. takanoi* of the same size class in the southeastern North Sea, but Asian shore crabs exceed the densities of *C. maenas* manifold.

Therefore, the total predation pressure of the Asian shore crab on the native community is nowadays much higher than that of the native *C. maenas* (Cornelius et al. 2021), with specific effects on the native community (Fig. 9.7). Such density-dependent effects of predatory nonnative crabs on native prey organisms are also known from other coasts (Griffen and Byers 2009, Gregory and Quijón 2011). These examples illustrate that a comparison of consumption effects of native and introduced predatory crabs is complex and can strongly depend on the respective crab traits and the composition and functioning of the recipient ecosystem.

Nonnative Crustaceans as Vectors for Parasite Introductions and Parasitic Crustaceans

Marine parasites can influence population and community dynamics by affecting host densities and their distributions as well as the interactions between hosts and their competitors and predators (Mouritsen and Poulin 2002, Blakeslee et al. 2020). Crustaceans, especially crabs, can have high parasite species richness and, therefore, serve as a vector for parasites when they are co-introduced together with their hosts into a new environment (see Chapter 5 in this volume). For example, the European shore crab *C. maenas* has become a well-studied model for disease introduction, and more than 72 pathogens have been identified (Bojko et al. 2021). Many of these parasites are lost when *C. maenas* invades a new environment (enemy release), but several pathogenic groups are

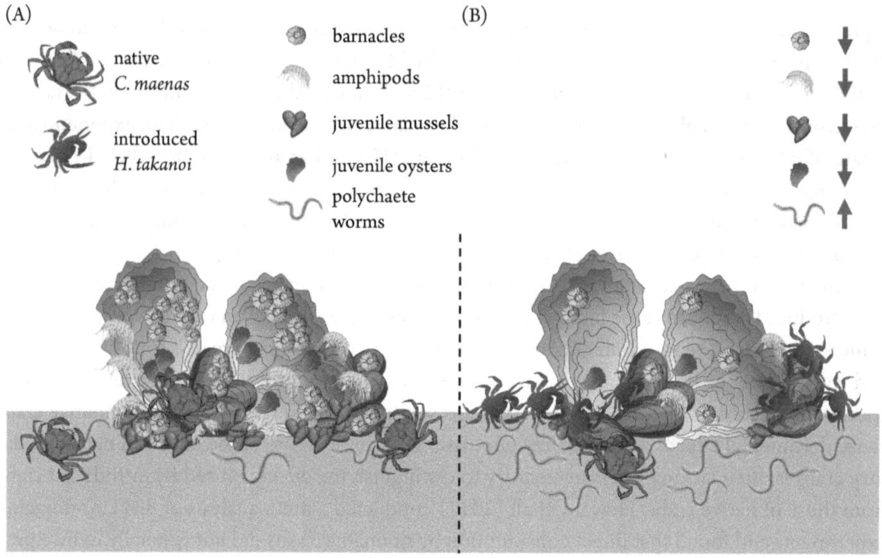

Fig. 9.7.
Schematic illustration of oyster reefs and densities of prey organisms in the southeastern North Sea before (A) and after (B) the introduction of Asian shore crabs *Hemigrapsus takanoi*. High densities of *H. takanoi* have caused higher predation pressure on epibenthic prey species such as barnacles, juvenile bivalves, and amphipods, while predation pressure on endobenthic polychaetes is reduced due to decreased densities of the native *Carcinus maenas*. Adapted from Cornelius et al. (2021), with permission from © Springer Nature.

regularly co-introduced with the shore crabs (parasite co-introduction with the host) and may then infect native hosts (parasite spillover; Goedknegt et al. 2016, Bojko et al. 2018). Additionally, introduced crustaceans can likely serve as hosts for native parasites and indirectly amplify the parasite load of native species (parasite spillback).

With the role as an introduction vector, crabs can influence parasite–host dynamics and disease occurrence in native ecosystems, while nonnative crustaceans can also be parasites themselves, and have been co-introduced with species of other taxonomic groups. Prominent examples are the copepod intestinal parasites *Mytilicola intestinalis* and *M. orientalis*, which have been co-introduced with bivalves to northwestern European coasts (Goedknegt et al. 2018). Both species occupy the gut of bivalves, where they attach to the gut wall with hook-like appendages. The introduction and effects of both species on native hosts exemplarily show how introduced parasites may affect native communities. *Mytilicola intestinalis* was introduced into the North Sea in the early 20th century and uses native blue mussels *Mytilus edulis* as host species. The parasite has been blamed for several mass mortalities of infested blue mussels because it restricts gut function and feeds on host tissue. Demann and Wegner (2019) revealed that mortality is not only a direct effect of the *M. intestinalis* infection of mussels. The infection of *M. edulis* with *M. intestinalis* also indirectly increases the susceptibility of mussels to secondary infection with *Vibrio* bacteria, which causes high mortality, especially at the elevated temperatures that are part of ocean environmental changes (Fox-Kemper et al. 2021). Thus, an increased introduction of nonnative parasites in combination with ocean warming may lead to increased mortality in native species, and this assumption is confirmed when considering the second copepod parasite species. The nonnative *M. orientalis* was co-introduced with Pacific oysters (*Magallana gigas*) to European waters in the 1960s and 70s, where the parasite was later also found in blue mussels (Fig. 9.8). In contrast to *M. intestinalis*, which does not infest

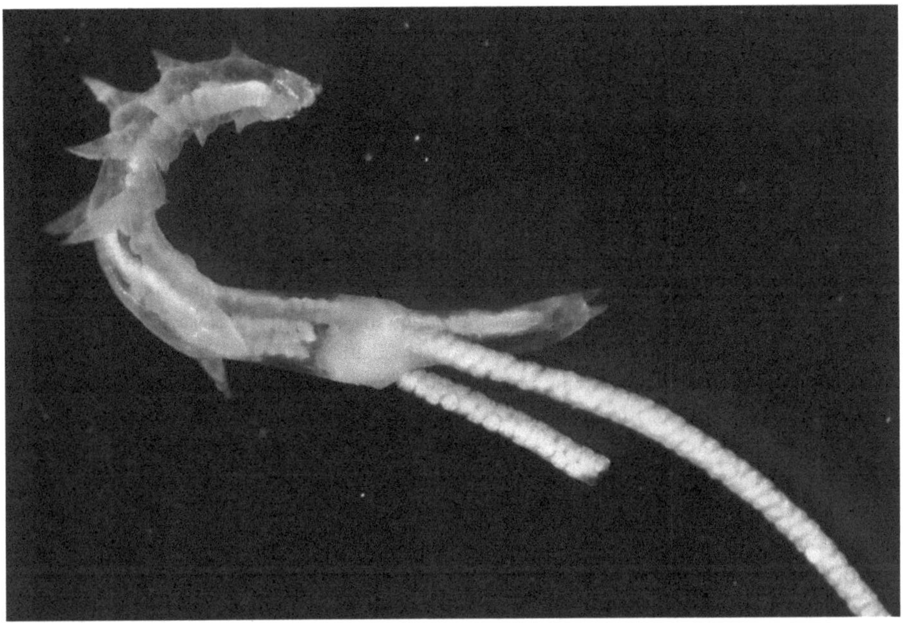

Fig. 9.8.
Nonnative intestinal parasitic copepod *Mytilicola orientalis* (female with egg sacs), which was co-introduced with Pacific oysters to North Sea coasts in the 1960s and 70s. The species spilled over to many native bivalves and causes direct and indirect detrimental effects on native hosts. Photo reproduced courtesy of Dagmar Lackschewitz.

hosts other than *M. edulis*, *M. orientalis* was additionally found in common cockles (*Cerastoderma edule*), Baltic tellins (*Limecola (Macoma) balthica*), and European flat oysters (*Ostrea edulis*). The high spillover frequency of *M. orientalis* suggests that it may cause strong parasite-mediated indirect impacts on native bivalve populations and, subsequently, on the ecosystem. In addition, it demonstrates that human-induced species introductions and ocean warming cause direct and indirect alterations of parasite–host interactions in native ecosystems with hitherto hardly considered consequences, in which crustaceans play a crucial role (Goedknegt et al. 2020).

CRUSTACEANS UNDER STRESS IN A COMMUNITY CONTEXT

Crustaceans number more than 50,000 recognized species worldwide (Thiel and Duffy 2007; Schram 2013) in diverse (Watling and Thiel 2015), often extreme (see case studies I and II above) habitats, from dry deserts to rivers and shallow intertidal ponds to the deep sea. Crustaceans encompass various feeding modes such as suspension feeders, deposit feeders, detritivores, and carnivores, and, additionally, symbiosis is found among the diverse lifestyles of crustaceans (e.g., Watling 2013, Chan and Høeg 2015, Baeza 2015). There are also multiple successful crustacean invaders in new habitats (see case study III above). Most benthic decapods have complex life histories that involve free-swimming larvae and settling stages (see case study II above, and Chang and Høeg [2015]). All these traits make crustaceans key species in food webs, and slight shifts in species abundances or performance may lead to strong ecosystem-wide changes (Paine 1992). Moreover, crustaceans can cause strong top-down control on aquatic macrophytes by grazing (Jormalainen 2015), and many crustaceans such as barnacles occupy living surfaces such as snails, bivalves, other crabs, and also macrophytes (e.g., algae and seagrass blades, algae holdfasts, and stipes, as well as seeds; Mejaes et al. 2015), resulting in interactions between the basibiont used as settlement substratum and the attached crustacean epibiont. Many of these, often very close, interactions are likely to change with climate change and will therefore lead to shifts in ecosystem composition and functioning.

In the face of global ocean changes, winners and losers can be identified across taxa (see examples highlighted above). Here, individual performance functions in response to one, or multiple, environmental or biotic stressors may lead to shifts in population sensitivities, ultimately with ecosystem-wide implications. Crustaceans might encompass some of the winners under, for example, acidification stress as they appear more tolerant than other marine taxa due to their strong ion regulatory capacities. For example, in the Western Baltic Sea, the common sea star *Asterias rubens* is strongly impacted by warming and acidification (e.g., Appelhans et al. 2012, 2014, Morón Lugo et al. 2020), while its direct crustacean competitors for mussel (*M. edulis*) prey, the shore crab *C. maenas*, and the Asian shore crabs *H. takanoi*, proved more robust to these environmental drivers (Appelhans et al. 2012, Morón Lugo et al. 2020). Adaptation to environmental changes is possible and has been demonstrated in several crustacean species. This will most likely lead to shifting impacts on mussel beds. Thus, the fate of a species or population not only depends on the sensitivities of all their life-history stages, and their potential for acclimation and adaptation to multiple changes, but will also be defined by the sensitivities of their direct (and indirect) competitors, their predators, or their prey.

Opportunistic and often stress-tolerant crustaceans are spreading globally and invade new habitats. Here, the newcomers often encounter less tolerant native species, which now experience direct interactions such as competition for similar food resources and space (Witte et al. 2010, Cornelius et al. 2021) or predation on their early life-history stages (Geburzi et al. 2018). However, general statements about the effects of introduced crustaceans on native species population dynamics and communities are difficult to make due to strong spatial and temporal variation in the direction and strength of interactions. Every new species in a native ecosystem can trigger a cascade of emerging direct and indirect species interactions that may change with time due to gradually changing abiotic factors and unpredictable, often random, events such as severe winters (Reise et al. 2017). Therefore, there remains considerable uncertainty about how new invaders, such as Pacific shore crabs *Hemigrapsus* spp. on both sides of the Atlantic Ocean, influence ecological shifts in the invaded marine systems under future environmental conditions (McDermott 1998, Morón Lugo et al. 2020, Nour et al. 2020, 2021). The combination of human-caused species translocations and ocean warming will likely be one of the most important stressors on the marine environment in the coming decades, causing significant changes in ecosystem functioning, especially in coastal zones throughout the world.

FUTURE DIRECTIONS

In the world's marine ecosystems, crustaceans are a very diverse group and can dominate some of the most extreme environments. Crustaceans have the capacities needed for short-term acclimation and long-term adaptation, such as efficient ion regulation and, often, short generation times. Population-specific differences in ocean acidification tolerance of, for example, barnacle crustaceans have already been demonstrated, but mechanisms explaining local adaptation are poorly understood. Therefore, long-term exposure experiments, involving multiple generations of species, will provide important insights into the abilities of crustaceans to "evolutionary" adapt to environmental change. Long-lived crustaceans, like the American lobster, might experience considerable climate change within their life span of 50–100 years, for which it may be crucial to investigate physiological or behavioral acclimation in an ocean change context. Here, a better understanding of mechanisms, including accumulation of heat shock proteins and epigenetic modifications, are important future tasks to assess the effects of the multitude of environmental change and their recurring extremes on crustaceans. Thus, linking acclimation and adaptation capacities to the regeneration time of a species will provide a more comprehensive understanding of the sensitivity of species to rapidly changing and fluctuating environmental shifts and their pulse stress events. Direct effects on single organisms, as assessed in mechanistic experimental approaches (single species, single driver, microcosms), are modulated by shifting biotic interactions (competition, predation, parasitism) in a community context. Thus, community responses to ocean change may only be understood if the apparent interactions of crustaceans with other species, their parasites, and with new invaders are considered under realistic climate change scenarios that encompass long-term community incubations under the combination of multiple environmental and realistic, albeit fluctuating, drivers. Finally, the ecosystem context might provide temporal and spatial refugia due to, for example, ocean acidification buffering (Kapsenberg and Cyronak 2019). In addition, environmental variability can lead to extremes but also provides temporal relaxation from stress. Considering this aspect of variability and the temporal (and spatial) occurrence of stress in the context of the stress memory of a species as well as an ecological memory of ecosystems over multiple environmental extremes will further contribute to our understanding of ongoing changes and potential future ecosystem shifts in the ocean.

CONCLUSIONS

Herein, we offer examples of several important drivers of ocean ecosystem change, such as warming and acidification and also shifting species compositions and their parasitic assemblages, and how these directly mediate cell to organismic physiology, their population- and species-wide performance windows, species interactions, and how such processes might result in changing ecosystems. It becomes evident that even by focusing on crustaceans, the complexity of biological responses to environmental change scales up with each level of biological organization from individuals to the population. At the ecosystem level, the daunting complexity of responses driven by competition, invasive species, and host–parasite interactions demonstrate the difficulty in making detailed predictions at the population and species level without a holistic experimental approach. However, this chapter reveals that at the species level, general predictions regarding single to multiple stressors are possible. For example, the necessity to survive in fluctuating environments resulted in the evolution of crustacean species with extraordinary osmo- and pH-regulatory capacities. This trait can be regarded as a unifying feature that pre-adapts certain species to abiotic changes induced by climate change (i.e., salinity, pH, and potentially also temperature), thereby facilitating the invasion of novel habitats. However, it remains uncertain how these crustaceans cope with further biotic stressors including competition, pathogens, and parasitism. Here, understanding the energetic expenses to sustain immunity, competition, and homeostasis in relation to available energy can be regarded as a challenging but promising approach for large-scale prediction of species' sensitivity to climate change.

REFERENCES

Aiuppa, A., M. Bitetto, A.L. Rizzo, F. Viveiros, P. Allard, M.L. Frezzotti, V. Valenti, and V. Zanon. 2020. The fumarolic CO_2 output from Pico do Fogo volcano (Cape Verde). Italian Journal of Geosciences 139:325–340.

Allen, G.J.P., P.L. Kuan, Y.C. Tseng, P.P. Hwang, A.R. Quijada-Rodriguez, and D. Weihrauch. 2020. Specialized adaptations allow vent-endemic crabs (*Xenograpsus testudinatus*) to thrive under extreme environmental hypercapnia. Scientific Reports 10:11720.

Anger, K. 1986. Changes of respiration and biomass of spider crab (*Hyas araneus*) larvae during starvation. Marine Biology 90:261–269.

Appelhans, Y.S., J. Thomsen, S. Opitz, C. Pansch, F. Melzner, and M. Wahl. 2014. Juvenile sea stars exposed to acidification decrease feeding and growth with no acclimation potential. Marine Ecology Progress Series 509:227–239.

Appelhans, Y.S., J. Thomsen, C. Pansch, F. Melzner, and M. Wahl. 2012. Sour times: Seawater acidification effects on growth, feeding behaviour and acid-base status of *Asterias rubens* and *Carcinus maenas*. Marine Ecology Progress Series 459:85–98.

Bącela-Spychalska, K., G.C.B. Poore, and M. Grabowski. 2020. Outlook: Crustaceans in the Anthropocene. Pages 464–491 *in* G.C.B. Poore and M. Thiel, editors. The natural history of the Crustacea, Volume 8: Evolution and biogeography. Oxford University Press, New York, USA.

Baeza, J.A. 2015. Crustaceans as symbionts: An overview of their diversity, host use and lifestyles. Pages 163–189 *in* M. Thiel and L. Watling, editors. The natural history of the Crustacea, Volume 2: Lifestyles and feeding biology. Oxford University Press, New York, USA.

Baillie, C.J., and J.H. Grabowski. 2018. Competitive and agonistic interactions between the invasive Asian shore crab and juvenile American lobster. Ecology 99:2067–2079.

Bakun, A., B.A. Black, S.J. Bograd, M. García-Reyes, A.J. Miller, R.R. Rykaczewski, and W.J. Sydeman. 2015. Anticipated effects of climate change on coastal upwelling ecosystems. Current Climate Change Reports 1:85–93.

Bednaršek, N., R. Ambrose, P. Calosi, R.K. Childers, R.A. Feely, S.Y. Litvin, W.C. Long, J.I. Spicer, J. Štrus, J. Taylor, F. Kessouri, M. Roethler, M. Sutula, and S.B. Weisberg. 2021. Synthesis of thresholds of ocean acidification impacts on decapods. Frontiers in Marine Science 8:651102.

Blakeslee, A.M.H., R.B. Barnard, K. Matheson, and C.H. McKenzie. 2020. Host-switching among crabs: Species introduction results in a new target host for native parasites. Marine Ecology Progress Series 636:91–106.

Bojko, J. 2020. The mitochondrial genome of UK (non-native) *Dikerogammarus haemobaphes* (Amphipoda: Gammaridae) informs upon *Dikerogammarus* evolution, invasions and associated microparasites. Hydrobiologia 847:229–242.

Bojko, J., A.L. Burgess, A.G. Baker, and C.H. Orr. 2021. Invasive non-native crustacean symbionts: Diversity and impact. Journal of Invertebrate Pathology 186:107482.

Bojko, J., P.D. Stebbing, A.M. Dunn, K.S. Bateman, F. Clark, R.C. Kerr, S. Stewart-Clark, Á. Johannesen, and G.D. Stentiford. 2018. Green crab *Carcinus maenas* symbiont profiles along a North Atlantic invasion route. Diseases of Aquatic Organisms 128:147–168.

Bopp, L., L. Resplandy, J.C. Orr, S.C. Doney, J.P. Dunne, M. Gehlen, P. Halloran, C. Heinze, T. Ilyina, R. Séférian, J. Tjiputra, and M. Vichi. 2013. Multiple stressors of ocean ecosystems in the 21st century: Projections with CMIP5 models. Biogeosciences 10:6225–6245.

Boron, W.F. 2004. Regulation of intracellular pH. Advances in Physiology Education 28:160–179.

Boyd, P.W., S. Collins, S. Dupont, K. Fabricius, J.P. Gattuso, J. Havenhand, D.A. Hutchins, U. Riebesell, M.S. Rintoul, M. Vichi, H. Biswas, A. Ciotti, K. Gao, M. Gehlen, C.L. Hurd, H. Kurihara, C.M. McGraw, J.M. Navarro, G.E. Nilsson, U. Passow, and H.O. Pörtner. 2018. Experimental strategies to assess the biological ramifications of multiple drivers of global ocean change—A review. Global Change Biology 24:2239–2261.

Brown, J.H., J.F. Gillooly, A.P. Allen, V.M. Savage, and G.B. West. 2004. Toward a metabolic theory of ecology. Ecology 85:1771–1789.

Buschbaum, C., D. Lackschewitz, and K. Reise. 2012. Nonnative macrobenthos in the Wadden Sea ecosystem. Ocean and Coastal Management 68:89–101.

Campbell, R., R. Baring, and S. Dittmann. 2019. Cracking the cuisine: Invasive European shore crabs (*Carcinus maenas*) select a menu of soft-shelled mussels over cockles. Journal of Experimental Marine Biology and Ecology 517:25–33.

Chan, B.K.K., and J.T. Høeg. 2015. Diversity of lifestyles, sexual systems and larval development pattern in sessile crustaceans. Pages 14–34 *in* M. Thiel and L. Watling, editors. The natural history of the Crustacea, Volume 2: Lifestyles and feeding biology. Oxford University Press, New York, USA.

Charmantier, G., C. Haond, J.H. Lignot, and M. Charmantier-Daures. 2001. Ecophysiological adaptation to salinity throughout a life cycle: A review in homarid lobsters. Journal of Experimental Biology 204:967–977.

Cieluch, U., K. Anger, F. Aujoulat, F. Buchholz, M. Charmantier-Daures, and G. Charmantier. 2004. Ontogeny of osmoregulatory structures and functions in the green crab *Carcinus maenas* (Crustacea, Decapoda). Journal of Experimental Biology 207:325–336.

Cieluch, U., K. Anger, M. Charmantier-Daures, and G. Charmantier. 2007. Osmoregulation and immunolocalization of Na+/K+-ATPase during the ontogeny of the mitten crab *Eriocheir sinensis* (Decapoda, Grapsoidea). Marine Ecology Progress Series 329:169–178.

Cornelius, A., K. Wagner, and C. Buschbaum. 2021. Prey preferences, consumption rates and predation effects of Asian shore crabs (*Hemigrapsus takanoi*) in comparison to native shore crabs (*Carcinus maenas*) in northwestern Europe. Marine Biodiversity 51:75.

Dawirs, R.R. 1987. Influence of limited starvation periods on growth and elemental composition (C,N,H) of *Carcinus maenas* (Decapoda, Portunidae) larvae reared in the laboratory. Marine Biology 93:543–549.

Dell, A.I., S. Pawar, and V. Savage. 2014. Temperature dependence of trophic interactions are driven by asymmetry of species responses and foraging strategy. The Journal of Animal Ecology 83:70–84.

Demann, F., and K.M. Wegner. 2019. Infection by invasive parasites increases susceptibility of native hosts to secondary infection via modulation of cellular immunity. The Journal of Animal Ecology 88:427–438.

Falconer, T.R.L., I.D. Marsden, J.V. Hill, and C.N. Glover. 2019. Does physiological tolerance to acute hypoxia and salinity change explain ecological niche in two intertidal crab species? Conservation physiology 7:coz086.

Faria, S.C., A. Bianchini, M.M. Lauer, A.L.R.L. Zimbardi, F. Tapella, M.C. Romero, and J.C. McNamara. 2020. Living on the edge: Physiological and kinetic trade-offs shape thermal tolerance in intertidal crabs from tropical to sub-antarctic South America. Frontiers in Physiology 11:312.

Faulkner, K.T., S. Clusella-Trullas, L.S. Peck, and S.L. Chown. 2014. Lack of coherence in the warming responses of marine crustaceans. Functional Ecology 28:895–903.

Feely, R.A., S.R. Alin, B. Carter, N. Bednaršek, B. Hales, F. Chan, T.M. Hill, B. Gaylord, E. Sanford, R.H. Byrne, C.L. Sabine, D. Greeley, and L. Juranek. 2016. Chemical and biological impacts of ocean acidification along the west coast of North America. Estuarine, Coastal and Shelf Science 183:260–270.

Fehsenfeld, S., R. Kiko, Y. Appelhans, D.W. Towle, M. Zimmer, and F. Melzner. 2011. Effects of elevated seawater pCO_2 on gene expression patterns in the gills of the green crab, *Carcinus maenas*. BMC Genomics 12:488.

Fehsenfeld, S., and D. Weihrauch. 2013. Differential acid-base regulation in various gills of the green crab *Carcinus maenas*: Effects of elevated environmental pCO_2. Comparative Biochemistry and Physiology Part A: Molecular and Integrative Physiology 164:54–65.

Fehsenfeld, S., and D. Weihrauch. 2017. Acid–base regulation in aquatic decapod crustaceans. Pages 151–191 *in* D. Weihrauch and M. O'Donnell, editors. Acid-base balance and nitrogen excretion in invertebrates: Mechanisms and strategies in various invertebrate groups with considerations of challenges caused by ocean acidification. Springer, Cham, Switzerland.

Florey, E., and G. Hoyle. 1976. The effects of temperature on a nerve-muscle system of the Hawaiian ghost crab, *Ocypode ceratophthalma* (Pallas). Journal of Comparative Physiology A: Neuroethology, Sensory, Neural, and Behavioral Physiology 110:14.

Foo, S.A., M. Byrne, E. Ricevuto, and M.C. Gambi. 2018. The carbon dioxide vents of Ischia, Italy, A natural system to assess impacts of ocean acidification on marine ecosystems: An overview of research and comparisons with other vent systems. Oceanography and Marine Biology: An Annual Review 56:237–310.

Fox-Kemper, B., H.T. Hewitt, C. Xiao, G. Aðalgeirsdóttir, S.S. Drijfhout, T.L. Edwards, N.R. Golledge, M. Hemer, R.E. Kopp, G. Krinner, A. Mix, D. Notz, S. Nowicki, I.S. Nurhati, L. Ruiz, J.-B. Sallée, A.B.A. Slangen, and Y. Yu. 2021. Ocean, cryosphere and sea level change. Pages 1211–1362 *in* V. Masson-Delmotte, P. Zhai, A. Pirani, S.L. Connors, C. Péan, S. Berger, N. Caud, Y. Chen, L. Goldfarb, M.I. Gomis, M. Huang, K. Leitzell, E. Lonnoy, J.B.R. Matthews, T.K. Maycock, T. Waterfield, O. Yelekçi, R. Yu, and B. Zhou, editors. Climate change 2021: The physical science basis. Contribution of working group I to the sixth assessment report of the intergovernmental panel on climate change. Cambridge University Press, Cambridge, UK.

Galil, B.S., S. Nehring, and V. Panov. 2008. Waterways as invasion highways—Impact of climate change and globalization. Pages 59–74 *in* W. Nentwig, editor. Ecological Studies, Volume 193: Biological invasions. Springer, Berlin, Heidelberg, Germany.

Gallardo, M., I. Rojas, K. Brokordt, G. Lovrich, V. Nuñez, K. Paschke, M. Thiel, and B. Yannicelli. 2019. Life on the edge: Incubation behaviour and physiological performance of squat lobsters in oxygen-minimum conditions. Marine Ecology Progress Series 623:51–70.

Gaonkar, C.A., and A.C. Anil. 2010. What do barnacle larvae feed on? Implications in biofouling ecology. Journal of the Marine Biological Association of the United Kingdom 90:1241–1247.

Gattuso, J.-P., and L. Hansson. 2011. Ocean acidification. Oxford University Press, New York, USA.

Geburzi, J.C., D. Brandis, and C. Buschbaum. 2018. Recruitment patterns, low cannibalism and reduced interspecific predation contribute to high invasion success of two Pacific crabs in northwestern Europe. Estuarine, Coastal and Shelf Science 200:460–472.

Geburzi, J.C., and M.L. McCarthy. 2018. How do they do it? Understanding the success of marine invasive species. Pages 109–124 *in* S. Jungblut, V. Liebich, and M. Bode, editors. YOUMARES 8—Oceans across boundaries: Learning from each other. Springer, Cham, Switzerland.

Gilbert, D., N.N. Rabalais, R.J. Díaz, and J. Zhang. 2010. Evidence for greater oxygen decline rates in the coastal ocean than in the open ocean. Biogeosciences 7:2283–2296.

Goedknegt, M.A., S. Bedolfe, J. Drent, J. van der Meer, and D.W. Thieltges. 2018. Impact of the invasive parasitic copepod *Mytilicola orientalis* on native blue mussels *Mytilus edulis* in the western European Wadden Sea. Marine Biology Research 14:497–507.

Goedknegt, M.A., C. Buschbaum, J. van der Meer, K.M. Wegner, and D.W. Thieltges. 2020. Introduced marine ecosystem engineer indirectly affects parasitism in native mussel hosts. Biological Invasions 22:3223–3237.

Goedknegt, M.A., M.E. Feis, K.M. Wegner, P.C. Luttikhuizen, C. Buschbaum, K. Camphuysen, J. van der Meer, and D.W. Thieltges. 2016. Parasites and marine invasions: Ecological and evolutionary perspectives. Journal of Sea Research 113:11–27.

González-Delgado, S., and J.C. Hernández. 2018. The importance of natural acidified systems in the study of ocean acidification: What have we learned? Advances in Marine Biology 80:57–99.

Grabowski, M., K. Bacela, and A. Konopacka. 2007. How to be an invasive gammarid (Amphipoda: Gammaroidea)—Comparison of life history traits. Hydrobiologia 590:75–84.

Gravinese, P.M., I.C. Enochs, D.P. Manzello, and R. van Woesik. 2018. Warming and pCO_2 effects on Florida stone crab larvae. Estuarine, Coastal and Shelf Science 204:193–201.

Gregory, G.J., and P.A. Quijón. 2011. The impact of a coastal invasive predator on infaunal communities: Assessing the roles of density and a native counterpart. Journal of Sea Research 66:181–186.

Griffen, B.D., I. Altman, J. Hurley, and H. Mosblack. 2011. Reduced fecundity by one invader in the presence of another: A potential mechanism leading to species replacement. Journal of Experimental Marine Biology and Ecology 406:6–13.

Griffen, B.D., and J.E. Byers. 2009. Community impacts of two invasive crabs: The interactive roles of density, prey recruitment, and indirect effects. Biological Invasions 11:927–940.

Griffen, B.D., T. Guy, and J.C. Buck. 2008. Inhibition between invasives: A newly introduced predator moderates the impacts of a previously established invasive predator. Journal of Animal Ecology 77:32–40.

Griffiths, J.R., M. Kadin, F.J.A. Nascimento, T. Tamelander, A. Tornroos, S. Bonaglia, E. Bonsdorff, V. Brüchert, A. Gårdmark, M. Järnström, and J. Kotta. 2017. The importance of benthic-pelagic coupling for marine ecosystem functioning in a changing world. Global Change Biology 23:2179–2196.

Grosholz, E.D., G.M. Ruiz, C.A. Dean, K.A. Shirley, J.L. Maron, and P.G. Connors. 2000. The impacts of a nonindigenous marine predator in a California bay. Ecology 81:1206–1224.

Gunderson, A.R., E.J. Armstrong, and J.H. Stillman. 2016. Multiple stressors in a changing world: The need for an improved perspective on physiological responses to the dynamic marine environment. Annual Review of Marine Science 8:357–378.

Halpern, B.S., M. Frazier, J. Afflerbach, J.S. Lowndes, F. Micheli, C. O'Hara, C. Scarborough, and K.A. Selkoe. 2019. Recent pace of change in human impact on the world's ocean. Scientific Reports 9:11609.

Hänfling, B., F. Edwards, and F. Gherardi. 2011. Invasive alien Crustacea: Dispersal, establishment, impact, and control. BioControl 56:573–595.

Henry, R., C. Lucu, H. Onken, and D. Weihrauch. 2012. Multiple functions of the crustacean gill: Osmotic/ionic regulation, acid-base balance, ammonia excretion, and bioaccumulation of toxic metals. Frontiers in Physiology 3:431.

Howard, B.R., T.W. Therriault, and I.M. Cote. 2017. Contrasting ecological impacts of native and non-native marine crabs: A global meta-analysis. Marine Ecology Progress Series 577:93–103.

Hu, M.Y., Y.-J. Guh, Y.-T. Shao, P.-L. Kuan, G.-L. Chen, J.-R. Lee, M.-S. Jeng, and Y.-C. Tseng. 2016. Strong ion regulatory abilities enable the crab *Xenograpsus testudinatus* to inhabit highly acidified marine vent systems. Frontiers in Physiology 7:14.

Hu, M.Y.A., W. Hagen, M.S. Jeng, and R. Saborowski. 2012. Metabolic energy demand and food utilization of the hydrothermal vent crab *Xenograpsus testudinatus* (Crustacea: Brachyura). Aquatic Biology 15:11–25.

Hughes, T.P., J.T. Kerry, S.R. Connolly, A.H. Baird, C.M. Eakin, S.F. Heron, A.S. Hoey, M.O. Hoogenboom, M. Jacobson, G. Liu, M.S. Pratchett, W. Skirving, and G. Torda. 2019. Ecological memory modifies the cumulative impact of recurrent climate extremes. Nature Climate Change 9:40–43.

Hurd, C.L., J. Beardall, S. Comeau, C.E. Cornwall, J.N. Havenhand, P.L. Munday, L.M. Parker, J.A. Raven, and C.M. McGraw. 2020. Ocean acidification as a multiple driver: How interactions between changing seawater carbonate parameters affect marine life. Marine and Freshwater Research 71:263–274.

Hutchinson, G. 1978. An introduction to population biology. Yale University Press, New Haven, Connecticut, USA and London, UK.

Jackson, M.C., S. Pawar, and G. Woodward. 2021. The temporal dynamics of multiple stressor effects: From individuals to ecosystems. Trends in Ecology and Evolution 36:402–410.

Jacobi, C.C., and K. Anger. 1985. Effect of temperature on respiration of larval stages of *Hyas araneus* and *H. coarctatus* (Decapoda, Majidae). Marine Ecology Progress Series 26:181–186.

Jeng, M.S., N.K. Ng, and P.K.L. Ng. 2004. Hydrothermal vent crabs feast on sea 'snow'. Nature 432:969.

Johnson M.J., L.M. Hennigs, Y. Sawall, C. Pansch, and M. Wall. 2021. Growth response of calcifying marine epibionts to biogenic pH fluctuations and global ocean acidification scenarios. Limnology and Oceanography 66:1125–1138.

Jormalainen, V. 2015. Grazers of macroalgae and higher plants. Pages502–534 in M. Thiel, and L. Watling, editors. The Natural History of The Crustacea, Volume 2: Lifestyles and Feeding Biology. Oxford University Press, New York, USA.

Kapsenberg, L., and T. Cyronak. 2019. Ocean acidification refugia in variable environments. Global Change Biology 25:3201–3214.

Klymasz-Swartz, A.K., G.J.P. Allen, J.R. Treberg, G.R. Yoon, A. Tripp, A.R. Quijada-Rodriguez, and D. Weihrauch. 2019. Impact of climate change on the American lobster (*Homarus americanus*): Physiological responses to combined exposure of elevated temperature and pCO_2. Comparative Biochemistry and Physiology Part A: Molecular and Integrative Physiology 235:202–210.

Köhn, E.E., S. Thomsen, D.L. Arévalo-Martínez, and T. Kanzow. 2017. Submesoscale CO_2 variability across an upwelling front off Peru. Ocean Science 13:1017–1033.

Kroeker, K.J., R.L. Kordas, R. Crim, I.E. Hendriks, L. Ramajo, G.S. Singh, C.M. Duarte, and J.P. Gattuso. 2013. Impacts of ocean acidification on marine organisms: Quantifying sensitivities and interaction with warming. Global Change Biology 19:1884–1896.

Lagerspetz, K.Y., and L.A. Vainio. 2006. Thermal behaviour of crustaceans. Biological Reviews of the Cambridge Philosophical Society 81:237–258.

Landschoff, J., D. Lackschewitz, K. Kesy, and K. Reise. 2013. Globalization pressure and habitat change: Pacific rocky shore crabs invade armored shorelines in the Atlantic Wadden Sea. Aquatic Invasions 8:77–87.

Langer, J.A.F., C.L. Meunier, U. Ecker, H.G. Horn, K. Schwenk, and M. Boersma. 2019. Acclimation and adaptation of the coastal calanoid copepod *Acartia tonsa* to ocean acidification: A long-term laboratory investigation. Marine Ecology Progress Series 619:35–51.

Lenz, M., B.A.P. da Gama, N.V. Gerner, J. Gobin, F. Groner, A. Harry, S.R. Jenkins, P. Kraufvelin, C. Mummelthei, J. Sareyka, E.A. Xavier, and M. Wahl. 2011. Non-native marine invertebrates are more tolerant towards environmental stress than taxonomically related native species: Results from a globally replicated study. Environmental Research 111:943–952.

Lima, F.P., and D.S. Wethey. 2012. Three decades of high-resolution coastal sea surface temperatures reveal more than warming. Nature Communications 3:704.

Lohrer, A.M., and R.B. Whitlatch. 2002. Interactions among aliens: Apparent replacement of one exotic species by another. Ecology 83:719–732.

Madeira, C., M.C. Leal, M.S. Diniz, H.N. Cabral, and C. Vinagre. 2018. Thermal stress and energy metabolism in two circumtropical decapod crustaceans: Responses to acute temperature events. Marine Environmental Research 141:148–158.

Martin, J.W., and T.A. Haney. 2005. Decapod crustaceans from hydrothermal vents and cold seeps: A review through 2005. Zoological Journal of the Linnean Society 145:445–522.

McDermott, J.J. 1998. The western Pacific brachyuran (*Hemigrapsus sanguineus*: Grapsidae), in its new habitat along the Atlantic coast of the United States: Geographic distribution and ecology. ICES Journal of Marine Science 55:289–298.

McMahon, B.R. 2001. Control of cardiovascular function and its evolution in Crustacea. Journal of Experimental Biology 204:923–932.

Meier, H.E.M., A. Höglund, K. Eilola, and E. Almroth-Rosell. 2017. Impact of accelerated future global mean sea level rise on hypoxia in the Baltic Sea. Climate Dynamics 49:163–172.

Mejaes, B.A., A.G.B. Poore, and M. Thiel. 2015. Crustaceans inhabiting domiciles excavated from macrophytes and stone. Pages 118–144 *in* M. Thiel and L. Watling, editors. The natural history of the Crustacea, Volume 2: Lifestyles and feeding biology. Oxford University Press, New York, USA.

Melzner, F., M.A. Gutowska, M. Langenbuch, S. Dupont, M. Lucassen, M.C. Thorndyke, M. Bleich, and H.O. Pörtner. 2009. Physiological basis for high CO_2 tolerance in marine ectothermic animals: Pre-adaptation through lifestyle and ontogeny? Biogeosciences 6:2313–2331.

Metzger, R., F.J. Sartoris, M. Langenbuch, and H.O. Pörtner. 2007. Influence of elevated CO_2 concentrations on thermal tolerance of the edible crab *Cancer pagurus*. Journal of Thermal Biology 32:144–151.

Miralles, L., A. Ibabe, M. González, E. Garcia-Vázquez, and Y.J. Borrell. 2021. "If you know the enemy and know yourself": Addressing the problem of biological invasions in ports through a new NIS invasion threat score, routine monitoring, and preventive action plans. Frontiers in Marine Science 8:633118.

Morón Lugo, S.C., M. Baumeister, O.M. Nour, F. Wolf, M. Stumpp, and C. Pansch. 2020. Warming and temperature variability determine the performance of two invertebrate predators. Scientific Reports 10:6780.

Mouritsen, K.N., and R. Poulin. 2002. Parasitism, community structure and biodiversity in intertidal ecosystems. Parasitology 124:S101–S117.

Nasrolahi, A., J. Havenhand, A.-L. Wrange, and C. Pansch. 2016. Population and life-stage specific sensitivities to temperature and salinity stress in barnacles. Scientific Reports 6:32263.

Nasrolahi, A., C. Pansch, M. Lenz, and M. Wahl. 2012. Being young in a changing world: How temperature and salinity changes interactively modify the performance of larval stages of the barnacle *Amphibalanus improvisus*. Marine Biology 159:331–340.

Nasrolahi, A., C. Pansch, M. Lenz, and M. Wahl. 2013. Temperature and salinity interactively impact early juvenile development: A bottleneck in barnacle ontogeny. Marine Biology 160:1109–1117.

Ng, P.K.L., D. Guinot, and P.J.F. Davie. 2008. Systema Brachyurorum: Part I. An annotated checklist of extant brachyuran crabs of the world. Raffles Bulletin of Zoology 17:1–286.

Nour, O.M., C. Pansch, M. Lenz, M. Wahl, C. Clemmesen, and M. Stumpp. 2021. Impaired larval development at low salinities could limit the spread of the non-native crab *Hemigrapsus takanoi* in the Baltic Sea. Aquatic Biology 30:85–99.

Nour, O.M., M. Stumpp, S.C.M. Lugo, F.R. Barboza, and C. Pansch. 2020. Population structure of the recent invader *Hemigrapsus takanoi* and prey size selection on Baltic Sea mussels. Aquatic Invasions 15:297–317.

Obert, B., M. Herlyn, and M. Grotjahn. 2007. First records of two crabs from the North West Pacific *Hemigrapsus sanguineus* and *H. takanoi* at the coast of Lower Saxony, Germany. Wadden Sea Newsletter 1:21–22.

O'Connor, N.J. 2018. Changes in population sizes of *Hemigrapsus sanguineus* (Asian shore crab) and resident crab species in Southeastern New England (2010–2016). Northeastern Naturalist 25:197–201.

Oliver, E.C.J., J.A. Benthuysen, S. Darmaraki, M.G. Donat, A.J. Hobday, N.J. Holbrook, R.W. Schlegel, and A.S. Gupta. 2021. Marine heatwaves. Annual Review of Marine Science 13:313–342.

Orr, J.A., R.D. Vinebrooke, M.C. Jackson, K.J. Kroeker, R.L. Kordas, C. Mantyka-Pringle, P.J. Van den Brink, F. De Laender, R. Stoks, M. Holmstrup, C.D. Matthaei, W.A. Monk, M.R. Penk, S. Leuzinger, R.B. Schäfer, and J.J. Piggott. 2020. Towards a unified study of multiple stressors: Divisions and common goals across research disciplines. Proceedings of the Royal Society B: Biological Sciences 287:20200421.

Paine, R.T. 1992. Food-web analysis through field measurement of per capita interaction strength. Nature 355:73–75.

Pane, E.F., and J.P. Barry. 2007. Extracellular acid-base regulation during short-term hypercapnia is effective in a shallow-water crab, but ineffective in a deep-sea crab. Marine Ecology Progress Series 334:1–9.

Pansch, C., G.S.I. Hattich, M.E. Heinrichs, A. Pansch, Z. Zagrodzka, and J.N. Havenhand. 2018a. Long-term exposure to acidification disrupts reproduction in a marine invertebrate. PLoS ONE 13:e0192036.

Pansch, C., and C. Hiebenthal. 2019. A new mesocosm system to study the effects of environmental variability on marine species and communities. Limnology and Oceanography Methods 17:145–162.

Pansch, C., A. Nasrolahi, Y.S. Appelhans, and M. Wahl. 2012. Impacts of ocean warming and acidification on the larval development of the barnacle *Amphibalanus improvisus*. Journal of Experimental Marine Biology and Ecology 420–421:48–55.

Pansch, C., A. Nasrolahi, Y.S. Appelhans, and M. Wahl. 2013a. Tolerance of juvenile barnacles (*Amphibalanus improvisus*) to warming and elevated pCO_2. Marine Biology (Berlin) 160:2023–2035.

Pansch, C., I. Schaub, J. Havenhand, and M. Wahl. 2014. Habitat traits and food availability determine the response of marine invertebrates to ocean acidification. Global Change Biology 20:765–777.

Pansch, C., P. Schlegel, and J. Havenhand. 2013b. Larval development of the barnacle *Amphibalanus improvisus* responds variably but robustly to near-future ocean acidification. ICES Journal of Marine Science 70:805–811.

Pansch, C., M. Scotti, F.R. Barboza, B. Al-Janabi, J. Brakel, E. Briski, B. Bucholz, M. Franz, M. Ito, F. Paiva, M. Saha, Y. Sawall, F. Weinberger, and M. Wahl. 2018b. Heat waves and their significance for a temperate benthic community: A near-natural experimental approach. Global Change Biology 24:4357–4367.

Pedersen, S.A., O.J. Håkedal, I. Salaberria, A. Tagliati, L.M. Gustavson, B.M. Jenssen, A.J. Olsen, and D. Altin. 2014. Multigenerational exposure to ocean acidification during food limitation reveals consequences for copepod scope for growth and vital rates. Environmental Science and Technology 48:12275–12284.

Pihl, L., S.P. Baden, and R.J. Diaz. 1991. Effects of periodic hypoxia on distribution of demersal fish and crustaceans. Marine Biology 108:349–360.

Popova, E., A. Yool, V. Byfield, K. Cochrane, A.C. Coward, S.S. Salim, M.A. Gasalla, S.A. Henson, A.J. Hobday, G.T. Pecl, W.H. Sauer, and M.J. Roberts. 2016. From global to regional and back again: Common climate stressors of marine ecosystems relevant for adaptation across five ocean warming hotspots. Global Change Biology 22:2038–2053.

Pörtner, H.O., and A.P. Farrell. 2008. Ecology physiology and climate change. Science 322: 690–692.

Quijada-Rodriguez, A.R., P.L. Kuan, P.H. Sung, M.T. Hsu, G.J.P. Allen, P.P. Hwang, Y.C. Tseng, and D. Weihrauch. 2021. Anthropogenic CO_2-mediated freshwater acidification limits survival, calcification, metabolism, and behaviour in stress-tolerant freshwater crustaceans. Biogeosciences 18:6287–6300.

Reise, K., C. Buschbaum, H. Büttger, and K.M. Wegner. 2017. Invading oysters and native mussels: From hostile takeover to compatible bedfellows. Ecosphere 8:e01949.

Reise, K., S. Olenin, and D.W. Thieltges. 2006. Are aliens threatening European aquatic coastal ecosystems? Helgoland Marine Research 60:77–83.

Reusch, T.B.H., J. Dierking, H.C. Andersson, and J.C. E. Bonsdorff, M. Casini, M. Czajkowski, B. Hasler, K. Hinsby, K. Hyytiäinen, and K. Johannesson, S. Jomaa, V. Jormalainen, H. Kuosa, S. Kurland, L. Laikre, B.R. MacKenzie, P. Margonski, F. Melzner, D. Oesterwind,H. Ojaveer, J.C. Refsgaard, A. Sandström, G. Schwarz, K. HTonderski, M. Winder, and M. Zandersen. 2018. The Baltic Sea as a time machine for the future coastal ocean. Science Advances 4:eaar8195.

Ries, J.B., A.L. Cohen, and D.C. McCorkle. 2009. Marine calcifiers exhibit mixed responses to CO_2-induced ocean acidification. Geology 37:1131–1134.

Ruiz, G.M., P. Fofonoff, A.H. Hines, and E.D. Grosholz. 1999. Non-indigenous species as stressors in estuarine and marine communities: Assessing invasion impacts and interactions. Limnology and Oceanography 44:950–972.

Schram, F.S. 2013. Comments on crustacean biodiversity and disparity of body plans. Pages 1–33 *in* L. Watling and M. Thiel, editors. The natural history of the Crustacea, Volume 1: Functional morphology and diversity. Oxford University Press, New York, USA.

Schramski, J.R., A.I. Dell, J.M. Grady, R.M. Sibly, and J.H. Brown. 2015. Metabolic theory predicts whole-ecosystem properties. Proceedings of the National Academy of Science USA 112:2617–2622.

Schiffer, M., L. Harms, M. Lucassen, F.C. Mark, H.-O. Pörtner, and D. Storch. 2014. Temperature tolerance of different larval stages of the spider crab *Hyas araneus* exposed to elevated seawater pCO_2. Frontiers in Zoology 11:87.

Schiffer, M., L. Harms, H. Pörtner, M. Lucassen, F.C. Mark, and D. Storch. 2013. Tolerance of *Hyas araneus* zoea I larvae to elevated seawater pCO_2 despite elevated metabolic costs. Marine Biology 160:1943–1953.

Schulz, K.G., S. Hartley, and B. Eyre. 2019. Upwelling amplifies ocean acidification on the East Australian Shelf: Implications for marine ecosystems. Frontiers in Marine Science 6:636.

Seabra, R., R. Varela, A.M. Santos, M. Gómez-Gesteira, C. Meneghesso, D.S. Wethey, and F.P. Lima. 2019. Reduced nearshore warming associated with eastern boundary upwelling systems. Frontiers in Marine Science 6:104.

Shields, J.D. 2019. Climate change enhances disease processes in crustaceans: Case studies in lobsters, crabs, and shrimps. Journal of Crustacean Biology 39:673–683.

Sinclair, B.J., K.E. Marshall, M.A. Sewell, D.L. Levesque, C.S. Willett, S. Slotsbo, Y.W. Dong, C.D.G. Harley, D.J. Marshall, B.S. Helmuth, and R.B. Huey. 2016. Can we predict ectotherm responses to climate change using thermal performance curves and body temperatures? Ecology Letters 19: 1372–1385.

Smale, D.A., T. Wernberg, E.C.J. Oliver, M. Thomsen, B.P. Harvey, S.C. Straub, M.T. Burrows, L.V. Alexander, J.A. Benthuysen, M.G. Donat, M. Feng, A.J. Hobday, N.J. Holbrook, S.E. Perkins-Kirkpatrick, H.A. Scannell, A. Sen Gupta, B.L. Payne, and P.J. Moore. 2019. Marine heatwaves threaten global biodiversity and the provision of ecosystem services. Nature Climate Change 9:306–312.

Small, D.P., P. Calosi, D. Boothroyd, S. Widdicombe, and J.I. Spicer. 2016. The sensitivity of the early benthic juvenile stage of the European lobster *Homarus gammarus* (L.) to elevated pCO_2 and temperature. Marine Biology 163:53.

Soliveres, S., R. Torices, and F.T. Maestre. 2012. Environmental conditions and biotic interactions acting together promote phylogenetic randomness in semi-arid plant communities: New methods help to avoid misleading conclusions. Journal of Vegetation Science 23:822–836.

Stasolla, G., E. Tricarico, and L. Vilizzi. 2021. Risk screening of the potential invasiveness of non-native marine crustacean decapods and barnacles in the Mediterranean Sea. Hydrobiologia 848: 1997–2009.

Stillman, J.H. 2002. Causes and consequences of thermal tolerance limits in rocky intertidal porcelain crabs, genus *Petrolisthes*. Integrative and Comparative Biology 42:790–796.

Stuart-Smith, R.D., G.J. Edgar, and A.E. Bates. 2017. Thermal limits to the geographic distributions of shallow-water marine species. Nature Ecology and Evolution 1:1846–1852.

Sunday, J.M., A.E. Bates, and N.K. Dulvy. 2012. Thermal tolerance and the global redistribution of animals. Nature Climate Change 2:686–690.

Sundell, K., A.-L. Wrange, P.R. Jonsson, and A. Blomberg. 2019. Osmoregulation in barnacles: An evolutionary perspective of potential mechanisms and future research directions. Frontiers in Physiology 10:877.

Swiney, K.M., W.C. Long, and R.J. Foy. 2017. Decreased pH and increased temperatures affect young-of-the-year red king crab (*Paralithodes camtschaticus*). ICES Journal of Marine Science 74:1191–1200.

Thiel, M., and J.E. Duffy, 2007. A primer in taxonomy, morphology and biology. Pages 3–28 *in* J.E. Duffy and M. Thiel, editors. Evolutionary ecology of social and sexual systems: Crustaceans as model organisms. Oxford University Press, New York, USA.

Thomsen, J., I. Casties, C. Pansch, A. Körtzinger, and F. Melzner. 2013. Food availability outweighs ocean acidification effects in juvenile *Mytilus edulis*: Laboratory and field experiments. Global Change Biology 19:1017–1027.

Thomsen, J., M. Gutowska, J. Saphörster, A. Heinemann, K. Trübenbach, J. Fietzke, C. Hiebenthal, A. Eisenhauer, A. Körtzinger, M. Wahl, and F. Melzner. 2010. Calcifying invertebrates succeed in a naturally CO_2-rich coastal habitat but are threatened by high levels of future acidification. Biogeosciences 7:3879–3891.

Thor, P., and S. Dupont. 2015. Transgenerational effects alleviate severe fecundity loss during ocean acidification in a ubiquitous planktonic copepod. Global Change Biology 21:2261–2271.

Trussell, G.C., P.J. Ewanchuk, and M.D. Bertness. 2002. Field evidence of trait-mediated indirect interactions in a rocky intertidal food web. Ecology Letters 5:241–245.

Ulloa, O., and S. Pantoja. 2009. The oxygen minimum zone of the eastern South Pacific. Deep Sea Research Part II: Topical Studies in Oceanography 56:987–991.

van den Brink, A., and S. Hutting. 2017. Clash of the crabs: Interspecific, inter-cohort competition between the native European green crab, *Carcinus maenas* and the exotic brush clawed crab *Hemigrapsus takanoi* on artificial oyster reefs. Journal of Sea Research 128:41–51.

Vaquer-Sunyer, R., and C.M. Duarte. 2008. Thresholds of hypoxia for marine biodiversity. Proceedings of the National Academy of Sciences of the United States of America 105:15452–15457.

Wahl, M., F.R. Barboza, B. Buchholz, S. Dobretsov, T. Guy-Haim, G. Rilov, R. Schuett, F. Wolf, J. Vajedsamiei, M. Yazdanpanah, and C. Pansch. 2021. Pulsed pressure: Fluctuating impacts of multifactorial environmental change on a temperate macroalgal community. Limnology and Oceanography 66:4210–4226.

Wahl, M., F.J. Werner, B. Buchholz, S. Raddatz, A. Graiff, B. Matthiessen, U. Karsten, C. Hiebenthal, J. Hamer, M. Ito, E. Gulzow, G. Rilov, and T. Guy-Haim. 2020. Season affects strength and direction of the interactive impacts of ocean warming and biotic stress in a coastal seaweed ecosystem. Limnology and Oceanography 65:807–827.

Walter, J., A. Jentsch, C. Beierkuhnlein, and J. Kreyling. 2013. Ecological stress memory and cross stress tolerance in plants in the face of climate extremes. Environmental and Experimental Botany 94:3–8.

Walther, K., K. Anger, and H.O. Pörtner. 2010. Effects of ocean acidification and warming on the larval development of the spider crab *Hyas araneus* from different latitudes (54 degrees vs. 79 degrees N). Marine Ecology Progress Series 417:159–170.

Wang, H.-Y., L.M. Tsang, F.P. Lima, R. Seabra, M. Ganmanee, G.A. Williams, and B.K.K. Chan. 2020. Spatial variation in thermal stress experienced by barnacles on rocky shores: The interplay between geographic variation, tidal cycles and microhabitat temperatures. Frontiers in Marine Science 7:553.

Waterman, T.H. 1960. The physiology of Crustacea, Volume 1. Academic Press New York, USA.

Watling, L. 2013. Feeding and digestive system. Pages 237–260 *in* L. Watling and M. Thiel, editors. The natural history of the Crustacea, Volume 1: Functional morphology and diversity. Oxford University Press, New York, USA.

Watling, L., and M. Thiel. 2015. The role of natural history in understanding the diversity of lifestyles in crustaceans. Pages 1–13 in M. Thiel and L. Watling, editors. The natural history of the Crustacea, Volume 2: Lifestyles and feeding biology. Oxford University Press, New York, USA.

Whiteley, N.M. 2011. Physiological and ecological responses of crustaceans to ocean acidification. Marine Ecology—Progress Series 430:257–271.

Whiteley, N.M., and E.W. Taylor. 2015. Responses to environmental stresses: Oxygen, temperature and pH. Pages 320–358 *in* E.S. Chang and M. Thiel, editors. The natural history of the Crustacea, Volume 4: Physiology. Oxford University Press, New York, USA.

Whiteley, N.M., E.W. Taylor, and A.J. El Haj. 1997. Seasonal and latitudinal adaptation to temperature in crustaceans. Journal of Thermal Biology 22:419–427.

Witte, S., C. Buschbaum, J.E.E. van Beusekom, and K. Reise. 2010. Does climatic warming explain why an introduced barnacle finally takes over after a lag of more than 50 years? Biological Invasions 12:3579–3589.

Wolf, F., K. Seebass, and C. Pansch. 2022. The role of recovery phases in mitigating the negative impacts of marine heatwaves on the sea star *Asterias rubens*. Frontiers in Marine Science 8:790241.

Wrange, A.L., C. Andre, T. Lundh, U. Lind, A. Blomberg, P.J. Jonsson, and J.N. Havenhand. 2014. Importance of plasticity and local adaptation for coping with changing salinity in coastal areas: A test case with barnacles in the Baltic Sea. BMC Evolutionary Biology 14:1–15.

Zeebe, R.E., A. Ridgwell, and J.C. Zachos. 2016. Anthropogenic carbon release rate unprecedented during the past 66 million years. Nature Geoscience 9:325–329.

Zenetos, A., S. Gofas, C. Morri, A. Rosso, D. Violanti, J.G. Raso, M.E. Çinar, A. Almogi-Labin, A.S. Ates, E. Azzurro, and E. Ballesteros. 2012. Alien species in the Mediterranean Sea by 2012. A contribution to the application of European Union's Marine Strategy Framework Directive (MSFD). Part 2. Introduction trends and pathways. Mediterranean Marine Science 13:328–352.

Fig. 1.3.
A male robber crab, *Birgus latro*, handling and killing a freshly caught red crab, *Gecarcoidea natalis*, on Christmas Island. Reproduced from Krieger et al. (2016), with permission from © Academia Sinica.

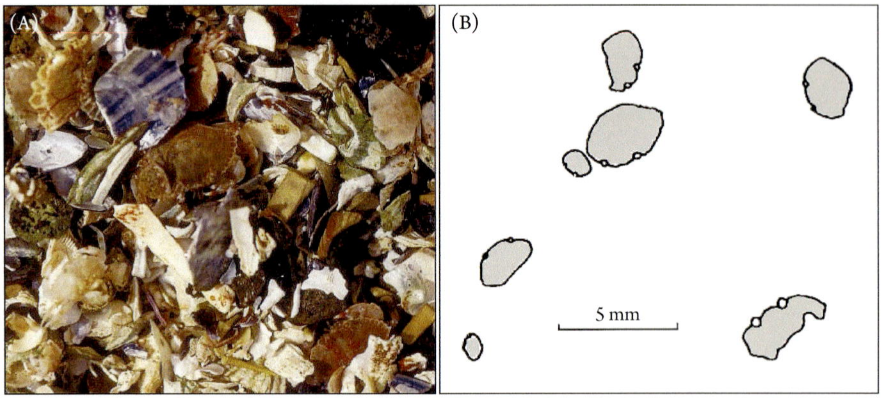

Fig. 1.5.
Juvenile rock crabs, *Cancer irroratus*, with cryptic coloration on shell hash. The cryptic individuals in (A) are outlined in (B). Adapted from Palma and Steneck (2001), with permission from © The American Society for Limnology and Oceanography.

Fig. 1.10.
Gray whale, *Eschrichtius robustus*, disturb sediments while foraging on benthic organisms. Tube-building ampeliscid amphipods are the major prey organisms of gray whales in the northern Bering Sea. Photo reproduced courtesy of Flip Nicklin, with permission from © Smithsonian Institution under Creative Commons Zero (CC0).

Fig. 2.1.
Male (bottom) and female (top) *Idotea balthica* on a blade of the brown alga *Fucus vesiculosus* taken through a microscope. Photo reproduced courtesy of Lars Gutow.

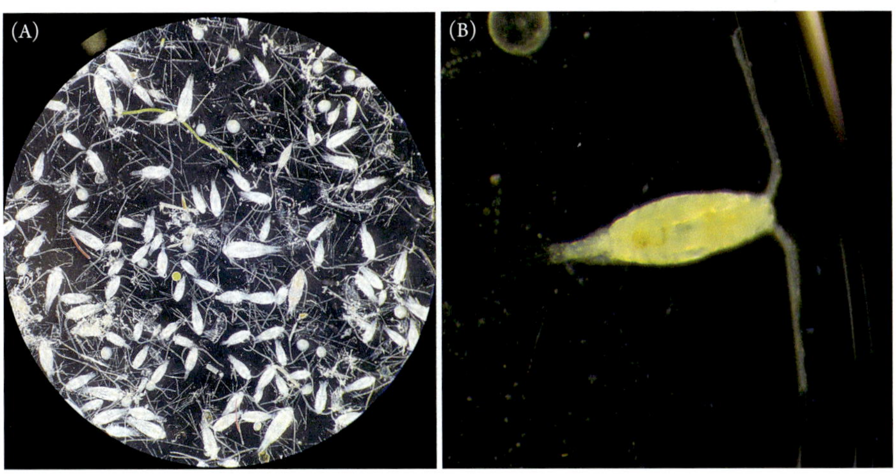

Fig. 2.2.
(A) Plankton tow sample under the microscope including a mixed copepod assemblage with many life stages, a long diatom chain in the upper middle, a single, round, centric diatom in the left center, and many pieces of microplastics; (B) Copepod *Acartia tonsa* on right side and centric diatom in upper left. Photos reproduced courtesy of Grace Saba.

Fig. 2.3. Spiny lobster, *Panulirus argus*, foraging in a red algae (*Laurencia* sp.) bed. Photo reproduced courtesy of Romuald Lipcius.

Fig. 2.4.
Adult blue crab (*Callinectes sapidus*) cannibalizing a young juvenile conspecific in a mesocosm with seagrass. Photo reproduced courtesy of Romuald Lipcius.

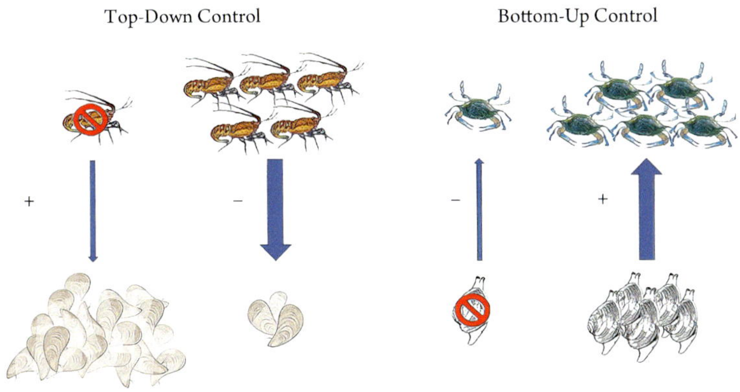

Fig. 2.5.
Diagram depicting top-down vs. bottom-up control in crustaceans. Example of top-down control by spiny lobsters (*Panulirus interruptus*) feeding on mussels (Robles et al. 1990), with left-most interaction showing few lobsters leading to many mussels (essentially a positive interaction), and second-from-left interaction showing many lobsters leading to few mussels (a negative interaction). Example of bottom-up control in blue crabs (*Callinectes sapidus*; Seitz et al. 2003), with second-from-right interaction showing few *Macoma* clams (*Macoma balthica*) leading to few crabs (essentially a negative interaction), and right-most interaction showing many clams leading to many crabs (a positive interaction). Figure reproduced courtesy of Taizhu Zhou.

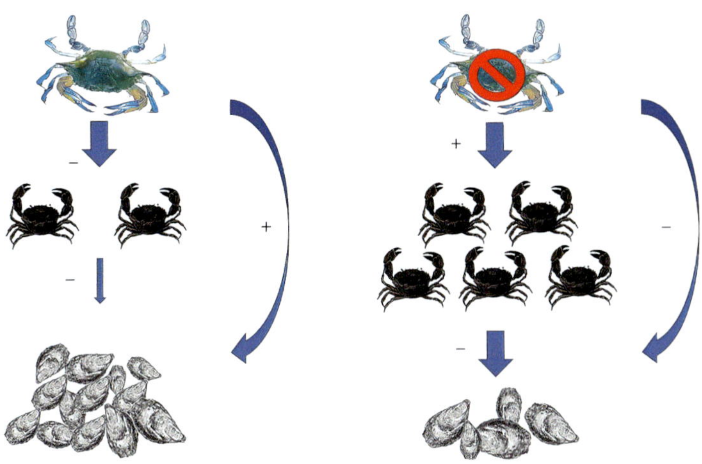

Fig. 2.7.
Trophic cascade involving blue crab (*Callinectes sapidus*), mud crab (*Panopeus herbstii*), and Eastern oyster (*Crassostrea virginica*) (Johnson et al. 2014). The left panel shows that when blue crabs are abundant, they greatly reduce mud crab densities (depicted with the negative sign next to the thick top arrow), which in turn exerts a minimal negative effect on oysters (depicted with the negative sign next to the thin bottom arrow), allowing oysters to proliferate in numerous, three-dimensional reefs (indirect positive effect depicted by the plus sign next to the curved arrow). In contrast, the right panel shows that in the absence of blue crabs, mud crabs occur in abundance (depicted by a positive sign next to the thick top arrow), and mud crabs exert a strong negative effect on oysters, resulting in fewer oysters (indirect negative curved arrow). Figure reproduced courtesy of Taizhu Zhou.

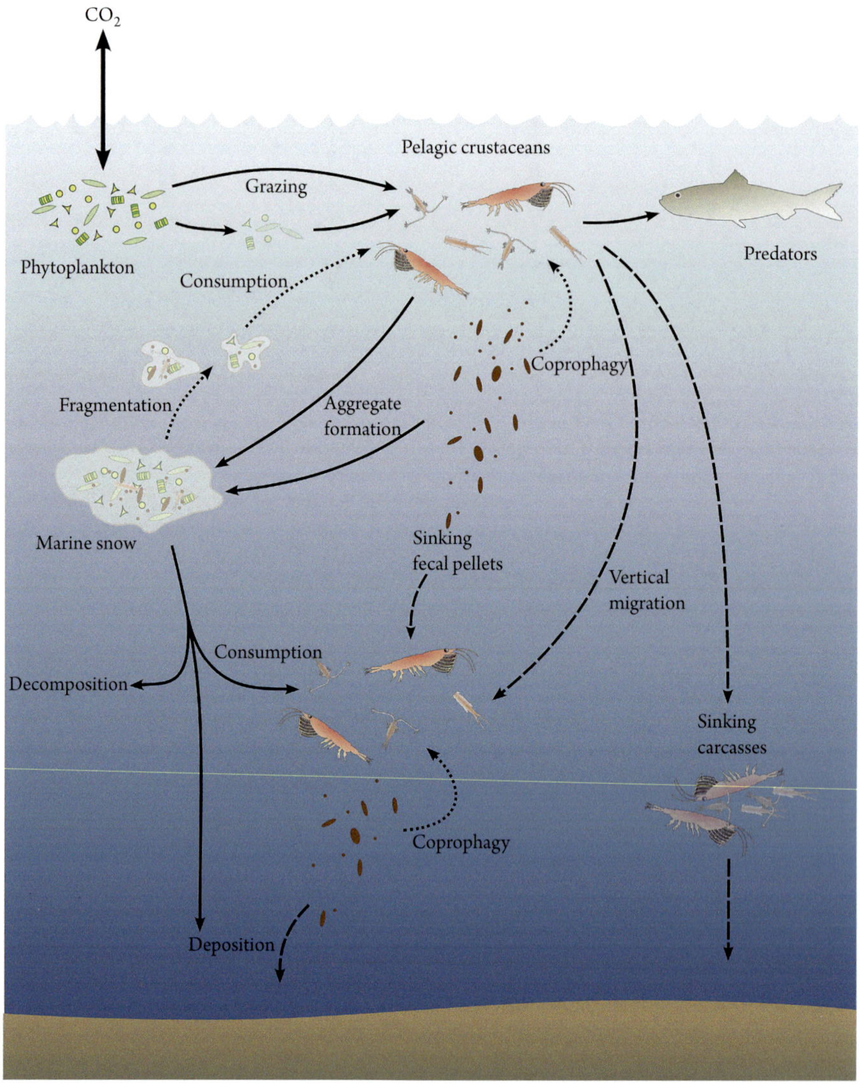

Fig. 3.6.
The role of pelagic crustaceans in oceanic carbon cycling (biological carbon pump). Crustaceans transfer carbon assimilated from atmospheric CO_2 to higher trophic levels by grazing on dead and live phytoplankton in the epipelagic zone. Production of fast-sinking fecal pellets and sinking carcasses of copepods and euphausiids, together with vertical migration, enhance the flux of detrital organic matter to depth (dashed arrows). At the same time, fragmentation and consumption of marine snow aggregates and fecal pellets by pelagic crustaceans attenuate the flux of detrital organic matter to depth (dotted arrows). Source: data from Turner (2015), inspired by the descriptions of Steinberg and Landry (2017).

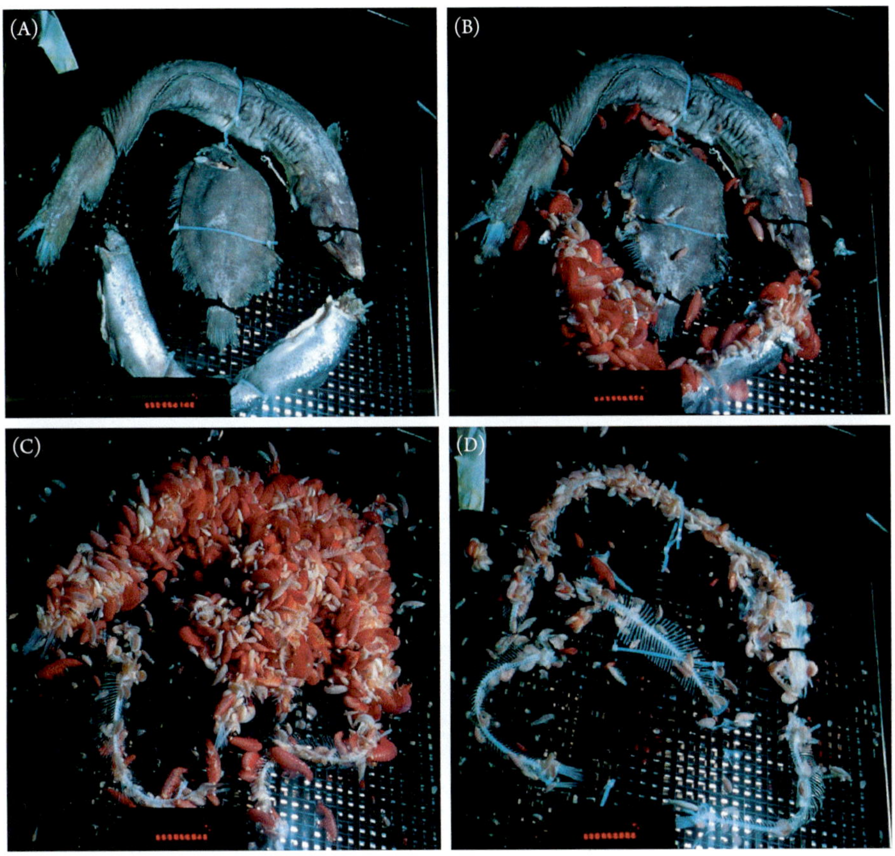

Fig. 3.7.
Processing of an experimental fish fall in the deep Arctic Ocean by aggregations of scavenging amphipods. The photographs show the state of the fish carcasses after (A) 0, (B) 6, (C) 10, and (D) 22.5 h of deployment, amphipods eventually cover entire carcasses (B, C). Reproduced from Premke et al. (2006), with permission from © Inter-Research Science Publisher.

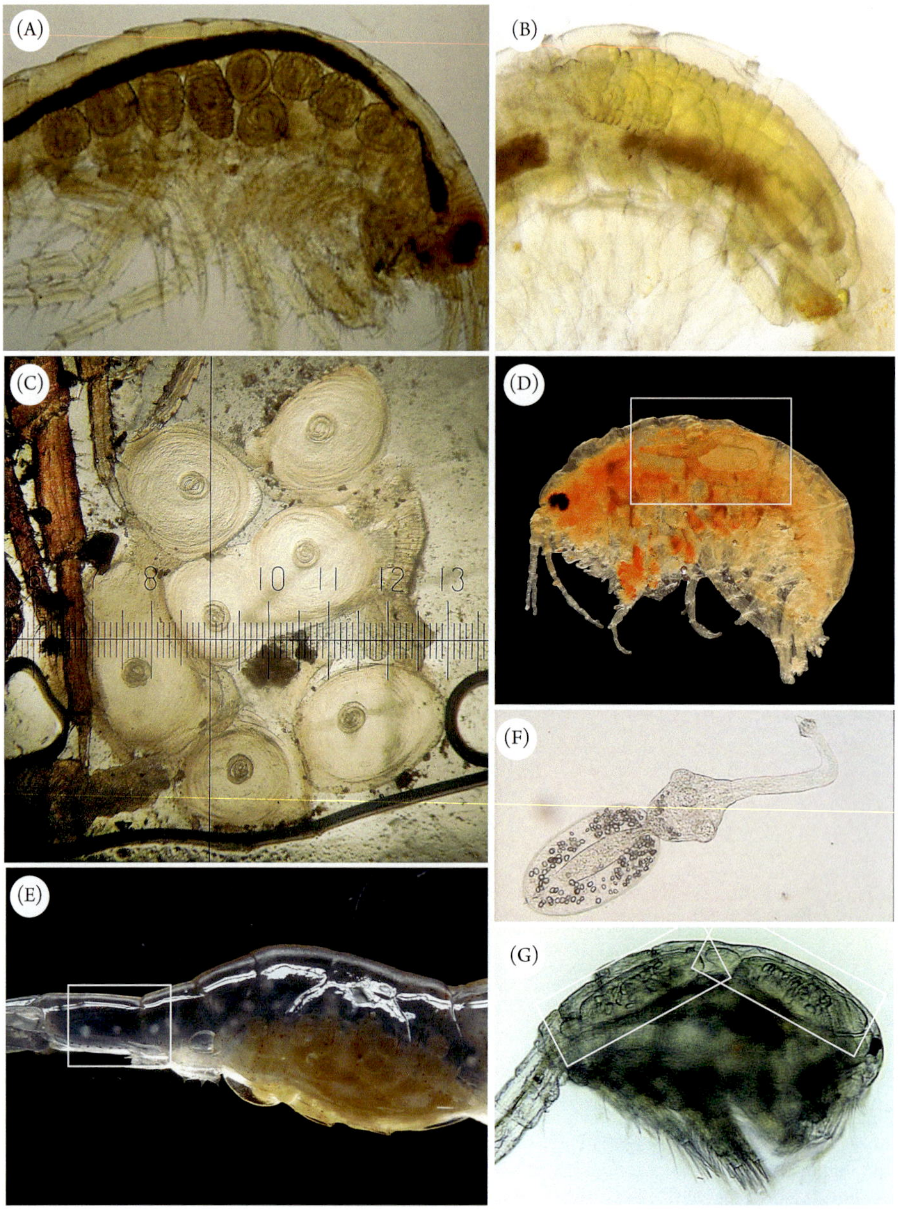

Fig. 5.1.
Various helminths in situ within their crustacean intermediate host. (A) Nine cystacanths of the acanthocephalan *Pomphorhynchus laevis* in the amphipod *Echinogammarus stammeri*, from Italy. Photo reproduced courtesy of B. Dezfuli. (B) Large cystacanth of the acanthocephalan *Acanthocephalus rhinensis* in the amphipod *Echinogammarus tibaldi*, from Italy. Photo reproduced courtesy of B. Dezfuli. (C) Encysted larvae of the nematode *Antechiniella septentrionalis* in the body cavity of the amphipod *Traskorchestia ditmari*, from Russia. Photo reproduced courtesy of N. Dokuchaev and E. Ivanova. (D) Juvenile nematode *Hedruris dratini* in the body cavity of the amphipod *Hyalella bonariensis*, from Argentina. Photo reproduced courtesy of E. Palumbo. (E) Metacercariae of the trematode *Microphallus szidati* (seen as white dots) in the body cavity of the freshwater shrimp *Palaemonetes argentinus*, from Argentina. Photo reproduced courtesy of M. Merlo. (F) Excysted cysticercoid of the cestode *Eurycestus avoceti* taken from the tissues of the brine shrimp *Artemia parthenogenetica*, from Spain. Photo reproduced courtesy of B. Georgiev and M. Sánchez. (G) Two procercoids of the cestode *Schistocephalus solidus* in the body cavity of the copepod *Macrocyclops albidus*, from Germany. Photo reproduced courtesy of D. Benesh.

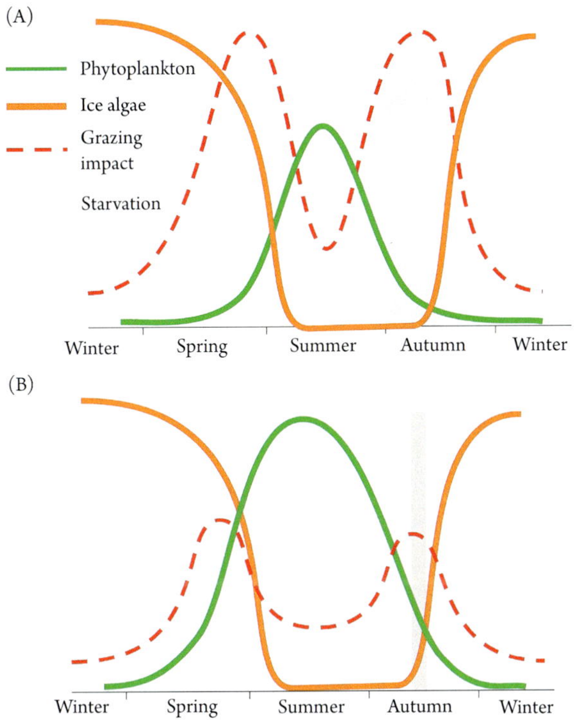

Fig. 6.2.
Graphical representation of seasonal changes in phytoplankton (green), ice algae (orange) and krill grazing impact (dashed line) during periods of (A) large and (B) small krill biomass. When the total krill biomass is large, autumn becomes the main bottleneck for population survival (indicated by the starvation period (gray bars)). Reproduced from Ryabov et al. (2017), with permission from © Springer Nature.

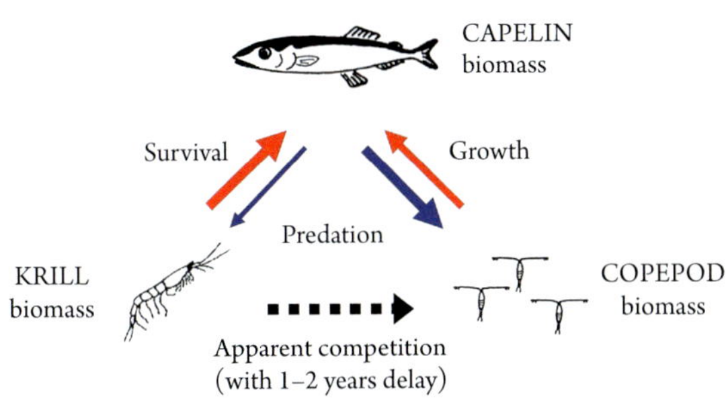

Fig. 6.6.
Schematic representation of apparent competition between krill and copepods in the Barents Sea mediated by their sharing of a predator, the capelin, based on inferences gained from state space statistical modeling of a 30+ year time series. Bottom-up (red) relationships between prey species and the capelin led to top-down control (blue). Arrow widths are approximately proportional to estimated effect sizes. Reproduced from Stige et al. (2018), with permission from © John Wiley & Sons.

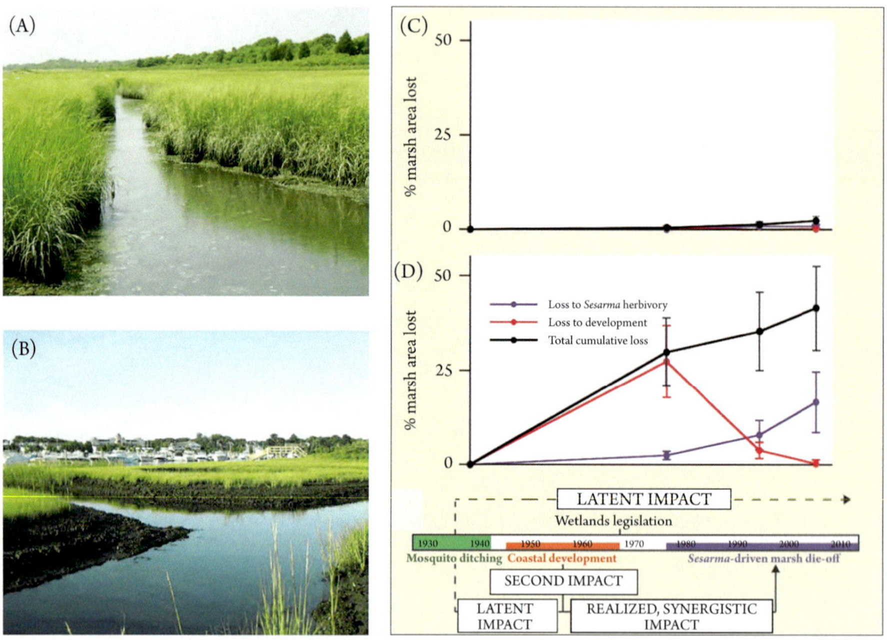

Fig. 8.3.
(A) Healthy *Spartina alterniflora* salt marsh, showing beneficial edge habitat formed by a tidal creek. (B) Marsh bank degradation due to herbivory by the crab *Sesarma reticulatum*. Both photographs from Cape Cod, Massachusetts, USA, reproduced from Altieri et al. (2012), with permission from © John Wiley & Sons. Also shown are time series of salt marsh habitat loss in (C) non-developed marshes, and (D) developed marshes in Cape Cod. Development was the chief agent of marsh loss until the mid-1970s, after which *Sesarma* herbivory accelerated due to overfishing of large predators. Reproduced from Coverdale et al. (2013), with permission from © John Wiley & Sons.

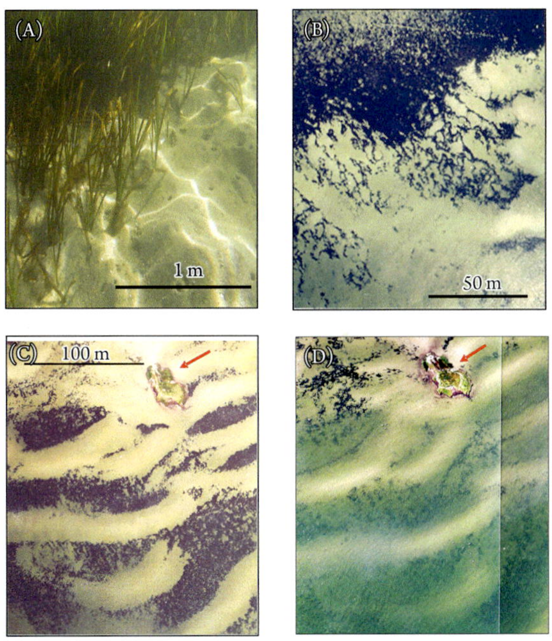

Fig. 8.5.
(A) Eelgrass (*Zostera marina*) shoots at a patch edge in sandy substratum at a depth of approximately 1.5 m in Mission Bay, California, USA. Small holes in sand are burrows of ghost shrimp *Neotrypaea californiensis*, which contributes to eelgrass patchiness via bioturbation. (B–D) Aerial photos of submerged eelgrass habitat in Chesapeake Bay, Virginia, USA, demonstrating different seascape configurations: (B) fragmented eelgrass along the edge of a continuous meadow; (C) large, isolated patches of eelgrass formed by current scouring; (D) the same seascape as shown in (C) after degradation (reductions in shoot density and biomass). Arrows point to a small island for orientation. Photos reproduced courtesy of K. Hovel.

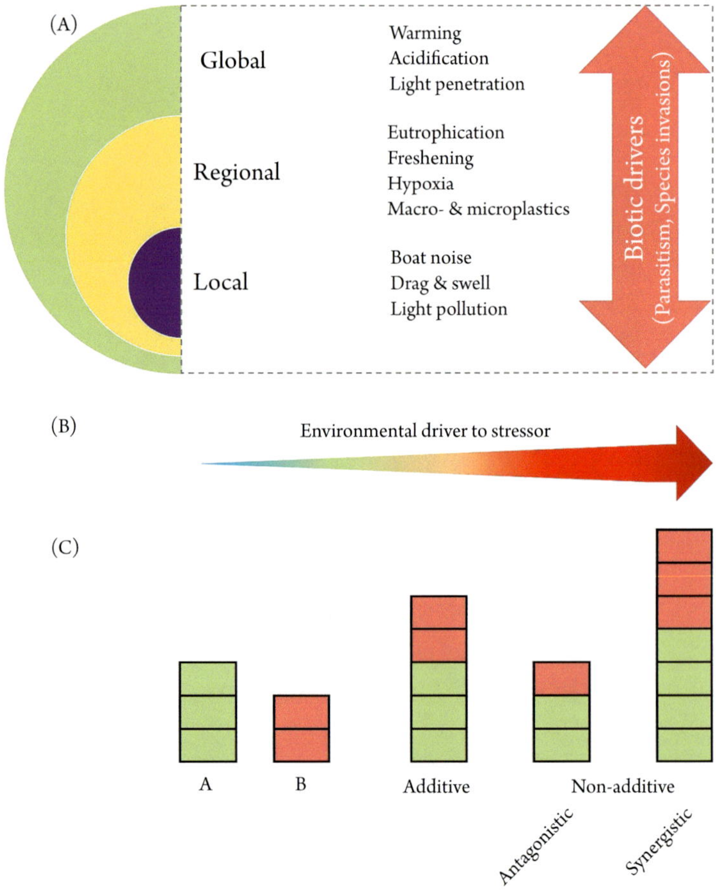

Fig. 9.1.
(A) Multiple abiotic drivers of ocean change that may be scaled from global to regional and local, and which may be influenced by the presence or absence of multiple biotic drivers. (B) Any environmental driver may turn into a stressor once tolerance limits of species and ecosystems are exceeded. (C) Abiotic and biotic drivers can act in an additive or non-additive (antagonistic, synergistic) way (exemplified as the interaction of one global environmental and one biotic driver).

Fig. 9.3.
Specimen of the vent-endemic crab *Xenograpsus testudinatus* thriving under extreme environmental hypercapnia in a shallow-water hydrothermal vent system off Kueishan Island (ROC Taiwan). Maps were illustrated in R, using the packages "ggOceanMapsData" and "ggOceanMaps." Photos reproduced courtesy of Dr. M.-S. Jeng.

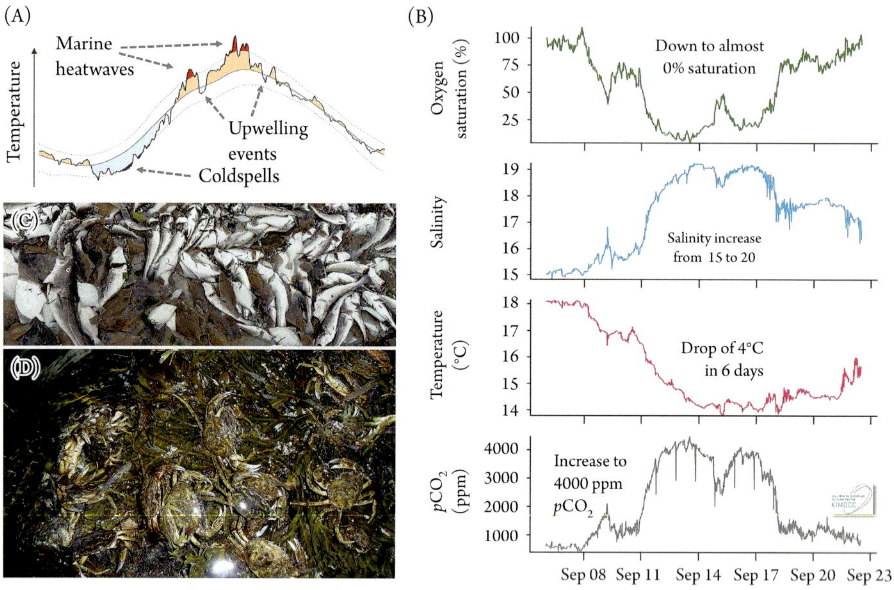

Fig. 9.4.
(A) Temperature course and marine heatwaves (dark shading above confidence interval) as recorded in the Western Baltic Sea Kiel Fjord in the year 2018 may be interrupted by (hypoxic) upwelling events, occasionally turning into cold spells (dark shading below confidence interval) of extremely low temperatures (F. Wolf, GEOMAR). (B) Seawater oxygen saturation, salinity, temperature, and pCO_2 were recorded at one meter depth in Kiel Fjord between September 6 and September 22, 2017, following an early-autumn upwelling event (plot in collaboration with M. Wall, GEOMAR, and C. Hiebenthal at KIMOCC GEOMAR; available at PANGAEA [https://doi.org/10.1594/PANGAEA.930979]). (C) Accumulations of dead fish washed ashore in September 2017 following the event. (D) *Carcinus maenas*, *Crangon crangon*, and fish aggregate on a stone in oxygenated surface water, inhabited by the bladderwrack *Fucus vesiculosus*, during a nocturnal upwelling event of hypoxic seawater in September 2020. Photos reproduced courtesy of Ivo Bobsien.

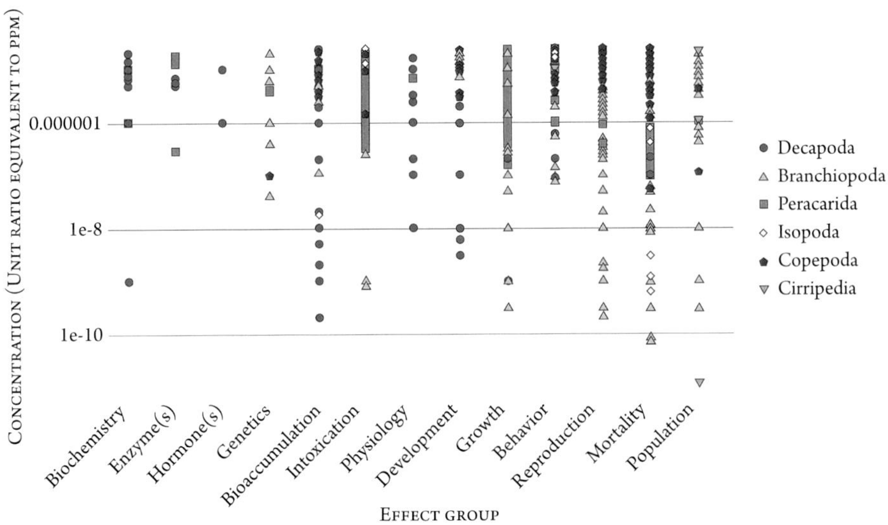

Fig. 11.1.
Articles relating to crustacean ecotoxicology in the US EPA ECOTOX Knowledgebase ordered by effect recorded and taxa. Graph represents ~900 species and more than 80,000 plottable records. Source: data from https://cfpub.epa.gov/ecotox/index.cfm (Oct 2023).

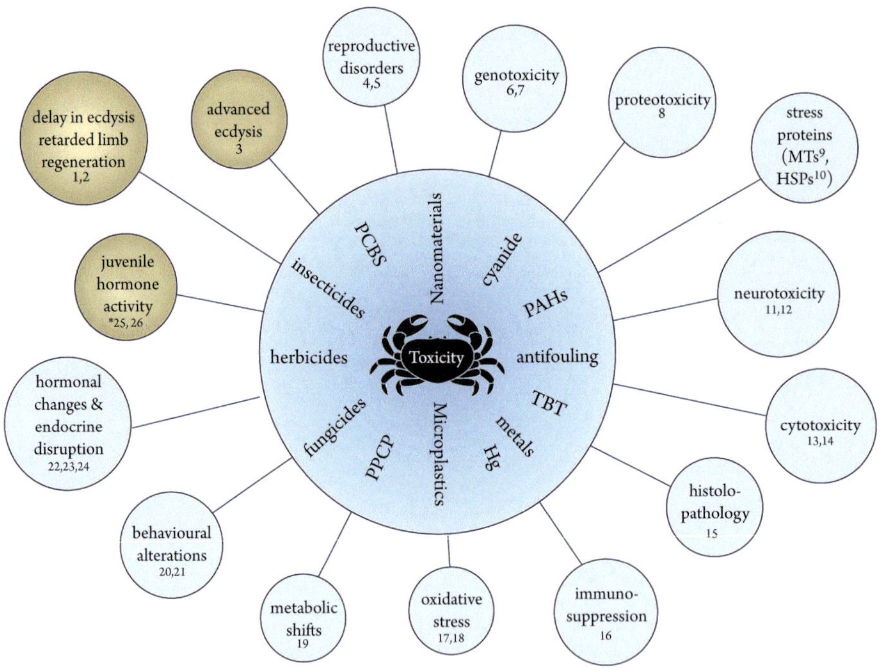

Fig. 11.3.

A non-exhaustive representation of classes of contaminants and physiological responses by crustaceans to them. General responses in brown, arthropod-specific responses in blue.(1 – Weis et al. 1987; 2 – Weis et al. 1992; 3 – Song et al. 2017; 4 – Sundelin and Eriksson 1998; 5 – Siegel and Wenner 1984; 6 – Klobučar et al. 2012; 7 – Pinheiro et al. 2013; 8 – Eckwert et al. 1997; 9 – Correi et al. 2002; 10 – Arts et al. 2004; 11 – Steevens and Benson 2000; 12 – Ben Kheder et al. 2017; 13 – Schultz and Kennedy 1976; 14 – Dell'Anno et al. 2008; 15 – Soegianto et al. 1999; 16 – Lorenzon et al. 2001; 17 – Lavarías et al. 2011; 18 – Ghedira et al. 2011; 19 – Bao et al. 2020; 20 – Chabenat et al. 2021; 21 – Krång and Dahlström 2006; 22 – Rodríguez et al. 2007; 23 – LeBlanc et al. 2007; 24 – Parmentier et al. 2019; 25 – Mu and LeBlanc 2004; 26 – Jaegers and Gismondi 2020). * that is, "Crustacean Juvenile Hormone" Methyl Farnesoate (Homola and Chang 1997).

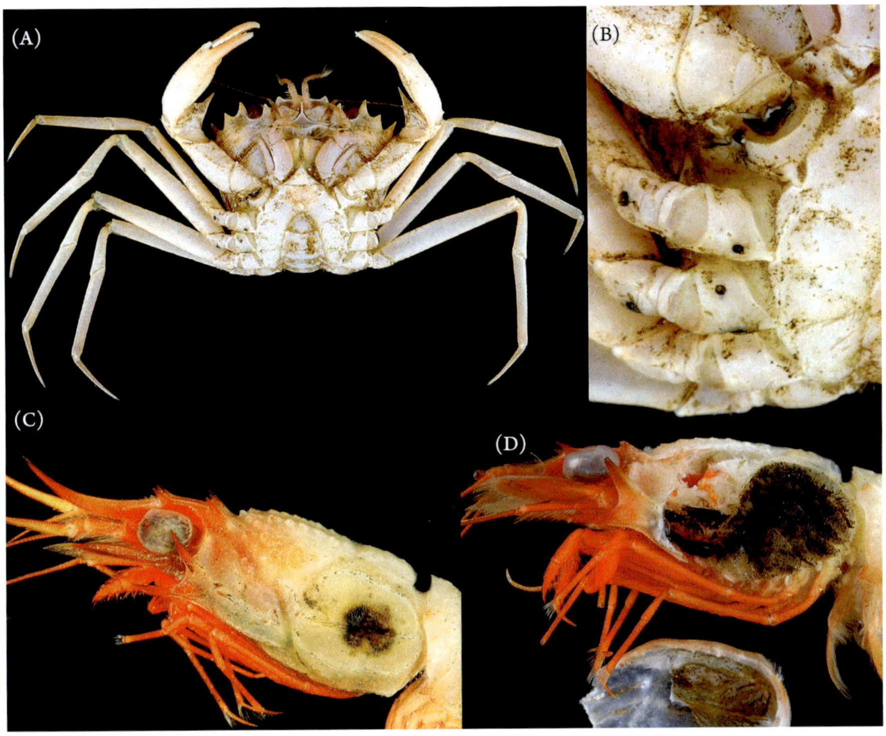

Fig. 11.5.
(A, B) Several dark lesions can be recognized on a deep-sea red crab, *Chaceon quinquedens*, collected about one and a half years after the Deepwater Horizon oil spill on a deep offshore bank in the Gulf of Mexico 49 km east of the spill site. (C, D) A lesion of the gill cavity wall (C) and gills covered by manganous precipitate (D) can be recognized on deep-sea shrimps, *Glyphocrangon longirostris*, collected one year after the Deepwater Horizon oil spill on a deep offshore bank in the Gulf of Mexico 110 km west–southwest of the spill site. Reproduced from Felder et al. (2014), with permission from © Oxford University Press.

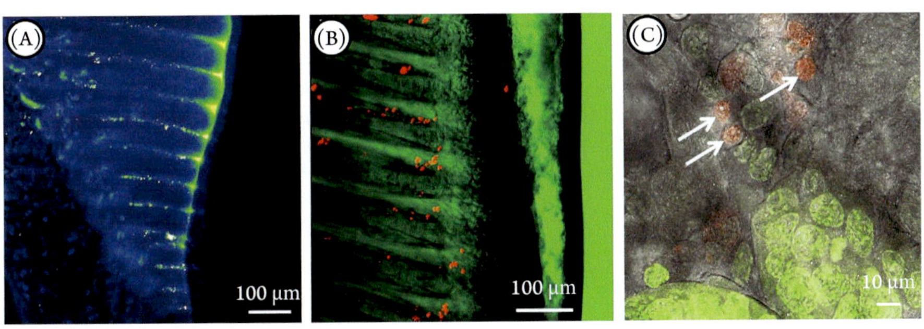

Fig. 11.6.
(A, B) Polystyrene microspheres (8 µm, respectively) visualized by Raman scattering as light shading (A) and dark dots (B) on the gill lamellae of the European shore crab, *Carcinus maenas*; scale bars = 100 µm. (C) Polystyrene microbeads emitting red fluorescence (0.1 µm, arrows) in the digestive gland of the sand shrimp, *Crangon crangon*. (A, B) reproduced from Watts et al. (2016), with permission from © The American Chemical Society; (C) reproduced from Korez et al. (2020), with permission from © Elsevier under Creative Commons license (CC BY).

Fig. 11.7.
(A) Accumulated plastic debris creates an obstacle for crabs on the beaches of the Cocos (Keeling) Islands. (B) A hermit crab inside a green bucket along the high tide line of South Island. (C) Accumulated plastic debris in the beach-back vegetation on West Island. (D) Crabs that became trapped and died inside a plastic drink bottle that washed up on Cocos. Reproduced from Lavers et al. (2020), with permission from © Elsevier.

Fig. 12.1.
Cave crayfish (*Procambarus orcinus* Hobbs and Means, 1972) showing distinctive cave adaptations, including loss of pigment (you can see through the carapace to the internal features), elongated sensory (see setae and antennules), and loss of eye (there is still an eye stalk as the crayfish molting mechanism is located there, but the eye is gone). Photo reproduced courtesy of C. Riley Nelson.

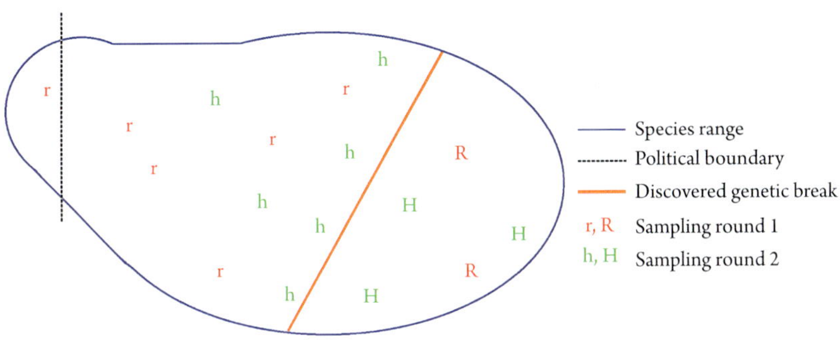

Fig. 12.2.
Sampling strategies (r's, h's for multiple rounds of sampling) for conservation genetics relative to political (dotted black line) and population subdivision (solid orange line) boundaries across a species' range (blue line). Capital letters represent one genetic subpopulation and lower case letters represent another.

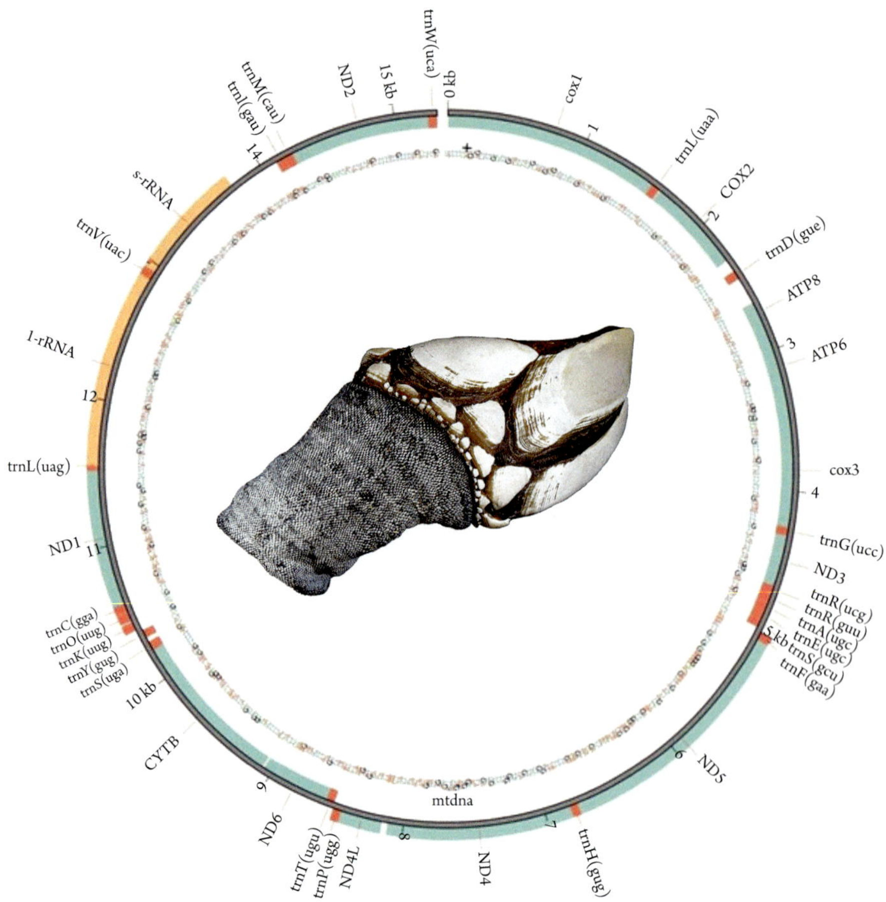

Fig. 12.4.
Mitochondrial genome (GenBank Accession CM029732) of *Pollicipes pollicipes*, the gooseneck barnacle, generated from a whole genome assembly GenBank Ref Seq: GCA_011947565.2 (Bernot et al. 2022). Transfer RNAs shown in pink, coding genes shown in green, and ribosomal subunits shown in yellow and labelled as rRNA, with an image of the species as the centerpiece (reproduced courtesy of Niklas Dreyer). Mitochondrial genome reproduced courtesy of Anton Zamyatin and Pavel Avdeyev.

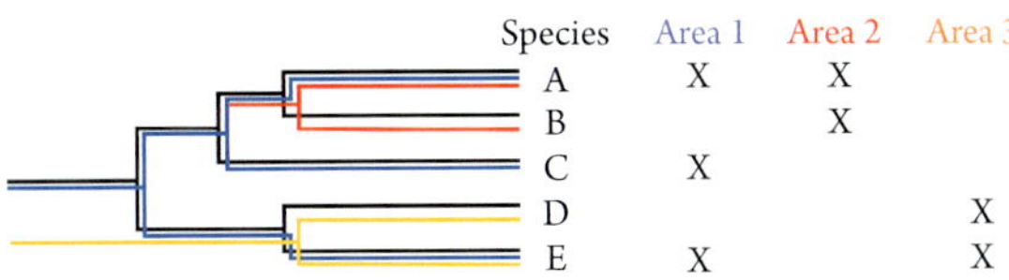

Fig. 12.5.
Phylogenetic diversity (PD) calculates the sum of branches leading to a given taxon and then the PD associated with a given area where that taxon is found.

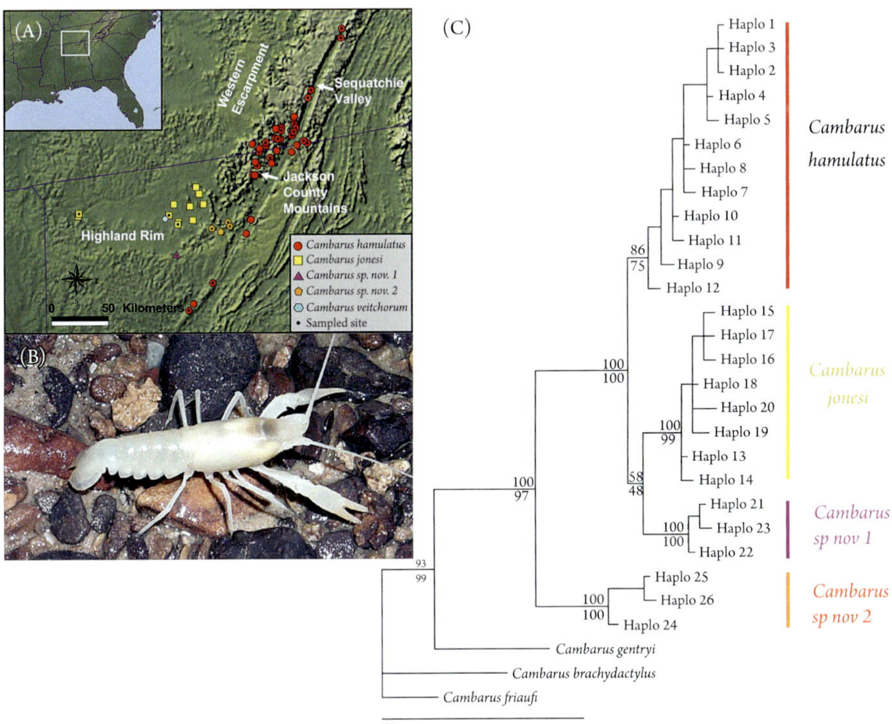

Fig. 12.6.
Geographically intensive and iterative sampling (B) of cave crayfish (A) from Appalachia allowed for an extensive and informative phylogenetic analysis (C) that resulted in the discovery of two new species and an effective conservation assessment of all four target species. Part (A) reproduced from Buhay et al. (2007), with permission from © Elsevier; parts (B) and (C) from Buhay and Crandall (2009), with permission from © Oxford University Press.

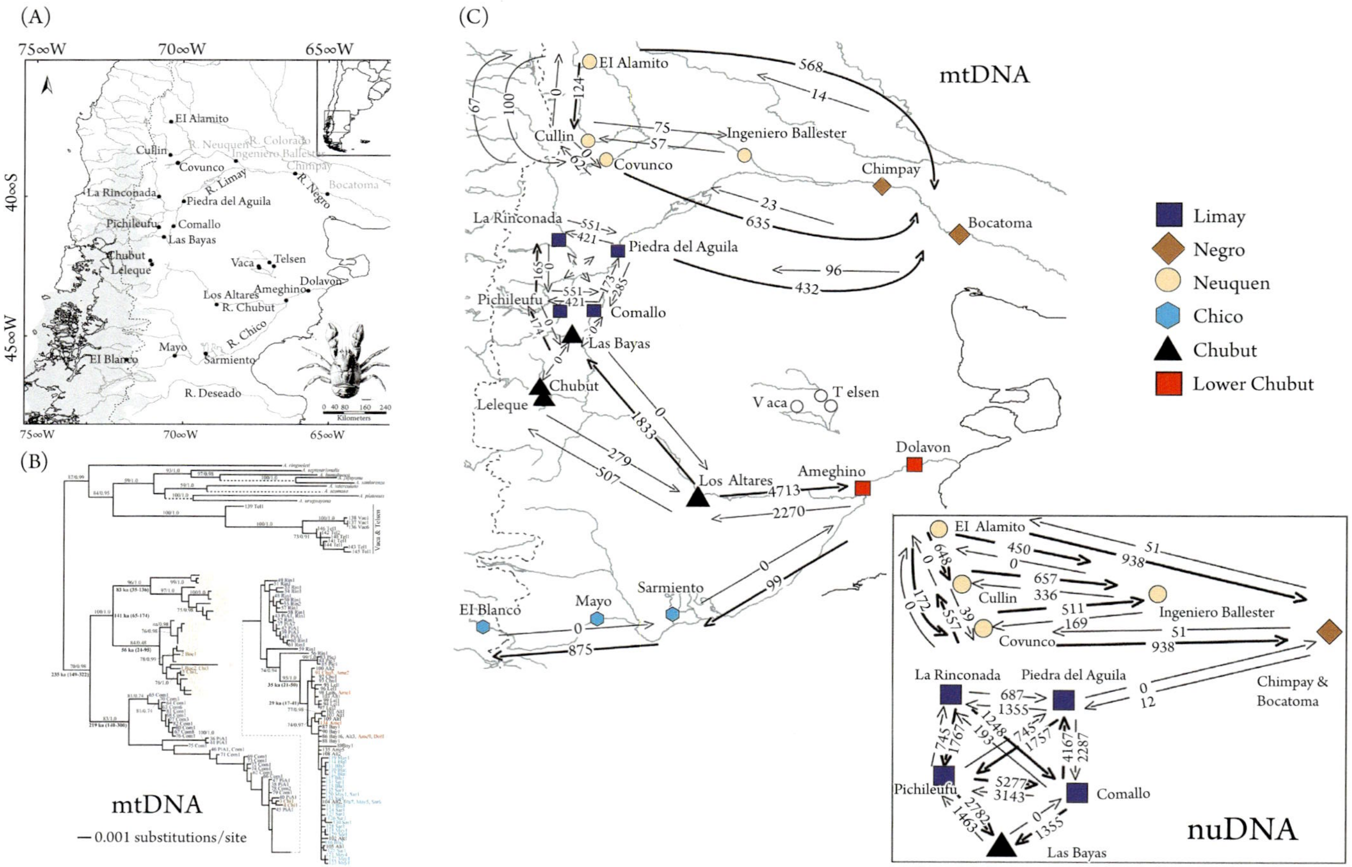

Fig. 12.7.
Phylogeographic study of *Aegla neuquensis* (A) using both mitochondrial (mtDNA) and nuclear (nuDNA) markers to estimate phylogenetic relationships among unique sequences (B) and different gene flow patterns for the mtDNA and nuDNA suggesting introgression (C). Reproduced from Barber et al. (2012) under a Creative Commons by Attribution License (CC BY).

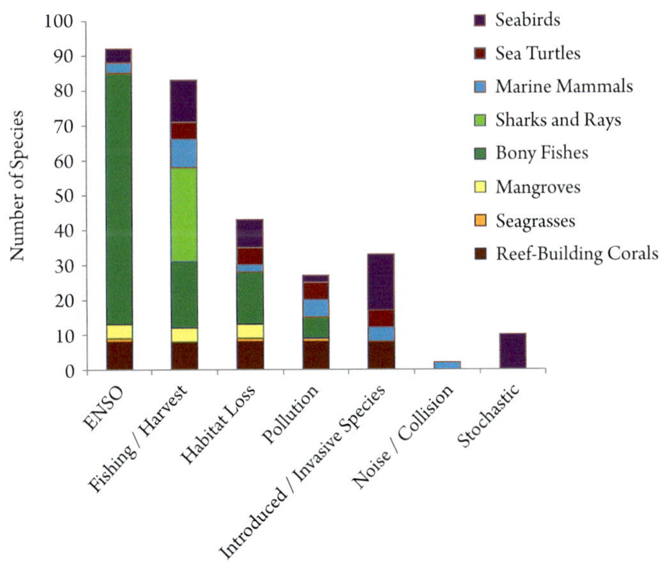

Fig. 13.3.
Major categories of threats to species listed in several threatened taxonomic categories in the Eastern Tropical Pacific including ENSO (El Niño–Southern Oscillation). Introduced species represent a significant risk to a wide range of threatened species, but are not the most important. Adapted from Polidoro et al. (2012), with permission from © Inter-Research Science Publishers.

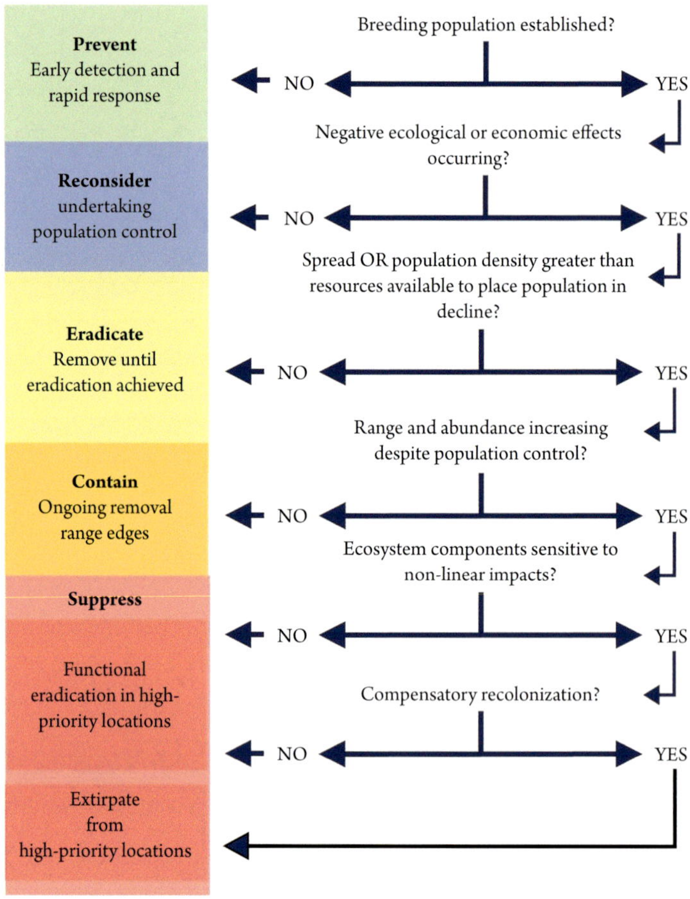

Fig. 13.5.
Detailed decision tree describing recommendations for invasive crustacean management given information about the current status of the invasion, the severity of impacts, magnitude of establishment and spread, available resources, ecosystem responses to invader density, population connectivity, likelihood of compensatory reproduction, and related attributes. Reproduced from Green and Grosholz (2021), with permission from © John Wiley & Sons.

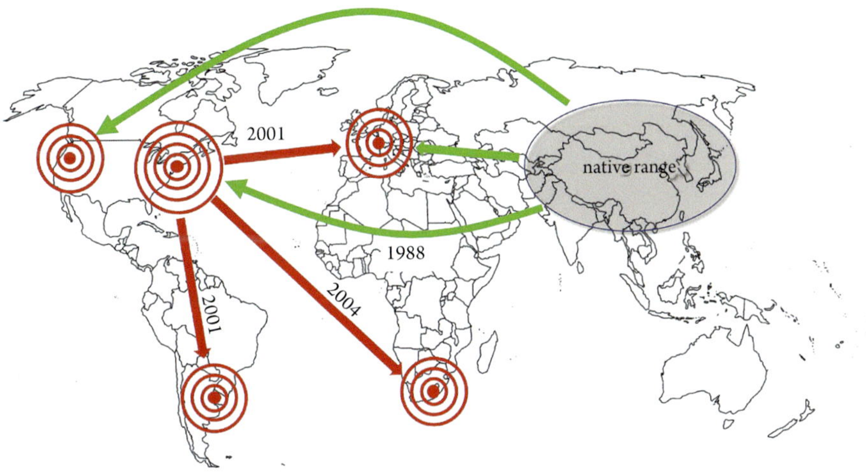

Fig. 13.6.
A map of the spread of an insect pest, the harlequin ladybird beetle *Harmonia axyridis* as an illustration of the "bridgehead effect" that can form the basis for new approaches to manage introductions via shipping vectors. The original pathways for the intentional introductions for biocontrol purposes (green arrows) show insects moved from the native range in Asia to Europe and North America. The subsequent unintentional invasions to South America, South Africa, and back to Europe (red arrows) demonstrate that the primary source of this global spread originated from the introduced populations in Eastern North America, and not the native Asian range. Thus, Eastern North America continues to function as a bridgehead region; therefore, this region should be the future focus of cooperative international actions. Reproduced from Ricciardi et al. (2021) under a Creative Commons License (CC BY).

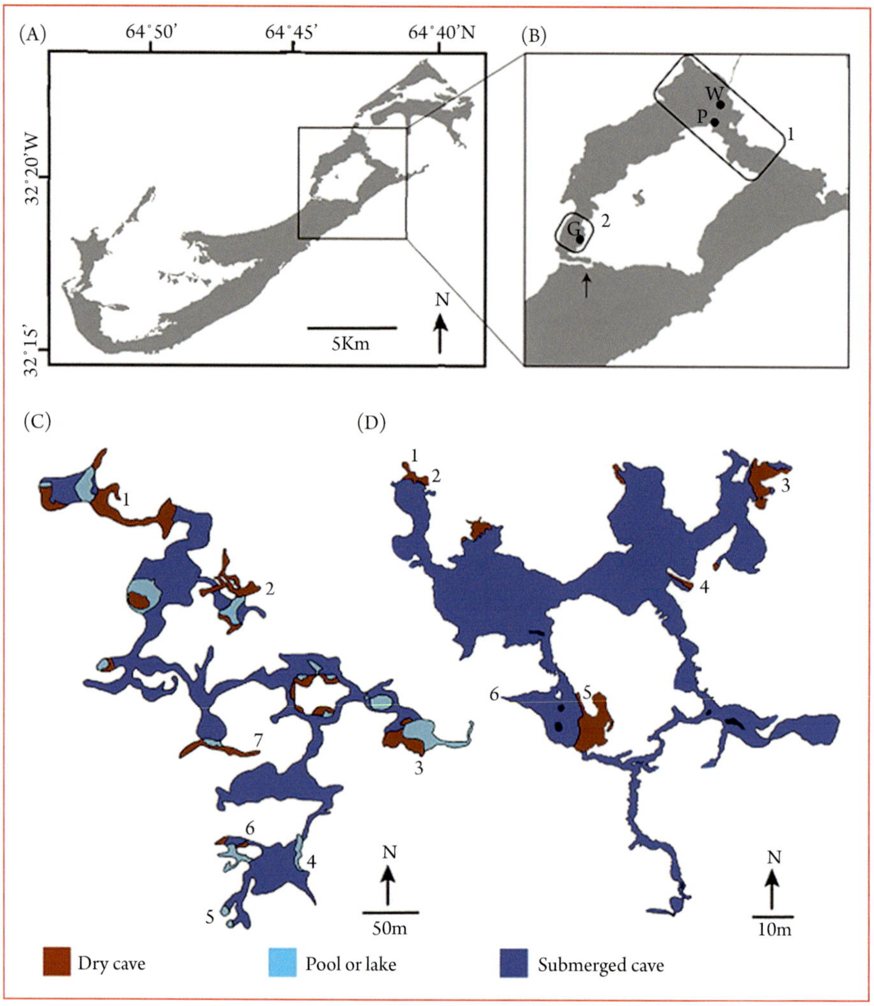

Fig. 14.6.
(A) Map of Bermuda in the North Atlantic Ocean. (B) Cave regions on opposite sides of Harrington Sound, a large inland body of water mostly encircled by land. The arrow indicates the only opening to the sea at Flatt's Inlet; rectangles show cave and karst regions in (1) the Walsingham Tract with the Walsingham Cave (W) and Palm Cave (P), and (2) Green Bay Cave System with the Green Bay Cave (G). (C) Map of Walsingham Cave System with cave entrances: (1) Wonderland, (2) Crystal, (3) Walsingham, (4) Deep Blue, (5) Vine, (6) Old Horse, (7) Fern Sink (modified from map by Robert Power). (D) Map of Palm Cave System with cave entrances: (1) Palm Slit, (2) Palm, (3) Strawmarket, (4) Sailor's Choice, (5) Myrtle Bank, (6) Cripple Gate. Modified from map by Jason Richards. Reproduced from Iliffe and Calderón-Gutiérrez (2021), with permission from © MDPI under Creative Commons by Attribution License (CC BY 4.0).

Fig. 14.7.
UW photo of divers and large speleothems in the Walsingham Cave System. Photo reproduced courtesy of Jill Heinerth.

Fig. 14.9.
(Top) Rock mound containing remnants of Wilkinson Quarry Cave at the center of the quarry floor (circa August 2017). (Bottom) After removal of the rock mound (May 2023). Photos reproduced courtesy of Thomas Iliffe.

Fig. 14.10.
Photo of century old, rusting bitumen barrels from a cave in the Walsingham Tract. Photo reproduced courtesy of Thomas Iliffe.

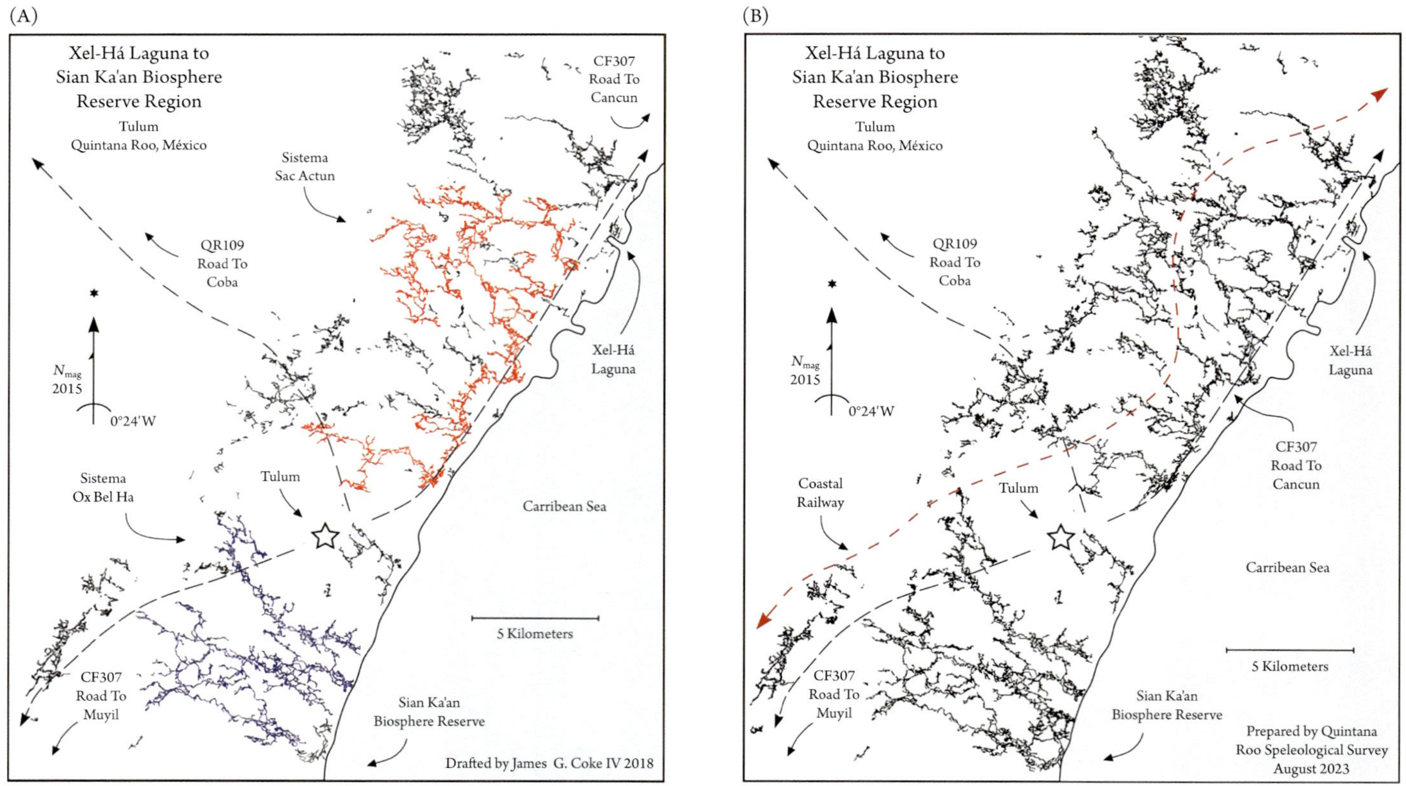

Fig. 14.13.
The Caribbean coast of Mexico's Yucatán Península in the state of Quinatana Roo has the world's longest underwater caves. (A) Sistema Sac Actun (red lines in upper center) is the world's longest submerged cave with a total extension of 372 km, while Sistema Ox Bel Ha (blue lines in lower left) is second at 319 km. Map from 2018 reproduced courtesy of James G. Coke IV. (B) The Tren Maya, a newly constructed coastal railway, raises concerns regarding conservation and protection of the underlying caves and anchialine aquifer. August 2023 map reproduced courtesy of James G. Coke IV.

Fig. 14.14.
Xibalbanus tulumensis is one of four species from the exclusively stygobiont class Remipedia in the Yucatán Península. Photo reproduced courtesy of Simon Richards.

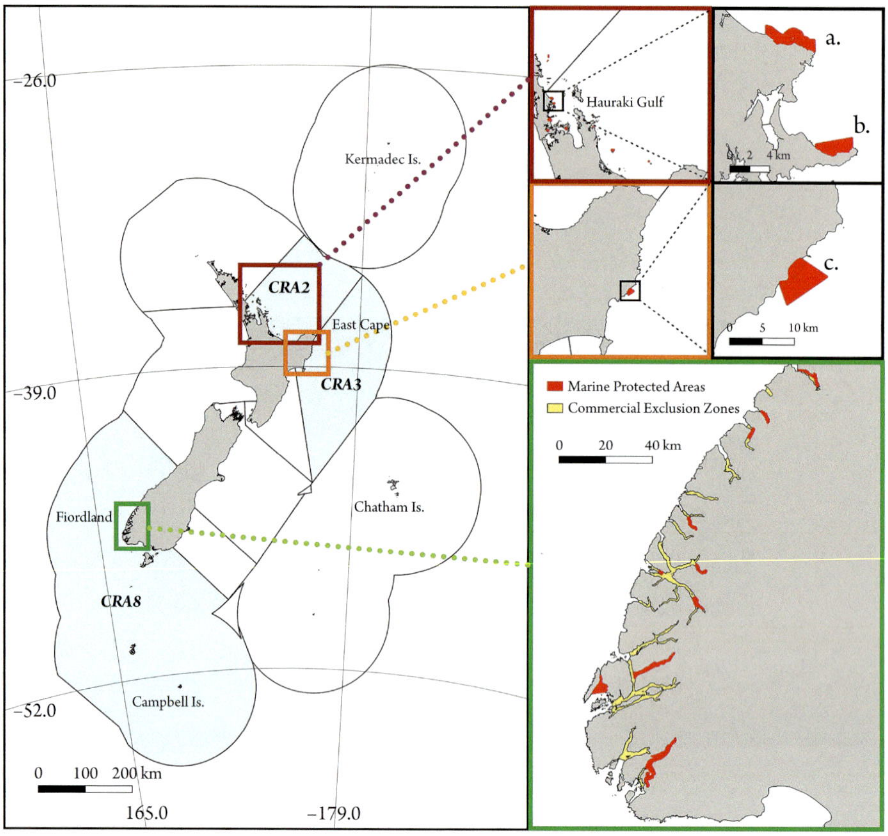

Fig. 15.2.
Key New Zealand red rock lobster regions described in this chapter: Northern New Zealand (upper box; *CRA2*), East Cape (central box; *CRA3*), and Fiordland (lower box; *CRA8*) quota management areas and marine protected areas: a. Cape Rodney to Okakari Point Marine Reserve, b. Tawharanui Marine Reserve, c. Te Tapuwae o Rongakako Marine Reserve. The Fiordland region incorporates two levels of protection: full protected marine protected areas (red shading) and commercial exclusion zones (yellow shading) where recreational fishing is still permitted.

Fig. 15.6.
Big Island of the Columbretes archipelago and *Palinurus elephas* in the Columbretes Islands marine reserve. Upper photo reproduced courtesy of Ben Stobart, SARDI, Australia; lower photo reproduced courtesy of David Diaz, CN-IEO-CSIC, Spain.

Fig. 15.7.
Part of a Roman mosaic from Pompeii (house n. 16, insula 2, Regio VIII; first century CE) that described predator–prey interactions between an octopus, a spiny lobster, and a moray eel (Soprintendenza Speciale per i Beni Archeologici di Napoli e Pompei).

Fig. 15.8.
(A) Map of the limits of the Columbretes Islands marine reserve (CIMR). Dots indicate locations where the fishery operated after the creation of the CIMR in 1990 (inner box, data from 2000–2008 shown). The CIMR was expanded north- and southward in 2008 (outer limits). Blue polygons are islands within the CIMR. Gray contour lines denote the 10–120 m isobaths. (B) Main island with lighthouse. Insert in (B): location of the CIMR (red circle) in the Western Mediterranean (© Google Earth, under fair use agreement).

Fig. 15.10.
European lobster *Homarus gammarus* (Linnaeus, 1758) with fresh exoskeleton and bright white denticles is a highly conspicuous animal when encountered outside its shelter. Photo reproduced courtesy of Erling Svensen.

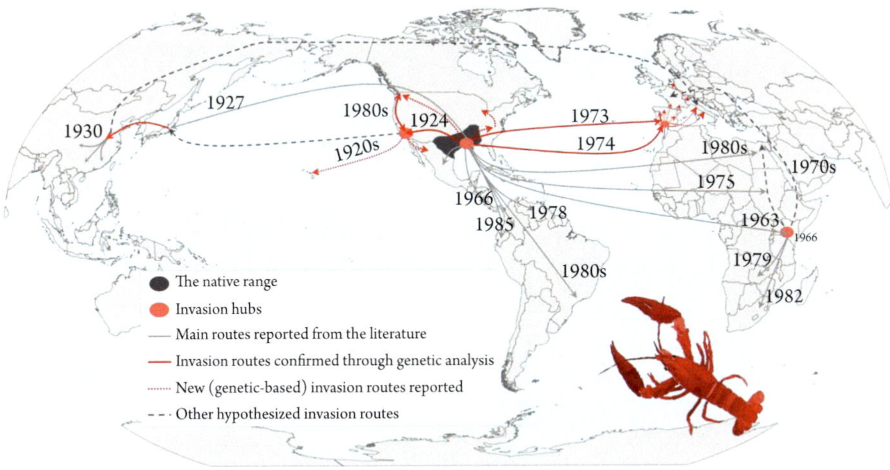

Fig. 16.1.
The global invasion routes of the red swamp crayfish, *Procambarus clarkii*, native of southern USA and north-eastern Mexico. Relevant invasion hubs, which usually function as recipients and sources of new invasions, are shown as red circles: Louisiana (in the native range), California, Kenya, and Spain. Reproduced from Oficialdegui et al. (2019), with permission from © John Wiley & Sons.

Fig. 16.2.
Two anostracan species protected under the USA Endangered Species Act. (A) *Branchinecta lynchi*; (B) *Branchinecta sandiegonensis*. Photos courtesy of ©Joel Sartore/ National Geographic Photo Ark.

Fig. 16.3.
(A) *Thaipotamon chulabhorn*, named in honor of the beloved Thai Princess Chulabhorn, and therefore protected under Thailand law; photo reproduced courtesy of the author. (B) *Lepidurus packardi* protected under the USA Endangered Species Act. Photo reproduced courtesy of J. Olesen.

Fig. 17.1.

(A–F) Selected *Bathynomus* models and figurines from Japan: (A) *B. giganteus* A. Milne Edwards, 1879, life size (Kitan Club). (B) *B. giganteus*, with fully articulated body parts (Bandai). (C) *B. doederleini* Ortmann, 1894 (Kaiyodo). (D–E) *B. giganteus* (Kaiyodo). (F) *B. giganteus* (T-Arts). (G–I) T-shirt designs: (G–H) Mantis shrimp designs (*Odontodactylus scyllarus* (Linnaeus, 1758)) from The Oatmeal. (I) Tourist souvenir promoting the famous dish, Singapore chilli crab (*Scylla* spp.). (J) *Wasser* by Giuseppe Arcimboldo (1566), oil on canvas (Kunsthistorisches Museum, Vienna) featuring (K) a penaeid prawn, (L) *Squilla mantis* Linnaeus, 1758, (M) *Cancer pagurus* Linnaeus, 1758 and *Astacus* sp.

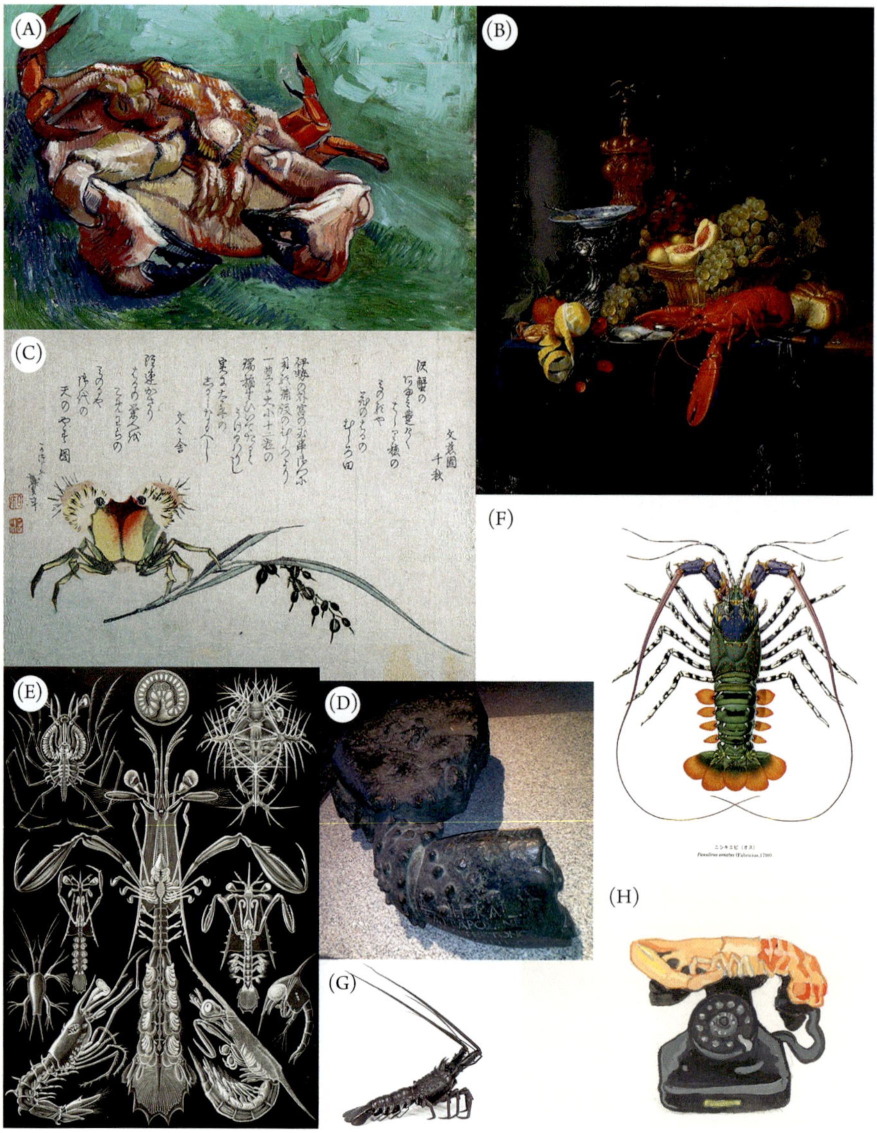

Fig. 17.2.

Crustaceans in fine art. (A) *Cancer pagurus* Linnaeus, 1758 in *A Crab on its Back* by Vincent Van Gogh (1887), oil on canvas (Van Gogh Museum, Amsterdam). (B) *Homarus gammarus* (Linnaeus, 1758) and *C. pagurus* in *Still Life with Golden Goblet* by Pieter de Ring (ca. 1640–1660), oil on canvas (Rijksmuseum, Leiden). (C) *Eriocheir* sp. in *Crab and Rice Plant* by 葛飾 北斎 (Katsushika Hokusai) (19th century). (D) Statue of *Eriphia verrucosa* (Forskål, 1775) in the collection of the Metropolitan Museum of Art, New York; formerly part of a set surrounding the base of the obelisk, Cleopatra's Needle. Photograph by Christopher Boyko. (E) Thoracostraca (Haeckel 1904: pl. 76). (F) *Panulirus ornatus* (Fabricius, 1798) (Sugiura and Asakura 2012). (G) Spiny lobster (Palinuridae) metal kusshin jizai (fully articulated) okimono model, unknown artist (Meiji period) (photo: ©Bonhams, used with permission). (H) Gouache on paper interpretation of Salvador Dalí's *Lobster Telephone* (*Homarus gammarus* (Linnaeus, 1758)) by Maria Spector (2009).

Fig. 17.3.

Crustaceans on stamps (identifications using current taxonomy). Line 1 (top; left to right): coral shrimp, *Stenopus tenuirostris* de Man, 1888; bumblebee shrimp, *Gnathophyllum americanum* Guérin-Méneville, 1855; anemone shrimp, *Ancylocaris brevicarpalis* Schenkel, 1902 (female); mantis shrimp, *Odontodactylus scyllarus* (Linnaeus, 1758). Line 2: anemone shrimp, *Ancylocaris brevicarpalis* Schenkel, 1902 (male); horned spider crab, *Schizophrys aspera* (H. Milne Edwards, 1831); fiddler crab, *Gelasimus vomeris* (McNeill, 1920); reticulated coral crab, *Trapezia septata* Dana, 1852; banded coral shrimp, *Stenopus hispidus* (Olivier, 1811); amphipod, *Amaryllis philatelica* Lowry and Stoddart 2002. Line 3: red nipper crab, *Geograpsus stormi* de Man, 1895; white-strip crab, *Circulium vitatum* (Ng and Davie, 2011); bright-eyed crab, *Chiromantes garfunkel* Davie and Ng, 2013; Kuhl's ghost crab, *Ocypode kuhlii* de Haan, 1835. Line 4: hairy crab, *Hirsutodynomene vespertilio* McLay and Ng, 2005; Cameron's spider crab, *Pleisticanthoides cameroni* Ng and Richer de Forges, 2012 (misidentified as *Dicranodromia danielae* Ng and McLay, 2005); small deep-water crab, *Carcinoplax nana* Guinot, 1989; Largo's spider crab, *Cyrtomaia largoi* Richer de Forges and Ng, 2007. Line 5: Stimpson's spider crab, *Oxypleurodon stimpsoni* Miers, 1886; *Cyrtomaia ericina* Guinot and Richer de Forges, 1982; giant barnacle, *Austromegabalanus psittacus* (Molina, 1782); hinged-beak shrimp, *Rhynchocinetes typus* H. Milne Edwards, 1837; hairy crab, *Romaleon setosum* (Molina, 1782); blackberry crab, *Homalaspis plana* (H. Milne Edwards, 1834). Line 6: pink shrimp, *Pandalus montagui* Leach, 1814; European lobster, *Homarus gammarus* (Linnaeus, 1758); brown crab, *Cancer pagurus* Linnaeus, 1758; blue-striped squat lobster, *Galathea strigosa* (Linnaeus, 1761). ©Posten Bring Group Norway, Design: Enzo Finger.

Fig. 17.4.

Crustaceans on coins. (A) Freshwater crab, *Potamon fluviatile* (Herbst, 1758) (Malta). (B) European green crab, *Carcinus maenas* (Linnaeus, 1758) (Iceland). (C) Mangrove crab, *Scylla serrata* (Forskål, 1775) (Seychelles). (D) Brown crab, *Cancer pagurus* (Linnaeus, 1758) (Guernsey). (E) Coconut crab, *Birgus latro* (Linnaeus, 1767) (Vanuatu). (F) Kuruma prawn, *Marsupenaeus japonicus* (Bate, 1888) (Mozambique). (G) Nordic shrimp, *Pandalus borealis* Krøyer, 1838 (Iceland). (H) Spiny lobster, *Panulirus* sp. (Turks and Caicos Islands). (I) Spiny lobster, *Panulirus* sp. (Cayman Islands). (J) Clawed lobster, *Homarus* sp. (San Marino). (Photos: ©Daniel Clements, used with permission, www.coinzoo.net).

Fig. 17.5.
Crustaceans in literature. (A) Mitten crab (*Eriocheir* sp.) featured in a depiction of the Japanese fable *The Crab and the Monkey* by 渓斎 英泉 (Keisai Eisen) (Edo period), woodblock print (Princeton University Art Museum). (B) Slavering brachyurans, loosely based on species of Ocypodidae and Gecarcinidae on the covers to *Night of the Crabs* (Smith 1989) and *Crabs on the Rampage* (Smith 1988). (C) Menacing giant lobster on the cover to *Fantastic* (July 1957 issue) (Valigursky 1957). (D) Vicious *Cancer* sp. and *Birgus latro* (Linnaeus, 1767) ripped their flesh on the covers to *All Man* (Nov 1959) and *Man's Conquest* (Nov 1956) issues (Collins and Hagenauer 2008). (E) A paguroid on the cover to *A House for Hermit Crab* (Carle 1987). (F) Cover to *Pagoo* (Holling 1957). (G) Crab-men on the cover of *Conan the Barbarian* #99 (Buscema and Chan 1979). (H) The titular monster (cf. Grapsoidea) from "I was trapped by Titano, the monster that time forgot!" (Leiber and Kirby 1960). (I) Two panels from the title pages of "They crawl by night!" (Keyes and Wolverton 1953). (J) The denouement of the story "Rough Riders!" featuring the attack of the *Cardisoma guanhumi* Latreille, 1828; from the original art (Kurtzman and Estrada 1953). (K) Panel from *The March of the Crabs* (de Pins 2015).

Fig. 17.6.
Crustaceans in music. (A) Half-man, half-lobster (*Homarus americanus* H. Milne Edwards, 1837) on the cover of the self-titled album by Nantucket (released 1978). (B) Truth in advertising on the cover of *A Giant Crab Comes Forth* by Giant Crab (released 1968). The title track is the same name as the album itself. (C) Cover of *Crab Tunes* by Noggins (released 1971); fittingly, the album contains six songs each named *Crab Tune* (#1–6). (D) Spiny lobsters in a crate on the cover of *A White Sport Coat and a Pink Crustacean* by Jimmy Buffet (released 1973). (E) Artist's rendition of a male "*Uca*" sp. (Ocypodidae) winning the race against another, indeterminate, crab on the cover of *Crab Race* by The Morwells (released 1978). (F) Giant anatomically inaccurate crab monster on the cover of *Crab Society North* by Stormtroopers of Death (released 2011). (G) A portunid crab covered in honey on the cover of aptly named album *Crab in Honey* by Fever the Ghost (released 2014). (H) Cover for the self-titled album by *Obey the Lobsters* (released 2017), although the image is a crayfish, not a lobster. (I) Album cover for *Attack of the Giant Purple Lobsters!* by Washington Dead Cats (released 2019), although the "purple lobsters" pictured are actually images of the lobster-like Ultraman villain Alien Baltan. (J) 1986 12" single rerelease of *Rock Lobster* by the B-52's; the original 1978 single release had no crustacean-themed cover art. (K) One of the pseudo-*Gecarcoidea natalis* Pocock, 1889 featured in the music video for Noisestorm's *Crab Rave* (released 2018). (L) Kazuki Kitagawa plays the Japanese giant spider crab (*Macrocheira kaempferi* Temminck, 1836) like a didgeridoo.

Fig. 17.7.
Crustaceans in film and television. (A) Ebirah, the giant lobster (Decapoda: cf. Nephropidae) who battled Godzilla in ゴジラ•エビラ•モスラ　南海の大決闘 (*Ebirah, Horror of the Deep*). (B) Zanika, a giant alien resembling a gall crab (Decapoda: Cryptochiridae) (from *Return of Ultraman* [1971]). (C) Crabgan, an alien resembling a giant crayfish (Decapoda: Astacoidea), shown here combined with Anemos, an alien resembling a large polychaete with which it forms a symbiotic relationship (from *Ultraman Gaia* [1998]). (D) Black Dome, a giant crab alien resembling a flying chocolate cake (from *Ultraman Leo* [1974]). (E) Giant *Cancer pagurus* animated by Ray Harryhausen from *The Mysterious Island*. (F) *Cancer pagurus* model (armature inside crab body) from *The Mysterious Island* as it looks today (Photo: ©Leonard Hawksmoor, used with permission). (G) Artist's conception of Onibaba from *Pacific Rim*. (H) Crab monster (*gen. et sp. indet.*) from Roger Corman's *Attack of the Crab Monsters*. (I) Sheldon J. Plankton (Copepoda, left) and Eugene H. Krabs (possible *Ocypode* sp. based on the eyestalks, right) from *SpongeBob SquarePants*. (J) Tamatoa (a fully aquatic coconut crab, *Birgus latro* (Linnaeus, 1767)) from Disney's *Moana*. (K) Sebastian, the hermit crab (Paguroidea) from Disney's *The Little Mermaid*.

Fig. 17.8.
Crustaceans in video games and toys. (A) Crawmerax from *Borderlands: The Secret Armory of General Knoxx*. (B) The Invincible Son of Crawmerax the Invincible from Borderlands 2. (C) Giant Enemy Crab from *Genji: Days of the Blade*. (D) Krabby from the *Pokémon* card game. (E) Mudcrab (Panopeidae) from *The Elder Scrolls* series. (F) Denizens of *Neo Aquarium – The King of Crustaceans*: *Chionoecetes opilio* (Fabricius, 1788), Paguroidea, *Ranina ranina* (Linnaeus, 1758), and *Panulirus* sp. (G) Characters from *King of Crabs*: *Pseudocarcinus gigas* (Lamarck, 1818), Cancridae sp., and a fiddler crab (Ocypodidae). (H) Not only crabs fight in *Fight Crab*: A *P. gigas* hefting a chainsaw versus an *Homarus americanus* H. Milne Edwards, 1837 brandishing a pair of pistols. (I) *Mantis*, card game with educational booklet on mantis shrimp.

Fig. 17.9.
Crustaceans as toys (A, B) and Cadbury "Yowie" figurines (C–I). (A) *Crazy Crustaceans* combination puzzle card set with educational facts. (B) Rubber toy mangrove crab, *Scylla serrata* (Forskål, 1775), apparently "cooked" (C) Mantis shrimp. (D) Freshwater crayfish "Yabby." (E) Spiny lobster. (F) Hairy hermit crab. (G) Sponge crab. (H) Blue swimmer crab. (I) Soldier crab.

Fig. 17.10.

Crustacean replicas from various Japanese manufacturers. (A) Assorted. (B) Spanner crab, *Ranina ranina* (Linnaeus, 1758) (Kaiyodo). (C) Ornate spiny lobster, *Panulirus ornatus* (Fabricius, 1798) (Kitan). (D) Australian giant crab, *Pseudocarcinus gigas* (Lamarck, 1818) (Kaiyodo).

Fig. 17.11.
Crustaceans as ornaments and paraphernalia. (A) Brass crab home ornament (*Scylla* sp.). (B) Refrigerator magnets. (C) Mantis shrimp cigarette lighter. (D–E) Various souvenirs and homewares. (F) *Bathynomus* tea infuser. (G) *Bathynomus* iPhone case. (H) Cigarette ash tray. (I) Kitchenware.

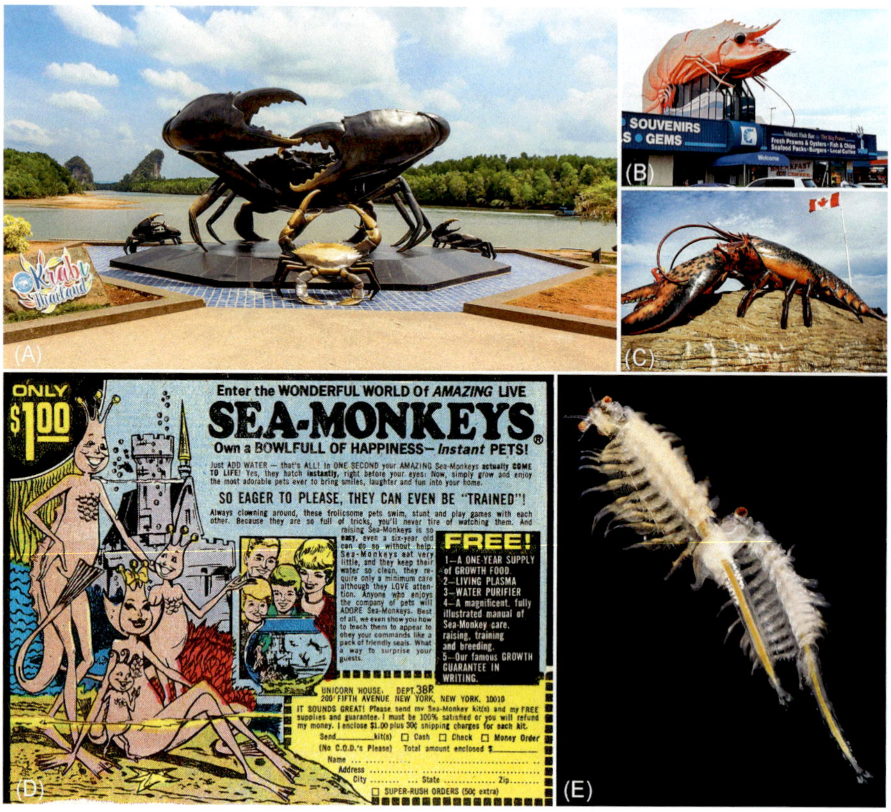

Fig. 17.12.

Crustaceans as public monuments (A–C); "Sea monkeys" (D–E). (A) Crab sculptures (*Scylla* sp.) at Krabi, Thailand. (B) The "Big Prawn" (Penaeidae), Ballina, Australia. (C) Lobster (*Homarus americanus* H. Milne Edwards, 1837), Shediac, Canada. (D) Advertisement for "Sea Monkeys" from 1970s *Archie and Me* comic book; art by Joe Orlando (1927–1998). (E) Live brine shrimp, *Artemia salina* (Linnaeus, 1758). (Photos: (A) Stefan Fussan, Wikimedia Commons, CC-BY-SA 3.0; (B) public domain; (C) Mike Young, Wikimedia Commons, CC-BY-SA 3.0; (E) Hans Hillewaert, Wikimedia Commons, CC-BY-SA-4.0).

10

EFFECTS OF ANTHROPOGENIC STRESSORS ON CRUSTACEAN ZOOPLANKTON IN LAKES

Dietmar Straile

Abstract

Crustacean zooplankton hold key roles in the pelagic food webs of lakes as they represent both important consumers of algae and prey for fish. Thus, stressors affecting both the bottom and top of the food web will cascade toward crustaceans, making them important indicators for anthropogenic stressors. This chapter focuses on three frequent stressors: changes in nutrient concentrations, climate warming, and neozoans. Stressor effects are mediated by the seasonality of stressor impact relative to other factors limiting zooplankton, and species traits, such as life cycle and anti-predator strategies of zooplankton. This chapter provides a background on seasonal dynamics and the diverse life cycle solutions and anti-predator strategies of zooplankton to exploit this seasonality. Using trait-based predictions, the responses of zooplankton to eutrophication and oligotrophication, warming, and invasive species are discussed with special focus on the crustacean community of Lake Constance. Results from other lakes will be used to analyze to which extent the responses of the Lake Constance crustaceans can be generalized. Changes in nutrient inflows can strongly change crustacean abundances in lakes and shift the relative importance of major groups. Warming has been shown to primarily shift the phenology of species, often without causing population declines or increases. Introductions or invasions of crustacean predators or competitors can change the dynamics of crustacean communities, often magnifying or reversing the effects of other stressors. The continuation of long-term studies will be crucial to improve our understanding of how the multitude of stressors and their interactions affect zooplankton dynamics and lake ecosystems.

Dietmar Straile, *Effects of Anthropogenic Stressors on Crustacean Zooplankton in Lakes*. In: *Ecology and Conservation*.
Edited by: Lars Gutow and Martin Thiel, Oxford University Press.
DOI: 10.1093/oso/9780197768242.003.0010

INTRODUCTION

Crustacean zooplankton play a central role in lake food webs, channeling the production of primary producers and of the microbial loop to higher trophic levels (Sterner 2009). Crustaceans control ecosystem functioning and services of lakes, such as maintenance of water quality through grazing of algae, and fish production (Hébert et al. 2017). According to its intermediate position in food webs, crustacean zooplankton dynamics are strongly influenced by bottom-up and top-down forces. Population growth rates may be affected by both the quantity and quality of food and by mortality induced by invertebrate and vertebrate predators, with predatory crustaceans contributing to the top-down control especially of smaller-sized herbivorous crustacean species. Due to bottom-up and top-down control of zooplankton populations, anthropogenic stressors affecting either prey or predators of crustacean zooplankton will have cascading effects on zooplankton dynamics.

Most freshwater crustacean plankton belong to the cladocerans and copepods (Dodson et al. 2010, Reid and Williamson 2010). Additionally, in some lakes, mysids are also important predators of other zooplankton. Four groups of cladocerans (Anomopoda, Ctenopoda, Haplopoda, and Onychopoda) and two groups of copepods (Calanoida, Cyclopoida) often account for an important share in plankton communities. Their positions in the food web range from filter-feeding species among the Anomopoda (e.g., *Daphnia* and *Bosmina* spp.) and Ctenopoda (e.g., *Diaphanosoma, Holopedium*), which feed predominantly on algae, to predatory species in the Haplopoda (*Leptodora*) and Onychopoda (e.g., *Bythotrephes*), which feed on rotifers and other crustacean plankton. Calanoid and cyclopoid copepods show life cycle omnivory with larval stages (nauplii) feeding on phytoplankton, whereas juvenile and adult stages switch at least partially to feed on ciliates, rotifers, and smaller crustaceans (Reid and Williamson 2010).

Crustacean populations in lakes have experienced multiple anthropogenic stressors during the last decades, affecting crustacean zooplankton across various spatial scales. Crustacean zooplankton in lakes all over the world have been affected by changes in nutrient concentrations (cultural eutrophication and oligotrophication; see, e.g., Straile and Geller 1998, Jeppesen et al. 2005), climate warming (e.g., Gyllström et al. 2005, Thackeray 2012, Wojtal-Frankiewicz 2012), and the intentional or unintentional introduction of fish or invertebrate species into lakes (e.g., Barbiero and Tuchman 2004, Eby et al. 2006). Regionally, marked acidic precipitation and/or poorly buffered soils exposed zooplankton populations to strong lake acidification, with declines in pH below five (Brett 1989). In softwater lakes of the Canadian Shield, the decline of calcium concentrations was associated with declines of calcium-rich *Daphnia* species (Jeziorski et al. 2012). Examples for lake-specific anthropogenic impacts include, for example, salt discharge causing species turnover within *Daphnia* communities in Onondaga Lake, New York (Hairston et al. 2005), and copper discharge eradicating the whole crustacean zooplankton in Lago d'Orta, Italy (Bonacina et al. 1988). Irrespective of the scale and type of a stressor, zooplankton populations have been shown to be sensitive indicators for these stressors because of their fast responses to changing environmental conditions and their central position in lake food webs (Jeppesen et al. 2011).

With the exception of stressors causing the disappearance of zooplankton, the effects of most stressors will be season-specific, mediated by, for example, seasonally varying nutrient inflow and warming rates, or by the life cycles of invasive species (Fig. 10.1). Furthermore, stressors interact with a highly seasonal environment in temperate lakes consisting of large gradients in temperature, food availability, and predation pressure. Consequently, the mutually depending effects of warming and food limitation on zooplankton can differ seasonally. Finally, zooplankton species have evolved specific traits, including life history and anti-predator strategies, which allow them to persist in the highly seasonal environments of temperate lakes. These traits, in combination with their seasonal niche, determine how strongly individual zooplankton species will be affected by anthropogenic stressors.

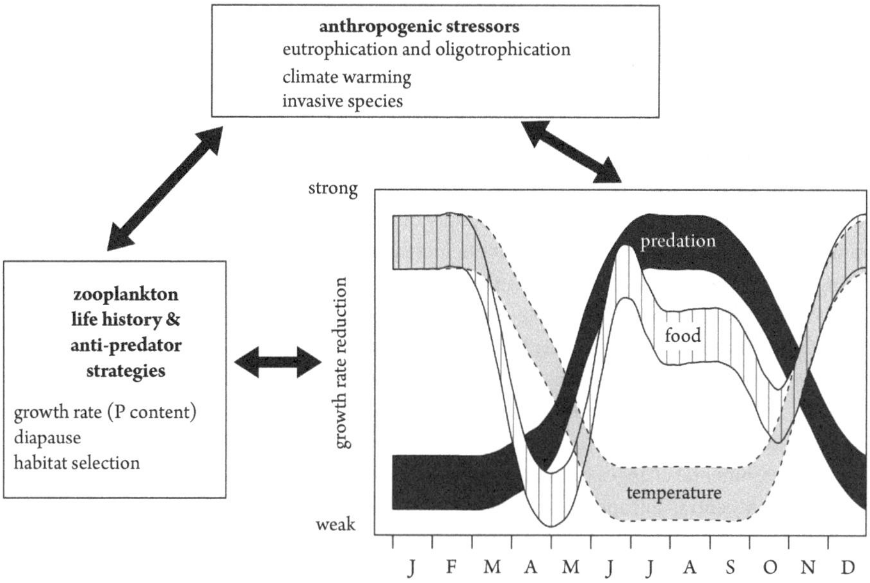

Fig. 10.1.
Interplay between anthropogenic stressors, species traits, and the seasonality of structuring forces in lakes. Anthropogenic stressors affect crustacean zooplankton species, which differ in their traits, for example, life cycle solutions and anti-predator strategies, in a highly seasonal environment in which food availability, predation, and temperature variably limit zooplankton growth rates throughout the season. Both zooplankton traits and the seasonal changes of limiting factors determine how strongly a specific anthropogenic stressor affects zooplankton populations.

Anthropogenic stressors might influence the seasonal occurrence patterns (phenologies) and traits of zooplankton populations with consequences for population abundances. For example, changes in phenologies resulting from an earlier start of favorable conditions or faster growth might result in population increases due to longer annual growth periods. If the phenologies of consumers and their food or predator organisms change differentially, there is a potential for food web disruptions (mismatch), resulting in a decline of the consumer population. Likewise, anthropogenic stressors might change the traits of crustacean populations because of plasticity and/or microevolutionary adaptation. These changes, in turn, might influence how crustacean populations respond to stressors.

The first section of this contribution will provide the background on seasonal forcing of zooplankton populations, including their strategies and adaptations to cope with this forcing, which is necessary to predict anthropogenic stressor effects on zooplankton. Then, the effects of three main stressors on crustacean zooplankton will be explored: changes in nutrient concentrations, climate warming, and invasive species. Throughout the chapter, large and deep Lake Constance will be used as a case study as this lake combines a long history of research and large environmental changes. The generality of the findings from this lake as well as lessons learned for crustacean plankton from other lakes will also be addressed.

LAKE CONSTANCE, ZOOPLANKTON SEASONAL FORCING, AND KEY TRAITS

The large and deep (z_{mean} = 101 m, area = 473 km^2) Upper Lake Constance, bordering Germany, Switzerland and Austria, is one of the best-studied lakes with regard to its ecology and zooplankton dynamics (Fig. 10.2A). Research on crustacean plankton in Lake Constance started in the 19th century (Leydig 1860, Weismann 1876, Hofer 1896). The first multi-year analysis of zooplankton seasonal dynamics dates back to the 1920s (Auerbach et al. 1924, 1926), and rather continuous zooplankton time series exist since the early 1950s (Kiefer and Muckle 1959, Einsle 1978, Straile and Geller 1998).

The mixing regime of Upper Lake Constance is warm-monomictic, with deep water mixing from December to March. Deep mixing strongly limits phytoplankton productivity resulting in low food availability for herbivorous crustaceans during this time of the year. With the onset of stratification, food availability for zooplankton increases rapidly due to enhanced light availability and high algal growth rates. The phytoplankton succession is typical for temperate lakes (Sommer et al. 1986), with abundant small edible algae during the spring bloom, which are replaced by less edible larger forms and a higher contribution of cyanobacteria in early summer. In early summer, predation by carnivorous crustaceans and juvenile fish, which overcome their gape limitation at this stage, also strongly increases. Thus crustacean zooplankton in Lake Constance need to adapt to a striking seasonality in food availability and predation pressure (Fig. 10.1).

The environmental changes occurring in Lake Constance are typically observed in many temperate lakes in central Europe and partially also in lakes around the globe. Historically, Lake Constance was considered a model system for deep oligotrophic lakes with total phosphorus concentrations below 7 μg/L, but it experienced severe eutrophication after World War II, with phosphorus concentration increasing to almost 90 μg/L from the 1950s to the late 1970s. Since then, the establishment of sewage plants in the catchment area, which started in the 1960s, reduced phosphorus concentrations again below 10 μg/L (Fig. 10.2B). Water temperatures in the lake increased during the last decades (Fig. 10.2C), including a step-wise increase during the late 1980s (Jochimsen et al. 2013). In addition, winter mixing and water temperature dynamics during spring have been shown to be influenced by the North Atlantic Oscillation (NAO) (Straile et al. 2003), a large-scale atmospheric circulation particularly affecting late winter and spring temperatures across Europe.

Zooplankton species evolved different strategies to cope with the seasonality in, for example, temperature constraints, food availability, and predation pressure. These strategies spread along a slow–fast continuum from species characterized by low growth and mortality rates to species with high growth and mortality rates (Kiørboe and Thomas 2020). In pelagic zooplankton, this continuum is associated with the reproductive mode of taxa (parthenogenetic vs. sexual reproduction), ontogenetic growth, and developmental rates. Fast growth is associated with a high tissue phosphorus content, as it requires fast protein synthesis and, consequently, high amounts of phosphorus-rich RNA (the "growth rate hypothesis"; Sterner and Elser 2002). While high phosphorus content, parthenogenetic reproduction, and the absence of a complex life cycle in cladocerans allows fast population growth, the complex life cycle of copepods, including metamorphosis and obligate sexual reproduction and lower phosphorus content, reduces population growth rates considerably. All other things being equal, the demographic costs of males halves the growth rate of a sexually reproducing species with equal sex ratio compared to a parthenogenetic species. Species with high growth rates often suffer from high mortality rates. However, species have evolved multiple strategies to reduce mortality due to predators, including life history and morphological and behavioral defenses. Behavioral defenses include the selection of habitat with reduced predation pressure, that is, residing at increased depth to hide from visual predators, and diel vertical migration.

A second major trait determining the population dynamics of planktonic crustacean taxa is the presence of diapause either via resting eggs (embryonic diapause) and/or the inclusion of a diapausing stage in the life cycle. This life cycle attribute is not related to the slow–fast continuum: while cladocerans and calanoid copepods produce resting eggs, cyclopoid copepods evolved the capacity to undergo diapause or developmental delays in various developmental stages, from CII

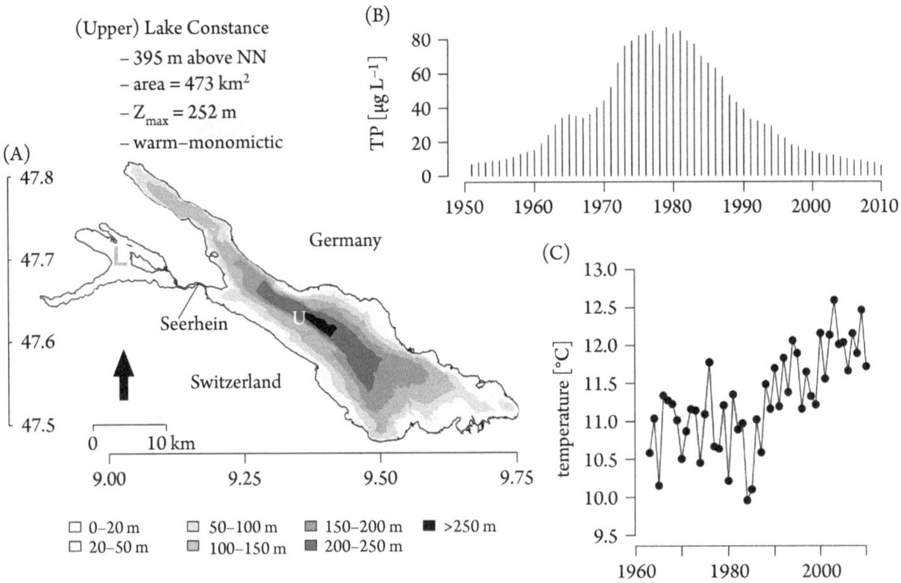

Fig. 10.2.
(A) Bathymetric map of Lake Constance showing Upper Lake Constance (U) and Lower Lake Constance (L) connected by the river Rhine (Seerhein), (B) Long-term development of total phosphorus concentrations measured during winter mixis (TP), and (C) annual mean water temperatures in the upper 5 m of Lake Constance. Source: data on TP and water temperatures provided by the Institute for Lake Research, Langenargen.

copepodids to adult females with fertilized eggs depending on species and population identity (Reid and Williamson 2010). Within these three groups, there is large variability in diapause behavior or investment in diapausing stages. For example, while the calanoid *Eudiaptomus gracilioides* produces resting eggs, its close relative *E. gracilis* does not (Santer et al. 2000). Likewise, populations of *Cyclops* spp. differ in life cycle duration and diapause behavior between lakes (Santer 1998) and even between different regions within Lake Constance (Seebens et al. 2013). Additionally, there is between-lake variability in *Daphnia* populations in how strongly they invest in resting egg production, suggesting selection on crustacean life cycle dynamics (Tessier and Cáceres 2004). Diapausing stages can be essential parts of the seasonal life cycle, that is, when a population is only able to complete the seasonal cycle via entering and leaving the diapause stage. Alternatively, a population may produce resting eggs during specific parts of the year but may also be able to complete the seasonal cycle and persist in the lake without the hatching of resting eggs. Thus, the diapause behavior may have an important impact on seasonal and long-term population dynamics depending on the allocation in diapausing stages.

Seasonal variability differed strongly between species of the pelagic crustacean community of Lake Constance (Fig. 10.3). At one extreme of the slow–fast continuum, cladoceran species with high P content and parthenogenetic growth show seasonal boom-bust cycles with rapid population increases of two to three orders of magnitude during a short period of two to three months (Fig. 10.3A–E), and population declines at various rates during the rest of the year. At the other extreme, low-P and complex life cycle calanoid copepods, such as *E. gracilis,* do not have the capacity for fast population growth, but their low mortality rates keep seasonal variability within a comparatively narrow range (Fig. 10.3F). The low growth rates and long generation times of

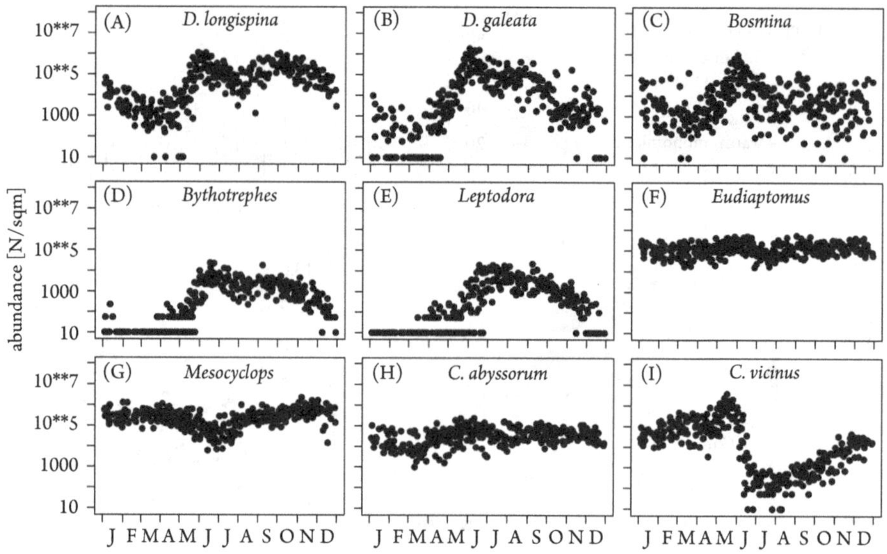

Fig. 10.3.
Seasonal dynamics of crustacean zooplankton in Lake Constance during the years 1986–1994. Samples with species abundances below the detection limit are plotted by default as 10 individuals per square meter. The species most often below the detection limit were the predatory cladocerans *Leptodora kindtii* (39% of samples) and *Bythotrephes longimanus* (30%), followed by *Daphnia galeata* (15%), the remaining cladocerans (*D. longispina*, 1.8%; *Bosmina* spp., 1.5%), and *Cyclops vicinus* (1.8%). *Eudiaptomus gracilis*, *Mesocyclops leuckartii*, and *Cyclops abyssorum* were found in each sample. Source: data from Ulrich Einsle, ISF Langenargen.

calanoid copepods precludes a strong numerical response to rather short periods of benign conditions as, for example, during the spring phytoplankton bloom. However, in the pelagic zone of Lake Constance, *E. gracilis* is the only crustacean without any resting stage and with continuous reproduction throughout the year, even during the winter season albeit with lower clutch sizes. As offspring is produced year round, seasonal abundance patterns are suggested to be not bottom-up but top-down controlled via predation on nauplii (Elster 1954, Seebens et al. 2007). All other copepod species use either a period of diapause or developmental retardation to complete their seasonal cycles. The two cyclopoid copepods *Mesocyclops leuckartii* and *Cyclops abyssorum* exhibit life cycles with developmental retardation mainly during their fifth copepodid stage. Development is retarded during winter in *M. leuckartii* but during summer in *C. abyssorum*. While all *M. leuckartii* copepodids perform the same strategy, a small fraction of the *C. abyssorum* population does not retard but continues its development during summer (Einsle 1964). However, both species were present in the plankton throughout the year (Fig. 10.3G–H) and had more than one generation in a season. In contrast, *Cyclops vicinus* reproduced only during the phytoplankton spring bloom, and performs a diapause in the fourth copepodid stage during summer in the sediment of the lake (Seebens et al. 2009). Entering diapause in June and emerging from diapause in the subsequent winter also resulted in large population variability (Fig. 10.3I). However, in contrast to cladocerans, large changes in population size of *C. vicinus* are not driven by population growth and decline, but by diapause and migration of specimens. Hence, the copepod species in eutrophic Lake Constance all exhibited different life cycle solutions to seasonal variability: continuous reproduction (*E. gracilis*), quiescence during summer (*C. abyssorum*) and winter (*M. leuckartii*), and a monocyclic life cycle with diapause (*C. vicinus*).

Life cycles of cladocerans are more similar among species, all characterized by rapid population increases based on parthenogenetic reproduction during spring or early summer, followed by subsequent population declines. However, these declines differ depending on mortality rates during the summer. For example, *Daphnia longispina* avoids intense fish predation via diel vertical migration (already documented in Lake Constance by Weismann [1876], Stich and Lampert [1981]) and population abundances remain high until autumn. In contrast, similar sized *D. galeata*, which does not migrate, decreases rapidly after its spring maximum due to high mortality caused by fish predation (Becker and Eckmann 1992). Likewise, small *Bosmina* species are not a preferred food item, whereas *Bythotrephes longimanus* is highly selected for by fish. However, for cladoceran species there is also variability in the degree of sexual production and the importance of resting eggs for overwintering. In Lake Constance, rare occurrences of *Bosmina* males were reported in the 19th century (Weismann 1878), but neither males nor resting eggs were reported thereafter. This observation is consistent with comparatively high *Bosmina* abundances throughout winter, suggesting that the populations can persist in the lake without the need for resting egg production. In contrast, the two predatory cladocerans, *Bythotrephes longimanus* and *Leptodora kindtii* switch more or less completely in autumn from production of subitaneous eggs to that of resting eggs. Concurrently, the numbers of males in the populations rise from negligible to above 50% for *Leptodora*, and almost 40% for *Bythotrephes*. Subsequently, the two species drop below detection level during the winter (Fig. 10.3C,G). The two *Daphnia* species occupy intermediate positions along the gradient from resting eggs to overwintering. *Daphnia longispina* is usually found year round in the plankton, where *D. galeata* is more often below the detection limit (Fig. 10.3A–B). The two species also differ in sexual allocation and the number of resting eggs produced: *D. galeata* produces more resting eggs, which is consistent with a stronger reliance on diapause for overwintering compared to *D. longispina*.

Sexual reproduction with the production of resting eggs involves demographic costs because energy will be channeled away from subitaneously developing eggs to males and resting eggs. Strong investment in resting eggs and/or males occurs in some species in autumn. For example, the percentage of *Bythotrephes* females carrying resting eggs increased during October from negligible to more than 80%. This investment in sexual reproduction will reduce population growth rates and contribute to autumnal population declines. However, the contribution of sexual reproduction to declines of cladoceran populations has not been thoroughly quantified. Likewise, the contribution of hatching from resting eggs to spring population growth has rarely been determined (Hairston et al. 2000). One exception is Saidenbach reservoir in Eastern Germany, where hatching from resting eggs has been estimated to account for < 1% of the total abundance and for only a tiny fraction of *Daphnia* population growth (Rother et al. 2010). The uncertainty regarding the actual size of overwintering populations, that is, population sizes below the detection limit, contributes to the difficulty in quantifying the contribution of hatching to population dynamics of cladoceran populations. As possibly both overwintering strategies (overwintering in an active stage vs. overwintering via diapause) might be differently affected by, for example, warming, the lack of knowledge in respect to the relative importance of overwintering strategies may hinder prediction of warming effects on future population dynamics.

Similar to life cycle strategies, predator avoidance strategies determine the susceptibility of crustacean zooplankton to anthropogenic environmental pressures. The different zooplankton species in Lake Constance occupy different depths during the summer period, resulting in differential exposure to visually hunting predators. Thereby, the depth selection of the species will also determine the food concentrations and the water temperatures they will experience and, thus, their susceptible to eutrophication and warming. For example, the two *Daphnia* species, *D. longispina* and *D. galeata*, occupy the most extreme positions within the depth gradient (Fig. 10.4), resulting in highly different birth and mortality rates during the summer season (Stich and Lampert 1981), and,

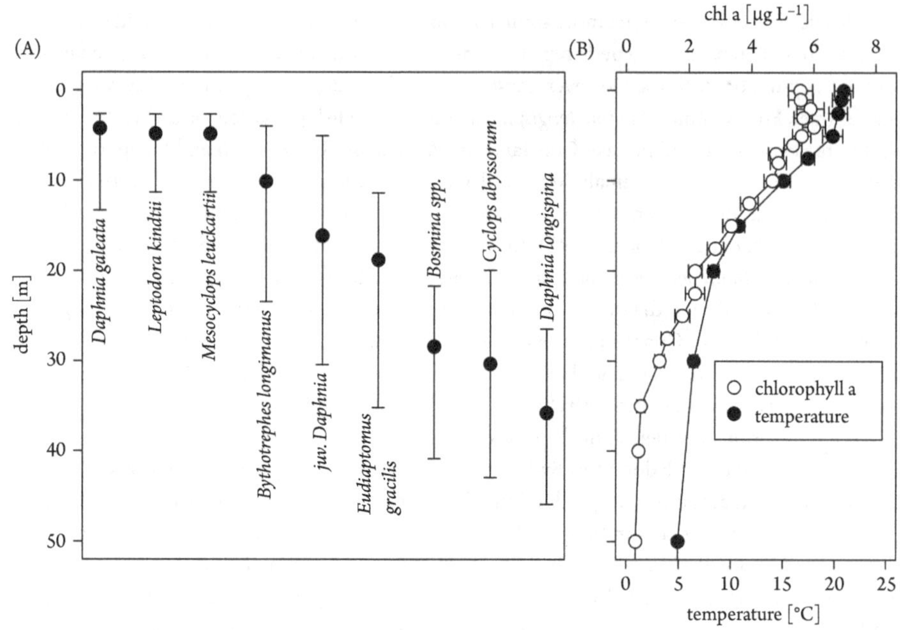

Fig. 10.4.
(A) Depth distribution (means of 20%, 50%, and 80% percentiles at individual sampling dates) of crustacean species in Lake Constance, and (B) the mean water temperatures and chlorophyll a concentrations at these depths during August 1986–1994. Note that, for example, *Daphnia longispina* performs diel vertical migrations and, thus, experiences higher food concentrations and water temperatures during night. Source: data from Ulrich Einsle, ISF Langenargen.

consequently, in different responses to eutrophication and oligotrophication (see below). Hence, species traits, including life cycle and anti-predator strategies, may strongly influence the responses of crustacean species to anthropogenic stressors.

THE INFLUENCE OF TROPHIC CHANGES ON CRUSTACEAN POPULATION DYNAMICS

Numerical Responses to Changes in Lake Trophic Status

Changes in the trophic status of a lake are likely to change birth and growth rates as well as the mortality rates of zooplankton populations in complex and species- and lake-specific ways. Birth and growth rates will increase with eutrophication because enhanced nutrient supply promotes primary productivity of algae and, consequently, crustacean food availability. However, changes in phytoplankton composition toward low-quality and toxic cyanobacteria might partially confound the nutrient–birth rate relationship. Mortality might increase when larger fish populations develop, or because of the disappearance of deep-water refugia when the hypolimnia of eutrophic lakes become anoxic.

In deep Lake Constance, phytoplankton biomass increased strongly with eutrophication (Güde and Straile 2016) as did the contribution of cyanobacteria to the phytoplankton, although their contribution to summer biomass rarely exceeded 15% (Isanta-Navarro et al. 2021). Likewise, fishery yields increased with eutrophication (Güde and Straile 2016), suggesting stronger predation pressures on zooplankton. Furthermore, the role of predation within the crustacean zooplankton assemblage might have changed with eutrophication, as predatory species such as the cladocerans *Bythotrephes* and *Leptodora* increased in abundance and the predatory copepod *C. vicinus* invaded the lake (see below).

Changes in population size thus depend on the species-specific variations in growth versus mortality rates within the zooplankton population. During the early phase of eutrophication in Lake Constance until the early 1950s, all species increased their population sizes (Straile and Geller 1998), suggesting that increases in birth rates outweighed elevated mortality rates. However, the increases in growth rates in response to eutrophication differed between species, resulting in shifts in dominance: the crustacean community changed from one dominated by calanoid copepods (*E. gracilis* and *Heterocope borealis*) in the 1920s to a community dominated by *Daphnia* (*D. longispina* and *D. galeata*) in the 1980s (Fig. 10.5). Likewise, the contribution of cyclopoid copepods increased with eutrophication as compared to calanoid copepods.

The changes in relative importance were accompanied by a shift in species composition. Increased food availability seems to have facilitated the invasion of new species in the pelagic zone of the lake. Most notably, the cladoceran *D. galeata* was first detected in plankton samples in 1952 and the cyclopoid copepod *C. vicinus* in 1956. Additionally, paleolimnological investigations suggest that in the 1940s a new *Bosmina* taxon, *B. longicornis kessleri*, had invaded the lake (Hofmann 1998). Other copepod species (*Acanthodiaptomus, Acanthocyclops*) also reached higher numbers with eutrophication in the eastern part of the lake, which is more shallow and also received higher nutrient inputs (Einsle 1988).

With oligotrophication, which resulted in a decline of phosphorus concentration to below 10 μg/L, that is, to values typically observed in the 1950s, zooplankton abundances started to decline again, and several changes in relative biomasses occurring with eutrophication were reversed. For example, the contribution of *Daphnia* to the crustacean biomasses decreased due to the decline of *D. galeata*. In contrast, the share of the calanoid copepod *Eudiaptomus* increased again, especially in the summer months (Fig. 10.5). *Diaphanosoma brachyurum*, which became extinct during eutrophication, reappeared in the lake (Stich 2004). Overall, the changes described here, that is, the potential of fast-growing cladocerans to exploit the enhanced resource availability caused by eutrophication better than copepods, were tightly related to the slow–fast gradient of zooplankton life histories.

The observed changes in the composition of the crustacean zooplankton community cannot be solely understood from species differences in birth rates; it is also necessary to consider differences in mortality rates. For example, *D. galeata* was able to invade Lake Constance with eutrophication although the species does not perform diel vertical migration to avoid fish predation (Stich and Lampert 1981). Large birth rates during the eutrophic period allowed *D. galeata* to invade and coexist with migrating *D. longispina*, despite much higher mortality rates compared to *D. longispina*. Stomach content analyses of the dominant planktivorous fish *Coregonus wartmanni* showed indeed that consumption of *D. galeata* is much higher than consumption of *D. longispina* (Becker and Eckmann 1992). With ongoing oligotrophication and declines in birth rates, high predation losses could not be sustained anymore, and *D. galeata* strongly declined in abundance (Straile 2015). Besides predation by fish, predation within the crustacean communities is also likely of importance. The invasive cyclopoid *C. vicinus* is suggested to have caused the extinction of the calanoid *Heterocope*, which disappeared shortly after the establishment of *C. vicinus* (see below) (Einsle 1983a). Likewise, the disappearance of *Diaphanosoma brachyurum* from the Lake Constance plankton in the 1960s has been suggested to be to the consequence of increased invertebrate predator abundances and predation pressure from *Bythotrephes longimanus* and/or *Leptodora kindtii* (Straile and Geller 1998). Changes in mortality rates were also likely important for differences in *E. gracilis*

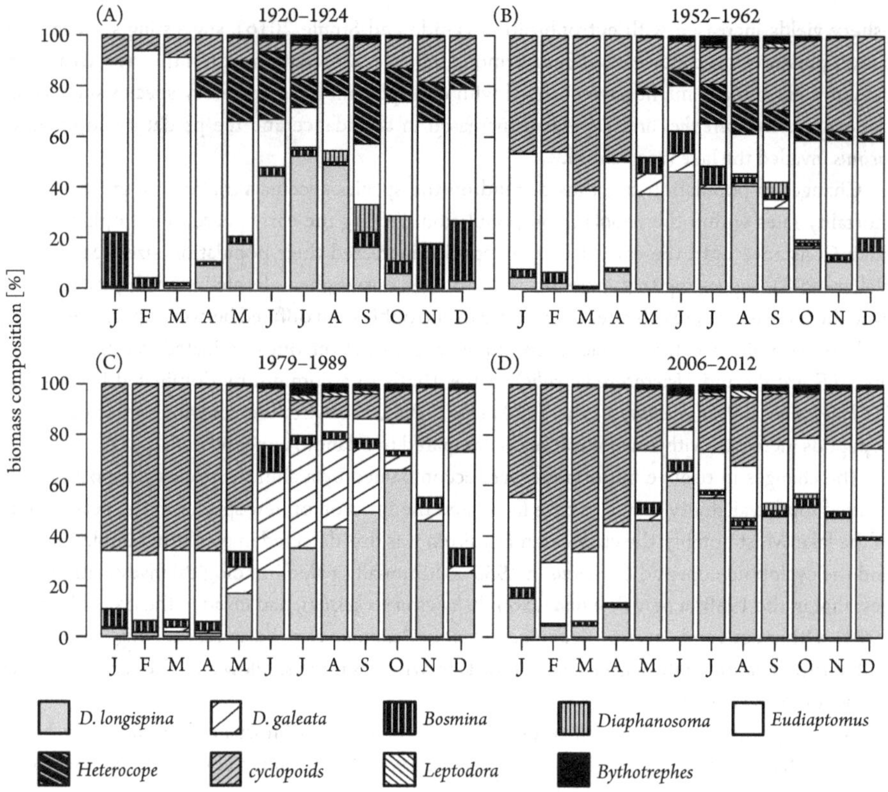

Fig. 10.5.
Biomass distribution among species of the Lake Constance crustacean zooplankton during periods of (A) ultra-oligotrophic, (B) oligotrophic, (C) eutrophic, and (D) again oligotrophic conditions after recovery. Source: data from Straile (2015).

seasonal dynamics between the oligotrophic and eutrophic periods (Seebens et al. 2007). While spring abundances strongly exceeded summer abundances during the eutrophic period, this difference was less pronounced during more oligotrophic years. Abundances of ovigerous females did not show large differences between eutrophic and oligotrophic years. Accordingly, these changes were likely due to differential nauplius survival (Seebens et al. 2007).

Similar changes in the relative abundances of cladocerans and of calanoid and cyclopoid copepods with changes in nutrient concentrations have also been observed in long-term data from other lakes (De Bernardi et al. 1988, Anneville et al. 2007) and are also evident from space-for-time substitutions (Jeppesen et al. 2011, Barbiero et al. 2019). The latter two studies examined crustacean composition in two very different systems: in the large oligotrophic Laurentian Great Lakes and in mesotrophic to eutrophic shallow Danish lakes (Fig. 10.6). In the oligotrophic Laurentian Great Lakes the contribution of calanoid copepods (various smaller species and *Limnocalanus macrurus*) to summer biomass surpasses 80% in the most oligotrophic Lake Superior, whereas in the least oligotrophic Lake Erie, the contribution of calanoid copepods (*Limnocalanus* and others) is less than 50% (Barbiero et al. 2019). Strong abundance declines with eutrophication were observed for the large calanoid species *Limnocalanus* in Lake Erie, similar to the strong declines and disappearance of the large calanoid *Heterocope* in Lake Constance and Lago Maggiore. This might suggest that large calanoids with univoltine life cycles seem to be particularly vulnerable to direct and indirect changes in trophic state.

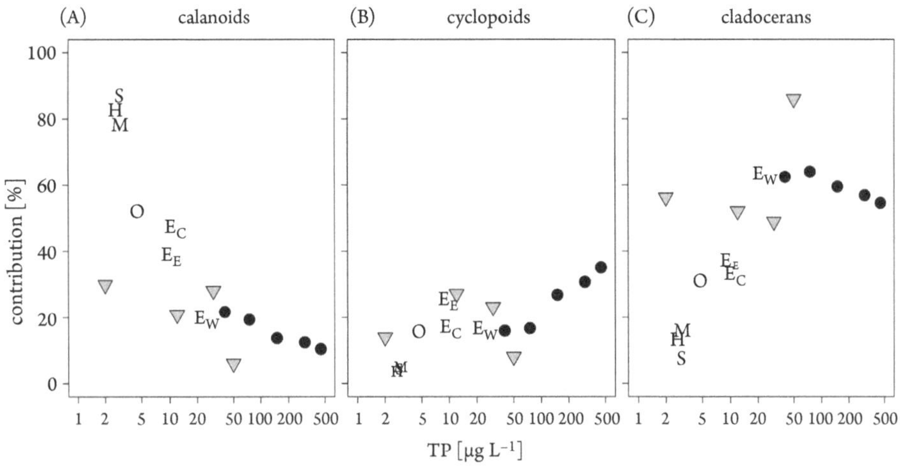

Fig. 10.6.
Contribution of (A) calanoid copepods, (B) cyclopoid copepods, and (C) cladocerans to total crustacean biomass in relation to total phosphorous (TP) concentrations in Danish lakes grouped according to TP concentrations (black circles, Jeppesen et al. 2011), various periods of Lake Constance (triangles, Fig. 10.5), and in the Laurentian Great Lakes (Lake Superior (S), Lake Michigan (M), Lake Huron (H), Lake Ontario (O), Central Lake Erie (E_C), Western Lake Erie (E_W), and Eastern Lake Erie (E_E); Barbiero et al. 2019). Note that for the Danish lakes averages are for the period May–October, whereas for Lake Constance and the Laurentinian Great Lakes August averages are shown.

In contrast, the contribution of cyclopoid copepods and *Daphnia* spp. increased from less than 10% in Lake Superior for both groups to 20% and 30%, respectively, in the Western Basin of Lake Erie (Fig. 10.6). The trophic status of shallow Danish lakes (< 50 µg L^{-1} – > 400 µg L^{-1}) strongly exceeds those of the Laurentian Great Lakes. However, in these shallow lakes the share of calanoid copepods also decreased with increasing phosphorus concentrations, while the share of cyclopoids increased. The share of cladocerans was high and on average between 50% and 60% in all Danish lakes. However, in hypereutrophic Danish lakes (phosphorus concentration > 100 µg L^{-1}), the share of cladocerans slightly decreased (Fig. 10.6), presumably due to a further increase of fish predation, especially on *Daphnia* spp., and a decline of edible algae and food quality (Jeppesen et al. 2003). Crustacean biomass composition changes observed during summer during the 20th century in Lake Constance (Fig. 10.5) also support the pattern derived from the Danish lakes and the Laurentian Great Lakes (Fig. 10.6), although the contribution of calanoids seems to be somewhat lower and the contribution of cladocerans higher in Lake Constance compared to the Danish and Laurentian lakes with similar TP values.

There are, however, also exceptions from this typically observed pattern of relative cladocerans and especially *Daphnia* increases and calanoid declines with eutrophication. A famous one is the response of *Daphnia* to eutrophication and recovery in Lake Washington, a 65 m deep lake close to Seattle, USA (Edmondson and Litt 1982). It is suggested that in Lake Washington, top-down control by a predatory crustacean, *Neomysis*, prevented the increase of the *Daphnia* population with eutrophication. Additionally, a mass development of filamentous cyanobacteria (*Oscillatoria*) during peak eutrophication presumably reduced food uptake and birth rates of *Daphnia*. Hence, only a reduction of phosphorus (with a subsequent decline of *Oscillatoria* concentrations) and a decrease of *Neomysis* due to increased fish predation, resulted in a strong increase of *Daphnia* numbers with oligotrophication, surpassing *Daphnia* numbers prior to eutrophication (Edmondson and Litt 1982). A potential interaction between the indirect effects of changes in nutrient concentrations and predation pressure was also suggested for Lago di Lugano in Switzerland/Italy (Lepori and Roberts 2017): improved hypolimnetic oxygen conditions with oligotrophication—providing a refuge from predation—resulted in an increase of large *D. longispina* and *D. galeata* relative to smaller *D. cucullata* in Lago Lugano. These examples show that changes in top-down control (related or unrelated to trophic changes) may in some lakes modify or even override the bottom-up effects of trophic changes on zooplankton dynamics.

Crustacean Phenology Responses to Changes in Lake Trophic Status

Changes in growth and mortality rates with eutrophication are predicted to also change the seasonal occurrence patterns of crustacean populations. According to the plankton succession model of the plankton ecology group (Sommer et al. 1986), the seasonality of zooplankton differs strongly between oligotrophic and eutrophic lakes: while in eutrophic lakes zooplankton seasonality is characterized by a spring and a summer maximum, only a summer maximum was predicted for oligotrophic lakes. Surprisingly, this prediction has rarely been tested. However, during the change from oligotrophic to eutrophic and back to oligotrophic conditions, this pattern was not detected in Lake Constance. Instead, in oligotrophic years a spring peak was observed, and its timing was mostly influenced by climatic conditions (Straile 2015). Furthermore, examination of the lakes studied in Sommer et al. (1986) suggests that the proposed difference between eutrophic and oligotrophic lakes pattern was at least partially confounded by regional climate differences, as in the 1970s and 1980s oligotrophic lakes were typically located at higher latitudes and/or higher altitudes than eutrophic lakes (Straile 2015).

Crustacean Trait Responses to Changes in Lake Trophic Status

Changes in the relative abundances of species ultimately result from interspecific trait differences, which determine how species respond to changes in food availability and predation pressure. However, there will likely also be changes in traits within species due to environmental changes. Such intraspecific changes in zooplankton traits in response to environmental stressors are poorly studied. In Lake Constance, higher food availability resulted in a more robust morphotype and an increase in body size in *M. leuckartii* from an average of 720 μm in 1921 to 880 μm in 1966. The morphological differences between the slender and the robust morphotypes were very pronounced so that the oligotrophic, slender morphotype was even described in the 1930s as a separate species (*Mesocyclops bodanicola*), which occurred in Lake Constance in the pelagic zone whereas the robust morphotype *M. leuckartii* occurred in the littoral zone of the lake. When the more robust type also became established in the pelagic during eutrophication, it was recognized that both forms represent different morphologies of *M. leuckartii* (Einsle 1968).

In addition, and partially as a consequence of changes in body size, the clutch size of crustacean zooplankton species also increased considerably with eutrophication in Lake Constance. For example, the maximum clutch size of both *E. gracilis* and *C. abyssorum* hardly exceeded 10 eggs per female during the 1920s but this doubled toward the 1960s (Einsle 1967). This trend was reversed with oligotrophication, and clutch sizes, for example, of *Eudiaptomus*, decreased again to levels below 10 eggs per female in recent years (unpublished data). Another important trait change was allocation in sexual reproduction in *D. longispina* as indicated by the presence of males. While during the 1930s, up to 20–30% of the autumn population consisted of males, this number decreased to less than 5% in the early 2000s (Jankowski and Straile 2003). A reduction in sexual allocation is also supported by the decrease and finally disappearance (since the 1970s) of *D. longispina* resting eggs in sediment layers of known age, the so-called egg bank. Since the 1970s, the resting egg bank is composed of resting eggs (ephippia) from *D. galeata* and *D. galeata* x *D. longispina* hybrids only (Weider et al. 1997, Jankowski and Straile 2003, Brede et al. 2009), although

D. longispina was present in the plankton in high numbers (Jankowski and Straile 2004). The lower allocation into males and the disappearance of resting eggs from the resting egg bank suggests that higher food availability with eutrophication might have increased the fitness of *Daphnia* genotypes with low investment into sexual reproduction and resting eggs as a means to overwinter.

Trait changes during eutrophication are evident in Lake Constance. However, it is unclear to what extent they result from phenotypic plasticity or represent microevolutionary change. The latter can be studied using resting eggs of species deposited in the egg bank. Unfortunately, genetic changes of *D. longispina* in Lake Constance cannot be studied using the egg bank because the species strongly reduced reproductive allocation in resting eggs (see above) so that since the 1970s resting eggs are hardly found anymore. However, the invasive *D. galeata* regularly produces resting eggs, which, after isolation, seem to be capable of hatching up to an age of approximately 50 years (Hairston et al. 1999, Isanta-Navarro et al. 2021). Studies based on allozymes, microsatellites, and genomic tools (Weider et al. 1997, Jankowski and Straile 2003, Brede et al. 2009, Cordellier et al. 2021) have identified changes in gene frequencies in *D. galeata* during eutrophication and oligotrophication. Furthermore, resurrection ecology approaches suggest microevolutionary change, both during eutrophication and oligotrophication: *D. galeata* hatched from resting eggs produced in the eutrophic period (1970s to 1990s) showed less growth rate reduction in the presence of cyanobacteria compared to *D. galeata* produced during both early eutrophication (1960s) and oligotrophic conditions (Hairston et al. 1999, Isanta-Navarro et al. 2021). Hence, *D. galeata* seems to have evolved an improved capacity to grow with low-quality and toxic food during eutrophication, but also seems to have rapidly lost this capacity after cyanobacteria concentrations declined in Lake Constance with oligotrophication (Isanta-Navarro et al. 2021).

In summary, changes in nutrient inflow have been shown to be a dominant cause of changes in crustacean population dynamics, whereas effects on crustacean phenology and intraspecific crustacean trait variability need to be studied more thoroughly. Nutrient enrichment favors species with the potential for fast growth rates most strongly, for example, cladocerans > cyclopoid copepods > calanoid copepods. However, this pattern can be modified in individual lakes by changes in predation pressure—related or unrelated to changes in nutrient inflows.

CLIMATE CHANGE RESPONSES

Seasonal Responses and Phenology Shifts

Similar to changes in trophic state, climate warming may influence crustacean population dynamics in a multitude of ways. Changes in temperature will inevitably affect the growth and development of all ectotherms and, hence, also crustaceans (Gillooly 2000). However, the impact of temperature changes on population dynamics will depend on the actual temperatures and on limitations by other factors. For example, during summer in temperate lakes, when temperatures are already close to optimum and food availability is reduced, a further temperature increase may not have a strong positive effect on growth rates, and may possibly even have a negative effect. In contrast, a huge impact of warming on population growth in temperate lakes is expected, especially during the phytoplankton spring bloom, when temperatures are still low and there is abundant food and reduced predation pressure (Fig. 10.1).

In addition to the direct effects of increasing temperatures, there is a multitude of indirect warming effects with respect to food availability and predation pressure on crustacean zooplankton. For example, warming will change the duration of ice cover and stratification, affecting the onset of the phytoplankton bloom and the length of the phytoplankton growth period in lakes (Gronchi et al. 2021), which will likely also lengthen the period of positive population growth of zooplankton.

However, winter or enhanced summer stratification may hamper upward mixing of nutrients in deeper lakes, thereby reducing primary productivity (Manca et al. 2015). Longer stratification periods combined with higher temperatures may result in longer periods of reduced oxygen concentrations in the hypolimnion (Jane et al. 2022). A reduced upward mixing of nutrients will affect food availability, while low oxygen concentrations in the hypolimnion reduce opportunities for zooplankton for hypolimnetic refuge from fish predation. Thus, warming might interact in multiple and complex ways with changes in nutrient inflow depending on lake characteristics.

Warming effects on crustacean populations can also be studied using a space-for-time substitution approach. For example, Gyllström et al. (2005) showed that in addition to changes in trophic state, climate influenced crustacean biomass and community structure in Europe across a large latitudinal range (38–68°N). Also, it is well established that diversity of crustacean zooplankton declines with increasing latitude, for example, in Canadian and Norwegian lakes (Patalas 1990, Hessen et al. 2007). In the Canadian lakes, the average number of species per lake declined from ~10 at latitudes with July air temperatures > 16°C to ≤ 5 species at latitudes with July air temperatures below 10°C. In addition, the regional species richness of cladocerans declined more strongly with increasing air temperature compared to that of calanoid copepods, which suggests that calanoid copepods perform better than cladocerans under conditions that allow only low population growth rates, for example, low temperatures and/or reduced food availability. However, these patterns need to be interpreted carefully, as the climatic gradient is almost always confounded by a nutrient gradient, that is, more northern lakes are often also less affected by eutrophication and, consequently, are more oligotrophic.

The main evidence for the effects of temperature change on zooplankton population dynamics comes from phenological studies. In mid and high latitude lakes, phenological responses are expected to be particularly strong during the spring period, when temperatures increase from below critical thermal limits for growth to optimum growth temperatures. Furthermore, food availability is high and predation pressure is low during spring, resulting in population dynamics being mostly controlled by temperature (Schalau et al. 2008). The effect of temperature on spring population growth is predicted to be particularly strong for cladoceran species, which display increases in population sizes by several orders of magnitude during spring (Fig. 10.2). Hence, rather small temperature-mediated changes in growth rates can have a strong effect on abundances and biomasses of species. In Lake Constance, warmer spring temperatures resulted in earlier onset of *Daphnia* growth and higher growth rates (Straile 2000). Consequently, in a year with warm spring temperatures (mean temperature in the upper 8 m in April and May: 11°C), *Daphnia* biomass in May was 200-fold higher than in years with cold spring temperatures (mean temperature: 7.5°C, Fig. 10.7A, B). Because of the earlier onset and acceleration of population growth due to warmer spring temperatures (Straile 2000), maximum spring abundances occurred roughly seven days earlier per decade during the period 1979–2007 (Straile et al. 2012). Due to the tight coupling of spring water temperatures with the status of the NAO, as indicated by the NAO index, *Daphnia* biomasses were also linked to this large-scale meteorological phenomenon, as were spring abundances and growth rates of the small cladoceran *Bosmina* (Straile and Müller 2010), that is, early spring biomasses and growth rates were higher in a positive phase, whereas early summer biomasses were lower in a positive phase of the NAO.

In contrast to cladocerans, the long generation times of copepods, which exceed the duration of the spring bloom, does not allow rapid multiplication of population sizes during one spring period. Hence, the effect of variations in spring temperatures on the development of biomass is expected to be smaller for copepods than for cladocerans. Nevertheless, warmer May temperatures also resulted in significantly higher May biomasses of the calanoid *Eudiaptomus* (Fig. 10.7d)—after the confounding effects of trophic changes had been accounted for analytically. Warmer temperatures

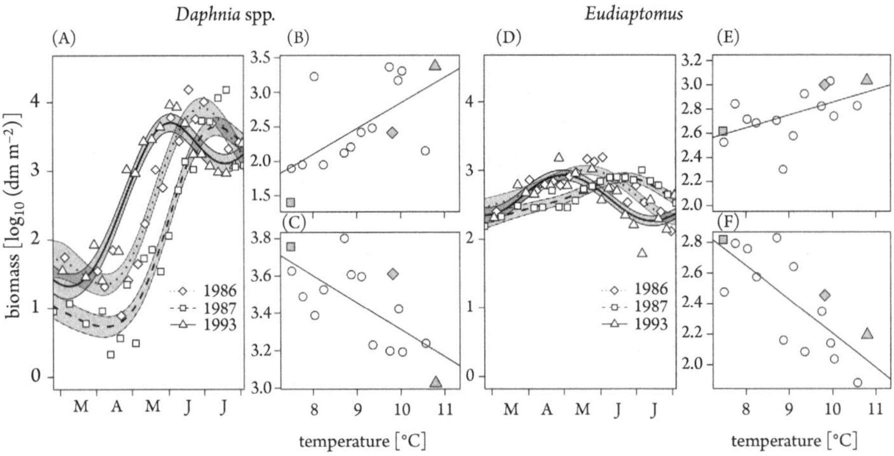

Fig. 10.7.
Spring biomass dynamics of *Daphnia* spp. (A–C) and *Eudiaptomus gracilis* (D–F) in Lake Constance years with cold (1987), average (1986), and warm (1993) April/May temperatures. Relationships of average May (B, E) and July (C, F) biomasses of the two species to average April/May temperatures during the period 1979–1994.

also result in accelerated development of copepods, suggesting the earlier appearance of certain developmental stages. Such an advancement has been demonstrated in Lake Constance for early copepodid stages of the continuously reproducing *E. gracilis* (Seebens et al. 2007) and of the monocyclic cyclopoid *C. vicinus* (Seebens et al. 2009).

Changes in phenology are of concern because they might result in a temporal mismatch in the occurrences of consumers and their prey, which in turn may have negative consequences for consumers. In Lake Constance, the timing of the onset of the spring bloom did not advance significantly (Peeters et al. 2007), which is in contrast to the advancement of, for example, *Daphnia* and *Eudiaptomus* phenology. Despite the differential responses of phytoplankton and crustaceans, we found no evidence that maximum biomass of crustaceans were high or lower in years with cold or warm springs, but a temporal shift of maximum biomasses (Fig. 10.7A, D). Hence, the Lake Constance spring dynamics provide evidence for a temporal decoupling of consumer and prey phenology, but only weak evidence for a disruption of the trophic links between the crustacean populations and their food resources with warming. However, for *C. vicinus* we found higher offspring survival and higher early copepodid abundances in years with an early onset of the phytoplankton bloom. Seebens et al. (2009) suggest that this positive effect of early bloom onset was due to a longer predator-free time period for development.

The examples of *Daphnia* and *Eudiaptomus* from Lake Constance show that spring temperatures trigger autogenous plankton succession and the demographic development of crustacean plankton beyond spring. As noted above, April/May temperatures influenced *Daphnia's* May biomass as well as average July biomass due to a seasonal shift in population dynamics. Similar seasonal shifts can be observed for the dynamics of *Eudiaptomus*. The seasonal minimum of CI copepodites in summer (June/July) occurs two to four weeks earlier after warm April temperatures compared to cold spring temperatures (Seebens et al. 2007). This suggests that spring temperatures act as a zeitgeber for further seasonal development influencing interannual variability of zooplankton well into the summer period, that is, differences in, for example, August abundances between study years do not result from differences in August temperatures, but likely depend on differences in vernal warming in the respective years.

The strong influence of water temperatures or climate indices such as the NAO, the Pacific Decadal Oscillation, or the East Atlantic Pattern on *Daphnia* spring dynamics has been documented in many lakes differing strongly in, for example, depth or trophic status (e.g., Straile and Adrian 2000, Winder and Schindler 2004a, Schindler et al. 2005, Manca et al. 2015, Liu et al. 2021). Similar to Lake Constance, interannual variability of the timing of the *Daphnia* spring maximum can be predicted from the timing when water temperatures in the upper 5 m of the water column surpasses 13°C in lakes as different as the shallow hypertrophic Müggelsee, Germany, and the mesotrophic Lake Washington, USA (Straile et al. 2012). Using a space-for-time substitution approach, this water temperature threshold successfully predicts the timing of the *Daphnia* maximum in 49 lakes/ponds of the northern hemisphere (Fig. 10.8). The strong influence of spring temperature is supported by modeling results, which suggest, however, that variability in trophic state and especially *Daphnia* mortality rates could also modify the temperature influence on *Daphnia* phenology (Schalau et al. 2008). For example, while the *Daphnia* peak in Lake Michigan was related to water temperature, it occurred later than predicted by the 13°C rule (Pothoven and Vanderploeg 2022). Unfortunately, the reason for this delayed phenology in Lake Michigan is unclear. Documenting and explaining deviations of *Daphnia* phenology from the 13°C rule might help to predict the response of *Daphnia* seasonal dynamics to warming in the presence of other drivers. Warmer spring temperatures also resulted in earlier population developments of other cladoceran species, for example, in Müggelsee, Lake Mjøsa, Norway, Lago Maggiore, Italy, and Bautzen reservoir, Germany (Straile and Adrian 2000, Benndorf et al. 2001, Tanentzap et al. 2020, Moe et al. 2022). Likewise, phenologies of copepod species were found to be related to temperature variability, for example, in Lake Washington, Lake Michigan, and Lake Mjøsa (Winder and Schindler 2004a, Moe et al. 2022, Pothoven and Vanderploeg 2022). Hence, there is strong evidence for temperate lakes that temperature increases results in a seasonal shift of crustacean population dynamics.

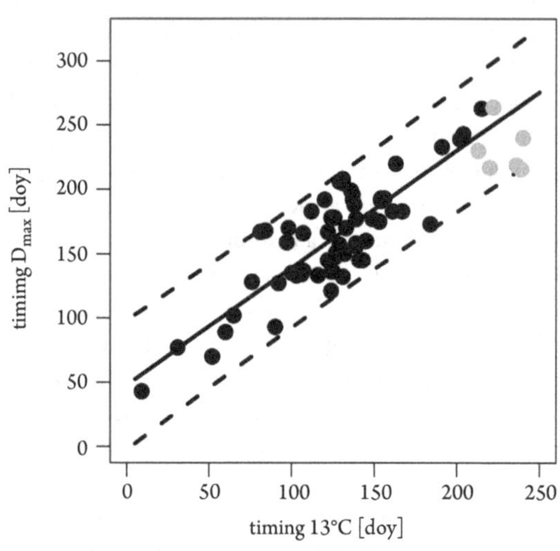

Fig. 10.8.
Relationship of the timing of the *Daphnia* maximum to the timing of the 13°C threshold in 47 lakes across the northern hemisphere (gray dots indicate lakes that did not warm up to 13°C and for which the day of maximum temperatures is shown). Lines shown are the regression fit ±95% confidence intervals. Source: data from Straile et al. (2012).

Effects of Climate Variability on Crustacean Abundances

In addition to changes in phenology, temperature variability, or warming, has also been related to population declines of increases of crustacean plankton. Such changes might occur via the indirect effects of warming, for example, when warming influences mixing dynamics in lakes, thereby changing nutrient availability for phytoplankton and thus primary productivity (Manca et al. 2015). Warming may also lengthen the overall seasonal growth period, resulting in higher annual average abundances. The effects of growth period lengthening are expected to be particularly pronounced in high altitude or high latitude lakes with short growth periods and, consequently, the potential for a large relative increase in the length of the ice-free period with warming. For example, in Green Lake 4, Southern Rocky Mountains, USA, the ice-free season increased from 82 to 134 days, which strongly increased average annual abundances of *Daphnia pulicaria* but not of *Hesperodiaptomus shoshone* (Loria et al. 2020). This difference in response between *Daphnia* and the calanoid copepod once more exemplifies the reduced growth potential of copepods relative to cladocerans.

Another cause for changes in overall abundances could be differential responses to temperature in crustacean zooplankton compared to its food or predator species, resulting in climate-induced food web disruptions. While there is abundant evidence for shifts in phenology (see above), there is not much evidence overall that these differences in response between, for example, crustacean populations and their food, actually resulted in declines or increases of crustacean populations (for a review, see Thackeray [2012]). However, a decline of *Daphnia* abundances in Lake Washington has been suggested to be due to a mismatch between the timing of the spring phytoplankton bloom, which advanced in Lake Washington, and the timing of the *Daphnia* peak, which did not advance (Winder and Schindler 2004b). Likewise, warming in Lago Maggiore resulted in changing phenologies, which are suggested to have caused a striking population increase in *Bythotrephes longimanus* of one order of magnitude, with cascading negative effects on *Daphnia* dynamics (Tanentzap et al. 2020). Unfortunately, these studies could not identify the special conditions in these lakes that made these crustacean species highly vulnerable to a mismatch. Hence, more work is needed to fully understand climate change effects on long-term crustacean population dynamics and to what extent they are indeed affected by climate-induced trophic mismatch.

Crustacean Trait Responses to Warming

Zooplankton populations are expected to adapt to climate warming both phenotypically and evolutionarily. Copepod life cycles are plastic, and conspecific populations may differ in the number of generations per year at different latitudes and mean temperatures. For example, the number of generations per year of the cyclopoid *Cyclops scutifer* decreased from one to two generations per year at latitudes $\leq 45°$N (mean July temperatures: 20°C) to 0.5 and 0.33 generations at a latitude of 82°N (mean July temperatures < 0°C) across Northern America (Elgmork 2004). As there was no systematic decline of copepod clutch size with latitude in these lakes, the response was most likely not mediated by food availability but a direct response to lower annual average temperatures and shorter ice-free seasons at the higher latitude. Hence, an increase in the number of generations with warming might be expected in some lakes, although the temperature range (> 20°C) studied by Elgmork (2004) was very large.

Warming might also change the overwintering strategy for *Daphnia*. In Lake Biwa, Japan, the number of *Daphnia* resting eggs in a sediment core decreased strongly from the 1980s onward, possibly because indirect effects of warming allowed *Daphnia* to overwinter without resting eggs (Tsugeki et al. 2009). This switch in reproductive allocation is similar to the one observed in Lake Constance in response to eutrophication, showing once more the similarity in response patterns to eutrophication and the indirect effects of warming. However, *Daphnia* may also respond directly to warmer temperatures. Experimental work on *Daphnia* clones established from resting eggs, which were isolated from a sediment core from Felbrigg Hall Lake, England, suggests that from the 1960s to the 2000s *Daphnia magna* evolved a higher capacity to cope with heat stress (Geerts et al. 2015).

In summary, climatic changes and variability have strongly influenced the seasonal dynamics of crustacean populations in temperate lakes, whereas effects on crustacean intraspecific trait variability, and long-term dynamics, that is, mediated via potential mismatches need to be studied more thoroughly. Indirect climatic effects operating via changes in nutrient supply or hypolimnetic refuge availability may strongly affect crustacean populations; however, such indirect effects also need more study. Similar to the effects of nutrient enrichment, seasonal warming favors species with the potential for fast growth rates most strongly, for example, cladocerans > cyclopoid copepods > calanoid copepods.

EFFECTS OF SPECIES INVASIONS ON CRUSTACEAN PLANKTON

Invasion of competitors and predators of zooplankton often have striking effects on native zooplankton populations (see Chapter 13 in this volume). Invasions via natural dispersal were often facilitated by modification of the recipient habitat by anthropogenic stressors, such as eutrophication, or via intentional or unintentional human transport of species. For example, invasions of *D. galeata* and *C. vicinus* in Lake Constance were facilitated by the eutrophication. Likewise, global warming is expected to facilitate the invasion of crustacean species from regions with higher temperatures. However, empirical evidence for climate-induced invasions of crustaceans in lakes is as yet lacking (Dexter and Bollens 2020).

With habitat modification, invasion of the pelagic zone of a lake may occur over small distances, for example, from nearby lakes or even within lakes from the littoral zone. Several zooplankton populations, which were able to establish a population in the pelagic zone of Upper Lake Constance following eutrophication, were previously present in the littoral zone of the lake or in Lower Lake Constance, a more eutrophic basin downstream of Upper Lake Constance. Also in Mondsee, Austria, the source of some species newly detected in the pelagic zone during the eutrophication period (the copepods *Acanthocyclops robustus* and *Thermocyclops oithonoides*, and the cladoceran *Chydorus sphaericus*) was the littoral zone of the lake (Nauwerck 1991). Unfortunately, there is no systematic review of such small-scale within-lake expansions of crustacean habitats, which are facilitated by ecosystem change.

Intentional transport of new species or unintentional transport (e.g., via ballast water) can occur over large distances. Regardless of the origin of the invading population, invasions can have strong effects on the population dynamics of native crustaceans, including population declines or even replacements, but also hybridization and introgression with native species. The natural invasion or the introduction of new fish species by humans into lakes provide classical examples of the effects of species invasions on prey communities. Brooks and Dodson (1965) showed that the establishment of alewifes (*Alosa pseudoharengus*) in several New England lakes resulted in population declines or the disappearance of large zooplankton species, such as *Daphnia* spp. and *Diaptomus* spp., while small zooplankton such as *Bosmina longirostris* or *Ceriodaphnia lacustris* increased in numbers. Similar changes were observed with the introduction of fish in previously fishless alpine lakes resulting in the eradication of populations of large zooplankton species (Eby et al. 2006). Likewise, introductions or invasions of invertebrates, such as *Dreissena* mussels (e.g., Bowen et al. 2022) or other zooplankton species, can strongly alter the dynamics of established zooplankton populations. For example, the introduction of the predatory crustacean *Mysis relicta* as fish food in Lake

Tahoe, USA, resulted in declines of *Bosmina* and *Daphnia* populations (Richards et al. 1991). Other examples include the accidental transport of the predatory cladoceran *Bythotrephes longimanus* via ballast water from Europe to North America and the introduction of the North American copepod *Skistodiaptomus pallidus* to New Zealand. Both *Bythotrephes* (Kerfoot et al. 2016, Barbiero and Tuchman 2004) and *Skistodiaptomus* (Duggan et al. 2014) invasions resulted in widespread structural changes in the zooplankton community.

The invasion of the cyclopoid copepod *C. vicinus* into Lake Constance during the eutrophication period has been suggested to have resulted in the local extinction of the calanoid copepod *Heterocope borealis*. The seasonal overlap of the monocyclic life cycles of the two species proved to be unfortunate for *Heterocope*: predatory C IV copepodid stages of *C. vicinus* emerged from diapause during autumn/winter, just before the hatching of the calanoid nauplii from resting eggs in late winter. Predation by *C. vicinus* imposed high mortality in *Heterocope* nauplii, driving the population to extinction within a few years (Einsle 1983a). Hence, the seasonal niches of both native and invasive species are crucial for the impact of invasive species. A similar rapid change in zooplankton has also been recorded in Lake Constance in recent years, presumably due to the habitat expansion of sticklebacks from the littoral zone to the pelagic zone (Ogorelec et al. 2022). This habitat expansion seems to have caused increasing predation pressure on the pelagic community, shifting the size structure toward smaller species. The small *D. cucullata,* which was previously found only occasionally in Upper Lake Constance, became dominant within a few years whereas the larger *D. longispina* declined in abundance.

Hybridization and subsequent backcrossing with resident species may be an important consequence of the invasion of new crustacean species. For example, the invasion of *D. galeata* in Lake Constance, presumably facilitated by eutrophication, resulted in hybridization with the native *D. longispina* and subsequent introgression (Weider and Stich 1992, Jankowski and Straile 2004, Brede et al. 2009), resulting in the appearance of new daphnid morphotypes (Einsle 1983b). Hybridization with native *Bosmina* is also assumed after the invasion of a new *Bosmina* species in Lake Constance (Hofmann 1998). Hybridization and introgression between species might have a long-lasting legacy preventing the recovery of the community composition after restoration of lakes. While the invasive *D. galeata* has declined strongly with oligotrophication in Lake Constance, hybrids still make up a large part of the *Daphnia* community. Likewise, *D. longispina* x *D. galeata* F1 hybrids tend to dominate the *Daphnia* communities in those European peri-alpine lakes that were most eutrophied in the past (Keller et al. 2008). Hence, species invasions might have long-lasting ecological and evolutionary consequences for native crustacean species.

In summary, invasive predators and competitors do have a strong potential to affect local crustacean populations with effects ranging from local extinctions to hybridization. Often nonhuman-mediated invasions are facilitated by other environmental stressors, and can magnify or reverse the effects of these stressors on crustaceans.

CONCLUSIONS AND FUTURE DIRECTIONS

Zooplankton populations in many lakes are affected by a variety of anthropogenic stressors. The most frequent anthropogenic stressors are changes in nutrient concentrations of lakes, global warming, and the invasion of neozoa. These stressors are strongly interlinked and operate via various processes on crustacean zooplankton. For example, both eutrophication and warming may change birth rates via changing food availability or increase predation pressure via reducing refuge space for zooplankton. Likewise, warming and nutrient change will facilitate invasions by nonnative species. This multitude of effects, which are often shared by stressors, will often make it difficult to disentangle the specific effects of the single stressors.

Anthropogenic stressors likely interact additively, synergistically, or antagonistically (Birk et al. 2020). Interactions of stressors are expected, for example, when warming and/or eutrophication affect the use of hypolimnetic refuge, which, in turn, may affect population growth rates more strongly in the presence of an invasive predator. In Canadian lakes, synergistic effects of the decline of calcium and of the invasion of *Bythotrephes longimanus* on *Daphnia* populations are expected as low calcium could impair the anti-predator defenses of *Daphnia* (Azan et al. 2015). Likewise, changes in food quality caused by eutrophication can affect the heat tolerance of zooplankton (Sarrazin and Sperfeld 2022) and thereby change its population dynamics. However, currently it is unknown to what extent such complex stressor interactions actually affect the growth or mortality rates of zooplankton in lakes.

In temperate lakes, zooplankton populations live in a highly seasonal environment in which they need to balance periods of positive population growth against periods of negative population growth in order to persist. Zooplankton species have evolved a huge diversity of life cycle strategies to exploit periods that enable positive population growth and anti-predator strategies allowing them to avoid or endure the periods of negative population growth. These strategies or traits predispose zooplankton species to be either less or strongly affected by anthropogenic stressors. Hence, the prediction of stressor effects requires information about the seasonality of environmental forcing on zooplankton and about the specific zooplankton traits. In this chapter, differences in the responses of crustacean populations to stressors have been linked to the species traits and their position along the growth rate–mortality continuum. *Daphnia* and calanoid copepods take opposite positions in this continuum, allowing calanoid copepods to dominate nutrient-poor cold lakes, whereas *Daphnia* are able to better exploit increased food availability and/or warmer conditions. While first trait databases for zooplankton have been established (Hébert et al. 2016), these data have rarely been linked systematically to changes in crustacean populations in response to environmental variations. Trait-specific analyses have been done for several other organism groups, and they could be an important step forward in predicting the responses of crustacean zooplankton to anthropogenic stressors. Furthermore, traits of zooplankton respond to stressors both phenotypically and genetically via microevolutionary change. Studying the role of adaptations for the persistence and response of zooplankton to stressors will be an important task for future studies. Improved possibilities for studying population genomics will allow the linking of changes in zooplankton growth or mortality rates to possible changes in genomic architecture, thereby improving our understanding of stressor effects on zooplankton dynamics.

All stressors may affect the phenology of zooplankton species as well as their maximum and mean abundances. The examples provided in this chapter tentatively suggest that—within the current range of stressor variations—shifts in phenology are of minor importance relative to changes in overall abundance in case of eutrophication. However, for warming the opposite might be the case, at least for those effects that do not operate via changes in primary production. Consequently, negative effects on zooplankton due to a mismatch of zooplankton with either prey or predator populations are not strongly supported. More data are needed to test the match-mismatch hypothesis rigorously and separately for different crustacean zooplankton species. The continuation of long-term monitoring programs on zooplankton populations at sufficient seasonal resolution will be crucial to improve our understanding of how the multitude of stressors and their interactions affect zooplankton and entire lake ecosystems.

ACKNOWLEDGMENTS

I thank the editors for inviting me to write this chapter and for many helpful comments on the manuscript. Thanks also to the Institute for Lake Research, Langenargen, Germany and Richard Barbiero Buffalo State University, USA for providing data.

REFERENCES

Anneville, O., J.C. Molinero, S. Souissi, G. Balvay, and D. Gerdeaux. 2007. Long-term changes in the copepod community of Lake Geneva. Journal of Plankton Research 29:i49–i59.

Auerbach, M., W. Maerker, and J. Schmalz. 1924. Hydrographisch-biologische Bodensee-Untersuchungen. I. Ergebnisse der Jahre 1920–1922. Archiv für Hydrobiologie Supplement 3:597–738.

Auerbach, M., W. Maerker, and J. Schmalz. 1926. Hydrographisch-biologische Bodensee-Untersuchungen II. Ergebnisse der Jahre 1923 und 1924 und Zusammenfassung 1920–1924. Verh.naturwiss.Ver.Karlsruhe 30:1–129.

Azan, S.S.E., S.E. Arnott, and N.D. Yan. 2015. A review of the effects of *Bythotrephes longimanus* and calcium decline on zooplankton communities—Can interactive effects be predicted? Environmental Reviews 23:395–413.

Barbiero, R.P., L.G. Rudstam, J.M. Watkins, and B.M. Lesht. 2019. A cross-lake comparison of crustacean zooplankton communities in the Laurentian Great Lakes, 1997–2016. Journal of Great Lakes Research 45:672–690.

Barbiero, R.P., and M.L. Tuchman. 2004. Changes in the crustacean communities of Lakes Michigan, Huron, and Erie following the invasion of the predatory cladoceran *Bythotrephes longimanus*. Canadian Journal of Fisheries and Aquatic Sciences 61:2111–2125.

Becker, M., and R. Eckmann. 1992. Plankton selection by pelagic european whitefish in Lake Constance: Dependency of season and time of day. Polskie Archiwum Hydrobiologii 39:393–402.

Benndorf, J., J. Kranich, T. Mehner, and A. Wagner. 2001. Temperature impact on the midsummer decline of *Daphnia galeata*: An analysis of long-term data from the biomanipulated Bautzen Reservoir (Germany). Freshwater Biology 46:199–211.

Birk, S., D. Chapman, L. Carvalho, B.M. Spears, H.E. Andersen, C. Argillier, S. Auer, A. Baattrup-Pedersen, L. Banin, M. Beklioğlu, E. Bondar-Kunze, A. Borja, P. Branco, T. Bucak, A.D. Buijse, A.C. Cardoso, R.M. Couture, F. Cremona, D. de Zwart, C.K. Feld, M.T. Ferreira, H. Feuchtmayr, M.O. Gessner, A. Gieswein, L. Globevnik, D. Graeber, W. Graf, C. Gutiérrez-Cánovas, J. Hanganu, U. Işkın, M. Järvinen, E. Jeppesen, N. Kotamäki, M. Kuijper, J.U. Lemm, S. Lu, A.L. Solheim, U. Mischke, S.J. Moe, P. Nõges, T. Nõges, S.J. Ormerod, Y. Panagopoulos, G. Phillips, L. Posthuma, S. Pouso, C. Prudhomme, K. Rankinen, J.J. Rasmussen, J. Richardson, A. Sagouis, J.M. Santos, R.B. Schäfer, R. Schinegger, S. Schmutz, S.C. Schneider, L. Schülting, P. Segurado, K. Stefanidis, B. Sures, S.J. Thackeray, J. Turunen, M.C. Uyarra, M. Venohr, P.C. von der Ohe, N. Willby, and D. Hering. 2020. Impacts of multiple stressors on freshwater biota across spatial scales and ecosystems. Nature Ecology and Evolution 4:1060–1068.

Bonacina, C., G. Bonomi, L. Barbanti, R. Mosello, and D. Ruggiu. 1988. Recovery of an industrially acidified, ammonium and heavy metals polluted lake (Lake Orta, N. Italy), due to the adoption of treatment plants. Verhandlungen der Internationalen Vereinigung für Theoretische und Angewandte Limnologie 23:535–544.

Bowen, K.L., W.J. Currie, H. Niblock, C.L. Ward, B. Metcalfe, K.M.D. Cuddington, T.B. Johnson, and M.A. Koops. 2022. Importance of long-term intensive monitoring programs for understanding multiple drivers influencing Lake Ontario zooplankton communities. Journal of Great Lakes Research 48:717–733.

Brede, N., C. Sandrock, D. Straile, P. Spaak, T. Jankowski, B. Streit, and K. Schwenk. 2009. The impact of human-made ecological changes on the genetic architecture of *Daphnia* species. Proceedings of the National Academy of Sciences of the United States of America 106:4758–4763.

Brett, M.T. 1989. Zooplankton communities and acidification processes (A review). Water, Air, and Soil Pollution 44:387–414.

Brooks, J.L., and S.I. Dodson. 1965. Predation, body size, and composition of plankton. Science 150:28–35.

Cordellier, M., M.W. Wojewodzic, M. Wessels, C. Kuster, and E. von Elert. 2021. Next-generation sequencing of DNA from resting eggs: Signatures of eutrophication in a lake's sediment. Zoology 145:125895.

De Bernardi, R., G. Giussani, M. Manca, and D. Ruggiu. 1988. Long-term dynamics of plankton communities in Lago Maggiore (N. Italy). Verhandlungen der Internationalen Vereinigung für Theoretische und Angewandte Limnologie 23:729–733.

Dexter, E., and S.M. Bollens. 2020. Zooplankton invasions in the early 21st century: A global survey of recent studies and recommendations for future research. Hydrobiologia 847:309–319.

Dodson, S.I., C.E. Cáceres, and D.C. Rogers. 2010. Cladocera and other Branchiopoda. Pages 773–827 *in* J.H. Thorp and A.P. Covich, editors. Ecology and classification of North American freshwater invertebrates, 3rd edition. Academic Press, Boston, MA, USA.

Duggan, I., M.W. Neale, K.V. Robinson, P. Verburg, and N.T.N. Watson. 2014. *Skistodiaptomus pallidus* (Copepoda: Diaptomidae) establishment in New Zealand natural lakes, and its effects on zooplankton community composition. Aquatic Invasions 9:195–202.

Eby, L., W. Roach, L. Crowder, and J. Stanford. 2006. Effects of stocking-up freshwater food webs. Trends in Ecology & Evolution 21:576–584.

Edmondson, W.T., and A.H. Litt. 1982. *Daphnia* in Lake Washington. Limnology and Oceanography 27:272–293.

Einsle, U. 1964. Die Gattung *Cyclops* s. str. im Bodensee. Archiv für Hydrobiologie 60:133–199.

Einsle, U. 1967. Über einige Auswirkungen der Eutrophierung des Bodensee-Obersees auf seine planktisch lebenden Copepodenpopulationen. Schweizerische Zeitschrift für Hydrologie 29:305–310.

Einsle, U. 1968. Die Gattung *Mesocyclops* im Bodensee. Archiv für Hydrobiologie 64:131–169.

Einsle, U. 1978. Qualitative und quantitative Änderungen im Crustaceenplankton des Bodensee-Obersees. Archiv für Hydrobiologie 82:300–315.

Einsle, U. 1983a. Long-term changes in planktonic associations of crustaceans in Lake Constance and adjacent waters and their effects on competitive situations. Hydrobiologia 106:127–134.

Einsle, U. 1983b. Die Entwicklung und Männchenbildung der *Daphnia*-Populationen im Bodensee-Obersee 1956–80. Schweizerische Zeitschrift für Hydrologie 45:321–332.

Einsle, U. 1988. The long-term dynamics of crustacean communities in Lake Constance (Obersee, 1962–1986). Schweizerische Zeitschrift für Hydrologie 50/2:136–165.

Elgmork, K. 2004. Life cycles of the freshwater, planktonic copepod *Cyclops scutifer* Sars on a north–south gradient in North America. Hydrobiologia 529:37–48.

Elster, H.J. 1954. Über die Populationsdynamik von *Eudiaptomus gracilis* Sars und *Heterocope borealis* Fischer im Bodensee. Archiv für Hydrobiologie Supplement 20:546–561.

Geerts, A.N., J. Vanoverbeke, B. Vanschoenwinkel, W. Van Doorslaer, H. Feuchtmayr, D. Atkinson, B. Moss, T.A. Davidson, C.D. Sayer, and L. De Meester. 2015. Rapid evolution of thermal tolerance in the water flea *Daphnia*. Nature Climate Change 5:665–668.

Gillooly, J.F. 2000. Effect of body size and temperature on generation time in zooplankton. Journal of Plankton Research 22:241–251.

Gronchi, E., K.D. Jöhnk, D. Straile, S. Diehl, and F. Peeters. 2021. Local and continental-scale controls of the onset of spring phytoplankton blooms: Conclusions from a proxy-based model. Global Change Biology 27:1976–1990.

Güde, H., and D. Straile. 2016. Bodensee: Ökologie und anthropogene Belastungen eines tiefen Voralpensees. Limnologie aktuell. Schweizerbart, Stuttgart, Germany.

Gyllström, M., L.A. Hansson, E. Jeppesen, F. Garcia-Criado, E. Gross, K. Irvine, T. Kairesalo, R. Kornijów, M.R. Miracle, M. Nykänen, T. Nõges, S. Romo, D. Stephen, E. van Donk, and B. Moss. 2005. The role of climate in shaping zooplankton communities of shallow lakes. Limnology and Oceanography 50:2008–2021.

Hairston Jr., N.G., A.M. Hansen, and W.R. Schaffner. 2000. The effect of diapause emergence on the seasonal dynamics of a zooplankton assemblage. Freshwater Biology 45:133–146.

Hairston Jr., N.G., C.M. Kearns, L.P. Demma, and S.W. Effler. 2005. Species-specific *Daphnia* phenotypes: A history of industrial pollution and pelagic ecosystem response. Ecology 86:1669–1678.

Hairston Jr., N.G., W. Lampert, C. Cáceres, C.L. Holtmeier, L.J. Weider, U. Gaedke, J.M. Fischer, J.A. Fox, and D.M. Post. 1999. Rapid evolution revealed by dormant eggs. Nature 401:446.

Hébert, M.P., B.E. Beisner, and R. Maranger. 2016. A compilation of quantitative functional traits for marine and freshwater crustacean zooplankton. Ecology 97:1081.

Hébert, M.-P., B.E. Beisner, and R. Maranger. 2017. Linking zooplankton communities to ecosystem functioning: Toward an effect-trait framework. Journal of Plankton Research 39:3–12.

Hessen, D.O., V. Bakkestuen, and B. Walseng. 2007. Energy input and zooplankton species richness. Ecography 30:749–758.

Hofer, B. 1896. Die Verbreitung der Tierwelt im Bodensee nebst vergleichenden Studien in einigen anderen Süßwasserbecken. Schriften des Vereins für Geschichte des Bodensees und seiner Umgebung 28:281–366.

Hofmann, W. 1998. The response of *Bosmina* (*Eubosmina*) to eutrophication of Upper Lake Constance: The subfossil record. Archiv für Hydrobiologie, Special Issues: Advances in Limnology 53:275–283.

Isanta-Navarro, J., N.G.J. Hairston, J. Beninde, A. Meyer, D. Straile, M. Möst, and D. Martin-Creuzburg. 2021. Reversed evolution of grazer resistance to cyanobacteria. Nature Communications 12:1945.

Jane, S.F., J.L. Mincer, M.P. Lau, A.S.L. Lewis, J.T. Stetler, and K.C. Rose. 2022. Longer duration of seasonal stratification contributes to widespread increases in lake hypoxia and anoxia. Global Change Biology 29:1009–1023.

Jankowski, T., and D. Straile. 2003. A comparison of egg-bank and long-term plankton dynamics of two *Daphnia* species, *D. hyalina* and *D. galeata*: Potentials and limits of reconstruction. Limnology and Oceanography 48:1948–1955.

Jankowski, T., and D. Straile. 2004. Allochronic differentiation among *Daphnia* species, hybrids and backcrosses: The importance of sexual reproduction for population dynamics and genetic architecture. Journal of Evolutionary Biology 17:312–321.

Jeppesen, E., J.P. Jensen, C. Jensen, B. Faafeng, D.O. Hessen, M. Søndergaard, T. Lauridsen, P. Brettum, and K. Christoffersen. 2003. The impact of nutrient state and lake depth on top-down control in the pelagic zone of lakes: A study of 466 lakes from the temperate zone to the arctic. Ecosystems 6:313–325.

Jeppesen, E., P. Nõges, T.A. Davidson, J. Haberman, T. Nõges, K. Blank, T.L. Lauridsen, M. Søndergaard, C. Sayer, R. Laugaste, L.S. Johansson, R. Bjerring, and S.L. Amsinck. 2011. Zooplankton as indicators in lakes: A scientific-based plea for including zooplankton in the ecological quality assessment of lakes according to the European Water Framework Directive (WFD). Hydrobiologia 676:279–297.

Jeppesen, E., M. Søndergaard, J.P. Jensen, K.E. Havens, O. Anneville, L. Carvalho, M.F. Coveney, R. Deneke, M.T. Dokulil, B. Foy, D. Gerdeaux, S.E. Hampton, S. Hilt, K. Kangur, J. Kohler, E.H.H.R. Lammens, T.L. Lauridsen, M. Manca, M.R. Miracle, B. Moss, P. Nõges, G. Persson, G. Phillips, R. Portielje, S. Romo, C.L. Schelske, D. Straile, I. Tatrai, E. Willén, and M. Winder. 2005. Lake responses to reduced nutrient loading—An analysis of contemporary long-term data from 35 case studies. Freshwater Biology 50:1747–1771.

Jeziorski, A., A.M. Paterson, and J.P. Smol. 2012. Changes since the onset of acid deposition among calcium-sensitive cladoceran taxa within softwater lakes of Ontario, Canada. Journal of Paleolimnology 48:323–337.

Jochimsen, M.C., R. Kümmerlin, and D. Straile. 2013. Compensatory dynamics and the stability of phytoplankton biomass during four decades of eutrophication and oligotrophication. Ecology Letters 16:81–99.

Keller, B., J. Wolinska, M. Manca, and P. Spaak. 2008. Spatial, environmental and anthropogenic effects on the taxon composition of hybridizing *Daphnia*. Philosophical Transactions of the Royal Society B: Biological Sciences 363:2943–2952.

Kerfoot, W.C., M.M. Hobmeier, F. Yousef, B.M. Lafrancois, R.P. Maki, and J.K. Hirsch. 2016. A plague of waterfleas (*Bythotrephes*): Impacts on microcrustacean community structure, seasonal biomass, and secondary production in a large inland-lake complex. Biological Invasions 18:1121–1145.

Kiefer, F., and R. Muckle. 1959. Beobachtungen am Crustaceenplankton des Überlinger Sees (Bodensee) 1952–1957 I. Das Auftreten der Arten im Jahresverlauf. Beiträge zur naturkundlichen Forschung in Südwestdeutschland 18:5–41.

Kiørboe, T., and M.K. Thomas. 2020. Heterotrophic eukaryotes show a slow-fast continuum, not a gleaner-exploiter trade-off. Proceedings of the National Academy of Sciences of the United States of America 117:24893–24899.

Lepori, F., and J.J. Roberts. 2017. Effects of internal phosphorus loadings and food-web structure on the recovery of a deep lake from eutrophication. Journal of Great Lakes Research 43:255–264.

Leydig, F. 1860. Naturgeschichte der Daphniden (Crustacea Cladocera). H. Laupp'sche Buchhandlung, Laupp & Siebeck, Tübingen, Germany.

Liu, X., G. Dur, S. Ban, Y. Sakai, S. Ohmae, and T. Morita. 2021. Quasi-decadal periodicities in growth and production of the copepod *Eodiaptomus japonicus* in Lake Biwa, Japan, related to the Arctic Oscillation. Limnology and Oceanography 66:3783–3795.

Loria, K.A., K.R. Christianson, and P.T.J. Johnson. 2020. Phenology of alpine zooplankton populations and the importance of lake ice-out. Journal of Plankton Research 42:727–741.

Manca, M., M. Rogora, and N. Salmaso. 2015. Inter-annual climate variability and zooplankton: Applying teleconnection indices to two deep subalpine lakes in Italy. Journal of Limnology 74:123–132.

Moe, S., A. Hobæk, J. Persson, B. Skjelbred, and J. Løvik. 2022. Shifted dynamics of plankton communities in a restored lake: Exploring the effects of climate change on phenology through four decades. Climate Research 86:125–143.

Nauwerck, A. 1991. Zooplankton changes in Mondsee. Verhandlungen der Internationalen Vereinigung für Theoretische und Angewandte Limnologie 24:974–979.

Ogorelec, Ž., A. Brinker, and D. Straile. 2022. Small but voracious: Invasive generalist consumes more zooplankton in winter than native planktivore. NeoBiota 78:71–97.

Patalas, K. 1990. Diversity of the zooplankton communities in Canadian shield lakes as a function of climate. Verhandlungen der Internationalen Vereinigung für Theoretische und Angewandte Limnologie 24:360–368.

Peeters, F., D. Straile, A. Lorke, and D.M. Livingstone. 2007. Earlier onset of the spring phytoplankton bloom in lakes of the temperate zone in a warmer climate. Global Change Biology 13:1898–1909.

Pothoven, S.A., and H.A. Vanderploeg. 2022. Variable changes in zooplankton phenology associated with the disappearance of the spring phytoplankton bloom in Lake Michigan. Freshwater Biology 67:365–377.

Reid, J.W., and C.E. Williamson. 2010. Copepoda. Pages 829–899 *in* J.H. Thorp and A.P. Covich, editors. Ecology and classification of North American freshwater invertebrates, 3rd edition. Academic Press, Boston, MA, USA.

Richards, K., C. Goldman, E. Byron, and C. Levitan. 1991. The mysids and lake trout of Lake Tahoe: A 25-year history of changes in the fertility, plankton, and fishery of an alpine lake. American Fisheries Society Symposium 9:30–38.

Rother, A., M. Pitsch, and S. Hülsmann. 2010. The importance of hatching from resting eggs for population dynamics and genetic composition of *Daphnia* in a deep reservoir. Freshwater Biology 55: 2319–2331.

Santer, B. 1998. Life cycle strategies of free-living copepods in fresh waters. Journal of Marine Systems 15:327–336.

Santer, B., E. Blohm-Sievers, C. Cáceres, and N.G.J. Hairston. 2000. Life-history variation in the coexisting freshwater copepods *Eudiaptomus gracilis* and *Eudiaptomus graciloides*. Archiv für Hydrobiologie 149:441–458.

Sarrazin, J., and E. Sperfeld. 2022. Food quality mediates responses of *Daphnia magna* life history traits and heat tolerance to elevated temperature. Freshwater Biology 67:1521–1531.

Schalau, K., K. Rinke, D. Straile, and F. Peeters. 2008. Temperature is the key factor explaining interannual variability of *Daphnia* development in spring: A modelling study. Oecologia 157:531–543.

Schindler, D.E., D.E. Rogers, M.D. Scheuerell, and C.A. Abrey. 2005. Effects of changing climate on zooplankton and juvenile sockeye salmon growth in southwestern Alaska. Ecology 86:198–209.

Seebens, H., U. Einsle, and D. Straile. 2009. Copepod life cycle adaptations and success in response to phytoplankton spring bloom phenology. Global Change Biology 15:1394–1404.

Seebens, H., U. Einsle, and D. Straile. 2013. Deviations from synchrony: Spatio-temporal variability of zooplankton community dynamics in a large lake. Journal of Plankton Research 35:22–32.

Seebens, H., D. Straile, R. Hoegg, H.B. Stich, and U. Einsle. 2007. Population dynamics of a freshwater calanoid copepod: Complex responses to changes in trophic status and climate variability. Limnology and Oceanography 52:2364–2372.

Sommer, U., Z.M. Gliwicz, W. Lampert, and A. Duncan. 1986. The PEG-model of seasonal succession of planktonic events in fresh waters. Archiv für Hydrobiologie 106:433–471.

Sterner, R. W. 2009. Role of zooplankton in aquatic ecosystems. Pages 678–688 *in* G.E. Likens, editor. Encyclopedia of inland waters. Academic Press, Cambridge, MA, USA.

Sterner, R.W., and J.J. Elser. 2002. Ecological Stoichiometry—The biology of elements from molecules to the biosphere. Princeton University Press, Princeton, New Jersey.

Stich, H.B. 2004. Back again: The reappearance of *Diaphanosoma brachyurum* in Lake Constance. Archiv für Hydrobiologie 159:423–431.

Stich, H.B., and W. Lampert. 1981. Predator evasion as an explanation of diurnal vertical migration by zooplankton. Nature 293:396–398.

Straile, D. 2000. Meteorological forcing of plankton dynamics in a large and deep continental European lake. Oecologia 122:44–50.

Straile, D. 2015. Zooplankton biomass dynamics in oligotrophic versus eutrophic conditions: A test of the PEG model. Freshwater Biology 60:174–183.

Straile, D., and R. Adrian. 2000. The North Atlantic Oscillation and plankton dynamics in two European lakes—Two variations on a general theme. Global Change Biology 6:663–670.

Straile, D., R. Adrian, and D.E. Schindler. 2012. Uniform temperature dependency in the phenology of a keystone herbivore in lakes of the Northern Hemisphere. PloS ONE 7:e45497.

Straile, D., and W. Geller. 1998. Crustacean zooplankton in Lake Constance from 1920 to 1995: Response to eutrophication and reoligotrophication. Archiv für Hydrobiologie, Special Issues: Advances in Limnology 53:255–274.

Straile, D., K.D. Jöhnk, and H. Rossknecht. 2003. Complex effects of winter warming on the physicochemical characteristics of a deep lake. Limnology and Oceanography 48:1432–1438.

Straile, D., and H. Müller. 2010. Response of *Bosmina* to climate variability and reduced nutrient loading in a large lake. Limnologica 40:92–96.

Tanentzap, A.J., G. Morabito, P. Volta, M. Rogora, N.D. Yan, and M. Manca. 2020. Climate warming restructures an aquatic food web over 28 years. Global Change Biology 26:6852–6866.

Tessier, A.J., and C.E. Cáceres. 2004. Differentiation in sex investment by clones and populations of *Daphnia*. Ecology Letters 7:695–703.

Thackeray, S.J. 2012. Mismatch revisited: What is trophic mismatching from the perspective of the plankton? Journal of Plankton Research 34:1001–1010.

Tsugeki, N.K., S. Ishida, and J. Urabe. 2009. Sedimentary records of reduction in resting egg production of *Daphnia galeata* in Lake Biwa during the 20th century: A possible effect of winter warming. Journal of Paleolimnology 42:155–165.

Weider, L.J., W. Lampert, M. Wessels, J.K. Colbourne, and P. Limburg. 1997. Long-term genetic shifts in a microcrustacean egg bank associated with anthopogenic changes in the Lake Constance ecosystem. Proceedings of the Royal Society B-Biological Sciences 264:1613–1618.

Weider, L.J., and H.B. Stich. 1992. Spatial and temporal heterogeneity of *Daphnia* in Lake Constance; intra- and interspecific comparisons. Limnology and Oceanography 37:1327–1334.

Weismann, A. 1876. Das Thierleben im Bodensee. Schriften des Vereins für Geschichte des Bodensees und seiner Umgebung 7:132–160.

Weismann, A. 1878. Ueber die Schmuckfarben der Daphnoiden. Zeitschrift für wissenschafltiche Zoologie 30:123–165.

Winder, M., and D.E. Schindler. 2004a. Climatic effects on the phenology of lake processes. Global Change Biology 10:1844–1856.

Winder, M., and D.E. Schindler. 2004b. Climate change uncouples trophic interactions in an aquatic ecosystem. Ecology 85:2100–2106.

Wojtal-Frankiewicz, A. 2012. The effects of global warming on *Daphnia* spp. population dynamics: A review. Aquatic Ecology 46:37–53.

11

IMPACTS OF CONTAMINANTS ON CRUSTACEANS IN THE FIELD

Alex T. Ford, and Thomas Knigge

Abstract

As with many groups of wildlife, crustaceans have seen considerable declines in their populations over the past century due to fisheries, habitat destruction, climate change, species invasions, and poor water quality. Many inorganic and organic pollutants can compromise the physiological performance of crustaceans, and it is yet unknown to what extent man-made chemicals can act as endocrine disruptors or interfere with chemical communication. Field monitoring programs have revealed examples of developmental abnormalities, including fluctuating asymmetry, limb development, intersexuality, and embryo aberrations. Ingested plastic debris may clog digestive organs, and entanglement in marine debris items often leads to severe injury or suffocation. Pathogens and disease are major sources of mortality in crustaceans both in the field and in aquaculture, and their impact is only just being realized. Some contaminants bioaccumulate in tissues, causing long-term intoxication whereas other contaminants are released in massive pulse events, such as oil spills, resulting in acute catastrophic effects. This chapter briefly summarizes the use of crustaceans in ecotoxicology and as bioindicators. We review the toxic effects of acute and chronic pollution and how it may negatively affect crustacean populations in the field. Based on notable declines of crustacean species over the last decades, either periodically or persistently, we assemble a body of evidence for the possible contribution of contaminants to this decline. Special attention is given to microplastics and developmental abnormalities, as well as diseases and parasites. We close this chapter with regards to open questions that should receive further attention for future research.

Alex T. Ford, and Thomas Knigge, *Impacts of Contaminants on Crustaceans in the Field*. In: *Ecology and Conservation*.
Edited by: Lars Gutow and Martin Thiel, Oxford University Press. © Oxford University Press (2025).
DOI: 10.1093/oso/9780197768242.003.0011

THE ROLE OF CRUSTACEANS IN ECOTOXICOLOGY AND ENVIRONMENTAL MONITORING

Crustaceans have been used extensively in international standard toxicity testing and environmental monitoring (LeBlanc 2007). Along with algae and certain fish species, they form the backbone of aquatic environmental acute and chronic toxicity testing and risk assessment (Nikinmaa 2014). The most commonly used crustaceans are parthenogenically reproducing species such as branchiopods (e.g., *Daphnia* sp. and *Artemia* sp.) as they are easy to culture and are short-lived (Nikinmaa 2014, Jepsen et al. 2020). Copepods (Maxillopoda) and amphipods (Malacostraca) also form a large component in laboratory-based research (Kwok et al. 2015, Podlesińska and Dąbrowska 2019) and offer opportunities to study sexual and direct embryo development. This means that researchers can undertake multigenerational studies that are too difficult with longer-living species. The US Environmental Protection Agency (EPA) Ecotoxicology Knowledgebase has over 1 million records from approximately 61,000 publications and nearly 14,000 species. Crustaceans make up nearly 1000 (7%) of the species entries (25 Orders and 125 Families) and ~10% of the records and publications (Fig. 11.1).

Larger crustaceans have more typically been used in field monitoring and are commonly associated with species of commercial importance or are studied alongside human health considerations (shellfish toxins) and zoonotic diseases. For example, crabs and shrimps have been used as indicator species for the monitoring of contamination in marine and freshwater environments throughout the world (EPA 2022).

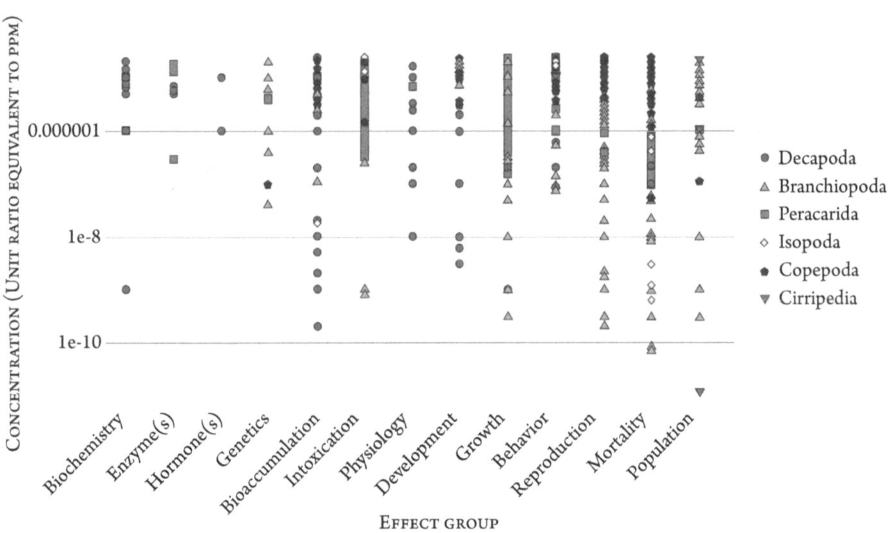

Fig. 11.1.
Articles relating to crustacean ecotoxicology in the US EPA ECOTOX Knowledgebase ordered by effect recorded and taxa. Graph represents ~900 species and more than 80,000 plottable records. Source: data from https://cfpub.epa.gov/ecotox/index.cfm (Oct 2023); see a color version of this figure in the centerfold.

Different Types of Pollution and Their Physiological Effects on Crustaceans

Crustaceans can encounter two major pollution scenarios in the field that will affect their population dynamics differently (Fig. 11.2):

(1) Repeated or continuous exposure to low concentrations of pollutants gradually released into the environment leads to long-term effects on growth, reproduction, and development. Sublethal impairment of reproduction and growth appear to be among the most pertinent consequences of chronic toxicity in crustaceans (Nimmo et al. 1981, Breteler et al. 1982) and may affect populations of sensitive species and alter community structures. Notably, the effects of chronic toxicity may be delayed and are often irreversible.

(2) Accidental pollution incidentally releases high concentrations of toxic substances into soils, lakes, rivers, or the sea. Large spills can be dramatic in their consequences and frequently led to local mass extinctions of many species. Smaller spills may happen more frequently, but may go unnoticed, although they may affect crustaceans in a similar manner. Despite the initially high mortality, more or less rapid recolonization has been reported for aquatic habitats although the pollutants may persist for decades. Sensitive species might disappear completely or return at significantly lower numbers and densities, while tolerant and opportunistic species may take over, for example, amphipods following oil or chemical spills (Dauvin 1987, ICPDR 2002, WWF 2002).

The effects of different pollutants have mostly been studied in the laboratory in specific combinations of contaminants and endpoints (Fig. 11.3). The situation in the field is much more

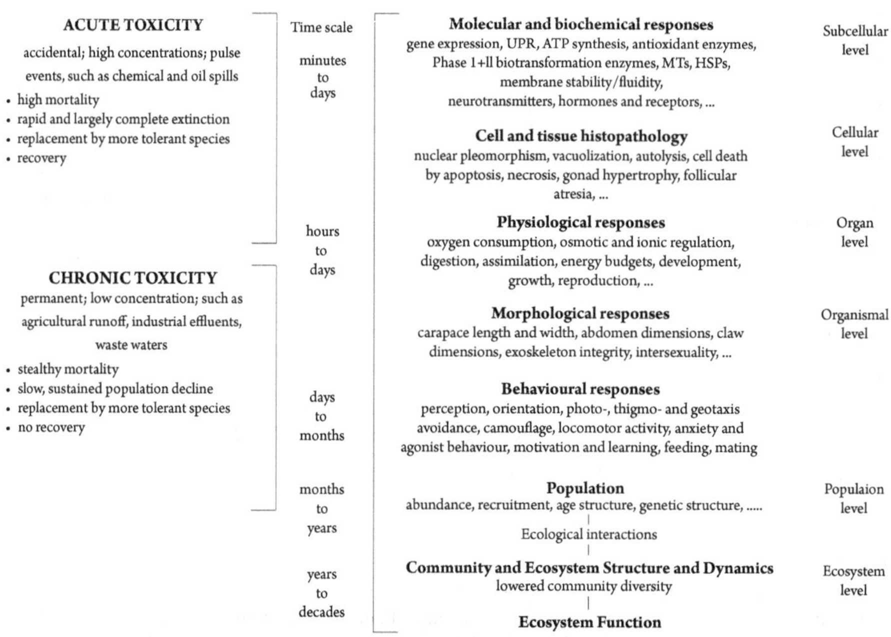

Fig. 11.2.
A chronological framework of acute and chronic toxicity with corresponding physiological effects at different levels of organization. UPR = unfolded protein response (ER stress), Phase I and II = mixed function oxygenases (notably cytochrome P450 enzymes) and biotransformation enzymes (superoxide dismutase, catalase, glutathione reductase, glutathione peroxidase, glutathione-transferase, etc.), MTs, Metallothioneines, HSPs, heat shock proteins. Source: data from Sheehan (1984) and Walker et al. (2012).

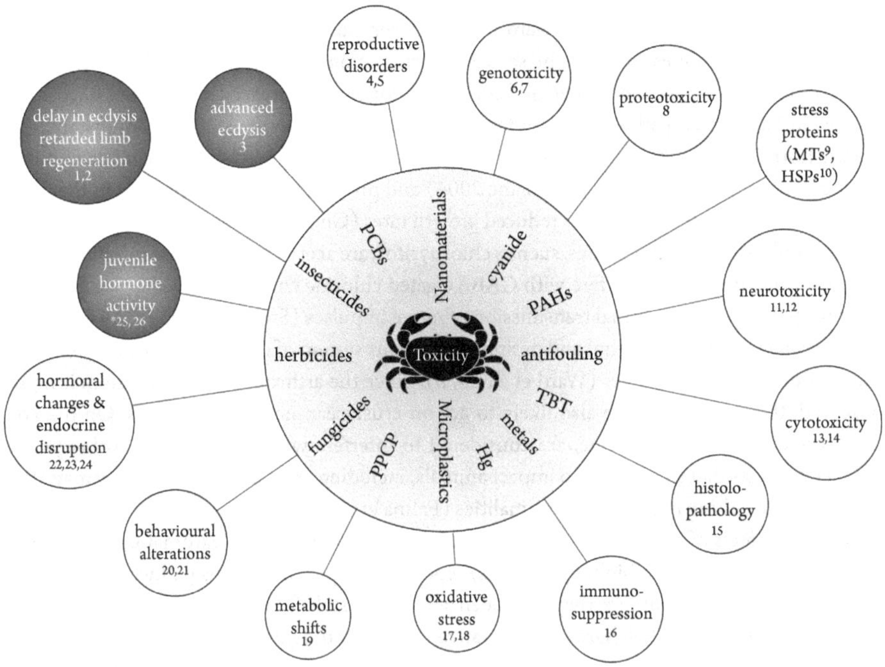

Fig. 11.3.
A non-exhaustive representation of classes of contaminants and physiological responses by crustaceans to them. General responses in light gray, arthropod-specific responses in dark gray. (1 – Weis et al. 1987; 2 – Weis et al. 1992; 3 – Song et al. 2017; 4 – Sundelin and Eriksson 1998; 5 – Siegel and Wenner 1984; 6 – Klobučar et al. 2012; 7 – Pinheiro et al. 2013; 8 – Eckwert et al. 1997; 9 – Correi et al. 2002; 10 – Arts et al. 2004; 11 – Steevens and Benson 2000; 12 – Ben Khedher et al. 2017; 13 – Schultz and Kennedy 1976; 14 – Dell'Anno et al. 2008; 15 – Soegianto et al. 1999; 16 – Lorenzon et al. 2001; 17 – Lavarías et al. 2011; 18 – Ghedira et al. 2011; 19 – Bao et al. 2020; 20 – Chabenat et al. 2021; 21 – Krång and Dahlström 2006; 22 – Rodríguez et al. 2007; 23 – LeBlanc 2007; 24 – Parmentier et al. 2019; 25 – Mu and LeBlanc 2004; 26 – Jaegers and Gismondi 2020). * that is, "Crustacean Juvenile Hormone" Methyl Farnesoate (Homola and Chang 1997). See a color version of this figure in the centerfold.

complex as the animals are exposed to several contaminants in mixtures that can act additively, synergistically, or antagonistically. Other stressors, such as temperature, oxygenation, or acidification can come into play, thus rendering the attribution of specific effects to contaminants difficult (Fig. 11.4). In addition, specific contaminants often result in complex toxicological effects across different levels of organization. For instance, mercury is considered neurotoxic, but in the marine copepod *Calanus finmarchicus* it also affected ATP synthesis, caused oxidative stress, and inhibited enzymes, or induced apoptosis and autophagy (Tollefsen et al. 2017). Metals specifically induce metallothioneins (Amiard et al. 2006), but other mechanisms of sequestration exist (Ahearn et al. 2004). They do, however, also cause proteotoxicity, cellular and tissue damage, immunosuppression, and metabolic shifts (Eckwert et al. 1997, Soegianto et al. 1999, Lorenzon et al. 2001, Bao et al. 2020). Organotins, such as tributyltin, primarily act as ATPase inhibitors, but they are also known to interact with the retinoid X receptor, thus putatively causing endocrine disruption (Parmentier et al. 2019).

Many insecticides are directed toward the insect nervous system, whereas insect growth regulators act on hormones involved in sexual determination and/or reproduction (i.e., methyl farnesoate) or molting (i.e., ecdysone) and corresponding signaling cascades (Sánchez-Bayo 2011) and may induce similar pathways in non-target crustaceans (Laufer et al. 2013). For instance, the juvenile hormone agonists pyriproxyfen and fenoxycarb inhibited ecdysteroid-regulated embryo development in daphnids (Mu and Leblanc 2004) and prolonged intermolt periods in the mysid shrimp *Neomysis integer*, resulting in reduced growth rates (Ghekiere et al. 2006).

Organophosphorus insecticides, such as chlorpyrifos, are acetylcholinesterase inhibitors, while organochlorine pesticides interfere with $GABA_A$-gated chloride channels as well as sodium channels, thus preventing the normal transmission of nerve impulses (Stenersen 2009). As such, these pesticides may impact the central and peripheral nervous system of crustaceans, leading to neurotoxicity and behavioral changes (Ward et al. 1976). Since the arthropod nervous system is highly conserved, these pesticides are also likely to act on crustacean non-target species. On the contrary, herbicides, such as atrazine, are considered to interfere with plant-specific pathways, but evidence suggests that they may also impact animals, including crustaceans, where they may affect development resulting in embryo abnormalities (Palma et al. 2009).

The toxic effects of organic pollutants, such as alkylphenols (APs), polyaromatic hydrocarbons (PAHs), or polychlorinated biphenyls (PCBs) reportedly impact numerous pathways and physiological functions. Among others, they have been held responsible for immunosuppression and have also been found to delay shell hardening, putatively via juvenoid hormone activity of APs (Laufer et al. 2013). Generally, PAHs are known to induce phase I and II biotransformation reactions that potentially increase the number of reactive oxygen species causing oxidative stress, which in turn may damage macromolecular cell components and require an increased aerobic catabolism (Rodrigues et al. 2013). Furthermore, the inhibition of acetylcholinesterase was observed (Rodrigues et al. 2013), which may negatively impact neuromuscular activity and affect locomotor or foraging behaviors (Dissanayake et al. 2010).

CHRONIC POLLUTION LEADING TO CRUSTACEAN POPULATION DECLINES.

Crustaceans, like many aquatic species, are faced with a broad range of contaminants. This includes industrial pollutants such as metals from mining, hydrocarbons from oil extraction, pesticides and herbicides from agriculture, chemical contaminants from everyday household products, personal care products, and pharmaceuticals as well as nitrates from sewage discharges and fertilizers (Fig. 11.4). Much of the crustacean declines, unless they are due to acute releases of contaminants, are hard to disentangle from other global issues such as overfishing and climate change.

The majority of crustacean species resides in marine environments, where they make up for about a quarter of the marine fauna (Costello et al. 2010), constituting nearly one gigaton of animal biomass (Bar-On et al. 2018). Although there is little evidence that these marine crustaceans are threatened by extinction, they are subject to various pressures, including pollution that, sporadically, contributes to massive declines.

Although merely one-sixth of all crustacean species occupy freshwater ecosystems (Balian et al. 2008, Covich et al. 2010), it is these freshwater crustaceans in particular that are seriously endangered (Cumberlidge et al. 2009, Richman et al. 2015, Reid et al. 2019). Cumberlidge et al. (2009) estimated that about two-thirds of the 1280 known species of freshwater crabs are at risk. A similar

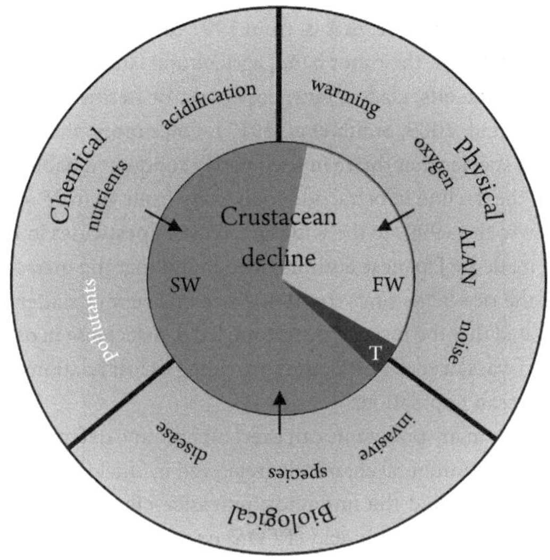

Fig. 11.4.
Combined action of physical, biological, and chemical stressors that may act on crustaceans in the field, presumably interacting in an additive or even synergistic manner, potentially causing the observed crustacean declines in the field (SW = seawater, FW = freshwater, T = terrestrial). Anthropogenic stressors, such as chemical pollution, and also noise or artificial light at night (ALAN), collectively contribute to the overall threats for crustacean populations and modulate their response capabilities.

---

picture was drawn by Gherardi (2011) and also Richman et al. (2015), who reported that about one-fourth to one-third of the 590 known freshwater crayfish species are threatened by extinction.

That pollution contributes to the threats that impact crustacean populations in the field has been reported by Bao et al. (2020), who state that the crayfish *Cambaroides dauricus* became an endangered species in northeastern China due to the pollution of the water bodies by heavy metals released by industries. European freshwater crayfish were massively decimated by crayfish plague introduced by invasive species about one century ago (Richman et al. 2015). However, the weakened endemic crayfish populations were further impacted by bad water quality and pollution in the second half of the 20th century, which the non-indigenous species tolerated better.

An example of how pollution and environmental changes are linked is given by environmental pollution from sulfuric and nitric acids, which in the poorly calcium-buffered boreal lakes led to acidification. This acidification was the primary cause for the decline of crustacean zooplankton, such as cladocerans and copepods, as well as crayfish in the Canadian and Scandinavian soft water lakes, as it affects the calcification of their cuticles (Edwards et al. 2016, Azan and Arnott 2018). Yet, increased acidification resulted mostly from massive air pollution (Grennfelt et al. 2020).

Change of land use also results in habitat loss that increases the extinction risk of many of the world's crustacean species (Cumberlidge et al. 2009). Agricultural or industrial land use, however, is followed by pollution from pesticides or heavy metals, which further threatens the animals. For instance, high mortality among freshwater crustaceans was demonstrated by Jergentz et al. (2004), who observed that the insecticides chlorpyrifos, endosulfan, and α-cypermethrin killed amphipods *Hyalella curvispina* and freshwater prawns *Macrobrachium borelli* in streams of the Argentinean pampas. Similarly, Leight et al. (2005) reported that grass shrimp *Palaemon* spp.

populations decreased within a decade, that is, from 1991 to 2000, in South Carolina tidal creeks receiving agricultural runoffs. On the other hand, agricultural land use may provide new habitats for crustaceans, such as ostracods, cladocerans, copepods, or freshwater crabs, as is the case for rice fields (Cumberlidge et al. 2009, Smith et al. 2017). Once more, the application of pesticides can affect the crustacean species that thrive in such newly conquered habitats. For example, Rossi et al. (2003) attributed the decline in ostracod biodiversity from 46 to 25 species over a period of 30 years from the 1960s to the 1990s to the widespread use of pesticides in Italian rice fields. Similar observations were made by Japanese authors, who found that the ostracod *Notodromas trulla*, a regular inhabitant of the rice fields since the 1940s, became very rare after 1970 (Okubo 2003). They further demonstrated that the insecticide fipronil led to a decrease in ostracod densities in experimental rice fields (Hayasaka et al. 2012), thus providing a clear relationship between pollution and the decline of crustacean populations.

The negative effects that many pollutants can exert on immune defense is a prominent example of the indirect action of environmental chemicals (reviewed in Shields and Huchin-Mian [2020]). Indeed, several studies have linked the immunosuppressive effects of environmental chemicals such as PCBs and perfluoroctanesulfonic acid (PFOS) or paper mill effluents and heavy metals to an increase in microsporidian infestation (Smith et al. 1995, Ford et al. 2006, Jacobson et al. 2010). Conversely, animals infested with parasites can be more susceptible to the effects of pollutants (Brown and Pascoe 1989, Lafferty et al. 2004).

Laufer et al. (2013) provided evidence that APs, as well as an insecticide with known juvenile hormone activity, delayed shell hardening in the lobsters of Long Island Sound, thus weakening their exoskeleton. They further concluded that prolonged cuticle sclerotization would facilitate microbial invasion and, thus, increase the prevalence of epizootic shell disease (ESD). In addition, PCBs, pesticides, metals, and PAHs from oil spills were found in the bay area (Shields 2013), which may have caused immunosuppression in crustaceans. Nevertheless, these contaminants alone were not responsible for the higher ESD prevalence, as contaminants were also present in parts of Long Island Sound where ESD was low (Shields 2013, Shields and Huchin-Mian 2020).

For a long time, pollution has been suspected of playing a role in the decline of amphipods of the genera *Diporeia/Monoporeia* in the Laurentian Great Lakes, the Baltic Sea, and the Sea of Okhotsk. Toxicological studies demonstrated the sensitivity of these amphipods to environmental chemicals (Landrum et al. 1991, Lukyanova et al. 2014, Löf et al. 2016). Notably, Ralston-Hooper et al. (2009) demonstrated that *Diporeia* sp. was several orders of magnitude more sensitive to the herbicide atrazine than the amphipod *Hyalella azteca*. Increasing concentrations of atrazine and its breakdown products were detected in Lake Michigan over time, suggesting a role of atrazine in the decline of *Diporeia*. Löf et al. (2016) provided evidence that sediments containing PAHs and PCBs reduced recruitment of *Monoporeia affinis* due to malformed embryos in the Baltic Sea, as did effluents from pulp mills and aluminum smelters (Sundelin and Eriksson 1998). Similarly, Blanchard et al. (2019) investigated the role of multiple stressors in the decline of *M. affinis* along the northeastern coast of Sakhalin Island in the Sea of Okhotsk. It appears plausible that PAH emissions from oil platforms and a benzene spill into the Amur River in 2005 have intensified the decline, as the nearshore populations showed lower densities than the offshore populations.

ACUTE/PULSE EVENTS IMPACTING CRUSTACEAN POPULATIONS

Acute events probably produced the most catastrophic and recognizable consequences for crustacean populations. The devastating effects of oil spills are most frequently reported and probably constitute the best-studied form of accidental pollution. Mining, chemical, and pesticide spills also cause dramatic losses of crustaceans. Although the devastating effects of acute spills are usually quite evident, they remain difficult to measure quantitatively. In most cases, reliable data sets before and after the spills are lacking. Furthermore, spills are often ephemeral and disappear from sight within days, so that the amounts of pollutants affecting the animals can only roughly be estimated.

Chemical and Pesticide Spills

In 1986, a fire destroyed a storehouse of Sandoz Ltd. near Basel (Switzerland) and released large quantities of organophosphorus insecticides, herbicides, and mercury containing fungicides into the Rhine River, causing massive kills of fish and benthic organisms (Giger 2009). At the same time the Sandoz accident happened, wastewater contaminated with atrazine from another accident at Ciba-Geigy was discharged into the Rhine. Toxicity testing with daphnids proved the insecticides to be harmful to crustaceans. As a matter of consequence, amphipods were found to be absent in the southern Upper Rhine after the spill and they were considerably reduced in numbers in the Middle Rhine (WWF 2002). Giger (2009), however, concluded that the damage was less extensive than might have been expected, as the Rhine was already chronically exposed to environmental chemicals and heavily engineered so that sensitive species had disappeared earlier. In this context, it is interesting to note that the invasion of the Rhine by the amphipod *Chelicorophium curvispinum* occurred shortly after native amphipod populations had been decimated by the Sandoz accident (Den Hartog et al. 1992); whether there is a causal relationship between acute pollution events and the subsequent appearance of invasive species deserves attention.

In the case of an insecticide spill, which consisted of a single pulse of the organophosphate chlorpyrifos into the Kennet River, a southern England tributary to the River Thames, the total abundance of macroinvertebrates was reduced along a 15 km stretch by 99.5% (Thompson et al. 2016). The previously dominating *Gammarus pulex* was most affected as it recovered at a much slower rate than the insect taxa.

Mining Spills

Mining activities often generate chronic water pollution that may profoundly change communities in rivers, lakes, estuaries, and coastlines. As with many other acute events, robust data on the effects of mining spills are scarce because faunal inventories from before the spills do not necessarily exist. In the case of the mining spills in the Tisza River system, which originates in the Carpathian Mountains (Romania) and is the largest tributary to the Danube River, inventories from the 1980s and data from several sampling campaigns in the 1990s provided a detailed overview of the macroinvertebrate taxa, including planktonic and benthic crustaceans. These data allowed for extensive follow-up studies after the heavy metal and cyanide spills that occurred in the year 2000 (ICPDR 2002, WWF 2002). Albeit being subject to chronic heavy metal pollution from the upstream mining activities, the Tisza River system was found to be rich in species due to a considerable diversity of aquatic habitats (WWF 2002). Several cladocearan species, for example, *Bosmina longirostris, Disparalona rostrata, Moina micrura,* and *Alona guttata,* as well as copepods, such as *Acanthocyclops robustus, Eucyclops serrulatus,* and *Eurytemora velox,* were listed among the zooplankton communities (ICPDR 2002). Crustaceans of the macroinvertebrate fauna that are characteristic for the Tisza River comprise amphipods, such as *Dikerogammarus villosus, Obesogammarus obesus, Gammarus balcanicus,* and *C. curvispinum* as well as the isopod *Jaera istri* and the crayfish *Pontastacus leptodactylus* (ICPDR 2002).

In January and March 2000, heavy rain and snow melts caused the dams of two mines located in the upper watershed of the Tisza River to overflow and break. In January, about 100,000 m^3 of heavy metal contaminated wastewater and up to 120 tons of cyanide were released into the Tisza

River, followed by another 100,000 m^3 of wastewater containing heavy metals and 20,000 tons of solid waste in March 2000. Concentrations of up to 32 mg cyanide·L^{-1} were measured in the Tisza tributaries. In the river stretches, which were closest to the source of the cyanide pollution, the plankton was completely extinguished, as exemplified by a 100% mortality of *Daphnia magna* in the cyanide plume (WWF 2002). As for the different crustacean populations, the report of the International Commission for the Protection of the Danube River (ICPDR 2002) stated that these were particularly affected by the mining spills. Further downstream, the cyanide pollution demonstrably destroyed up to 80% of the cladoceran and copepod species (ICPDR 2002). Furthermore, populations of two malacostracan species, the amphipod *C. curvispinum* and the isopod *J. istri*, were completely eradicated. Overall, the mining spills diminished the existing crustacean populations by more than half, resulting in very poor crustacean fauna in the Tisza River system following the spills (WWF 2002). Interestingly, one and a half years after the mining spills, the crustacean fauna had completely recovered (ICPDR 2002). Not only have *C. curvispinum* and *J. istri* recolonized the river stretches of the Tisza, where they were present before the spill, but the Danube crayfish *P. leptodactylus* was also found in the middle stretches of the Tisza River where it was found before the spill.

Oil Spills

Oil spills have repeatedly affected crustacean populations. However, oil spills are quite variable, as different types of crude oil and other forms of oil, like fuel oil, are spilled in various quantities (Suchanek 1993, Jewett et al. 1999). These may get more or less dispersed in the water column or could sink to the ground, where they may harm subtidal communities. Oil spills may occur offshore or close to the coast where they are more likely to affect the benthic communities.

Several studies from the 1970s, 1980s, and 1990s provide estimates as well as detailed biodiversity analyses reflecting the losses of crustaceans related to major oil spills (reviewed in NRC [2003]). For instance, Krebs and Burns (1977) reported on the effects of a fuel oil spill in Buzzards Bay, Massachusetts, that affected the salt marshes and drastically reduced the population of burrowing fiddler crabs *Uca pugnax* over a period of at least seven years, and Sanders et al. (1980) found mass mortality of ampeliscid amphipods, which, however, recolonized the area more or less quickly. In 1996, the North Cape oil spill off Cape Cod, Massachusetts, released almost 3000 tons of light heating oil, which dispersed in the water column, thereby releasing various PAHs. The National Oceanic and Atmospheric Administration (NOAA 1998) estimated losses of 9 million mostly juvenile lobsters as well as 7.6 million rock and hermit crabs. Mortality of amphipods in these oil spills could be correlated with the presence of PAHs that persisted for more than half a year in the sediment (Ho et al. 1999). Several authors identified amphipod species as being particularly affected by the oil spills (Sanders et al. 1980, Dauvin 1987, Jewett et al. 1999). This may be particularly true for infaunal amphipods, if their burrows get clogged by heavy oil and because the fine sediments of the intertidal zone may get impregnated by oil residues, which particularly exposes them to oil-derived PAHs (Cabioch et al. 1978, Sanders et al. 1980, Jewett et al. 1999). Eventually, crustaceans being equipped with different P450 cytochromes (CYP) and lower CYP1A activity seem to detoxify organic pollutants, such as PAHs, at lower rates than fish (Livingstone 1998, Koenig et al. 2012). This may render them more susceptible to PAHs from oil spills.

The reduction in crustacean populations following oil spills is initially driven by the high acute mortality (Cabioch et al. 1978, Suchanek 1993), but long-term effects due to impaired recruitment of *U. pugnax* and peracarid crustaceans (amphipods, cumaceans, isopods, mysids, tanaids) have been demonstrated (Krebs and Burns 1977, Dauvin and Gentil 1990). Similarly, Mathews et al. (1993) observed collapses of prawn populations in the Persian Gulf following crude oil spills during the Gulf War, which caused a reduction of 99% of eggs, larvae, and postlarvae, and resulted in a 73% reduction of total shrimp biomass. Amphipods needed about six years to recover from the *Exxon Valdez* oil spill in Prince William Sound, Alaska (Jewett et al. 1999) and 15 years from the *Amoco Cadiz* oil spill along the shorelines of Brittany to attain former densities (Dauvin 1998).

A more recent study reported on the impacts of the Deepwater Horizon oil spill on subtidal crustacean communities in seaweed beds of the Gulf of Mexico (Felder et al. 2014). That study profited from pre-spill crustacean inventories, which allowed a more precise comparison with the situation after the spill. Regular surveys of two subtidal banks between 100 and 300 km from the well where the spill occurred revealed a rich diversity of benthic decapod assemblages in the seaweed beds before 2010. These assemblages declined significantly after the oil spill; crustacean diversity was reduced by 29% and 42%, respectively, and species richness by 9% and 12%, respectively. Interestingly, the impact was more pronounced on the bank further away from the oil spill, presumably because the assemblages of the bank closer to the spill had already been modified by discharges from the Mississippi River. The reduction in diversity and species richness of the decapod assemblages of both seaweed beds was associated with a remarkable decline of the species that formerly dominated the benthic crustacean communities. By contrast, more resilient species increased their relative abundance fivefold. The authors could attribute the declines to precipitates on the gills that hindered respiratory gas exchange as well as the degradation of the seaweed cover (Fig. 11.5).

BIOACCUMULATION AND TROPHIC TRANSFER OF CONTAMINANTS

In their great variety of life forms, crustaceans occupy different positions in the food web, representing different trophic levels and feeding strategies. Hence, they play a central role in the trophic transfer from primary producers to higher-level consumers and, consequently, also for the biomagnification of various pollutants across the food chain (Jakimska et al. 2011, Piscia et al. 2019). Crabs, as well, may feed on plant material and, therefore, be involved in the trophic transfer of pollutants from flora to fauna (Martínez-Colón et al. 2021). At higher levels of the trophic pyramid, secondary and tertiary consumers such as decapod crustaceans biomagnify pollutants along the food chain. As predators and scavengers, they can present very high concentrations of contaminants (Alava 2020, Kim 2020).

Not only do crustaceans represent different trophic levels and feeding strategies, they also differ considerably in their capacity to deal with metals and organic pollutants as well as microplastics and nanoparticles (Rainbow and Luoma 2011, Alava 2020, Kim 2020).

Metals

Many crustaceans demonstrate a high capacity to accumulate metals. Knigge and Köhler (2000) measured up to ~500 mg of lead per kg dry weight in woodlice *Porcellio scaber* exposed to leaf litter contaminated with high levels of lead and Jones and Hopkin (1998) found >1000 mg zinc per kg dry weight in *P. scaber* from sites close to smelting works at Avonmouth, UK. Isopods are known to sequester metals in an undissolved form within lysosome-derived granules in S- and B-cells (Schill and Köhler 2004, Kampe et al. 2018). Interestingly, studies by Schill and Köhler (2004) as well as Arts et al. (2004) suggest different kinetics of uptake, accumulation, and loss of metals in two species of woodlice from the vicinity of the Avonmouth smelting works, with higher ingestion and egestion rates of zinc in *Oniscus asellus*, which stores zinc in B-cells, whereas *P. scaber* accumulates zinc at a higher rate due to a long-lasting sequestration in the S-cells of the digestive gland. In contrast to *O. asellus*, *P. scaber* seems to reduce zinc ingestion by selective feeding (Arts et al. 2004, Schill and Köhler 2004). Similar mechanisms of metal sequestration are described for barnacles, amphipods, and shrimp (Rainbow and Luoma 2011). According to these authors, the barnacle *Amphibalanus amphitrite* takes up zinc at high rates and detoxifies it irreversibly as zinc pyrophosphate granules with practically no excretion, thus accumulating zinc to remarkably high concentrations. The amphipod *Orchestia gammarellus* equally accumulates zinc as zinc-rich granules in the embryonic and fibrillar cells of the digestive gland, from which part of it is lost with the dying cells. In contrast, the decapod *Palaemon serratus* seems to lack irreversible storage detoxification and regulates the zinc body burden by increased excretion to a maximal limit after which the regulation breaks down. Some of the zinc may also be reversibly detoxified by metallothioneins. In addition, there seem to be remarkable differences between metals. In the river shrimp *Palaemon varians* fed with field-contaminated *Nereis diversicolor*, zinc did not accumulate, probably due to excretion, but copper did and significantly induced detoxification by metallothionein-like proteins (Rainbow and Smith 2013). The large

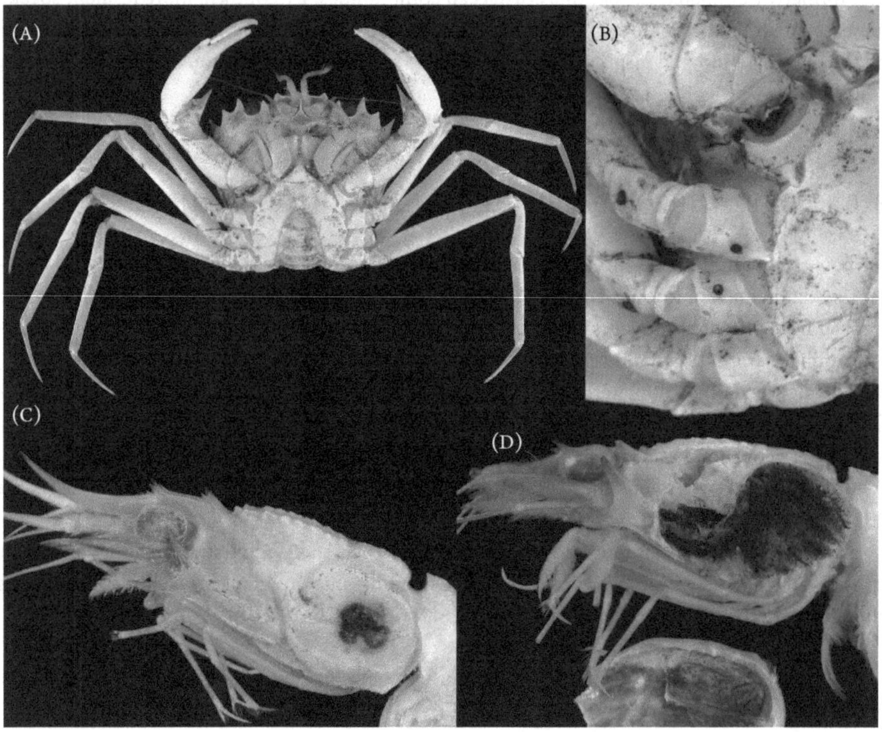

Fig. 11.5.
(A, B) Several dark lesions can be recognized on a deep-sea red crab, *Chaceon quinquedens*, collected about one and a half years after the Deepwater Horizon oil spill on a deep offshore bank in the Gulf of Mexico 49 km east of the spill site. (C, D) A lesion of the gill cavity wall (C) and gills covered by manganous precipitate (D) can be recognized on deep-sea shrimps, *Glyphocrangon longirostris*, collected one year after the Deepwater Horizon oil spill on a deep offshore bank in the Gulf of Mexico 110 km west-southwest of the spill site. Reproduced from Felder et al. (2014), with permission from © Oxford University Press; see a color version of this figure in the centerfold.

diversity in lifestyle and feeding strategies among crustaceans likely also leads to differences in the digestive system (e.g., Štrus et al. 2019) and, consequently, to fundamental differences in the uptake, detoxification, sequestration, and excretion of metals in the various groups of crustaceans.

Organic Compounds

Besides metals, crustaceans also accumulate organic pollutants to remarkably high concentrations. For instance, the amphipods *Gammarus tigrinus* and *Cryptorchestia cavimana* as well as the isopod *Asellus aquaticus* accumulated lower polychlorinated biphenyl congeners (PCB 28, 52, 101) relative to the sediment concentration to much higher concentrations than eel, despite the eel having a 17 times higher lipid content than the peracarids (van der Oost et al. 1988). The higher chlorinated PCBs (PCB 138, 153, 180), however, were found to be less biomagnified in crustaceans. The authors attributed this to the slower uptake kinetics for the higher chlorinated congeners that did not allow the short-lived amphipods and isopods to reach an equilibrium between uptake and elimination. In another field study, Kim (2020) investigated the bioaccumulation and trophic transfer of organochlorine pesticides comprising hexachlorobenzenes (HCBs), hexachlorocyclohexanes (HCHs), chlorodanes (CHLs), and dichloro-diphenyl-trichloroethane (DDT) in the calanoid copepods *Paracalanus* spp. and *Acartia* spp. and the grapsid crab *Hemigrapsus penicillatus*. In contrast to van der Oost et al. (1988), who found that the more water-soluble lower-chlorinated PCBs were highly enriched in crustaceans due to more rapid uptake kinetics and, putatively, uptake via both ventilation and food, Kim (2020) reported that organochlorine pesticides with n-octanol-water partition coefficients (K_{ow}) lower than log K_{ow} 5, that is, HCHs, had low bioaccumulation factors (BAFs) (~17-fold in copepods and ~15.5-fold in crabs) and showed no trophic transfer. On the other hand, HCBs were bioaccumulated somewhat less and CHLs slightly more than a 100-fold in copepods, and about 350- and 500-fold in crabs with trophic transfer factors between copepods and crabs of about 4.5. The highest bioaccumulation was detected for DDT, which was enriched more than 1500-fold in copepods and more than 4000-fold in *H. penicillatus*, but with a lower trophic transfer factor of about 2.5.

A high capacity of crustaceans (mantis shrimp, *Oratosquilla oratoria*, a penaeid prawn, *Penaeus chinensis*, and two portunid crab species, *Portunus pelagicus* and *P. trituberculatus*) to bioaccumulate organic pollutants was also reported by Du et al. (2021) for perfluoroalkyl substances (PFASs), ranging between 12.2 $ng{\cdot}g^{-1}$ dw in crabs to 22.4 $ng{\cdot}g^{-1}$ dw in mantis shrimp with perfluorooctane-sulfonic acid (PFOS) accounting for more than one-third of the accumulated PFASs. The PFOS and the perfluorooctanoic acid also showed the highest BAFs (> 5000, i.e., very bioaccumulative) among the PFASs. Biomagnification of PFOS was also the highest among the PFASs and particularly prominent in crabs. For the sum of the seven most important PFASs, portunid crabs showed biomagnification when feeding on the gastropod *Reishia clavigera* and mantis shrimps when feeding on crabs and/or penaeid prawns, thus demonstrating the role of crustaceans in the trophic transfer of organic pollutants through the food chain (Du et al. 2021). Similarly, Munoz et al. (2019) found that PFOS accounted for about half to two-thirds of all PFASs detected in copepods in the Gironde estuary, France. These authors also reported varying BAFs for copepods, mysids, and shrimps (*Palaemon longirostris* and *Crangon crangon*) ranging from log BAF 1.9 to 5 for the different PFASs analyzed, that is, BAFs greater than 5000 for certain compounds, such as PFOS or L-perfluorooctane sulfonamide. Interestingly, Munoz et al. (2019) calculated biomagnification factors greater than one for copepods and mysids, indicating that biomagnification already occurs at the trophic level of zooplankton. Yet another study found BAFs beyond 5000 for perfluorooctanoic acid and PFOS in crabs (*Ocypode stimpsoni* and *Clibanarius infraspinatus*) in southeastern coastal waters of China (Dai and Zheng 2019). Based on the results of these studies, the overall picture of an important bioaccumulation of PFASs in crustacean species at different levels of the food web is confirmed.

PLASTICS AND MICROPLASTICS

Plastics are now ubiquitous around the planet and are considered a major threat to aquatic life through entanglement, restricted feeding through ingestion, and their toxic components (MacLeod et al. 2021). Plastic production is still thought to be rising and concerns are that larger pieces of plastics can break down into smaller components of microplastics and nanoplastics whereby they can cause harm through a range of mechanisms from mechanical obstruction to toxic effects. Plastics are known to contain over 10,000 different chemical additives, of which over 2000 are known to have neurological, immunocompromising, and endocrine-disrupting effects (Wiesinger et al. 2021). There are now numerous studies whereby crustacean species have been known to harbor microplastics (Cole et al. 2013, Watts et al. 2014, 2016, Welden et al. 2018; Fig. 11.6). For example, Welden et al. (2018) observed over 40% of spider crabs *Maja squinado* had microplastics in their tissues, and the abundance of plastics was inversely proportional to the distance from the land. Field data for the bioaccumulation and trophic transfer of microplastics are, however, scarce, and inconclusive (Carbery et al. 2018, Miller et al. 2020). There has been considerable debate as to whether the presence of microplastics in tissues are indicative of harmful effects, with several reviews and meta-analyses indicating that many laboratory-based studies are not conducted with environmentally relevant concentrations of plastic particles (Burns and Boxall 2018, Adam et al. 2021, Ogonowski et al. 2023).

Importantly, micro- and nanoplastics are known to be vectors for metals and organic pollutants bound to their surfaces. For example, Chua et al. (2014) demonstrated that polybrominated diphenyl ethers (PBDEs) derived from microplastic particles were assimilated by the marine amphipod *Allorchestes compressa*. Crustaceans might also have very effective mechanisms to filter out small (non-nutritious) particles (Saborowski et al. 2019, Korez et al. 2020). Crustaceans themselves contribute to plastic breakdown into smaller particles. For example, Antarctic krill, as well as some amphipods, are known to break down microplastics into smaller plastic particles (Dawson et al. 2018, Mateos-Cárdenas et al. 2020). One study that has clearly shown harm

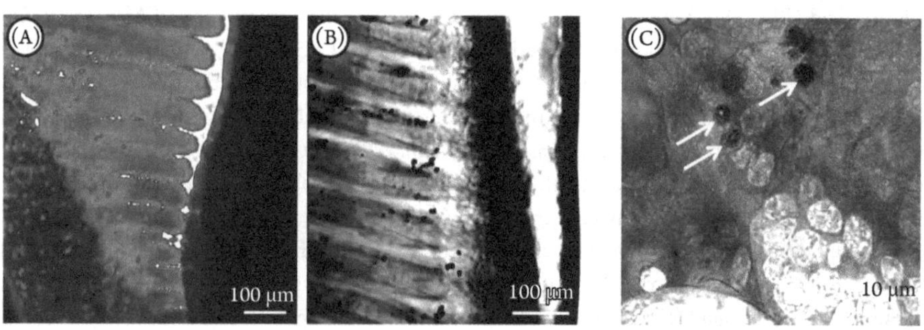

Fig. 11.6.
(A, B) Polystyrene microspheres (8 μm, respectively) visualized by Raman scattering as light shading (A) and dark dots (B) on the gill lamellae of the European shore crab, *Carcinus maenas*; scale bars = 100 μm. (C) Polystyrene microbeads emitting red fluorescence (0.1 μm, arrows) in the digestive gland of the sand shrimp, *Crangon crangon*. (A, B) reproduced from Watts et al. (2016), with permission from © The American Chemical Society; (C) reproduced from Korez et al. (2020), with permission from © Elsevier under Creative Commons license (CC BY); see a color version of this figure in the centerfold.

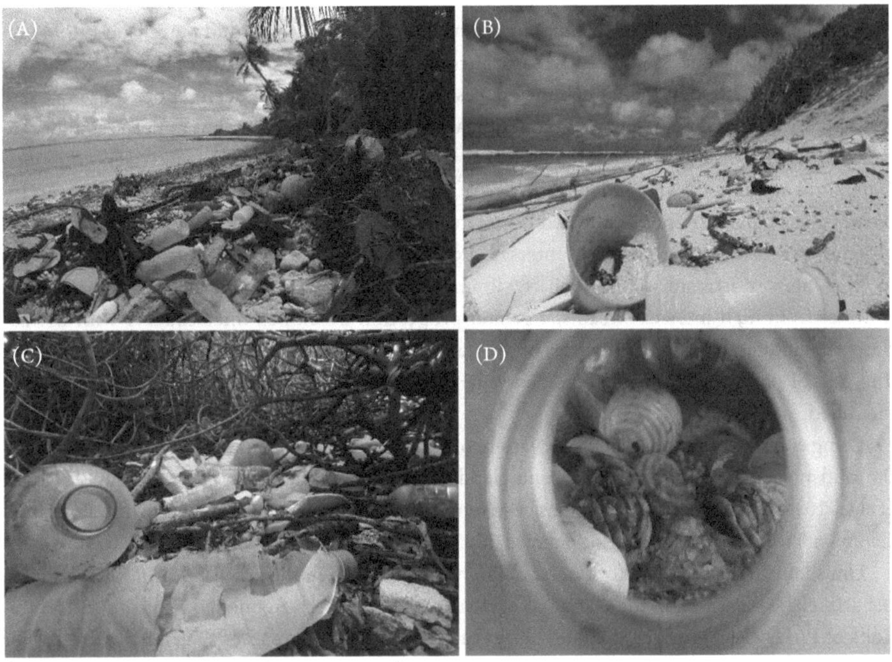

Fig. 11.7.
(A) Accumulated plastic debris creates an obstacle for crabs on the beaches of the Cocos (Keeling) Islands. (B) A hermit crab inside a green bucket along the high tide line of South Island. (C) Accumulated plastic debris in the beach-back vegetation on West Island. (D) Crabs that became trapped and died inside a plastic drink bottle that washed up on Cocos. Reproduced from Lavers et al. (2020), with permission from © Elsevier; see a color version of this figure in the centerfold.

to populations of crustaceans is the entrapment of hermit crabs in beach litter (Fig. 11.7). Lavers et al. (2020) estimated that over half a million hermit crabs had become entrapped in plastic litter each year on two tropical islands, adversely affecting the populations and local ecosystem.

DEVELOPMENTAL ABNORMALITIES IN THE FIELD

Intersexuality

One of the most common forms of developmental abnormalities reported in the field for crustaceans is intersexuality, reviewed in Ford (2012). Intersexuality is the abnormal condition where gonochoristic species display characteristics of being male and female (Ford 2012). Some crustaceans are simultaneous and sequential hermaphrodites, meaning they may go through stages of appearing intersex naturally (Ford 2008). Reporting of intersexuality in wildlife became prominent during the late 1990s with the discovery that a large variety of chemicals can act like hormones and disrupt the functioning of the endocrine systems (Colborn et al. 1996, Ford et al. 2006). It has been speculated that these endocrine-disrupting chemicals cause some of the observed incidences of intersexuality in crustaceans. However, thus far there has been no concrete evidence from the field despite a large body of research reporting higher incidences of intersex specimens in polluted vs. unpolluted (i.e., reference) locations. For example, Moore and Stevenson (1994) reported a high prevalence of intersex in harpacticoid copepods from the Firth of Forth, UK, though no causal agent was identified. High incidences of intersexuality have been reported in male and female freshwater crabs from streams contaminated with arsenic in Japan, but none in reference locations (Takahashi et al. 2000). Jungmann et al. (2004) found that some German rivers contained *Gammarus* amphipods with a high incidence of intersexuality and others with low incidences. Interestingly, when amphipods collected from streams with a low prevalence of intersexuality were caged in streams with high levels of intersex, a greater proportion became intersexed than if kept in "low-intersex" stream water. Similarly, if amphipods were kept under laboratory conditions in water from high-intersex streams, a greater proportion developed intersexuality with the authors concluding that some properties of the water were the causal factor (Jungmann et al. 2004). The role of parasites could not confidently be ruled out or any specific chemical contaminant identified.

Using crayfish as a model species, a variety of studies have demonstrated how mechanical ablation and molecular silencing of androgenic gland hormone genes can result in intersexuality (Rosen et al. 2010). Similarly, intersexuality has been reported as being very prevalent within the amphipod *Corophium volutator* (Barbeau and Grecian 2003, McCurdy et al. 2004).

Unusually high incidences of intersexuality were observed in the crayfish *P. leptodactylus* from the artificial Lake Pinok in the Ostrava-Karviná coal basin, Czech Republic, with high concentrations of PAHs (Mazurová et al. 2008). Few studies report frequencies of intersex that are higher than 10% (e.g., Moore and Stevenson 1991, Barbeau and Grecian 2003, Ford et al. 2004, Levy et al. 2020; for a comprehensive overview, see also the review of Grilo and Rosa [2017]). The 18% of *P. leptodactylus* in PAH-contaminated Lake Pinok supports the idea of environmental chemicals being able to induce intersex in crustaceans, although specific habitat characteristics and evolutionary drivers as described by Levy et al. (2020) for the Australian red claw crayfish *Cherax quadricarinatus* may play their part as well. More strikingly, Moore and Stevenson (1991) found 91% intersex among 30 individuals of the harpacticoid copepod *Paramphiascella hyperborea* close to the outfall of a sewage treatment plant, an extraordinarily high number given that previous records of intersex in harpacticoid copepods were extremely rare (Moore and Stevenson 1991). This suggests that environmental chemicals from the sewage treatment plant induced intersex in the copepods at this site.

Overall, the body of evidence gathered by the aforementioned studies strongly suggests that pollutants are responsible for above-normal frequencies of intersex, although the mechanism(s) by which they act on sex determination and the chemical nature of the pollutants causing intersex remain elusive. Despite the diversity in sex-determining mechanisms in crustaceans, two mechanisms, however, could be involved in the disruption of normal sex determination (Knigge et al. 2021): first, sexual differentiation in malacostracans is brought about by an insulin-like androgenic gland hormone (AGH), which is produced by the androgenic gland (Charniaux-Cotton 1954, Hasegawa et al. 1987, Sagi et al. 1997). The AGH induces masculinization, while the absence of the androgenic gland or a lack of AGH results in feminization (Levy and Sagi 2020). Hence, any environmental chemical that interferes with the synthesis and secretion of AGH or any putative upstream regulation of the androgenic gland via neuropeptides of the crustacean hyperglycemic hormone-superfamily (Ford et al. 2003, Ford 2008), such as the gonad-inhibiting hormone (Tang et al. 2021), could affect sex determination during critical windows of development. Second, methyl farnesoate, a juvenile hormone-like sesquiterpenoid, is involved in sex determination in branchiopods (Olmstead and LeBlanc 2002) by inducing the doublesex gene (dsx1) responsible for male sex differentiation (Kato et al. 2011). Because sex determination in *Daphnia* spp. depends on environmental cues that are likely to be linked to methyl farnesoate via neuroendocrine signaling pathways (Toyota et al. 2015), environmental pollutants that interfere with the neuroendocrine axis, such as psychotropic drugs released into the environment, could potentially disturb this mechanism of sex determination (Camp et al. 2020).

Gynandromorphism, where individuals develop bilaterally into a male and female phenotype, has been reported in several crabs and lobsters (Chace and Moore 1959, Farmer 1972, Taylor 1986; Fig. 11.8) and is speculated to be caused by pollution (Ford 2012, Stentiford 2012). However, the frequencies of such abnormalities are rare in the wild and possibly due to genetic and sporadic changes in gene expression during early embryonic development (Ford 2012, Scholtz 2020). Studies in the laboratory with *Daphnia* have shown that pesticides known to increase the frequency of males can also increase the likelihood of gynandromorph phenotypes, thus making the hypothesis that exogenous contaminants inducing gynandromorphs theoretically possible in the wild (Olmstead and LeBlanc 2007).

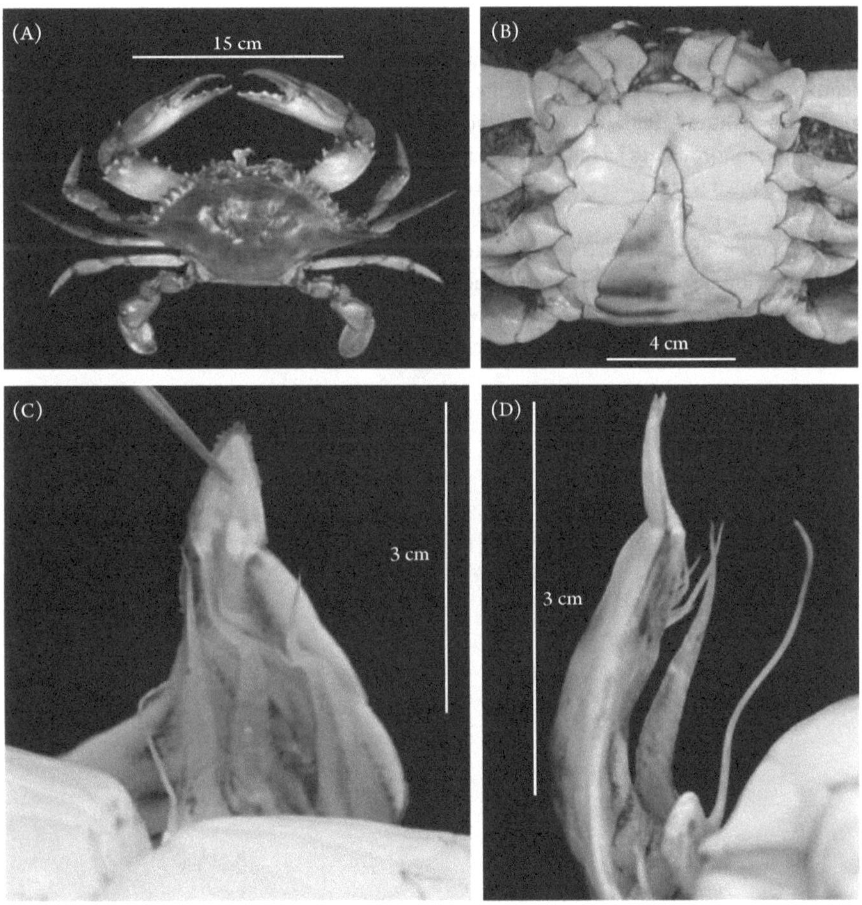

Fig. 11.8.
Bilateral gynandromorphism in the blue crab (*Callinectes sapidus*). (A) Posterior view of a conserved blue crab individual. (B) Anterior view of the individual with bilateral gynandromorphism, displaying female (left) and male (right) characteristics. (C) Telson inner view, displaying male and female sexual organs. (D) Telson lateral view, displaying male and female sexual organs. Photos reproduced courtesy of Karen Reed from the specimen catalog number USNM1182904, in Smithsonian National Museum of Natural History under Creative Commons Zero (CC 0).

Morphological Abnormalities

Although an extensive body of literature describes morphological abnormalities (as reviewed in Scholtz [2020, chapter 4]), these are not frequent in crustaceans. Most abnormalities seem to be a result of wound healing following damage caused to the exoskeleton (Scholtz 2020). Developmental abnormalities, other than those related to intersexuality mentioned above, are more likely to be associated with pollution. For example, several studies have shown that estuarine amphipods collected and maintained on oil- and PCB-contaminated sediments develop abnormal hairs along their oostegites (brood plates), which correlates with reductions in brood sizes (Borowsky et al. 1993; Fig. 11.9). Similarly, intersex amphipods are also known to develop malformed setae on their oostegites (Ford et al. 2003). The hairs are under the control of a "putative" ovarian hormone and thus might indicate a type of endocrine disruption (Ford et al. 2003). Béguer et al. (2008) observed high proportions (40%) of skeletal abnormalities in *Palaemon* shrimp in the Gironde estuary, France, which was speculated to be due to heavy metal pollution. These abnormalities included malformed cephalothorax and rostrum, and to a lesser extent scaphocerites and uropods (including dissymmetry). Similarly, de Oliveira Dias (1999) reported large numbers of morphological abnormalities (30%) of *Acartia lilljeborgi* (Copepoda, Crustacea) in the Espírito Santo Bay, Brazil. These abnormalities included "intestinal prolapse and partial rupture of the chitinous carapace accompanied with a mass extrusion of cell material and protoplasm" and were suggested to be due to urbanized pollution.

Limb abnormalities have also been reported in crustaceans although, again, the majority of these have been attributed to the regenerative capacity of crustaceans and the abnormal signaling in stem cells (Scholtz 2020). Weis et al. (1992) reviewed the effects of pollution on molting and limb regeneration in crustaceans and highlighted several reports where a range of contaminants have caused

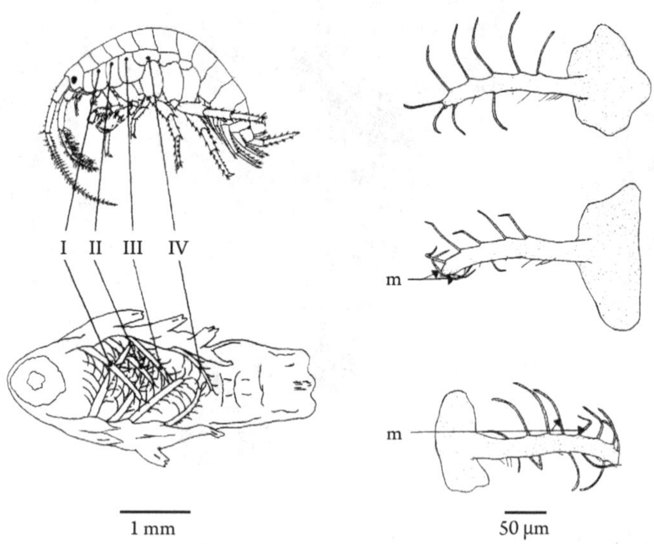

Fig. 11.9.
(Left) Female *Melita nitida* (Amphipoda) in lateral and ventral view displaying oostegites (I–IV). (Right) Pair IV ostegites removed and displaying normal (top) and abnormal/misaligned setae (middle and bottom). Reproduced from Borowsky et al. (1993), with permission from © Elsevier.

limb/claw abnormalities and defects in chitin formation. For example, tributyltin exposure to *Uca pugilator* causes abnormal limb bending in the chela in males and abnormal curvature of the dactyl (Weis et al. 1987). All things considered, proof for pollutants causing morphological abnormalities is scarce, but cannot be excluded (Scholtz 2020).

Fluctuating Asymmetry

Fluctuating asymmetry (FA) has often been used to measure the effects of anthropogenic disturbance and is defined as the subtle, random deviations from the expected bilateral symmetry (Palmer and Strobeck, 1986). In crustaceans, FA has been used to measure the effects of various environmental stressors from urbanized vs. non-urbanized beaches (Lezcano et al. 2015, Frota et al. 2019), contaminated vs. non-contaminated harbors (Li 2002), through to metals (Godet et al. 2012), oil pollution (Barca-Bravo et al. 2008), and radiation (Fuller et al. 2017).

DISEASE/PARASITES

Environmental contaminants are known to increase disease through immunosuppression but also decrease some diseases/parasites through variable susceptibility and toxicity between species and broken linkages in complex parasitic life cycles (Williams and MacKenzie 2003). Crustaceans can harbor many different diseases and parasites, but also themselves be parasitic on other species (including other crustaceans) (Nowak et al. 2020). The significantly higher incidences of intersexuality in the marine amphipod *Marinogammarus marinus* from industrially polluted locations in Scotland compared to reference locations (Ford et al. 2004, 2006) may have also been related to a novel coinfection by two phylogenetically distinct parasites (Microsporidia and Paramyxea) known to induce feminization in amphipod hosts (Guler et al. 2012, Short et al. 2012a). This led the authors to speculate that the pollution had somehow directly or indirectly led to an increase in feminizing parasites or inhibited the natural ability of these parasites to feminize their hosts, leaving them intersex. Interestingly, the authors also observed a higher frequency of specimens that were intersex internally but not visibly intersex externally. These specimens displaying intersex internally were also more prevalent in industrially contaminated locations but were negative for feminizing parasites (Short et al. 2012b). Transcriptomic analysis of intersex specimens believed to be induced with and without feminizing parasites invoked similar gene changes, which highlighted that the transcriptional fingerprints of feminization and/or de-masculinization were similar with the two phenotypes (Short et al. 2014).

There is considerable evidence in the literature that large-scale die-offs of crabs and lobsters in coastal areas suffering from algal blooms and low oxygen have been attributed to disease and increased disease susceptibility (Mullen et al. 2004, Stentiford and Shields 2005, Wang 2011). Following a mass mortality of an estimated 11 million lobsters along the coast of Long Island Sound, USA, in 1999, Mullen et al. (2004) could not detect significant amounts of trace elements, PAHs, or PCBs in lobster tissues and the study failed to isolate any bacteria from the hemolymph and hepatopancreas. What the study did observe was considerable infection from a protozoan parasite throughout the nervous tissues of impacted lobsters, leading the authors to conclude that this parasite was the primary cause of mass mortality in 1999. A mass mortality of multiple crab and lobster species off the northeast coast of England in 2021, stretching over 70 km, was speculated to have been caused by a novel pathogen, although there was no conclusive proof (Henderson et al. 2023).

FUTURE DIRECTIONS

Although considerable knowledge about the toxicity of xenobiotics to crustaceans has been acquired, a couple of questions require increased research efforts.

(1) Endocrine disruption is a long-standing issue; it has only occasionally been demonstrated that environmental chemicals do interfere with crustacean hormone signaling (Knigge et al. 2021; Fig. 11.1). It is, however, likely that many of the more than 30,000 environmental chemicals would display endocrine-disrupting effects. Indeed, so-called insect growth regulators have even been specifically designed to interfere with juvenoid hormone synthesis and molting in insects (LeBlanc 2007, Rodríguez et al. 2007). Their action on crustaceans stands to reason. On the other hand, only a limited number of studies have investigated neuroendocrine disruption by environmental chemicals, and little attention has been given to pollutant-related impairment of hormonal sex determination specific to crustaceans.

(2) Sensitive sublethal behavioral endpoints are useful to assess the effects of chemicals at very low concentrations (Hellou 2011, Melvin and Wilson 2013), and several studies have investigated the effects of pollutants on locomotor and feeding activity, or agonistic behavior and camouflage (Dissanayake et al. 2009, Chabenat et al. 2021, Kohler et al. 2021). Caution is needed, however, when correlating behavioral changes with chemical exposure as other stressors may alter behavioral programs as well. It is important to establish the consequences of these behavioral changes at the individual and population level when they happen to take place in natural environments.

(3) Many crustaceans are active in the dark and/or their visual orientation may be hampered by turbid waters, which is why crustaceans rely heavily on chemoreception for foraging, sensing of predators and congeners, or for habitat selection (Dodson et al. 1994, Breithaupt and Thiel 2010). Little is known about how environmental chemicals interfere with the reception and processing of chemical information in a crustacean's environment. Recently, Kozma et al. (2020) revealed a highly complex inventory of chemical sensors on the antennules of different decapod crustaceans. A better understanding of the interactions of this multitude of receptors, excitatory ion channels, and signaling pathways is needed in order to evaluate how pollutants may interfere with olfactory behavior. Given the importance of chemoreception in crustaceans and the plausibility of its chemical perturbation (Olsén 2010), it is astonishing how little attention has been given to the effects of environmental pollution on olfactory pathways.

(4) This chapter has mainly focused on chemical pollution. However, other types of pollution affect crustaceans too (Fig. 11.4). For instance, ship noise was found to affect foraging, mating, and anti-predator behavior as well as feeding activity and color change in *Carcinus maenas* (Wale et al. 2013, Carter et al. 2020, Rising et al. 2022). Furthermore, benthic crustaceans are sensitive to vibrations generated by drilling, piling, and dredging that are propagated via the seafloor (Roberts et al. 2016, Aimon et al. 2021). Hence there is reason for concern that anthropogenic underwater noise and substratum vibration may affect crustacean physiology.

(5) Circadian rhythms in crustaceans are mediated by light (Sullivan et al. 2009, Strauss and Dircksen 2010) and control many behavioral activities, such as diel vertical migration of zooplankton (Dawidowicz and Pijanowska 2018), foraging, reproduction, and retreat into shelter. Artificial light at night was found to reduce the diel vertical migration of *Daphnia* spp. (Moore et al. 2000) and to disturb nocturnal activities (Fallaci et al. 1999, Duarte et al. 2019, Torres et al. 2020, Lynn et al. 2021).

(6) Finally, a large proportion of field studies have focused on female fecundity as a measure of water quality in crustaceans. Eggs and/or embryos can be assessed for developmental abnormalities and whether they have been fertilized (Sundelin and Eriksson 1998, Ford et al. 2003). Yet the effects of environmental contaminants on male fertility have been very much overlooked (Lewis and Ford 2012). Indeed, Yang et al. (2008) recorded significantly lower sperm counts in amphipods collected from locations receiving industrial contaminants such as metals, PAHs, and PCBs. Botelho et al. (2021), in a 10-year monitoring study with *Marinogammarus marinus*, recorded consistently low sperm counts in amphipods from a marine protected area in southern England receiving regular sewage discharges and in the proximity to historical coastal landfills. These studies indicate that we need a better understanding of how populations of crustaceans may be impacted by environmental contaminants.

SUMMARY AND CONCLUSIONS

Many studies have investigated how inorganic and organic pollutants act on crustaceans. Thus, a considerable knowledge base about the toxic effects of a great variety of environmental chemicals on crustaceans has been built over decades. The results of this research strongly suggest that crustaceans are susceptible to many of the thousands of pollutants released into nature by human activities, albeit many questions remain to be answered. Obviously, pollutants do not occur singly in the field but as complex mixtures, the interactions of which are very difficult to predict. In addition, toxic chemicals and environmental pollution of all sorts do not act alone, but may interact either additively or even synergistically, with natural stressors, such as temperature, osmotic concentration, acidity, oxygenation, or infectious diseases and parasitism as well as prey–predator interactions and nutrient availability (van Straalen 2003, Holmstrup et al. 2010). As a matter of consequence, it is complicated to separate anthropogenic from natural stressors so as to evaluate the risk of pollution for crustacean populations in natural environments. A recent case of mysteriously high mortality of crustaceans off the northeast coast of England has demonstrated once more the difficulties in trying to disentangle causes for mass mortality of crustacean species and to identify pollution as a threat for crustacean populations in the field (Henderson et al. 2023).

Apart from obviously disastrous events of mass mortality or high incidences of morphological modifications, such as intersex, that might be correlated to pollution on a spatial or temporal scale, it must be assumed that many of the adverse effects of chronic pollution take place at a low threshold that remains invisible for most of the time but is likely to weaken field populations in their capacity to resist other stressors. Eventually, this might have negative effects on crustaceans, such as population decreases in the field, without this being directly assignable to environmental pollution. Nevertheless, there is a consensus among most scientists working in the field that pollution has threatened and keeps threatening crustaceans in the wild.

REFERENCES

Adam, V., A. von Wyl, and B. Nowack. 2021. Probabilistic environmental risk assessment of microplastics in marine habitats. Aquatic Toxicology 230:105689.

Ahearn, G.A., P.K. Mandal, and A. Mandal. 2004. Mechanisms of heavy-metal sequestration and detoxification in crustaceans: A review. Journal of Comparative Physiology B 174:439–452.

Aimon, C., S.D. Simpson, R.A. Hazelwood, R. Bruintjes, and M.A. Urbina. 2021. Anthropogenic underwater vibrations are sensed and stressful for the shore crab *Carcinus maenas*. Environmental Pollution 285:117148.

Alava, J.J. 2020. Modeling the bioaccumulation and biomagnification potential of microplastics in a cetacean foodweb of the Northeastern Pacific: A prospective tool to assess the risk exposure to plastic particles. Frontiers in Marine Science 7:566101.

Amiard, J.C., C. Amiard-Triquet, S. Barka, J. Pellerin, and P.S. Rainbow. 2006. Metallothioneins in aquatic invertebrates: Their role in metal detoxification and their use as biomarkers. Aquatic Toxicology 76:160–202.

Arts, M.J., R.O. Schill, T. Knigge, H. Eckwert, J.E. Kammenga, and H.R. Köhler. 2004. Stress proteins (hsp70, hsp60) induced in isopods and nematodes by field exposure to metals in a gradient near Avonmouth, UK. Ecotoxicology 13:739–755.

Azan, S.S.A., and S.E. Arnott. 2018. The impact of calcium decline on population growth rates of crustacean zooplankton in Canadian Shield lakes. Limnology and Oceanography 63:602–616.

Balian, E.V., H. Segers, C. Lévèque, and K. Martens. 2008. The freshwater animal diversity assessment: An overview of the results. Hydrobiologia 595:627–637.

Bao, J., Y. Xing, C. Feng, S. Kou, H. Jiang, and X. Li. 2020. Acute and sub-chronic effects of copper on survival, respiratory metabolism, and metal accumulation in *Cambaroides dauricus*. Scientific Reports 10:16700.

Barbeau, M.A., and L.A. Grecian. 2003. Occurrence of intersexuality in the amphipod *Corophium volutator* (Pallas) in the Upper Bay of Fundy, Canada. Crustaceana 76:665–679.

Barca-Bravo, S., M.J. Servia, F. Cobo, and M.A. Gonzalez. 2008. The effect of human use of sandy beaches on developmental stability of *Talitrus saltator* (Montagu, 1808) (Crustacea, Amphipoda). A study on fluctuating asymmetry. Marine Ecology 29:91–98.

Bar-On, Y.M., R. Phillips, and R. Milo. 2018. The biomass distribution on Earth. Proceedings of the National Academy of Sciences 115:6506–6511.

Béguer, M., S. Pasquaud, P. Noël, M. Girardin, and P. Boët. 2008. First description of heavy skeletal deformations in *Palaemon* shrimp populations of European estuaries: The case of the Gironde (France). Hydrobiologia 607:225–229.

Blanchard, A., N. Demchenko, L. Aerts, S. Yazvenko, V. Ivin, I. Shcherbakov, and H. Melton. 2019. Prey biomass dynamics in gray whale feeding areas adjacent to northeastern Sakhalin (the Sea of Okhotsk), Russia, 2001–2015. Marine Environmental Research 145:123–136.

Borowsky, B., P. Aitken-Ander, and J.T. Tanacredi. 1993. The effects of low doses of waste crankcase oil on *Melita nitida* Smith (Crustacea: Amphipoda). Journal of Experimental Marine Biology and Ecology 166:39–46.

Botelho, M.T., N. Fuller, M. Vannuci-Silva, G. Yang, K. Richardson, and A.T. Ford. 2021. Unusual male size vs sperm count relationships in a coastal marine amphipod indicate reproductive impairment by unknown toxicants. Aquatic Toxicology 233:105793.

Breithaupt, T., and M. Thiel. 2010. Chemical communication in crustaceans. Springer, New York, USA.

Breteler, R.J., J.W. Williams, and R.L. Buhl. 1982. Measurement of chronic toxicity using the opossum shrimp *Mysidopsis bahia*. Pages 189–194 *in* M.D. Morgan, editor. Ecology of Mysidacea. Developments in hydrobiology, Volume 10. Springer, Dordrecht, The Netherlands.

Brown, A., and D. Pascoe. 1989. Parasitism and host sensitivity to cadmium: An acanthocephalan infection of the freshwater amphipod *Gammarus pulex*. Journal of Applied Ecology 26:473–487.

Burns, E.E., and A.B. Boxall. 2018. Microplastics in the aquatic environment: Evidence for or against adverse impacts and major knowledge gaps. Environmental Toxicology and Chemistry 37: 2776–2796.

Cabioch, L., J.C. Dauvin, and F. Gentil. 1978. Preliminary observations on pollution of the sea bed and disturbance of sub-littoral communities in northern Brittany by oil from the Amoco Cadiz. Marine Pollution Bulletin 9:303–307.

Camp, A.A., J. Yun, S.A. Chambers, M.H. Haeba, and G.A. LeBlanc. 2020. Involvement of glutamate and serotonin transmitter systems in male sex determination in *Daphnia pulex*. Journal of Insect Physiology 121:104015.

Carbery, M., W. O'Connor, and T. Palanisami. 2018. Trophic transfer of microplastics and mixed contaminants in the marine food web and implications for human health. Environment International 115:400–409.

Carter, E.E., T. Tregenza, and M. Stevens. 2020. Ship noise inhibits colour change, camouflage, and anti-predator behaviour in shore crabs. Current Biology 30:R191–R214.

Chabenat, A., C. Bellanger, and T. Knigge. 2021. Effects of environmental antidepressants on colour change and locomotor behaviour in juvenile shore crabs, *Carcinus maenas*. Aquatic Toxicology 234:105808.

Chace, F.A. Jr., and G.M. Moore. 1959. A bicolored gynandromorph of the lobster, *Homarus americanus*. The Biological Bulletin 116:226–231.

Charniaux-Cotton, H. 1954. Discovery in an amphipod crustacean (*Orchestia gammarella*) of an endocrine gland responsible for the differentiation of primary and secondary male sex characteristics. Comptes Rendus Hebdomadaires des Seances de L'academie des Sciences 239:780–782.

Chua, E.M., J. Shimeta, D. Nugegoda, P.D. Morrison, and B.O. Clarke. 2014. Assimilation of polybrominated diphenyl ethers from microplastics by the marine amphipod, *Allorchestes compressa*. Environmental Science and Technology 48:8127–8134.

Colborn, T., D. Dumanoski, and J.P. Myers. 1996. Our stolen future: Are we threatening our fertility, intelligence, and survival? A scientific detective story. Dutton, New York, USA.

Cole, M., P. Lindeque, E. Fileman, C. Halsband, R. Goodhead, J. Moger, and T.S. Galloway. 2013. Microplastic ingestion by zooplankton. Environmental Science and Technology 47:6646–6655.

Correi, A.D., D.R. Livingstone, and M.H. Costa. 2002. Effects of water-borne copper on metallothionein and lipid peroxidation in the marine amphipod *Gammarus locusta*. Marine Environmental Research 54:357–360.

Costello, M., M. Coll, R. Danovaro, P. Halpin, H. Ojaveer, and P. Miloslavich. 2010. A census of marine biodiversity knowledge, resources, and future challenges. PloS ONE 5:E12110.

Covich, A.P., J.H. Thorp, and D.C. Rogers. 2010. Introduction to the subphylum Crustacea. Pages 695–723 *in* J.H. Thorp and A.P. Covich, editors. Ecology and classification of North American freshwater invertebrates, 2nd edition. Academic Press, San Diego, California, USA.

Cumberlidge, N., P.K.L. Ng, D.C.J. Yeo, C. Magalhães, M.R. Campos, F. Alvarez, T. Naruse, S.R. Daniels, L.J. Esser, F.Y.K. Attipoe, F.-L. Clotilde-Ba, W. Darwall, A. McIvor, J.E.M. Baillie, B. Collen, and M. Ram. 2009. Freshwater crabs and the biodiversity crisis: Importance, threats, status, and conservation challenges. Biological Conservation 142:1665–1673.

Dai, Z., and F. Zheng. 2019. Distribution and bioaccumulation of perfluoroalkyl acids in Xiamen coastal waters. Journal of Chemistry 36:2612853.

Dauvin, J.C. 1987. Evolution à long terme (1978–1986) des populations d'amphipodes des sables fins de la Pierre Noire (Baie de Morlaix, Manche Occidentale) après la catastrophe de l'Amoco Cadiz. Marine Environmental Research 21:247–273.

Dauvin, J.C. 1998. The fine sand *Abra alba* community of the Bay of Morlaix twenty years after the Amoco Cadiz oil spill. Marine Pollution Bulletin 36:669–676.

Dauvin, J.C., and F. Gentil. 1990. Conditions of the peracarid populations of subtidal communities in Northern Brittany ten years after the Amoco Cadiz oil spill. Marine Pollution Bulletin 21:123–130.

Dawidowicz, P., and J. Pijanowska. 2018. Diel vertical migration of aquatic crustaceans—Adaptive role, underlying mechanisms, and ecosystem consequences. Pages 232–257 *in* G. Wellborn and M. Thiel, editors. The natural history of the Crustacea, Volume 5: Life histories. Oxford University Press, New York, USA.

Dawson, A.L., S. Kawaguchi, C.K. King, K.A. Townsend, R. King, W.M. Huston, and S.M. Bengtson Nash. 2018. Turning microplastics into nanoplastics through digestive fragmentation by Antarctic krill. Nature Communications 9:1001.

Dell'Anno, A., F. Raffaelli, R. Danovaro, L. Nanetti, A. Vignini, C. Moroni, and L. Mazzanti. 2008. Cytotoxic effects induced by hexachlorobenzene in *Squilla mantis* (L.) (Crustacea, Stomatopoda). Environmental Toxicology 23:9–14.

de Oliveira Dias, C. 1999. Morphological abnormalities of *Acartia lilljeborgi* (Copepoda, Crustacea) in the Espírito Santo Bay (ES Brazil). Hydrobiologia 394:249–251.

Den Hartog, C., F.W.B. Van Den Brink, and G. van Der Velde. 1992. Why was the invasion of the Rhine by *Corophium curvispinum* and *Corbicula* species so successful? Journal of Natural History 26: 1121–1129.

Dissanayake, A., T.S. Galloway, and M.B. Jones. 2009. Physiological condition and intraspecific agonistic behaviour in *Carcinus maenas* (Crustacea: Decapoda). Journal of Experimental Marine Biology and Ecology 375:57–63.

Dissanayake, A., C. Piggott, C. Baldwin, and K.A. Sloman. 2010. Elucidating cellular and behavioural effects of contaminant impact (polycyclic aromatic hydrocarbons, PAHs) in both laboratory-exposed and field-collected shore crabs, *Carcinus maenas* (Crustacea: Decapoda). Marine Environmental Research 70:368–373.

Dodson, S.I., T.A. Crowl, B.L. Peckarsky, L.B. Kats, A.P. Covich, and J.M. Culp. 1994. Non-visual communication in freshwater benthos: An overview. Journal of the North American Benthological Society 13:268–282.

Du, D., Y. Lu, Y. Zhou, Q. Li, M. Zhang, G. Han, H. Cui, and E. Jeppesen. 2021. Bioaccumulation, trophic transfer and biomagnification of perfluoroalkyl acids (PFAAs) in the marine food web of the South China Sea. Journal of Hazardous Materials 405:124681.

Duarte, C., D. Quintanilla-Ahumada, C. Anguita, P.H. Manríquez, S. Widdicombe, J. Pulgar, E.A. Silva-Rodríguez, C. Miranda, K. Manríquez, and P.A. Quijón. 2019. Artificial light pollution at night (ALAN) disrupts the distribution and circadian rhythm of a sandy beach isopod. Environmental Pollution 248:565–573.

Eckwert, H., G. Alberti, and H.R. Köhler. 1997. The induction of stress proteins (hsp) in *Oniscus asellus* (Isopoda) as a molecular marker of multiple heavy metal exposure: I. Principles and toxicological assessment. Ecotoxicology 6:249–262.

Edwards, B., D. Jackson, and K. Somers. 2016. Effects of declining calcium availability on the survival, growth and calcium content of a freshwater crayfish, *Orconectes virilis*. Freshwater Biology 61:914–922.

EPA. 2022. ECOTOX Knowledgebase, last update of the database: Sep 15, 2022. United States Environmental Protection Agency. Accessed November 25, 2023. [https://cfpub.epa.gov/ecotox/].

Fallaci, M., A. Aloia, M. Audoglio, I. Colombini, F. Scapini, and L. Chelazzi. 1999. Differences in behavioral strategies between two sympatric talitrids (Amphipoda) inhabiting an exposed sandy beach of the French Atlantic Coast. Estuarine, Coastal and Shelf Science 48:469–482.

Farmer, A.S. 1972. A bilateral gynandromorph of *Nephrops norvegicus* (Decapoda: Nephropidae). Marine Biology 15:344–349.

Felder, D.L., B.P. Thoma, W.E. Schmidt, T. Sauvage, S.L. Self-Krayesky, A. Chistoserdov, H.D. Bracken-Grissom, and S. Fredericq. 2014. Seaweeds and decapod crustaceans on Gulf deep banks after the Macondo oil spill. BioScience 64:808–819.

Ford, A.T., 2008. Can you feminise a crustacean? Aquatic Toxicology 88:316–321.

Ford, A.T. 2012. Intersexuality in Crustacea: An environmental issue? Aquatic Toxicology 108:125–129.

Ford, A.T., T.F. Fernandes, S.A. Rider, P.A. Read, C.D. Robinson, and I.M. Davies. 2003. Measuring sublethal impacts of pollution on reproductive output of marine Crustacea. Marine Ecology Progress Series 265:303–309.

Ford, A.T., T.F. Fernandes, S.A. Rider, P.A. Read, C.D. Robinson, and I.M. Davies. 2004. Endocrine disruption in a marine amphipod? Field observations of intersexuality and de-masculinisation. Marine Environmental Research 58:169–173.

Ford, A.T., T.F. Fernandes, C.D. Robinson, I.M. Davies, and P.A. Read. 2006. Can industrial pollution cause intersexuality in the amphipod, *Echinogammarus marinus*? Marine Pollution Bulletin 53:100–106.

Frota, G.P., T.M.B. Cabrini, and R.S. Cardoso. 2019. Fluctuating asymmetry of two crustacean species on fourteen sandy beaches of Rio de Janeiro State. Estuarine, Coastal and Shelf Science 223:138–146.

Fuller, N., J.T. Smith, L.L. Nagorskaya, D.I. Gudkov, and A.T. Ford. 2017. Does Chernobyl-derived radiation impact the developmental stability of *Asellus aquaticus* 30 years on? Science of the Total Environment 576:242–250.

Ghedira, J., J. Jebali, M. Banni, L. Chouba, H. Boussetta, J. López-Barea, and J. Alhama. 2011. Use of oxidative stress biomarkers in *Carcinus maenas* to assess littoral zone contamination in Tunisia. Aquatic Biology 14:87–98.

Ghekiere, A., T. Verslycke, N. Fockedey, and C.R. Janssen. 2006. Non-target effects of the insecticide methoprene on molting in the estuarine crustacean *Neomysis integer* (Crustacea: Mysidacea). Journal of Experimental Marine Biology and Ecology 332:226–234.

Gherardi, F. 2011. Towards a sustainable human use of freshwater crayfish (Crustacea, Decapoda, Astacidea). Knowledge and Management of Aquatic Ecosystems 401:02.

Giger, W. 2009. The Rhine red, the fish dead—The 1986 Schweizerhalle disaster, a retrospect and long-term impact assessment. Environmental Science and Pollution Research International 16:98–111.

Godet, J.P., S. Demuynck, C. Waterlot, S. Lemière, C. Souty-Grosset, F. Douay, A. Leprêtre, and C. Pruvot. 2012. Fluctuating asymmetry analysis on *Porcellio scaber* (Crustacea, Isopoda) populations living under metals-contaminated woody habitats. Ecological Indicators 23:130–139.

Grennfelt, P., A. Engleryd, M. Forsius, Ø. Hov, H. Rodhe, and E. Cowlin. 2020. Acid rain and air pollution: 50 years of progress in environmental science and policy. Ambio 49:849–864.

Grilo, T.F., and R. Rosa. 2017. Intersexuality in aquatic invertebrates: Prevalence and causes. Science of the Total Environment 592:714–728.

Guler, Y., S. Short, P. Kile, and A.T. Ford. 2012. Integrating field and laboratory evidence for environmental sex determination in the amphipod, *Echinogammarus marinus*. Marine Biology 159:2885–2890.

Hasegawa, Y., K. Haino-Fukushima, and Y. Katakura. 1987. Isolation and properties of androgenic gland hormone from the terrestrial isopod, *Armadillidium vulgare*. General and Comparative Endocrinology 67:101–110.

Hayasaka, D., T. Korenaga, K. Suzuki, F. Saito, F. Sánchez-Bayo, and K. Goka. 2012. Cumulative ecological impacts of two successive annual treatments of imidacloprid and fipronil on aquatic communities of paddy mesocosms. Ecotoxicology and Environmental Safety 80:355–362.

Hellou, J. 2011. Behavioural ecotoxicology, an "early warning" signal to assess environmental quality. Environmental Science and Pollution Research 18:1–11.

Henderson, G., E. Bresnan, J. Brooke, K. Davidson, M. Dearnaley, M. Fitzsimons, A. Ford, T. Galloway, C. Halsall, T. Horton, and M. Inall. 2023. Independent expert assessment of unusual crustacean mortality in the north-east of England in 2021 and 2022. DEFRA report, UK. Accessed November 5, 2023 [https://www.gov.uk/government/publications/assessment-of-unusual-crustacean-mortality-in-the-north-east-of-england-in-2021-and-2022].

Ho, K., L. Patton, J. Latimer, R. Pruell, M. Pelletier, R. McKinney, and S. Jayaraman. 1999. The chemistry and toxicity of sediment affected by oil from the North Cape spilled into Rhode Island Sound. Marine Pollution Bulletin 38:314–323.

Holmstrup, M., A.-M. Bindesbol, G.J. Oostingh, A. Duschl, V. Scheil, H.-R. Köhler, S. Loureiro, A.M.V.M. Soares, A.L.G. Ferreira, C. Kienle, A. Gerhardt, R. Laskowski, P. Kramarz, M. Bayley, C. Svendsen, and D. Spurgeon. 2010. Interactions between effects of environmental chemicals and natural stressors: A review. Science of the Total Environment 408:3746–3762.

Homola, E., and E.S. Chang. 1997. Methyl farnesoate: Crustacean juvenile hormone in search of functions. Comparative Biochemistry and Physiology Part B: Biochemistry and Molecular Biology 117: 347–356.

ICPDR. 2002. Joint Danube survey: Investigation of the Tisza River and its tributaries. Accessed November 5, 2023. [https://www.icpdr.org/main/sites/default/files/jds-itr_report.pdf].

Jacobson, T., K. Holmström, G. Yang, A.T. Ford, U. Berger, and B. Sundelin. 2010. Perfluoroctane sulfonate accumulation and parasite infestation in a field population of the amphipod *Monoporeia affinis* after microcosm exposure. Aquatic Toxicology 98:99–106.

Jaegers, J., and E. Gismondi. 2020. Gammarid exposure to pyriproxyfen and/or cadmium: What effects on the methylfarnesoate signalling pathway? Environmental Science and Pollution Research 27:31330–31338.

Jakimska, A., P. Konieczka, K. Skóra, and J. Namieśnik. 2011. Bioaccumulation of metals in tissues of marine animals, part II: Metal concentrations in animal tissues. Polish Journal of Environmental Studies 20:1127–1146.

Jepsen, P.M., K. Syberg, G. Drillet, and B.W. Hansen. 2020. Planktonic crustacean culture: Live planktonic crustaceans as live feed for finfish and shrimps in aquaculture. Pages 341–366 *in* G. Lovrich and M.

Thiel, editors. The natural history of the Crustacea, Volume 9: Fisheries and aquaculture. Oxford University Press, New York, USA.

Jergentz, S., P. Pessacq, H. Mugni, C. Bonetto, and R. Schulz. 2004. Linking *in situ* bioassays and population dynamics of macroinvertebrates to assess agricultural contamination in streams of the Argentine pampa. Ecotoxicology and Environmental Safety 59:133–141.

Jewett, S.C., T.A. Dean, R.O. Smith, and A. Blanchard. 1999. 'Exxon Valdez' oil spill: Impacts and recovery in the soft-bottom benthic community in and adjacent to eelgrass beds. Marine Ecology Progress Series 185:59–83.

Jones, D.T., and S.P. Hopkin. 1998. Reduced survival and body size in the terrestrial isopod *Porcellio scaber* from a metal-polluted environment. Environmental Pollution 99:215–223.

Jungmann, D., V. Ladewig, K.-U. Ludwichowski, P. Petzsch, and R. Nagel. 2004. Intersexuality in *Gammarus fossarum* (Koch)—A common inducible phenomenon? Archives in Hydrobiology 159:511–529.

Kampe, S., R. Kaegi, K. Schlich, C. Wasmuth, H. Hollert, and C. Schlechtriem. 2018. Silver nanoparticles in sewage sludge: Bioavailability of sulfidized silver to the terrestrial isopod *Porcellio scaber*. Environmental Toxicology and Chemistry 37:1606–1613.

Kato, Y., K. Kobayashi, H. Watanabe, and T. Iguchi. 2011. Environmental sex determination in the branchiopod crustacean *Daphnia magna*: Deep conservation of a doublesex gene in the sex-determining pathway. PLoS Genetics 7:e1001345.

Khedher, S.B., Z. Haouas, and H. Boussetta. 2017. Cholinesterase activity and histopatological changes in the Mediterranean crab, *Carcinus maenas*, exposed to environmental contaminants. International Journal of Aquatic Biology 5:348–359.

Kim, S.-K. 2020. Trophic transfer of organochlorine pesticides through food-chain in coastal marine ecosystem. Environmental Engineering Research 25:43–51.

Klobučar, G.I., O. Malev, M. Šrut, A. Štambuk, S. Lorenzon, Ž. Cvetković, E.A. Ferrero, and I. Maguire. 2012. Genotoxicity monitoring of freshwater environments using caged crayfish (*Astacus leptodactylus*). Chemosphere 87:62–67.

Knigge, T., and H.R. Köhler. 2000. Lead impact on nutrition, energy reserves, respiration and stress protein (hsp 70) level in *Porcellio scaber* (Isopoda) populations differently preconditioned in their habitats. Environmental Pollution 108:209–217.

Knigge, T., G.A. LeBlanc, and A.T. Ford. 2021. A crab is not a fish: Unique aspects of the crustacean endocrine system and considerations for endocrine toxicology. Frontiers in Endocrinology 12:587608.

Koenig, S., P. Fernández, and M. Solé. 2012. Differences in cytochrome P450 enzyme activities between fish and crustacea: Relationship with the bioaccumulation patterns of polychlorobiphenyls (PCBs). Aquatic Toxicology 108:11–17.

Kohler, S.A., M.O. Parker, and A.T. Ford. 2021. High-throughput screening of psychotropic compounds: Impacts on swimming behaviours in *Artemia franciscana*. Toxics 9:64.

Korez, Š., L. Gutow, and R. Saborowski. 2020. Coping with the "dirt": Brown shrimp and the microplastic threat. Zoology 143:125848.

Kozma, M.T., H. Ngo-Vu, Y.Y. Wong, N.S. Shukla, S.D. Pawar, A. Senatore, M. Schmidt, and C.D. Derby. 2020. Comparison of transcriptomes from two chemosensory organs in four decapod crustaceans reveals hundreds of candidate chemoreceptor proteins. PLoS ONE 15:e0230266.

Krång, A.S., and M. Dahlström. 2006. Effects of a candidate antifouling compound (medetomidine) on pheromone induced mate search in the amphipod *Corophium volutator*. Marine Pollution Bulletin 52:1776–1783.

Krebs, C.T., and K.A. Burns. 1977. Long-term effects of an oil spill on populations of the salt-marsh crab *Uca pugnax*. Science 197:484–487.

Kwok, K.W., S. Souissi, G. Dur, E.J. Won, and J.S. Lee. 2015. Copepods as references species in estuarine and marine waters. Pages 281–308 *in* C. Amiard-Triquet, J.-C. Amiard, and C. Mouneyrac, editors. Aquatic ecotoxicology: Advancing tools for dealing with emerging risks. Academic Press, Cambridge, USA.

Lafferty, K.D., J.W. Porter, and S.E. Ford. 2004. Are diseases increasing in the ocean? Annual Review of Ecology, Evolution, and Systematics 35:31–54.

Landrum, P., B. Eadie, and W. Faust. 1991. Toxicokinetics and toxicity of a mixture of sediment-associated polycyclic aromatic hydrocarbons to the amphipod *Diporeia* sp. Environmental Toxicology and Chemistry 10:35–46.

Laufer, H., M. Chen, B. Baclaski, J.M. Bobbitt, J.D. Stuart, Y. Zuo, and M.W. Jacobs. 2013. Multiple factors in marine environments affecting lobster survival, development, and growth, with emphasis on alkylphenols: A perspective. Canadian Journal of Fisheries and Aquatic Sciences 70:1588–1600.

Lavarías S, H. Heras, N. Pedrini, H. Tournier, and M Ansaldo. 2011. Antioxidant response and oxidative stress levels in *Macrobrachium borellii* (Crustacea: Palaemonidae) exposed to the water-soluble fraction of petroleum. Comparative Biochemistry and Physiology Part C: Toxicology and Pharmacology 153:415–421.

Lavers, J.L., P.B. Sharp, S. Stuckenbrock, and A.L. Bond. 2020. Entrapment in plastic debris endangers hermit crabs. Journal of Hazardous Materials 387:121703.

LeBlanc, G.A. 2007. Crustacean endocrine toxicology: A review. Ecotoxicology 16:61–81.

Leight, A.K., G.I. Scott, M.H. Fulton, and J.W. Daugomah. 2005. Long term monitoring of grass shrimp *Palaemonetes* spp. Population metrics at sites with agricultural runoff influences. Integrative and Comparative Biology 45:143–150.

Levy, T., and A. Sagi. 2020. The "IAG-Switch"—A key controlling element in decapod crustacean sex differentiation. Frontiers in Endocrinology 11:651.

Levy, T., T. Ventura, G. De Leo, N. Grinshpan, F.A. Abu Abayed, R. Manor, A. Savaya, M.Y. Sklarz, V. Chalifa-Caspi, D. Mishmar, and A. Sagi. 2020. Two homogametic genotypes—One crayfish: On the consequences of intersexuality. iScience 23:101652.

Lewis, C., and A.T. Ford. 2012. Infertility in male aquatic invertebrates: A review. Aquatic Toxicology 120:79–89.

Lezcano, A.H., M.L.R. Quiroga, A.L. Liberoff, and S. van der Molen. 2015. Marine pollution effects on the southern surf crab *Ovalipes trimaculatus* (Crustacea: Brachyura: Polybiidae) in Patagonia Argentina. Marine Pollution Bulletin 91:524–529.

Li, M.H. 2002. Fluctuating asymmetry and intersexuality in the shore crab *Grapsus albolineatus* near a coastal landfill site in northern Taiwan. Bulletin of Marine Science 70:75–88.

Livingstone, D.R. 1998. The fate of organic xenobiotics in aquatic ecosystems: Quantitative and qualitative differences in biotransformation by invertebrates and fish. Comparative Biochemistry and Physiology Part A: Molecular and Integrative Physiology 120:43–49.

Löf, M., B. Sundelin, C. Bandh, and E. Gorokhova. 2016. Embryo aberrations in the amphipod *Monoporeia affinis* as indicator of toxic pollutants in sediment, a field evaluation. Ecological Indicators 60:18–30.

Lorenzon, S., M. Francese, V.J. Smith, and E.A. Ferrero. 2001. Heavy metals affect the circulating haemocyte number in the shrimp *Palaemon elegans*. Fish and Shellfish Immunology 11:459–472.

Lukyanova, O.N., E.V. Zhuravel, D.N. Chulchekov, and O.V. Podgurskaya. 2014. An assessment of the ecological state of the northeastern Sakhalin Island shelf: Hydrochemical analysis and bioassay. Russian Journal of Marine Biology 40:539–547.

Lynn, K.D., D. Quintanilla-Ahumada, C. Anguita, S. Widdicombe, J. Pulgar, P.H. Manríquez, P.A. Quijón, and C. Duarte. 2021. Disruption and recovery: Artificial light at night (ALAN) alter the activity and feeding behavior of sandy beach amphipods from Atlantic Canada. Science of the Total Environment 780:146568.

MacLeod, M., H.P.H. Arp, M.B. Tekman, and A. Jahnke. 2021. The global threat from plastic pollution. Science 373:61–65.

Martínez-Colón, M., H. Alegría, A. Huber, H. Kubra-Gul, and P. Kurt-Karakus. 2021. Bioaccumulation and biomagnification of potential toxic elements (PTEs): An *Avicennia germinans–Uca rapax* trophic transfer story from Jobos Bay, Puerto Rico. Ecological Indicators 121:107038.

Mateos-Cárdenas, A., J. O'Halloran, F.N. van Pelt, and M.A. Jansen. 2020. Rapid fragmentation of microplastics by the freshwater amphipod *Gammarus duebeni* (Lillj.). Scientific Reports 10:1–12.

Mathews, C., S. Kedidi, N. Fita, A. Al-Yahya, and K. Al-Rasheed. 1993. Preliminary assessment of the effects of the 1991 Gulf War on Saudi Arabian prawn stocks. Marine Pollution Bulletin 27: 251–271.

Mazurová, E., K. Hilscherová, V. Jálová, H.R. Köhler, R. Triebskorn, J.P. Giesy, and L. Bláha. 2008. Endocrine effects of contaminated sediments on the freshwater snail *Potamopyrgus antipodarum in vivo* and in the cell bioassays *in vitro*. Aquatic Toxicology 89:172–179.

Melvin, S.D., and S.P. Wilson. 2013. The utility of behavioral studies for aquatic toxicology testing: A meta-analysis. Chemosphere 93:2217–2223.

McCurdy, D.G., M.R. Forbes, S.P. Logan, M.T. Kopec, and S.I. Mautner. 2004. The functional significance of intersexes in the intertidal amphipod *Corophium volutator*. Journal of Crustacean Biology 24: 261–265.

Miller, M., M. Hamann, and F. Kroon. 2020. Bioaccumulation and biomagnification of microplastics in marine organisms: A review and meta-analysis of current data. PLoS ONE 15:e0240792.

Moore, C.G., and J.M. Stevenson. 1991. The occurrence of intersexuality in harpacticoid copepods and its relationship with pollution. Marine Pollution Bulletin 22:72–74.

Moore, C.G., and J.M. Stevenson. 1994. Intersexuality in benthic harpacticoid copepods in the Firth of Forth, Scotland. Journal of Natural History 28:1213–1230.

Moore, M.V., S.M. Pierce, H.M. Walsh, S.K. Kvalvik, and J.D. Lim. 2000. Urban light pollution alters the diel vertical migration of *Daphnia*. Internationale Vereinigung für Theoretische und Angewandte Limnologie: Verhandlungen 27:1–4.

Mu, X., and G.A. Leblanc. 2004. Cross communication between signaling pathways: Juvenoid hormones modulate ecdysteroid activity in a crustacean. Journal of Experimental Zoology Part A: Ecological Genetics and Physiology 301:793–801.

Mullen, T.E., S. Russell, M.T. Tucker, J.L. Maratea, C. Koerting, L. Hinckley, S. De Guise, S. Frasca, R.A. French Jr., T.G. Burrage, and C. Perkins. 2004. Paramoebiasis associated with mass mortality of american lobster *Homarus americanus* in Long Island Sound, USA. Journal of Aquatic Animal Health 16:29–38.

Munoz, G., H. Budzinski, M. Babut, J. Lobry, J. Selleslagh, N. Tapie, and P. Labadie. 2019. Temporal variations of perfluoroalkyl substances partitioning between surface water, suspended sediment, and biota in a macrotidal estuary. Chemosphere 233:319–326.

Nikinmaa, M. 2014. An introduction to aquatic toxicology. Academic Press, Cambridge, USA.

Nimmo, D.R., T.L. Hamaker, E. Matthews, and J.C. Moore. 1981. An overview of the acute and chronic effects of first and second generation pesticides on an estuarine mysid. Pages 3–20 *in* J. Vernberg, A. Calabrese, F.P. Thurberg, and W.B. Vernberg, editors. Biological monitoring of marine pollutants. Academic Press, Cambridge, USA.

NOAA (National Oceanic and Atmospheric Administration). 1998. Damage assessment center. Damage assessment and restoration plan for the North Cape oil spill. Rhode Island Department of Environmental Protection and U.S. Department of the Interior. Silver Spring, Maryland, USA.

NRC. 2003. National Research Council (US) Committee on oil in the sea: Inputs, fates, and effects. The National Academies of Sciences, National Academies Press, Washington D.C., USA.

Nowak, B.F., M.B. Martin, and S. Boltaña. 2020. Parasitic crustaceans. Pages 401–434 in G. Lovrich and M. Thiel, editors. The natural history of the Crustacea, Volume 9: Fisheries and aquaculture. Oxford University Press, New York, USA.

Okubo, I. 2003. The freshwater ostracod *Notodromas monacha* (O.F. Müller, 1776). Bulletin of the Biogeographical Society of Japan 58:43–48.

Olmstead, A.W., and G.A. LeBlanc. 2002. Juvenoid hormone methyl farnesoate is a sex determinant in the crustacean *Daphnia magna*. Journal of Experimental Zoology 293:736–739.

Olmstead, A.W., and G.A. LeBlanc. 2007. The environmental-endocrine basis of gynandromorphism (intersex) in a crustacean. International Journal of Biological Sciences 3:77.

Ogonowski, M., M. Wagner, B. Rogell, M. Haave, and A. Lusher. 2023. Microplastics could be marginally more hazardous than natural suspended solids—A meta-analysis. Ecotoxicology and Environmental Safety 264:115406.

Olsén, K.H. 2010. Effects of pollutants on olfactory mediated behaviors in fish and crustaceans. Pages 507–529 *in* T. Breithaupt and M. Thiel, editors. Chemical communication in crustaceans. Springer, New York, USA.

Palma, P., V.L. Palma, C. Matos, R.M. Fernandes, A. Bohn, A.M. Soares, and I.R. Barbosa. 2009. Effects of atrazine and endosulfan sulphate on the ecdysteroid system of *Daphnia magna*. Chemosphere 74:676–681.

Palmer, A.R., and C. Strobeck. 1986. Fluctuating asymmetry: Measurement, analysis, patterns. Annual Review of Ecology and Systematics 17:391–421.

Parmentier, K.F.V., Y. Verhaegen, B.P. De Witte, S. Hoffman, D.H.R. Delbare, P.M. Roose, K.D.E. Hylland, T. Burgeot, G.J. Smagghe, and K. Cooreman. 2019. Tributyltin: A bottom-up regulator of the *Crangon crangon* population? Frontiers in Marine Science 6:633.

Pinheiro, M.A., L.F. Duarte, T.R. Toledo, M.L. Adam, and R.A. Torres. 2013. Habitat monitoring and genotoxicity in *Ucides cordatus* (Crustacea: Ucididae), as tools to manage a mangrove reserve in southeastern Brazil. Environmental Monitoring and Assessment 185:8273–8285.

Piscia, R., M. Mazzoni, R. Bettinetti, R. Caroni, D. Cicala, and M.M. Manca. 2019. Stable isotope analysis and persistent organic pollutants in crustacean zooplankton: The role of size and seasonality. Water 11:1490.

Podlesińska, W., and H. Dąbrowska. 2019. Amphipods in estuarine and marine quality assessment—A review. Oceanologia 61:179–196.

Rainbow, P., and S. Luoma. 2011. Metal toxicity, uptake and bioaccumulation in aquatic invertebrates—Modelling zinc in crustaceans. Aquatic Toxicology 105:455–465.

Rainbow, P., and B. Smith. 2013. Accumulation and detoxification of copper and zinc by the decapod crustacean *Palaemonetes varians* from diets of field-contaminated polychaetes *Nereis diversicolor*. Journal of Experimental Marine Biology and Ecology 449:312–320.

Ralston-Hooper, K., J. Hardy, L. Hahn, H. Ochoa-Acuña, L. Lee, R. Mollenhauer, and M. Sepúlveda. 2009. Acute and chronic toxicity of atrazine and its metabolites deethylatrazine and deisopropylatrazine on aquatic organisms. Ecotoxicology 18:899–905.

Reid, A.J., A.K. Carlson, I.F. Creed, E.J. Eliason, P.A. Gell, P.T.J. Johnson, K.A. Kidd, T.J. MacCormack, J.D. Olden, S.J. Ormerod, J.P. Smol, W.W. Taylor, K. Tockner, J.C. Vermaire, D. Dudgeon, and S. Cooke. 2019. Emerging threats and persistent conservation challenges for freshwater biodiversity. Biological Reviews 94:849–873.

Richman, N.I., M. Böhm, S.B. Adams, F. Alvarez, E.A. Bergey, J.J.S. Bunn, Q. Burnham, J. Cordeiro, J. Coughran, K.A. Crandall, K.L. Dawkins, R.J. DiStefano, N.E. Doran, L. Edsman, A.G. Eversole, L. Füreder, J.M. Furse, F. Gherardi, P. Hamr, D.M. Holdich, P. Horwitz, K. Johnston, C.M. Jones, J.P.G. Jones, R.L. Jones, T.G. Jones, T. Kawai, S. Lawler, M. López-Mejía, R.M. Miller, C. Pedraza-Lara, J.D. Reynolds, A.M.M. Richardson, M.B. Schultz, G.A. Schuster, P.J. Sibley, C. Souty-Grosset, C.A. Taylor, R.F. Thoma, J. Walls, T.S. Walsh, and B. Collen. 2015. Multiple drivers of decline in the global status of freshwater crayfish (Decapoda: Astacidea). Philosophical Transactions of The Royal Society B: Biological Sciences 370:20140060.

Rising, K., J. Hardege, T. Tregenza, and M. Stevens. 2022. Anthropogenic noise may impair the mating behaviour of the shore crab *Carcinus maenas*. PLoS ONE 17:e0276889.

Roberts, L., S. Cheesman, M. Elliott, and T. Breithaupt. 2016. Sensitivity of *Pagurus bernhardus* (L.) to substrate-borne vibration and anthropogenic noise. Journal of Experimental Marine Biology and Ecology 474:185–194.

Rodrigues, A.P., K.K. Lehtonen, L. Guilhermino, and L. Guimarães. 2013. Exposure of *Carcinus maenas* to waterborne fluoranthene: Accumulation and multibiomarker responses. Science of the Total Environment 443:454–463.

Rodríguez, E.M., D.A. Medesani, and M. Fingerman. 2007. Endocrine disruption in crustaceans due to pollutants: A review. Comparative Biochemistry and Physiology Part A: Molecular and Integrative Physiology 146:661–671.

Rosen, O., R. Manor, S. Weil, O. Gafni, A. Linial, E.D. Aflalo, T. Ventura, and A. Sagi, 2010. A sexual shift induced by silencing of a single insulin-like gene in crayfish: Ovarian upregulation and testicular degeneration. PLoS ONE 5:e15281.

Rossi, V., G. Benassi, M. Veneri, C. Bellavere, P. Menozzi, A. Moroni, and K.G. McKenzie. 2003. Ostracoda of the Italian ricefields thirty years on: New synthesis and hypothesis. Journal of Limnology 62:1–8.

Saborowski, R., E. Paulischkis, and L. Gutow. 2019. How to get rid of ingested microplastic fibers? A straightforward approach of the Atlantic ditch shrimp *Palaemon varians*. Environmental Pollution 254:113068.

Sagi, A., E. Snir, and I. Khalaila. 1997. Sexual differentiation in decapod crustaceans: Role of the androgenic gland. Invertebrate Reproduction and Development 31:55–61.

Sánchez-Bayo, F. 2011. Insecticides mode of action in relation to their toxicity to non-target organisms. Journal of Environmental and Analytical Toxicology S4:1–9.

Sanders, H.L., J.F. Grassle, L.S. Hampson, L.S. Morse, S. Garner-Price, and C. Jones. 1980. Anatomy of an oil spill: Long-term effects from the grounding of the barge Florida off West Falmouth, Massachusetts. Journal of Marine Research 38:265–380.

Schill, R.O., and H.R. Köhler. 2004. Energy reserves and metal-storage granules in the hepatopancreas of *Oniscus asellus* and *Porcellio scaber* (Isopoda) from a metal gradient at Avonmouth, UK. Ecotoxicology 13:787–796.

Scholtz, G. 2020. Duplicated, twisted, and in the wrong place: Patterns of malformation in crustaceans. Pages 112–142 in K. Anger, S. Harzsch, and M. Thiel, editors. The natural history of the Crustacea, Volume 7: Developmental biology and larval ecology. Oxford University Press, New York, USA.

Schultz, T.W., and J.R. Kennedy. 1976. Cytotoxic effects of the herbicide 3-amino-1,2,4-triazole on *Daphnia pulex* (Crustacea: Cladocera). Biological Bulletin 151:370–385.

Sheehan, P.S. 1984. Effects on individuals and populations. Pages 23–54 *in* P.J. Sheehan, D.R. Miller, G.C. Butler, and P. Bourdeau, editors. Effects of pollutants at the ecosystem level. SCOPE, John Wiley and Sons Ltd., Hoboken, New Jersey, USA.

Shields, J.D. 2013. Complex etiologies of emerging diseases in lobsters (*Homarus americanus*) from Long Island Sound. Canadian Journal of Fisheries and Aquatic Sciences 70:1576–1587.

Shields, J.D., and J.P. Huchin-Mian. 2020. Ecological factors in the emergence of pathogens in commercially important crustaceans. Pages 367–399 *in* G. Lovrich and M. Thiel, editors. The natural history of the Crustacea, Volume 9: Fisheries and aquaculture. Oxford University Press, New York, USA.

Short, S., Y. Guler, G. Yang, P. Kille, and A.T. Ford. 2012a. Paramyxean–microsporidian co-infection in amphipods: Is the consensus that Microsporidia can feminise their hosts presumptive? International Journal for Parasitology 42:683–691.

Short, S., G. Yang, P. Kille, and A.T. Ford. 2012b. A widespread and distinctive form of amphipod intersexuality not induced by known feminising parasites. Sexual Development 6:320–324.

Short, S., G. Yang, Y. Guler, A. Green Etxabe, P. Kille, and A.T. Ford. 2014. Crustacean intersexuality is feminization without demasculinization: Implications for environmental toxicology. Environmental Science and Technology 48:13520–13529.

Siegel, P.R., and A.M. Wenner. 1984. Abnormal reproduction of the sand crab *Emerita analoga* in the vicinity of a nuclear generating station in Southern California. Marine Biology 80:341–345.

Smith, R.J., D. Zhai, S. Savatenalinton, T. Kamiya, and N. Yu. 2017. A review of rice field ostracods (Crustacea) with a checklist of species. Journal of Limnology 77:1.

Smith, V.J., R.J. Swindlehurst, P.A. Johnston, and A.D. Vethaak. 1995. Disturbance of host defence capability in the common shrimp, *Crangon crangon*, by exposure to harbour dredge spoils. Aquatic Toxicology 32:43–58.

Soegianto, A., M. Charmantier-Daures, J.-P. Trilles, and G. Charmantier. 1999. Impact of cadmium on the structure of gills and epipodites of the shrimp *Penaeus japonicas* (Crustacea: Decapoda). Aquatic Living Resources 12:57–70.

Song, Y., D.L. Villeneuve, K. Toyota, T. Iguchi, and K.E. Tollefsen. 2017. Ecdysone receptor agonism leading to lethal molting disruption in arthropods: Review and adverse outcome pathway development. Environmental Science and Technology 51:4142–4157.

Stenersen, J. 2009. Chemical pesticides: Mode of action and toxicology. CRC Press, Boca Raton, Florida, USA.

Stentiford, G.D. 2012. Histological intersex (ovotestis) in the European lobster *Homarus gammarus* and a commentary on its potential mechanistic basis. Diseases of Aquatic Organisms 100:185–190.

Stentiford, G.D., and J.D. Shields. 2005. A review of the parasitic dinoflagellates *Hematodinium* species and *Hematodinium*-like infections in marine crustaceans. Diseases of Aquatic Organisms 66:47–70.

Steevens, J.A., and W.H. Benson. 2000. Interactions of chlorpyrifos and methyl mercury: A mechanistic approach to assess chemical mixtures. Marine Environmental Research. 50:113–117.

Strauss, J., and H. Dircksen. 2010. Circadian clocks in crustaceans: Identified neuronal and cellular systems. Frontiers in Biosciences 15:1040–1074.

Štrus, J., N. Žnidaršič, P. Mrak, U. Bogataj, and G. Vogt. 2019. Structure, function and development of the digestive system in malacostracan crustaceans and adaptation to different lifestyles. Cell and Tissue Research 377:415–443.

Suchanek, T. 1993. Oil impacts on marine invertebrate populations and communities. American Zoologist 33:510–523.

Sullivan, J.M., M.C. Genco, E.D. Marlow, J.L. Benton, B.S. Beltz, and D.C. Sandeman. 2009. Brain photoreceptor pathways contributing to circadian rhythmicity in crayfish. Chronobiology International 26:1136–1168.

Sundelin, B., and A.-K. Eriksson. 1998. Malformations in embryos of the deposit-feeding amphipod *Monoporeia affinis* in the Baltic Sea. Marine Ecology Progress Series 171:165–180.

Takahashi, T., A. Araki, Y. Nomura, M. Koga, and K. Arizono. 2000. The occurrence of dual-gender imposex in Japanese freshwater crab. Journal of Health Science 46:376–379.

Tang, M., Z. Lu, S.V. Babu, G. Yang, Y. Li, Z. Xu, G. Pan, Z. Qin, and L. Lin. 2021. The regulatory relationships between the gonad-inhibiting hormone and insulin-like androgenic gland hormone-binding protein genes in the eyestalk-androgenic gland-testis axis of *Macrobrachium rosenbergii*. Frontiers in Marine Sciences 8:775191.

Taylor, D.M. 1986. A bilateral gynandromorph of the snow crab, *Chionoecetes opilio*, from Newfoundland, Canada. Crustaceana 51:309–312.

Thompson, M., C. Bankier, T. Bell, A. Dumbrell, C. Gray, M. Ledger, K. Lehmann, B.A. McKew, C.D. Sayer, F. Shelley, M. Trimmer, S.L. Warren, and G. Woodward. 2016. Gene-to-ecosystem impacts of a catastrophic pesticide spill: Testing a multilevel bioassessment approach in a river ecosystem. Freshwater Biology 61:2037–2050.

Tollefsen, K.E., Y. Song, T. Høgåsen, I.B. Øverjordet, D. Altin, and B.H. Hansen. 2017. Mortality and transcriptional effects of inorganic mercury in the marine copepod *Calanus finmarchicus*. Journal of Toxicology and Environmental Health Sciences, Part A 80:845–861.

Torres, D., S. Tidau, S. Jenkins, and T. Davies. 2020. Artificial sky-glow disrupts celestial migration at night. Current Biology 30:R696–R697.

Toyota, K., H. Miyakawa, K. Yamaguchi, S. Shigenobu, Y. Ogino, N. Tatarazako, S. Miyagawa, and T. Iguchi. 2015. NMDA receptor activation upstream of methyl farnesoate signaling for short day-induced male offspring production in the water flea, *Daphnia pulex*. BMC Genomics 16:186.

Van der Oost, R., H. Heida, and A. Opperhuizen. 1988. Polychlorinated biphenyl congeners in sediments, plankton, molluscs, crustaceans, and eel in a freshwater lake: Implications of using reference chemicals and indicator organisms in bioaccumulation studies. Archives of Environmental Contamination and Toxicology 17:721–729.

Van Straalen, N.M. 2003. Ecotoxicology becomes stress ecology. Environmental Science and Technology 37:324A–330A.

Wale, M.A., S.D. Simpson, and A.N. Radford. 2013. Noise negatively affects foraging and antipredator behaviour in shore crabs. Animal Behaviour 86:111–118.

Walker, C.H., R.M. Sibly, S.P. Hopkin, and D.B. Peakall. 2012. Principles of ecotoxicology, 4th edition. CRC Press, Boca Raton, Florida, USA.

Wang, W. 2011. Bacterial diseases of crabs: A review. Journal of Invertebrate Pathology 106:18–26.

Ward, D.V., B.L. Howes, and D.F. Ludwig. 1976. Interactive effects of predation pressure and insecticide (temefos) toxicity on populations of the marsh fiddler crab *Uca pugnax*. Marine Biology 35:119–126.

Watts, A.J., C. Lewis, R.M. Goodhead, S.J. Beckett, J. Moger, C.R. Tyler, and T.S. Galloway. 2014. Uptake and retention of microplastics by the shore crab *Carcinus maenas*. Environmental Science and Technology 48:8823–8830.

Watts, A.J.R., M.A. Urbina, R. Goodhead, J. Moger, C. Lewis, and T.S. Galloway. 2016. Effect of microplastic on the gills of the shore crab *Carcinus maenas*. Environmental Science and Technology 50: 5364–5369.

Weis, J.S., A. Cristini, and K. Ranga Rao. 1992. Effects of pollutants on molting and regeneration in Crustacea. American Zoologist 32:495–500.

Weis, J.S., J. Gottlieb, and J. Kwiatkowski. 1987. Tributyltin retards regeneration and produces deformities of limbs in the fiddler crab *Uca pugilator*. Archives of Environmental Contamination and Toxicology 16:321–326.

Welden, N.A., B. Abylkhani, and L.M. Howarth. 2018. The effects of trophic transfer and environmental factors on microplastic uptake by plaice, *Pleuronectes platessa*, and spider crab, *Maja squinado*. Environmental Pollution 239:351–358.

Wiesinger, H., Z. Wang, and S. Hellweg. 2021. Deep dive into plastic monomers, additives, and processing aids. Environmental Science and Technology 55:9339–9351.

Williams, H.H., and K. MacKenzie. 2003. Marine parasites as pollution indicators: An update. Parasitology 126:S27–41.

WWF. 2002. The ecological effects of mining spills in the Tisza River System in 2000. Vienna, Italy. Accessed November 5, 2023. [https://wwfeu.awsassets.panda.org/downloads/Tisza_Cyanide_Report.pdf].

Yang, G., P. Kille, and A.T. Ford. 2008. Infertility in a marine crustacean: Have we been ignoring pollution impacts on male invertebrates? Aquatic Toxicology 88:81–87.

CONSERVATION GENETICS OF CRUSTACEANS

Keith A. Crandall

Abstract

With over 42,000 species of known crustaceans inhabiting mainly marine, but also freshwater and even terrestrial habitats, they are often key members of these diverse habitats and associated ecosystems. However, many crustaceans are either endangered or their endangered status is simply unknown because we lack critical information about these important species. Conservation genetics offer efficient tools, both molecular and analytical, to gain keen insights into population dynamics, population sizes, species boundaries, and habitat associations to inform conservation efforts based on either ecosystems and/or species. This chapter covers key concepts in conservation genetics as applied to crustacean biodiversity and provides examples of the application of these concepts in ongoing crustacean conservation efforts as case studies. We explore informative reference databases for geographic, taxonomic, and conservation information for crustaceans. We discuss sampling strategies as applied to conservation genetics and explore a diversity of molecular markers from mitochondrial DNA to anchored hybrid enrichment. We also highlight key analytical approaches for estimating population genetic parameters that help inform conservation decisions such as gene flow, genetic diversity, and phylogenetic diversity estimates. Finally, we demonstrate the application of some of these concepts through the exploration of case studies based on our own conservation genetics work in crustaceans. This chapter highlights not only the useful tools and a diversity of data types used in conservation genetics, but the extreme need for the application of these tools, data, and associated insights to the world's crustacean diversity that is severely understudied with respect to conservation.

Keith A. Crandall, *Conservation Genetics of Crustaceans*. In: *Ecology and Conservation*. Edited by: Lars Gutow and Martin Thiel, Oxford University Press. © Oxford University Press (2025). DOI: 10.1093/oso/9780197768242.003.0012

INTRODUCTION

Crustaceans are often keystone species in a diversity of habitats (including marine, freshwater, cave, and even terrestrial) (e.g., Gaedke and Straile 1998) and can serve as bioindicators for the relative health of such habitats (e.g., de Almeida Rodrigues et al. 2021). While certain crustaceans are invasive species (Weis 2010), many more crustaceans are considered endangered, especially those from particularly sensitive habitats that have been heavily impacted by human activities such as caves and freshwater. Much of conservation biology and the associated legal protective efforts (e.g., the Endangered Species Act of the USA) are focused on *species*. Yet, for many crustaceans, we have a fundamental lack of information about species status. Many crustacean groups have seen an increase in taxonomic activity in recent years, greatly facilitated by the incorporation of molecular systematics to aid in identifying species boundaries (species delimitation) and species assignment (e.g., Crivellaro et al. 2018, Rodríguez-Flores et al. 2019, Lovrenčić et al. 2020). Despite this increasing rate of species descriptions, many species remain undescribed and therefore particularly susceptible to factors impacting their endangered status. Furthermore, for those species that are described, very few have been assessed for conservation status at regional, national, or international levels.

Molecular tools can be highly effective in a diversity of applications for conservation purposes, including delimiting species, estimating effective population sizes, inferring changes in population size through evolutionary history, measuring gene flow and migration patterns, and even ranking geographic areas for relative conservation priorities based on phylogenetic diversity concepts. In this chapter, we discuss endangered crustaceans and their habitats, review the state of crustacean conservation assessment, and describe some key molecular tools and then analytical tools applied to conservation efforts with some demonstrations of their utility through crustacean case studies. Throughout the chapter, we hope to introduce readers to a diversity of informatics tools and molecular genetic tools for application in conservation genetics, highlight the amazing diversity of the Crustacea, and emphasize the dire need for more conservation efforts for these important community members across a diversity of habitats. We hope this chapter might motivate and set the foundation for future work in crustacean conservation genetics and the conservation of these amazing animals and their diverse habitats.

ENDANGERED CRUSTACEANS AND THEIR HABITATS

While most crustaceans are marine, there are important and particularly endangered groups of species that inhabit freshwater and cave environments. There are even terrestrial crustaceans, including amphipods, oniscidean isopods, and decapods (Richardson and Araujo 2015). This is without considering the phylogenetic reality that the Hexapoda (insects and relatives) are crustaceans (i.e., Pancrustacea; Regier et al. 2005, Giribet and Edgecombe 2019, Bracken-Grissom and Wolfe 2020, von Reumont and Edgecombe 2020). Thus, the crustaceans are a highly successful group having established themselves in a diversity of habitats and many of these habitats are suffering from increasing endangerment due to human activities both locally and globally.

Marine

The vast majority of crustaceans are found in the marine environment. While once thought of as a homogeneous environment that allowed for global gene flow within species across great distances, especially with larval dispersal, many species can have very restricted distributions. Indeed, marine crustaceans show a diversity of gene flow and phylogeographic patterns from panmictic to isolation by distance, to small-isolated populations, depending on the species (Ewers-Saucedo and Wares 2020). Thus, measuring gene flow and identifying patterns of population subdivision becomes a key component of conservation biology where genetic markers are essential to inform such questions. For example, De Wit et al. (2020) used RADSeq (see below) single nucleotide polymorphism data to measure population subdivision in the isopod *Idotea balthica* from the Baltic Sea. They sampled this target isopod from 31 locations throughout the Baltic Sea, with 45% of their sites located in marine protected areas. They used the genetic data to integrate with oceanographic current models and estimated life history traits. Using these combined data, they showed population structure on scales of hundreds of kilometers across the Baltic Sea, with reduced genetic diversity along its salinity gradient. Similar studies have been done on the same species using a variety of molecular markers and sampling strategies (e.g., Wares 2001, Wares et al. 2007, Panova et al. 2017). Additionally, conservation genetics is highly useful for measuring dispersal events, either natural or human aided (often resulting in the introduction of exotic species to new marine environments causing substantial disruption of the native community; see Carlton 2011).

The marine environment is under substantial human pressure, from fisheries to pollution to global warming, all impacting species' distributions and causing endangerment (Duarte et al. 2020). While fish and marine mammals receive the bulk of the attention with respect to marine conservation, marine crustaceans are just as impacted by such habitat modifications and overfishing. For example, Rumbold et al. (2018) studied diversity and seasonal variation in peracarids between a natural rocky shore and a man-made and impacted harbor showing substantial community differences in composition and abundance of the targeted peracarids. Yet our understanding of even the basics of marine crustacean diversity is very limited. The marine environment is particularly difficult to sample due to the vastness of the habitat. Yet, recent advances in environmental DNA (eDNA; see below) sampling approaches have shown that there is much diversity in the marine crustaceans yet to be discovered (Bezeng and van der Bank 2019, Rossel and Martínez Arbizu 2019, Arfianti and Costello 2020) and, when sampled, eDNA from marine environments shows that arthropods (overwhelmingly made up of the crustaceans) are often the dominant taxon (Nguyen et al. 2020).

Freshwater

While 97% of the Earth's water is in the oceans, the remaining 3% of freshwater is essential for human life. Of course, not only humans rely on freshwater: many crustacean species are endemic to freshwater environments, which receive similar impacts from human usage but often to a much more detrimental degree due to freshwater being an extremely limited and often highly localized resource (Reid et al. 2019, Dudgeon 2020). Many crustaceans are highly endemic to localized headwaters and/or springs in freshwater systems making them even more susceptible to habitat degradation or loss. For example, Hudson et al. (2021) document direct human activities such as industrial extraction sites for mining and petroleum and the compounded effects of climate change differentially impacted widely distributed and range-restricted endemic freshwater crab species from Columbia. On average, more attention has been paid to freshwater than marine crustaceans, especially at regional levels, with respect to conservation efforts. This is nicely exemplified by the US Endangered Species List (FWS 2021), where only 28 crustacean species are listed, and all are freshwater taxa (crayfish, shrimp, isopods, amphipods, and fairy shrimp). This conservation focus on freshwater crustaceans is not unique to the USA, with an international assessment of endangerment provided by the International Union for Conservation (IUCN), where the IUCN Red List of endangered species reports 3189 crustaceans (out of some 52,000+ known species; see Chapter 16 in this volume), of which 26% are considered to be endangered in some form, and 83.6% are from freshwater environments (IUCN SSC 2021). Thus, the freshwater habitat is of particular interest and focus for crustacean conservation, and genetic techniques have been highly successfully applied in this environment for species delimitation, gene flow/migration estimates, historical population size estimates, and even eDNA techniques (Deiner et al. 2015, Fernández et al. 2018, Harper et al. 2019, Gold et al. 2020).

Caves

Caves represent a unique environment both with respect to evolution (Porter and Crandall 2003, Stern et al. 2017, Porter and Sumner-Rooney 2018) and conservation (Bishop et al. 2020, Iannella et al. 2021). Nearly half of the 28 crustaceans listed by the US Endangered Species Act are cave species. Cave crustaceans often have unique adaptations to the cave environment (Fig. 12.1), including lack of pigmentation, loss of eyes, elongated sensory apparati (setae, antennae, etc.). Cave crustaceans are particularly difficult to sample due to limited access to the water within a given cave system, and often there is a lack of understanding of water flow and connectedness within and among geographically proximal caves and cave systems. Yet these cave environments lead to highly locally endemic fauna that often result in these species being highly endangered as they are often restricted to single caves or cave systems (see Chapter 14 in this volume). One cave environment, the anchialine caves, is particularly unique in that they are inland bodies of freshwater with a subterranean connection to marine water resulting in density-stratified water, with the freshwater near the surface, the saltwater below, and a brackish interface between the two. As a consequence of this unique habitat (Calderón-Gutiérrez et al. 2018), anchialine caves have very distinctive and endemic crustacean fauna (Pérez-Moreno et al. 2016), including the relatively recently discovered remipeds, which are so unique they were placed in their own class (Yager 1981). Often considered an evolutionary "dead end," cave environments, to the contrary, allow for speciation at rates comparable to other crustacean habitats (Stern et al. 2017), yet many of these species are often critically endangered and considered endangered at a much higher rate than their non-cave counterparts. These subterranean environments are heavily impacted by human activities and often overlooked in conservation policies (Mammola et al. 2019).

Fig. 12.1.
Cave crayfish (*Procambarus orcinus* Hobbs and Means, 1972) showing distinctive cave adaptations, including loss of pigment (you can see through the carapace to the internal features), elongated sensory (see setae and antennules), and loss of eye (there is still an eye stalk as the crayfish molting mechanism is located there, but the eye is gone). Photo reproduced courtesy of C. Riley Nelson; see a color version of this figure in the centerfold.

Terrestrial

The terrestrial environment is an interesting consideration with respect to crustacean diversity and conservation. Crustaceans are traditionally considered to be highly diverse in aquatic environments with limited and yet highly interesting evolutionary lineages invading the terrestrial habitat (Richardson and Araujo 2015). However, evolutionarily, the Hexapoda (insects) are a subgroup of crustaceans and arguably the most successful terrestrial organisms from a species diversity perspective (Regier et al. 2005). Restricting ourselves to the traditional concept of crustaceans, at least five major lineages have successfully adapted to the terrestrial habitat, including Isopoda, Amphipoda, Astacida, Anomura, and Brachyura (Hansson et al. 2011). This transition from an aquatic to a terrestrial lifestyle opens the door to exciting investigations into evolutionary adaptations associated with such transitions, including around osmoregulation, communication, vision, behavior, and reproduction (e.g., Ayari et al. 2021). Perhaps the most famous example of a terrestrial crustacean is the coconut crab (*Birgus latro*)—the largest terrestrial crustacean—with its unique communication approach through various tapping frequencies (Oka et al. 2019). The genome of the coconut crab was recently sequenced and compared to two marine crustaceans, the ornate spiny lobster, *Panulirus ornatus* and the red king crab *Paralithodes camtschaticus*, and showed a proliferation of genes associated with visual, respiratory, olfactory, and cytoskeletal systems in the terrestrial transition of the coconut crab (Veldsman et al. 2021). Krieger et al. (2021) examined multiple crustacean lineages and their molecular adaptations to terrestrialization across Isopoda, Brachyura, and Anomala to show distinct olfactory pathway evolution and morphological transformation across these different lineages. These recent genomic and morphological studies highlight the uniqueness of the terrestrial crustacean lineages and provide tools (see the "Genomics/Transcriptomics" section below) for advancing conservation genetics. These crustaceans have unique conservation issues typically associated with terrestrial habitats, including deforestation, alteration of drainage patterns, and pollution (Cumberlidge et al. 2009).

RESOURCES FOR TRACKING ENDANGERED STATUS

Critical to conservation is the ability to consistently identify endangered species and targeted habitats. While much of conservation biology is beyond the scope of science (politics, economics, capacity, etc.), conservation genetics can provide critical information to help infuse strong science into conservation decisions. Genetic information provides critical insights relative to endangered species status criteria. Unfortunately, these criteria often vary across geographic scales and often do not consider entire distributions and populations of endangered species. Thus, effective conservation genetics rely on extensive sampling throughout the species' range, regardless of political boundaries, to identify genetic constraints on gene flow, resulting in population subdivision and aiding in the identification of distinct populations and new species.

Local, Regional, National, and International Criteria and Sampling Strategies

Critical elements of local, regional, national, and international criteria for defining "endangered species" typically focus on the prospects of species survival being impacted by modification or loss of habitat, overutilization (overfishing, etc.), disease, predation, etc. However, how one demonstrates such impact is typically left to the science, and such standards are often phrased in terms of "small population size" or "limited distribution." Many crustaceans have very localized species distributions (e.g., cave species limited to a single cave or stream species inhabiting a single freshwater spring). Crustaceans can be difficult to directly assess for population size because researchers often have limited access to the target species (e.g., hydrothermal vent species in deep ocean habitats, cave species where only a small fraction of the water is accessible for sampling). Conservation genetics can offer tools for estimating population sizes, both historical and current (see below and Crandall et al. [1999], Buhay and Crandall [2005]). However, such approaches require proper sampling strategies. Such sampling strategies need to take into account the entire species range, regardless of political boundaries. They also need to consider variation at multiple genetic loci (see below). Importantly, sampling strategies for conservation genetics should also be iterative. Conservation genetics can often uncover hidden genetic variation in certain parts of a species range that may require more extended sampling in a follow-on targeted sampling effort to fully understand boundaries of genetic breaks and accurately measure population subdivision and estimates of gene flow and effective population sizes. For example, consider a species that has a given geographic range (Fig. 12.2, blue outline). Local/regional conservation efforts can be hampered by focusing just on their jurisdiction. In our example, the black line represents a political boundary that cuts through the distribution, leaving a very small portion of the species distribution in one jurisdiction and the remainder in another jurisdiction. The jurisdiction with the limited portion of the range might consider this species "endangered" because the "population" is small within that jurisdiction, but such a political decision ignores the biological reality of a species range that spans jurisdictions. Thus, it is important for conservation efforts to consider the species/habitats regardless of political boundaries.

Next, let's use the same example to consider sampling strategies. Suppose you initiate a sampling effort across the entire range of the species (Fig. 12.2, red r's). You analyze the genetic data from this sampling effort and identify a potential partitioning of genetic variation (Fig. 12.2, capital R vs. lowercase r). However, the initial sampling is sparse due to the effort to sample throughout the range. To better define the break in genetic diversity (Fig. 12.2, orange line), additional sampling effort is required (Fig. 12.2, green h with capitalization again indicating subpopulation membership). Thus, species boundaries, political boundaries, and sampling strategies are all important to consider in conservation genetic studies and in using conservation genetic information to make recommendations on endangered species status.

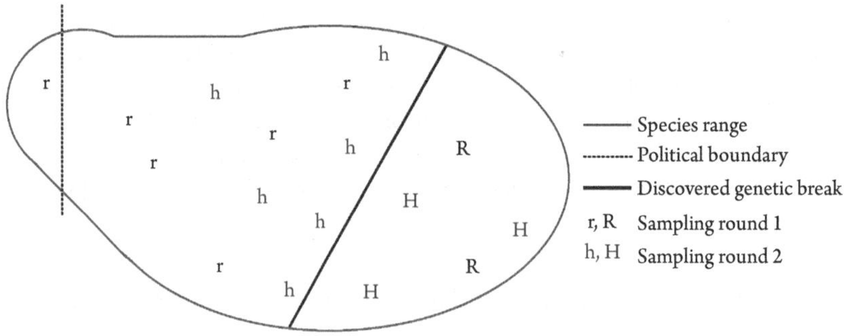

Fig. 12.2.
Sampling strategies (r's, h's for multiple rounds of sampling) for conservation genetics relative to political (dotted black line) and population subdivision (solid black line) boundaries across a species' range (gray line). Capital letters represent one genetic subpopulation and lower case letters represent another. See a color version of this figure in the centerfold.

Importance of Taxonomic Framework (WoRMS)

Note that the previous section focused on knowing a species and its associated range. This is critical information for any conservation genetic study. Yet such information is often difficult to access as species descriptions are often one-off studies in a diversity of journals published around the world. There have been some exceptional recent efforts to organize crustacean taxonomic information (e.g., Martin and Davis 2001, Ng et al. 2008, De Grave et al. 2009, McLaughlin et al. 2010, Crandall and De Grave 2017) and include this information in a searchable database, namely the World Registry of Marine Species (WoRMS; Costello et al. 2013). This taxonomic information is critical foundational information for any conservation study focused on species as the taxonomy provides the organizational framework for species-centric conservation efforts. Additionally, a web-based taxonomy is dynamic, expertly curated, and electronically searchable rather than in hard copy journals or static PDF files, making this taxonomic information exceptionally helpful for organizing conservation efforts.

IUCN, Criteria, and the Importance of Genetics

Given the taxonomic foundation of well-defined and geographically articulated species, we can then begin the assessment process. Note, however, that for many (most) crustaceans and certainly most invertebrates, "conservation assessment" needs to start with taxonomy. Conservation assessment and taxonomy can go hand in hand as authors describe new species. Some authors have explicitly incorporated conservation assessments into the descriptions of new species (e.g., Jara et al. 2018) with explicit criteria for assessment. There is no better time for such an assessment as most species descriptions have information on habitat, species range, perhaps even species numbers, ecological impactors of habitat, etc. No doubt such information could be extended with additional study, but an initial conservation assessment is critical and should be done at the time of new species descriptions.

We discussed earlier that differing agencies (local, regional, national, and international) can use different criteria (and even different parts of a species' distribution) to make inferences concerning conservation. For consistency and broadest applicability, the International Union for Conservation of Nature (IUCN) has well-established and globally utilized "Red List" criteria for assessing conservation status (IUCN 2021). The IUCN capitalizes on expert knowledge through Specialist Groups for conducting Red List assessments of individual species against the Red List criteria. Specialist groups (IUCN SSC 2021) are organized according to taxonomic groups from particular habitats (e.g., freshwater crustaceans, cave invertebrates, Atlantic Islands invertebrates) and also by cross-cutting themes such as climate change, conservation genetics, phylogenetic diversity, etc. The taxonomic groups are particularly focused on species-level Red List assessments, which can capitalize on information and methodologies from the cross-cutting themes. It is, however, notable that there is not a marine crustacean specialist group! In fact, a query of the IUCN Red List shows only 3189 species of the 52,000+ crustacean species have been assessed (Fig. 12.3), and of these, 36% are considered "Data Deficient" (see also Chapter 16 in this volume). Interestingly, 26% of the crustaceans assessed are considered threatened to some degree. Thus, the crustacean community has significant work to do in terms of assessing conservation status for the breadth of crustacean diversity. It is also notable that while WoRMS is an outstanding taxonomic database that includes a field for the IUCN Red List category, and the IUCN Red List is based on species taxonomy, these two databases do not currently share information. This is a significant issue for the limited taxonomic and conservation expertise in the community to have to redo their efforts independently for each database. Thus, these electronic resources need to better harmonize and share data to help advance science, conservation, and save effort from experts with limited time to devote to such initiatives.

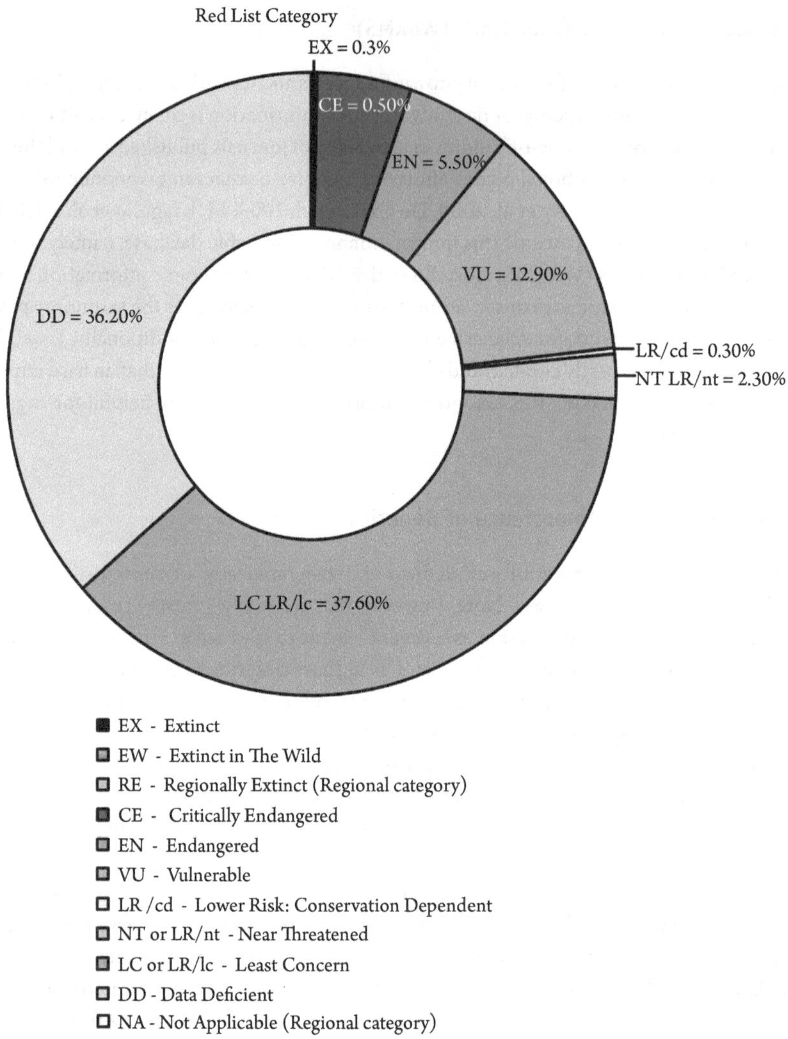

Fig. 12.3.
IUCN Red List endangered status for crustaceans (Malacostraca [3019], Maxillopoda [73], Branchiopoda [47], Hexanauplia [37], and Ostracoda [13]). Source: data from IUCN SSC (2021).

Genetic data can effectively be used on Red List assessments in a variety of ways. Many of the IUCN Red List criteria have to do with a "reduction in population size." Population sizes (both historical and current) can be effectively estimated from population genetic data by estimating different diversity parameters (Crandall et al. 1999). For example, pairwise genetic distance can be used to estimate current population sizes with the aid of Tajima's π (Tajima 1983) and compared to a historical estimator of genetic diversity, such as Watterson's θ (Watterson 1975), to measure changes in population sizes over evolutionary timescales. For example, Buhay and Crandall (2005) assessed these two measures of genetic diversity for a group of cave crayfish from the genus *Orconectes* and found that most of the species examined were both stable and had surprisingly large population sizes, except for *Orconectes australis australis*, which showed a clear decrease in current compared to historical population size. Interestingly, both surface species examined had significant population size reductions despite having double or more the effective population sizes (Buhay and Crandall 2005). The application of such methods requires careful consideration of sampling approaches, yet such information can be highly useful for conservation biology (Hare et al. 2011). Such population size estimates can be directly integrated into the IUCN Red List assessments. Another criterion for the Red List (often coupled with population size) is population subdivision (fragmentation). Conducting a population genetic survey of target species is an ideal way to measure population fragmentation for such assessments. We cover methods associated with both of these concepts below.

Endangered Habitats

While some government conservation effort is focused on species (e.g., USA Endangered Species Act; hence the importance of a robust taxonomy), conservation efforts by some governments and NGOs (e.g., WWF, Conservation International, The Nature Conservancy) are often focused on habitats. Often, the hope is that by preserving rare and endangered habitats, one protects a wide variety of species within such habitats. There has been significant effort to document conservation "hotspots" around the world (e.g., Myers et al. 2000, Sechrest et al. 2002). Ironically, these hotspots are often defined by species counts for given geographic areas/habitats. Unfortunately, they are often based on species counts of large vertebrate groups (e.g., mammals or birds), terrestrial plants, and occasionally butterflies. This is due to the extensive taxonomic knowledge of these groups compared to, for example, crustaceans. Thus, certain conservation planning and efforts are missing entire clades of biodiversity and often entire associated habitats.

Conservation priorities can go beyond simple species counts and also take into account phylogenetic diversity (PD) to aid in ranking areas for conservation, including taking into account concepts of complementarity of geographic areas that maximize PD conservation (Faith 1992, 2018, Owen et al. 2019). Such approaches put a higher weight on unique evolutionary lineages and allow one to avoid requiring a robust taxonomy and differences in "splitting versus lumping" in taxonomies among different organisms. Such approaches have been used effectively in crustaceans to rank conservation habitats in South America and Australia (Whiting et al. 2000, Pérez-Losada et al. 2009), providing *very* different rank orders for conservation compared to mammals, birds, frogs, etc. Indeed, PD measures (Evolutionary Diversity—ED) have been effectively combined with the IUCN Red Listings (Global Endangerment—GE) to produce EDGE priorities for conservation (Isaac et al. 2007, Gumbs et al. 2018). These EDGE rankings are now part of the IUCN Red List prioritizations for conservation and have been effectively used to set species conservation priorities in freshwater crayfish (Owen et al. 2015).

DATA TYPES FOR CONSERVATION GENETICS

As outlined above, genetics can play a key role in a wide variety of areas of conservation from defining species, to estimating population sizes, to identifying population subdivision (fragmentation), to estimating PD, all of which is relevant information for conserving species. Here, we review some of the molecular tools for measuring genetic diversity in populations and lay out their relative strengths and weaknesses.

mtDNA

Mitochondrial DNA (mtDNA) has been the workhorse of conservation genetics and phylogeography since the late 1980s (Moritz 1994). It has the advantage of having highly variable regions flanked by conserved regions and multiple identical copies in a given organism, making it an excellent target for polymerase chain reaction (PCR) amplification and sequencing. The higher levels of variation relative to many nuclear genetic markers make it exceptionally useful for conservation genetics as typical target loci (e.g., the 16S, COI genes) (Fig. 12.4).

Indeed, the cytochrome oxidase I (COI) gene has become the standard for DNA barcoding (Hebert et al. 2003). Mitochondrial DNA is maternally inherited as a single locus. This has two implications for downstream analyses and interpretation. First, in measuring gene flow and population

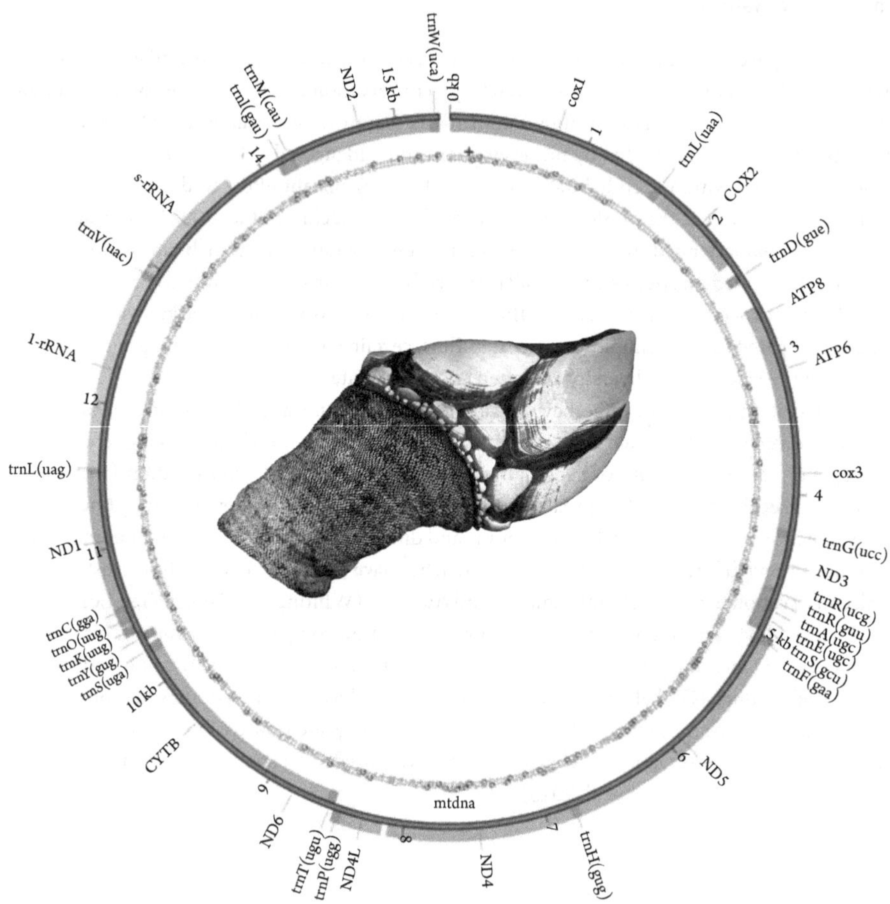

Fig. 12.4.
Mitochondrial genome (GenBank Accession CM029732) of *Pollicipes pollicipes*, the gooseneck barnacle, generated from a whole genome assembly GenBank Ref Seq: GCA_011947565.2 (Bernot et al. 2022). Transfer RNAs shown in dark shading, coding genes shown in light shading, and ribosomal subunits shown in light shading and labeled as rRNA, with an image of the species as the centerpiece (reproduced courtesy of Niklas Dreyer). Mitochondrial genome reproduced courtesy of Anton Zamyatin and Pavel Avdeyev; see a color version of this figure in the centerfold.

subdivision, if there is sex-biased dispersal, this single genetic marker will only pick up the female side of things. To get the whole picture, nuclear markers would need to be added. This is especially important in the context of invasive species with some degree of hybridization making reliance on mtDNA alone problematic for identifying native vs. invasive species and hybrids (Perry et al. 2002). Second, as described above, much of the inference with genetics is based, one way or another, on phylogenetic inference. Yet phylogenetic inference is far superior with multiple informative genetic loci. Because mtDNA is inherited as a single unit with no recombination, it is effectively a single locus. So even studies based on whole mitochondrial genomes are still single-locus-based phylogenetic analyses and would be better served by broader genomic sampling. Mitochondrial DNA also has potential issues with pseudogenes, nuclear copies of parts of the mtDNA that can get amplified by the PCR primers instead of the mitochondrial versions of the targeted genes (Song et al. 2008). Yet mtDNA has been highly informative and continues to be a critical marker for conservation genetics. Mitochondrial and nuclear genes can be specifically targeted to take advantage of high throughput sequencing (HTS) using a targeted amplicon sequencing approach that provides greater depth of coverage and larger numbers of simultaneously sequenced loci for larger numbers of individuals at a reduced cost compared to traditional Sanger sequencing approaches (Bybee et al. 2011). A recent special issue of the *Journal of Crustacean Biology* highlights several studies producing high-quality mitochondrial genomes from a diversity of crustacean species (e.g., Baeza et al. 2022, Collins et al. 2022, Cronin et al. 2022, O'Brien et al. 2022) as well as an informative synopsis of bioinformatics tools for mitochondrial genome assembly and annotation (Baeza 2022).

Microsatellites

Microsatellite markers are tandemly repeated short DNA sequences (typically two or three nucleotides in length) that tend to be highly variable (more variable than mtDNA). They are exceptional at characterizing fine-scale genetic diversity and even paternity relationships among organisms. The downside of microsatellites is that they are generally not transferable across species (certainly not without ascertainment bias) and therefore require a significant amount of lab effort to identify variable microsatellite markers that are applicable to each species of interest. Nevertheless, microsatellite markers have been developed for many commercially important crustacean species such as *Penaeus* (Meehan et al. 2003), *Homarus* (Tam and Kornfield 1996), *Xiphopenaeus* (Kerkhove et al. 2019), *Pollicipes* (Parrondo et al. 2022), *Portunus* (Yap et al. 2002), and *Cardisoma* using HTS approaches to mine microsatellite markers (Gomes et al. 2019). They have also been effectively used for conservation genetics of crustaceans (e.g., Bartholomei-Santos et al. 2011, Iorgu et al. 2011).

RADSeq

Restriction site-associated DNA sequencing (RADSeq) takes advantage of classic restriction enzyme cut sites historically used in population genetics. Traditionally, polymorphisms at restriction cut sites would be characterized and used to infer population structure. With RADSeq, however, these restrictions cut sites serve as anchors for HTS. This has the advantage of sampling loci throughout the genome in a more cost-effective way than full genome sequencing. Additionally, it does not require prior genome sequence information and is, therefore, broadly applicable even to new study systems. RADSeq takes full advantage of HTS allowing for accurate inference of polymorphisms due to depth of coverage at given sites. A disadvantage of RADSeq is that users do not get locus-specific information. So, while the approach is highly effective for measuring population genetic polymorphism, it is harder to assign that polymorphism to functional variation. Nevertheless, RADSeq has been highly successfully used in a variety of evolutionary, ecological, and conservation studies (Andrews et al. 2016), including studies focused on crustaceans (e.g., Schwentner et al. 2020).

Anchored Hybrid Probes

Anchored hybrid enrichment (AHE) is another molecular technology that takes full advantage of HTS but targets specific loci. Thus, it has the advantage of potentially providing functional information along with a multi-locus genome-wide survey of genetic variation spanning both shallow and deep phylogenetic relationships. The basic approach is to align genome/transcriptome data and design target probes in highly conserved regions that flank more variable regions (Lemmon et al. 2012). Thus, in the same way that conserved PCR primers for mtDNA flanking variable regions allows for effective variation surveys, the AHE probes target conserved regions to hybridize and sequence through variable regions. Well-designed probe sets allow for broad phylogenetic spread of genetic variation (because of both conserved and variable regions), targeting single-copy gene regions that are distributed throughout the genome. The drawback of this approach is that it does require substantial genomic/transcriptome information for the targeted group of organisms upon which to design the probe set. The only crustacean group we are aware of that has AHE probes designed are the decapod crustaceans, including specific probe sets for (1) all decapods (including 410 loci surveying over 86,000 nucleotides), (2) Dendrobranchiata, (3) Astacidea, (4) Achelata, (5) Anomura, and (6) Brachyura (Wolfe et al. 2019). Our group is currently developing AHE sets for both copepods and barnacles as well. An interesting example of the application of the AHE probes for the freshwater crayfish was a recent focus on a species complex where the Astacidea AHE probe set of Wolfe et al. (2019) was used to estimate a species-level phylogeny for the genus *Lacunicambarus* (Glon et al. 2022). The study demonstrated the effectiveness of AHE probes across this species complex, the successful amplification from previously collected specimens (museum specimens preserved in 70% EtOH), and provided a direct comparison of the resulting phylogeny relative to mitochondrial loci. The AHE probe tree had much higher and uniformly higher bootstrap values compared to the mitochondrial tree and, importantly, provided inference from multiple, independent genetic loci.

Genomics/Transcriptomics

Genomic and transcriptomic resources in the crustaceans are fairly limited. Despite "phylogenomics" in the titles of various recent papers, such studies tend to be focused on transcriptomics. Transcriptomics has the advantage of again surveying variation across the genome for many, many single-copy loci and thereby providing informative phylogenetic information. But transcriptomics and genomics are both much more expensive and more geared toward deeper-level phylogenetics. Transcriptomics, in particular, has been the centerpiece data for the ongoing debate of the crustacean–hexapod relationships (von Reumont et al. 2012, Giribet and Edgecombe 2019). There is greater promise for full genomic sequence data providing a diversity of insights for conservation biology and there is a significant effort moving forward to sequence at least one genome from every species on earth as part of the Earth BioGenome Project (Lewin et al. 2018, Blaxter et al. 2022, Lewin et al. 2022). This effort, coupled with the diversity of transcriptome-based phylogenetic studies, provides the raw data from which effective probes can be designed for targeted conservation genetic efforts for both functional genomics and understanding genetic diversity within and among species.

eDNA

Environmental DNA (eDNA) is a novel survey technique for identifying organismal diversity within and among communities by simply extracting DNA from environmental samples (soil, water, corals, etc.), sequencing using either a metabarcoding approach (Valentini et al. 2016) that targets specific loci or a shotgun metagenomic approach that produces short DNA sequencing reads, and then maps these reads to a reference database for taxonomic assignment. Metabarcoding attempts to use conserved primers across a great taxonomic diversity to characterize and quantify the relative abundance of different taxa in a given community. However, it is still a PCR-based approach and therefore susceptible to PCR bias, differential amplification, and other problems. Metagenomics, on the other hand, directly targets the DNA from a sample, but is biased by the reference databases available for taxonomic assignment (Nasko et al. 2018). Metabarcoding has been used more in studies focused on animal taxa (Nguyen et al. 2020), whereas the metagenomic approach has been used more for characterizing microbial communities where there are better reference genomes available. However, with projects such as the Earth BioGenome Project, it is conceivable that metagenomics could be applied to organismal conservation work in the future. Nevertheless, the metagenomic approach can be highly useful in examining pathogenic microbes impacting conservation efforts for endangered crustaceans.

ANALYTICAL INSIGHTS FOR CONSERVATION

There is an ever-growing list of computational approaches to analyzing the diversity of data that impacts conservation biology and conservation genetics. Therefore, a comprehensive review of analytical approaches is simply beyond the scope of this chapter. Nevertheless, we hope to provide readers with both some standard and some newer approaches for data analyses relevant to the data types discussed above. Note that as the data volumes grow due to our abilities to capture both environmental data (through air quality measures, water quality measures, satellite data, etc.) and genetic data (HTS, transcriptomics, metagenomics, genomics) as well as novel data types included in analyses (specific ecological data, morphological data, metabolomics, proteomics, etc.), our ability to conduct meaningful and integrated analyses becomes strained. Therefore, it is critical that effective conservation biologists in today's world have significant data science skills. There are a number of courses/workshops available around the world that are specifically focused on sharpening data science skills with applications in conservation biology. Much of conservation genetic analyses we would traditionally do with independent pieces of software (phylogenetics, diversity calculations, etc.) can now be performed in R (a scripting language heavily used in statistical analyses), capitalizing on a number of R packages for such analyses. Additionally, because of the diversity and volume of data types and the often-unstructured nature of much of these data, machine-learning techniques are being employed in ever more powerful ways for data integration and analysis (Schrider and Kern 2018). Again, it is beyond the scope of this chapter to get into such techniques, but readers are encouraged to explore such options as they embark on conservation genetic analyses. We direct the readers to the Missouri Botanical Garden's The Global Change Conservation Lab (http://www.earthskysea.org/software/), where they link to a diversity of R tools for data access (including the IUCN Red List), cleaning, GIS information, etc. Likewise, for phylogenetic and population genetic analyses, popular R packages include pegas and ape (https://github.com/emmanuelparadis/pegas) (Paradis 2010, Paradis and Schliep 2019).

Species Status

One of the main efforts in crustacean conservation genetics is simply to identify and delimit species and species boundaries. While phylogenetics can provide useful information on units of conservation (Crandall et al. 2000), species delimitation in the genomic era can incorporate the increasing volume of genetic data more effectively using a multi-species coalescent framework (Rannala 2015). Such approaches have been used extensively in crustacean research (e.g., Stokkan et al. 2018, Ramirez et al. 2021). Yet recently, Sukumaran et al. (2021) argued that the multi-species coalescent is limited in cases where there is within-species population structure. Instead, they argue that aspects of the speciation process itself should be incorporated into speciation-based delimitation that distinguishes species and population boundaries using full probabilistic models. As this approach is very new, it is yet to be seen how it will be adopted by the research community.

Gene Flow

Gene flow (migration rate, population substructure) is a key concept and parameter in conservation genetics as it speaks to the concept of population fragmentation. Gene flow and species delimitation are intimately linked (Jackson et al. 2017) as it is the prolonged lack of gene flow that often determines species boundaries (via a variety of mechanisms). Measurements of gene flow can be obtained from a variety of software packages, which make different assumptions about the underlying populations. Thus, users must be careful to match assumptions to the analytical methods employed (Ravinet et al. 2017). F-statistics, originally defined by Sewall Wright (1965), have been used to partition genetic diversity across individuals (I), subdivisions (S), and the total population (T). Thus, F_{ST} is a measure of genetic differentiation over subpopulations with respect to the total population, where higher levels of F_{ST} suggest more structured populations. F_{ST} has been used extensively in conservation genetics and has even seen a resurgence in whole-genome scanning approaches to identify regions of local adaptation (Matthey-Doret and Whitlock 2019). Yet there are other approaches for more accurately measuring population substructure from genetic data (Pearse and Crandall 2004).

Effective Population Sizes

Effective population size is similarly complicated and has multiple approaches for estimation with varying assumptions (Crandall et al. 1999, Hare et al. 2011). Inbreeding effective size measures longer-term evolutionary effective size by incorporating gene genealogies as part of the estimation process, whereas variance effective size measures over contemporary timescales by taking into account allele frequency information instead of gene genealogies. Both estimates converge as populations are large and panmictic, but can be misleading if population substructure exists and is not taken into account. However, in conservation genetics, we are often dealing with small, structured populations requiring special emphasis on method choice for estimating effective population size. Regardless, the effective population size is typically very distinct from the census size of a population (the raw count of the number of individuals). One can think of the variance effective size as the number of those individuals contributing variants to the next generation, whereas inbreeding effective size looks more at the ancestral population that gave rise to the current observed population.

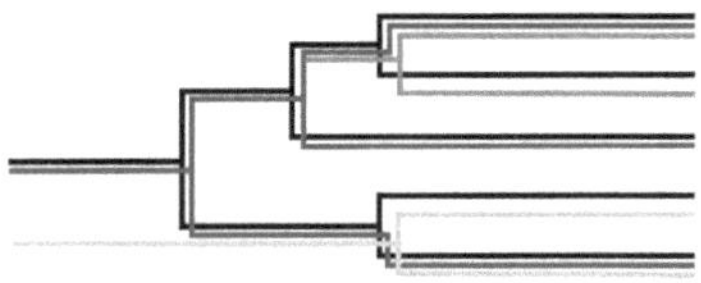

Species	Area 1	Area 2	Area 3
A	X	X	
B		X	
C	X		
D			X
E	X		X

Fig. 12.5.
Phylogenetic diversity (PD) calculates the sum of branches leading to a given taxon and then the PD associated with a given area where that taxon is found. See a color version of this figure in the centerfold.

Conservation Priorities for Species and Habitats (Phylogenetic Diversity)

Phylogenetic diversity calculates the branch length sums of given taxa and/or areas within which taxa occur. Faith (1992) originally described the concept of PD to account for distinctiveness of phylogenetic lineages in addition to simple species counts. PD is relatively easy to compute and has seen a resurgence in application as a principal measure of diversity in microbiome studies (Armstrong et al. 2021). The measure has the added advantage of being less reliant on "species" (counts) per se as it counts accumulated branch lengths of any terminal samples rather than strictly on species. Protecting PD has the added potential benefit of promoting ecosystem stability (Cadotte et al. 2012) through preservation of ecosystem functions (Srivastava et al. 2012). Areas for conservation can be effectively ranked by relative and complimentary PD accumulation in a given area (Fig. 12.5). For example, Area 1 would receive the highest priority rank for conservation efforts because it would preserve the greatest amount of PD. Such analyses have been effectively used in freshwater crayfish to rank bioregions in Australia (Whiting et al. 2000) and in freshwater anomuran crabs to rank bioregions in South America (Pérez-Losada et al. 2009).

CASE STUDIES

To illustrate a number of the tools and concepts presented in this chapter, two case studies are presented below. Note that these are mainly drawn from our own work as we are most familiar with these examples, but such tools and concepts have been effectively and broadly applied in a variety of studies to advance crustacean conservation efforts.

Cave Crayfish—Species Status and Effective Population Size

In the investigation of conservation genetics of a group of endangered species of cave crayfish, Buhay et al. (2007) used the iterative and extensive sampling approach described above coupled with sequencing of nuclear and mitochondrial DNA to estimate genetic diversity and test species boundaries. Their resulting phylogenetic estimate clearly identified two clades that corresponded to known species of cave crayfish (*Cambarus hamulatus* and *C. jonesi*) (Fig. 12.6). The extensive sampling also allowed for the discovery of two very distinct evolutionary lineages (high PD) corresponding to two new species which were subsequently described as *C. speleocoopi* (sp. nov. 2) and *C. laconensis* (sp. nov. 1) after even more extensive follow-up sampling for these two new clades (Buhay and Crandall 2009). The original study was also able to estimate genetic diversity (both historical and current levels with conversions to effective population sizes) and showed that three of the four target species had a decrease in genetic diversity and effective population size, with *C. laconensis* being the only species to show a relatively stable population size. This information was

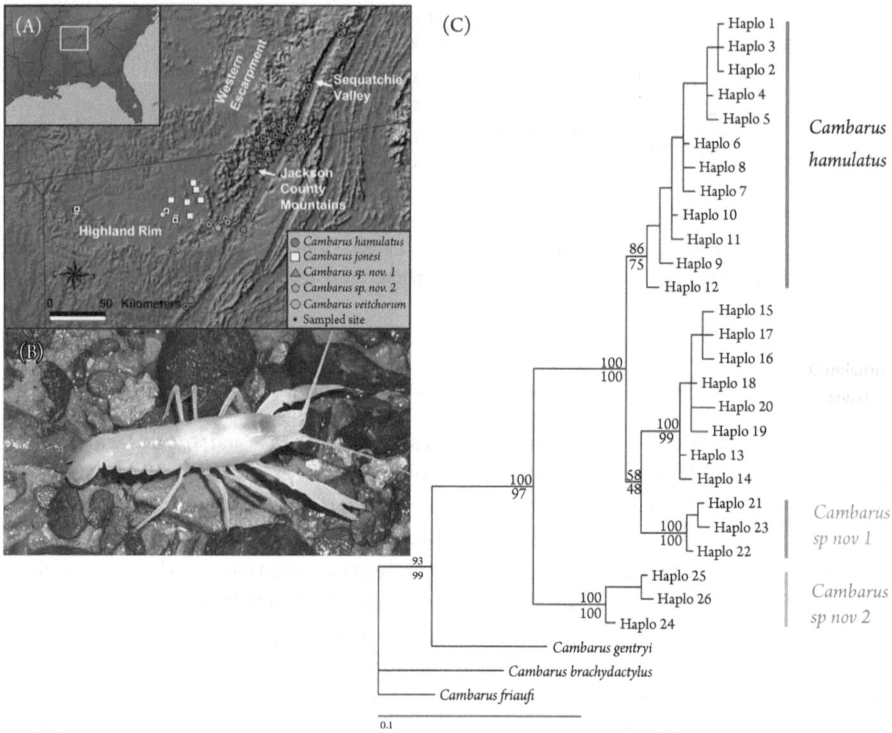

Fig. 12.6.
Geographically intensive and iterative sampling (B) of cave crayfish (A) from Appalachia allowed for an extensive and informative phylogenetic analysis (C) that resulted in the discovery of two new species and an effective conservation assessment of all four target species. Part (A) reproduced from Buhay et al. (2007), with permission from © Elsevier; parts (B) and (C) from Buhay and Crandall (2009), with permission from © Oxford University Press; see a color version of this figure in the centerfold.

then used to assess the conservation status for these species using the IUCN Red List criteria, resulting in a Vulnerable status assessment for all species except *C. hamulatus,* which was assessed as Near Threatened. With subsequent study (Richman et al. 2015), these listings have been updated to *C. speleocoopi* Endangered, *C. hamulatus* Least Concern, *C. jonesi* Vulnerable, and *C. laconensis* Critically Endangered.

Freshwater Crabs—Patterns of Gene Flow

The anomuran freshwater crab group *Aegla* are the sole members of the family Aeglidae and are phylogenetically related to the Hippoidea, Lomisoidea, and Chirostyloidea (Bracken-Grissom et al. 2013). The genus currently contains 93 species (WoRMS 2022), including four species described in 2021, many with help from COI sequence data to distinguish new species from congeners, for example, *Aegla buenoi* (Marçal et al. 2021). The species have been assessed using the IUCN Red List criteria, resulting in nearly 70% of the species having some kind of endangered status and 16 species considered Critically Endangered (Santos et al. 2017). We studied one of the non-endangered species, *Aegla neuquensis,* to test hypotheses of the phylogeographic impact of Patagonian glaciation cycles (Barber et al. 2012). Again, with extensive sampling including 295 individuals from throughout the species range (Fig. 12.7A) focusing on all the major river drainages in the species' distribution, we sequenced both nuclear and mitochondrial DNA to infer the evolutionary history of this group. Our data allowed us to again identify unique lineages that appear to be highly distinct and require further investigation (see the Vaca and Telsen lineages in Fig. 12.7B, which cluster with other species). Additionally, our phylogeny coupled with more extensive population genetic analyses suggested gene flow from the Negro to the Chico-Chubut drainage as a recent colonization event (supported by both the short branch lengths in Fig. 12.7B and the gene flow analysis in Fig. 12.7C). The population genetic analyses suggested that gene flow was facilitated by paleolakes during the Patagonian glaciation, with subsequent fragmentation following the disappearance of these paleolakes. The mitochondrial DNA and nuclear DNA showed different phylogeographic patterns (demonstrating the importance of surveying both mtDNA and nuclear DNA), suggesting introgression, with each genetic component showing distinct phylogeographic patterns (for details, see Barber et al. [2012]). Interestingly, these glacially impacted patterns of gene flow were very distinct from those of a species studied on the Chilean side of the Andes, *Aegla alacalufi*. Here, we found isolated refugia with recent spread across previously glaciated areas regardless of river drainage structure (Xu et al. 2009).

FUTURE DIRECTIONS

It is an exciting time to be working on crustacean conservation genetics. The genomics era has ushered in outstanding tools that provide the ultimate in genetic resolution: DNA sequence data. Additionally, we can now also explore and identify key genes important for species fitness and survival, especially in changing environments. These new molecular tools provide exceptional genetic and genomic insights into conservation efforts. However, there is a great need for bioinformatic tools that allow the full utility of such high-resolution data. Additionally, a new generation of conservation geneticist is needed, one with substantial data science skills. Of course, effective conservation efforts require the integration of a diversity of data. Data integration and metadata analysis also require significant abilities in managing large quantities of data. An emphasis on diverse data and data science skills will allow for more comprehensive and effective evaluations and decisions to enhance the conservation efforts for crustaceans. Of course, it is often difficult for single individuals to have the molecular skills necessary to generate meaningful molecular data, field skills to effectively collect samples for molecular analyses and the associated phenotypic and ecological data, and bioinformatic skills to effectively analyze these diverse data sets. Consequently, impactful conservation often requires a team of experts with diverse and complementary skills and insights, especially including organismal insights that can help make sense of patterns observed in the molecular data or identify when the molecular data and bioinformatic analyses are leading researchers astray with either poor quality data, analyses, or interpretations.

CONCLUSIONS

Genetic information can be critical to assessment, management, and prioritization of endangered species and habitats. Such information is most powerful in a robust taxonomic context with associated ecological and phenotypic data. New omics technologies are advancing both the molecular and analytical tools for making inferences from genetic/genomic data to aid in our conservation efforts. Crustaceans, in particular, are both an important target for conservation efforts and in desperate need of conservation efforts. Freshwater species are often restricted to highly human-impacted environments and often serve as indicator species for the entire freshwater habitat.

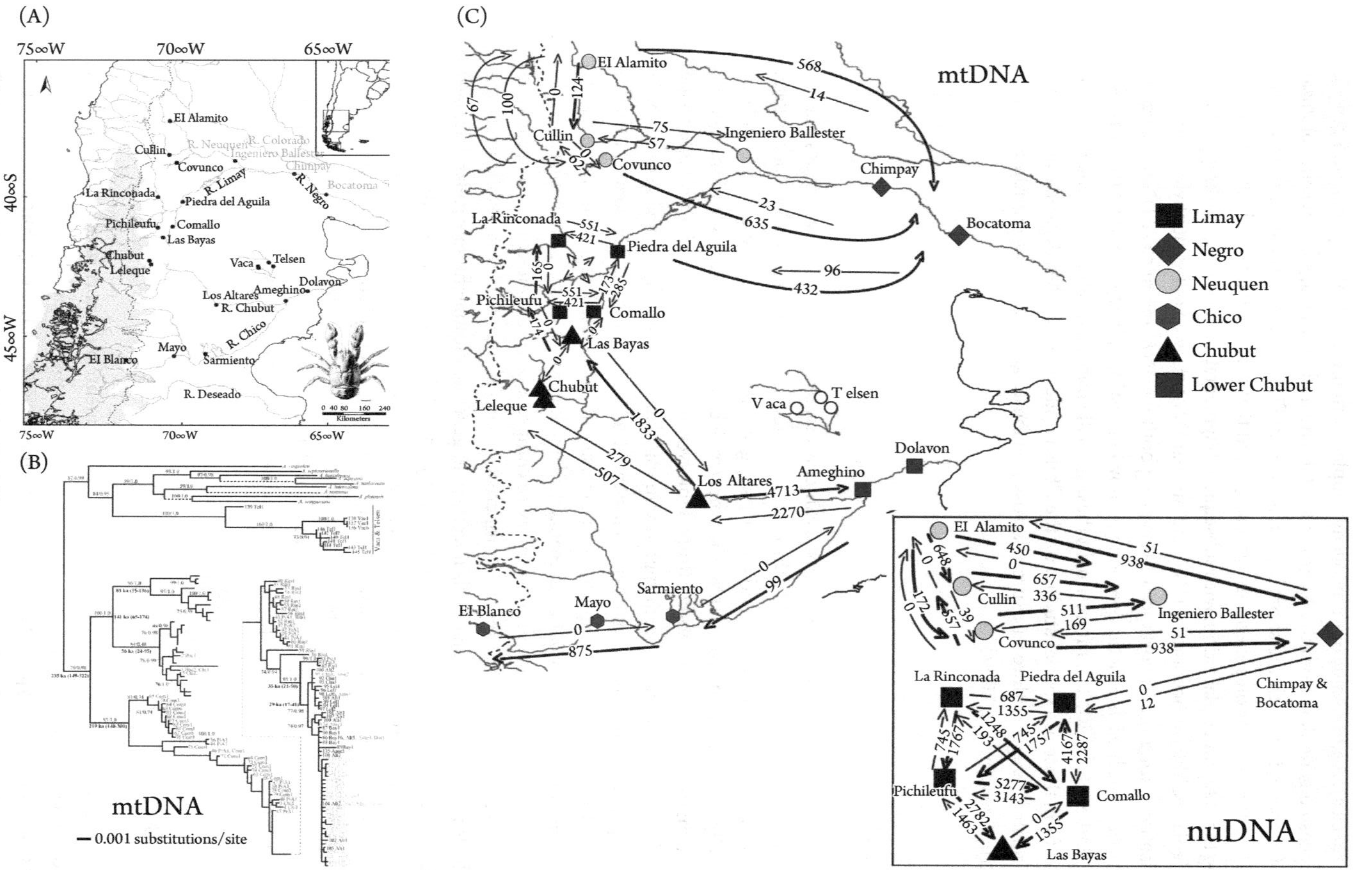

Fig. 12.7.
Phylogeographic study of *Aegla neuquensis* (A) using both mitochondrial (mtDNA) and nuclear (nuDNA) markers to estimate phylogenetic relationships among unique sequences (B) and different gene flow patterns for the mtDNA and nuDNA suggesting introgression (C). Reproduced from Barber et al. (2012) under a Creative Commons by Attribution License (CC BY); see a color version of this figure in the centerfold.

Marine species show extraordinary diversity and are often key members of impacted communities such as coral reefs and deep-sea vents. They are keystone species upon which many of the marine mammals, so often the focus of conservation efforts, rely. Given the lack of effort by organizations such as IUCN to assess marine crustaceans, it is time for the crustacean community to take up this effort and provide the foundational information for broader conservation efforts of our imperiled marine ecosystems. It is an exciting time to study conservation genetics of crustaceans and contribute to conservation efforts around the world of such an amazing section of the global biodiversity.

ACKNOWLEDGMENTS

I thank the editors of this volume for their patience and editing of this chapter. This chapter represents the last 30 years of efforts to attempt to understand a small fraction of the wonderful world of crustaceans. Along the way, I have had amazing students, collaborators, and mentors around the world in crustacean biology, molecular biology, and bioinformatics. Thus, throughout the paper, I have used "we" to acknowledge that this work stems from group knowledge learned over decades from some fantastic colleagues. Thanks to all of you for making this such a fun field for me.

REFERENCES

Andrews, K.R., J.M. Good, M.R. Miller, G. Luikart, and P.A. Hohenlohe. 2016. Harnessing the power of RADseq for ecological and evolutionary genomics. Nature Reviews Genetics 17:81–92.

Arfianti, T., and M.J. Costello. 2020. Global biogeography of marine amphipod crustaceans: Latitude, regionalization, and beta diversity. Marine Ecology Progress Series 638:83–94.

Armstrong, G.W., K. Cantrell, S. Huang, D. McDonald, N. Haiminen, A.P. Carrieri, Q. Zhu, A. Gonzalez, I. McGrath, K. Beck, D. Hakim, A.S. Havulinna, G. Méric, T. Niiranen, L. Lahti, V. Salomaa, M. Jain, M. Inouye, A.D. Swafford, H.-C. Kim, L. Parida, Y. Vázquez-Baeza, and R. Knight. 2021. Efficient computation of Faith's phylogenetic diversity with applications in characterizing microbiomes. Genome Research 31:2131–2137.

Ayari, A., C. Ghemari, and K. Nasri-Ammar. 2021. Reproductive adaption as a survival strategy to life in an arid environment: The terrestrial crustacean *Hemilepistus reaumurii* as a model. Zoologischer Anzeiger 294:10–19.

Baeza, J.A. 2022. An introduction to the special section on crustacean mitochondrial genomics: Improving the assembly, annotation, and characterization of mitochondrial genomes using user-friendly and open-access bioinformatics tools, with decapod crustaceans as an example. Journal of Crustacean Biology 42:ruac012.

Baeza, J.A., A.M. Baker, and H. Liu. 2022. Genome survey sequencing of the long-legged spiny lobster *Panulirus longipes* (A. Milne-Edwards, 1868) (Decapoda: Achelata: Palinuridae): Improved mitochondrial genome annotation, nuclear repetitive elements classification, and SSR marker discovery. Journal of Crustacean Biology 42:ruac006.

Barber, B.R., J. Xu, M. Pérez-Losada, C.G. Jara, and K.A. Crandall. 2012. Conflicting evolutionary patterns due to mitochondrial introgression and multilocus phylogeography of the Patagonian freshwater crab *Aegla neuquensis*. PLoS ONE 7:e37105.

Bartholomei-Santos, M.L., P.A. Roratto, and S. Santos. 2011. High genetic differentiation of *Aegla longirostri* (Crustacea, Decapoda, Anomura) populations in southern Brazil revealed by multi-loci microsatellite analysis. Genetics and Molecular Research 10:4133–4146.

Bernot, J.P., P. Avdeyev, A. Zamyatin, N. Dreyer, N. Alexeev, M. Pérez-Losada, and K.A. Crandall. 2022. Chromosome-level genome assembly, annotation, and phylogenomics of the gooseneck barnacle *Pollicipes pollicipes*. Giga Science 11:giac021.

Bezeng, B.S., and H.F. van der Bank. 2019. DNA barcoding of southern African crustaceans reveals a mix of invasive species and potential cryptic diversity. PLoS ONE 14:e0222047.

Bishop, R.E., W. Humphreys, and D. Jaume. 2020. Subterranean and anchialine waters. Pages 331–358 *in* G.C.B. Poore and M. Thiel, editors. The natural history of the Crustacea, Volume 8: Evolution and biogeography. Oxford University Press, New York, USA.

Blaxter, M., J.M. Archibald, A.K. Childers, J.A. Coddington, K.A. Crandall, F. Di Palma, R. Durbin, S.V. Edwards, J.A.M. Graves, K.J. Hackett, N. Hall, E.D. Jarvis, R.N. Johnson, E.K. Karlsson, W.J. Kress, S. Kuraku, M.K.N. Lawniczak, K. Lindblad-Toh, J.V. Lopez, N.A. Moran, G.E. Robinson, O.A. Ryder, B. Shapiro, P.S. Soltis, T. Warnow, G. Zhang, and H.A. Lewin. 2022. Why sequence all eukaryotes? Proceedings of the National Academy of Sciences of the United States of America 119:e2115636118.

Bracken-Grissom, H.D., M.E. Cannon, P. Cabezas, R.M. Feldmann, C.E. Schweitzer, S.T. Ahyong, D.L. Felder, R. Lemaitre, and K.A. Crandall. 2013. A comprehensive and integrative reconstruction of evolutionary history for Anomura (Crustacea: Decapoda). BMC Evolutionary Biology 13:1–29.

Bracken-Grissom, H., and J.M. Wolfe. 2020. Phylogeny with an emphasis on Crustacea. Pages 80–104 *in* G.C.B. Poore and M. Thiel, editors. The natural history of the Crustacea, Volume 8: Evolution and biogeography. Oxford University Press, New York, USA.

Buhay, J.E., and K.A. Crandall. 2005. Subterranean phylogeography of freshwater crayfishes shows extensive gene flow and surprisingly large population sizes. Molecular Ecology 14:4259–4273.

Buhay, J.E., and K.A. Crandall. 2009. Taxonomic revision of cave crayfish in the genus *Cambarus*, subgenus *Aviticambarus* (Decapoda: Cambaridae) with descriptions of two new species, *C. speleocoopi* and *C. laconensis*, endemic to Alabama, U.S.A. Journal of Crustacean Biology 29:121–134.

Buhay, J.E., G. Moni, N. Mann, and K.A. Crandall. 2007. Molecular taxonomy in the dark: Evolutionary history, phylogeography, and diversity of cave crayfish in the subgenus *Aviticambarus*, genus *Cambarus*. Molecular Phylogenetics and Evolution 42:435–448.

Bybee, S.M., H. Bracken-Grissom, B.D. Haynes, R.A. Hermansen, R.L. Byers, M.J. Clement, J.A. Udall, E.R. Wilcox, and K.A. Crandall. 2011. Targeted amplicon sequencing (TAS): A scalable next-gen approach to multilocus, multitaxa phylogenetics. Genome Biology and Evolution 3:1312–1323.

Cadotte, M.W., R. Dinnage, and D. Tilman. 2012. Phylogenetic diversity promotes ecosystem stability. Ecology. 93:S223–S233.

Calderón-Gutiérrez, F., C.A. Sánchez-Ortiz, and L. Huato-Soberanis. 2018. Ecological patterns in anchialine caves. PLoS ONE 13:e0202909.

Carlton, J.T. 2011. The global dispersal of marine and estuarine crustaceans. Pages 3–23 *in* B.S. Galil, P.F. Clark, and J.T. Carlton, editors. In the wrong place—Alien marine crustaceans: Distribution, biology and impacts. Springer, Dordrecht, The Netherlands.

Collins, S.B., H.D. Bracken-Grissom, and J.A. Baeza. 2022. The complete mitochondrial genome of the pink shrimp *Farfantepenaeus duorarum* (Burkenroad, 1939) (Decapoda: Dendrobranchiata: Penaeidae). Journal of Crustacean Biology 42:ruac007.

Costello, M.J., P. Bouchet, G. Boxshall, K. Fauchald, D. Gordon, B.W. Hoeksema, G.C.B. Poore, R.W.M. van Soest, S. Stöhr, T.C. Walter, B. Vanhoorne, W. Decock, and W. Appeltans. 2013. Global coordination and standardisation in marine biodiversity through the World Register of Marine Species (WoRMS) and related databases. PLoS ONE 8:e51629.

Crandall, K.A., O.R. Bininda-Emonds, G.M. Mace, and R.K. Wayne. 2000. Considering evolutionary processes in conservation biology. Trends in Ecology and Evolution 15:290–295.

Crandall, K.A., and S. De Grave. 2017. An updated classification of the freshwater crayfishes (Decapoda: Astacidea) of the world, with a complete species list. Journal of Crustacean Biology 37:615–653.

Crandall, K.A., D. Posada, and D. Vasco. 1999. Effective population sizes: Missing measures and missing concepts. Animal Conservation Forum 2:317–319.

Crivellaro, M.S., B.L. Zimmermann, M.L. Bartholomei-Santos, K.A. Crandall, M. Pérez-Losada, G. Bond-Buckup, and S Santos. 2018. Looks can be deceiving: Species delimitation reveals hidden diversity in the freshwater crab *Aegla longirostri* (Decapoda: Anomura). Zoological Journal of the Linnean Society 182:24–37.

Cronin, T.J., S.J.M. Jones, and J.A. Baeza. 2022. The complete mitochondrial genome of the spot prawn, *Pandalus platyceros* Brandt in von Middendorf, 1851 (Decapoda: Caridea: Pandalidae), assembled from linked-reads sequencing. Journal of Crustacean Biology 42:ruac003.

Cumberlidge, N., P.K.L. Ng, D.C.J. Yeo, C. Magalhães, M.R. Campos, F. Alvarez, T. Naruse, S.R. Daniels, L.J. Esser, F.Y.K. Attipoe, F.-L. Clotilde-Ba, W. Darwall, A. McIvor, J.E.M. Baillie, B. Collen, and M. Ram. 2009. Freshwater crabs and the biodiversity crisis: Importance, threats, status, and conservation challenges. Biological Conservation 142:1665–1673.

de Almeida Rodrigues, P., R.G. Ferrari, L.S. Kato, R.A. Hauser-Davis, and C.A. Conte-Junior. 2021. A systematic review on metal dynamics and marine toxicity risk assessment using crustaceans as bioindicators. Biological Trace Element Research 200:881–903.

De Grave, S., N.D. Pentcheff, S.T. Ahyong, T.-Y. Chan, K.A. Crandall, P.C. Dworschak, D.L. Felder, R.M. Feldmann, C.H.J.M. Fransen, L.Y.D. Goulding, R. Lemaitre, M.E.Y. Low, J.W. Martin, P.K.L. Ng, C.E. Schweitzer, S.H. Tan, D. Tshudy, and R. Wetzer. 2009. A classification of living and fossil genera of decapod crustaceans. Raffles Bulletin of Zoology, Supplement 21:1–109.

Deiner, K., J.-C. Walser, E. Mächler, and F. Altermatt. 2015. Choice of capture and extraction methods affect detection of freshwater biodiversity from environmental DNA. Biological Conservation 183:53–63.

De Wit, P., P.R. Jonsson, R.T. Pereyra, M. Panova, C. André, and K. Johannesson. 2020. Spatial genetic structure in a crustacean herbivore highlights the need for local considerations in Baltic Sea biodiversity management. Evolutionary Applications 13:974–990.

Duarte, C.M., S. Agusti, E. Barbier, G.L. Britten, J.C. Castilla, J.-P. Gattuso, R.W. Fulweiler, T.P. Hughes, N. Knowlton, C.E. Lovelock, H.K. Lotze, M. Predragovic, E. Poloczanska, C. Roberts, and B. Worm. 2020. Rebuilding marine life. Nature 580:39–51.

Dudgeon, D. 2020. Freshwater biodiversity. Cambridge University Press, Cambridge, UK.

Ewers-Saucedo, C., and J.P. Wares. 2020. Population connectivity and phylogeography of crustaceans. Pages 440–463 *in* G.C.B. Poore and M. Thiel, editors. The natural history of the Crustacea, Volume 8: Evolution and biogeography. Oxford University Press, New York, USA.

Faith, D.P. 1992. Conservation evaluation and phylogenetic diversity. Biological Conservation 61:1–10.

Faith, D.P. 2018. Phylogenetic diversity and conservation evaluation: Perspectives on multiple values, indices, and scales of application. Pages 1–26 *in* R.A. Scherson and D.P. Faith, editors. Phylogenetic diversity: Applications and challenges in biodiversity science. Springer, Cham, Switzerland.

Fernández, S., S. Rodríguez, J.L. Martínez, Y.J. Borrell, A. Ardura, and E. García-Vázquez. 2018. Evaluating freshwater macroinvertebrates from eDNA metabarcoding: A river Nalón case study. PLoS ONE 13:e0201741.

FWS (U.S. Fish and Wildlife Service). 2021. Endangered species. Accessed August 24, 2022. [https://www.fws.gov/endangered/].

Gaedke, U., and D. Straile. 1998. Daphnids: Keystone species for the pelagic food web structure and energy flow. A body size-related analysis linking seasonal changes at the population and ecosystem levels. Advances in Limnology 53:587–610.

Giribet, G., and G.D. Edgecombe. 2019. The phylogeny and evolutionary history of arthropods. Current Biology 29:R592–R602.

Glon, M.G., M.B. Broe, K.A. Crandall, M. Daly, S. Kong, R.F. Thoma, and J.V. Freudenstein. 2022. Anchored hybrid enrichment resolves the phylogeny of *Lacunicambarus* Hobbs, 1969 (Decapoda: Astacidea: Cambaridae). Journal of Crustacean Biology 42:ruab073.

Gold, Z., A.R. Wall, E.E. Curd, R.P. Kelly, N.D. Pentcheff, L. Ripma, P.H. Barber, and R. Wetzer. 2020. eDNA metabarcoding bioassessment of endangered fairy shrimp (*Branchinecta* spp.). Conservation Genetics Resources 12:685–690.

Gomes, S.O., I.G.B. Souza, M.F. Santos, G.R. Silva, M. Albrecht, A.S. McKinley, P. Bentzen, and F.M. Diniz. 2019. Discovery of novel NGS-mined microsatellite markers and an exploratory analysis of genetic differentiation between two Western Atlantic populations of *Cardisoma guanhumi* Latreille, 1825 (Decapoda: Brachyura: Gecarcinidae). Journal of Crustacean Biology 39:181–185.

Gumbs, R., C.L. Gray, O.R. Wearn, and N.R. Owen. 2018. Tetrapods on the EDGE: Overcoming data limitations to identify phylogenetic conservation priorities. PLoS ONE 13:e0194680.

Hansson, B.S., S. Harzsch, M. Knaden, and M. Stensmyr. 2011. The neural and behavioral basis of chemical communication in terrestrial crustaceans. Pages 149–173 *in* T. Breithaupt and M. Thiel, editors. Chemical communication in crustaceans. Springer, New York, USA.

Hare, M.P., L. Nunney, M.K. Schwartz, D.E. Ruzzante, M. Burford, R.S. Waples, K. Ruegg, and F. Palstra. 2011. Understanding and estimating effective population size for practical application in marine species management. Conservation Biology 25:438–449.

Harper, L.R., A.S. Buxton, H.C. Rees, K. Bruce, R. Brys, D. Halfmaerten, D.S. Read, H.V. Watson, C.D. Sayer, E.P. Jones, V. Priestley, E. Mächler, C. Múrria, S. Garcés-Pastor, C. Medupin, K. Burgess, G. Benson, N. Boonham, R.A. Griffiths, L. Lawson Handley, and B. Hänfling. 2019. Prospects and challenges of environmental DNA (eDNA) monitoring in freshwater ponds. Hydrobiologia 826:25–41.

Hebert, P.D.N., A. Cywinska, S.L. Ball, and J.R. deWaard. 2003. Biological identifications through DNA barcodes. Proceedings of the Royal Society B: Biological Sciences 270:313–321.

Hudson, D.M., G. Phillips, C.A. Lasso, and M.R Campos. 2021. Threats to endemic Colombian freshwater crabs (Decapoda: Pseudothelphusidae, Trichodactylidae) associated with climate change and human-mediated activities. Pages 425–444 in T. Kawai and D.C. Rogers, editors. Recent advances in freshwater crustacean biodiversity and conservation. CRC Press, Boca Raton, Florida, USA.

Iannella, M., B. Fiasca, T. Di Lorenzo, M. Di Cicco, M. Biondi, S. Mammola, and D.M.P. Galassi. 2021. Getting the "most out of the hotspot" for practical conservation of groundwater biodiversity. Global Ecology and Conservation. 31:e01844.

Iorgu, E.I., O.P. Popa, A.-M. Petrescu, and L.O. Popa. 2011. Cross-amplification of microsatellite loci in the endangered stone-crayfish *Austropotamobius torrentium* (Crustacea: Decapoda). Knowledge and Management of Aquatic Ecosystems 401:08.

Isaac, N.J.B., S.T. Turvey, B. Collen, C. Waterman, and J.E.M. Baillie. 2007. Mammals on the EDGE: Conservation priorities based on threat and phylogeny. PLoS ONE 2:e296.

IUCN (International Union for Conservation of Nature). 2021. The IUCN red list of threatened species. Accessed August 24, 2022. [https://www.iucnredlist.org/].

IUCN SSC (International Union for Conservation of Nature, SSC Groups). 2021. Our Work. Accessed August 24, 2022. [https://www.iucn.org/ssc-groups/].

Jackson, N.D., B.C. Carstens, A.E. Morales, and B.C. O'Meara. 2017. Species delimitation with gene flow. Systematic Biology 66:799–812.

Jara, C.G., M. Pérez-Losada, and K.A. Crandall. 2018. *Aegla chilota*, new species of anomuran freshwater crab from Chiloé Island, western Patagonia. Nauplius 26:e2018029.

Kerkhove, T.R., B. Hellemans, M. De Troch, A. De Backer, and F.A. Volckaert. 2019. Isolation and characterisation of 14 novel microsatellite markers through next generation sequencing for the commercial Atlantic seabob shrimp *Xiphopenaeus kroyeri*. Molecular Biology Reports 46:6565–6569.

Krieger, J., M.K. Hörnig, M. Kenning, B.S. Hansson, and S. Harzsch. 2021. More than one way to smell ashore—Evolution of the olfactory pathway in terrestrial malacostracan crustaceans. Arthropod Structure and Development 60:101022.

Lemmon, A.R., S.A. Emme, and E.M. Lemmon. 2012. Anchored hybrid enrichment for massively high-throughput phylogenomics. Systematic Biology 61:727–744.

Lewin, H.A., G.E. Robinson, W.J. Kress, W.J. Baker, J. Coddington, K.A. Crandall, R. Durbin, S.V. Edwards, F. Forest, M.T.P. Gilbert, M.M. Goldstein, I.V. Grigoriev, K.J. Hackett, D. Haussler, E.D. Jarvis, W.E. Johnson, A. Patrinos, S. Richards, J.C. Castilla-Rubio, M.-A. van Sluys, P.S. Soltis, X. Xu, H. Yang, and G. Zhang. 2018. Earth BioGenome Project: Sequencing life for the future of life. Proceedings of the National Academy of Sciences of the United States of America 115:4325–4333.

Lewin, H.A., S. Richards, A.E. Lieberman, M.L. Allende, J.M. Archibald, M. Bálint, K.B. Barker, B. Baumgartner, K. Belov, G. Bertorelle, M.L. Blaxter, J. Cai, N.D. Caperello, K. Carlson, J.C. Castilla-Rubio, S.-M. Chaw, L. Chen, A.K. Childers, J.A. Coddington, D.A. Conde, M. Corominas, K.A. Crandall, A.J. Crawford, F. DiPalma, R. Durbin, T.E. Ebenezer, S.V. Edwards, O. Fedrigo, P. Flicek, G. Formenti, R.A. Gibbs, M.T.P. Gilbert, M.M. Goldstein, J.M. Graves, H.T. Greely, I.V. Grigoriev, K.J. Hackett, N. Hall, D. Haussler, K.M. Helgen, C.J. Hogg, S. Isobe, K.S. Jakobsen, A. Janke, E.D. Jarvis, W.E. Johnson, S.J.M. Jones, E.K. Karlsson, P.J. Kersey, J.-H. Kim, W.J. Kress, S. Kuraku, M.K.N. Lawniczak, J.H. Leebens-Mack, X. Li, K. Lindblad-Toh, X. Liu, J.V. Lopez, T. Marques-Bonet, S.

Mazard, J.A.K. Mazet, C.J. Mazzoni, E.W. Myers, R.J. O'Neill, S. Paez, H. Park, G.E. Robinson, C. Roquet, O.A. Ryder, J.S.M. Sabir, H.B. Shaffer, T.M. Shank, J.S. Sherkow, P.S. Soltis, B. Tang, L. Tedersoo, M. Uliano-Silva, K. Wang, X. Wei, R. Wetzer, J.L. Wilson, X. Xu, H. Yang, A.D. Yoder, and G. Zhang. 2022. The Earth BioGenome Project 2020: Starting the clock. Proceedings of the National Academy of Sciences of the United States of America 119:e2115635118.

Lovrenčić, L., L. Bonassin, L.L. Boštjančić, M. Podnar, M. Jelić, G. Klobučar, M. Jaklič, V. Slavevska-Stamenković, J. Hinić, and I. Maguire. 2020. New insights into the genetic diversity of the stone crayfish: Taxonomic and conservation implications. BMC Evolutionary Biology 20:146.

Mammola, S., P. Cardoso, D.C. Culver, L. Deharveng, R.L. Ferreira, C. Fišer, D.M.P. Galassi, C. Griebler, S. Halse, W.F. Humphreys, M. Isaia, F. Malard, A. Martinez, O.T. Moldovan, M.L. Niemiller, M. Pavlek, A.S.P.S. Reboleira, M. Souza-Silva, E.C. Teeling, J.J. Wynne, and M. Zagmajster. 2019. Scientists' warning on the conservation of subterranean ecosystems. BioScience 69:641–650.

Marçal, I.C., F.P. Páez, L. Souza-Shibatta, S.H. Sofia, and G.M. Teixeira. 2021. *Aegla buenoi* n. sp. (Decapoda: Anomura): First record of aeglid crab from Cinzas River basin, Brazil. Zootaxa 5005:291–303.

Martin, J.W., and G.E. Davis. 2001. An updated classification of the recent Crustacea. Natural History Museum of Los Angeles County, Los Angeles, USA.

Matthey-Doret, R., and M.C. Whitlock. 2019. Background selection and FST: Consequences for detecting local adaptation. Molecular Ecology 28:3902–3914.

McLaughlin, P.A., T. Komai, R. Lemaitre, and D.L. Rahayu. 2010. Annotated checklist of anomuran decapod crustaceans of the world (exclusive of the Kiwaoidea and families Chirostylidae and Galatheidae of the Galatheoidea) Part I. Lithodoidea, Lomisoidea and Paguroidea. Raffles Bulletin of Zoology 23:131–137.

Meehan, D., Z. Xu, G. Zuniga, and A. Alcivar-Warren. 2003. High frequency and large number of polymorphic microsatellites in cultured shrimp, *Penaeus* (*Litopenaeus*) *vannamei* [Crustacea: Decapoda]. Marine Biotechnology 5:311–330.

Moritz, C. 1994. Applications of mitochondrial DNA analysis in conservation: A critical review. Molecular Ecology 3:401–411.

Myers, N., R.A. Mittermeier, C.G. Mittermeier, G.A. da Fonseca, and J. Kent. 2000. Biodiversity hotspots for conservation priorities. Nature 403:853–858.

Nasko, D.J., S. Koren, A.M. Phillippy, and T.J. Treangen. 2018. RefSeq database growth influences the accuracy of k-mer-based lowest common ancestor species identification. Genome Biology 19:165.

Ng, P.K.L., D. Guinot, and P.J.F. Davie. 2008. Systema Brachyurorum: Part I. An annotated checklist of extant brachyuran crabs of the world. Raffles Bulletin of Zoology 17:1–286.

Nguyen, B.N., E.W. Shen, J. Seemann, A.M.S. Correa, J.L. O'Donnell, A.H. Altieri, N. Knowlton, K.A. Crandall, S.P. Egan, W.O. McMillan, and M. Leray. 2020. Environmental DNA survey captures patterns of fish and invertebrate diversity across a tropical seascape. Scientific Reports 10:6729.

O'Brien, C., H.D. Bracken-Grissom, and J.A. Baeza. 2022. The complete mitochondrial genome of the Atlantic ghost crab *Ocypode quadrata* (Fabricius, 1787) (Brachyura: Ocypodidae: Ocypodinae). Journal of Crustacean Biology 42:ruac005.

Oka, S.-I., N. Kobayashi, T. Sato, K. Ueda, and M. Yamagishi. 2019. Sound production in the coconut crab, the largest terrestrial crustacean. Zoology 137:125710.

Owen, C.L., H. Bracken-Grissom, D. Stern, and K.A. Crandall. 2015. A synthetic phylogeny of freshwater crayfish: Insights for conservation. Philosophical Transactions of the Royal Society B Biological Sciences 370:20140009.

Owen, N.R., R. Gumbs, C.L. Gray, and D.P. Faith. 2019. Global conservation of phylogenetic diversity captures more than just functional diversity. Nature Communications 10:859.

Panova, M., A. Nygren, P.R. Jonsson, and S. Leidenberger. 2017. A molecular phylogeny of the north-east Atlantic species of the genus *Idotea* (Isopoda) with focus on the Baltic Sea. Zoologica Scripta 46:188–199.

Paradis, E. 2010. pegas: An R package for population genetics with an integrated-modular approach. Bioinformatics 26:419–420.

Paradis, E., and K. Schliep. 2019. ape 5.0: An environment for modern phylogenetics and evolutionary analyses in R. Bioinformatics 35:526–528.

Parrondo, M., P. Morán, M. Ballenghien, J.L. Acuña, A. Aguión, J. Arrontes, J. Chiss, T. Cruz, J.N. Fernandes, L. García-Flórez, and E. García-Vázquez. 2022. Chaotic genetic patchiness in the highly valued Atlantic stalked barnacle *Pollicipes pollicipes* from the Iberian Peninsula: Implications for fisheries management. Frontiers in Marine Science 9:801780.

Pearse, D.E., and K.A. Crandall. 2004. Beyond FST: Analysis of population genetic data for conservation. Conservation Genetics 5:585–602.

Pérez-Losada, M., G. Bond-Buckup, C.G. Jara, and K.A. Crandall. 2009. Conservation assessment of southern South American freshwater ecoregions on the basis of the distribution and genetic diversity of crabs from the genus *Aegla*. Conservation Biology 23:692–702.

Pérez-Moreno, J.L., T.M. Iliffe, and H.D. Bracken-Grissom. 2016. Life in the underworld: Anchialine cave biology in the era of speleogenomics. International Journal of Speleology 45:5.

Perry, W.L., D.M. Lodge, and J.L Feder. 2002. Importance of hybridization between indigenous and nonindigenous freshwater species: An overlooked threat to North American biodiversity. Systematic Biology 51:255–275.

Porter, M.L., and K.A. Crandall. 2003. Lost along the way: The significance of evolution in reverse. Trends in Ecology and Evolution 18:541–547

Porter, M.L., and L. Sumner-Rooney. 2018. Evolution in the dark: Unifying our understanding of eye loss. Integrative and Comparative Biology 58:367–371.

Ramirez, J.L., L. Simbine, C.G. Marques, E. Zelada-Mázmela, L.E. Reyes-Flores, A.S. López, J. Gusmão, C. Tavares, P.M. Galetti, and P.D. Freitas. 2021. DNA barcoding of Penaeidae (Decapoda; Crustacea): Non-distance-based species delimitation of the most economically important shrimp family. Diversity 13:460.

Rannala, B. 2015. The art and science of species delimitation. Current Zoology 61:846–853.

Ravinet, M., R. Faria, R.K. Butlin, J. Galindo, N. Bierne, M. Rafajlović, M.A.F. Noor, B. Mehlig, and A.M. Westram. 2017. Interpreting the genomic landscape of speciation: A road map for finding barriers to gene flow. Journal of Evolutionary Biology 30:1450–1477.

Regier, J.C., J.W. Shultz, and R.E. Kambic. 2005. Pancrustacean phylogeny: Hexapods are terrestrial crustaceans and maxillopods are not monophyletic. Proceedings of the Royal Society B: Biological Sciences 272:395–401.

Reid, A.J., A.K. Carlson, I.F. Creed, E.J. Eliason, P.A. Gell, P.T.J. Johnson, K.A. Kidd, T.J. MacCormack, J.D. Olden, S.J. Ormerod, J.P. Smol, W.W. Taylor, K. Tockner, J.C. Vermaire, D. Dudgeon, and S.J. Cooke. 2019. Emerging threats and persistent conservation challenges for freshwater biodiversity. Biological Reviews 94:849–873.

Richardson, A., and P.B. Araujo. 2015. Lifestyles of terrestrial crustaceans. Pages 299–336 *in* M. Thiel and L. Watling, editors. The natural history of the Crustacea. Volume 2: Lifestyles and feeding biology. Oxford University Press, New York, USA.

Richman, N.I., M. Böhm, S.B. Adams, F. Alvarez, E.A. Bergey, J.J.S. Bunn, Q. Burnham, J. Cordeiro, J. Coughran, K.A. Crandall, K.L. Dawkins, R.J. DiStefano, N.E. Doran, L. Edsman, A.G. Eversole, L. Füreder, J.M. Furse, F. Gherardi, P. Hamr, D.M. Holdich, P. Horwitz, K. Johnston, C.M. Jones, J.P.G. Jones, R.L. Jones, T.G. Jones, T. Kawai, S. Lawler, M. López-Mejía, R.M. Miller, C. Pedraza-Lara, J.D. Reynolds, A.M.M. Richardson, M.B. Schultz, G.A. Schuster, P.J. Sibley, C. Souty-Grosset, C.A. Taylor, R.F. Thoma, J. Walls, T.S. Walsh, and B. Collen. 2015. Multiple drivers of decline in the global status of freshwater crayfish (Decapoda: Astacidea). Philosophical Transactions of the Royal Society B: Biological Sciences 370:20140060.

Rodríguez-Flores, P.C., A. Machordom, P. Abelló, J.A. Cuesta, and E. Macpherson. 2019. Species delimitation and multi-locus species tree solve an old taxonomic problem for European squat lobsters of the genus *Munida* Leach, 1820. Marine Biodiversity 49:1751–1773.

Rossel, S., and P. Martínez Arbizu. 2019. Revealing higher than expected diversity of Harpacticoida (Crustacea: Copepoda) in the North Sea using MALDI-TOF MS and molecular barcoding. Scientific Reports 9:9182.

Rumbold, C., S. Obenat, S.N. Velazquez, B. Gancedo, and E. Spivak. 2018. Seasonal variation of peracarid assemblages in natural and artificial marine environments of the Southwestern Atlantic Ocean. Marine Biodiversity 48:1743–1754.

Santos, S., G. Bond-Buckup, A.S. Gonçalves, M.L. Bartholomei-Santos, L. Buckup, and C.G. Jara. 2017. Diversity and conservation status of *Aegla* spp. (Anomura, Aeglidae): An update. Nauplius 25:e2017011.

Schrider, D.R., and A.D. Kern. 2018. Supervised machine learning for population genetics: A new paradigm. Trends in Genetics 34:301–312.

Schwentner, M., G. Giribet, D.J. Combosch, and B.V. Timms. 2020. Genetic differentiation in mountain-dwelling clam shrimp, *Paralimnadia* (Crustacea: Branchiopoda: Spinicaudata), in eastern Australia. Invertebrate Systematics 34:88–100.

Sechrest, W., T.M. Brooks, G.A.B. da Fonseca, W.R. Konstant, R.A. Mittermeier, A. Purvis, A.B. Rylands, and J.L. Gittleman. 2002. Hotspots and the conservation of evolutionary history. Proceedings of the National Academy of Sciences of the United States of America 99:2067–2071.

Song, H., J.E. Buhay, M.F. Whiting, and K.A. Crandall. 2008. Many species in one: DNA barcoding overestimates the number of species when nuclear mitochondrial pseudogenes are coamplified. Proceedings of the National Academy of Sciences of the United States of America 105:13486–13491.

Srivastava, D.S., M.W. Cadotte, A.A.M. MacDonald, R.G. Marushia, and N. Mirotchnick. 2012. Phylogenetic diversity and the functioning of ecosystems. Ecology Letters 15:637–648.

Stern, D.B., J. Breinholt, C. Pedraza-Lara, M. López-Mejía, C.L. Owen, H. Bracken-Grissom, J.W. Fetzner Jr., and K.A. Crandall. 2017. Phylogenetic evidence from freshwater crayfishes that cave adaptation is not an evolutionary dead-end. Evolution 71:2522–2532.

Stokkan, M., J.A. Jurado-Rivera, P. Oromí, C. Juan, D. Jaume, and J. Pons. 2018. Species delimitation and mitogenome phylogenetics in the subterranean genus *Pseudoniphargus* (Crustacea: Amphipoda). Molecular Phylogenetics and Evolution 127:988–999.

Sukumaran, J., M.T. Holder, and L.L. Knowles. 2021. Incorporating the speciation process into species delimitation. PLoS Computational Biology 17:e1008924.

Tajima, F. 1983. Evolutionary relationship of DNA sequences in finite populations. Genetics 105: 437–460.

Tam, Y.K., and I. Kornfield. 1996. Characterization of microsatellite markers in *Homarus* (Crustacea, Decapoda). Molecular Marine Biology and Biotechnology 5:230–238.

Valentini, A., P. Taberlet, C. Miaud, R. Civade, J. Herder, P.F. Thomsen, E. Bellemain, A. Besnard, E. Coissac, F. Boyer, C. Gaboriaud, P. Jean, N. Poulet, N. Roset, G.H. Copp, P. Geniez, D. Pont, C. Argillier, J.-M. Baudoin, T. Peroux, A.J. Crivelli, A. Olivier, M. Acqueberge, M. Le Brun, P.R. Møller, E. Willerslev, and T. Dejean. 2016. Next-generation monitoring of aquatic biodiversity using environmental DNA metabarcoding. Molecular Ecology 25:929–942.

Veldsman, W.P., K.Y. Ma, J.H.L. Hui, T.F. Chan, J.A. Baeza, J. Qin, and K.H. Chu. 2021. Comparative genomics of the coconut crab and other decapod crustaceans: Exploring the molecular basis of terrestrial adaptation. BMC Evolutionary Biology 22:313.

von Reumont, B.M., and G.D. Edgecombe. 2020. Crustaceans and insect origins. Pages 105–120 *in* G.C.B. Poore and M. Thiel, editors. The natural history of the Crustacea, Volume 8: Evolution and biogeography. Oxford University Press, New York, USA.

von Reumont, B.M., R.A. Jenner, M.A. Wills, E. Dell'ampio, G. Pass, I. Ebersberger, B. Meyer, S. Koenemann, T.M. Iliffe, A. Stamatakis, O. Niehuis, K. Meusemann, and B. Misof. 2012. Pancrustacean phylogeny in the light of new phylogenomic data: Support for Remipedia as the possible sister group of Hexapoda. Molecular Biology and Evolution 29:1031–1045.

Wares, J.P., 2001. Intraspecific variation and geographic isolation in *Idotea balthica* (Isopoda: Valvifera). Journal of Crustacean Biology 21:1007–1013.

Wares, J.P., S. Daley, R. Wetzer, and R.J. Toonen. 2007. An evaluation of cryptic lineages of *Idotea balthica* (Isopoda: Idoteidae): Morphology and microsatellites. Journal of Crustacean Biology 27:643–648.

Watterson, G.A. 1975. On the number of segregating sites in genetical models without recombination. Theoretical Population Biology 7:256–276.

Weis, J.S. 2010. The role of behavior in the success of invasive crustaceans. Marine and Freshwater Behaviour and Physiology 43:83–98.

Whiting, A.S., S.H. Lawler, and P. Horwitz. 2000. Biogeographic regionalization of Australia: Assigning conservation priorities based on endemic freshwater crayfish phylogenetics. Animal Conservation Forum 3:155–163.

Wolfe, J.M., J.W. Breinholt, K.A. Crandall, A.R. Lemmon, E.M. Lemmon, L.E. Timm, M.E. Siddall, and H.D. Bracken-Grissom. 2019. A phylogenomic framework, evolutionary timeline and genomic resources for comparative studies of decapod crustaceans. Proceedings of the Royal Society B: Biological Sciences 286:20190079.

WoRMS (World Register of Marine Species). 2022. *Aegla* Leach, 1821. Accessed August 24, 2022. [https://www.marinespecies.org/aphia.php?p=taxdetails&id=366970].

Wright, S. 1965. The interpretation of population structure by F-statistics with special regard to systems of mating. Evolution 19:395–420.

Xu, J., M. Pérez-Losada, C.G. Jara, and K.A. Crandall. 2009. Pleistocene glaciation leaves deep signature on the freshwater crab *Aegla alacalufi* in Chilean Patagonia. Molecular Ecology 18:904–918.

Yager, J. 1981. Remipedia, a new class of Crustacea from a marine cave in The Bahamas. Journal of Crustacean Biology 1:328–333.

Yap, E.S., E. Sezmis, J.A. Chaplin, I.C. Potter, and P.B.S. Spencer. 2002. Isolation and characterization of microsatellite loci in *Portunus pelagicus* (Crustacea: Portunidae). Molecular Ecology Notes. 2:30–32.

CAUSES AND CONSEQUENCES OF INVASIONS BY NONNATIVE CRUSTACEANS

Edwin Grosholz

Abstract

This chapter provides a review and synthesis of the causes and consequences of crustacean invasions in aquatic, estuarine, and marine systems. The discussion covers both human-mediated and nonhuman-mediated vectors that have moved large numbers of invasive crustaceans around the planet. Historically important vectors such as aquaculture have given way to ballast water and hull fouling, while other vectors, such as live bait and the ornamental aquarium trade, grow in importance. Recipient communities also influence invasion success, and this prompts a discussion of the impacts of biotic resistance, release from native enemies, and the evolutionary history of predator and prey. Human impacts such as fisheries exploitation, contaminants, and a changing climate are also discussed, as well as growing human economies and current and future invasions. Future invasions may also include efforts to save endangered species such as "assisted colonization" or to create new fisheries. Crustacean invasions have affected native species and ecosystem processes as well as commercial fisheries and human infrastructure. They also show nonrandom taxonomic patterns, which are discussed in their role as vectors of introduction for nonnative pathogens and parasites. Case studies are presented of some of the world's worst invasive crustaceans in marine, estuarine, and freshwater systems including amphipods, cladocerans, copepods, decapods, isopods, and mysids. Finally, the damaging consequences of these invasions have necessitated costly management approaches to mitigate their impacts, and the chapter finishes with a summary of current and future options for managing invasive crustaceans in a changing world.

Edwin Grosholz, *Causes and Consequences of Invasions by Nonnative Crustaceans*. In: *Ecology and Conservation*.
Edited by: Lars Gutow and Martin Thiel, Oxford University Press.
DOI: 10.1093/oso/9780197768242.003.0013

INTRODUCTION

Crustaceans are the most successful invasive taxon in aquatic, estuarine, and marine systems (Ricciardi 2011, Ruiz et al. 2011). In the estuarine and marine waters of North America, fully 28% of nonnative species are crustaceans, which makes it the largest group of invaders (Ruiz et al. 2011). In the freshwater ecosystems of Europe, 53% of nonnative species are Crustacea, which are the largest group of invaders (Hänfling et al. 2011). Their successful spread across the globe has been mediated by human activities like commercial shipping that includes both ballast water and ship hulls. However, other vectors include the construction of canals and waterways, and economic activities including aquarium and aquaculture sales. As for most introduced species, the impacts of the majority of crustacean invaders are largely unknown; however, many dozens of studies have documented serious impacts on native ecosystems, economic activity, and human health. Their impacts can occur as the result of burrowing, biofouling, and otherwise damaging natural and human infrastructure; their consequences for numerous fisheries have been well documented, as has their ability to be a vector for nonnative parasites and pathogens.

The crustacean groups that have produced the most established invaders are the Amphipoda, Isopoda, Decapoda, and Copepoda in marine and estuarine systems, and similarly in freshwater systems, but with the addition of Cladocera and Mysida (Fig. 13.1). Many studies have documented the intercontinental transfers of multi-species groups, such as the damaging invasions of the North American Great Lakes by planktonic Ponto–Caspian species, and the invasion of North American crayfish into continental Europe. In the process, some individual species have reached almost global distribution, such as the European green crab *Carcinus maenas,* the red swamp crayfish *Procambarus clarkii,* and the barnacles *Amphibalanus improvisus, A. eburneus,* and *Balanus glandula.*

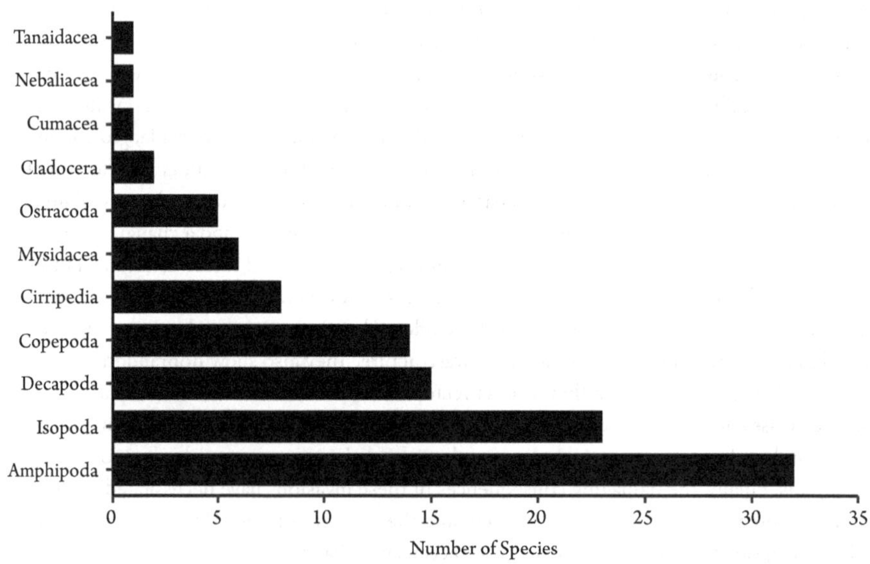

Fig. 13.1.
The numbers of crustacean species in estuarine and marine systems that are nonnative invaders in North American waters. For each taxonomic group the numbers represent species verified as nonnative and that have documented established populations. Source: data from Ruiz et al. (2011).

The reasons for the success of nonnative crustaceans vary considerably by taxa, which include a broad range of life histories from entirely benthic to holoplanktonic. However, most invaders are also tolerant of a wide range of environmental conditions, have a relatively broad/generalist diet, and a life history with short generation times and high rates of reproduction. Many holoplanktonic taxa are frequently introduced in ballast water tanks due to their ability to tolerate those conditions or, in some cases, to produce resistant stages that can better manage those conditions.

PATHWAYS OF INVASION

Human-Mediated Pathways

The most important vector for introductions of nonnative crustaceans are ship-based vectors, including ballast water and hull fouling (Ruiz et al. 1997). These have been documented to be the most important vectors for invasions generally and include larger-bodied crustaceans with dispersing larval stages (e.g., decapods) as well as smaller holoplanktonic species (e.g., copepods) (Minchin et al. 2009). There are also somewhat less important vectors including the live bait trade, in which crayfish and shrimp (primarily) are used as bait for larger fishery species (Keller and Lodge 2007). Movements of species as live bait can often result in further spread well after the initial establishment (Lapointe et al. 2016). Many studies have documented the movement of invasive crayfish species within regions as the result of bait-related activities (Loureiro et al. 2015). Various other human-mediated pathways include movements of harbor-based traps, escape from education or research facilities, dispersal with marine construction equipment, and the movement of sediments and sands for channel engineering (Ruiz et al. 1997).

Aquaculture was historically a significant source of invasions for many taxa including crustaceans, and was likely responsible for many invasive species associated with shellfish, macroalgae, and other species distributed for fisheries enhancement. The role of aquaculture has changed dramatically as a vector, largely due to increased oversight by management agencies, and shifts to using larval and juvenile stages, rather than adults, to transfer fishery species (Williams et al. 2013). That study found that changes in aquaculture importation practices over the past decade have greatly reduced the opportunities for introduction of crustacean "hitchhikers."

By comparison, movement of species via the ornamental trade (largely for residential aquariums) continues to be a significant source of new invasions (Williams et al. 2014). Unintentional aquarium releases are now considered to be among the most rapidly growing vectors for the introduction of nonnative aquatic species. Unfortunately, this vector continues to receive comparatively less attention in both the science and policy realm, despite the fact that the sales of aquarium and ornamental species are growing by 14% annually, with most of this going to developing countries (Padilla and Williams 2004). Aquarium species are generally imported as large, healthy adults, and consequently have a greater probability of surviving and becoming established if conditions permit. Although only a few significant crustacean invasions have occurred through this vector, such as the marbled crayfish *Procambarus virginalis* in Europe (Chucholl and Wendler 2017, Scheers et al. 2021), this is likely to be an increasingly important vector as sales of ornamental species from developing countries increase.

Movement of crustaceans has also been greatly facilitated by waterways, canals, and other conveyances (Galil et al. 2008). These generally do not represent natural waterways and can include canals to move water for human consumption, for irrigation of agricultural commodities, or for creating new shipping channels. Canal systems in Europe helped to facilitate the spread of highly invasive Ponto–Caspian species to the North Sea and Baltic drainages (Hänfling et al. 2011). The engineering of existing river systems for navigation and channel modification has

Table 13.1. The numbers of non-indigenous species (NIS) relative to the number of all species in the freshwater sections of the Rhine River showing that 51.1% of all NIS taxa are crustaceans and a remarkable 76.7% of all crustacean species are nonnative. Source: data from Leuven et al. (2009).

Taxon	Total number of species in 2000 (IKSR 2002)[a]	Number of NIS in 2005	Percent NIS per taxon of total number of NIS	Percent NIS of total number of species per taxon
Acarina	6	1	2.2	16.7
Bryozoa	6	1	2.2	16.7
Hydrozoa	2	1	2.2	50.0
Hirudinea	14	1	2.2	7.1
Trichoptera	58	1	2.2	1.7
Tricladida	6	2	4.4	33.3
Annelida	28	5	11.1	17.9
Mollusca	45	10	22.2	22.2
Crustacea	30	23	51.1	76.7
Total	195	45	100	23.1

[a] See Leuven et al. (2009).

also contributed to the spread of invasive crustaceans including many Ponto–Caspian and North American invaders throughout the Rhine River system (Leuven et al. 2009). This has likely contributed to the very high percentage (76.7%) of crustacean species in the Rhine being nonnative (Table 13.1).

The creation of large shipping channels has contributed substantially to the spread of invasive species, including the Suez Canal linking the Red Sea to the Mediterranean Sea, and the Panama Canal linking the Caribbean with the Pacific Ocean (Galil 2011). The Suez Canal is perhaps the most spectacular example, in which dispersal of hundreds of species from the Red Sea to the Mediterranean via the Suez Canal occurred as part of the "Lessepsian migration." This has likely facilitated the invasion of at least 100 nonnative Red Sea crustaceans into the Mediterranean Sea (Galil 2011).

Another emerging vector, known as "assisted colonization," in which organisms are translocated outside their native range for conservation purposes, can result in intentional introductions into novel environments (Brodie et al. 2021). This well-intended approach ensures that threatened/endangered species, which may not be able to disperse to new regions, would become established in more tolerable habitats to create new populations. Assisted colonization for conservation purposes has not yet been proposed for marine crustaceans, since these are likely able to disperse to new areas in response to threats like increasing sea surface temperatures, although this has been attempted within geographically distinct populations of the southern rock lobster *Jasus edwardsii* in Australia (Green et al. 2010). The unintended consequences of introductions via assisted colonization are unknown, although there are clear risks and rewards to this approach (Ricciardi and Simberloff 2009, 2021, Schlaepfer et al. 2009).

Intentional introductions for other positive outcomes include introducing species into previously unoccupied areas to create/enhance fisheries opportunities. These include the well-intended intentional introduction of the red king crab (*Paralithodes camtschaticus*) and snow crab (*Chionoecetes opilio*) into the Barents Sea region, which have raised concerns about impacts on both benthic habitats and competing commercial fisheries (Falk-Petersen et al. 2011, Kaiser et al. 2018). Introduced in the 1960s, the red king crab is a generalist benthic consumer, and studies have implicated these crabs in declines of bivalves, echinoderms, and polychaetes (Fuhrmann et al. 2017). However, the lack of long-term data and conflicting impacts of other factors such as bottom-trawling have prevented clear conclusions regarding their impacts (Falk-Petersen et al. 2011). The invasive snow crab is also capable of significant impacts on benthic habitats, potentially competing for food with native spider crabs (*Hyas araneus*) and hermit crabs (*Pagurus pubescens*) as the result of substantial diet overlap. However, studies to date suggest the evidence for significant impacts has been so far equivocal (Gebruk et al. 2020). Furthermore, broader economic impacts of snow crabs on other commercial fisheries in the southern Barents Sea such as cod and haddock, which share substantial diet overlap, have also been equally uncertain, in part due to higher overall productivity in this region associated with climate warming (Dvoretsky and Dvoretsky 2015).

Nonhuman-Mediated Pathways

Many crustaceans have long-lived larvae and, consequently, have the potential to disperse over comparatively long distances (Shanks 2009). Many examples of episodic recruitment well outside of the established range may be assisted by climate-driven phenomena. Several species of portunid crabs native to the Caribbean Sea, including *Callinectes bocourti*, *C. marginatus*, and *C. exasperatus*, have been documented as adults from the southeast Atlantic region in South Carolina, Georgia, and Florida. In western North America, adult *Callinectes arcuatus*, which are native to Mexico and further south, have been found repeatedly and abundantly in Southern California estuaries, although they have failed to become established. Another portunid crab, *Portunus xantusii*, dramatically expanded its range from Southern California nearly 400 km northward as the result of an unusual ocean warming event in 2016 (Sadowski et al. 2018). This same unusual warming event also resulted in substantial northward range expansions of several other crustaceans including the pelagic red crab *Pleuroncodes planipes* and two barnacles, *Megabalanus californicus* and *Paraconcavus pacificus* (Sanford et al. 2019).

Although many invasive crustaceans have been documented to expand their range over large distances in comparatively short periods of time, the rate of range expansion can vary dramatically even for the same species in different invaded regions. As an example, *Carcinus maenas* has expanded its range very slowly in South Australia and Tasmania relative to the rapid rate of spread in eastern North America (Thresher et al. 2003). Also, some crabs introduced into San Francisco Bay, California, including *Rhithropanopeus harrisii* and *Eriocheir sinensis*, have barely expanded their range at all along the California coast since their initial introductions in 1937 for *R. harrisii* (Peterson 2006) and 1992 for *E. sinensis* (Hanson and Sytsma 2008).

Range expansions may also occur as the result of unusual events that can transport organisms over much greater distances than they might themselves traverse. Processes such as rafting of organisms attached to floating debris have the potential to move species across great distances. Over 300 species of crustaceans have been described as rafting on floating objects (Thiel and Gutow 2005), and the consequences of rafting for establishing new populations or connecting previously isolated populations are substantial (Thiel and Haye 2006). Recent studies of an epic tsunami event in Japan in 2011 documented the trans-Pacific movement of debris from eastern Japan to the West Coast of North America on which nearly 300 species including > 30 species of crustaceans were described

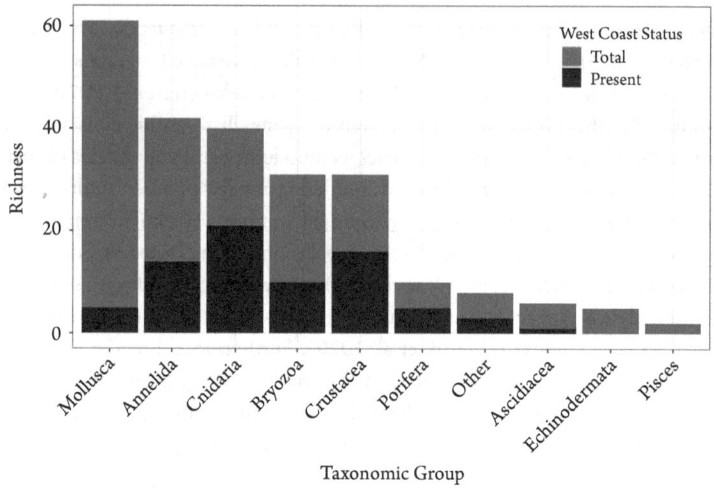

Fig. 13.2.
Total number of species (richness) for each taxon including Crustacea (> 30 spp.) identified from floating debris associated with the massive 2011 tsunami in Japan arriving on the coast of western North America from Japan over a six-year period. This includes species already present due to natural pan-Pacific distribution or as the result of prior introductions (black bars). Reproduced from Carlton et al. (2017), with permission from © Science.

over a multi-year period (Fig. 13.2; Carlton et al. 2017). This study suggests that even rare seismic events that spawn rafting of debris over large distances can contribute substantially to new invasions. Even for this well-studied event, the actual number of species that have become established through reproduction or otherwise is uncertain. The lack of taxonomic knowledge for many taxa complicates interpreting whether these species are actually novel in the introduced region (Carlton et al. 2017). These kinds of events make clear the need for better systematic descriptions of crustaceans and other invertebrate faunas worldwide.

Climate change is likely to change future patterns of species introductions and expansions in a few respects. First, there will likely be a change in shipping and climate matching of source regions associated with climate-driven warming of ocean surface waters. The future sources for nonnative crustaceans will likely be different from the current sources due to the variable warming patterns of different ocean basins and coastlines that will influence climate matching. Also, these sources will likely be influenced by the changing economic fortunes of different countries and the resulting influence on shipping activity (Seebens et al. 2016, Sardain et al. 2019). Among countries worldwide, the most important invasion vectors differ between high-income countries and low-income countries and, unfortunately, most countries do not have much capacity to prevent or to manage new invasions (Early et al. 2016).

Climate change is also likely to facilitate newly emerging shipping routes through the Arctic Ocean (trans-Arctic) linking polar and cold-temperate regions that had previously been isolated from one another (Chan et al. 2019). The volume of trans-Arctic shipping traffic is increasing rapidly, with a projected average annual increase of 20% per year over the next 25 years (Miller and Ruiz 2014). Climate change projections also apply to the Antarctic, and although shipping vectors are less likely to change here, changes in the circulation patterns of the Southern Ocean and the expected warming of surface waters may result in new opportunities for invasion (McCarthy et al. 2019).

OTHER FACTORS FACILITATING INVASIONS

Many other factors besides species characteristics and pathway details influence the likelihood of invader establishment, their impact on native ecosystems, or the degree of secondary spread to areas beyond the initial introduction. Among the multiple factors responsible for determining the success or failure of invasions is the ability of native communities to prevent or slow invasions generally referred to as "biotic resistance" (Elton 1958). Recent meta-analyses have investigated what factors influence the strength of biotic resistance in freshwater systems (Alofs and Jackson 2014). These studies included crustacean invaders such as the cladoceran *Daphnia lumholtzi*, the amphipods *Dikerogammarus villosus*, *Echinogammarus ischnus*, and *Gammarus tigrinus*, and the crayfish *Pacifastacus leniusculus* and *Faxonius rusticus*. Similar meta-analyses have also examined the strength of biotic resistance in estuarine and marine systems and included the invasive crab *Petrolisthes armatus* (Kimbro et al. 2013). Both meta-analyses examined how competition with, and predation by, native species influenced the strength of biotic resistance in these habitats. Overall, the effects of native consumers on invasive species (consumptive biotic resistance) were relatively similar in freshwater and estuarine/marine systems, while the effects of native competitors on invaders (competitive biotic resistance) were relatively weaker in freshwater systems. Studies of estuarine/marine systems also found that biotic resistance had a greater impact on invading consumers, but a weaker impact on invading primary producers (Kimbro et al. 2013).

Other factors can also affect invasion success, such as the presence of "enemies" including parasites, pathogens, and consumers. The absence of these enemies in their introduced range, relative to their native range, is one explanation for invasion success and is known as the enemy release hypothesis (Torchin and Mitchell 2004). This hypothesis suggests that introduced species are more successful when they are "released" from the parasites, pathogens, predators, and competitors that are left behind in their native range. Of course, this assumes that the enemies in the new range are less able to reduce invasion success, either because these enemies fail to recognize the invader or are unable to overcome their defenses (Keane and Crawley 2002). Studies of 26 introduced species, which included three crabs, *Cancer novaezelandiae*, *Carcinus maenas*, and *Hemigrapsus sanguineus*, found that introduced populations had fewer species of parasites as well as lower overall prevalence relative to populations in the native range (Torchin et al. 2003). This reduced parasite load is proposed to contribute to the invasion success of these species.

Another hypothesis, known as the novel weapons hypothesis, suggests that invaders with morphological or chemical weapons that are evolutionarily novel in the introduced range are more likely to be successful (Callaway and Ridenour 2004). The parasites and pathogens that invaders transmit can also be considered novel weapons. Few have been more influential in facilitating invasion than the diseases affecting native crayfish in Europe (Holdich et al. 2009, Peeler et al. 2011). Invasion by the North American signal crayfish *Pacifastacus leniusculus* and the associate transfer of the "crayfish plague" (*Aphanomyces astaci*) resulted in substantial mortalities of native European crayfish (*Astacus astacus* and *Astacus fluviatilis*) and likely increased the success of the invader.

Other hypotheses have been suggested to contribute to invasion success such as the degree of naivety of predators and prey (Sih et al. 2010). Nonnative predators may experience the advantage of novelty associated with the naivety of native prey who fail to recognize the new predators as a threat. But these same nonnative predators may also suffer from a novelty disadvantage, because they may be naive to their new prey and fail to recognize them as a suitable food. Examples of predator naivety include the invasion of the southern Atlantic coast of the US by the invasive crab *Petrolisthes armatus*. Studies have shown that despite the high value of this species as prey for the abundant native crab *Panopeus herbstii*, this native crab has been slow to learn to consume the invader, which may contribute to the success of this invasion (Hostert et al. 2019).

Fisheries overharvest that reduces or removes native predators, which had previously limited invader establishment, can also contribute to the increased impact and spread of invasive species, including notable crustacean invaders. Studies have shown that extensive recreational fishing in New England has reduced native, higher trophic level consumers. These fisheries-related reductions in top predators have resulted in increased abundances of smaller crabs including the invasive *Carcinus maenas* and the native crab *Sesarma reticulatum*. Increased abundances of *C. maenas* and *S. reticulatum* have had substantial impacts on marsh vegetation through burrowing activities in many Massachusetts salt marshes (Altieri et al. 2012, Coverdale et al. 2013).

The nature of the habitat can also influence the success of invasions. Artificial habitats such as docks, piers, and pilings can provide novel habitat that is particularly amenable to colonization and establishment of nonnative fouling species such as tunicates, bryozoans, hydroids, bivalves, sponges, and barnacles, which can also provide habitat for many smaller crustaceans including amphipods, isopods, and tanaids (Bulleri and Chapman 2010). In many cases, these artificial habitats are less accessible to benthic predators because they are often floating or suspended far off the substratum. Studies in Australia demonstrated that barnacles such as *Megabalanus coccopoma* were significantly more abundant on artificial structures relative to surrounding habitat (Dafforn et al. 2009). Similar studies of artificial substrata in Brazil found high densities of introduced barnacles, including *Amphibalanus reticulatus* (Neves et al. 2007). Experimental studies in Chile showed high densities of the invasive tunicate *Ciona intestinalis* on artificial structures, but not in natural benthic habitats (Dumont et al. 2011).

Nonnative species themselves can create novel habitat that disproportionately favors high abundances of nonnative species including invasive crustaceans. The nonnative bryozoan *Watersipora* spp. (formerly *Watersipora subtorquata*) creates artificial reefs in many areas, which studies in California have shown to support high densities of invasive caprellid amphipods including *Caprella mutica* (Sellheim et al. 2010). *Watersipora* spp. has also been shown to enhance the abundance of invasive crustaceans when it colonizes ship hulls and, thus, may facilitate ship-based dispersal of invasive species including barnacles (*Amphibalanus amphitrite*) and amphipods (*Corophium* spp.) (Floerl et al. 2004).

Pollution from water-born contaminants can also result in conditions that favor the establishment and spread of nonnative species in a wide range of habitats from offshore sites in New England, USA, to estuarine locations in southeastern Australia and the western US (Carman et al. 2007, Piola and Johnston 2008, Crooks et al. 2011). These studies generally focus on sessile, attached fouling taxa including barnacles. Studies of copper pollution in New South Wales, Australia, showed the nonnative barnacle *Amphibalanus amphitrite* was facilitated by elevated levels of copper that simultaneously reduced the abundances of native barnacles *Balanus trigonus* (Piola and Johnston 2008). Subsequent laboratory studies showed that this was the result of the invader having a higher tolerance of copper toxicity (Piola and Johnston 2009). Similar studies of copper pollution in San Francisco Bay, California, showed that abundances of the introduced barnacle *Amphibalanus improvisus* were significantly increased under high contaminant conditions (Crooks et al. 2011).

INVADER IMPACTS

The spread and establishment of nonnative species threaten ecosystem structure and function worldwide, and the rates of invasion and their impacts continue to increase in frequency and geographic extent (Mack et al. 2000). There are numerous documented impacts of invaders on native species and ecosystems, including predation-mediated extirpations of native prey (Witte et al. 1992), exclusion of native competitors, and changes to ecosystem processes and nutrient cycling

(D'Antonio and Vitousek 1992). These changes can often have substantial impacts on human populations through interruptions of nature-derived goods and services as well as substantial human health concerns (Diagne et al. 2021).

Taxonomic Patterns of Invasion

The type of impacts that introduced crustaceans create is often a function of the types of species that successfully become established. Recent summaries have documented 108 crustacean species in estuarine and marine habitats in North America. The majority of these are amphipods, which make up 30% of these documented invasions. Other groups that are also common in the lists of invaders are isopods (21%), decapods (14%), and copepods (13%) (Ruiz et al. 2011). Even within more restricted taxonomic groups, the numbers of invaders are highly nonrandom (Karatayev et al. 2009). Within nonnative crabs, which are among the best-studied and most easily identified group of crustacean invaders, there are strong biases in invasion rates across families. In terms of documented establishment, spread, and impacts, crab introductions are limited to a relatively small number of families. Of the 20 families of anomuran crabs, only crabs in the Lithodidae and Porcellanidae have resulted in established introductions. Similarly, out of the 98 families of brachyuran crabs, there are only 18 families for which established invasions have been documented. Families like the Leucosiidae, Majidae, Pilumnidae, Panopeidae, Porcellanidae, Portunidae, Varunidae, and Xanthidae have multiple species that have become successfully established outside their native range, with the Portunidae being the most frequently established (Grosholz 2010).

Impacts on Native Ecosystems

The types and range of impacts of invading species are numerous and can include various impacts on native species and communities (Grosholz 2002). Consumers like decapods are likely to have the most significant impacts as this class of invader, along with ecosystem engineers, generally have the largest ecosystem impacts. Individual native species, including threatened and endangered ones, are the most at risk. However, unlike many documented extinctions in freshwater systems, global extinction in marine systems is comparatively rare outside of marine mammals and birds (Carlton et al. 1999, IUCN 2021). It is likely that the taxa that crustaceans most likely impact, such as fish and other invertebrates, may have been overlooked, and there have been few such listings in these taxa (Polidoro et al. 2012). Certainly, there have been many more documented extinctions in aquatic freshwater systems, particularly in lakes and isolated water bodies, but these have not been the result of invasive crustaceans (Gurevitch and Padilla 2004; Fig. 13.3).

Impacts on Habitats

Among the most potentially destructive species are ecosystem engineers that can degrade or remove habitat that other species depend on. Among these are species that burrow into structures such as foundation species, or benthic habitats. This includes the boring sphaeromatid isopod *Sphaeroma quoianum* that has caused substantial erosion of creek banks in the salt marsh creeks where it has become common on the US West Coast (Talley et al. 2001). Another example of a boring isopod affecting a key foundation species is *Sphaeroma terebrans*, which can substantially influence the growth and survival of red mangroves from Florida to Panama (Davidson et al. 2016). Recent studies have also shown that this species has substantially damaged at least three mangrove species in Hainan, China (Xin et al. 2020). The potential impacts of these boring isopods may become even greater as these increasingly threatened mangroves, which uniquely define an important

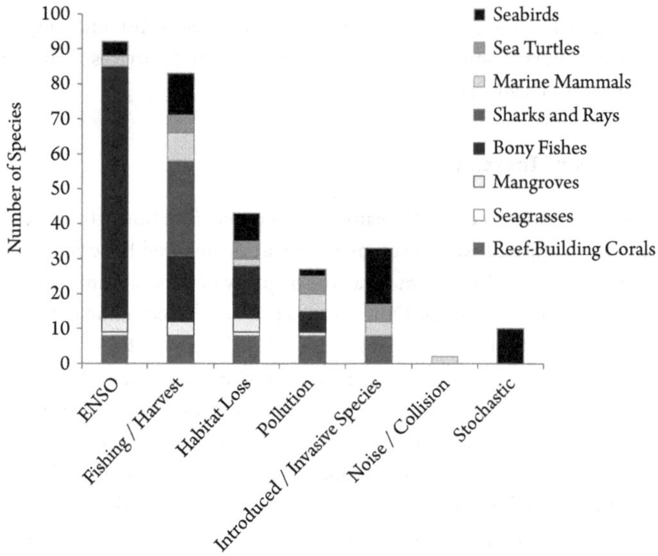

Fig. 13.3.
Major categories of threats to species listed in several threatened taxonomic categories in the Eastern Tropical Pacific including ENSO (El Niño-Southern Oscillation). Introduced species represent a significant risk to a wide range of threatened species, but are not the most important. Adapted from Polidoro et al. (2012), with permission from © Inter-Research Science Publishers; see a color version of this figure in the centerfold.

coastal habitat, suffer from harvest, habitat degradation, and other stress factors. Another important invader that can also be an ecosystem engineer is the Chinese mitten crab *Eriocheir sinensis*, which invaded the western United States in the 1990s, where it has been shown to burrow into the walls of earthen canals and create burrow networks that substantially weaken surrounding structures both natural and human-created (Dittel and Epifanio 2009). Similar concerns exist in western Europe where *E. sinensis* was introduced in the early 20th century and can have similar impacts on canals and other waterway structures (Herborg et al. 2003). Additional examples include the impacts of red swamp crayfish on benthos and water quality and via its prodigious burrowing activities (Gallardo et al. 2016).

Consequences for Disease

Work on ballast water shipments has documented that crustaceans in ballast water tanks. including copepods (some nonnative), can harbor infectious strains of the serious human pathogen *Vibrio cholerae*, which is the infectious agent responsible for human cholera. These studies have shown that the transport of ballast water has brought cholera to the Chesapeake Bay region and likely other harbors as well (Ruiz et al. 2000). Other human pathogens are known to be vectored by invasive crustaceans such as the Asian lung fluke *Paragonimus westermani*, for which species of *Eriocheir* have been shown to be second intermediate hosts (Banzai et al. 2021). Luckily, this parasite is currently not found in populations in Europe or North America, although significant concern still remains.

What we know about the role of human-mediated vectors in the global movement of macroorganisms is largely based on observed translocations and biogeographical studies (Pagenkopp-Lohan et al. 2020). Consequently, microorganism translocations and the magnitude and scale of their impacts are likely to be seriously underestimated (Pagenkopp-Lohan et al. 2020). Some examples do stand out, such as the "crayfish plague," which is the fungal-like disease *Aphanomyces astaci* that has accompanied introductions of crayfish from North America into European waters. This pathogen is lethal to several species of native crayfish despite being apparently harmless to North American crayfish, and this has likely contributed to the loss of native Eurasian crayfish from much of their range.

The most well-studied species in terms of parasites and pathogens are the European green crab *C. maenas* and the Chinese mitten crab *E. sinensis*, which have been shown to carry and transmit multiple pathogens and parasites of concern. For *E. sinensis*, in addition to the Asian lung fluke (above), they are also a host for the white spot syndrome virus (Pagenkopp-Lohan et al. 2020). Surveys of *C. maenas* found 67 taxa, the most for any crustacean, including ostreid herpesvirus and the bitter crab disease agent *Hematodinium perezi* (Bojko et al. 2021). Of these, the most common were helminths, protists, and, to a lesser extent, bacteria and microsporidia (Bojko et al. 2021). Although not all these symbionts were disease agents, the potential for moving pathogenic symbionts is significant. In general, detection of nonnative parasites and other small-bodied organisms is generally much more difficult compared to most free-living hosts (Pagenkopp-Lohan et al. 2020).

Newly emerging genetic methods suggest that for planktonic adult organisms (e.g., copepods, dinoflagellates) or resting stages of parasites, infected individuals may be comparatively common in the ballast water of commercial ships (Pagenkopp-Lohan et al. 2016), but the extent to which these parasites successfully establish is uncertain. Ship hulls can also be important vectors for disease transmission via invasive crustaceans (Georgiades et al. 2021). For example, parasitic invertebrates can also be translocated by barnacles, including the castrating isopod *Hemioniscus balani* (Davidson et al. 2016). Human pathogens, such as the species *Vibrio cholerae*, have also been found in the surface biofilms of vessel hulls (Shikuma and Hadfield 2010). In European coastal waters, intentional introduction of the invasive Pacific oyster (*Magallana gigas*) inadvertently resulted in the introduction of the parasitic copepod *Mytilicola orientalis* that impacts growth and production of native blue mussels (*Mytilus edulis*) and other native bivalves (Goedknegt et al. 2018). Overall, there is a vast potential for crustacean invaders to move marine parasites and pathogens around the globe (Table 13.2). Unfortunately, few of these invasions have been well documented, so much more work on parasites and pathogens is needed.

Climate Change Consequences

The impacts of climate change interact with many other stressors, including overharvest, contaminants, and habitat loss. Climate change may also modify the impacts of introduced species as gradients in temperature, salinity, and other variables are altered. There is experimental evidence that invaders in freshwater as well as in estuarine and marine systems are more resilient to several categories of climate change, particularly marine invaders (Sorte et al. 2012). Although these studies examine physiological tolerance levels of invaders relative to natives, there are few studies of species interactions such as competition, predation, or parasitism between native and introduced species under climate change. Some work has attempted to look at physiological performance and the limits of introduced and native predatory–prey interactions, which is providing some insights into how climate change, particularly how changes in temperature and salinity regimes, may influence the impacts of invasive predator–prey dynamics (Cheng et al. 2017).

Table 13.2. A listing of diseases vectored by crustaceans identified by The World Organization for Animal Health. These are parasites and pathogens that would be likely to cause mass mortalities if they were to become established in a new region. Source: data from Pagenkopp-Lohan et al. (2020).

Disease Name	Disease-Causing Agent
Acute hepatopancreatic necrosis disease	Unidentified bacteria
Crayfish plague	*Aphanomyces astaci*
Necrotizing hepatopancreatitis	*Hepatobacter penaei*
Infection with infectious hypodermal and hematopoietic necrosis virus	Hypodermal and hematopoietic necrosis virus
Infection with infectious myonecrosis virus	Myonecrosis virus
White tail disease	
Infection with Taura syndrome virus	*Macrobrachium rosenbergii*
Infection with white spot syndrome virus	nodavirus Taura syndrome virus
Infection with white spot syndrome virus	White spot syndrome virus
Infection with yellow head virus genotype 1	Yellow head virus genotype 1

As arctic habitats continue to warm, we may see additional invasions similar to the red king crab *Paralithodes camtschaticus,* which was intentionally introduced off the Norwegian coast (Falk-Petersen et al. 2011). Studies of feeding rates (see above) suggest that introduced red king crabs could significantly affect native scallop populations in the future in this region. As a large benthic consumer, red king crabs can potentially have substantial impacts on benthic ecosystems and reduced benthic diversity has been associated with this invasion (Falk-Petersen et al. 2011).

EXAMPLES OF HIGH-IMPACT INVASIVE CRUSTACEANS

Several crustaceans are listed among the world's 100 worst invaders (Lowe et al. 2000), including the Chinese mitten crab *E. sinensis,* the European green crab *C. maenas,* and the fishhook water flea *Cercopagis pengoi* (ISSG 2015). While these species clearly deserve special recognition for their substantial and widely distributed impacts, many others have been found to be similarly problematic. While these lists are not definitive in terms of the spatial extent or the magnitude of the impact on native habitats, the world's 100 worst invasive species are continually cited by the public press, and species on this list are often overemphasized relative to others not on the list. The species listed in Table 13.3 may or may not be candidates for future lists, but all certainly have a significant body of information documenting substantial negative impacts.

Freshwater Invaders

Amphipoda

Several amphipods have been introduced to North America from the Ponto–Caspian region. and several of these species have also resulted in declines in native invertebrates in the inland waters of Europe. In the North American Great Lakes, *Echinogammarus ischnus* has largely replaced the native gammarid *Gammarus fasciatus* on rocky substrata (Barrett et al. 2017). Interactions between native and exotic amphipods commonly involve intraguild predation, which can occur during molting

Table 13.3. Examples of crustacean invaders of greatest concern based on reported impacts and distribution. Columns for each species show the taxonomic identity and current genus and species (subject to change), the geographic region where it is native, one or more geographic regions where that species has become established as nonnative, and a short description of the primary impacts.

	Taxa	Native Region	Introduced Regions	Impacts
Amphipoda	*Chelicorophium curvispinum*	Ponto–Caspian basin	Baltic, North Sea drainages, Europe	native Amphipoda, benthos
	Dikerogammarus villosus	Ponto–Caspian basin	Baltic, North Sea drainages, Europe	native Amphipoda, benthos
	Echinogammarus ischnus	Ponto–Caspian basin	Baltic, North Sea, Europe, North American Great Lakes	native Amphipoda, benthos
Cladocera	*Cercopagis pengoi*	Ponto–Caspian basin	North American Great Lakes	native cladocera and fish
	Bythotrephes longimanus	Eurasia	North American Great Lakes	native cladocera and fish
Copepoda	*Pseudodiaptomus forbesi*	South China Sea	Northeast Pacific	native copepods and fish
	Mytilicola orientalis	North West Pacific	Northeast Pacific	commercial shellfish
Decapoda	*Carcinus maenas*	Europe	North America, Australia, Argentina, South Africa, Japan	native bivalves, smaller crabs
	Eriocheir sinensis	China	Europe, North America	native shrimp, fishes, channel banks
	Pacifastacus leniusculus	North America	Asia, Europe, western US	native crayfish, disease
	Procambarus clarkii	North America	Mediterranean, Africa, Asia, Europe, North America	native crayfish, ecosystems, disease
	Faxonius rusticus	US	North American Great Lakes, interior US	native crayfish
Mysida	*Hemimysis anomala*	Ponto–Caspian basin	Baltic and North Sea drainages, North American Great Lakes	native zooplankton
Isopoda	*Sphaeroma quoianum*	Australasia	Western North America	marsh banks, channels

periods and can determine which species dominates. The most effective invaders are often ones with high reproductive output and consumption rates (Dick et al. 2013).

Another noteworthy example is *Dikerogammarus villosus* (Platvoet and Dick 2000, Dodd et al. 2014), an effective predator known as "killer shrimp," which is larger than many other amphipods in these habitats. Lab experiments have shown that it consumes a wide variety of invertebrates and rapidly reduces and eliminates many other crustaceans. As it expanded its range into northwestern Europe, it largely replaced native and other nonnative amphipods.

Another Ponto–Caspian species, a filter-feeding corophiid amphipod *Chelicorophium curvispinum* (*Corophium curvispinum*), invaded the Rhine River in the 1980s and within a few years had colonized hundreds of kilometers of benthos at very high densities (> 100,000 individuals m^{-2}), where it altered bottom habitats, thus reducing the diversity of benthic species as it created galleries of mud tubes on hard substrata (van den Brink et al. 1993, van Riel et al. 2006). In the process of colonizing rocky substrata and consuming suspended food particles, *C. curvispinum* nearly extirpated populations of filter-feeding caddisflies *Hydropsyche* and the zebra mussel from the lower Rhine River (van den Brink et al. 1993).

It is important to note that in some drainages like the Rhine River, there have been multiple amphipod invasions, so assessing the impact of a single species is difficult. In addition to *C. curvispinum*, other key invaders also include *Dikerogammarus villosus*, *Echinogammarus ischnus*, and several other nonnative *Gammarus* species including *Gammarus tigrinus* (van Riel et al. 2006). High levels of cannibalism and intraguild predation have resulted in substantial declines of both native and nonnative invasive species in this habitat (Hänfling et al. 2011).

Cladocera

The fishhook waterflea *Cercopagis pengoi* is a consumer of other zooplankton and has had impacts on other planktivores. In the North American Great Lakes, this has included planktivorous fish such as alewife (*Alosa pseudoharengus*) and rainbow smelt (*Osmerus mordax*) (Bushnoe et al. 2003). It possesses a long caudal spine, which allows it to evade consumption by larger consumers, including the same planktivorous fish it competes with for food (Vanderploeg et al. 2002). *Cercopagis pengoi* establishment also fouls fishing gear, with impacts on commercial fishing more broadly. The Baltic Sea is another region where *C. pengoi* has had substantial impacts on zooplankton and fish communities (Holopainen et al. 2016). For example, stable isotope analyses have demonstrated a significant shift in $\delta^{15}N$ for juvenile herring in the presence of *C. pengoi*, which it competes with for zooplankton. These studies suggest that due to competition for food with native species, this invader has caused a substantial shift in the Baltic Sea food web (Gorokhova et al. 2005).

The introduction of another cladoceran, the Eurasian spiny waterflea *Bythotrephes longimanus*, has resulted in substantial reductions in both species richness and abundance of zooplankton in lakes it invaded in Europe and North America. Its impact has been largely due to competition and predation on native cladocerans. *Bythotrephes longimanus* has a long tail spine that also deters feeding by planktivores and allows this species to dominate zooplankton assemblages. It is a significant consumer of smaller zooplankton, including other cladocerans as well as copepods and rotifers, and it also competes for food with larval fishes (Vanderploeg et al. 2002).

Decapoda

The red swamp crayfish *Procambarus clarkii* is native to Gulf coastal plains from the Florida panhandle to Mexico and the southern Mississippi River (Gallardo et al. 2016). It has had significant impacts on benthic habitats by increasing sediment resuspension. Its activities have created rapid shifts from a clear water, macrophyte-dominated state to a turbid, phytoplankton-dominated state.

This has resulted in significant reductions in populations of submerged macrophytes in lakes within just a few years of invasion (Souty-Grosseta et al. 2016). Turbidity and nutrient concentration have increased substantially in the presence of omnivorous invaders like *P. clarkii* (Gallardo et al. 2016). This species has also invaded Europe and novel ranges in US, where it was transported for fisheries and food, and it continues to change habitats and to threaten native crayfish in many countries (Gallardo et al. 2016).

Another important crayfish invader is the rusty crayfish *Faxonius rusticus*. Although native to a small, Midwestern region in the US, rusty crayfish have been spread widely across North America via bait movements. They have had large impacts on native crayfish such as *Faxonius virilis* and other native invertebrates across much of this introduced region (Twardochleb et al. 2013). Also, an effective omnivore, this species can decrease macrophyte cover and biomass, thus altering ecosystem function and nutrient cycling. Its overall impact on the biodiversity of these systems has been substantial (Lodge et al. 2012).

A third crayfish species that has had significant impacts is the signal crayfish *Pacifastacus leniusculus*. The impacts of the introduction of *P. leniusculus* to Europe have been substantial, including greatly reducing aquatic macrophyte abundance relative to native crayfish. This has resulted in more substantially altered habitat structure, with potential consequences for invertebrate and fish distributions (Lodge et al. 2012). This species, along with others like the red swamp crayfish, has been responsible for the introduction and spread of the crayfish plague, *Aphanomyces astaci*, which is a fungus-like organism also not native to Eurasian freshwaters (see above). This pathogen is lethal for several native Eurasian crayfish, but not so for North American crayfish, whose invasions, together with predation and habitat change, have resulted in the displacement of native Eurasian crayfish from the majority of their historical range (Lodge et al. 2000).

Mysida

Several genera of mysids including *Mysis*, *Hemimysis*, *Limnomysis*, and *Neomysis* have been introduced widely in many regions and have had a large impact on native communities and local ecosystems. Unlike other crustacean groups, many mysids have been intentionally introduced to provide supplementary food sources for sport fishes in North American and European lakes. An unintended but frequent consequence of many of these introductions has been the substantial reductions in the abundance of large native zooplankton. For example, the introduction of the bloody red mysid *Hemimysis anomala* has had substantial negative impacts on native zooplankton in many areas of Europe (Dick et al. 2013). Studies have shown that this mysid is a more voracious consumer of zooplankton than native consumers. The magnitude of impacts on field populations of native zooplankton can be estimated from laboratory measures of the relative functional response of invaders vs. native consumers (Dick et al. 2017). The relative impact of invasive mysids is generally greater on cladocerans compared with copepods (Fig. 13.4) and can result in declines in the growth and abundance of pelagic fishes as well as increased rates of parasitism on local fishes.

Estuarine and Marine Invaders

Decapoda

The European green crab *Carcinus maenas* is among the most widely distributed invasive crustaceans and has established populations on both coasts of North America, in southern South America, Australia, South Africa, and Japan. As a generalist predator, *C. maenas* feeds on a wide range of taxa, but particularly bivalve mollusks, polychaetes, and small crustaceans including other

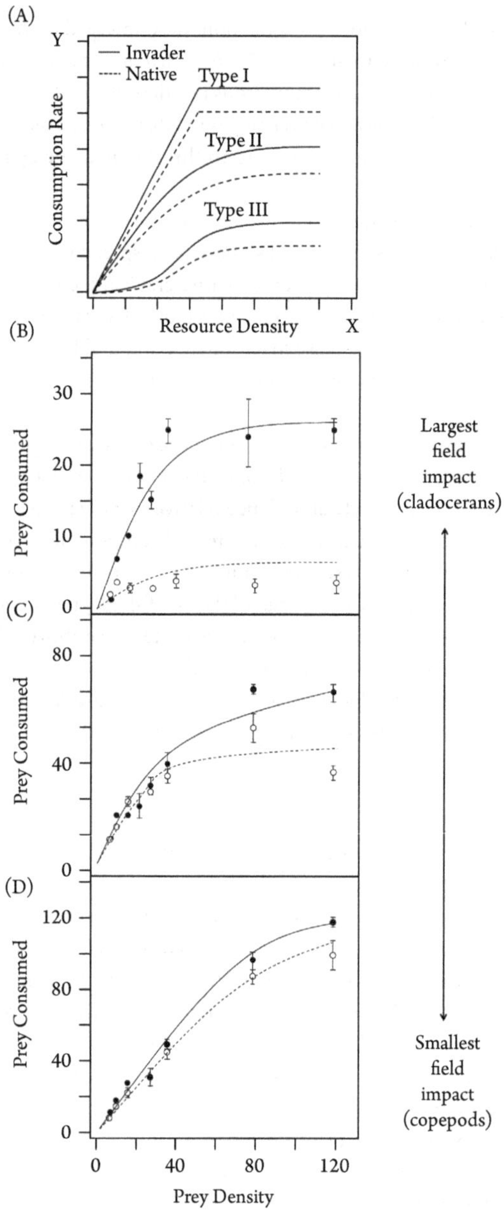

Fig. 13.4.
The functional response (per capita prey consumption as prey density increases) of an invasive mysid consumer *Hemimysis anomala* (solid circles) vs. a native mysid *Mysis salemaai* (open circles) showing a positive relationship between difference in functional responses between native and invasive species in the lab and impacts in the field: prey species in (B) and (C) are cladocerans and in (D) copepods. Reproduced from Dick et al. (2017), with permission from © John Wiley & Sons.

crabs. Studies of the introduction of the European green crab in the western US following its introduction circa 1990 demonstrated that this invader reduced small native clams in a central California bay (*Nutricola* spp.) by >90% (Grosholz et al. 2000) and created similar reductions in populations of smaller native shore crabs (*Hemigrapsus oregonensis*) (deRivera et al. 2011). Subsequent studies showed that predation by *C. maenas* on native clams reduced their competitive impact on nonnative gem clams *Gemma gemma*. The invasive gem clam had been introduced at least thirty years earlier but remained restricted in its distribution due to competition with native clams. Thus, one of the consequences of green crab predation was a substantial increase in the abundance and distribution of *Gemma gemma* thirty years after its introduction (Grosholz 2005).

The impacts of *C. maenas* on commercial fisheries have also been substantial. Green crabs were strongly implicated in the destruction of the soft-shell clam (*Mya arenaria*) fishery on the East Coast of the United States and Canada in the mid-20th century (Glude 1955). The impacts to the soft-shell fishery have continued, resulting in a 75% decline over the past forty years concurrent with increases in green crab abundance (Beal et al. 2016). Green crabs also threaten populations of other commercially important bivalves such as scallops *Argopecten irradians* and northern quahogs *Mercenaria mercenaria* (Walton and Walton 2001), and these crabs have been responsible for millions of dollars of losses annually for commercial clam fisheries in the eastern US (Grosholz et al. 2011). In eastern Maine, recent work has shown that as much as 99% of commercial clams have disappeared largely due to green crab predation (Beal et al. 2018).

Green crabs can also have significant impacts on local habitats through their foraging activities. As they dig for clams in soft-sediment habitats, the crabs uproot beds of eelgrass, which protects small marine animals like young salmon and provides feeding areas for migrating birds. Within a year of Maine's 2012 green crab spike, more than half of the eelgrass (*Zostera marina*) cover in Casco Bay disappeared (Neckles 2015). Green crabs have also been shown to have substantial negative impacts on the abundance of eelgrass in Newfoundland (Matheson et al. 2016) and British Columbia (Howard et al. 2019), Canada.

The Chinese mitten crab *E. sinensis* is another crab with substantial impacts in several regions. It invaded Europe in the early 1900s (Herborg et al. 2003), and during this period of high abundance, they impacted native fishes, damaged fishing nets, and caused losses of up to 80 million Euros. Mitten crabs also burrowed into the earthen walls of channels and levees, which has caused additional damage (Herborg et al. 2005). Mitten crabs have invaded the Atlantic, Pacific, and Gulf coasts of the United States as well. In San Francisco Bay, there have been substantial economic impacts due to the mitten crab invasion (Rudnick et al. 2003). Mitten crabs moved dozens of miles up the river system that feeds the bay. Crabs entered pumping facilities in large numbers that are crucial for moving water for irrigation and drinking water elsewhere in the state. Very expensive emergency efforts were required to remove and exclude crabs from water pumping facilities, where mitten crabs had clogged screens and damaged fish, including some endangered species that would otherwise be salvaged (Rudnick et al. 2003). During periods of high abundance, mitten crabs have also affected shrimp trawling by clogging nets, as well as hook-and-line fishing by stealing bait from hooks.

The porcelain crab *Petrolisthes elongatus*, a New Zealand native, has significantly impacted benthic communities on Australian shores. Studies have found that the biomass of *P. elongatus* was positively correlated with overall native species abundance and biomass, which highlights its dependence on the same cobble refugia as many native species (Gribben et al. 2015). However, additional analyses showed that community structure was strongly affected, with sessile gastropods (chitons and limpets) positively correlated with *P. elongatus* biomass. There were clear and consistent relationships between habitat availability and the biomass of *P. elongatus*, and negative relationships with the abundances of native species, including mobile gastropods and native crabs (*Brachynotus spinosus* and *Paragrapsus gaimardii*), and *P. elongatus* biomass, most likely due to space competition (Gribben et al. 2015).

Copepoda

The invasive copepod *Pseudodiaptomus forbesi,* native to the Chinese region of Asia, first invaded the lower Columbia River (Washington and Oregon, USA) in 2002. A model of food web dynamics based upon a 12-year time-series of plankton community and environmental data from the Columbia River showed that plankton communities in the lower Columbia River were strongly impacted by the copepod *P. forbesi* at multiple trophic levels (Dexter et al. 2020). Impacts were observed for different stages with copepodite stages negatively affecting *Daphnia* and cyclopoid copepods and nauplii stages negatively affecting ciliates and autotrophs.

In San Francisco Bay, multiple nonnative copepod species have been introduced in addition to *P. forbesi,* which complicates interpretations of the impacts of these invaders. Species including *Limnoithona tetraspina, Tortanus tridextrilobatus,* and *Oithona davisae* have resulted in dramatic changes in the zooplankton community. Among the most consequential impacts is that the zooplankton assemblage has shifted from larger-bodied species that are more easily captured by juvenile fishes to smaller, quicker species that are less easily captured. These changes have likely contributed to the dramatic declines in fish populations (pelagic organism decline) documented in San Francisco Bay over the past few decades, with consequences for the entire ecosystem (Kimmerer 2004). These introductions certainly make the case that in many highly invaded systems it is very difficult to connect declines in native species or ecosystem functions with any single invader species. Studies of multiple invasive species are rare, but more are desperately needed.

MANAGING INVASIVE CRUSTACEANS

Several tools are available for managing invasive crustaceans as they are for most other aquatic and marine invasions. The most cost-effective and efficient use of management resources is prevention of new invasions, which relies on identifying and managing the invasion vectors most responsible for these introductions. Ultimately, the management decision will be a function of how long the species has become established, how widely distributed it has become, how accessible it is to removal methods, and what resources are at risk from its invasion.

Control of nonnative species typically involves removal of invaders from the invaded habitat to mitigate their ecological and economic impacts (Perrings 2005, Epanchin-Niell and Hastings 2010). Population control generally requires large quantities of resources and effort, which defines and often limits the possible scale of control efforts (Zavaleta et al. 2001). The likelihood of the success of programs aimed at complete eradication is a function of the size of the invading population and its geographic extent (Simberloff 2009).

Mitigating the invasion impacts is particularly problematic in marine and freshwater habitats because the rate of establishment and spread of invaders often outstrips the available resources. Aquatic habitats cover more than 70% of the planet, and many of these organisms disperse broadly, propelled by water- and wind-driven currents (Shanks 2009). Additionally, humans have limited access to aquatic environments because of depth restrictions. Consequently, many aquatic invaders can rapidly achieve a broad distribution before they are detected and determined to be established.

For most marine and freshwater invaders, complete eradication is not a viable option (Green and Grosholz 2021). Recent studies have demonstrated the existence of overcompensatory reproduction for some taxa in response to removal, thus risking program failure and loss of valuable management resources (Grosholz et al. 2021). In this case, population suppression and containment are advised to be the primary goals of these management programs.

New studies have developed methods for "functional eradication," which involves identifying thresholds for target densities for suppression that would allow recovery of native species or other desired ecosystem functions (Green and Grosholz 2021). Examples of these threshold-based

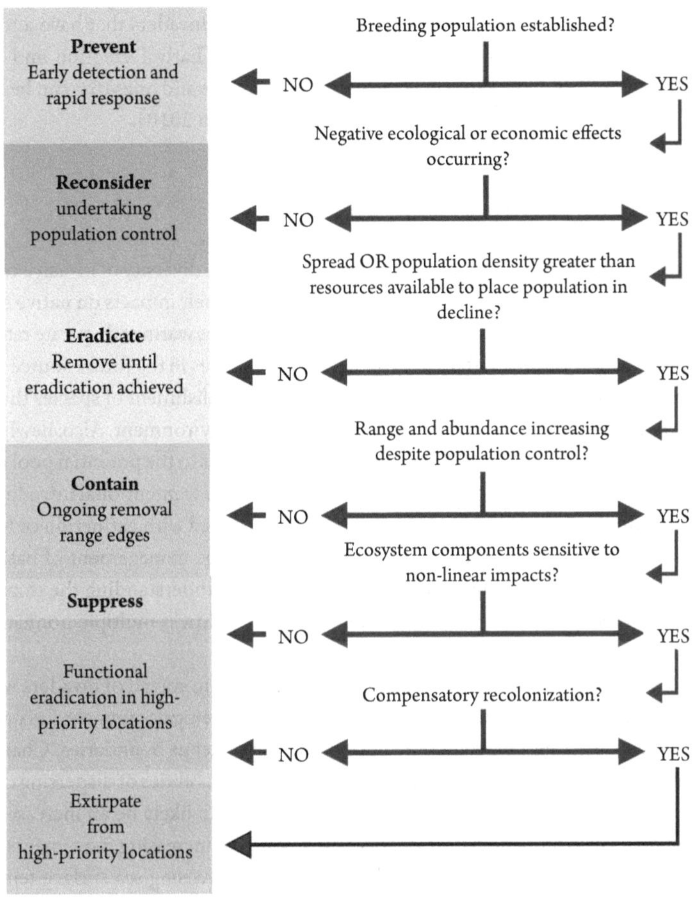

Fig. 13.5.
Detailed decision tree describing recommendations for invasive crustacean management given information about the current status of the invasion, the severity of impacts, magnitude of establishment and spread, available resources, ecosystem responses to invader density, population connectivity, likelihood of compensatory reproduction, and related attributes. Reproduced from Green and Grosholz (2021), with permission from © John Wiley & Sons; see a color version of this figure in the centerfold.

targets have been developed for well-studied crustacean species like the rusty crayfish (*Faxonius rusticus*) and the European green crab (*C. maenas*) for which sufficient data allow the determination of suppression thresholds (Green and Grosholz 2021). This approach can be integrated into a more comprehensive decision process that includes other types of information: (1) how well established and how widely spread the invasion is, (2) the severity of impacts and level of concern about these, (3) how sensitive the ecosystem is to these impacts, (4) life history attributes that describe how connected the population is, and (5) how susceptible the species might be to overcompensatory reproduction and related issues. An example of this type of decision process or "decision tree" is shown in Fig. 13.5.

The management of invasive crustaceans has also moved toward greater emphasis on detecting invasions early in the process. It is well known that the success of invasion management is directly related to how early the intervention can be undertaken, regardless of whether it is eradication, suppression, or containment. Programs for early detection of invasions and a subsequent rapid follow-up response have been developed for many high priority invaders that have a track record of producing damaging invasions. These programs, known as "Early Detection and Rapid Response," are among the most cost-effective management strategies and ones that can be reasonable implemented with modest annual surveys (Pysek and Richardson 2010).

FUTURE CHALLENGES AND OPPORTUNITIES

Climate change and growing economies will continue to affect the diversity of invasive crustaceans, the landscape of vectors that move them around the planet, and their impacts on native ecosystems and human economies. As the climate and sea surface temperatures warm at disparate rates depending on the Continent and coastline, there will be significant changes in the future source regions for any particular location. These new sources may result in the establishment of species that were not previously viewed as being able to establish in the surrounding environment. Also, newly emerging economies will contribute a greater number and volume of species to the potential pool of invaders associated with ornamental and aquaculture industries. Finally, new intentional introductions that will likely result from increasing pressure to save species threatened with extinction or to maintain fisheries that have been overharvested, will continue to challenge management of native ecosystems. Because most systems already contain multiple invaders, understanding the impacts of any future invaders will require investigations that simultaneously address multiple nonnative species in each system.

Changing economies and climates will also influence the future spread of invaders without human assistance. Increasing rates of range expansions are already being documented as ocean basins warm and previously existing biogeographic boundaries fail to act as boundaries. Changes in upwelling rates and coastal currents will also likely contribute to the abilities of dispersing crustaceans to reach novel regions and to become established. Also, there will likely be an increase in rates of introduction of novel parasites and pathogens accompanying the increasing movements of species. The parasites and pathogens themselves are also reacting to increasing sea surface temperatures, which may further exacerbate their impacts.

The future of managing invasive crustaceans, both benthic and holoplanktonic, will involve a change in focus from reactive responses once an invader has been established to proactive prevention of invasions by managing invasion vectors. The real future success will involve new strategies for international cooperation. For widely dispersing invaders, such as many crustaceans with broadly dispersing larvae or adult stages, even comprehensive actions of a single country may be swamped by the inactivity of neighboring countries, and therefore broad regional cooperation will likely be needed.

A particularly useful new approach is to identify and address areas where invaders have already become established, known as "bridgehead regions," and from where new invasions have been shown to originate (Ricciardi et al. 2021). This may allow cooperative international management to focus more resources on shipping vectors that originate from bridgehead regions and spend less on shipping vectors from regions that pose less of a risk (Fig. 13.6). Cooperative international management that can focus resources on vectors from these "bridgehead regions" will likely be an emerging and highly productive approach to reduce the threat of new invasions.

CONCLUSIONS

Crustaceans are the most commonly introduced species in aquatic and estuarine/marine ecosystems worldwide. Most invaders are members of the Amphipoda, Isopoda, Decapoda, and Copepoda in estuarine and marine systems, as well as Cladocera and Mysida in freshwater systems. Human activities are most responsible for the introduction of nonnative crustaceans

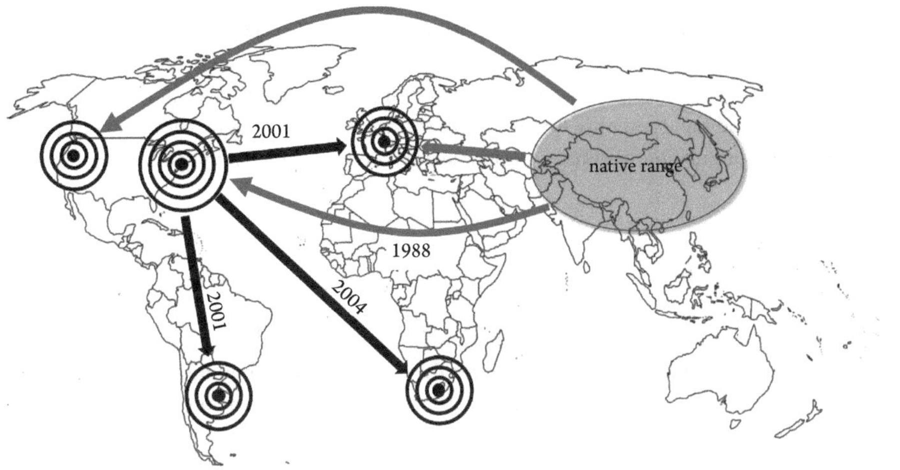

Fig. 13.6.
A map of the spread of an insect pest, the harlequin ladybird beetle *Harmonia axyridis* as an illustration of the "bridgehead effect" that can form the basis for new approaches to manage introductions via shipping vectors. The original pathways for the intentional introductions for biocontrol purposes (gray arrows) show insects moved from the native range in Asia to Europe and North America. The subsequent unintentional invasions to South America, South Africa, and back to Europe (black arrows) demonstrate that the primary source of this global spread originated from the introduced populations in Eastern North America, and not the native Asian range. Thus, Eastern North America continues to function as a bridgehead region; therefore, this region should be the future focus of cooperative international actions. Reproduced from Ricciardi et al. (2021) under a Creative Commons License (CC BY); see a color version of this figure in the centerfold.

around the planet with shipping vectors, ballast water, and hull fouling being the most important in terms of numbers of established invaders. Aquaculture has historically been important, but has become less so as the methods of shipment have changed. At the same time, the aquarium and pet trade have increased in importance as these industries have increased the number of species and volume and will continue to do so as these developing economies grow. Rates of invasion via nonhuman mechanisms such as range expansion have been increased by climate change and associated warming of sea surface temperatures. Oceanographic phenomena such as persistent parcels of warm ocean waters have permitted dispersal and survival of species hundreds of kilometers beyond their current native range. Many aspects of recipient communities also influence crustacean invasions, including biotic resistance of native species, presence of natural enemies, and the evolutionary history of natives and invaders. Other environmental factors, including fisheries exploitation, pollution, and prior invasions, can also strongly influence native communities, in many cases making them more susceptible to new invasions.

The impacts of these nonnative Crustacea are far-reaching and depend considerably on the taxa and the region of introduction. Among the most common impacts are other invertebrate taxa that they either consume or compete with for resources. Nonnative crustaceans can affect primary production and nutrient cycling through their consumption, or indirect impacts on phytoplankton and benthic macrophytes and macroalgae. Other impacts include compromising structural aspects of the habitat including features like marsh edges and channel walls through their burrowing and foraging activities. Crustacean invaders may also be vectors for pathogens and parasites that have had substantial impacts on native species, commercial fisheries, and human health. Climate change is likely to affect the future movement of new parasites and pathogens and also likely exacerbate their impacts. Efforts to mitigate these impacts via management programs typically aimed at reducing invader populations have increasingly moved away from reactive strategies of eradication, containment, and suppression toward approaches that emphasize prevention and better use of existing management resources. Effective approaches are now in place that focus on early detection and rapid response for new invasions, as well as new approaches to manage more effectively with population targets for invader suppression and avoiding pitfalls associated with eradication.

ACKNOWLEDGMENTS

I would like to thank Martin Thiel, Lars Gutow, and Miguel Penna for the opportunity to contribute a chapter to this volume. I am also thankful for their editorial contributions, which have greatly improved this manuscript. Many thanks to the authors and publishers who provided permission to use the figures and tables used herein.

REFERENCES

Alofs, K.M., and D.A. Jackson. 2014. Meta-analysis suggests biotic resistance in freshwater environments is driven by consumption rather than competition. Ecology 95:3259–3270.

Altieri, A.H., M.D. Bertness, T.C. Coverdale, N.C. Herrmann, and C. Angelini. 2012. A trophic cascade triggers collapse of a salt-marsh ecosystem with intensive recreational fishing. Ecology 93:1402–1410.

Banzai, A., H. Sugiyama, M. Hasegawa, Y. Morishima, and Y. Kawakami. 2021. *Paragonimus westermani* metacercariae in two freshwater crab species in Kagoshima Prefecture, Japan, as a possible source of infection in wild boars and sika deer. Journal of Veterinary Medical Science 83:412–418.

Barrett, K.B., J.M. Haynes, and D.I. Warton. 2017. Thirty years of change in a benthic macroinvertebrate community of southwestern Lake Ontario after invasion by four Ponto-Caspian species. Freshwater Science 36:90–102.

Beal, B.F., D. Nault, A. Hannah, P. Thayer, H. Leighton, and B. Ellis. 2016. Comparative large-scale field trials along the Maine coast to assess management options to enhance populations of the commercially important softshell clam, *Mya arenaria* L. Journal of Shellfish Research 35:711–727.

Beal, B.F., C.R. Coffin, S.F. Randall, C.A. Goodenow, K.E. Pepperman, B.W. Ellis, C.B. Jourdet, and G.C. Protopopescu. 2018. Spatial variability in recruitment of an infaunal bivalve: Experimental effects of predator exclusion on the softshell clam (*Mya arenaria* L.) along three tidal estuaries in southern Maine, USA. Journal of Shellfish Research 37:1–27.

Bojko, J., A.L. Burgess, A.G. Baker, and C.H. Orr. 2021. Invasive non-native crustacean symbionts: Diversity and impact. Journal of Invertebrate Pathology 186:107482.

Brodie, J.F., S. Lieberman, A. Moehrenschlager, K.H. Redford, J.P. Rodríguez, M. Schwartz, P.J. Seddon, and J.E.M. Watson. 2021. Global policy for assisted colonization of species. Science 372:456–458.

Bulleri, F., and M.G. Chapman. 2010. The introduction of coastal infrastructure as a driver of change in marine environments. Journal of Applied Ecology 47:26–35.

Bushnoe, T.M., D.M. Warner, L.G. Rudstam, and E.L. Mills. 2003. *Cercopagis pengoi* as a new prey item for alewife (*Alosa pseudoharengus*) and rainbow smelt (*Osmerus mordax*) in Lake Ontario. Journal of Great Lakes Research 29:205–212.

Callaway, R.M., and W.M. Ridenour. 2004. Novel weapons: Invasive success and the evolution of increased competitive ability. Frontiers in Ecology and the Environment 2:436–443.

Carlton, J.T., J.B. Geller, M.L. Reaka-Kudla, and E.A. Norse. 1999. Historical extinctions in the sea. Annual Review of Ecology and Systematics 30:515–538.

Carlton, J.T., J.W. Chapman, J.B. Geller, J.A. Miller, D.A. Carlton, M.I. McCuller, N.C. Treneman, B.P. Steves, and G.M. Ruiz. 2017. Tsunami-driven rafting: Transoceanic species dispersal and implications for marine biogeography. Science 357:1402–1406.

Carman, M.R., S.G. Bullard, and J.P. Donnelly. 2007. Water quality, nitrogen pollution, and ascidian diversity in coastal waters of southern Massachusetts, USA. Journal of Experimental Marine Biology and Ecology 342:175–178.

Chan, F.T., K. Stanislawczyk, A.C. Sneekes, A. Dvoretsky, S. Gollasch, D. Minchin, M. David, A. Jelmert, J. Albretsen, and S.A. Bailey. 2019. Climate change opens new frontiers for marine species in the Arctic: Current trends and future invasion risks. Global Change Biology 25:25–38.

Cheng, B.S., L.M. Komoroske, and E.D. Grosholz. 2017. Invasive predator and native prey response under climate change: Integrating physiological and environmental data. Functional Ecology 31:642–652.

Chucholl, C., and F. Wendler. 2017. Positive selection of beautiful invaders: Long-term persistence and bio-invasion risk of freshwater crayfish in the pet trade. Biological Invasions 19:197–208.

Coverdale, T.C., E.E. Axelman, C.P. Brisson, E.W. Young, A.H. Altieri, and M.D. Bertness. 2013. New England salt marsh recovery: Opportunistic colonization of an invasive species and its non-consumptive effects. PLoS ONE 8:e73823.

Crooks, J.A., A.L. Chang, and G.M. Ruiz. 2011. Aquatic pollution increases the relative success of invasive species. Biological Invasions 13:165–176.

Dafforn, K.A., E.L. Johnston, and T.M. Glasby. 2009. Shallow moving structures promote marine invader dominance. Biofouling 25:277–287.

D'Antonio, C.M., and P.M. Vitousek. 1992. Biological invasions by exotic grasses, the grass/fire cycle, and global change. Annual Review of Ecology and Systematics 23:63–87.

Davidson, T.M., G.M. Ruiz, and M.E. Torchin. 2016. Boring crustaceans shape the land–sea interface in brackish Caribbean mangroves. Ecosphere 7:e01430.

deRivera, C.E., E.D. Grosholz, and G.M. Ruiz. 2011. Multiple and long-term effects of an introduced predatory crab. Marine Ecology Progress Series 429:145–155.

Dexter, E., S.L. Katz, S.M. Bollens, G. Rollwagen-Bollens, and S.E. Hampton. 2020. Modeling the trophic impacts of invasive zooplankton in a highly invaded river. PLoS ONE 15:e0243002.

Diagne, C., B. Leroy, A.-C. Vaissiere, R.E Gozlan, D. Roiz, I. Jaric, J.-M. Salles, C.J.A. Bradshaw, and F. Courchamp. 2021. High and rising economic costs of biological invasions worldwide. Nature 592:571–576.

Dick, J.T.A., K. Gallagher, S. Avlijas, H.C. Clarke, S.E. Lewis, S. Leung, D. Minchin, J. Caffrey, M.E. Alexander, C. Maguire, C. Harrod, N. Reid, N.R. Haddaway, K.D. Farnsworth, M. Penk, and A. Ricciardi. 2013. Ecological impacts of an invasive predator explained and predicted by comparative functional responses. Biological Invasions 15:837–846.

Dick, J.T.A., C. Laverty, J.J. Lennon, D. Barrios-O'Neill, P.J. Mensink, J.R. Britton, V. Medoc, P. Boets, M.E. Alexander, N.G. Taylor, A.M. Dunn, M.J. Hatcher, P.J. Rosewarne, S. Crookes, H.J. MacIsaac, M. Xu, A. Ricciardi, R.J. Wasserman, B.R. Ellender, O.L.F. Weyl, F.E. Lucy, P.B. Banks, J.A. Dodd, C. MacNeil, M.R. Penk, D.C. Aldridge, and J.M. Caffrey. 2017. Invader relative impact potential: A new metric to understand and predict the ecological impacts of existing, emerging and future invasive alien species. Journal of Applied Ecology 54:1259–1267.

Dittel, A.I., and C.E. Epifanio. 2009. Invasion biology of the Chinese mitten crab *Eriochier sinensis*: A brief review. Journal of Experimental Marine Biology and Ecology 374:79–92.

Dodd, J.A., J.T.A. Dick, M.E. Alexander, C. MacNeil, A.M. Dunn, and D.C. Aldridge. 2014. Predicting the ecological impacts of a new freshwater invader: Functional responses and prey selectivity of the 'killer shrimp', *Dikerogammarus villosus*, compared to the native *Gammarus pulex*. Freshwater Biology 59:337–352.

Dumont, C.P., C.F. Gaymer, and M. Thiel. 2011. Predation contributes to invasion resistance of benthic communities against the non-indigenous tunicate *Ciona intestinalis*. Biological Invasions 13:2023–2034.

Dvoretsky, A.G., and V.G Dvoretsky. 2015. Commercial fish and shellfish in the Barents Sea: Have introduced crab species affected the population trajectories of commercial fish? Reviews in Fish Biology and Fisheries 25:297–322.

Early, R., B.A. Bradley, J.S. Dukes, J.J. Lawler, J.D. Olden, D.M. Blumenthal, C.M. D'Antonio, P. Gonzalez, E.D. Grosholz, I. Ibanez, L.P. Miller, C.J.B. Sorte, and A.J. Tatem. 2016. Invasive species in the 21st century: Global threats and national response capabilities. Nature Communications 7:12485.

Elton, C.S. 1958. The ecology of invasions by plants and animals. Methuen, London, UK.

Epanchin-Niell, R.S., and A. Hastings. 2010. Controlling established invaders: Integrating economics and spread dynamics to determine optimal management. Ecology Letters 13:528–541.

Falk-Petersen, J., P. Renaud, and N. Anisimova. 2011. Establishment and ecosystem effects of the alien invasive red king crab (*Paralithodes camtschaticus*) in the Barents Sea – a review. ICES Journal of Marine Science 68:479–488.

Floerl, O., T.K. Pool, and G.J. Inglis. 2004. Positive interactions between nonindigenous species facilitate transport by human vectors. Ecological Applications 14:1724–1736.

Fuhrmann, M.M., T. Pedersen, and E.M. Nilssen. 2017. Trophic niche of the invasive red king crab *Paralithodes camtschaticus* in a benthic food web. Marine Ecology Progress Series 565:113–129.

Galil, B.S. 2011. The alien crustaceans in the Mediterranean Sea: An historical review. Pages 377–401 *in* B.S. Galil, P.F. Clark, and J.T. Carlton, editors. In the wrong place – alien marine crustaceans: Distribution, biology and impacts, Volume 6: Invading nature – Springer series in invasion ecology. Springer, Dordrecht, The Netherlands.

Galil, B.S., S. Nehring, and V. Panov. 2008. Waterways as invasion highways—Impact of climate change and globalization. Pages 59–74 *in* W. Nentwig, editor. Ecological studies, Volume 193: Biological invasions. Springer, Berlin, Heidelberg, Germany.

Gallardo, B., M. Clavero, M.I. Sanchez, and M. Vila. 2016. Global ecological impacts of invasive species in aquatic ecosystems. Global Change Biology 22:151–163.

Gebruk, A., A.K. Zalota, P. Dgebuadze, Y. Ermilova, V.A. Spiridonov, N. Shabalin, L.A. Henry, S.F. Henley, and V.O. Mokievsky. 2020. Trophic niches of benthic crustaceans in the Pechora Sea suggest that the invasive snow crab *Chionoecetes opilio* could be an important competitor. Polar Biology 44: 57–71.

Georgiades, E., C. Scianni, I. Davidson, M.N. Tamburri, M.R. First, G. Ruiz, K. Ellard, M. Deveney, and D. Kluza. 2021. The role of vessel biofouling in the translocation of marine pathogens: Management considerations and challenges. Frontiers in Marine Science 8:660125.

Goedknegt, M.A., S. Bedolfe, J. Drent, J. van der Meer, and D.W. Thieltges. 2018. Impact of the invasive parasitic copepod *Mytilicola orientalis* on native blue mussels *Mytilus edulis* in the western European Wadden Sea. Marine Biology Research 14:497–507.

Glude, J.B. 1955. The effects of temperature and predators on the abundance of the softshell clam, *Mya arenaria*, in New England. Transactions of the American Fisheries Society 84:13–26.

Gorokhova, E., S. Hansson, H. Hoglander, and C.M. Andersen. 2005. Stable isotopes show food web changes after invasion by the predatory cladoceran *Cercopagis pengoi* in a Baltic Sea Bay. Oecologia 143:251–259.

Green, B.S., C. Gardner, A. Linnane, and P.J. Hawthorne. 2010. The good, the bad and the recovery in an assisted migration. PLoS ONE 5:e14160

Green, S.J., and E.D. Grosholz. 2021. Functional eradication as a framework for invasive species control. Frontiers in Ecology and the Environment 19:98–107.

Gribben, P.E., M. Simpson, and J.T. Wright. 2015. Relationships between an invasive crab, habitat availability and intertidal community structure at biogeographic scales. Marine Environmental Research 110:124–131.

Grosholz, E.D. 2002. Ecological and evolutionary consequences of coastal invasions. Trends in Ecology and Evolution 17:22–27.

Grosholz, E.D. 2005. Recent biological invasion may hasten invasional meltdown by accelerating historical introductions. Proceedings of the National Academy of Sciences USA. 102:1088–1091.

Grosholz, E.D. 2010. Crabs. Pages 125–128 *in* D. Simberloff and M. Rejmánek, editors. Encyclopedia of invasive introduced species. University of California Press, Berkeley, CA, USA.

Grosholz, E.D., S. Lovell, E. Besedin, and M. Katz. 2011. Modeling the impacts of the European green crab on commercial shellfisheries. Ecological Applications 21:915–924.

Grosholz, E.D., G.M. Ruiz, C.A. Dean, K.A. Shirley, J.L. Maron, and P.G. Connors. 2000. The impacts of a nonindigenous marine predator in a California bay. Ecology 81:1206–1224.

Grosholz, E.D., G. Ashton, M. Bradley, C. Brown, L. Ceballos-Osuna, A. Chang, C. de Rivera, J. Gonzalez, M. Heineke, M. Marraffini, L. McCann, E. Pollard, I. Pritchard, G. Ruiz, B. Turner, and C. Tepolt. 2021.

Stage-specific overcompensation, the hydra effect and the failure to eradicate an invasive predator. Proceedings of the National Academy of Sciences USA 118:e2003955118.

Gurevitch, J., and D.K. Padilla. 2004. Are invasive species a major cause of extinctions? Trends in Ecology and Evolution 19:470–474.

Hänfling, B., F. Edwards, and F. Gherardi. 2011. Invasive alien Crustacea: Dispersal, establishment, impact and control. BioControl 56:573–595.

Hanson, E., and M. Sytsma. 2008. The potential for mitten crab *Eriocheir sinensis* H. Milne Edwards, 1853 (Crustacea: Brachyura) invasion of Pacific Northwest and Alaskan estuaries. Biological Invasions 10:603–614.

Herborg L., S.P. Rushton, A. Clare, and M.G. Bentley. 2003. Spread of the Chinese mitten crab (*Eriocheir sinensis* H. Milne Edwards) in continental Europe: Analysis of a historical data set. Hydrobiologia 503:21–28.

Herborg, L.M, S.P. Rushton, A.S. Clare, and M.G. Bentley. 2005. The invasion of the Chinese mitten crab (*Eriocheir sinensis*) in the United Kingdom and its comparison to continental Europe. Biological Invasions 7:959–968.

Holdich, D.M., J.D. Reynolds, C. Souty-Grosset, and P.J. Sibley. 2009. A review of the ever increasing threat to European crayfish from non-indigenous crayfish species. Knowledge and Management of Aquatic Ecosystems 394–395:1.

Holopainen, R., M. Lehtiniemi, H.E.M. Meier, J. Albertsson, E. Gorokhova, J. Kotta, and M. Viitasalo. 2016. Impacts of changing climate on the non-indigenous invertebrates in the northern Baltic Sea by end of the twenty-first century. Biological Invasions 18:3015–3032.

Hostert, L.E., L.M. Pintor, and J.E. Byers. 2019. Sex, size, and prey caloric value affect diet specialization and consumption of an invasive prey by a native predator. Current Zoology 65:499–507.

Howard, B.R., F.T. Francis, I.M. Cote, and T.W. Therriault. 2019. Habitat alteration by invasive European green crab (*Carcinus maenas*) causes eelgrass loss in British Columbia, Canada. Biological Invasions 21:3607–3618.

ISSG (Invasive Species Specialist Group) 2015. The Global Invasive Species Database. Version 2015.1. Accessed August 25, 2021. [https://www.iucngisd.org/gisd/].

IUCN 2021. The IUCN Red List of threatened species. Version 2021-2. [https://www.iucnredlist.org].

Kaiser, B.A., M. Kourantidou, and L. Fernandez. 2018. A case for the commons: The snow crab in the Barents. Journal of Environmental Management 210:338–348.

Karatayev A.Y., L.E. Burlakova, D.K. Padilla, S.E. Mastitsky, and S. Olenin. 2009. Invaders are not a random selection of species. Biological Invasions 11:2009–2019.

Keane, R.M., and M.J. Crawley. 2002. Exotic plant invasions and the enemy release hypothesis. Trends in Ecology and Evolution 17:164–170.

Keller, R.P., and D.M. Lodge. 2007. Species invasions from commerce in live aquatic organisms: Problems and possible solutions. BioScience 57:428–436.

Kimbro, D.L., B.S. Cheng, and E.D. Grosholz. 2013. Biotic resistance in coastal environments. Ecology Letters 16:821–833.

Kimmerer, W. 2004. Open water processes of the San Francisco Estuary: From physical forcing to biological responses. San Francisco Estuary and Watershed Science 2:1.

Lapointe, N.W.R., P.L. Fuller, M. Neilson, B.R. Murphy, and P.L. Angermeier. 2016. Pathways of fish invasions in the Mid-Atlantic region of the United States. Management of Biological Invasions 7:221–233.

Leuven, R.S.E.W., G. van der Velde, I. Baijens, J. Snijders, C. van der Zwart, H.J.R. Lenders, A. bij de Vaate. 2009. The river Rhine: A global highway for dispersal of aquatic invasive species. Biological Invasions 11:1989–2008.

Lodge, D.M., C.A. Taylor, D.M. Holdich, and J. Skurdal. 2000. Nonindigenous crayfishes threaten North American freshwater biodiversity: Lessons from Europe. Fisheries 25:7–20.

Lodge, D.M., A. Deines, F. Gherardi, D.C.J. Yeo, T. Arcella, A.K. Baldridge, M.A. Barnes, W.L. Chadderton, J.L. Feder, C.A. Gantz, G.W. Howard, C.L. Jerde, B.W. Peters, J.A. Peters, L.W. Sargent, C.R. Turner, M.E. Wittmann, and Y. Zeng. 2012. Global introductions of crayfishes: Evaluating the impact of species invasions on ecosystem services. Annual Review of Ecology, Evolution and Systematics 43:449–472.

Loureiro, T.G., P.M. Anastácio, S.L.S. Bueno, P.B. Araujo, C. Souty-Grosset, and M.P. Almerão. 2015. Distribution, introduction pathway, and invasion risk analysis of the North American crayfish *Procambarus clarkii* (Decapoda: Cambaridae) in Southeast Brazil. Journal of Crustacean Biology 35:88–96.

Lowe, S., M. Browne, S. Boudjelas, and M. De Poorter. 2000. 100 of the world's worst invasive alien species: A selection for the Global Invasive Species Database. The Invasive Species Specialist Group (ISSG) of the World Conservation Union (IUCN), Auckland, New Zealand.

Mack, R.N., D. Simberloff, W.M. Lonsdale, H. Evans, M. Clout, and F.A. Bazzaz. 2000. Biotic invasions: Causes, epidemiology, global consequences, and control. Ecological Applications 10:689–710.

Matheson, K., C.H. McKenzie, R.S. Gregory, D.A. Robichaud, I.R. Bradbury, P.V.R. Snelgrove, and G.A. Rose. 2016. Linking eelgrass decline and impacts on associated fish communities to European green crab *Carcinus maenas* invasion. Marine Ecology Progress Series 548:31–45.

McCarthy, A., L.S. Peck, K.A. Hughes, and D.C. Aldridge. 2019. Antarctica: The final frontier for marine biological invasions. Global Change Biology 25:2221–2241.

Minchin, D., S. Gollasch, A.N. Cohen, C.L. Hewitt, and S. Olenin. 2009. Characterizing vectors of marine invasion. Pages 109–116 *in* G. Rilov and J.A. Crooks, editors. Biological invasions in marine ecosystems, Volume 204: Ecological Studies (Analysis and Synthesis). Springer, Berlin, Heidelberg, Germany.

Miller, A.M., and G.M. Ruiz. 2014. Arctic shipping and marine invaders. Nature Climate Change 4:413–416.

Neckles, H. 2015. Loss of eelgrass in Casco Bay, Maine, linked to green crab disturbance. Northeastern Naturalist 22:478–500.

Neves, C.S., R.M. Rocha, F.B. Pitombo, and J.J. Roper. 2007. Use of artificial substrata by introduced and cryptogenic marine species in Paranagua Bay, southern Brazil. Biofouling 23:319–330.

Padilla, D.K., and S.L. Williams. 2004. Beyond ballast water: Aquarium and ornamental trades as sources of invasive species in aquatic ecosystems. Frontiers in Ecology and the Environment 2:131–138.

Pagenkopp-Lohan, K.M., R.C. Fleischer, K.J. Carney, K.K. Holzer, and G.M. Ruiz. 2016. Amplicon-based pyrosequencing reveals high diversity of protistan parasites in ships' ballast water: Implications for biogeography and infectious diseases. Microbial Ecology 71:530–542.

Pagenkopp-Lohan, K.M., G. Ruiz, and M. Torchin. 2020. Invasions can drive marine disease dynamics. Pages 115–138 *in* D.C. Behringer, B.R. Silliman, and K.D. Lafferty, editors. Marine disease ecology. Oxford University Press, New York, USA.

Peeler, E.J., B.C. Oidtmann, P.J. Midtlyng, L. Miossec, and R.E. Gozlan. 2011. Non-native aquatic animals introductions have driven disease emergence in Europe. Biological Invasions 13:1291–1303.

Perrings, C. 2005. Mitigation and adaptation strategies for the control of biological invasions. Ecological Economics 52:315–325.

Peterson, C. 2006. Range expansion in the northeast Pacific by an estuary mud crab—A molecular study. Biological Invasions 8: 565–576.

Piola, R.F., and E.L. Johnston. 2008. Pollution reduces native diversity and increases invader dominance in marine hard-substrate communities. Diversity and Distributions 14:329–342.

Piola, R.F., and E.L. Johnston. 2009. Comparing differential tolerance of native and non-indigenous marine species to metal pollution using novel assay techniques. Environmental Pollution 157:2853–2864.

Platvoet, D., and J.T. Dick. 2000. Invading predatory crustacean *Dikerogammarus villosus* eliminates both native and exotic species. Proceeding of the Royal Society B – Biological Sciences 267:977–983.

Polidoro, B.A., T. Brooks, K.E. Carpenter, G.J. Edgar, S. Henderson, J. Sanciangco, and D.R. Robertson. 2012. Patterns of extinction risk and threat for marine vertebrates and habitat-forming species in the Tropical Eastern Pacific. Marine Ecology Progress Series 448:93–104.

Pysek, P., and D.M. Richardson. 2010. Invasive species, environmental change and management, and health. Annual Review of Environment and Resources 35:25–55.

Ricciardi, A. 2011. Crustaceans (other). Pages 135–137 *in* D. Simberloff and M. Rejmánek, editors. Encyclopedia of biological invasions. University of California Press, Berkeley and Los Angeles, USA.

Ricciardi, A., and D. Simberloff. 2009. Assisted colonization: Good intentions and dubious risk assessment. Trends in Ecology and Evolution 24:476–477.

Ricciardi, A., and D. Simberloff. 2021. Assisted colonization risk assessment. Science 372:925.

Ricciardi, A., J.C. Iacarella, D.C. Aldridge, T.M. Blackburn, J.T. Carlton, J.A. Catford, J.T.A. Dick, P.E. Hulme, J.M. Jeschke, A.M. Liebhold, J.L. Lockwood, H.J. MacIsaac, L.A. Meyerson, P. Pyšek, D.M. Richardson, G.M. Ruiz, D. Simberloff, M. Vila, and D.A. Wardle. 2021. Four priority areas to advance invasion science in the face of rapid environmental change. Environmental Reviews 29:119–141.

Rudnick, D.A, K. Hieb, K.F. Grimmer, and V.H. Resh. 2003. Patterns and processes of biological invasion: The Chinese mitten crab in San Francisco Bay. Basic and Applied Ecology 4:249–262.

Ruiz, G., J. Carlton, E. Grosholz, and A.H. Hines. 1997. Global invasions of marine and estuarine habitats by non-indigenous species: Mechanisms, extent and consequences. American Zoologist 37:621–632.

Ruiz, G., P. Fofonoff, B. Steves, and A. Dahlstrom. 2011. Marine crustacean invasions in North America. A synthesis of historical records and documented impacts. Pages 215–250 *in* B.S. Galil, P.F. Clark, and J.T. Carlton, editors. In the wrong place – alien marine crustaceans: Distribution, biology and impacts, Volume 6: Invading Nature – Springer Series in Invasion Ecology. Springer, Dordrecht, The Netherlands.

Ruiz, G.M., T.K. Rawlings, F.C. Dobbs, L.A. Drake, T. Mullady, A. Huq, and R.R. Colwell. 2000. Global spread of microorganisms by ships – ballast water discharged from vessels harbours a cocktail of potential pathogens. Nature 408:49–50.

Sadowski, J.S., J.A. Gonzalez, S.I. Lonhart, R. Jeppesen, T.M. Grimes, and E.D. Grosholz. 2018. Temperature induced range expansion of a subtropical crab along the California coast. Marine Ecology 39:e12528.

Sanford, E., J.L. Sones, M. Garcia-Reyes, J.H.R Goddard, and J.L. Largier. 2019. Widespread shifts in the coastal biota of northern California during the 2014–2016 marine heatwaves. Scientific Reports 9:1–14.

Sardain, A., E. Sardain, and B. Leung. 2019. Global forecasts of shipping traffic and biological invasions to 2050. Nature Sustainability 2:274–282.

Scheers, K., R. Brys, T. Abeel, D. Halfmaerten, S. Neyrinck, and T. Adriaens. 2021. The invasive parthenogenetic marbled crayfish *Procambarus virginalis* Lyko, 2017 gets foothold in Belgium. Bioinvasions Records 10:326–340.

Schlaepfer, M.A., W.D. Helenbrook, K.B. Searing, and K.T. Shoemaker. 2009. Assisted colonization: Evaluating contrasting management actions (and values) in the face of uncertainty. Trends in Ecology and Evolution 24:471–472.

Seebens, H., N. Schwartz, P.J. Schupp, and B. Blasius. 2016. Predicting the spread of marine species introduced by global shipping. Proceedings of the National Academy of Sciences USA 113:5646–5651.

Sellheim, K., J.J. Stachowicz, and R.C. Coates. 2010. Effects of a nonnative habitat-forming species on mobile and sessile epifaunal communities. Marine Ecology Progress Series 398:69–80.

Shanks, A.L. 2009. Pelagic larval duration and dispersal distance revisited. Biological Bulletin 216:373–385.

Shikuma, N.J., and M.G. Hadfield. 2010. Marine biofilms on submerged surfaces are a reservoir for *Escherichia coli* and *Vibrio cholerae*. Biofouling 26:39–46.

Sih, A., D.I. Bolnick, B. Luttbeg, J.L. Orrock, S.D. Peacor, L.M. Pintor, E. Preisser, J.S. Rehage, and J.R. Vonesh. 2010. Predator-prey naïveté, antipredator behavior, and the ecology of predator invasions. Oikos 119:610–621.

Simberloff, D. 2009. We can eliminate invasions or live with them: Successful management projects. Biological Invasions 11:149–157.

Sorte, C., I. Ibanez, D. Blumenthal, N. Molinari, L. Miller, E. Grosholz, J. Diez, C. D'Antonio, J. Olden, S. Jones, and J. Dukes. 2012. Poised to prosper? A cross-system comparison of climate change effects on native and non-native species performance. Ecology Letters 16:261–270.

Souty-Grosseta, C., P.M. Anastáciob, L. Aquilonic, F. Banhab, J. Choquera, C. Chucholl, and E. Tricarico. 2016. The red swamp crayfish *Procambarus clarkii* in Europe: Impacts on aquatic ecosystems and human well-being. Limnologica 58:78–93.

Talley, T.S., J.A. Crooks, and L.A. Levin. 2001. Habitat utilization and alteration by the invasive burrowing isopod, *Sphaeroma quoyanum*, in California salt marshes. Marine Biology 138:561–573.

Thiel, M., and L. Gutow. 2005. The ecology of rafting the marine environment. II. The rafting organisms and community. Oceanography and Marine Biology: An Annual Review 43:279–418.

Thiel, M., and P.A. Haye. 2006. The ecology of rafting in the marine environment. III. Biogeographical and evolutionary consequences. Oceanography and Marine Biology: An Annual Review 44:323–429.

Thresher, R., C. Proctor, G. Ruiz, R. Gurney, C. MacKinnon, W. Walton, L. Rodriguez, and N. Bax. 2003. Invasion dynamics of the European shore crab, *Carcinus maenas,* in Australia. Marine Biology 142:867–876.

Torchin, M.E., and C.E. Mitchell. 2004. Parasites, pathogens, and invasions by plants and animals. Frontiers in Ecology and the Environment 2:183–190.

Torchin, M.E., K.D. Lafferty, A.P. Dobson, V.J. McKenzie, and A.M. Kuris. 2003. Introduced species and their missing parasites. Nature 421:628–630.

Twardochleb, L.A., J.D. Olden, and E.R. Larson. 2013. A global meta-analysis of the ecological impacts of nonnative crayfish. Freshwater Science 32:1367–1382.

van den Brink, F.W.B., G. van der Velde, and A. bij de Vaate. 1993. Ecological aspects, explosive range extension and impact of a mass invader, *Corophium curvispinum* Sars, 1895 (Crustacea: Amphipoda), in the Lower Rhine (The Netherlands). Oecologia 93:224–232.

van Riel, M.C., G. van der Velde, S. Rajagopal, S. Marguillier, F. Dehairs, and A. bij de Vaate. 2006. Trophic relationships in the Rhine food web during invasion and after establishment of the Ponto-Caspian invader *Dikerogammarus villosus.* Hydrobiologia 565:39–58.

Vanderploeg, H.A., T.F. Nalepa, D.J. Jude, E.L. Mills, K.T. Holeck, J.R. Liebig, I.A. Grigorovich, and H. Ojaveer. 2002. Dispersal and emerging ecological impacts of Ponto-Caspian species in the Laurentian Great Lakes. Canadian Journal of Fisheries and Aquatic Sciences 59:1209–1228.

Walton, W.C., and W.C. Walton. 2001. Problems, predators, and perception: Management of quahog (hard clam), *Mercenaria mercenaria,* stock enhancement programs in southern New England. Journal of Shellfish Research 20:127–134.

Williams, S.L., I.C. Davidson, J.R. Pasari, G.V. Ashton, J.T. Carlton, R.E. Crafton, R.E. Fontana, E.D. Grosholz, A.W. Miller, G.M. Ruiz, and C.J. Zabin. 2013. Managing multiple vectors for marine invasions in an increasingly connected world. BioScience 63:952–966.

Williams, S.L., R.E. Crafton, R.E. Fontana, E.D. Grosholz, G. Ha, J.R. Pasari, and C.J. Zabin. 2014. A vector analysis of marine ornamental species in California. Management of Biological Invasions 6:13–29.

Witte, F., T. Goldschmidt, J. Wanink, M. van Oijen, K. Goudswaard, E. Witte-Maas, and N. Bouton. 1992. The destruction of an endemic species flock: Quantitative data on the decline of the haplochromine cichlids of Lake Victoria. Environmental Biology of Fishes 34:1–28.

Xin, K., Z. Xie, C. Zhong, N. Sheng, C. Gao, and X. Xiao. 2020. Damage caused by *Sphaeroma* to mangrove forests in Hainan, Dongzhaigang, China. Journal of Coastal Research 36:1197–1203.

Zavaleta, E.S., R.J. Hobbs, and H.A. Mooney. 2001. Viewing invasive species removal in a whole-ecosystem context. Trends in Ecology and Evolution 16:454–459.

14

HABITAT-BASED CONSERVATION OF ANCHIALINE CAVE CRUSTACEANS

Thomas M. Iliffe, Alejandro Martínez, Fernando Alvarez, and William F. Humphreys

Abstract

Anchialine ecology is a relatively new endeavor, dating to the early 1980s with the application of technical cave diving methodology to scientific investigations of underwater cave ecosystems. The first major discoveries involved the new class Remipedia from blue holes in the Bahamas and the new order Mictacea from Bermuda caves. Anchialine stygobiont crustaceans include representatives from the Copepoda, Ostracoda, Procarididea, Caridea, Anomura, Isopoda, Amphipoda, Ingolfiellida, Thermosbaenacea, Tanaidacea, Mictacea, Bochusacea, Cumacea, Stygiomysida, Mysida, and Remipedia. Common threats to anchialine habitats are construction and mining activities, sewage or wastewater disposal, solid waste dumps, overpumping of groundwater, disturbance of salinity-stratified water layers by divers, invasive species, and climate change. Principal conservation initiatives involve the creation of parks and nature reserves, enacting of laws and regulations to protect anchialine species and associated caves, education of the public on the importance of cave habitats, and addition of cave organisms to endangered species lists. This chapter focuses on strategies to protect threatened subterranean habitats with unique endemic crustacean populations. Case studies from Bermuda, Lanzarote (Canary Islands), Quintana Roo (Mexico), and Western Australia were reviewed. The occurrence of exceptional crustacean assemblages with valuable ecological function has led to the classification and protection of valuable habitats as management units and reconition of keynote crustacean species including the mictacean *Mictocaris halope* (Bermuda), squat lobster *Munidopsis polymorpha* (Lanzarote), thermosbaenacean *Halosbaena tulki* (Western Australia), and remipedes *Kumonga exleyi* (Western Australia), *Xibalbanus tulumensis* (Yucatán), and *Morlockia ondinae* (Lanzarote). Adequate protection of anchialine crustaceans is essential in maintaining unique components of crustacean biodiversity.

Thomas M. Iliffe et al., *Habitat-Based Conservation of Anchialine Cave Crustaceans*. In: *Ecology and Conservation*.
Edited by: Lars Gutow and Martin Thiel, Oxford University Press. © Oxford University Press (2025).
DOI: 10.1093/oso/9780197768242.003.0014

INTRODUCTION

Anchialine habitats were originally defined based on distinctive shrimp found in coastal, landlocked pools (Holthuis 1973). Such salinity-stratified pools occur in volcanic and limestone terrain from tropical and subtropical regions around the world. Subsequent scuba diving explorations led to the discovery of extensive submerged cave systems inhabited by a diverse cave-adapted fauna dominated by crustaceans (Iliffe et al. 1983, 1984b). New, higher crustacean taxa discovered and described from anchialine caves include the class Remipedia (Yager 1981) and the orders Mictacea (Bowman et al. 1985) and Bochusacea (Gutu and Iliffe 1998), as well as many novel families, genera, and species.

Due to their proximity to the coast, anchialine caves and groundwater have come under severe pressure from tourist, residential, and commercial development. Threats to the anchialine habitat arise from sewage and wastewater disposal, solid waste dumps, rock quarries, construction, invasive species, and climate change (Mammola et al. 2019, Wynne et al. 2021). Unfortunately, numerous caves were destroyed or polluted before they could be explored or scientifically investigated (see, e.g., Iliffe 1979, Gonzalez et al. 2020, Iliffe and Calderón-Gutiérrez 2021). Since many anchialine taxa occur in just a single cave or a few adjacent caves, demolition or pollution can result in the extinction of numerous, yet unknown species. Therefore, it is critical that proactive steps are taken to explore, scientifically investigate, protect, and preserve the anchialine caves and groundwater (Mammola et al. 2022). The aim of this chapter is to present case studies from four globally significant regions (Bermuda, the Canary Islands, Western Australia, and the Yucatán Península; Fig. 14.1) with a discussion of their anchialine caves and groundwater, notable crustacean fauna (Appendix 14.1), environmental threats, and conservation action plans.

WESTERN AUSTRALIA

Bundera Sinkhole, Australia (22° 24′ 44.5″ S, 113° 45′ 54.6″ E)

The fauna of Bundera Sinkhole forms a significant component of the Outstanding Universal Value of the Ningaloo Coast World Heritage Area (inscribed in June 2011). Bundera Sinkhole

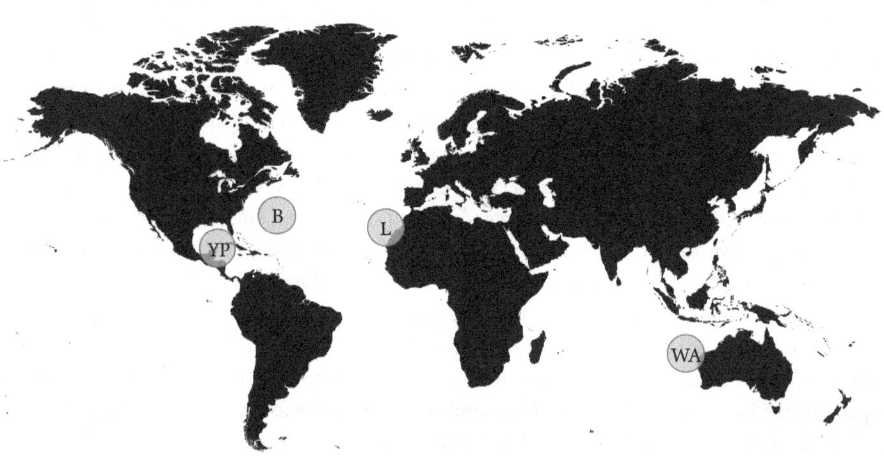

Fig. 14.1.
World map with the locations of the four study sites: Yucatán Península (YP), Bermuda (B), Lanzarote (L), and Western Australia (WA). Note: the selected world map omitted the Antarctica continent to improve the visual scale for the points of interest (original figure under CC license Wikimedia commons).

(Fig. 14.2) is a global hotspot of subterranean biodiversity supporting more than 16 stygobiont species (i.e., obligate groundwater fauna) of which 10 species are endemic to Bundera, with 9 being crustaceans (Appendix 14.1). Noteworthy are the stygiobiont crustaceans with a full Tethyan distribution (Stock 1993) belonging to genera or higher taxa also known from Lanzarote (Carnary Islands) or the Caribbean, especially the remipede *Kumonga exleyi* (Yager and Humphreys 1996), ostracod *Welesia kornickeri* (Danielopol et al. 2000), thermosbaenacean *Halosbaena tulki* Poore and Humphreys 1992, and isopod *Bunderanthura bundera* Poore and Humphreys 2013. The site (termed the Cape Range Remipede Community, Carnarvon Basin) is listed as a threatened ecological community under Western Australia legislation [CR B) ii)—Critically Endangered] and as vulnerable under the Environment Protection and Biodiversity Conservation Act 1999 (Commonwealth of Australia legislation).

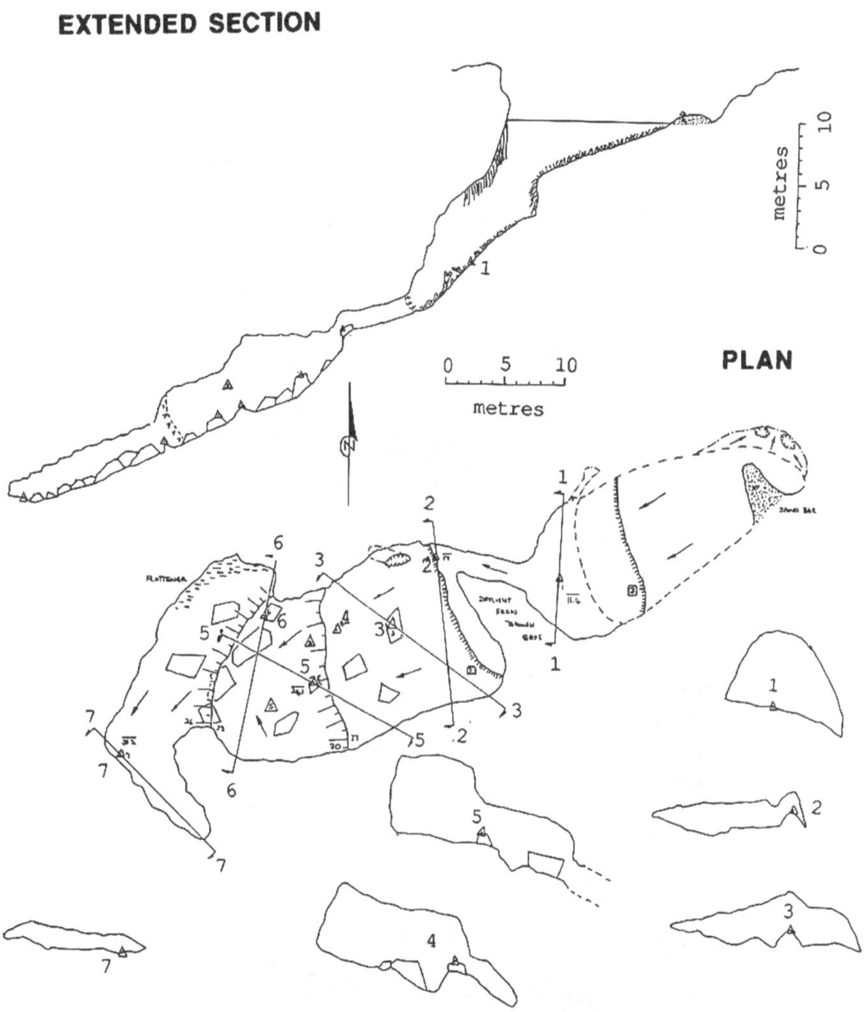

Fig. 14.2.
Plan and extended section map with cross-sectional views of Bundera Sinkhole. This anchialine cave on the Cape Range peninsula of northwestern Australia is 33 m deep, with 19 and 32 g/L haloclines at 6.1 and 15.9 m, respectively (Humphreys 1999). Reproduced from Poole and Warren (1995), with permission from © The Royal Society of Western Australia.

Threats to Bundera Sinkhole's Remipede community include disturbance and mixing of the water layers by divers, dumping of rubbish or toxic substances (Fig. 14.3), introduction of exotic species of fish, and eutrophication or pollution. Any long-term decline in groundwater quality or aquifer water levels would also be a serious concern. Considering the fragile nature of this ecosystem, swimming and diving are strictly prohibited.

Clades of the thermosbaenacean *Halosbaena* (Fig. 14.4), which occur throughout the Tethyan realm (Poore and Humphreys 1992), are found widely in the Pilbara region of Australia and on

Fig. 14.3.
Removal of a truck that crashed into Bundera Sinkhole in Western Australia. Only a small oil spill needed clean-up. Photo reproduced courtesy of Bill Humphreys.

Fig. 14.4.
Halosbaena tulki is the only described member of the order Thermosbaenacea known from the Southern Hemisphere. It is widespread in ground waters of the Cape Range peninsula. Photo reproduced courtesy of D. Elford for Western Australian Museum.

Christmas Island (Page et al. 2016). *Halosbaena* inhabits deep groundwater (Exmouth town water supply), caves, and sinkholes, as well as the basal gravels of episodic watercourses of the Pilbara plateau and coastal plain. Concern that borefield drawdown of the Exmouth town water supply would deplete the contained stygofauna (Humphreys 1995) was eased when intensive test pumping showed no entrainment of stygofauna into the pumped water (Humphreys 1996a, 1996b). This indicates that stygofauna actively move against the current at the periphery of the entraining flow.

Australian Arid Zone Aquifers

The Pilbara in the north of Western Australia (Fig. 14.5) is an arid region with abundant fresh groundwater, inhabited by a globally related, highly diverse stygofauna comprising about 350 species, of which ~80% are crustaceans, and with up to 54 stygobiont species occurring in one borehole (Guzik et al. 2011, Halse et al. 2014). Calcrete aquifers separated by about 350 km in the Fortescue River valley contain two of the four species known globally of the order Spelaeogriphacea, the others occurring in sandstone in caves on top of Table Mountain, South Africa, and the karstic western Mato Grosso do Sul, Brazil, both sites being in conservation areas. These sites all occur in geological formations of Cretaceous age or older (145–66 Ma BP), predating the breakup of SW Gondwana, which rifted over 40 Ma from the Early Jurassic into the Early Cretaceous (178–127 Ma; Muir et al. 2020).

Both Australian sites are subjected to the impact of large-scale water pumping or are downstream of major opencast mining operations. In the Murchison region, about 800 km south, three clades of chiltoniid amphipods comprise at least 13 morphospecies (Cooper et al. 2007) and seven genera among the different calcrete aquifers (Bradford et al. 2010) with three sympatric genera occurring in a single calcrete (King et al. 2012).

The stygobiont isopod genus *Haloniscus* (Philosciidae) includes at least 13 morphospecies, mostly distributed among different calcrete aquifers of the Murchison district and the Ngalia Basin, Northern Territory, in the Australian arid zone (Taiti and Humphreys 2001, Cooper et al. 2008, Guzik et al. 2019). Eyed species occur in mound springs of South Australia fed by the Great Artesian Basin aquifer (Stringer et al. 2019).

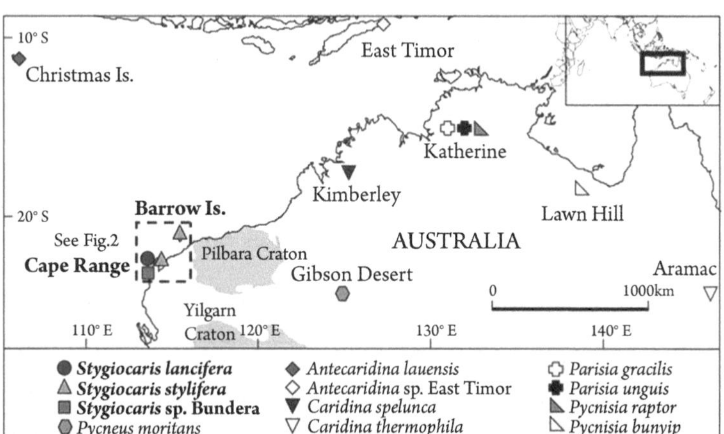

Fig. 14.5.
Map of collection locations of subterranean atyids in the Australian region. Reproduced from Page et al. (2008) under Creative Commons by Attribution License (CC BY).

Half of the stygobiont species in the Pilbara have ranges <680 km^2, so development projects involving extensive water drawdown may deplete stygobiont populations or threaten the survival of species with small ranges (Halse et al. 2014). Stygobionts inhabiting calcrete aquifers are vulnerable to several threatening processes.

Groundwater calcrete deposits are typically ~8–10 m thick (Jacobson and Arakel 1986), so reduced rainfall, perhaps due to climate change, or groundwater pumping will likely result in many calcrete aquifers drying up due to the water table falling below the level of the calcrete. The habitat itself is vulnerable because calcrete is a resource that is widely used as road fill, but it is used more intensively to neutralize acids in mineral processing in this mineraliferous region and, to this end, vast areas are mined, with one mine in the Murchison extracting calcrete in an area of ~40 km^2 over several decades. Uranium also concentrates from the water in groundwater calcretes, and uranium itself becomes the resource targeted (Cameron 1984).

Barrow Island lies off the northwest coast of Australia but was connected to the mainland during the Pleistocene low seas. Numerous oilfield-related boreholes dot the karst surface, penetrating the major freshwater lens atop the seawater intrusion. The lens supports diverse stygofauna comprising at least 63 species, of which 55 are crustaceans (Humphreys 2000, Humphreys et al. 2013, Larson et al. 2013). The lens is subjected to water extraction to support oilfield operations and receives contaminants arising from petroleum exploration and development, and from the production oilfield (Humphreys 2002): oil entering karst systems, while being invisible, is persistent (Maksimovich and Khmurchik 2015).

Christmas Island in the northern Indian Ocean is an ancient basaltic volcano of 135 km^2 with a karst limestone capping (Grimes and Humphreys 1998), of which 85 km^2 is a Commonwealth of Australia National Park. The island is buffered from anthropogenic impacts (save climate change) by its remoteness, forest cover, and status as a national park, despite which the island's ecology is suffering from past phosphate mining and, especially, invasive species (Beeton et al. 2010). Christmas Island is notable for the presence of 14 species of land crabs, especially the mass migration of tens of millions of red crabs, *Gecarcoidea natalis,* between rainforest and ocean associated with breeding (Shaw and Kelly 2013), as well as a large population of the coconut crab, *Birgus latro.*

The groundwaters of Christmas Island support stygofauna of great biogeographical and evolutionary significance, possibly relicts of the Mesozoic Tethys Sea 225 to 65 million years ago (Namiotko et al. 2004, Humphreys et al. 2009, James et al. 2019). The coastal fringe supports a few endemic stygobionts comprising the thermosbaenacean *Halosbaena* (Fig. 14.5), the shrimp *Metabetaeus,* and the thaumatocyprid ostracod *Humphreysella* (Humphreys and Eberhard 2001, Humphreys and Danielopol 2006, Kornicker et al. 2006, Humphreys 2014, James et al. 2019). The freshwaters are inhabited by the endemic cytherurid ostracod *Microceratina martensi,* the first record of a complete specimen of the genus (Namiotko et al. 2004), and diverse darwinulid ostracods including *Isabenula* and *Vestalenula* (Rossetti et al. 2011). The evolutionary and biogeographical significance of some endemic species was emphasized by James et al. (2019). For example, the ostracods referred to above and *Halosbaena* and the cave shrimp *Procaris* (Bruce and Davie 2006) represent lineages of great antiquity and remarkably odd biogeography (Page et al. 2018).

In the monsoonal Kimberley region of Western Australia, an emergent Devonian limestone reef forms the limestone ranges, including the Napier, Oscar, and Ningbing Ranges (Playford et al. 2009). The area is rich in lead-zinc ores typical of Mississippi Valley-type deposits (Leach et al. 2010) underlying the limestone ranges. The karst features include caves and tunnels, which support some ancient relictual crustaceans including the Western Australian endemic isopod family Tainisopidae (Wilson and Ponder 1992), with a related genus in the Pilbara (Wilson 2003). The area is increasingly being incorporated within conservation reserves and indigenous protected areas.

Conservation for Western Australia

The faunas of many calcrete aquifers are listed as a threatened ecological community (DBCA 2021), including freshwater subterranean aquifers of Barrow Island, the Bungaroo hyporheic, and the stygofaunal communities of the Western Fortescue Plains freshwater aquifer. Forty-four discrete calcrete aquifer systems in the Gascoyne and Murchison regions of Western Australia are classified as Priority 1 by the DBCA (2021). As such they are rated as possibly threatened ecological communities that do not meet survey criteria or that are not adequately defined but are ranked in the highest priority for survey and/or definition of the community, and evaluation of conservation status, so that consideration can be given to their declaration as threatened ecological communities.

The Devonian reefs in the Kimberley region are increasingly being incorporated within conservation reserves under the umbrella of the Balili Conservation Park (Devonian Reef Conservation Park). About 63% of Christmas Island's land area is incorporated within the Christmas Island National Park run by the Australian National Parks, an Australian Commonwealth agency. The Western Australian Government has a subterranean Fauna Management Plan "to maintain stygofauna and troglofauna representation, diversity, viability and ecological function at the species, population and assemblage level." The legislation provides for the protection of subterranean fauna through three Acts of State and Commonwealth parliaments, namely the Wildlife Conservation Act 1950 (Western Australia), the Environmental Protection Act 1986 (Western Australia), and the Environment Protection and Biodiversity Conservation Act 1999 (Commonwealth of Australia).

To implement this legislation, the Environmental Protection Authority of Western Australia developed two key documents to guide the assessment process for activities that could potentially impact subterranean fauna (EPAWA 2016, 2021). These documents provide advice to proponents and the public on the EPAWA's minimum requirements for environmental impact assessment and management of subterranean fauna.

BERMUDA

Geology and Caves

Bermuda is a group of oceanic islands atop a seamount in the Northwestern Atlantic, lying 1000 km east of North Carolina, United States, at 32° 20′N, 64° 45′W (Fig. 14.6A). Warm Gulf Stream water passing near Bermuda produces the island's subtropical climate and allows for the presence of the northernmost coral reefs in the Atlantic. The islands of Bermuda are on the southeastern margin of a shallow, reef-rimmed platform atop a long-extinct, Meso-Cenozoic volcanic seamount (Vogt and Jung 2007). The Bermuda volcano is exceptional in that it is formed from material rising from deep in the transition zone between the upper and lower mantle, rather than from hotspots along a crack between tectonic plates (Mazza et al. 2019).

A 15–30-m-thick layer of Pleistocene age limestone caps the top of the Bermuda seamount. The main islands have a core of older limestone, fronted on the seaward sides by prominent ridges of younger limestone. During periods of higher sea level, the Bermuda platform was submerged, allowing production of reef-derived carbonate sand. As glacial sea levels fell, reaching −134 m at the Last Glacial Maximum (Lambeck et al. 2014), winds blew loose sand on the exposed platform top into large dunes. Rainwater percolating through the sand dunes lithified them into the eolian limestone that covers the island today. This limestone has undergone extensive weathering and solution to produce a characteristic karst landscape (Bretz 1960, Palmer et al. 1977) with more than 150 caves.

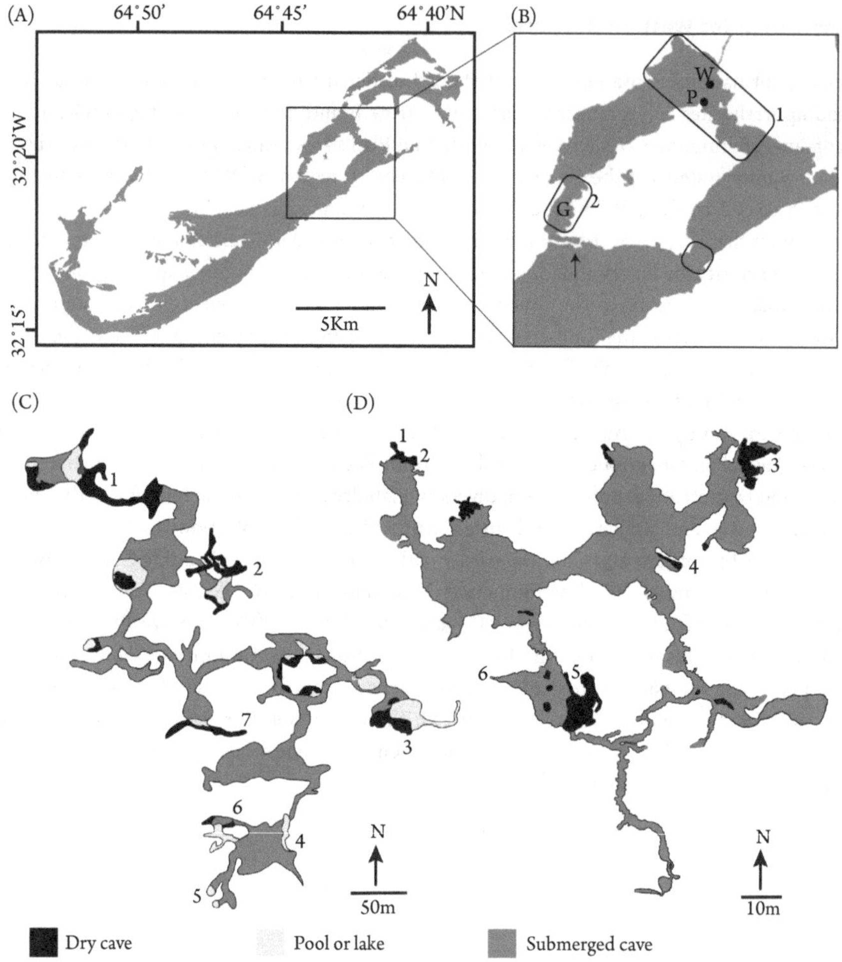

Fig. 14.6.
(A) Map of Bermuda in the North Atlantic Ocean. (B) Cave regions on opposite sides of Harrington Sound, a large inland body of water mostly encircled by land. The arrow indicates the only opening to the sea at Flatt's Inlet; rectangles show cave and karst regions in (1) the Walsingham Tract with the Walsingham Cave (W) and Palm Cave (P), and (2) Green Bay Cave System with the Green Bay Cave (G). (C) Map of Walsingham Cave System with cave entrances: (1) Wonderland, (2) Crystal, (3) Walsingham, (4) Deep Blue, (5) Vine, (6) Old Horse, (7) Fern Sink (modified from map by Robert Power). (D) Map of Palm Cave System with cave entrances: (1) Palm Slit, (2) Palm, (3) Strawmarket, (4) Sailor's Choice, (5) Myrtle Bank, (6) Cripple Gate. Modified from map by Jason Richards. Reproduced from Iliffe and Calderón-Gutiérrez (2021), with permission from © MDPI under Creative Commons by Attribution License (CC BY 4.0); see a color version of this figure in the centerfold.

Caves developed as water flowed downward through cracks and joints in the limestone until reaching an impermeable layer. This point occurs either at sea level, where non-saturated freshwater meets seawater that is already saturated with calcium carbonate, or at the contact point between soluble limestone and insoluble volcanic basalt (~30 m below present sea level). Most of Bermuda's caves contain rubble mounds formed by roof collapse. In this way, caves are migrating upward from the level where they were initially formed. Eventually, roof collapse produces the sinkhole entrances characteristic of Bermuda caves (Chandler 2006).

Most caves developed in early Pleistocene age Walsingham limestone, the oldest of Bermuda's five limestone formations (Vacher et al. 1989, Vacher and Rowe 1997). This rock is especially dense and durable, resulting from the tight cementation of the original carbonate sand grains (Rowe 2020).

The Walsingham limestone formation lies along the narrow strip of land separating Harrington Sound and Castle Harbor (Fig. 14.6A, B). Harrington Sound is a large inland body of saltwater, 3.5 km long and 2.3 km wide, surrounded by land with only a small channel connecting it to the sea. Due to its restricted connection to the sea, tides in Harrington Sound are damped, occurring 2½ to 3 hours later, with 30% of the range of comparable open ocean tides (Morris et al. 1977). Because of this delay, the tides in Harrington Sound are out of sync with those in the ocean or in Castle Harbor, and the pressure differential alternately pushes or pulls water through submerged cave passages, the flow changing direction every six hours. This tidal flushing was likely a prominent factor in forming caves specifically in those areas where the border between Harrington Sound and the sea is the narrowest.

Bermuda's longest cave is the submerged Green Bay Cave with two entrances and 2+ km of surveyed passages. Large diameter tunnels at 18 m depth cross beneath the peninsula separating Harrington Sound from Bermuda's North Shore, while smaller passageways form an anastomosing complex. Massive stalactites and stalagmites in the underwater cave provide clear proof that the cave was dry and air-filled for long periods during the Ice Ages (Fig. 14.7).

The Walsingham (7 entrances, 1.3 km long) and Palm (8 entrances, 1.5 km) Cave Systems carry tidal flow between Harrington Sound and Castle Harbour (Iliffe and Calderón-Gutiérrez 2021; Fig. 14.6B, C). Crystal and Fantasy Caves in the Walsingham System have been operated as major tourist attractions for nearly a century. A floating pontoon bridge allows visitors to cross the large anchialine lake in Crystal Cave, while a sightseeing path runs along the edge of a lake in Fantasy Cave.

Fig. 14.7.
UW photo of divers and large speleothems in the Walsingham Cave System. Photo reproduced courtesy of Jill Heinerth; see a color version of this figure in the centerfold.

Cave diving explorations (Fig. 14.7) have been ongoing in Bermuda since the early 1980s and have interconnected many formerly isolated cave pools into large networks of submerged cave passages (Iliffe 2003). Although divers have explored many of Bermuda's anchialine caves, likely numerous others remain to be discovered. Until now, all known underwater cave passages lie beneath the island and do not extend past the coastline or under the seafloor.

Anchialine Cave Habitat

Bermuda's anchialine cave habitats contain brackish to marine waters with dampened and delayed tides. Except at coastal cave entrances where tidally exchanging water enters and leaves on approximately six-hour cycles, water currents are negligible. The cave water column is characterized by gradually increasing salinity and warming water temperature instead of the well-defined halo/thermoclines observed in anchialine caves from Yucatán (Mexico) or the Bahamas. Dissolved oxygen levels decline to about 3.5 mg/l, while the pH of approximately 7.6 stabilizes in deeper water. During winter, colder water sinks to depressions on the cave floor where it persists over the summer. In summer, warmer and less dense water accumulates in domes on the cave ceiling where the normally dark walls of the cave are bleached white (Iliffe 2003).

The caves have low productivity, high levels of endemism, and truncated, but diverse trophic linkages. The fauna includes suspension feeders, detritivores, and a few predators with heterotrophic marine microbes and macroinvertebrates (Iliffe et al. 2000, 2020). Biological zonation is evident with increasing distances farther into the Green Bay Cave from the submerged entrance on the coast of Harrington Sound. Close to the entrance, brightly colored sponges, hydroids, tunicates, and other encrusting organisms cover the walls and cave ceiling. Because of lack of light, decreasing tidal currents, and particulate matter suspended in the water, the density of these organisms declines with distance into the cave. In the much clearer, low particulate waters of the deep cave interior, cave-limited (stygobiont) species predominate. These latter species are specially adapted to life in caves, thus lacking or with only reduced eyes and pigmentation, elongated body, and appendages, and enhanced non-visual sensory receptors.

Bermuda caves represent a global hotspot for anchialine stygobiont species (Iliffe and Calderón-Gutiérrez 2021). The fauna from Bermuda's Walsingham track includes 78 cave-dwelling species, of which 66 are crustaceans: 20 species of Copepoda (of which 16 are endemic), 19 Ostracoda (17 endemic), 7 Amphipoda (5 endemic), 5 Cumacea (2 endemic), 5 Decapoda (3 endemic), 4 Isopoda (4 endemic), 2 Mysidacea (2 endemic), 2 Tanaidacea (1 endemic), 1 Mictacea (1 endemic), and 1 Ingolfiellida (1 endemic).

Many of Bermuda's anchialine crustaceans are noteworthy in terms of their taxonomic novelty and/or their biogeographic affinity to related species on opposite sides of oceans, or even opposite sides of the Earth. *Mictocaris halope* (Bowman and Iliffe 1985; Fig. 14.8) is the sole species in the peracrid crustacean order Mictacea. Its relationships are unclear but may include ties to the deep-sea, freshwater cave, and anchialine orders Bochusacea and Spelaeogriphacea (Poore 2015).

The stygiobiont halocyprid ostracod genus *Spelaeoecia* includes seven species from the Bahamas, two from Cuba, and one each from Jamaica, the Yucatán Península, and Bermuda. The anchialine misophrioid copepod genus *Speleophria* includes two species from Western Australia, and a single species from each of the Yucatán Península (Mexico), the Balearic Islands, Croatia, Palau, and Bermuda. The decapod genus *Procaris* inhabits anchialine caves and pools in Hawaii (mid-Pacific), Christmas Island (Indian Ocean), Yucatán Península (Caribbean Sea), Ascension Island (South Atlantic), and Bermuda (North Atlantic). Such biogeographic affinities suggest ancient origins followed by vicariant dispersal through plate tectonics and seafloor spreading (Iliffe et al. 1983).

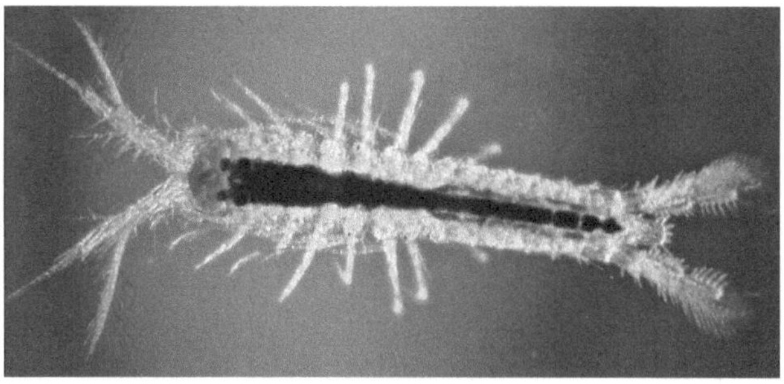

Fig. 14.8.
Mictocaris halope Bowman and Iliffe, 1985 from the monotypic peracarid crustacean order Mictacea known only from anchialine caves in Bermuda. Photo reproduced courtesy of Peter Parks, under Creative Commons license Wikimedia commons.

Several noteworthy stygiobiont crustaceans characteristically inhabiting anchialine caves from the Caribbean, East Atlantic, or Mediterranean regions are absent from Bermuda, despite extensive exploration. The reasons why Remipedia, Stygiomysida, and Thermosbaenacea do not occur in Bermuda's caves are likely significant but are as yet unknown.

Threats to Bermuda Caves and Cave Fauna

Bermuda is among the world's most densely populated small island nations with 1309 inhabitants per km^{-2} in 2017 (Ritchie 2019). As such, issues of (1) construction and quarrying, (2) water pollution, (3) trash disposal, (4) vandalism, and (5) climate change threaten Bermuda's caves and anchialine fauna. Most of Bermuda's cave species have extremely restricted distributions, with many known only from a single cave system. Thus, pollution or destruction of a single cave could result in the extinction of multiple endemic species. Numerous Bermuda caves have likely disappeared due to quarry and construction activities before they were explored or studied, considering that two large limestone quarries and a hotel/golf course lie at the north and south ends of the Walsingham region where most Bermuda caves are located.

Government Quarry Cave was discovered during quarrying operations in 1969 but was destroyed when quarrying resumed in the mid-1980s. The cave contained two pools, one of which was initially explored to a complex fissure system reaching depths of 24 m, making it one of the deepest underwater caves in Bermuda. However, before diving exploration could continue, large amounts of refuse and debris were bulldozed into the cave pool to fill it before the quarrying away of the overlying rock (Iliffe et al. 1984a). Because of this dumping, water in the Government Quarry Cave and several adjacent caves turned anoxic and sulfurous. Geological drilling operations at Government Quarry also encountered a large cave at 18 m below sea level. The drill bit subsequently dropped to 33 m where volcanic bedrock was recovered. This would imply that caves in the Government Quarry area are the deepest in Bermuda and have been developed at the island's limestone basalt interface, averaging about 30 m below the present sea level.

Wilkinson Quarry Cave (Fig. 14.9), which lacked a natural entrance, was uncovered during blasting operations in 2002 (Darrell 2020). Its existence became widely known when a quarry worker took a large and freshly broken stalagmite into a local bar. Soon thereafter, Bermuda Government officials inspected the cave, finding an inclined fissure descending a steep flowstone slope to a tidal, clear water pool. The cave was damaged by the quarrying operations when the artificial

Fig. 14.9.
(Top) Rock mound containing remnants of Wilkinson Quarry Cave at the center of the quarry floor (circa August 2017). (Bottom) After removal of the rock mound (May 2023). Photos reproduced courtesy of Thomas Iliffe; see a color version of this figure in the centerfold.

entrance was opened and by blasting, which shattered large columns and split apart two-meter-thick flowstone slabs. When going to the cave a week later, the Government Conservation Officer saw large stalactites, which were intact on his first visit, were now lying smashed on the cave floor. Exploratory cave dives found the flowstone floored fissure continued to 15 m water depth, before coming back up at another fissure that ended in an unstable air chamber, choked off by fresh collapse. During the dives, biological collections identified three species of endemic crustaceans listed as critically endangered by the International Union for Conservation of Nature and covered under the Bermuda Protected Species Act. While biologists and government environmental officials argued that the cave is of exceptional environmental and ecological importance, consultants hired by quarry management proposed the "demolition of an unstable mound with a subterranean void in Wilkinson Quarry" since the cave in it was small, had been structurally compromised by blasting, and was ecologically or aesthetically not significant. Ultimately, the Bermuda Development

Applications Board approved the destruction of the cave and the removal of the surrounding bedrock to the level of the lower quarry floor. The fate of the cave's endemic crustaceans is uncertain since access to the quarry is restricted, and it is unknown what, if anything, will remain of the cave after quarrying is completed.

Due to the lack of central sewage treatment for most of Bermuda, septic tanks are required for developments in cave areas (Government of Bermuda 2022). Effluent from the tanks flows into sealed boreholes, reaching depths of 12 to 33.5 m below sea level, where most anchialine caves occur. Without scientifically trained cave divers, the impact of sewage pollution cannot be assessed.

Multiple caves in Bermuda have been used as dumps for solid and liquid waste. Bassett's Cave on the former US Naval Air Station Annex, and known since at least 1779 based on historical signatures on the cave wall, was used to dump trash, raw sewage, rusted drums spilling an unidentified white powder, and waste motor oil. A 60 to 120-cm-thick layer of oil floats on an anchialine cave lake, fluctuating in response to the tides. The volume of oil in the cave is estimated to be ~3500 m^3, based on an average oil thickness of 80 cm (SET 2014).

Bitumen Cave (Fig. 14.10) on the grounds of the former Castle Harbor Hotel contains hundreds of disintegrating bitumen barrels with tar still covering their interior. The barrels are suspended precariously on a steep slope, 15 to 20 m above an anchialine lake (Iliffe and Calderón-Gutiérrez 2021). Sears Cave in Smith's Parish is a large sinkhole, the bottom of which is filled with domestic trash. The Audubon Society now owns the cave, which harbors several rare species of endemic cave ferns and fungi (Iliffe 1979, Bermuda Audubon Society 2022).

Climate change is already affecting anchialine caves due to an increasing frequency and severity of hurricanes and storms, with their storm surge and intense rains altering haloclines and salinity. On the medium- to long-term time horizon, ocean acidification enhances limestone dissolution, while unpredictably rapid sea level rise can submerge or destroy coastal anchialine habitats. Projected sea levels could reach 2 m or more in 2100. A 2 m rise would reduce Bermuda's land area by 14% (Glasspool 2008) and result in substantial alternation to the suitable habitat for anchialine fauna.

Fig. 14.10.
Photo of century old, rusting bitumen barrels from a cave in the Walsingham Tract. Photo reproduced courtesy of Thomas Iliffe; see a color version of this figure in the centerfold.

Table 14.1 Taxonomic table listing the number of anchialine species in the major crustacean taxa for each region. The number of endemic species are noted in parentheses.

Taxa	W Australia	Bermuda	Lanzarote	Yucatán
Remipedia	1(1)		2(2)	4(4)
Ostracoda	4(3)	13(12)	4(4)	4(4)
Copepoda	8(8)	20(19)	11(10)	13(13)
Mictacea		1(1)		
Thermosbaenacea	1(1)		1(1)	1(1)
Mysida		2(2)	1(1)	1(1)
Stygiomysida				2(1)
Cumacea		2(2)	1(1)	
Tanaidacea		1(1)		
Amphipoda	2(2)	7(7)	3(3)	8(8)
Spelaeogriphacea	2(2)			
Ingolfiellida		1(1)		
Isopoda	2(2)	4(4)	1(1)	7(7)
Anomura			1(1)	
Caridea	2(2)	5(3)		16(12)
Total	**22(21)**	**56(52)**	**25 (24)**	**56(51)**

Conservation Measures

Government planning regulations, protected species laws, and a cave species management plan endeavor to protect the caves, cave species, and overall cave habitat, respectively. The Planning Act 1974 states: "The protection of caves shall take precedence over all other planning considerations." The Protected Species Act 2003 lists 22 cave-dwelling species (of which 20 are crustaceans) for legal protection. The 2014 "Management Plan for Bermuda's Critically Endangered Cave Fauna" was intended to protect cave habitat through legislation, raising public awareness, comprehensive mapping of caves, identifying and managing point source pollution, monitoring cave air and water, examining the potential for hatchery breeding, facilitating ecological research, and undertaking the active restoration of impacted caves (Glasspool 2014; Table 14.1).

A major research objective would be the development of optimal breeding and rearing techniques for the more abundant cave species. The Department of Conservation Services Hatchery would provide facilities for these studies. Restoration of impacted caves involves identifying point sources of pollution through dye tracing or mapping the spatial distribution of contaminants, then managing pollution levels so they do not exceed tolerance levels. It would also involve the removal of trash from dumping sites within caves. Long-term monitoring of water quality within caves would be carried out with the use of dissolved oxygen, pH, temperature, salinity, and/or nutrient data loggers to ensure that environmental quality is maintained at levels necessary to support the cave fauna (Glasspool 2014). Unfortunately, since there are currently no resident cave diving scientists in Bermuda, measurements and monitoring of crustacean populations in the caves are not possible.

Although Bermuda has some of the world's more robust legislation intended to safeguard the anchialine cave fauna, these laws have not been adequately enforced, allowing the destruction, contamination, and permanent alteration of caves to occur. Other problems include the lack of funding for the 2014 Cave Fauna Management Plan and uncontrolled pollution of anchialine caves by cesspit seepage and solid waste disposal.

The best way to ensure the protection of a cave is to own it. Five private and government nature reserves and parks in the Walsingham karst region of Bermuda have been established to own and protect several of the island's most important and biodiverse caves including the Walsingham and the Palm Cave Systems.

LANZAROTE, CANARY ISLANDS, SPAIN

The island of Lanzarote, located 160 km off the Eastern Atlantic coast of North Africa, is one of the oldest of the Canary Islands. The low elevation, scarce annual rainfall, and prevalence of porous volcanic rocks along much of the island's coastline have favored marine infiltration inland and the resulting salinization of the island's aquifer (Martínez et al. 2016). This saline groundwater is accessible through wells, pools, and caves, but it has been primarily explored through a 22,000 years old, 7.8 km long lava tube cave (Carracedo et al. 2003). As this cave extends down the slopes of La Corona volcano, it first reaches sea level and saltwater pools at Cueva de los Lagos. On the coast, the naturally illuminated Jameos del Agua (29°09′26″N, 13°25′46″W) has been developed into a popular tourist attraction. The last 1800 meters of this cave, known as Túnel de la Atlántida (Fig. 14.11), extends past the coastline and under the seafloor where it is flooded by seawater. It hosts one of the richest anchialine ecosystems known in the world, with up to 40 endemic species described so far (Martínez and Gonzalez 2019). Although La Corona's anchialine assemblages consist mostly of stygobiont crustaceans, there are also highly adapted species of annelids (Martínez et al. 2013, 2017, Gonzalez et al. 2018, Worsaae et al. 2019) and two endemic flatworms (Gobert et al. 2019, Martínez et al. 2019a). These include remarkable species such as the only known galatheoidean cave crab *Munidopsis polymorpha* Koelbel 1892 (Fig. 14.12), the only described remipedes from the Eastern Atlantic *Morlockia ondinae* (Garcia-Valdecasas 1984) and *M. atlantida* (Koenemann et al. 2009), as well as highly specialized lineages of thermosbaenaceans, ostracods, amphipods, and copepods (Bowman and Iliffe 1986, Boxshall and Iliffe 1990, Jaume and Boxshall 1996).

The Canarian Remipedes and the Phylogenetic Importance of Stygobionts in Lanzarote

While most anchialine lineages are unequivocally of marine origin, many of them have been classified among genera, families, and even higher taxonomic ranks that are restricted to caves distributed across broad geographic ranges (Iliffe et al. 1984b, Wilkens et al. 2009). Such disjunct global distributions have been interpreted as the result of vicariance events driven by plate tectonics during the breakup of Pangea and the opening of the Atlantic Ocean (Stock 1993, Iliffe and Kornicker 2009, Jurado-Rivera et al. 2017, Wilson et al. 2019).

Remipedes are the most emblematic anchialine lineage since they represent an ancient clade, probably sister to all terrestrial hexapods (Lozano-Fernandez et al. 2019). Lanzarote is among the few world localities in which more than one remipede species occur sympatrically, along with Cottage Pond in Turks and Caicos, Dan's Cave in the Bahamas, and Cenote Crustacean in the Yucatán Península, Mexico (Schram et al. 1986, Yager 1987, Koenemann et al. 2004, Neiber et al. 2012, Gonzalez et al. 2020). However, it is remarkable that in Lanzarote, these two sympatric species are phylogenetically closely related and morphologically very similar, presumably feeding on similar prey and exploiting similar niches (Koenemann et al. 2009).

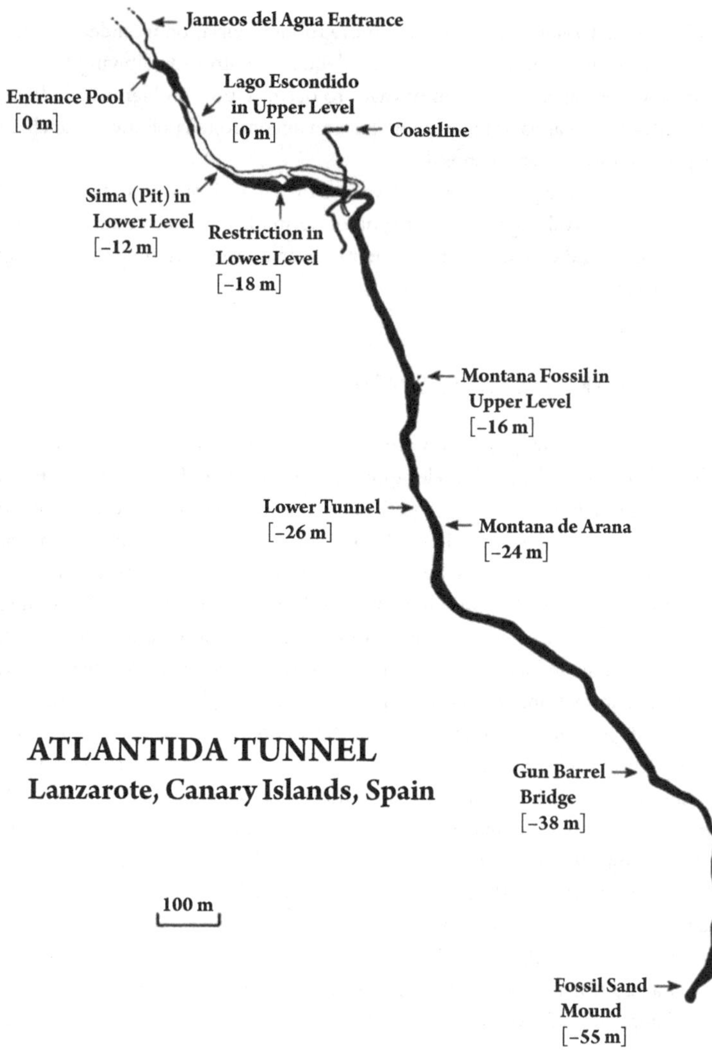

Fig. 14.11.
Map of the Túnel de la Atlántida anchialine lava cave on Lanzarote, Canary Islands. Map by Sheck Exley, reproduced from Kornicker and Iliffe (1995), with permission from © The Smithsonian Institution under Creative Commons Zero (CC 0).

Little ecological information is available for the two Canarian remipede species. They seem to occur in low density along the cave, with 1–2 specimens observed on most dives. They are more common in the sub-seafloor Túnel de la Atlántida, although they also can be found in the farthest inland waters of the cave system at Cueva de los Lagos or, more rarely, in the naturally illuminated Jameos del Agua lake. The rarity of these two species, as well as their restricted distribution, has been used as a strong argument to prohibit recreational diving in La Corona lava tube, limiting access to the cave for scientific expeditions.

The two Canarian species are sister groups and form a clade with two Caribbean species, *Morlockia williamsi* from Little Bahama Bank and *M. emersoni* from the Dominican Republic. A fifth, still undescribed species, has recently been reported from the Turks and Caicos Islands

(Gonzalez et al. 2020). The genus *Morlockia*, therefore, exhibits an Amphiatlantic disjunct distribution, along with the thermosbaenacean *Halosbaena*, the isopod *Curassanthura*, the ostracod *Humphreysella*, the misophrioids *Expansophria* and *Speleophriopsis*, and the harpacticoid *Neoechinophora* (Page et al. 2018, Boxshall and Jaume 1999).

The Jameíto and Related Deep-Sea Stygobionts

A possible deep-sea origin for anchialine stygobites offers an alternative to a Tethyan origin, with dispersal occurring through deep water crevices or cave-like habitats (Iliffe 1990, Danielopol et al. 1996). This hypothesis was inspired by the discovery of many anchialine species that belong to entirely deep-sea lineages as well as by the similarities between cave and deep-sea environments, for example, lightless, no photosynthesis, and stable conditions (Hart et al. 1985, Martínez García et al. 2009, Gonzalez et al. 2017). Several species from La Corona lava tube have close deep-sea relatives such as the annelids *Speleobregma lanzaroteum*, *Fauveliopsis jameoaquensis*, and *Gesiella jameensis*, the thaumatocyprid ostracods *Humphreysella wilkensi* and *H. phalanx*, and a variety of misophrioid copepods (Bertelsen 1983, Pettibone 1985, Ohtsuka et al. 1993, Kornicker and Iliffe 1995, Boxshall and Jaume 1999, Boxshall and Jaume 2000, Karanovic and Brandão 2012, Martínez et al. 2013, Gonzalez et al. 2017, 2018).

A charismatic example of a cave animal from a deep-sea lineage is *Munidopsis polymorpha*, a small, blind, pale squat lobster known exclusively from La Corona lava tube (Fig. 14.12), while other members of the genus inhabit hydrothermal vents (Ahyong et al. 2011). In addition to its evolutionary importance, *M. polymorpha* is a cultural icon to the island's inhabitants. Likely, the species was already known by the Majos people, who populated the island before the arrival of the Europeans. After the Spanish conquered the archipelago, it became well known to the inhabitants of the northern part of Lanzarote, who referred to *M. polymorpha* as the "grillos blancos" (i.e., white crickets). Their reputation increased with the opening of the Jameos del Agua touristic center in 1979, when this species became an important element for merchandising and touristic promotion under the name of "jameíto." Today, *M. polymorpha* is considered the island's symbolic animal, and a little *Munidopsis* called "Ale" has been adopted as the mascot of the UNESCO Geopark of Lanzarote and the Chinijo Islands. Therefore, *M. polymorpha* represents an intriguing animal that integrates culture with natural heritage (Martínez et al. 2019b).

Fig. 14.12.
The squat lobster *Munidopsis polymorpha* inhabits the Atlántida anchialine lava tube on Lanzarote, Canary Islands. Photo reproduced courtesy of Thomas Iliffe.

Munidopsis polymorpha is classified as endangered on the Spanish National List of Protected Species. The population of the species suffered a major decline in the 1980s, probably in relation to the construction and opening of the Jameos del Agua as a touristic center. This decline was probably due to the pollution of the Jameos del Agua lake, which contained the largest population, since visitors to the cave used the lake as a wishing well in which to toss good luck coins (Martínez et al. 2020). Fortunately, continuous monitoring and clean-up efforts led by the local government and the administration of the Jameos del Agua have improved the situation, safeguarding the presence of an abundant population of *M. polymorpha* in the cave.

Munidopsis polymorpha is an opportunistic species with large populations in the Jameos del Agua anchialine lake and the entrance of Túnel de la Atlántida where primary production and algal growth are stimulated by natural skylights or artificial lights. However, a few individuals have also been observed in totally dark sections of Cueva de los Lagos and Túnel de la Atlántida at least as far as the site known as "Montaña de Arena" (i.e., sand mountain). Despite their high abundance, these populations exhibit a relatively low genetic diversity (Cabezas et al. 2008, 2012). Outside the cave, the species has been recorded in wells of Los Cocoteros salt works, as well as in the Charcos de Luis anchialine pools near the town of Órzola (Martínez et al. 2019b).

Threats and Conservation Measures

Anchialine habitats of Lanzarote are susceptible to negative anthropogenic impact mainly derived from recreational uses of the anchialine environments, but also from other potentially harmful activities such as unregulated animal collections for scientific or display purposes, as well as the uncontrolled use of the land especially in the areas near the lava field of La Corona (Martínez et al. 2019b). The Spanish Government has been working together with the Government of the Canary Islands and the Cabildo de Lanzarote to protect *Munidopsis polymorpha* and *Morlockia ondinae*, along with the emblematic anchialine community living in the island. This has been achieved by laws to directly protect these two species but also by defining protected terrains around their habitats. Any anthropogenic access to those terrains has the potential to introduce contamination, ultimately impacting the anchialine ecosystems (Wynne et al. 2021). Fortunately, Lanzarote has adopted a sustainable touristic development since the 1970s and several areas of the island, including La Corona lava field and Jameos del Agua, have integrated in the so-called Red Canaria de Espacios Protegidos since the 1990s. This network of protected spaces regulates human activities and prohibits any major alteration. The efforts for conservation reached a milestone with the declaration of Lanzarote as a UNESCO Biosphere Reserve in 1993 and a UNESCO Geopark in 2015 (Martínez et al. 2019b). Today, these institutions work together with the local government and a team of scientists to ensure a better understanding of the anchialine ecosystems of the island that guarantee its preservation.

Locally, the inland-most sections of La Corona lava tube and nearby anchialine pools are threatened by inappropriate recreational activities, vandalism, and dumping of trash and garbage in the cave, as well as unintended pollution of the water with sunscreen and other cosmetic products. Fortunately, the Jameos del Agua touristic center protects a major section of La Corona lava tube from such activities but exposes the ecosystem to the tossing of coins into the main anchialine lake by the visitors, despite the displayed signs forbidding such activity (Brito et al. 2009). Consequently, heavy metals have accumulated at the bottom of the lake over the years, favoring the co-selection of bacteria carrying antibiotic resistant genes (Martínez et al. 2020). The impact of toxic heavy metals in those coins, however, has been reduced by periodical cleaning of the lake by trained divers. The occasional dropping of random objects by tourists during the evening musical events and the periodic festival may also have a detrimental effect. However, to date, no obvious signs of stress have been observed in the cave species. Scientists are working with the local authorities and the administration of the touristic center to understand the cumulative effects of these disturbances and minimize any long-term impacts.

THE YUCATáN PENíNSULA, MEXICO

The calcareous nature of the Yucatán Península (YP) has created a landscape with very few superficial bodies of water and a very large karst aquifer of ~165,000 km^2 in Mexico, Guatemala, and Belize (Bauer-Gottwein et al. 2011). The northern portion of the YP, comprising the Mexican States of Campeche, Yucatán, and Quintana Roo, and the island of Cozumel is unique in that it has thousands of sinkholes known locally as "cenotes." In many cases, cenotes connect to large open conduits, forming networks of flooded caves that can only be explored by divers. Three main areas with anchialine caves can be identified within this region: (1) the Ring of Cenotes area around the city of Merida in the northwestern portion of the peninsula, where hundreds of cenotes and their associated flooded caves have developed, (2) the cave area of the "Riviera Maya," along the Caribbean coast of the YP, where some of the longest flooded caves in the world have been discovered, and (3) the flooded caves of the island of Cozumel. Even though 2000 km of dry and underwater passageways have been explored and mapped (QRSS 2022; Fig. 14.13), only a small fraction of these vast cave systems has been surveyed biologically.

A rich stygobiont fauna has been described from the YP comprising 61 species: sponges (6 spp.), mollusks (1 sp.), crustaceans (51 spp.), echinoderms (1 sp.), and fishes (2 spp.) (Álvarez and Iliffe 2008, Álvarez et al. 2015, Gómez and Calderón-Gutiérrez 2020), of which 95% are endemic to the region. Within the crustaceans, several genera are also endemic to the YP, such as *Xibalbanus* (Remipedia), *Tulumella* (Thermosbaenacea), *Mayaweckelia* and *Tuluweckelia* (Amphipoda), *Creaseriella* and *Yucatalana* (Isopoda), and *Anchialocaris*, *Creaseria*, and *Yagerocaris* (Decapoda), giving the region a special zoogeographic and evolutionary relevance. In contrast, other genera have representatives in the Greater Caribbean, Europe, and various oceanic islands (e.g., *Curassanthura*, *Procaris*, *Typhlatya*, *Triacanthoneus*, and *Barbouria*), producing a rich mixture of lineages. New anchialine species have continuously been described from these caves in the last two decades, marking a trend that suggests that many more new species will be described in the years to come.

We generally refer to the "Yucatán Península" as including the island of Cozumel, which is separated from the mainland by an 18 km wide, 400 m deep channel; however, it is important to consider that the two geographic areas have almost completely different anchialine faunas. Only 4 species out the 61 recorded so far for the region are shared between the two areas: the amphipod *Bahadzia bozanici* (Holsinger 1992), the cirolanid isopod *Metacirolana mayana* (Bowman 1987), the alpheid shrimp *Yagerocaris cozumel* (Kensley, 1988), and the barbourid shrimp *Barbouria cubensis* (von Martens 1872). The island of Cozumel is an oceanic island that has never been in contact with the mainland (Spaw 1978, Ward 1985), and so the anchialine faunas of both regions have been acquired independently. Crustaceans compose the most diverse group on the island, including several genera that are absent from the mainland: *Agostocaris*, *Procaris*, *Anchialocaris*, and *Parhippolyte*.

Comments on Selected Taxa

The case of the Mexican remipedes (Fig. 14.14), which has been instrumental in several groundbreaking studies on the phylogeny of arthropods (e.g., von Reumont et al. 2012, Lozano-Fernandez et al. 2019), the discovery of the first venomous crustacean (von Reumont et al. 2014) and in

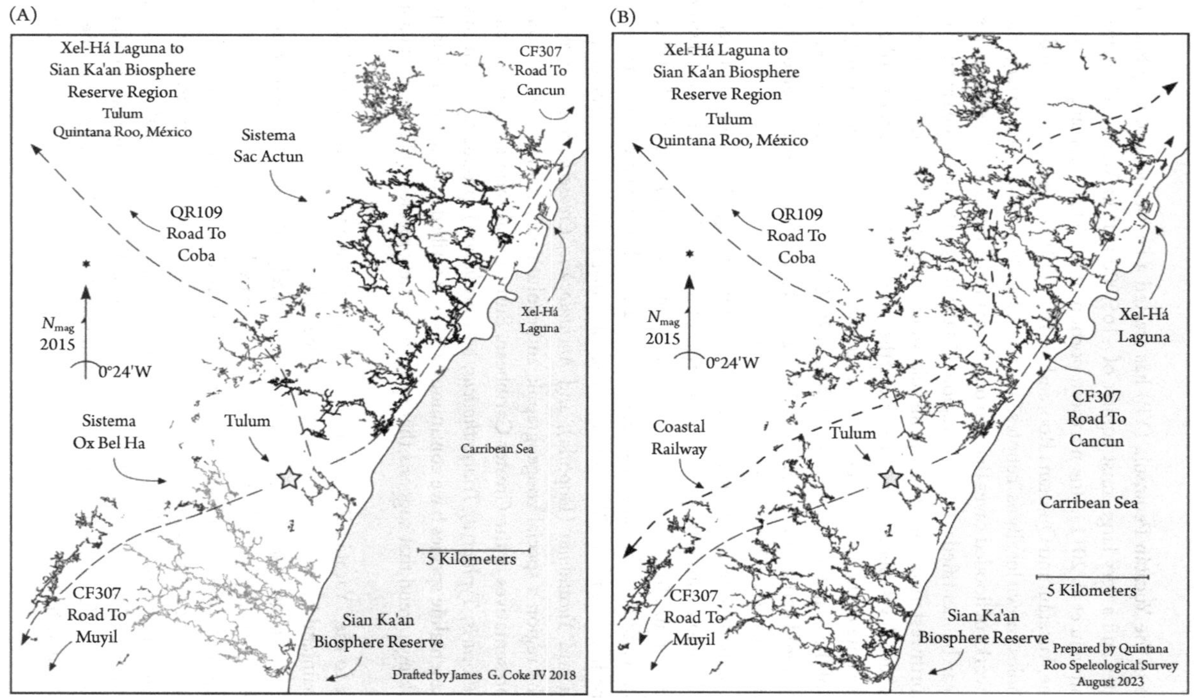

Fig. 14.13.
The Caribbean coast of Mexico's Yucatán Península in the state of Quinatana Roo has the world's longest underwater caves. (A) Sistema Sac Actun (darker lines in upper center) is the world's longest submerged cave with a total length of 372 km, while Sistema Ox Bel Ha (light lines in lower left) is second at 319 km. Map from 2018 reproduced courtesy of James G. Coke IV. (B) The Tren Maya, a newly constructed coastal railway, raises concerns regarding conservation and protection of the underlying caves and anchialine aquifer. August 2023 map reproduced courtesy of James G. Coke IV; see a color version of this figure in the centerfold.

Fig. 14.14.
Xibalbanus tulumensis is one of four species from the exclusively stygobiont class Remipedia in the Yucatán Península. Photo reproduced courtesy of Simon Richards; see a color version of this figure in the centerfold.

the study of their neural system (e.g., Stemme et al. 2015), among other topics, is particularly promising. Two species occur in this area: *Xibalbanus tulumensis* (Yager 1987), in caves from Puerto Morelos to Tulum, and *X. fuchscockburni* (Neiber et al. 2012), known only from Cenote Crustacea south from Puerto Morelos. *Xibalbanus tulumensis* has been the species of choice in these studies because it can be found in more than 10 cave systems, it has a relatively high natural abundance, and, in many cases, there is easy access to the caves inhabited by this species. However, the distribution area of the two remipedes is exactly that of the so-called Riviera Maya, which is a 100 km long, 15 km wide band of coastline that goes from Puerto Morelos in the north to Tulum in the south. In addition to remipedes, there are other species endemic to the area such as the alpheid shrimp *Triacanthoneus akumalensis* (Alvarez et al. 2012) and the hippolytid shrimp *Calliasmata nohochi* (Escobar-Briones et al. 1997).

The Riviera Maya is the most important tourist corridor in Mexico and Latin America, with thousands of hotel rooms, dozens of aquatic parks and tourist attractions, and two of the fastest-growing cities in Mexico, Cancún and Playa del Carmen. Consequently, the area is being rapidly transformed and the pressure on the environment is enormous. The aquifer is starting to show alarming levels of contaminants in sinkholes and their associated caves due to the rapid development. Several pollutants have been detected for some time in the Riviera Maya groundwater, such as hydrocarbons (Medina-Moreno et al. 2014), human coliform bacteria (Leal-Bautista et al. 2013), and other pathogens and pharmaceuticals (Leal-Bautista et al. 2011). Recent data show that fuel from cars and road paving with asphalt are two major sources of pollutants for the aquifer in the area (León-Borges and Lizardi-Jiménez 2017).

The third remipede species from the YP is *Xibalbanus cozumelensis* Olesen et al. 2017, known only from Cueva Quebrada in Cozumel. Although morphologically like *X. tulumensis*, it is genetically distinct (Olesen et al. 2017). No other reports exist on its distribution on the island, so it can be considered a microendemic. Cozumel is an island where tourism is the main activity and where rapid development of tourist infrastructure is occurring. Due to poor public services and sanitary regulations, the aquifer now shows high levels of total dissolved solids and fecal coliforms (Contreras-Tax et al. 2021).

Conservation Challenges

An explosive development is now occurring in the YP fueled by prosperous agricultural and tourist industries, involving an intense growth of urban areas with the resulting changes in land use and over-exploitation of resources, especially water. The negative impact of pollutants on the aquifer has been known for a decade (Hernández-Terrones et al. 2011, Metcalfe et al. 2011). About 30% of the northern section of the YP, mainly in the State of Yucatán, but also in Campeche and Quintana Roo, has been deforested to give way to agriculture and livestock farming, facilitating the infiltration of pesticides, fertilizers, and other pollutants into the aquifer (Polanco-Rodríguez et al. 2018). The aquifer around the city of Merida, where the Ring of Cenotes is, now shows important levels of contamination by organochlorine pesticides, which have low decay rates and easily bioaccumulate (Polanco-Rodríguez et al. 2019). The Mexican federal government is constructing a railway across the YP that has been highly controversial because of the lack of proper environmental impact and engineering studies; consequently, there is concern this project can cause cave collapse and groundwater pollution (Ayala Azcárraga et al. 2022).

The stygobiotic fauna present in the northwestern section of the YP, including the Ring of Cenotes, is diverse and can be locally abundant. A series of caves south from the city of Merida can be regarded as diversity hotspots with 12 or more anchialine species each, including mysids (*Antromysis cenotensis, Stygiomysis holthuisi, S. cokei*), ostracods (*Humphreysella mexicana*), atyids (*Typhlatya mitchelli, T. pearsei*), amphipods (*Mayaweckelia cenoticola*), isopods (*Creaseriella anops, Haptolana bowmani, Curassanthura yucatanensis*), palaemonids (*Creaseria morleyi*), and two species of fish (*Typhlias pearsei, Ophisternon infernale*). Like the Riviera Maya, this area with dozens of cenotes attracts hundreds of tourists, with a clear increment during the summer months, putting considerable pressure on the biological systems and ultimately degrading the sites (Blanco-Gregory et al. 2019).

COMPARISONS

Flooded caves harboring anchialine fauna typically form in karst areas near or on the coastline in a porous limestone matrix, although they can also develop in lava tubes or in cracks produced by geologic faults in other types of rock. They often have open entrances due to the collapse of the cave's ceiling or to superficial fractures. These entrance sites to the underground aquifer are locally known as "cenotes" in Yucatán, "blue holes" in the Bahamas and Belize, or "jameos" in Lanzarote.

In contrast to these caves, anchialine pools in Hawaii and elsewhere in the Indo-South Pacific region occur near the coast and are mostly or fully exposed to daylight and open air. Fluctuating salinities and water levels occur periodically with the tides resulting in mixing of marine and groundwater. Characteristic fauna includes a variety of obligate anchialine shrimp. Benthic microbial communities, algae, filter feeders, detritivores, scavengers, and primary and secondary consumers are also present (Sakihara et al. 2020).

A third type of anchialine system has been discovered by cave divers using mixed gas rebreathers or forms of technical diving to penetrate sub-seafloor caves for considerable distances and/or depths to reach previously unexplored areas. Pockets of marine water separated by chemoclines and/or thermoclines from tidal circulation were found to contain typical anchialine fauna including remipedes, ostracods, etc. (Ballou et al. 2021). During past periods of glacially lowered sea level, such caves would have been on land and been classified as anchialine.

Similarities in Habitat and Fauna

The interior of anchialine caves lack light. Therefore, photosynthesis is absent, they have limited nutrient resources, and are density-stratified into discrete layers separated by one or two well-defined halo/thermo/chemoclines. Beneath the lower halocline, the water is of fully marine salinity. Owing to a longer residence time and lack of strong currents, particulate matter is generally absent from the water column and the water tends to be exceptionally clear. Despite this, the majority of anchialine crustaceans live suspended in the water as either plankton or nekton.

Anchialine faunas are characterized by a diverse biota dominated by endemic crustaceans, often representing novel higher taxa such as Remipedia, Mictacea, and Bochusacea. In addition, anchialine stygobionts include Ostracoda (Thaumatocyprididae and Podocopida), Copepoda (Platycopioida, Misophrioida, Cyclopoida, Harpacticoida, and Calanoida, especially Epacteriscidae and Ridgewayiidae), Thermosbaenacea, Isopoda (Anthuroidea, Asellota, and Cirolanidae), Amphipoda (Bogidiellida, Crangonyctidira, Hadziidae, and Niphargidae), and Caridea (Agostocarididae, Alpheidae, Atyidae, Hippolytidae, Palaemonidae, and Procarididae).

Anchialine organisms are primarily stygobionts, lacking eyes and pigment, while possessing enhanced sensory mechanisms to find food and mates in the absence of light, or physiological adaptation for lower metabolic rates in a food-poor environment. Even within a cave, anchialine species are commonly limited to discrete water masses and regions of the cave as determined by salinity, dissolved oxygen levels, food resources, presence of prey or predators, etc. Since many anchialine species are known only from a single or a few adjacent caves, they are highly susceptible to potential extinction.

Regional Importance of Threats

Water pollution and habitat destruction are two of the greatest threats (Table 14.2), especially in Bermuda, which is a small island with a high population density. Disposal of wastewater into the ground has led to high nutrient levels and depletion of dissolved oxygen from the cave water. Under such circumstances, anoxic conditions and a buildup of black hydrogen sulfide precipitate produces dark clouds in the water. Furthermore, untold numbers of Bermuda caves have been obliterated in limestone quarries or through the construction of hotels and golf courses. In Yucatán, highways have collapsed into underlying voids, creating sinkholes that were ultimately filled. Sinkholes accessing the groundwater have been used as garbage dumps in Bermuda and Yucatán.

Water extraction from the aquifer for human, industrial, or agricultural use is another important threat to the fragile underground ecosystem. In coastal areas, excessive freshwater extraction allows for saltwater intrusions that can extend for many kilometers inland, thus modifying the entire ecosystem. Saltwater intrusions in coastal aquifers are a worldwide problem. However, it becomes especially critical in the case of anchialine caves and their associated fauna since the necessary freshwater layer to create the anchialine conditions is depleted. Coupled with water extraction, the

Table 14.2 Overview qualitatively ranking common threats in each region. Scores: + threat present, ++ serious threat, +++ threat endangering integrity and persistence of cave ecosystems, ? not known well.

Main threats	Western Australia	Bermuda	Lanzarote	Yucatán
Habitat destruction (Construction)	+	+++	?	+
Water extraction	?	+	?	++
Pollution (dissolved contaminants)	?	++	+	++
Pollution (solid)	?	++	+	?
Tourism	?	++	++	++
Climate change	?	++	+	++

following threat is the release of wastewater. After the extraction, the disposal of byproducts derived from anthropogenic activities, often directly to the same aquifer from which water was extracted, becomes a primary source of contamination. Due to the natural high connectivity that has been found in groundwater and, more so, in anchialine caves, every contamination event has far-reaching consequences.

Popular tourist attractions that are constantly visited have been disturbed or modified to different extents. Tourists visiting show caves in Bermuda and Lanzarote commonly drop copper coins into saltwater pools where they corrode, releasing toxic copper ions into the water. In Yucatán, eyed fish from surface pools have learned to follow tourist divers into the dark cave interior where the fish use the diver's lights to gobble up all visible organisms from the water column. In such caves, the anchialine fauna is now virtually absent.

Climate change and sea level rise is a medium- to long-term threat that can potentially submerge those caves closest to the coast, thereby drastically altering the ecosystem. More frequent and intense hurricanes and storm surges also raise the water table and flush anchialine fauna from the cave or kill them by changing the salinity.

Regional Approach to Conservation Strategies

Considering the fragility of the anchialine environment and, consequently, that of its peculiar fauna, the conservation strategies must be specifically designed for each site and region. Strategies should preserve areas that contain flooded caves, limiting access to reasonable levels and regulating water extraction and wastewater disposal. Land development needs to be regulated, and authorization of permits must be restricted to preserve surface karst and underlying groundwater.

One of the best ways to protect the anchialine ecosystem is by declaring anchialine organisms as endangered species that benefit from existing endangered species laws and regulations. In Bermuda, for example, 20 anchialine stygobionts have been officially named as protected species and the 2014 "Management Plan for Bermuda's Critically Endangered Cave Fauna" is designed to protect the entire cave habitat (Table 14.3). The plan sets priorities to (1) prevent extinction, (2) prevent significant population decline, and (3) promote species recovery. It declares the caves as "critical habitat," initiates a cave mapping program, evaluates the potential for breeding of cave species, determines tolerance to contaminants, identifies point sources of pollution, implements long-term monitoring, facilitates ongoing research, undertakes cave restoration, expands education and public awareness programs, and improves information sharing.

FUTURE DIRECTIONS

The anchialine environment and its highly specialized fauna, composed mainly of crustacean species, poses many questions about its origin, evolution, and functioning. How were the first anchialine caves colonized, and how has their fauna survived through millions of years despite their apparent fragility? It has been pointed out that anchialine caves have served as refugia through geologic time (Iliffe et al. 1983, 1984b, Jurado-Rivera et al. 2017, van Hengstum et al. 2019), but how can this be if coastlines, where these ecosystems primarily form, are constantly changing? How are these caves colonized, and how is the anchialine fauna dispersing to occupy new anchialine habitats?

Regarding the evolution of anchialine fauna, perhaps one of the most intriguing questions to pursue in the future is that related to the role of Remipedia in the evolution of arthropods (von Reumont and Burmester 2010, von Reumont et al. 2012, Stemme et al. 2013). Phylogenetic studies in the last decade have found remipedes to be the link between Crustacea and

Table 14.3 Implementation of the Bermuda Cave Fauna Management Plan. DCS–Dept. of Conservation Services, AG–Attorney General, DEP–Dept. of Environmental Protection. Priority 1: an action that must be taken to prevent extinction or to prevent the species from declining irreversibly. Priority 2: an action that must be taken to prevent a significant decline in the species population/habitat quality, or some other significant negative impact short of extinction. Priority 3: all other actions necessary to provide for full recovery of the species. Source: data from Glasspool (2014).

Priority	Task Description	Task Duration	Responsible Party
1	1. Protect the cave habitat by declaring the caves as "critical habitat" under the Protected Species Act 2003.	4 months	DCS, AG's Chambers
2	2. Undertake a comprehensive mapping initiative of the Islands' cave system and integrate into the GIS framework.	10 years	DCS
2	3. Evaluate the potential for ex-situ/hatchery breeding of representative cave species and conduct experiments to determine their tolerance to known contaminants of the cave system.	10 years	DCS/hatchery
1	4. Identify point sources of pollution and implement the necessary procedures to manage these so that they do not exceed species tolerance levels. If necessary, explore and implement alternative technologies for sewage disposal.	10 years	DCS
2	5. Establish baseline levels of air and water in the cave systems and implement long-term monitoring of these.	10 years	DEP, DCS
3	6. Facilitate ongoing research to gain a better understanding of the baseline population status, the distribution of the cave species, the hydrography of the cave system, and the establishment of a monitoring system.	20 years	DCS
3	7. Undertake active restoration of suitable caves.	5 years	DCS
3	8. Expand current education and public awareness programs.	20 years	DCS, Ministry of Education
3	9. Improve information sharing between key stakeholders.	6 months	DCS, IT

Hexapoda (see Volume 8, Chapters 4 and 8, respectively, in this book series; Bracken-Grissom and Wolfe 2020, von Reumont and Edgecombe 2020). They show mixed characters of both major groups in their genetic makeup, nervous system, and internal biochemistry, among various aspects, suggesting that forms closely related to Remipedia could be the ancestors of one of the most abundant and diverse groups in the planet, the insects.

Ecologically, the crustacean fauna from anchialine caves display unique behaviors as they must survive in a nutrient-poor habitat. Besides exhibiting several morphological adaptations in feeding appendages (e.g., Atyidae, Agostocarididae, and Macromaxillocarididae), they have developed intricate food webs and unique adaptations. Recently, Brankovits et al. (2017) described a novel chemosynthetic route by which atyid shrimp may use methane to obtain energy and food, constituting a process that could shape the entire anchialine community. It is now relevant to ask how widespread is this process in anchialine habitats? But also, other major discoveries have come from anchialine crustaceans, such as the description of the first venomous crustacean of the world, the remipede *Xibalbanus tulumensis* (von Reumont et al. 2014). Such discoveries represent only the starting point for future research on their ecological significance.

To explore and eventually answer questions like these, studies will have to integrate varied techniques in geology, genetics, zoology, chemistry, microbiology, hydrology, geochemistry, and other disciplines. The complexity of the functioning and interactions within the anchialine ecosystem will require new interdisciplinary approaches and international collaborations.

Cave Exploration and Biodiversity Surveys

Cave exploration techniques have improved constantly with better and safer diving equipment. Combined with increasing exploring capabilities, we now have sophisticated cave mapping, positioning systems, and GIS software to precisely locate the position of species, and advanced measuring instruments to characterize the water column physically and chemically. The interconnection and interdependence between the surface and the anchialine cave environment have been established (e.g., Brankovits et al. 2017, Adams et al. 2020). We have made good progress in recognizing the major groups of organisms, and more than several hundred anchialine cave species have been described. Crustaceans dominate in terms of number of species and abundance in this environment. Very important taxa, such as the Remipedia, that have helped us understand the relationships among the major groups within Arthropoda, or the atyid shrimps of the genus *Typhlatya*, which are an essential link in a novel chemosynthetic route, come from anchialine caves. Exploration, faunistic studies, and conservation of the crustacean fauna should continue to be the primary objectives of anchialine cave research.

Outreach

During research expeditions, invited lectures are often presented to school children and to the general public that emphasize the importance of caves and cave animals, why cave ecosystems are threatened, and what conservation measures need to be taken. These lectures have been highly impactful. For example, after one lecture, a young schoolboy asked if he could have friends who will attend his upcoming birthday party donate money to our cave research project rather than to give him gifts. This was one of the greatest compliments I have ever received!

Press releases describing the results of our research expeditions have produced numerous internationally distributed newspaper and magazine articles. Internet podcast interviews allow for connection to a wider audience through audio broadcasts. Lectures during major conferences, workshops, or invited presentations further disseminate the results of this research, while discussion of cave research or even field trips to caves have been included in university classes ranging from introductory biology, ecology, evolution, and tropical ecology, to scientific diving.

Our cave research has also been prominently featured on internationally broadcast documentary films from Netflix, the BBC, PBS, National Geographic, Discovery Channel, History Channel, German Public Broadcasting, French Public Broadcasting, and others. With a viewership of many millions of people worldwide, there is no better way to reach a broader audience than through television.

SUMMARY AND CONCLUSIONS

The Crustacea dominates anchialine ecosystems worldwide in terms of taxonomic diversity and relative abundance. Numerous new higher crustacean taxa have been discovered from anchialine habitats including the class Remipedia, orders Mictacea and Bochusacea, as well as many new families, genera, and species. It is common for anchialine species to be found only in a single cave or cave system but at the same time have closely related taxa occurring in anchialine habitats on opposite sides of oceans or even opposite sides of the Earth. Remipedes, with their 8 families, 12 genera, and 30 species, provide an excellent example of such an anomalous biogeographic distribution. Most remipedes, 21 species, are from the Lucayan Archipelago (Bahamas and Turks and Caicos Islands), while six species live in the Caribbean (Belize, Cuba, Dominican Republic, and the Yucatán Península of Mexico), two species are from Lanzarote, Canary Islands, and a single species is from Western Australia. Only two caves outside the Western Atlantic region are inhabited by remipedes: Túnel de la Atlántida volcanic cave, Lanzarote, and the Bundera Sinkhole, Western Australia. Three remipedes from the Bahamas and Belize are found at significant depth or distance inside sub-seafloor caves. Other than caves on the mainland of the Yucatán Península and Bundera Sinkhole on the Cape Range peninsula of Western Australia, all other remipedes inhabit caves on or associated with islands. However, regardless of where in the world remipedes occur, they are found in caves, under total darkness, close to or below haloclines, in higher salinity seawater, and inhabited by characteristic stygofauna.

Anchialine habitats are exceptionally fragile environments with a unique crustacean fauna and are susceptible to activities that disturb the earth or groundwater. Because anchialine caves exist below the surface and, in the most part, are accessible only to highly trained and skilled cave divers, they are relatively unknown or, at best, poorly appreciated by the general public, conservation scientists, and government officials. The result is that most anchialine cave habitats are not effectively protected.

Appendix 14.1. Taxonomic table listing the crustacean genera and number of species in that genus for each site, number of endemic taxa are in parentheses.

Family	Genus	W Australia	Bermuda	Lanzarote	Yucatán
Order Remipedia					
Kumongidae	*Kumonga*	1(1)			
Morlockiidae	*Morlockia*			2(2)	
Xibalbaniidae	*Xibalbanus*				4(4)
Order Ostracoda					
Bairdiidae	*Aponesidea*		1(1)		
Candonidae	*Dolerocypria*		1(1)		
Cytheruidae	*Cytherella*		1(1)		
	Microceratina	1(1)			
Deeveyidae	*Spelaeoecia*		1(1)		1(1)

(Continued)

Appendix 14.1. Continued

Family	Genus	W Australia	Bermuda	Lanzarote	Yucatán
Halocypridae	*Metapolycope*		1(1)		
	Micropolycope		2(2)		
Philomedidae	*Pseudophilomedes*		1(1)		
Polycopidae	*Eupolycope*			1(1)	
	Polycopissa		1(1)		
	Pseudopolycope				1(1)
Pontocyprididae	*Iliffeoecia*		1		
	Kareloecia		1(1)		
	Thomontocypris		1(1)		
Sarsiellidae	*Eusarsiella*		1(1)	1(1)	
Thaumatocyprididae	*Humphreysella*	1		2(2)	2(2)
	Welesina	1(1)			
Paracyprididae	*Phlyctenophora*	1(1)			
Order Copepoda					
Ameiridae	*Nitokra*	1(1)			
Arietellidae	*Paramisophria*			1(1)	
Cyclopidae	*Diacyclops*				2(2)
	Halicyclops	1(1)			
	Mesocyclops				2(2)
	Prehendocyclops				3(3)
Epacteriscidae	*Balinella*				1(1)
	Bunderia	1(1)			
	Enantiosis		1(1)		
	Enantronia			1(1)	
	Epacteriscus		1(1)		
	Erebonectes		1(1)		
Halicyclopidae	*Halicyclops*		2(2)		1(1)
Laophontidae	*Onychocamptus*	1(1)			
Misophriidae	*Dimisophria*			1(1)	
Palpophriidae	*Palpophria*			1(1)	
Peltidiidae	*Eupelte*		1(1)		
Platycopiidae	*Antrisocopia*		1(1)		
	Nanocopia		1(1)		
Pseudocyclopiidae	*Exumella*		1(1)		1(1)
	Paracyclopia	1(1)	1(1)		
	Stygocyclopia	1(1)		1(1)	
Ridgewayiidae	*Ridgewayia*		1(1)		
	Stygoridgewayia	1(1)			
Schminkepinellidae	*Muceddina*			1(1)	
Smirnovipinidae	*Oromiina*			1(1)	
Speleoithonidae	*Speleoithona*		1(1)		
Speleophriidae	*Boxshallia*			1	

Family	Genus	W Australia	Bermuda	Lanzarote	Yucatán
	Expansophria			1(1)	
	Mexicophria				1(1)
	Speleophria	1(1)	1(1)		1(1)
	Speleophriopsis		1(1)	1(1)	
Stephidae	*Miostephos*		1		
	Stephos				1(1)
Superornatiremidae	*Intercrusia*		1(1)		
	Neoechinophora		3(3)	1(1)	
	Superornatiremis		1(1)		
Order Mictacea					
Mictocarididae	*Mictocaris*		1(1)		
Order Thermosbaenacea					
Halosbaenidae	*Halosbaena*	1(1)		1(1)	
Tulumellidae	*Tulumella*				1(1)
Order Mysida					
Mysidae	*Antromysis*				1(1)
	Bermudamysis		1(1)		
	Heteromysoides			1(1)	
	Platyops		1(1)		
Order Stygiomysida					
Stygiomysidae	*Stygiomysis*				2(1)
Order Cumacea					
Bodotriidae	*Speleocuma*			1(1)	
Nannastacidae	*Cumella*		2(2)		
Order Tanaidacea					
Apseudidae	*Apseudes*		1(1)		
Order Amphipoda					
Bogidiellidae	*Bermudagidiella*		1(1)		
	Bogidiella			1(1)	
Hadziidae	*Bahadzia*				2(2)
	Hadzia	2(2)	1(1)		
	Mayaweckelia				3(3)
	Tuluweckelia				1(1)
Hyalellidae	*Hyalella*				2(2)
Hyalidae	*Parhyale*			1(1)	
Liljeborgiidae	*Idunella*		1(1)		
Pardaliscidae	*Spelaeonicippe*			1(1)	
Phoxocephalidae	*Cocoharpinia*		1(1)		
Podoceridae	*Podobothrus*		1(1)		
Pseudoniphargidae	*Pseudoniphargus*		2(2)		

(Continued)

Appendix 14.1. Continued

Family	Genus	W Australia	Bermuda	Lanzarote	Yucatán
Order Spelaeogriphacea					
Spelaeogriphidae	*Mangkurtu*	2(2)			
Order Ingolfiellida					
Ingolfiellidae	*Ingolfiella*		1(1)		
Order Isopoda					
Atlantasellidae	*Atlantasellus*		1(1)		
Cirolanidae	*Arubolana*		1(1)		
	Cirolana				2(2)
	Creaseriella				1(1)
	Haptolana	1(1)			1(1)
	Metacirolana				1(1)
	Yucatalana				1(1)
Leptanthuridae	*Bunderanthura*	1(1)			
	Curassanthura		1(1)	1(1)	1(1)
Stenetriidae	*Stenobermuda*		1(1)		
Order Anomura					
Galatheidae	*Munidopsis*			1(1)	
Order Caridea					
Agostocarididae	*Agostocaris*				2(2)
Alpheidae	*Bermudacaris*		1(1)		
	Triacanthoneus				1(1)
	Yagerocaris				1(1)
Anchialocarididae	*Anchialocaris*				1(1)
Atyidae	*Jonga*				1
	Stygiocaris	2(2)			
	Typhlatya		1(1)		4(4)
Barbouriidae	*Barbouria*		1		1
	Parhippolyte		1		2
Hippolytidae	*Calliasmata*				1(1)
Palaemonidae	*Creaseria*				1(1)
Procarididae	*Procaris*		1(1)		1(1)
	Totals	**22(21)**	**56(52)**	**25(24)**	**56(51)**

ACKNOWLEDGMENTS

The final crustacean checklist is available on the World Register of marine Cave Species (WoRCS, http://www.marinespecies.org/worcs/, last accessed March 1, 2024), a thematic species database of WoRMS that aims at creating a comprehensive taxonomic and ecological database of species from marine caves worldwide. FA acknowledges the financial support obtained through grants A1-S-32846 from CONAHCYT and DGAPA-PAPIIT-IN206523 from UNAM.

REFERENCES

Adams, R.E., T.M. Iliffe, and J.B. West. 2020. Identifying tree roots in the caves of Quintana Roo, Mexico as a step toward ecological insights and improved conservation. Plants, People, Planet 2:133–139.

Ahyong, S.T., N. Andreakis, and J. Taylor. 2011. Mitochondrial phylogeny of the deep-sea squat lobsters, Munidopsidae (Galatheoidea). Zoologischer Anzeiger—A Journal of Comparative Zoology 250:367–377.

Álvarez, F., and T.M. Iliffe. 2008. Fauna anquihalina de Yucatán. Pages 379–418 *in* F. Alvarez and G. Rodríguez-Almaraz, editors. Crustáceos de México: Estado actual de su Conocimiento. Universidad Autónoma de Nuevo León, Nuevo León, México.

Álvarez, F., T.M. Iliffe, S. Benitez, D. Brankovits, and J.L. Villalobos. 2015. New records of anchialine fauna from the Yucatan Peninsula, Mexico. Check List 11:1–10.

Alvarez, F., T.M. Iliffe, B. Gonzalez, and J.L. Villalobos. 2012. *Triacanthoneus akumalensis*, a new species of alpheid shrimp (Crustacea: Caridea: Alpheidae) from an anchialine cave in Quintana Roo, Mexico. Zootaxa 3154:61–68.

Ayala Azcárraga, S.C., M.F. Lases Hernández, E. Monroy Ríos, I. Sosa Aranda, F. Calderón Gutiérrez, R. Pacheco Muñoz, S.D. Acosta Sinesio, J.D. Graf Pérez, and A. Merio Galeazzi. 2022. Análisis de la manifestación de impacto ambiental modalidad regional (MIA-R) del proyecto "Tren Maya Tramo 5 Sur". Clave del proyecto 23QR2022V0020. Consulta Pública, Mayo, 2022. Accessed December 12, 2023. [https://miatrenmaya.wordpress.com/2022/06/18/nuestras-observaciones-de-la-manifestacion-de-impacto-ambiental-del-proyecto-tren-maya-tramo-5-sur/].

Ballou, L., T.M. Iliffe, B. Kakuk, B.C. Gonzalez, K.J. Osborn, K. Worsaae, K. Meland, K. Broad, H. Bracken-Grissom, and J. Olesen. 2021. Monsters in the dark: Systematics and biogeography of the stygobitic genus *Godzillius* (Crustacea: Remipedia) from the Lucayan Archipelago. European Journal of Taxonomy 751:115–139.

Bauer-Gottwein, P., B.R.N. Gondwe, G. Charvet, L.E. Marín, M. Reboyedo-Vieyra, and G. Merediz-Alonso. 2011. The Yucatan Peninsula karst aquifer, Mexico. Hydrogeology Journal 19:507–524.

Beeton, R.J.S., A. Burbidge, G. Grigg, P. Harrison, R. How, W. Humphreys, N. McKenzie, and J. Woinarski. 2010. Final report of the Christmas Island Expert Working Group to Minister for the Department of Environment Protection, Heritage and the Arts. Queensland, Australia.

Bermuda Audubon Society. 2022. Sears Cave. Accessed December 6, 2023. [https://audubon.bm/reserve/sears-cave/].

Bertelsen, R.D. 1983. *Speleobregma lanzaroteum*, a new genus and species of Scalibregmatidae (Polychaeta) from a marine cave in the Canary Islands. Proceedings of the Biological Society of Washington 99:375–379.

Blanco-Gregory, R., F. Enseñat-Soberanis, and J.A. Modragón-Mejía. 2019. La capacidad de carga psicosocial del turista: Instrumento de medición para el desarrollo sostenible de la turistificación de los cenotes. Cuadernos de Turismo 43:169–186.

Bowman, T.E. 1987. *Bahalana mayana*, a new troglobitic cirolanid isopod from Cozumel Island and the Yucatan Peninsula, Mexico. Proceedings of the Biological Society of Washington 100:659–663.

Bowman, T.E., and T.M. Iliffe. 1985. *Mictocaris halope*, a new unusual peracaridan crustacean from marine caves on Bermuda. Journal of Crustacean Biology 5:58–73.

Bowman, T.E., and T.M. Iliffe. 1986. *Halosbaena fortunata*, a new thermosbaenacean crustacean from the Jameos del Agua marine lava cave, Lanzarote, Canary Islands. Stygologia 2:84–89.

Bowman, T.E., S.P. Garner, R.R. Hessler, T.M. Iliffe, and H.L. Sanders. 1985. Mictacea, a new order of Crustacea Peracarida. Journal of Crustacean Biology 5:74–78.

Boxshall, G.A., and T.M. Iliffe. 1990. Three new species of misophrioid copepods from oceanic islands. Journal of Natural History 24:595–613.

Boxshall, G., and D. Jaume. 1999. On the origin of misophrioid copepods from anchialine caves. Crustaceana 72:957–963.

Boxshall, G.A., and D. Jaume. 2000. Discoveries of cave misophrioids (Crustacea: Copepoda) shed new light on the origin of anchialine faunas. Zoologischer Anzeiger 239:1–19.

Bracken-Grissom, H., and J.M. Wolfe. 2020. The pancrustacean conundrum: A conflicted phylogeny with emphasis on Crustacea. Pages 80–104 *in* G.C.B. Poore and M. Thiel, editors. The natural history of the Crustacea, Volume 8: Evolution and biogeography. Oxford University Press, New York, USA.

Bradford, T., M. Adams, W.F. Humphreys, A.D. Austin, and S.J.B. Cooper. 2010. DNA barcoding of stygofauna uncovers cryptic amphipod diversity in a calcrete aquifer in Western Australia's arid zone. Molecular Ecology Resources 10:41–50.

Brankovits, D., J. Pohlman, H. Niemann, M. Leigh, M.C. Leewis, K. Becker, T. Iliffe, F. Alvarez, M. Lehmann, and B. Phillips. 2017. Methane- and dissolved organic carbon-fueled microbial loop supports a tropical subterranean estuary ecosystem. Nature Communications 8:1835.

Bretz, J.H. 1960. Bermuda: A partially drowned, late mature Pleistocene karst. Geological Society of America Bulletin 71:1729–1754.

Brito, M.D.C., A. Martínez, and J. Núñez. 2009. Changes in the stygobiont polychaete community of the Jameos del Agua, Lanzarote, as a result of bioturbation by the echiurid *Bonellia viridis*. Marine Biodiversity 39:183–188.

Bruce, A.J., and P.J.F. Davie. 2006. A new anchialine shrimp of the genus *Procaris* from Christmas Island: The first occurrence of the Procarididae in the Indian Ocean (Crustacea: Decapoda: Caridea). Zootaxa 1238:23–33.

Cabezas, P., F. Alda, E. Macpherson, and A. Machordom. 2012. Genetic characterization of the endangered and endemic anchialine squat lobster *Munidopsis polymorpha* from Lanzarote (Canary Islands): Management implications. ICES Journal of Marine Science 69:1030–1037.

Cabezas, P., P. Bloor, I. Acevedo, C. Toledo, M. Calvo, E. Macpherson, and A. Machordom. 2008. Development and characterization of microsatellite markers for the endangered anchialine squat lobster *Munidopsis polymorpha*. Conservation Genetics 10:673–676.

Cameron, E. 1984. The Yeelirrie calcrete uranium deposit, Western Australia, in surficial uranium deposits. International Atomic Energy Agency IAEA–TECDOC–322 16:157–164.

Carracedo, J.C., B. Singer, B. Jicha, H. Guillou, E. Rodríguez Badiola, J. Meco, F.J. Pérez Torrado, D. Gimeno, S. Socorro, and A. Láinez. 2003. La erupción y el tubo volcánico del Volcán Corona (Lanzarote, Islas Canarias). Estudios Geológicos 59:277–302.

Chandler, R. 2006. Bermuda's caves. Bermuda Zoological Society, Bermuda. Accessed December 12, 2023. [https://conservation-services.squarespace.com/s/Bermudas-Caves.pdf]

Contreras-Tax, A.F., O. Frausto-Martínez, and W.D. Uhu-Yam. 2021. Groundwater quality in a karstic zone of the island of Cozumel, Quintana Roo, Mexico. Tropical and Subtropical Agroecosystems 24:27.

Cooper, S.J.B, J.H. Bradbury, K.M. Saint, R. Leys, A.D. Austin, and W.F. Humphreys. 2007. Subterranean archipelago in the Australian arid zone: Mitochondrial DNA phylogeography of amphipods from central Western Australia. Molecular Ecology 16:1533–1544.

Cooper, S.J.B., K.M. Saint, S. Taiti, A.D Austin, and W.F. Humphreys. 2008. Subterranean archipelago II: Mitochondrial DNA phylogeography of stygobitic isopods (Oniscidea: *Haloniscus*) from the Yilgarn region of Western Australia. Invertebrate Systematics 22:195–206.

Danielopol, D.L., A. Baltanás, and G. Bonaduce. 1996. The darkness syndrome in subsurface-shallow and deep-sea dwelling Ostracoda (Crustacea). Biosystematics and Ecology Series 11:123–143.

Danielopol, D.L., A. Baltanás, and W.F. Humphreys. 2000. *Danielopolina kornickeri* sp. n. (Ostracoda, Thaumatocypridoidea) from a Western Australian anchialine cave: Morphology and evolution. Zoologica Scripta 29:1–16.

Darrell, M. 2020. Bermuda National Trust objection to proposed demolition of unstable mound with subterranean void in Wilkinson Quarry, Hamilton Parish, Letter to Mrs. Victoria Pereira, Director of Planning, Hamilton, Bermuda. Accessed December 12, 2023. [https://bnt.bm/wilkinson-quarry-objection/].

DBCA (Department of Biodiversity, Conservation and Attractions). 2021. Priority ecological communities for Western Australia, version 32. Species and Communities Program, Department of Biodiversity, Conservation and Attractions, 15 July 2021. Accessed December 6, 2023. [https://www.dbca.wa.gov.au/wildlife-and-ecosystems/threatened-ecological-communities/list-threatened-ecological-communities].

EPAWA (Environmental Protection Authority of Western Australia). 2016. Environmental factor guideline: Subterranean fauna. Environmental Protection Authority of Western Australia, Joondalup, Western Australia. Accessed November 25, 2024. [https://www.epa.wa.gov.au/sites/default/files/Policies_and_Guidance/Guideline-Subterranean-Fauna-131216_3.pdf].

EPAWA (Environmental Protection Authority of Western Australia). 2021. Technical Guidance - Subterranean fauna surveys for environmental impact assessment. Environmental Protection Authority of Western Australia, Joondalup, Western Australia. Accessed November 25, 2024 [https://www.epa.wa.gov.au/sites/default/files/Policies_and_Guidance/Technical Guidance - Subterranean fauna surveys for EIA.pdf].

Escobar-Briones, E., M.E. Camacho, and J. Alcocer. 1997. *Calliasmata nohochi*, new species (Decapoda: Caridea: Hippolytidae), from anchialine cave systems in continental Quintana Roo, Mexico. Journal of Crustacean Biology 17:733–744.

Garcia-Valdecasas, A. 1984. *Morlockiidae* new family of Remipedia (Crustacea) from Lanzarote (Canary Islands). Eos 60:329–333.

Glasspool, A.F. 2008. The impact of climate change on Bermuda. The Bermuda National Trust, Hamilton, Bermuda.

Glasspool, A.F. 2014. Management plan for Bermuda's critically endangered cave fauna. Bermuda Department of Conservation Services, Hamilton, Bermuda.

Gobert, S., P. Reygel, and T. Artois. 2019. *Schizorhynchia* (Platyhelminthes Rhabdocoela) of Lanzarote (Canary Islands), with the description of eight new species. Marine Biodiversity 49: 2089–2107.

Gómez, P., and F. Calderón-Gutiérrez. 2020. Anchialine cave-dwelling sponge fauna (Porifera) from la Quebrada, Mexico, with the description of the first Mexican stygobiont sponges. Zootaxa 4803:125–151.

Gonzalez, B.C., A. Martínez, E. Borda, T.M. Iliffe, D. Fontaneto, and K. Worsaae. 2017. Genetic spatial structure of an anchialine cave annelid indicates connectivity within—but not between—islands of the Great Bahama Bank. Molecular Phylogenetics and Evolution 109:259–270.

Gonzalez, B.C., A. Martínez, J. Olesen, S.B. Truskey, L. Ballou, M. Allentoft-Larsen, J. Daniels, P. Heinerth, M. Parrish, N. Manco, J. Ward, T.M. Iliffe, K.J. Osborn, and K. Worsaae. 2020. Anchialine biodiversity in the Turks and Caicos Islands: New discoveries and current faunal composition. International Journal of Speleology 49:71–86.

Gonzalez, B.C., K. Worsaae, D. Fontaneto, and A. Martínez. 2018. Anophthalmia and elongation of body appendages in cave scale worms (Annelida: Aphroditiformia). Zoologica Scripta 47:106–121.

Government of Bermuda. 2022. How to design a septic tank. Accessed December 12, 2023. [https://www.gov.bm/how-design-septic-tank].

Grimes, K.G., and W.F. Humphreys. 1998. Christmas Island cave studies. Helictite 36:17–18.

Gutu, G., and T.M. Iliffe, 1998. Description of a new hirsutiid (n.g., n.sp.) and reassignment of this family from order Mictacea to the new order, Bochusacea (Crustacea, Peracrida). Travaux du Muséum d'Histoire Naturelle "Grigore Antipa" 40:93–120.

Guzik, M.T., A.D. Austin, S.J.B. Cooper, M.S. Harvey, W.F. Humphreys, T. Bradford, S.M. Eberhard, R.A. King, R. Leys, K.A. Murihead, and M. Tomlinson. 2011. Is the Australian subterranean fauna uniquely diverse? Invertebrate Systematics 24:407–418.

Guzik, M.T., D.N. Stringer, N.P. Murphy, S.J.B. Cooper, S. Taiti, R.A. King, W.F. Humphreys, and A.D. Austin. 2019. Molecular phylogenetic analysis of Australian arid zone oniscidean isopods (Crustacea: *Haloniscus*) reveals strong regional endemicity and new putative species. Invertebrate Systematics 33:556–574.

Halse, S.A., M.D. Scanlon, J.S. Cocking, H.J. Barron, J.B. Richardson, and S.M. Eberhard. 2014. Pilbara stygofauna: Deep groundwater of an arid landscape contains globally significant radiation of biodiversity. Records of the Western Australian Museum 78:443–483.

Hart Jr., C.W., R.B. Manning, and T.M. Iliffe. 1985. The fauna of Atlantic marine caves: Evidence of dispersal by sea floor spreading while maintaining ties to deep waters. Proceedings of the Biological Society of Washington 98:288–292.

Hernández-Terrones, L., M. Rebolledo-Vieyra, M. Merino-Ibarra, M. Soto, A. Le-Cossec, and E. Monroy-Ríos. 2011. Groundwater pollution in a karstic region (NE Yucatan): Baseline nutrient content and flux to coastal ecosystems. Water, Air, and Soil Pollution 218:517–528.

Holsinger, J.R. 1992. Two new species of the subterranean amphipod genus *Bahadzia* (Hadziidae) from the Yucatan Peninsula region of southern Mexico, with an analysis of phylogeny and biogeography of the genus. Stygologia 7:85–105.

Holthuis, L.B. 1973. Caridean shrimps found in land-locked saltwater pools at four Indo-West Pacific localities (Sinai Peninsula, Funafuti Atoll, Maui and Hawaii Islands), with the description of one new genus and four new species. Zoologische Verhandelingen 128:1–48.

Humphreys, W.F. 1995. Subterranean fauna. Pages 45–47 *in* Consultative Environmental Review: Extensions to Exmouth Water Supply Borefield, CER, June 1995. Water Authority of Western Australia, Perth, Australia.

Humphreys, W.F. 1996a. An examination of the likely impacts of the Exmouth borefield extension on the subterranean fauna. Appendix D: Subterranean fauna. *In* Muir Environmental. 1996. Extensions to Exmouth Water Supply Borefield Consultative Environmental Review. Report No. WP 225. Water Resources Directorate. Water Resources Planning Branch, Water Corporation of Western Australia, Perth, Australia.

Humphreys, W.F. 1996b. Appendix 3: Stygofauna and energy entrainment in the Exmouth borefield. In Supplementary investigation of the effect of public water supply abstraction on the stygofauna and aquifer of the Cape Range. An addendum to Extensions to Exmouth Water Supply Borefield, CER, June 1995. Water Corporation of Western Australia, Perth, Australia.

Humphreys, W.F. 1999. Physico-chemical profile and energy fixation in Bundera Sinkhole, an anchialine remiped habitat in north-western Australia. Journal of the Royal Society of Western Australia. 82:89.

Humphreys, W.F. 2000. The hypogean fauna of the Cape Range Peninsula and Barrow Island, northwestern Australia. Pages 581–601 *in* H. Wilkens, D.C. Culver, and W.F. Humphreys, editors. Ecosystems of the world, Volume 30: Subterranean ecosystems. Elsevier, Amsterdam, The Netherlands.

Humphreys, W.F. 2002. The subterranean fauna of Barrow Island, northwestern Australia, and its environment. Mémoires de Biospéologie—International Journal of Subterranean Biology 28:107–127.

Humphreys, W.F. 2014. Subterranean fauna of Christmas Island: Habitats and salient features. Raffles Bulletin of Zoology, 30:29–44.

Humphreys, W.F., and D.L. Danielopol. 2006. Danielopolina (Ostracoda, Thaumatocyprididae) on Christmas Island, Indian Ocean, a sea mount island. Crustaceana 78:1339–1352.

Humphreys, W.F., and S. Eberhard. 2001. Subterranean fauna of Christmas Island, Indian Ocean. Helictite 37:59–74.

Humphreys, W.F., L.S. Kornicker, and D.L. Danielopol. 2009. On the origin of *Danielopolina baltanasi* sp. n. (Ostracoda, Thaumatocypridoidea) from three anchialine caves on Christmas Island, a seamount in the Indian Ocean. Crustaceana 82:1177–1203.

Humphreys, G., J. Alexander, M.S. Harvey, and W.F. Humphreys. 2013. The subterranean fauna of Barrow Island, Northwestern Australia: 10 years on. Records of the Western Australian Museum 83:145–158.

Iliffe, T.M. 1979. Bermuda's caves: A non-renewable resource. Environmental Conservation 6:181–186.

Iliffe, T.M. 1990. Crevicular dispersal of marine faunas. Mémoires de Biospéologie 17:93–96.

Iliffe, T.M. 2000. Anchialine cave ecology. Pages 59–76 *in* H. Wilkens, D.C. Culver, and W.F. Humphreys, editors. Ecosystems of the world, Volume 30: Subterranean ecosystems. Elsevier Science, Amsterdam, The Netherlands.

Iliffe, T.M. 2003. Submarine caves and cave biology of Bermuda. NSS News 2003:217–224.

Iliffe, T.M., and F. Calderón-Gutiérrez. 2021. Bermuda's Walsingham caves: A global hotspot for anchialine stygobionts. Diversity 13:352.

Iliffe, T.M., and L. Kornicker. 2009. Worldwide diving discoveries of living fossil animals from the depths of anchialine and marine caves. Pages 269–280 *in* M.A. Lang, I.G. Macintyre, and K. Rützler, editors. Smithsonian Institution Scholarly Press, Washington, USA.

Iliffe, T.M., C.W. Hart, and R.B. Manning. 1983. Biogeography and the caves of Bermuda. Nature 302:141–142.

Iliffe, T.M., T.D. Jickells, and M.S. Brewer. 1984a. Organic pollution of an inland marine cave from Bermuda. Marine Environmental Research 12:173–189.

Iliffe, T.M., H. Wilkens, J. Parzefall, and D. Williams 1984b. Marine lava cave fauna: Composition, biogeography and origins. Science 225:309–311.

Iliffe, T.M., D. Brankovits, V. Gerovasileiou, B. Gonzalez, A. Martínez García, and D.A. Keith. 2020. SM1.1 Anchialine caves. Page 92 *in* D.A. Keith, J.R. Ferrer-Paris, E. Nicholson, and R.T. Kingsford, editors.) The IUCN Global Ecosystem Typology 2.0: Descriptive profiles for biomes and ecosystem functional groups. IUCN, Gland, Switzerland.

Jacobson, G., and A.V. Arakel. 1986. Calcrete aquifers in the Australian arid zone. Pages 515–523 *in* Proceedings of the International Conference on Groundwater Systems Under Stress. Australian Water Resources Council, Brisbane, Australia.

James, D.J., P.T. Green, W.F. Humphreys, and J.C.Z. Woinarski. 2019. Endemic species of Christmas Island, Indian Ocean. Records of the Western Australian Museum 34:55–114.

Jaume, D., and G.A. Boxshall. 1996 Two new genera of cyclopinid copepods (Crustacea) from anchihaline caves on western Mediterranean and eastern Atlantic islands. Zoological Journal of the Linnean Society 117:283–304.

Jurado-Rivera, J.A., J. Pons, F. Alvarez, A. Botello, W.F. Humphreys, T.J. Page, T.M. Iliffe, E. Willassen, K. Meland, C. Juan, and D. Jaume. 2017. Phylogenetic evidence that both ancient vicariance and dispersal have contributed to the biogeographic patterns of anchialine cave shrimps. Scientific Reports 7:2852.

Karanovic, I., and S.N. Brandão. 2012. On the genus *Thaumatoconcha* Kornicker and Sohn (Halocyprida) with description of two new species from Southern Ocean deep sea. Helgoland Marine Research 66:275–294.

Kensley, B. 1988. New species and records of cave shrimps from the Yucatan Peninsula (Decapoda: Agostocarididae and Hippolytidae). Journal of Crustacean Biology 8:688–699.

King, R.A., T. Bradford, A.D. Austin, W.F. Humphreys, and S.J.B. Cooper. 2012. Divergent molecular lineages and not-so-cryptic species: The first descriptions of stygobitic chiltoniid amphipods (Talitroidea: Chiltoniidae) from Western Australia. Journal of Crustacean Biology 32:465–488.

Koenemann, S., A. Bloechl, A. Martínez, T.M. Iliffe, M. Hoenemann, and P. Oromí. 2009. A new, disjunct species of *Speleonectes* (Remipedia, Crustacea) from the Canary Islands. Marine Biodiversity 39:215–225.

Koenemann, S., T.M. Iliffe, and J. Yager. 2004. *Kaloketos pilosus*, a new genus and species of Remipedia (Crustacea) from the Turks and Caicos Islands. Zootaxa 618:1–12.

Kornicker, L.S., and T.M. Iliffe. 1995. Ostracoda (Halocypridina, Cladocopina) from an anchialine lava tube in Lanzarote, Canary Islands. Smithsonian Contributions to Zoology 568:1–32.

Kornicker, L.S., D.L. Danielopol, and W.F. Humphreys. 2006. Description of the anchialine ostracode *Danielopolina* sp. cf. *D. kornickeri* from Christmas Island, Indian Ocean. Crustaceana 79:77–88.

Lambeck, K., H. Rouby, A. Purcell, Y. Sun, and M. Sambridge. 2014. Sea level and global ice volumes from the Last Glacial Maximum to the Holocene. Proceedings of the National Academy of Sciences 111:15296–15303.

Larson, H.L., R. Foster, W.F. Humphreys, and M.I. Stevens. 2013. A new species of the blind cave gudgeon *Milyeringa* (Gobioidei, Eleotridae, Butinae) from Barrow Island, Western Australia, with a redescription of *M. veritas* Whitley. Zootaxa 3616:135–150.

Leach, D.L., R.D. Taylor, D.L. Fey, S.F. Diehl, and R.W. Saltus. 2010. A Deposit Model for Mississippi Valley-type lead-zinc ores. Chapter A of Mineral Deposit Models for Resource Assessment. Scientific Investigations Report 2010–5070–A. U.S. Department of the Interior, U.S. Geological Survey, USA.

Leal-Bautista, R.M., G. Hernández-Zárate, N.A. Jaime, G. Cuevas, and G. Velázquez-Oliman. 2011. Pathogens and pharmaceuticals pollutants as indicators of contamination at the northeastern aquifer of Quintana Roo. Tropical and Subtropical Agroecosystems 13:211–219.

Leal-Bautista, R.M., M. Lenczewski, C. Morgan, A. Gahala, and J.E. McLain. 2013. Assessing fecal contamination in groundwater from the Tulum region, Quintana Roo, Mexico. Journal of Environmental Protection 4:1272–1279.

León-Borges, J.A., and M.A. Lizardi-Jiménez. 2017. Hydrocarbon pollution in underwater sinkholes of the Mexican Caribbean caused by tourism and asphalt: Historical data series and cluster analysis. Tourism Management 63:179–186.

Lozano-Fernandez, J., M. Giacomelli, J.F. Fleming, A. Chen, J. Vinther, P.F. Thomsen, H. Glenner, F. Palero, D.A. Legg, T.M. Iliffe, D. Pisani, and J. Olesen. 2019. Pancrustacean evolution illuminated by taxon-rich genomic-scale data sets with an expanded remipede sampling. Genome Biology and Evolution 11:2055–2070.

Maksimovich, N.G., and V.T. Khmurchik. 2015. Remediation of oil-polluted groundwater aquifers at karst region. Pages 417–419 *in* G. Lollino, M. Arattano, M. Rinaldi, O. Giustolisi, J.-C. Marechal, and G.E. Grant, editors. Engineering geology for society and territory, Volume 3. Springer International Publishing, Switzerland.

Mammola, S., P. Cardoso, D.C. Culver, L. Deharveng, R.L. Ferreira, C. Fišer, D.M.P. Galassi. C. Griebler, S. Halse, W.F. Humphreys, M. Isaia, F. Malard, A. Martinez, O.T. Moldovan, M.L. Niemiller, M. Pavlek, A.S.P.S. Reboleira, M. Souza-Silva, E.C. Teeling, J.J. Wynne, and M. Zagmajster. 2019. Scientists' warning on the conservation of subterranean ecosystems. BioScience 69:641–650.

Mammola, S., M.B. Meierhofer, P.A.V. Borges, R. Colado, D.C. Culver, L. Deharveng, T. Delić, T. Di Lorenzo, T. Dražina, R.L. Ferreira, B. Fiasca, C. Fišer, D.M.P. Galassi, L. Garzoli, V. Gerovasileiou, C. Griebler, S. Halse, F.G. Howarth, M. Isaia, J.S. Johnson, A. Komerički, A. Martínez, F. Milano, O.T. Moldovan, V. Nanni, G. Nicolosi, M.L. Niemiller, S. Pallarés, M. Pavlek, E. Piano, T. Pipan, D. Sanchez-Fernandez, A. Santangeli, S.I. Schmidt, J.J. Wynne, M. Zagmajster, V. Zakšek, and P. Cardoso. 2022. Towards evidence-based conservation of subterranean ecosystems. Biological Reviews 97:1476–1510.

Martínez, A., and B.C. Gonzalez. 2019. Research in anchialine caves: Anchialine lava environments in Lanzarote as case of study. Pages 399–414 *in* O.T. Moldovan, .Ľ Kováč, and S. Halse, editors. Cave ecology. Springer Verlag, Cham, Switzerland.

Martínez, A., M. Di Domenico, and K. Worsaae. 2013. Evolution of cave *Axiokebuita* and *Speleobregma* (Scalibregmatidae, Annelida). Zoologica Scripta 42:623–636.

Martínez, A., A. Di Cesare, N. Mari-Mena, G. García-Gómez, A. Garcia-Herrero, D. Fontaneto, and E.M. Eckert. 2020. Tossed 'good luck' coins as vectors for anthropogenic pollution into aquatic environment. Environmental Pollution 259:113800.

Martínez, A., M. Di Domenico, F. Leasi, M. Curini-Galletti, M.A. Todaro, M.D. Zotto, S. Gobert, T. Artois, J. Norenburg, K.M. Jörger, J. Núñez, D. Fontaneto, and K. Worsaae. 2019a. Patterns of diversity and endemism of soft-bodied meiofauna in an oceanic island, Lanzarote, Canary Islands. Marine Biodiversity 49:2033–2055.

Martínez, A., G. García-Gómez, A. García-Herrero, Á.D. Cesare, G. Corno, R. Herrera, L. Moro, E. Eckert, and D. Fontaneto. 2019b. Lanzarote and Chinijo Islands: An anchialine UNESCO Global Geopark. Pages 109–121 *in* E. Mateo, J. Martínez-Frías, and J. Vegas, editors. Lanzarote and Chinijo Islands geopark: From Earth to space. Springer International Publishing, Cham, Switzerland.

Martínez, A., B.C. Gonzalez, K. Worsaae, H. Wilkens, P. Oromí, T.M. Iliffe, and K. Worsaae. 2016. Guide to the anchialine ecosystems of Jameos del Agua and Túnel de la Atlántida. Medio Ambiente, Cabildo de Lanzarote, Arrecife, Lanzarote, Spain.

Martínez, A., K. Kvindebjerg, T.M. Iliffe, and K. Worsaae. 2017. Evolution of cave suspension feeding in Protodrilidae (Annelida). Zoologica Scripta 46:214–226.

Martínez García, A., A.M. Palmero, M. Carmen Brito, J. Núñez, and K. Worsaae. 2009. Anchialine fauna of the Corona lava tube (Lanzarote, Canary Islands): Diversity, endemism and distribution. Marine Biodiversity 39:169–182.

Mazza, S.E., E. Gazel, M. Bizimis, R. Moucha, P. Béguelin, E.A. Johnson, R.J. McAleer, and A.V. Sobolev. 2019. Sampling the volatile-rich transition zone beneath Bermuda. Nature 569:398–403.

Medina-Moreno, S.A., A. Jiménez-González, A. Gutierrez-Rojas, and M.A. Lizardi-Jiménez. 2014. Hydrocarbon pollution studies of underwater sinkholes along Quintana Roo as a function of tourism development in the Mexican Caribbean. Revista Mexicana de Ingeniería Química 13:509–516.

Metcalfe, C.D., P.A. Beddows, G. Gold Bouchot, T.L. Metcalfe, L. Hongxia, and H. Van Lavieren. 2011. Contaminants in the coastal karst aquifer system along the Caribbean coast of the Yucatan Peninsula. Environmental Pollution 159:991–997.

Morris, B., J. Barnes, F. Brown, and J. Markham. 1977. The Bermuda marine environment, Volume 1. Bermuda Biological Station Special Report 15:1–120.

Muir, R.M., E.M. Bordy, R. Mundil, and D. Frei. 2020. Recalibrating the breakup history of SW Gondwana: U–Pb radioisotopic age constraints from the southern Cape of South Africa. Gondwana Research 84:177–193.

Namiotko, T., K. Wouters, D.L. Danielopol, and W.F. Humphreys. 2004. On the origin and evolution of a new anchialine stygobitic *Microceratina* species (Crustacea, Ostracoda) from Christmas Island (Indian Ocean). Journal of Micropalaeontology 23:49–60.

Neiber, M.T., F.C. Hansen, T.M. Iliffe, B. Gonzalez, and S. Koenemann. 2012. Molecular taxonomy of *Speleonectes fuchscockburni*, a new pseudocryptic species of Remipedia (Crustacea) from an anchialine cave system on the Yucatán Peninsula, Quintana Roo, Mexico. Zootaxa 3190:31–46.

Ohtsuka, S., A. Fosshagen, and T.M. Iliffe. 1993. Two new species of *Paramisophria* (Copepoda, Calanoida, Arietellidae) from anchialine caves on the Canary and Galápagos Islands. Sarsia 78:57–67.

Olesen, J., K. Meland, H. Glenner, P. Van Hengstum, and T.M. Iliffe. 2017. *Xibalbanus cozumelensis*, a new species of Remipedia (Crustacea) from Cozumel, Mexico, and a molecular phylogeny of *Xibalbanus* on the Yucatán Peninsula. European Journal of Taxonomy 316:1–27.

Page, T.J., W.F. Humphreys, and J.M. Hughes. 2008. Shrimps down under: Evolutionary relationships of subterranean crustaceans from Western Australia (Decapoda: Atyidae: *Stygiocaris*). PLoS ONE 3:e1618.

Page, T.J., J.M. Hughes, K.M. Real, M.I. Stevens, R.A. King, and W.F. Humphreys. 2016. Allegory of the cave crustacean: Systematic and biogeographic reality of *Halosbaena* (Peracarida: Thermosbaenacea) sought with molecular data at multiple scales. Marine Biodiversity 48:1185–1202.

Page, T.J., M.I. Stevens, M. Adams, R. Foster, A. Velasco-Castrillón, and W.F. Humphreys. 2018. Multiple molecular markers reinforce the systematic framework of unique Australian cave fishes (*Milyeringa*: Gobioidei). Australian Journal of Zoology 66:115–127.

Palmer, A.N., M.V. Palmer, and J.M. Queen. 1977. Geology and origin of the caves of Bermuda. Pages 336–339 *in* Proceedings of the seventh international congress of speleology. International Union of Speleology British Cave Research Association, Sheffield, UK.

Pettibone, M. 1985. Polychaete worms from a cave in the Bahamas and from experimental wood panels in deep water of the North Atlantic (Polynoidae, Macellicephalinae, Harmothoinae). Proceedings of the Biological Society of Washington 98:127–149.

Playford, P.E., R.M. Hocking, and A.E. Cockbain. 2009. Devonian reef complexes of the Canning Basin, Western Australia. Geological Survey of Western Australia Bulletin 145:1–444.

Polanco-Rodríguez, A.G., T.V. Magaña-Castro, J. Cetz-Iuit, and R. Quintal-López. 2019. Uso de agroquímicos cancerígenos en la región agrícola de Yucatán, México. Centro Agrícola 46:72–83.

Polanco-Rodríguez, A.G., M.I. Riba-López, A. DelValls-Casillas, J.A. Araujo-León, and S. Datta Banik. 2018. Impact of pesticides in karst groundwater. Review of recent trends in Yucatan, Mexico. Groundwater for Sustainable Development 7:20–29.

Poole, A., and D. Warren. 1995. Bundera Sinkhole. Accessed December 12, 2023. [http://www.cavediving.net.au/docs/cape-range/6C-28.jpg].

Poore, G.C.B. 2015. Orders Bochusacea, Mictacea and Spelaeogriphacea. Pages 77–92 *in* C. von Vaupel Lein, editor. Treatise on zoology—Anatomy, taxonomy, biology. The Crustacea, Volume 5. Brill, Leiden, The Netherlands.

Poore, G.C.B., and W.F. Humphreys. 1992. First record of Thermosbaenacea (Crustacea) from the Southern Hemisphere: A new species from a cave in tropical Western Australia. Invertebrate Taxonomy 6:719–725.

Poore, G.C.B., and W.F. Humphreys. 2013. *Bunderanthura bundera* gen. et sp. nov. from Western Australia, first anchialine Leptanthuridae (Isopoda) from the Southern Hemisphere. Records of the Western Australia Museum 28:21–29.

QRSS. 2022. Quintana Roo Speleological Survey. Accessed December 12, 2023. [https://legacy.caves.org/project/qrss/qrss.htm].

Ritchie, H. 2019. Which countries are most densely populated? Accessed December 12, 2023. [https://ourworldindata.org/most-densely-populated-countries].

Rossetti, G., R.L. Pinto, and K. Martens. 2011. Description of a new genus and two new species of Darwinulidae (Crustacea, Ostracoda), from Christmas Island (Indian Ocean) with some considerations on the morphological evolution of ancient asexuals. Belgian Journal of Zoology 141:55–74.

Rowe, M.P. 2020. Bermuda geology. Accessed December 12, 2023. [https://bermudageology.com/].

Sakihara, T.S., M. Lamson, B. Seidel, D.M. Sedar, S. Santos, J. Havird, T.M. Iliffe, and D.A. Keith. 2020. Anchialine pools. Page 92 *in* D.A. Keith, J.R. Ferrer-Paris, E. Nicholson, and R.T. Kingsford, editors. The IUCN global ecosystem typology 2.0: Descriptive profiles for biomes and ecosystem functional groups. International Union for Conservation of Nature, Gland, Switzerland.

Schram, F.R., J. Yager, and M.J. Emerson. 1986. Remipedia. Part I. Systematics. San Diego Society of Natural History Memoir 15:4–60.

SET (Southern Environmental Technologies). 2014. Bassett's Cave survey report, Morgan's Point, Bermuda, August 12, 2014, submitted to Langan Engineering and Environmental Services Inc. Accessed December 12, 2023. [https://southernenvironmental.us/projects/].

Shaw, A.K., and K.A. Kelly. 2013. Linking El Niño, local rainfall, and migration timing in a tropical migratory species. Global Change Biology 19:3283–3290.

Spaw, R.H. 1978. Late Pleistocene carbonate bank deposition: Cozumel Island, Quintana Roo, Mexico. Gulf Coast Association Geological Society 28:601–619.

Stemme, T., T.M. Iliffe, and G. Bicker. 2015. Olfactory pathway in *Xibalbanus tulumensis*: Remipedian hemiellipsoid body as homologue of hexapod mushroom body. Cell and Tissue Research 363:635–648.

Stemme, T., T.M. Iliffe, B.M. von Reumont, S. Koenemann, S. Harzsch, and G. Bicker. 2013. Serotonin-immunoreactive neurons in the ventral nerve cord of Remipedia (Crustacea): Support for a sister group relationship of Remipedia and Hexapoda? BMC Evolutionary Biology 13:1–15.

Stock, J.H. 1993. Some remarkable distribution patterns in stygobiont Amphipoda. Journal of Natural History 27:807–819.

Stringer, D.N., R.A. King, S. Taiti, M.T. Guzik, S.J.B. Cooper, and A.D. Austin. 2019. Systematics of *Haloniscus* Chilton, 1920 (Isopoda: Oniscidea: Philosciidae), with description of four new species from threatened Great Artesian Basin springs in South Australia. Journal of Crustacean Biology 39:651–668.

Taiti, S., and W.F. Humphreys. 2001. New aquatic Oniscidea (Crustacea, Isopoda) from groundwater calcretes of Western Australia. Pages 133–151 *in* W.F. Humphreys and M.S. Harvey, editors. Subterranean biology in Australia 2000, supplement no. 64. Records of the Western Australian Museum, Australia.

Vacher, H.L., and M.P. Rowe. 1997. Geology and hydrogeology of Bermuda. Pages 35–90 *in* H.L. Vacher and T. Quinn, editors. Geology and hydrogeology of carbonate islands, Volume 54: Developments in sedimentology. Elsevier, Amsterdam, The Netherlands.

Vacher, H.L., M. Rowe, and P. Garrett. 1989. The geological map of Bermuda. Bermuda Ministry of Works and Engineering. Oxford Cartographers, London. Public Works Department, Bermuda, UK.

van Hengstum, P.J., J.N. Cresswell, G.A. Milne, and T.M. Iliffe. 2019. Development of anchialine cave habitats and karst subterranean estuaries since the last ice age. Scientific Reports 9:11907.

Vogt, P., and W. Jung. 2007. Origin of the Bermuda volcanoes and the Bermuda Rise: History, observations, models, and puzzles. Special Papers-Geological Society of America 430:553.

von Martens, E. 1872. Über Cubanische Crustaceen. Archiv für Naturgeschichte 38:77–147.

von Reumont, B.M., and T. Burmester. 2010. Remipedia and the evolution of hexapods. Pages 1–6 *in* Encyclopedia of Life Sciences. John Wiley & Sons, Chichester, UK.

von Reumont, B.M., and G.D. Edgecombe. 2020. Crustaceans and insect origins. Pages 105–120 *in* G.C.B. Poore and M. Thiel, editors. The natural history of the Crustacea, Volume 8: Evolution and biogeography. Oxford University Press, New York, USA.

von Reumont, B.M., A. Blanke, S. Richter, F. Alvarez, C. Bleidorn, and R.A. Jenner. 2014. The first venomous crustacean revealed by transcriptomics and functional morphology: Remipede venom glands express a unique toxin cocktail dominated by enzymes and a neurotoxin. Molecular Biology and Evolution 31:48–58.

von Reumont, B.M., R.A. Jenner, M.A. Wills, E. Dell'Ampio, G. Pass, I. Ebersberger, B. Meyer, S. Koenemann, T.M. Iliffe, A. Stamakis, O. Niehaus, K. Meusemann, and B. Misof. 2012. Pancrustacean phylogeny in the light of new phylogenomic data: Support for Remipedia as the possible sister group of Hexapoda. Molecular Biology and Evolution 29:1031–1045.

Ward, W.C. 1985. Quaternary geology of northeastern Yucatan Peninsula. Pages 23–53 *in* W.C. Ward, E. Weidie, and W. Back, editors. Geology and hydrogeology of the Yucatan Peninsula. New Orleans Geological Society, New Orleans, USA.

Wilkens, H., T.M. Iliffe, P. Oromí, A. Martínez, T.N. Tysall, and S. Koenemann. 2009. The Corona lava tube, Lanzarote: Geology, habitat diversity and biogeography. Marine Biodiversity 39:155–167.

Wilson, G.D.F. 2003. A new genus of Tainisopidae fam. nov. (Crustacea: Isopoda) from the Pilbara, Western Australia. Zootaxa 245:1–20.

Wilson, G.D.F., and W.F. Ponder. 1992. Extraordinary new subterranean isopods (Peracarida: Crustacea) from the Kimberley region, Western Australia. Records of the Australian Museum 44:279–298.

Wilson, K., B.C. Gonzalez, A. Kerbl, S. Holdflod, J.T. Jørgensen, M. Armenteros, T.M. Iliffe, and A. Martínez. 2019. Diversity and evolution of the stygobitic *Speleonerilla* nom. nov. (Nerillidae, Annelida) with description of three new species from anchialine caves in the Caribbean and Lanzarote. Marine Biodiversity 49:2167–2192.

Worsaae, K., M.D. Mikkelsen, and A. Martínez. 2019. Description of six new species of *Mesonerilla* (Nerillidae, Annelida) and an emended description of *M. intermedia* Wilke, 1953, from marine and cave environments. Marine Biodiversity 49:2141–2165.

Wynne, J.J., F.G. Howarth, S. Mammola, R.L. Ferreira, P. Cardoso, T. Di Lorenzo, D.M.P. Galassi, R.A. Medellin, B.W. Miller, D. Sánchez-Fernández, M.E. Bichuette, J. Biswas, C.W. BlackEagle, C. Boonyanusith, I.R. Amorim, P.A. Vieira Borges, P.J. Boston, R.N. Cal, N. Cheeptham, L. Deharveng, D. Eme, A. Faille, D. Fenolio, C. Fišer, Ž. Fišer, S.M. 'Ohukani'ōhi'a Gon III, F. Goudarzi, C. Griebler, S. Halse, H. Hoch, E. Kale, A.D. Katz, L'. Kováč, T.M. Lilley, S. Manchi, R. Manenti, A. Martínez, M.B. Meierhofer, A.Z. Miller, O.T. Moldovan, M.L. Niemiller, S.B. Peck, T. Giovannini Pellegrini, T. Pipan, C.M. Phillips-Lander, C. Poot, P.A. Racey, A. Sendra, W.A. Shear, M. Souza Silva, S. Taiti, M. Tian, M.P. Venarsky, S. Yancovic Pakarati, M. Zagmajster, and Y. Zhao. 2021. A conservation roadmap for the subterranean biome. Conservation Letters 14:e12834.

Yager, J. 1981. A new class of Crustacea from a marine cave in the Bahamas. Journal of Crustacean Biology 1:328–333.

Yager, J. 1987. *Cryptocorinetes haptodiscus*, new genus, new species, and *Speleonectes benjamini*, new species, of remipede crustaceans from anchialine caves in the Bahamas, with remarks on distribution and ecology. Proceedings of the Biological Society of Washington 100:302–320.

Yager, J. and W.F. Humphreys. 1996. *Lasionectes exleyi*, sp. nov., the first remipede crustacean recorded from Australia and the Indian Ocean, with a key to the world species. Invertebrate Taxonomy 10:171–187.

15

RECOVERY OF CRUSTACEAN POPULATIONS IN PROTECTED AREAS

Even Moland, Raquel Goñi, Benjamin J. Hanns, and Nick T. Shears

Abstract

Crustaceans are targeted in fisheries worldwide. Although populations have shown remarkable resilience in the face of centuries of exploitation, they often exist in depleted states. Clawed and spiny lobsters have thus been the focus of long-term monitoring and scientific study aimed at gaining a firm understanding of the effects of harvesting and conservation on populations. In marine protected areas (MPAs), areas where fishing pressure has been eliminated or reduced, population effects have been unequivocally demonstrated. This chapter provides an overview of conservation efforts and studies that have contributed to increased understanding of such effects. Three well-studied species and conservation cases are showcased, namely those of *Jasus edwardsii* (New Zealand), *Palinurus elephas* (Mallorca), and *Homarus gammarus* (Norway). In these cases, long-term scientific monitoring and fisheries data have provided information for evaluation of the effects of MPAs and pointed to caveats regarding their design and integration with other fisheries management regulations. The chapter focuses on commercially exploited, large-bodied decapod crustacean species to highlight the relevance of protected areas as harvest refugia, and functional hotspots that sustain resources for adjacent fishing grounds. Appropriately scaled applications of MPAs throughout the marine realm are now deemed essential supplements to traditional fisheries management to achieve sustainability in global fisheries and reach sustainable development goals.

INTRODUCTION

Wild decapod crustacean fisheries are growing faster than fisheries of any other major group globally (Boenish et al. 2022). Throughout the world, decapod crustaceans are highly sought-after target species in recreational, artisanal, and commercial fisheries, and high-value species often show

Even Moland et al., *Recovery of Crustacean Populations in Protected Areas*. In: *Ecology and Conservation*.
Edited by: Lars Gutow and Martin Thiel, Oxford University Press. © Oxford University Press (2025).
DOI: 10.1093/oso/9780197768242.003.0015

clear signs of overharvesting (Spanier and Lavalli 1998, Guzman and Tewfik 2004, Melville-Smith and de Lestang 2006, Agnalt et al. 2007, Kleiven et al. 2012, 2022, LaScala-Gruenewald et al. 2021). A progressive decline in size at maturity over time has been documented for both sexes in the western rock lobster (*Panulirus cygnus*), which forms the basis of a valuable fishery in Western Australia (Melville-Smith and de Lestang 2006). At present, the Norway lobster (*Nephrops norvegicus*) is one of the most important demersal species in the Northeast Atlantic, subject to an intensive bottom trawl fishery with increased landings over the past 50 years (Haynes et al. 2016). In general, decapod crustaceans such as spiny and clawed lobsters have shown rapid population recovery within marine protected areas (MPAs) in settings ranging from tropical to temperate (Table 15.1) (for a recent review, see Spanier 2024). Likewise, while less-iconic crab species are less studied in conservation settings, they tend to show similar responses in fishery closures from Thailand to Alaska (Table 15.1).

Decapod crustaceans are versatile study subjects and models in fisheries science, and iconic species came to pioneer the early marine conservation literature. There are multiple ways to study them, including:

- Population ecology. They are suitable for capture-mark-recapture studies as they are readily caught by non-destructive gear (traps) with minimal risk of damage to the ecosystem, or to the crustacean being caught. Although they shed their exoskeletons (molt—see section on life history below), they retain conventional tags (e.g., T-bar tags, streamer tags) when these are inserted in the dorsal or ventral musculature between cephalotorax and abdomen (e.g., Kelly and MacDiarmid 2003, Agnalt et al. 2007, Fernández-Chacón et al. 2020, 2021).
- Movement and space use. Their exoskeletons make them resilient to handling and facilitate attachment of external tags (e.g., acoustic transmitters) without adverse effects to the animal. For a comprehensive review of telemetry-based studies on decapod crustaceans, see Florko et al. (2021).
- Fecundity, parentage studies, and sexual selection. As females carry and brood their eggs underneath the abdomen (outside their bodies), eggs might be sampled from individuals with no adverse effects to the animal, and on multiple occasions (e.g., Sørdalen et al. 2018, 2020).
- Importantly, their pelagic larval phase resembles that of many marine fish species where a lot has been learned in the realm of biophysical modeling (Kough et al. 2013) and its coupling to population genetics (Segura-García et al. 2019).

Harvesting of coastal crustaceans has occurred since prehistoric times, overlapping with the entire history of human utilization of marine populations for consumption. Consequently, crustaceans are deeply engrained in human lore and culture. Lobster utilization by worldwide human populations was comprehensively reviewed by Spanier et al. (2015). Written records suggest that coastal lobster populations were dense, also in the presence of abundant groundfish predators. As human harvesting slowly transitioned from coastal gleaning and trapping by simple means, technological development in fisheries, boats, and navigational aids eventually led to overfishing and collapse of coastal ecosystems worldwide (Jackson et al. 2001). Fishing is well established as the driver having had the most impact on biodiversity in marine systems in the past 50 years (IPBES, 2019). However, the magnitude of crustacean population reductions and consequences for ecosystem function were poorly understood until the effects of rebounds were studied in MPAs (e.g., Shears and Babcock 2002, 2003).

Table 15.1. Crustaceans for which population recovery in MPAs have been evaluated in the peer-reviewed literature.* Effect = population, fishery, or ecosystem effects of protection as reported in the peer-reviewed literature. Region = part of the species' distribution range where effects have been evaluated. CPUE = catch per unit effort. WPUE = weight per unit effort.

Family	Species	Common name	Effect	Region	Reference
Cancridae	*Cancer magister*	Dungeness crab	CPUE increase, size increase	North Pacific	Frid et al. 2016
	Cancer pagurus	Brown crab, edible crab	Size increase, abundance increase	East Atlantic	Hoskin et al. 2011, Sheehan et al. 2013
Lithodidae	*Paralithodes camtschaticus*	Red king crab, Kamchatka crab, Alaskan king crab	Size increase, abundance increase, spawning biomass increase**	Bering Sea (Bristol Bay)	Kruse et al. 2010
Nephropidae	*Homarus americanus*	American lobster	Male lobster size increase, survival increase	NW Atlantic	Rowe 2002
	Homarus gammarus	European lobster	CPUE increase, WPUE increase, size increase, survival increase, body growth difference, mating behavior difference, relaxed harvest selection, trait rescue, biomass spillover, egg production increase	East Atlantic, NE Atlantic (Skagerrak)	Hoskin et al. 2011, Moland et al. 2013a, b, Howarth et al. 2017, Sørdalen et al. 2018, 2020, 2022, Thorbjørnsen et al. 2018. Stewart et al. 2020, Kraufvelin et al. 2023
	Nephrops norvegicus	Norway lobster	CPUE increase, size increase, abundance increase, biomass increase	NE Atlantic (Kattegat), Mediterranean	Sköld et al. 2022, Vigo et al. 2023

(Continued)

Table 15.1. Continued

Family	Species	Common name	Effect	Region	Reference
Palinuridae	*Jasus edwardsii*	Southern rock lobster, red rock lobster, spiny rock lobster, crayfish, kōura papatea	Density increase, biomass increase, fecundity increase, recruitment subsidy, kelp recovery (trophic cascade)	South Australia, New Zealand	Kelly et al. 2000, Shears and Babcock 2002, 2003, Shears et al. 2006, Barrett et al. 2009
	Jasus lalandii	Cape rock lobster, west coast rock lobster	Abundance difference, size difference	South Atlantic	Mayfield et al. 2005
	Palinurus elephas	European spiny lobster, crayfish, cray, crawfish, common spiny lobster	Density increase, size increase, net contribution to adjacent fishery through spillover, increase in reproductive output	Mediterranean	Goñi et al. 2006, Goñi et al. 2010, Díaz et al. 2011, Follesa et al. 2011, Cau et al. 2019
	Panulirus argus	Caribbean spiny lobster	Size increase, legal-sized lobsters abundance increase, spillover	Caribbean	Cox and Hunt 2005, Ley-Cooper et al. 2014
	Panulirus cygnus	Western rock lobster	Density increase, biomass increase, egg production increase, density dependence (increased natural mortality, stunted growth)	Western Australia	Babcock et al. 2007, Tuffley and de Lestang 2024

(Continued)

Table 15.1. Continued

Family	Species	Common name	Effect	Region	Reference
	Panulirus interruptus	California spiny lobster	CPUE increase, size increase, abundance gradient, fishery benefits through spillover, kelp recovery (trophic cascade)	East Pacific, Mexico to Southern California	Kay et al. 2012a, b, Eisaguirre et al. 2020, Lenihan et al. 2021
	Panulirus penicillatus	Red lobster, pronghorn spiny lobster, coral cray	No effect (noncompliance assumed)	East Pacific	Buglass et al. 2018
Portunidae	*Necora puber*	Velvet crab, fighter crab, lady crab	CPUE increase	East Atlantic	Howarth et al. 2017
	Scylla serrata	Black crab, mud crab, mangrove crab	CPUE increase, size increase, spillover	Indo-Pacific	Pillans et al. 2005, Jones et al. 2017
Scyllaridae	*Scyllarides latus*	Mediterranean slipper lobster	Abundance difference, density difference, size difference	Mediterranean	Miller et al. 2023

*Reference list is not exhaustive. **Effects throughout fishing area from rebuilding efforts including large areas closed to trawling.

Crustacean Life History Traits Facilitating Spatial Management

Large-bodied decapod crustaceans share a suite of life history characteristics that have a bearing on their population responses to harvesting and protection. Body growth is slow and, as for all crustaceans, their growth is intermittent and consists of a series of consecutive "molt cycles," involving shedding of their exoskeletons. Growth is conferred by hydrostatic inflation of the new, soft exoskeleton (one size up) prior to its hardening (Dillaman et al. 2013, Webster 2015). Because of the molting process, crustacean age estimation is challenging (but see Sheehy et al. 1999, Gnanalingam et al. 2019, Huntsberger et al. 2020). However, longevity of large-bodied decapod crustaceans can span several decades (e.g., Sheehy et al. 1999). Natural mortality is usually low (e.g., Frisch and Hobbs 2012, Moland et al. 2013b), and species reach considerable size (and age) prior to sexual maturity.

Sperm transfer is direct, and females may store sufficient sperm in their internal sperm pouch (from one or multiple matings) to perform multiple egg-extrusions (the equivalent of spawning) during which eggs are fertilized as they are extruded through the oviduct and pass the opening of the sperm pouch. Complex sexual dynamics and behavior in lobsters (MacDiarmid and Butler 1999, Sørdalen et al. 2018) confer traits that are susceptible to harvest selection (Sørdalen et al. 2020). Eggs are glued to the swimming legs (pleopods) below the abdomen, where they are carried externally by the brooding female until hatching. This represents an opportunity for females to position hatching larvae into a water mass that might confer retention or dispersal of offspring immediately after hatching. Fecundity is limited by the available volume below the abdomen, and generally follows a linear increase with increasing body size (e.g., Goñi et al. 2003, Agnalt 2008).

The planktonic larval phase in large-bodied decapod crustaceans represents considerable potential for dispersal. Although the European lobster *H. gammarus* has a pelagic larval duration (PLD) lasting from three to eight weeks, larvae might be retained locally through positioning in water masses subject to retention or limited transport (Øresland and Ulmestrand 2013). Spiny lobsters, such as *Panulirus cygnus*, have a highly adapted larval phase in which the drifting phyllosoma larvae will be transported over several hundreds of kilometers of open ocean before settling in nursery habitat as the more actively swimming puerulus larval stage over a PLD that might last up to 11 months (see Kolbusz et al. [2021] and references therein). At the extreme end of the scale, *Jasus edwardsii* has a PLD lasting up to 24 months (see below). To date, genetic structuring of metapopulations for species at both ends of the PLD range point to a high degree of connectivity (e.g., Ellis et al. 2017, Segura-García et al. 2019, Moksnes and Jonsson 2020).

Below, we provide descriptions of three well-studied species and conservation cases, namely those of *J. edwardsii* (New Zealand), *Palinurus elephas* (Mallorca) and *H. gammarus* (Norway). In these cases, long-term scientific monitoring and fisheries data have provided information for evaluation of the effects of MPAs and pointed to caveats regarding their design and integration with other fisheries management regulations.

CASE STUDY 1: *JASUS EDWARDSII* AND THE BENEFITS OF MARINE PROTECTED AREAS IN NEW ZEALAND

The New Zealand Spiny Lobster, *Jasus edwardsii* (Hutton, 1875)

In Aotearoa New Zealand, four species of spiny lobster occur, with the most common and well-known species being the red rock lobster *J. edwardsii* (Fig. 15.1; colloquially: crayfish, Māori: kōura or kōura papatea) (Booth 2006). The packhorse or green rock lobster, *Sagmariasus verreauxi*, is regionally common in northern New Zealand, and the other two species, the deep-water lobster *Projasus parkeri* and the tropical species *Panulirus versicolor*, are less common/conspicuous (Booth 2006).

Jasus is a pan-Southern Hemisphere genus of rock lobster with an exceptionally wide latitudinal distribution from ~23°S (Pollock, 1986) to the sub-Antarctic islands (Booth and Webber 2001, Booth 2006). *Jasus edwardsii* is widespread throughout much of Australasia. In New Zealand, *J. edwardsii* is present from the Three Kings Islands (Manawatāwhi) in the north, to the Auckland Islands (Motu Maha/Maungahuka) in the south, and the Chatham Islands (Rēkohu/Wharekauri) in the east (Fig. 15.2; Booth and Ovenden 2000, Booth 2008).

Jasus edwardsii has a PLD of 12–24 months (Booth 1994), which may be the longest for any rock lobster or benthic marine organism (Bradbury and Snelgrove 2001, Pollock 1995). Coupled with the limited swimming ability of their pueruli, successful recruitment is heavily reliant on retention mechanisms, which keep phyllosoma close to the coast. As a result, the

Fig. 15.1.
Large male *Jasus edwardsii* found in the open during the day within the Leigh marine reserve, northern New Zealand. Photo reproduced courtesy of Nick Shears.

large-scale spatial distribution of lobsters is closely connected to key coastal and oceanic processes, which impact retention or export of larval production. For example, modeling and pueruli collections have suggested larvae off the east coast of the lower North Island of New Zealand become entrained in the Wairarapa/East Cape eddy system leading to high recruitment south of East Cape (Booth 1994, Chiswell and Booth 2008), while north of East Cape, most larval production is suggested to be exported southward into the Wairarapa/East Cape eddy (Chiswell and Booth, 2008). In addition, the genetic homogeneity between Australian and New Zealand populations (Ovenden et al. 1992), and feasible transport times via known oceanic currents (Chiswell et al. 2003), suggests Australian sourced phyllosoma may in part supplement Fiordland populations.

Adult *J. edwardsii* are mostly nocturnal and predominantly reef-associated but will utilize adjacent deep soft sediment environments (MacDiarmid 1991). The species has a varied and broad natural diet primarily consisting of mollusks, echinoderms and other invertebrates (Booth 2006). Historically, *J. edwardsii* were considered an important predator in coastal ecosystems (MacDiarmid et al. 2013, Eddy et al. 2014). Their ecological importance is evident within marine reserves where the recovery of *J. edwardsii* has had a large influence on ecosystem structure and dynamics (Babcock et al. 1999, 2010, Shears and Babcock 2002, 2003, 2004, 2007, Langlois et al. 2005, Shears et al., 2006, 2008, Salomon et al. 2008, 2010). The most notable ecological effect of *J. edwardsii* is their role in controlling sea urchin populations and, indirectly, kelp forests. In particular, large *J. edwardsii* are highly effective at feeding on sea urchins (Shears and Babcock 2003, Ling et al. 2009, Ling and Johnson 2012, Peleg et al. 2023).

New Zealand Lobster Fishery

Across Aotearoa New Zealand the indigenous Māori people have long valued kōura (Paulin 2007), with the presence of lobster remains in ancient middens indicating lobster harvest well before the 18th century (Leach and Anderson 1979). The first recorded "commercial sale" of *J. edwardsii* occurred in 1769 in Tolaga Bay between local Māori and Joseph Banks. This interaction was fortuitously illustrated by Tupaia (Fig. 15.3), the Tahitian navigator/high priest/translator on Captain James Cook's maiden trip to New Zealand (Salmond 2003). The exchange of lobsters played an important part in the early interactions between Europeans and Māori. From Joseph Banks' journal: "But above all the luxuries we met with the lobster or sea crawfish must not be forgot. . . of

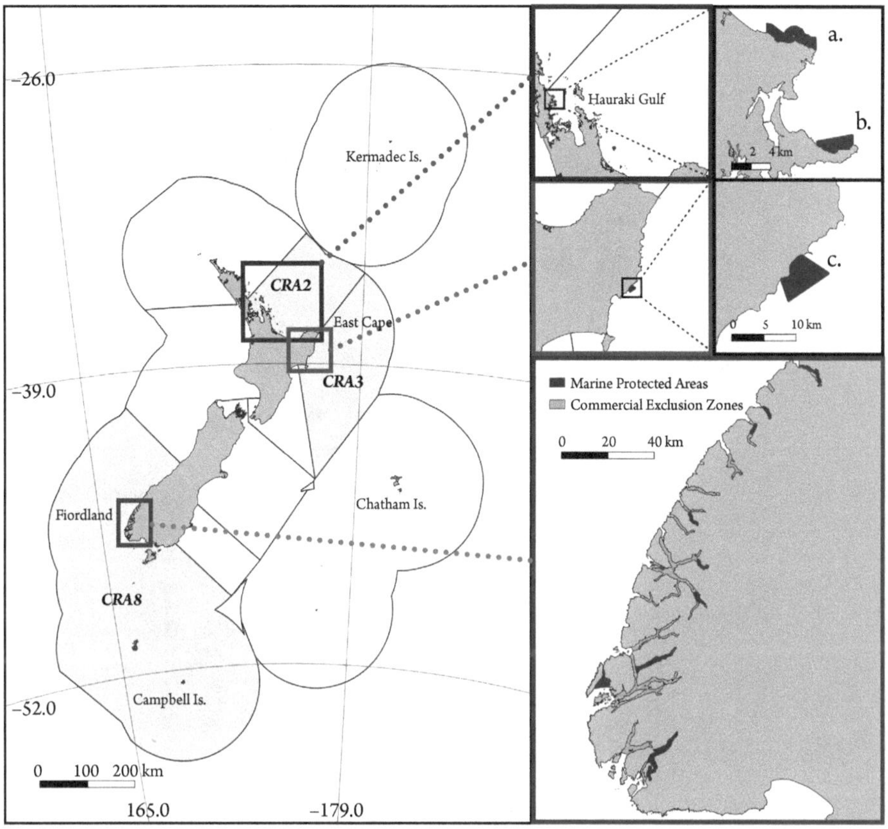

Fig. 15.2.
Key New Zealand red rock lobster regions described in this chapter: Northern New Zealand (upper box; *CRA2*), East Cape (central box; *CRA3*), and Fiordland (lower box; *CRA8*) quota management areas and marine protected areas: a. Cape Rodney to Okakari Point Marine Reserve, b. Tawharanui Marine Reserve, c. Te Tapuwae o Rongakako Marine Reserve. The Fiordland region incorporates two levels of protection: full protected marine protected areas (dark shading) and commercial exclusion zones (light shading) where recreational fishing is still permitted. See a color version of this figure in the centerfold.

them we bought great quantities of the native everywhere to the northward, who catch them by diving near the shore, feeling first with their feet till they find out where they lie" (Beaglehole 1963).

Despite the abundance of lobster at that time and the relative ease of harvest, a trade in lobster was rarely established with early European settlers (Annala 1983). This was in part due to the rapid degradation of lobster flesh once dead, and an inability to salt the flesh to preserve it for export (Arbuckle 1971). By 1883, refrigeration had become more widespread, and frozen fish products could be more confidently exported. By the 1930s a lobster fishery had developed in the Hauraki Gulf region supplying domestic and international markets (MacDiarmid et al. 2016). With the development of true industrialized fishing practices in the 1950s (Smith 2013), the first boom in New Zealand's lobster fishery occurred in 1956 in Fiordland, resulting in dramatic increases in catch as fishers from the east coast based themselves on the remote southwest coast (Gibbs 2008). This was followed by the lobster fishery boom in the Chatham Islands in 1966, peaking in 1968 at a yield of over 6000 tonnes, with one vessel landing 2 tonnes in a single day, contributing to the maximum annual national catch of 10,600 tonnes (Annala and King 1983).

Fig. 15.3.
Drawings illustrative of Captain Cook's first voyage to New Zealand, 1768–1770. The only known drawing of Tupaia's (Māori high priest and navigator), sharing a lobster with Captain Cook's party member. Reproduced courtesy of British Library, Public Domain Mark 1.0.

Typical of many commercial lobster fisheries, the New Zealand lobster fishery was characterized by heavy exploitation of new areas, marked by extremely high initial catch rates followed by dramatic declines (Booth 2006). Alongside the explosive growth of intensive commercial fishing, most lobster fisheries have seen dramatic declines over the "statistical era" (circa mid-1930s–present day) (MacDiarmid et al. 2016). Since the late 1980s national commercial landings have remained relatively stable, fluctuating between 3000 and 4000 tonnes per annum (Jeffs et al. 2013), but fishing effort has had to steadily increase (Booth 2008).

The New Zealand lobster fisheries have been managed by the quota management system, since April 1, 1990 (New Zealand Fisheries 2020). Prior to this, more rudimentary management strategies were implemented, such as minimum legal sizes, which were first regulated in 1950, and escape gap regulations on fishing gear to limit the occurrence of handling damage on undersize lobsters (Arbuckle 1971). Two species of lobsters are commercially harvested in New Zealand waters, *J. edwardsii* and *S. verreauxi*, and are managed as separate fisheries. *Sagmariasus verreauxi* is only caught in the northern-most part of the country and the overall catch is small in comparison to *J. edwardsii* (~1.5% of national *J. edwardsii* landings), which is fished throughout New Zealand's territorial waters (New Zealand Fisheries 2020). The New Zealand fishery for *J. edwardsii* is broken up into 10 quota management areas (QMAs) (Fig. 15.2). The boundaries between QMAs are not based on biology or ecology, but rather a means of fishery management. Since 2001, stock assessments have been carried out at the QMA level with the total allowable catch (TAC) for each area set annually. The TAC includes recreational fishing and customary Māori uses, with the remainder allocated to the commercial fishery as total allowable commercial catch (TACC). Commercial takings and TACCs vary substantially between the QMAs, reflecting natural dissimilarities in recruitment (Fig. 15.4). For example, the northeastern New Zealand QMA, CRA2, has largely been characterized by catch rates below that of the East Cape, CRA3, and the Fiordland, CRA8, QMAs. The TACC is set and adjusted independently for each QMA to ensure that stocks are maintained at or above sustainability limits.

Jasus edwardsii and New Zealand Marine Protected Areas

New Zealand demonstrated an early commitment to marine conservation through the drafting of the Marine Reserves Act 1971, followed shortly after by the establishment of one of the world's first no-take MPAs, the Cape Rodney–Okakari Point (Leigh) Marine Reserve, in 1977 (Ballantine and Gordon 1979). The Marine Reserves Act 1971 had the original purpose of protecting marine organisms and the environment to allow their scientific study (Ballantine 2014). With a total of 44 marine reserves now in place in New Zealand's territorial waters, this has allowed for extensive research into *J. edwardsii* and how they and their associated ecosystems respond to protection from fishing.

Jasus edwardsii has generally responded positively to protection in MPAs throughout New Zealand (Freeman et al. 2012b). In MPAs located in areas that have historically supported *J. edwardsii* populations, lobster size and abundance generally increased to levels well above those in adjacent fished areas. In addition, *J. edwardsii* within MPAs have been shown to be healthier and grow faster (Freeman and MacDiarmid 2009, Freeman et al. 2012a; see also Sørdalen et al. 2022). The absence of recovery in the Poor Knights Island marine reserve suggests positive responses to MPAs are likely context-dependent (Balemi and Shears 2023). While trajectories in long-term data suggest that the degree to which lobster populations are effectively protected from fishing within MPAs is variable, Freeman et al. (2012b) proposed that reserve populations may be more resilient than fished populations to fluctuation in larval supply due to large residual populations of legal-sized animals. However, the extent of this resilience is ultimately dependent on the MPA's ability to effectively protect adult *J. edwardsii* from fishing in the adjacent waters. Marine reserve size, regional variation in recruitment, and number of effective MPAs within a region can impact an MPA's overall efficacy at conserving *J. edwardsii*. Below, we contrast the response of *J. edwardsii* populations to MPA protection in three different regions of New Zealand, which provide contrasting lessons for the management and conservation of lobsters.

Lobster Declines in Northern MPAs

Extensive research in northeastern New Zealand from the 1990s and early 2000s has provided valuable insights into the complexities of *J. edwardsii* behavior (MacDiarmid 1994, Kelly et al. 1999), movement (MacDiarmid 1991, MacDiarmid et al. 1991, Kelly 2001, Kelly and MacDiarmid 2003), and reproduction (MacDiarmid 1989a, b). This research has helped demonstrate how MPAs can promote the recovery of *J. edwardsii* populations (Kelly et al. 2000) and contribute to spillover fisheries (Kelly et al. 2002). Protected areas such as the Leigh and Tawharanui marine reserves have been characterized by higher abundances and greater mean individual sizes, including higher abundances and biomasses of legal-sized lobsters (Kelly et al. 2000, Shears et al. 2006). In addition

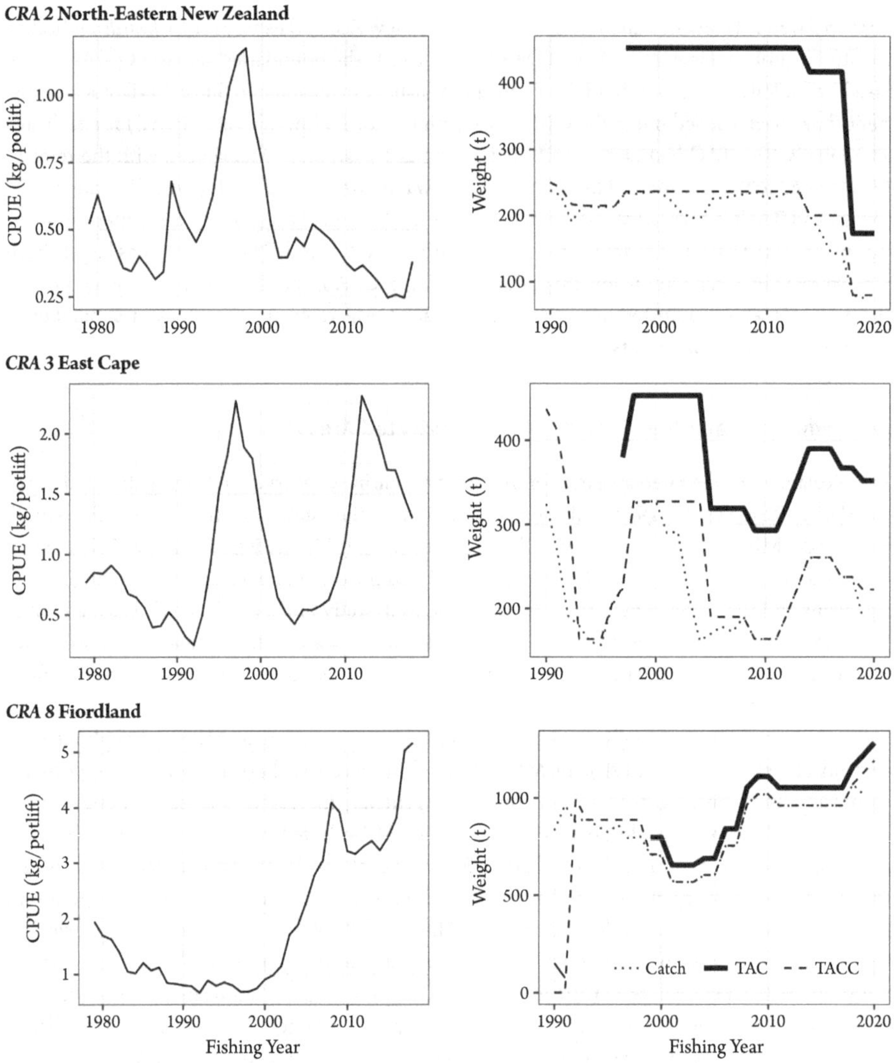

Fig. 15.4.
New Zealand lobster (*Jasus edwardsii*) quota management area *CRA2*, *CRA3*, and *CRA8* catch per unit effort (CPUE, kg/potlift) and total catch weight, total allowable catch (TAC), and total allowable commercial catch (TACC) information for the years 1990 to 2020. *Source*: data from New Zealand Fisheries (2020) November Plenary.

to demonstrating the direct impacts of restricting fishing activity, recovery of lobsters within protected areas has also been coupled with the resurgence of kelp forests (*Ecklonia radiata*) on areas of reef previously characterized by sea urchin (*Evechinus chloroticus*) barrens (Shears and Babcock 2002). It has been demonstrated that lobster predation on sea urchins indirectly promotes kelp forest recovery, persistence, and stability via density- and behavior-driven trophic cascades (Cole and Keuskamp 1998, Shears and Babcock 2003, Spyksma et al. 2017, Peleg et al. 2023). Similarly, Eisaguirre et al. (2020) showed that the California spiny lobster (*Panulirus interruptus*) contributed to recovery of understory kelps (several species) and Giant kelp (*Macrocystis pyrifera*) in Northern Channel Islands MPAs through predation on purple urchins (*Strongylocentrotus purpuratus*).

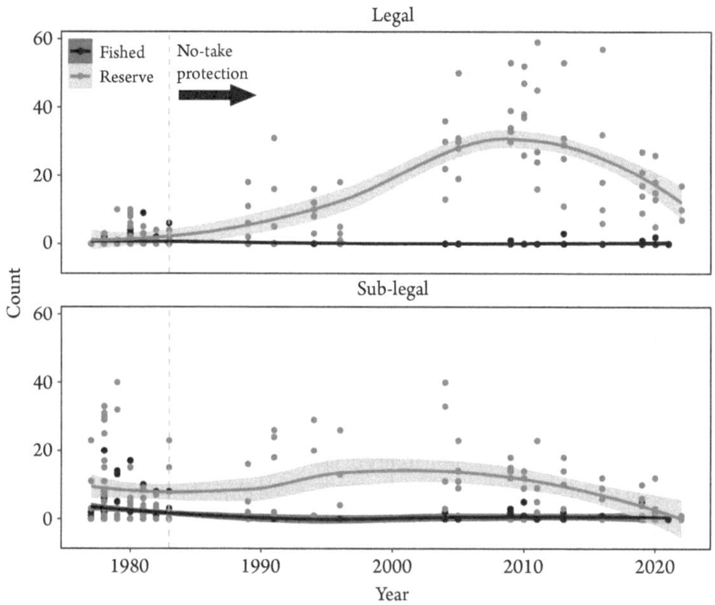

Fig. 15.5.
Numbers of legal and sub-legal sized *Jasus edwardsii* across permanent transects (500 m^2) monitored within the Tawharanui Marine Reserve area and the adjacent non-protected fished area.

Despite clear evidence for recovery of *J. edwardsii* within northeastern New Zealand MPAs, long-term monitoring data have indicated large declines in the abundance and biomass within multiple marine reserves across the region in recent decades (LaScala-Gruenewald et al. 2021; Fig. 15.5). These declines coincide with reported declines in catch per unit effort in the surrounding fishery (see Fig. 15.4, CRA2; Webber et al. 2018, LaScala-Gruenewald et al. 2021, Hanns and Shears 2023). Declines in catch rates in the fishery correspond with a period of consistently low recruitment since the mid-1990s. While this may account for some of the decline observed in MPAs, the high rate of decline in MPAs suggests these reserves are not effectively protecting adult lobsters and implies that these protected populations are being directly impacted by fishing (LaScala-Gruenewald et al. 2021, Hanns and Shears 2023).

Movement studies within New Zealand marine reserves have shown that adult *J. edwardsii* seasonally move from shallow reefs to deeper patch reefs or sand habitat for foraging and reproduction (Kelly 2001, Freeman et al. 2009). Within small northeastern New Zealand marine reserves such as Leigh, lobsters have been found to frequently cross reserve offshore boundaries while undertaking seasonal offshore movements. Sampling of commercial catch between 1995 and 1997 on the reserve's offshore boundaries found fishers were, on average, obtaining catches of equal size and value to other coastal areas but that catch composition and catch per unit effort (CPUE) was highly variable, reflecting seasonality in movement (Kelly et al. 2002). Following these findings, Kelly and MacDiarmid (2003) issued the pertinent warning that sustained capture of lobsters undertaking seasonal offshore movements would reduce protected population size and growth. While Kelly et al. (2002) indicated that MPA implementation did not have a negative effect on the livelihood of local fishers, the subsequent declines within the northeastern New Zealand MPAs have suggested that the offshore extent of the MPAs is not sufficient to protect populations of adult *J.*

edwardsii. The Hauraki Gulf Marine Spatial Plan (Department of Conservation and Fisheries New Zealand 2021) has recognized the plight of lobsters in the Leigh (Goat Island) Marine Reserve and proposed extending the offshore boundary from 800 m to 3 km offshore. The proposed extension aims to mitigate the effects of fishing on the offshore boundary and increase resilience within the protected population to fluctuations in rates of recruitment.

Gisborne—Historically Most Important Fishery?

For comparison, the Te Tapuwae o Rongokako Marine Reserve (TTR) situated on New Zealand's East Cape has an offshore boundary approximately 5 km from the shore (Fig. 15.2). Tagging studies within TTR have revealed that lobster do not move beyond the offshore boundary and generally only cross longshore boundaries where continuous reef habitat also crosses (Freeman et al. 2009). Freeman et al. (2009) found abundances of lobster on reef increased with distance from the longshore boundaries, suggesting that the probability of fishing mortality decreased toward the centre of the MPA. Due to TTR's large size and far offshore extent, the impacts of fishing are largely confined to lobsters occupying reef habitat near the MPA's longshore boundaries. As a result, the lobster population across the remainder of the MPA has rebuilt and maintains old growth with minimal impact from the surrounding fishery (Freeman et al. 2012b). In addition, tag recapture information has revealed that *J. edwardsii* within the TTR tend to grow faster than those in the fished open areas (Freeman et al. 2012a), while incidences of the infection "tail fan necrosis" decreased toward the MPA's center (Freeman and MacDiarmid 2009). Both results were attributed to the negative effects of handling of sub-legal-sized individuals in the fishery.

However, given the long PLD of *J. edwardsii*, meaningful protection must be extended and replicated at large spatial scales for MPAs to disproportionately contribute to the larval pool. At the time of writing, TTR is the sole MPA on ~285 km of the East Cape coastline and one of only two along ~655 km of coastline along the southern east coast of New Zealand's North Island. It is therefore unlikely for MPA protection to have had a meaningful impact across the wider East Cape region. Across the East Cape lobster fishery, the absence of a meaningful MPA impact is reflected in large fluctuations in CPUE and a general downward trend in commercial landings (Fig. 15.4). Although this fishery has previously been described as a "fisheries management success story" due to what were considered as effective and innovative fisheries management strategies (Breen and Kendrick, 1997), subsequent instability and declines in the fishery strongly implies that fisheries measures alone cannot effectively insure against recruitment variability.

The Fiordland Marine Area—A Successful Integrated Approach

In 2005, a large-scale network of MPAs and commercial exclusion zones was implemented across the Fiordland Marine Area (Te Moana o Atawhenua), southern New Zealand, a geographically expansive coastal area adjacent to the 1.2 million ha Fiordland National Park. The inner regions of 11 fjords were closed to commercial fishing (46,002 ha; 59% of the Fiordland Marine Area) and a network of 10 MPAs was nested within the commercial exclusion zones (Wing et al. 2004). Prior to these efforts the fishery also experienced consecutive years of reduced TACs (Bentley et al. 2003). The impact of these combined fisheries measures and MPA implementation provides a positive example of large-scale marine management integrating MPAs and fisheries management strategies (Wing and Jack 2014). In the years following, lobster populations recovered both inside and outside the MPAs. This was evident in large increases in commercial CPUE (Fig. 15.4), which were followed by increases to the regional TACC and increases in recreational limits within some fjords (Starr 2010). As of 2020, this is the most productive lobster fishery in New Zealand (New

Zealand Fisheries 2020). Spatial modeling of New Zealand *J. edwardsii* larval sources and sinks suggests that most of the Fiordland recruits are sourced from within the region (Chiswell and Booth, 2008). In Fiordland, Jack and Wing (2010) indicated the reproductive output per unit area of *J. edwardsii* within MPAs was as much as eight times higher than in the surrounding fished areas. This suggests the regional increases in *J. edwardsii* are consistent with the buildup of broodstock within the Fiordland MPA network (Jack and Wing 2013). The Fiordland MPA network provides a critical example of how large-scale protection of mature, fecund subpopulations of *J. edwarsii* in refuges distributed across an expansive coastal area can help insure against regional population decline.

Conclusions

The contrasting responses of *J. edwardsii* to MPAs over the Northern, East Cape, and Fiordland regions of New Zealand presents important lessons for the management and conservation of this species. To effectively protect *J. edwardsii*, MPAs must be large enough to encompass seasonal movement, but for MPA implementation to be meaningful at larger regional spatial scales, protection must be extended and replicated. Effective conservation of spawning biomass within a network of well-designed MPAs can improve an area's resilience to recruitment uncertainties. The northern New Zealand example shows how MPA effectiveness is compromised when seasonal movements take lobster beyond offshore boundaries. The Gisborne example indicates that even large effective MPAs will have minimal impact at the fishery scale. When brood stocks are effectively protected, populations within MPAs can disproportionately contribute to the larval pool. The Fiordland example provides a valuable example of the power of integrating fisheries (i.e., TAC reduction) and conservation (i.e., MPA implementation) management strategies to both rebuild a region's population and provide insurance against recruitment variability.

CASE STUDY 2: *PALINURUS ELEPHAS* AND THE BENEFITS OF THE MARINE RESERVE OF THE COLUMBRETES ISLANDS (NW MEDITERRANEAN)

The European Spiny Lobster *Palinurus elephas* (Fabricius, 1787)

The European spiny lobster *Palinurus elephas* (Fabricius, 1787) is a temperate, deep-water spiny lobster species widely distributed in the Northeast Atlantic and the Western Mediterranean Sea (Holthuis 1991). It is the most abundant and accessible spiny lobster in the region and a traditionally preferred target of fisheries off Ireland, the UK, France, Portugal, Spain, Italy, Croacia, Greece, Tunisia, Morocco, and adjacent Mediterranean waters (Goñi and Latruite 2005).

Palinurus elephas (Fig. 15.6) lives between the shore and 200 m depth on rocky and coralliferous substrata where micro-caves and natural protective holes are numerous (Ceccaldi and Latrouite 2000). Tagging studies indicate that adult movement is restricted, with most individuals moving < 5 km, although there are reports of recaptures up to 70 km away (Goñi and Latrouite 2005). Inshore–offshore migrations linked to reproduction and feeding have been reported for both Atlantic and Mediterranean populations and have been inferred for the Western Mediterranean population based on seasonal bathymetric changes in size and sex structure (Goñi et al. 2001a).

Palinurus elephas is a long-lived (maximum estimated age 28.3 yr for females and 23.1 yr for males) (Díaz et al. 2016), slow-growing species that reproduces once a year (Goñi and Latrouite 2005). In the Mediterranean, individuals mature at an age of 3 to 4 yr (~80 mm carapace length, CL), and the embryo incubation period extends from September to February (Goñi et al. 2003).

Fig. 15.6.
Big Island of the Columbretes archipelago and *Palinurus elephas* in the Columbretes Islands marine reserve. Upper photo reproduced courtesy of Ben Stobart, SARDI, Australia; lower photo reproduced courtesy of David Diaz, CN-IEO-CSIC, Spain; see a color version of this figure in the centerfold.

Palinurus elephas is a generalist, opportunistic feeder. While mollusks and sea urchins are the most important prey in the diet of the species, other prey such as decapod crustaceans, ophiuroids, or coralline algae are also consumed in certain areas (Goñi et al. 2001b). In the Western Mediterranean, postlarvae settle during the spring/summer in holes and in crevices at 5–15 m depth, although they have also been observed down to 35 m depth in late summer. Early juveniles leave shelters to forage at night (Díaz et al. 2001). Adult *P. elephas* are solitary, in pairs, or in small groups preferentially at the base of rocks or boulders over gravel or maerl beds. Like juveniles, they are primarily active at night; their movements are generally limited and seem motivated for foraging and reproduction.

The high unit value of the species makes the modern fisheries economically feasible despite low yields and is believed to be the prime reason for the overfished status of their populations. However, the lack of reliable historical catch and effort data prevents testing the hypothesis that increased value compensated a putative long-term decline in the Western Mediterranean. Published declines of CPUE in Atlantic fisheries and widespread concern for the lack of proper assessment of its status led to the classification of *P. elephas* by the International Union for Conservation of Nature as Threatened Vulnerable under criteria A2bd (Goñi 2014).

Western Mediterranean Fisheries of *Palinurus elephas*

Among the edible spiny lobsters in the Mediterranean, *P. elephas* has been historically considered an excellent food, as portrayed in a 3rd-century Roman mosaic from Pompeii (Fig. 15.7). Its yields

Fig. 15.7.
Part of a Roman mosaic from Pompeii (house n. 16, insula 2, Regio VIII; first century CE) that described predator-prey interactions between an octopus, a spiny lobster, and a moray eel (Soprintendenza Speciale per i Beni Archeologici di Napoli e Pompei). See a color version of this figure in the centerfold.

started to decrease during Roman times and the deterioration of the stocks worsened in the 20th century with the intensification of effort and introduction of modern technologies (Spanier et al. 2015). Although later reports of overfishing appeared as early as the 1930s, the widespread decline of *P. elephas* fisheries may be traced back to the period 1960–1980, when a major change in the exploitation strategy took place. Fishing effort increased dramatically as pots and diving were replaced by trammel nets, hauling gear, and other technological advances, which are significantly more efficient and resulted in further population declines (Goñi and Latrouite 2005). All circumstantial evidence points the blame at the great efficiency of nets in catching lobsters, the poor selectivity of trammel nets relative to pots, and excessive fishing effort.

Recent studies indicate that trammel nets catch a greater proportion of large lobsters than pots and that their negative impact on lobster habitats and benthic communities is far greater (e.g., Mallol et al. 2014, Enrichettia et al. 2019). Moreover, at-sea effort control of fisheries is difficult, and it is easy for the limits of net length per boat to be exceeded. Additionally, while pots had to be hauled every day to restock the bait, trammel nets are left in the water for two or several days to bait the net with fishes entangled during the first day. The Mediterranean population is considered overfished, and it is still targeted by artisanal boats in many areas, especially in archipelagos and islands.

Unfortunately, irregular reporting by most countries, and the mistaken inclusion of *Palinurus mauritanicus* or other species (e.g., *H. gammarus* in Greece or *Panulirus regius* in Morocco) in some records renders many historic data sets mostly useless (Goñi and Latrouite 2005). Also, in fisheries where *P. elephas* is sold locally (e.g., southern Europe and the Mediterranean) a potentially significant proportion of the catch may go unrecorded. A reconstruction of Corsican landings (1950–2008) suggested that 16 times more lobster were landed than are reflected in the data reported to the United Nations Food and Agriculture Organization (FAO) (Harper and Zeller 2011).

Currently, there are several measures in place for the management of this species, including: prohibition of landing ovigerous females, minimum size limits (90 mm CL), restrictions on gear type and number of nets per boat, and closed seasons covering the egg-bearing period, ranging up to seven months in the Balearic Islands (NW Mediterranean), which appears to be the most effective management measure for the species. Our knowledge is insufficient to discriminate between increased fishing mortality due to greater fishing effort and the changes in size- and sex-related catchability associated with trammel netting as the factors responsible for the depleted status of *P. elephas* fisheries. As a high-value commodity, *P. elephas* continues to be pursued despite low yields because rapidly growing prices maintain viable fisheries. Therefore, some fishermen are requesting managers impose more restrictions to help rebuild stocks, and additional measures, such as permanent closures of fractions of the fishing grounds or of spawning grounds, may provide the solution.

Palinurus elephas in the Columbretes Islands Marine Reserve

The Columbretes Islands marine reserve archipelago (CIMR) is located in the Western Mediterranean, 30 miles from the east coast of the Spanish peninsula (between 39°51' and 39°55'N and 0°40' and 0°42'E) (Fig. 15.8). Known by Greeks and Romans, Pliny the Elder (AD 23–79) writer, naturalist, and naval commander of the early Roman Empire mentioned Columbretes Islands in his treaty Naturalis Historia (Plinio il Vecchio 77–78). The first modern published reference was that of Captain Smyth (1831) whose primary studies were dedicated to the terrestrial fauna and flora as well as the birds of the archipelago.

The great distance that separates the Columbretes Islands from the Spanish mainland coast, together with the lack of contaminant sources in the islands, determines the high quality of the local marine waters. The transparency of the water allows the presence of erect algae (e.g., *Laminaria*) at considerable depth (90 m). Benthic communities harbor maerl beds, algal covers that reach great depths, and meadows of the seagrass *Cymodocea nodosa*, banks of the Mediterranean stony coral *Cladocora caespitosa*, the fan shell *Pinna nobilis*, the red gorgonian *Paramuricea clavata*, and species of greatest commercial importance, such as the red lobster, *P. elephas*.

Palinurus elephas is a highly appreciated resource of the archipelago; knowledge about its extraction exists from the 19th century. At that time, lobster fishermen visited the islands from mainland harbors in small coastal boats with lateen sails, catching on average 3000 kg by boat per season (Templado and Calvo 2002). The CIMR was established in 1990 and has become an emblematic marine reserve due to its strict protection, extent (55 km^2), and empowering legislation that prohibits all commercial bottom fishing and is well enforced.

In the archipelago, adult *P. elephas* inhabit rocky and coralliferous habitats from the shore to the circa-littoral stratum, especially from 30–90 m depth. Postlarvae and juveniles settle in very shallow waters (< 10 m) but are more abundant at depths > 20 m (Díaz et al. 2001).

Evidence of Fisheries Benefits from the Reserve

The prohibition of fishing in the reserve (1990–present) offered an exceptional opportunity to advance the knowledge of processes and benefits of no-take areas for benthic commercial species of restricted movements, such as *P. elephas*. A long-term study monitored the evolution of *P. elephas* abundance, biomass, egg production potential, and spillover to the fishery (Goñi et al. 2001a, 2003, 2006, 2008, 2010, Díaz et al. 2011, 2016).

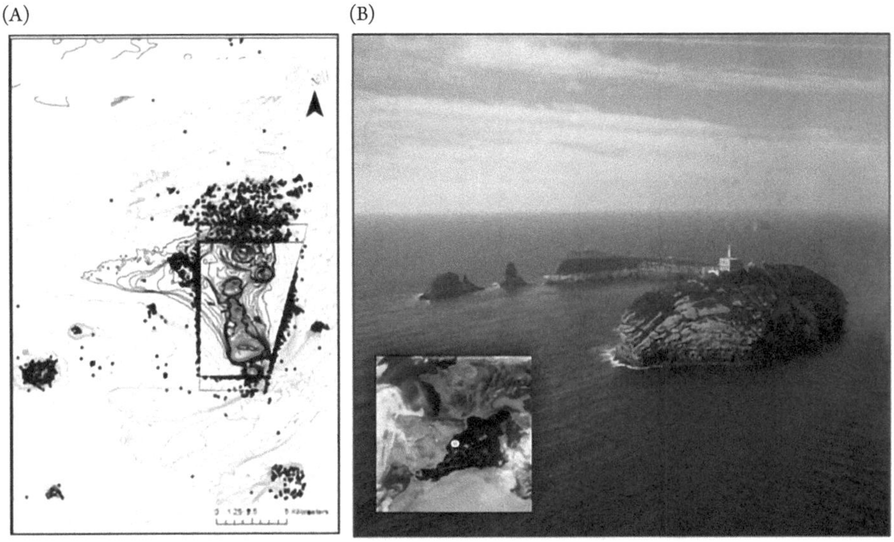

Fig. 15.8.
(A) Map of the limits of the Columbretes Islands marine reserve (CIMR). Dots indicate locations where the fishery operated after the creation of the CIMR in 1990 (inner box, data from 2000–2008 shown). The CIMR was expanded north- and southward in 2008 (outer limits). Dark polygons are islands within the CIMR. Gray contour lines denote the 10–120 m isobaths. (B) Main island with lighthouse. Insert in (B): location of the CIMR (white circle) in the Western Mediterranean (© Google Earth, under fair use agreement). See a color version of this figure in the centerfold.

Biomass Export: Spillover of Lobsters from the Reserve to the Adjacent Fishing Grounds

The first investigation into the effects of the reserve on the adjacent *P. elephas* fishing grounds was conducted 9 to 12 years after "no-take" was implemented in the reserve (Goñi et al. 2006). Catch and effort data were collected onboard commercial lobster fishing boats operating in the region and inside the reserve. The CPUE of *P. elephas* in the fishery outside the CIMR showed a significant nonlinear decline with distance from the center of the CIMR, with a depression at the boundary followed by a plateau, indicating spillover from the MPA to the adjacent fishing grounds. This depression was caused by local depletion associated with concentration of fishing effort at the CIMR boundary (Fig. 15.9), while the plateau suggested that lobster export from the CIMR was sufficient to maintain stable catch rates up to 1500 m from the CIMR boundary. Commercial catch and effort data were combined to estimate CPUE, which declined linearly with distance away from the reserve limits. Analyses of recaptures of lobsters tagged and released inside the reserve indicate that the density gradient was caused by lobsters leaving the reserve. Fishing effort concentration along its boundaries (see Fig. 15.8A) and high exploitation rates in the local fishery limited the spatial extent of spillover (Fig. 15.9).

Despite abundant evidence of spillover in the MPA literature, Goñi et al. (2010) was the first study to quantify the number and biomass of individuals spilling over annually from an MPA to the fishing grounds and their contribution to the local lobster catch. Based on a decade (1997–2007) of tag recapture data of *P. elephas* from the CIMR, the estimated annual emigration probabilities were 3.7% for females and 6.7% for males. During the period of 8 to 17 years of protection, harvested spillover offset the loss of yield resulting from the reduction of fishing grounds set aside in the CIMR, producing a mean annual net benefit of 10% of the catch by weight.

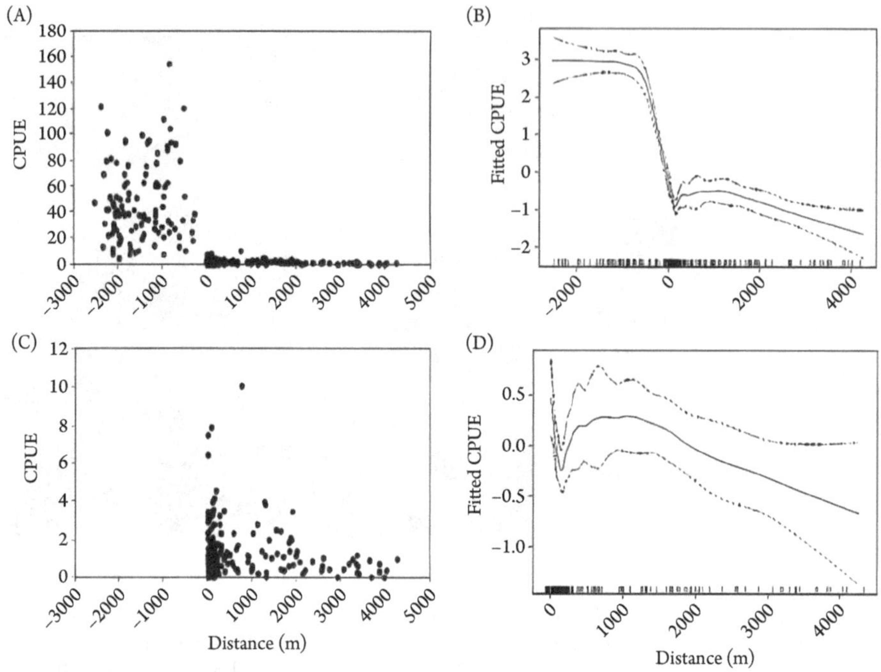

Fig. 15.9.
(A) Catch per unit effort (CPUE) data inside the CIMR (−2000 to 0 m). (C) CPUE data inside the CIMR (0–4000 m). (B, D) Fitted mean *P. elephas* CPUE as a function of distance of fishing sets to the CIMR boundary derived from a generalized additive model. Models incorporate logarithmic link, gamma variance function, smooth variable distance, linear predictor depth and factor side (in (B) only). Inside tick marks on the x-axis indicate individual observations. Dashed lines indicate two standard errors. Reproduced from Goñi et al. (2006), with permission from © Inter-Research Science Publisher.

Although the number of lobsters spilling over annually did not quite make up for the loss of fishing grounds in the CIMR, it did so in weight because the mean size of lobsters emigrating from the CIMR was larger than the mean size of lobsters outside. The potential of net emigration to offset the loss of fishing grounds in the CIMR was high: it was estimated that 7% of the lobsters residing in the Columbretes MPA emigrated every year to the adjacent fishing grounds, and that 90% of them were harvested within one year. The resulting spillover contributed 31%–43% of the annual local catch. This benefit was primarily due to lobsters growing larger in the MPA before emigrating. Nomadic or home range movements of individuals near MPA boundaries, seasonal migrations and density-dependent movements facilitated spillover from the CIMR to the fishery.

Egg/Larvae Export from the Reserve

Increased recruitment from eggs and larvae exported from the reserve could produce much greater benefits for exploited populations than spillover of adults. However, the link between local egg production and regional recruitment can currently not be resolved. Therefore, the first step to evaluate the potential was to quantify the contribution of the reserve to the regional reproductive output. Díaz et al. (2011) assessed egg production of *P. elephas* in the CIMR and its contribution to regional reproductive output from year 10 to year 19 of protection. Results showed that: (1) mean female fecundity in the CIMR increased by 41% over the 10-year study, (2) at the end of the study, mature females were on average 20 times more abundant, and egg production per unit area was 30 times greater in the MPA than in nearby fished areas, and (3) at that time, the MPA occupying 18% of the regional lobster habitats was responsible for over 80% of the regional egg production. Overall, it was estimated that after nearly two decades of protection of *P. elephas* in the CIMR, the MPA raised the regional egg production six times. When compared to a net benefit through biomass spillover from the MPA to the regional lobster fishery of +11% of the annual catch, a six-fold increase in egg production suggests that egg and larval export have far greater potential to support exploited populations than spillover.

Mature Reserve as a Proxy for the Unfished Condition of Exploited Species

In the absence of a historical baseline, long-term no-take MPAs can provide a reference to the unfished condition of exploited species of limited mobility. Díaz et al. (2016) document the recovery of *P. elephas* in the CIMR and use it as a baseline to assess the stock status in exploited grounds in the region. Lobster indices of density and biomass in the CIMR continued to increase after 25 years of protection, a period close to the species' lifespan. CPUE in weight more than doubled from years 10 to 25 without fishing in the CIMR, while the increase of CPUE in numbers was only 20%. This faster increase in biomass reflected higher individual growth inside the MPA. A highly depleted status of exploited populations outside CIMR was inferred from the demographic structure, and from CPUEs with "fishery:CIMR" ratios below 0.05. In the CIMR, the size class of maximum egg production increased over time, while the sizes remained stable in fished areas. The contribution of small lobsters (< 90 mm CL) to overall egg production was only 1.5% in the CIMR but 30% in the fished areas. Mortality estimated from recent size compositions was three to four times higher in fished areas than inside the CIMR, resulting in a reduction in predicted spawning potential per recruit close to 20% of the unfished level. This experiment emphasizes the value of long-term no-take areas as reference laboratories for investigating fishing effects.

CASE STUDY 3: *HOMARUS GAMMARUS* AND THE BENEFITS OF LOBSTER RESERVES IN COASTAL SKAGERRAK, NORWAY

The European Lobster *Homarus gammarus* (Linnaeus, 1758)

The European lobster *H. gammarus* (Linnaeus, 1758) (Fig. 15.10) is a large, long-lived sexually dimorphic decapod crustacean of ecological and commercial importance, distributed from northern Norway to Morocco in North Africa, including the Mediterranean (Triantafyllidis et al. 2005, Pavičić et al. 2020). The species is considered a nocturnal animal and generally spends light hours solitary inside shelters on rocky bottoms (Mehrtens et al. 2005, Moland et al. 2011a). European lobsters rarely move more than a few kilometers for periods up to multiple years (Dannevig 1936, Smith et al. 2001, Agnalt et al. 2007, Huserbråten et al. 2013) and typically live within limited shorter-term (< 1 yr) home ranges (~0.02 km^2) once they settle into a suitable habitat (Moland et al. 2011b, Skerritt et al. 2015). European lobsters are primarily associated with hard bottom habitats from 0–60 m (Galparasoro et al. 2009) and some individuals tend to spend the winter (or the period with minimum temperatures) in the deepest habitat available within their home range (Moland et al. 2011a, b). Based on gut content analyses, the major components of *H. gammarus*' diet include malacostracan crustaceans, gastropods, and polychaete annelids, although remnants of bivalves and echinoderms can also be distinguished (Hallbäck and Warén 1972).

European lobster longevity may potentially span several decades, during which large specimens may grow to weights up to 10 kg. The average age of large (150 to 170 mm carapace length, CL) male and female specimens has been estimated to be as high as 31 and 54 years, respectively, based on accumulation of lipofuscin in the eyestalk ganglion (Sheehy et al. 1999). The same study estimated that the 80–89.9 mm CL size class contained males that were 3.5 to 12.9 years old and females that were 4.1 to 13.0 years old. Growth studies conducted in southern Norway have estimated average

Fig. 15.10.
European lobster *Homarus gammarus* (Linnaeus, 1758) with fresh exoskeleton and bright white denticles is a highly conspicuous animal when encountered outside its shelter. Photo reproduced courtesy of Erling Svensen; see a color version of this figure in the centerfold.

growth increments of ~7 mm CL at each molt for females, with a two-year molt cycle being the predominant pattern (Agnalt et al. 2007). Based on data from wild-caught females, size at maturity is usually ~80 mm CL. Fecundity at maturation is ~5000 eggs, while a large female will carry ~40,000 eggs at 150 mm CL (Agnalt 2008). Ovigerous (= egg-bearing) females brood their eggs for 9–11 months, depending on the accumulation of degree-days (a function of time and temperature required to complete embryo development). The European lobster has a PLD lasting from three to eight weeks during which larvae undergo the first four molt cycles. Although the potential for dispersal is thus considerable, larvae might be retained locally through positioning in water masses subject to retention or limited transport (Øresland and Ulmestrand 2013).

Spatial genetic structure throughout the species' range in Europe is limited, with only a weak differentiation between larger regions of the Swedish Skagerrak and the Atlantic areas to the west, with the population in Norway being a mix of these two (Ellis et al. 2017). Similar work along the Skagerrak coast from Sweden to southern Norway suggest high connectivity among sampled populations (Huserbråten et al. 2013). However, population structure is evident within the species' distribution range, with implication for source–sink dynamics, management, and conservation (Pavičić et al. 2020).

The Coastal Fisheries of Homarus gammarus in Skagerrak

The trap-based fishery for European lobsters along the southern Norway coast has a unique history and background as the technique was originally introduced by Dutch sailors in 1660–1670. At this time, Dutch trade in Norway was primarily for lumber used in construction and as fuel on the Continent. However, traders took note of the abundance of lobsters and soon established a highly profitable industry based on live transport of lobsters from Norway to the capitals of Europe (Knutsen et al. 2022). This incurred a status change for the species as currency and tradable commodity in coastal communities, which also saw the early beginnings of landing statistics. Over the period covering officially reported catch statistics, the official catches of lobster have ranged between 500 and 700 tonnes annually, with a peak of 1300 tonnes in 1932 (Dannevig 1936, Kleiven et. al 2012, Kleiven et al. 2022). However, effects of the long-term intensive harvesting were noted

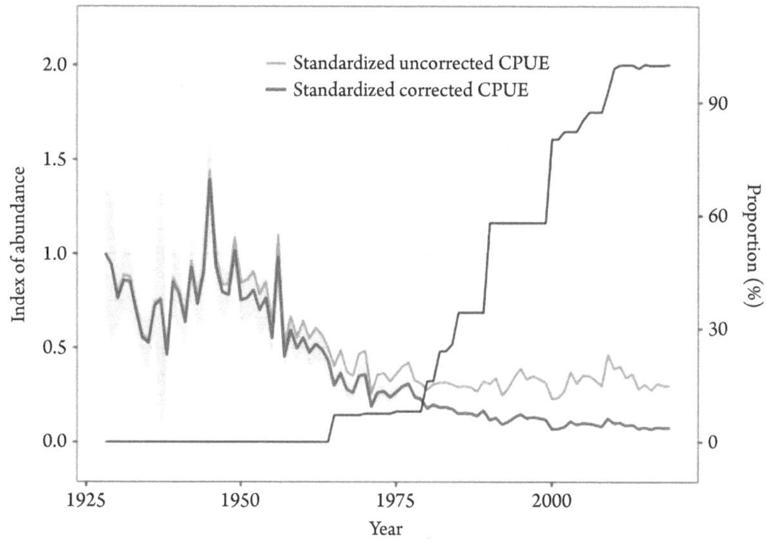

Fig. 15.11.
Standardized uncorrected catch per unit effort (CPUE): mean lobsters $trap^{-1}$ day^{-1} as reported by a selection of professional lobster fishers to the Norwegian Institute of Marine Research from 1928 to 2020. Corrected CPUE is adjusted for "technological creep": the transition from one- to two-chambered traps (black line). Error band: ±1 SE. Adapted from Kleiven et al. (2022), with permission from © Nature under Creative Commons License 4.0 (CC BY 4.0).

by the early 1800s, when the abundance of large lobsters was said to be in decline (Breton 1835). Over the past few decades, the population has collapsed due to overfishing and has remained at historically low levels since 1970. Recently, a study incorporated the intermittent development of gear effectiveness, "technological creep," in the fishery when analyzing the catch trend as reported from volunteer fishers since 1928 (see below) and showed that across nine decades the southern Norway *H. gammarus* lobster population likely declined by more than 90% (Kleiven et al. 2022).

In the year 2000, the official catches (50 tonnes) were estimated at less than 10% of the historical average. The historically low population size was confirmed by Norway's Institute of Marine Research's extensive time series index on lobster CPUE dating back to 1928, based on standardized logbook reports from commercial fishers (number of lobsters/trap/24 h) (Fig. 15.11). For example, before 1950, CPUE was 0.15–0.20 lobsters $trap^{-1}$ day^{-1}, compared to a CPUE of less than 0.1 lobsters $trap^{-1}$ day^{-1} over the last decades (Agnalt et al. 2007). In southern Norway, lobsters can only be fished from October 1 until November 30, whereas on the west coast the fishing season is extended to December 31. Commercial fishers may use up to 100 traps per day, whereas recreational fishers may use 10 traps per day. In the Skagerrak, the species is targeted in an intense trap fishery, dominated by recreational fishers (Kleiven et al. 2012, 2022) exerting high levels of fishing mortality on legal lobsters (Wiig et al. 2013). Management measures have been revised or adopted in recent years: closed season, gear restriction, and minimum (25 cm) and maximum (32 cm, in Skagerrak only) total length legal-size limits (corresponding to ~90 mm and ~115 mm carapace length, respectively), protection of incubating females, escape vents for undersized individuals, and the rigging of traps with biodegradable cotton thread weak links to prevent ghost fishing if lost.

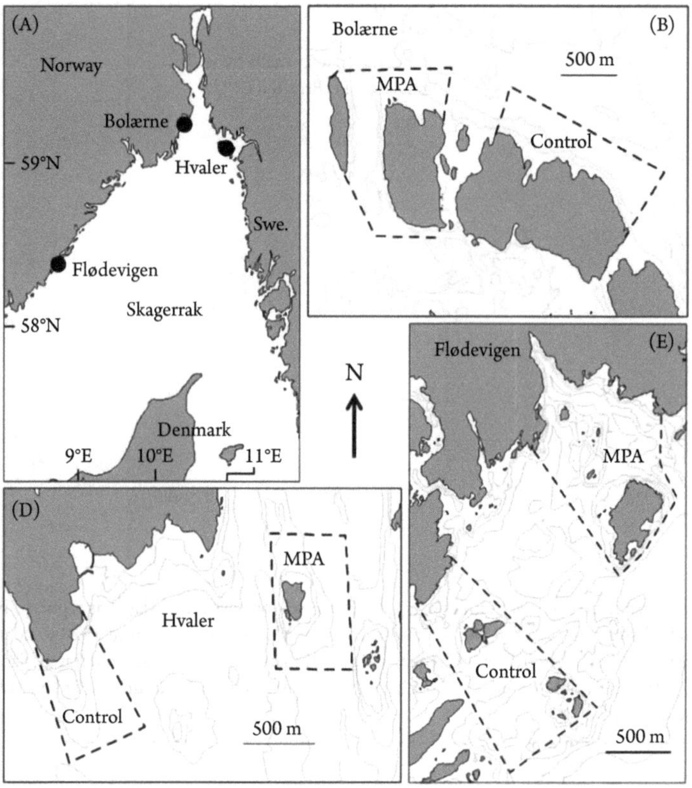

Fig. 15.12.
The original lobster reserves in Skagerrak. (A) Established in 2006 and adjacent control areas subject to long-term monitoring; in Vestfold (B), Agder (C), and Østfold (D) counties. Adapted from Moland et al. (2013a), with permission from © Royal Society Publishing under Creative Commons by Attribution License (CC BY 3.0).

Introduction and Demonstration of Marine Conservation Through Lobster Reserves

Based on comprehensive consultations with local fishers and managers and inspired by early experiences with lobster protection in Sweden (Moland et al. 2013b), candidate areas for protection of lobsters were designated in 2004. These were implemented as partially protected areas (PPAs) in 2006, effectively banning the capture of lobsters through a bylaw prohibiting the use of fixed gear (Pettersen et al. 2009).

Long-term monitoring of three PPAs and adjacent control areas (Fig. 15.12) resulted in rigorous scientific before–after control–impact (BACI) evaluation of lobster population effects pertaining to density, growth, demography, behavior, and phenotypic diversity (reviewed by Moland et al. [2021]). As effects from protection manifested, the lobster reserves received high public attention and were soon considered a credible supplement to traditional fisheries management. To date, more than 50 lobster reserves have been implemented by local municipalities in Norway, and over a 20-year period lobster PPAs have become an important tool for local fisheries management (reviewed by Knutsen et al. [2022]).

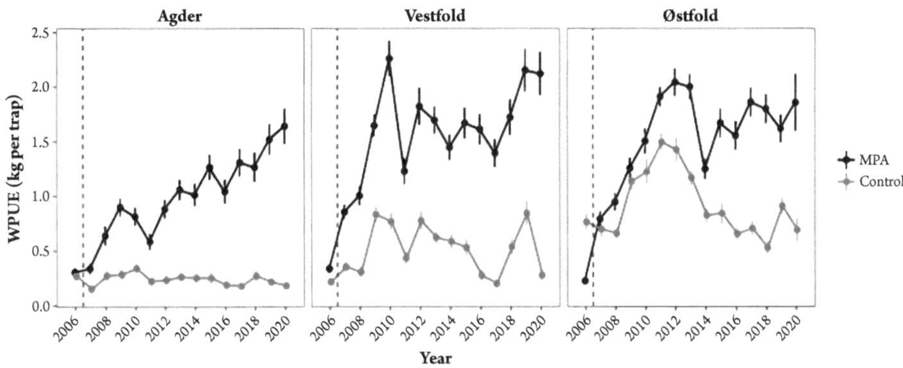

Fig. 15.13.
Weight per unit effort (WPUE; lobster in kg per trap day^{-1}) in the lobster MPAs (black) and control areas (gray) from 2006 to 2020 in Agder (Flødevigen), Vestfold (Bolærne), and Østfold (Kvernskjær) counties. In total, 19,512 lobsters were captured, marked, and released in the study areas throughout the period, with a sex ratio of 33:67 (female:male). Adapted from Knutsen et al. (2022), with permission from © Elsevier under Creative Commons by Attribution License 4.0 (CC BY 4.0).

Changes in Population Density

The long-term monitoring of the Norwegian lobster PPAs is based on CPUE as an indicator of abundance or population density in PPAs versus control areas. Moland et al. (2013a) showed a 245% average increase in CPUE for lobster over the first four years of protection, versus an 87% increase in control areas. Recently, Fernández-Chacón et al. (2021) applied a robust design model to the capture-recapture time series data to estimate annual abundances (corrected for imperfect detection) in the same areas (2006–2014). The results indicated that abundance levels increased after protection, albeit highly variably among PPAs. It also confirmed that the CPUE-based estimates for the first four years were indeed a realistic approximation of the average abundance increase. Knutsen et al. (2022) presented the long-term development in CPUE as weight per unit effort (WPUE; lobster in kg per trap day^{-1}), which suggests an increase of more than 500% in the lobster reserves relative to control areas from 2006 to 2020 (Fig. 15.13).

Long-Term Effects on Demography and Reproductive Potential

Population monitoring with capture-recapture allowed for modeling of vital rates in longer term studies. Moland et al. (2013b) found a positive linear trend in survival with body size in lobster in a Swedish PPA (Kåvra), with an additive effect of sex, implying that females have lower natural mortality than males. Catchability (in research traps) was found to be higher in males in the same data, while the capture-recapture modeling applied to the data revealed that although caught less, and thus observed less than males in catches, the female population had higher survival, most likely due to lower growth rates and less risk-prone behavior (Biro et al. 2014, Biro and Sampson 2015). The same sex-specific differences in lobster survival and catchability were found in a recent study using capture-recapture data collected in the Norwegian Skagerrak (Fernández-Chacón et al. 2021), confirming this pattern for the species. In this case, they also reported from one PPA that survival increased by 125.2% and 78.5% after protection for large-sized (> 25 cm) males and females, respectively.

Rebuilding of age and size structures are hallmarks of lowered harvest pressure incurring lowered mortality in any harvested animal population (see Francis et al. [2007], and references

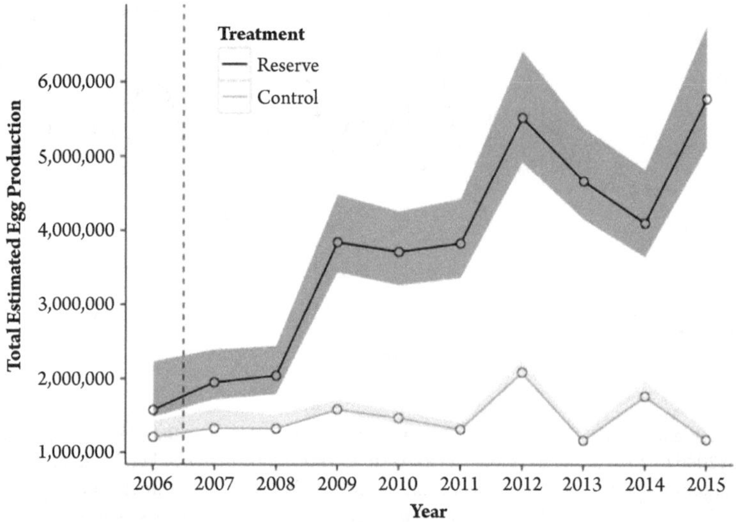

Fig. 15.14.
Average reproductive potential in *H. gammarus* in three partially protected areas—lobster reserves and control areas. Dashed vertical line denotes the time of implementation of protection in the PPAs. Error bands are 95% confidence limits. Reproduced from Calef (2016), with permission.

therein). In southern Norway, at the onset of the monitoring time series, lobsters were managed with a 24 cm total length (TL) minimum legal size (MLS). Harvested populations were thus characterized by few larger individuals. An increased average body size (13%) of lobsters in PPAs relative to controls was evident after four years of protection from the annual lobstering season (Moland et al. 2013b), an effect somewhat dampened by introduction of a 25 cm MLS in 2008. Thorbjørnsen et al. (2018) presented "before" (2006) and "after" (2014) data on the demography of lobsters in the three original PPAs, demonstrating a substantial widening of the size structure in the protected populations compared to the harvested before state. By 2014, the average size increase in PPAs was 15% compared to 1% in control areas. More recently, Fernández-Chacón et al. (2020) analyzed body size data of tagged lobsters sampled from 2004 to 2015 in all PPA–control area pairs and reported a rapid change in the size structure characterized by an increase of the large-sized fraction of the population with a subsequent increase in mean size. Overall, the diversity of sizes tended to increase inside protected areas.

Simultaneous increase in average body size and population size ought to confer increase of the reproductive potential, that is, the potential total egg production in a protected population. Calef (2016) explored this relationship and the contrast between lobster reserves and control areas, considering the intermittent growth of lobsters and the best available knowledge on relationships between fecundity and spawning frequency with increasing body size. The models indicated a more than three-fold increase of total egg production on average in the PPAs during the first eight years of protection (2006–2015) (Fig. 15.14). This estimate is similar to the greater than three-fold increase in egg production estimated after 10 years of protection in Vinga, a Swedish no-take MPA (Kraufvelin et al. 2023). In contrast, the control areas were subject to a slight decrease in total estimated egg production (−2.8%), notwithstanding an increase of minimum legal length from 24 to 25 cm TL, and a ban on landing and trading ovigerous females, both introduced in 2008.

Harvest Selection on Phenotypes

Fernández-Chacón et al. (2020) provided size-dependent survival estimates of lobsters in both MPA and control areas in Skagerrak, together with the shape of the relationship between survival and total body length. They documented significant negative relationships between survival and body size in the control areas but not in the MPAs, where the effect of body size was predominantly positive, and concluded that MPAs are a viable management approach for protecting against fisheries-induced selection, through the spatial refuge in both size-dependence and overall mortality.

MPAs, in combination with control areas, can also be useful as field laboratories for understanding how fishing can affect mating systems of species. For instance, as lobsters have become more numerous and grown to larger sizes in the MPAs (Fernández-Chacón et al. 2020, 2021), the opportunity and scope for mate choice should also increase. In a parental assignment study conducted in Flødevigen, Sørdalen et al. (2018) found a female preference for males with larger body sizes, but that the relative size differences between females and males of mated pairs (the size-assortative mating pattern) were significantly larger in the reserve compared to mated pairs in the control area. They also found positive selection differentials on male body and claw size inside the MPA, but not in the control area, implying that size has less influence on male mating success in fished areas, most likely because fisheries-induced selection weakens female choice and the competition between males. Furthermore, estimations of sexual selection gradients on male traits uncovered selection to be acting strongest on relative claw size (claw size relative to body size), rather than on absolute claw and body size. This trait was later linked to an increased risk of being captured in the fishery (Moland et al. 2019), and the finding that male lobsters have up to 8% larger claws inside MPAs compared to similar-sized males in fished areas (Sørdalen et al. 2020). In sum, these findings strongly suggest that MPAs can be an effective means to preserve functional mating patterns, as well as restore and maintain diversity in sexually selected traits.

In a recent seminal study, Sørdalen et al. (2022) demonstrated increased growth rates in protected lobster relative to those measured in the fished control sites. The effect—measured as increase in body length per molt—was more pronounced in females. The authors interpreted this as an effect of relaxed harvest selection in the protected sites, contrasted by an effect of fishing out of fast-growing phenotypes in the fished control sites. The authors speculated that males respond similarly, but the effect was masked by measurements of body length, which do not capture size increase manifest in claw size. Capturing this would have required a time series of weight measurements.

Spillover and Interaction with Surrounding Fisheries

The initially implemented MPAs were small (~0.5–1.1 km^2), and since the onset of the research capture and tagging of lobster, recoveries have been reported from fishers operating outside of the MPAs or in adjacent control areas. Huserbråten et al. (2013) quantified the emigration from MPAs and found that during the initial six years of protection, 4.7% of tagged individuals emigrated from MPA sites. Both Huserbråten et al. (2013) and Thorbjørnsen et al. (2018) reported a highly left-skewed distribution of movement behaviors, with most individuals recaptured (in research traps) close to their tagging location. However, movement up to 24 km away from tagging location was reported (range: 34 m to 24.67 km) by fishers recovering tagged lobsters.

As described above, benefits of MPAs have been reported as net contribution to spiny lobster *P. elephas* fishery catches through spillover (Goñi et al. 2010). In an effort to test for a similar pattern in *H. gammarus* in year 0 to year 8 of protection, Thorbjørnsen et al. (2018) compared cases of emigration and subsequent recovery by fishers for MPAs and control areas from 2006 to 2014.

While there was no difference in movement rates between MPAs and control areas, the lobster emigrating from MPAs to fished areas were significantly larger (1.19 cm TL, $p < 0.005$) than lobsters emigrating from control areas.

Nillos Kleiven et al. (2019) used experimental lobster fishing with randomized trap deployments in a wider area inside and around a lobster MPA before and after implementation. The aim was to test for an effect gradient from inside to far outside the MPA in a before-after-gradient design (Ellis and Schneider 1997). After four years, lobster abundance measured by CPUE increased significantly by a factor of 2.6 inside the MPA, with a decreasing gradient toward the border and into the fished areas outside. CPUE values in fished areas further away (3 km) remained like values before MPA establishment. However, a depression of population density developed at the MPA border compared to before data. Using the effort data, the depression in density could be explained by the increased fishing effort close to the reserve border. The study thus demonstrated the effect of "fishing the line" (see Kellner et al. 2007), possibly even penetrating into the MPA, and the need for effort control in areas surrounding MPAs (see also LaScala-Gruenewald et al. 2021). Nillos Kleiven et al. (2019) found a ~80% increase of effort (trap numbers) in the total area and pinpointed fishing hotspots with a three-fold increase in trap density. The highest increase was near the borders, indicating a shift in the preferred fishing areas.

COMMON PATTERNS AND LESSONS LEARNED FROM EXISTING EXAMPLES

Harvested species that are slow growing and late maturing are vulnerable to over-exploitation. Concomitantly, studies evaluating the effects of protection usually report rapid population responses to lowered fishing mortality when the fishing pressure is heavy and long-lasting (Jaco and Steele 2020). In large decapod crustaceans, such responses can be sustained in the longer term (e.g., Goñi et al. 2010, Fernández-Chacón et al. 2020), subject to fluctuation (e.g., Babcock et al. 2010), or even eroded if the protected areas are too small, or if fishing effort remains high at the borders. Small marine reserves do not provide a safeguard against overfishing if species flux out of the areas within which they are protected, for example, during seasonal migrations or other intrinsically motivated movement that crosses borders (e.g., LaScala-Gruenewald et al. 2021). Outside fishing pressure can thus penetrate marine reserve borders and deplete putatively protected populations (e.g., Nillos Kleiven et al. 2019). Such examples underscore the need for adaptive design of reserves and for viewing such measures in synergy with the overall management regime (Guenther et al. 2015).

Negative results are rare in the peer-reviewed literature on crustacean conservation. However, lacking effects of protection reported in the spiny lobster *Panulirus penicillatus* in the Galapagos was attributed to poaching or noncompliance due to weak enforcement (Buglass et al. 2018). Partially protected areas where effects of destructive gear (e.g., scallop dredging, bottom trawling) have been removed have shown positive indirect effects on *Cancer pagurus* in Lyme Bay, UK (Sheehan et al. 2013), although Davies et al. (2021) found no difference over time for exploited invertebrates there. Recently, Tuffley and de Lestang (2024) demonstrated increased natural mortality and stunted growth in western rock lobster *Panulirus cygnus*, as putative density-dependent effects of protection from fishing in a MPA.

Social behavior of large decapod crustaceans ranges from solitary to gregarious, and confinement to the seafloor confers site fidelity and affinity to certain habitat features. Large decapod crustaceans are commonly harvested with traps, although trammel nets are used in some areas (see Case Study 2, above). Wiig et al. (2013) showed that in a fishery where fishers adjust their behavior throughout the open season, large swathes of the available habitat can be subject to intensive harvest pressure, incurring high rates of fishing mortality of legally landed individuals. In cases such as these, the innate and highly variable susceptibility to harvest (the catchability) is the primary driver for the fate of an individual over the duration of a fishing season. However, catchability can be affected by both morphological and behavioral traits (Moland et al. 2019) and thus incur harvest selection on both. Well-managed fisheries seek to alleviate this through harvest rules and technical measures that secure long-term sustainability and maintenance of ecosystem function (e.g., Barrett et al. 2009). Importantly, conservation and recovery of large-bodied crustaceans have conferred recovery of their ecological roles—for example, as predators on sea urchins. This is manifest in studies demonstrating resurgence of kelp forests in New Zealand (Cole and Keuskamp 1998, Shears and Babcock 2003, Spyksma et al. 2017) and California (Eisaguirre et al. 2020) MPAs. These examples show that rebuilding of crustacean abundance and size structure in protected states allow their functional roles to be fully realized in the ecosystem. Crustaceans' contribution to functional redundancy supports the notion that marine reserves are "functional hotspots" with the ability to mitigate and promote adaptation to climate change (Roberts et al. 2017).

FUTURE OF CRUSTACEAN CONSERVATION: NETWORKS AND CLIMATE REFUGIA?

The best available science (including the work summarized herein) confirms the usefulness of harvest refugia in the form of MPAs in rebuilding natural population densities and conserving old growth age structure, as well as providing the benefit of natural population dynamics (e.g., Sørdalen et al. 2020). Sedentary lifestyles contrasted by life histories tuned to long-distance dispersal and high connectivity seem to fit with the benefits of MPA networks (Olsen et al. 2013, Pérez-Ruzafa et al. 2017, Moksnes and Jonsson 2020). In all regions presented and discussed herein (New Zealand, Skagerrak/Norway, Western Mediterranean), networks of areas in which lobsters are afforded protection have been implemented. The designation and implementation of these MPAs is likely a result of the knowledge accrued based on these very examples. Considering anthropogenic global warming and its effect on variability in marine populations, it seems worthwhile to start designing reserve networks that take environmental change in the present and on the horizon into account. In addition to the increased pressure from a surge in crustacean fisheries (Boenish et al. 2022), resilience of crustacean populations is likely to be affected by impacts of incremental ocean warming (e.g., Linnane et al. 2024) and increased frequency of marine heat waves (e.g., Caputi et al. 2019, Oppenheim et al. 2019), as well as ocean acidification (e.g., Keppel et al. 2012, Tai et al. 2021). The buffering of such impacts will benefit from MPAs and networks being placed in areas that are likely to act as climate refugia (Davis et al. 2021).

REFERENCES

Agnalt, A.L. 2008. Fecundity of the European lobster (*Homarus gammarus*) off southwestern Norway after stock enhancement: Do cultured females produce as many eggs as wild females? ICES Journal of Marine Science 65:164–170.

Agnalt, A.L., T.S. Kristiansen, and K.E. Jørstad. 2007. Growth, reproductive cycle, and movement of berried European lobsters (*Homarus gammarus*) in a local stock off southwestern Norway. ICES Journal of Marine Science 64:288–297.

Annala, J.H. 1983. New Zealand rock lobsters: Biology and fishery. New Zealand Ministry of Agriculture and Fisheries, New Zealand.

Annala, J. and M. King. 1983. The 1963–73 New Zealand rock lobster landings by statistical area. New Zealand Ministry of Agriculture and Fisheries, New Zealand.

Arbuckle, G. 1971. The development of the crayfishing industry. Chatham Islands in perspective: A socio-economic review. Hicks Smith, Wellington, New Zealand.

Babcock, R.C., S. Kelly, N.T. Shears, J.W. Walker, and T.J. Willis. 1999. Changes in community structure in temperate marine reserves. Marine Ecology Progress Series 189:125–134.

Babcock, R.C., J.C. Phillips, M. Lourey, and G. Clapin. 2007. Increased density, biomass and egg production in an unfished population of Western Rock Lobster (*Panulirus cygnus*) at Rottnest Island, Western Australia. Marine and Freshwater Research 58:286–292.

Babcock, R.C., N.T. Shears, A.C. Alcala, N.S. Barrett, G.J. Edgar, K.D. Lafferty, T. McClanahan, and G. Russ. 2010. Decadal trends in marine reserves reveal differential rates of change in direct and indirect effects. Proceedings of the National Academy of Science U.S.A. 107:18256–18261.

Balemi, C.A., and N.T. Shears. 2023. Emergence of the subtropical sea urchin *Centrostephanus rodgersii* as a threat to kelp forest ecosystems in northern New Zealand. Frontiers in Marine Science 10: 224067.

Ballantine, B. 2014. Fifty years on: Lessons from marine reserves in New Zealand and principles for worldwide network. Biological Conservation 176:297–307.

Ballantine, W., and D. Gordon. 1979. New Zealand's first marine reserve, Cape Rodney to Okakari point, Leigh. Biological Conservation 15:273–280.

Barrett, N.S., C. Buxton, and G.J. Edgar. 2009. Changes in in vertebrate and macroalgal populations in Tasmanian marine reserves in the decade following protection. Journal of Experimental Marine Biology and Ecology 370:104–119.

Beaglehole, J.C. 1963. The *Endeavour* journal of Joseph Banks, 1768–1771. Trustees of the Public Library of New South Wales, New South Wales, Australia.

Bentley, N., P.A. Breen, P.J. Starr, and I. Bay. 2003. Design and evaluation of a revised management decision rule for red rock lobster fisheries (*Jasus edwardsii*) in CRA 7 and CRA 8. New Zealand Fisheries Assessment Report 30:1–43.

Biro, P.A., B. Adriaenssens, and P. Sampson. 2014. Individual and sex-specific differences in intrinsic growth rate covary with consistent individual differences in behaviour. The Journal of Animal Ecology 83:1186–1195.

Biro, P.A., and P. Sampson. 2015. Fishing directly selects on growth rate via behaviour: Implications of growth-selection that is independent of size. Proceedings of the Royal Society B: Biological Sciences 282:20142283.

Boenish, R., J.P. Kritzer, K. Kleisner, R.S. Steneck, K.M. Werner, W. Zhu, F. Schram, D. Rader, W. Cheung, J. Ingles, Y. Tian, and J. Mimikakis. 2022. The global rise of crustacean fisheries. Frontiers in Ecology and the Environment 20:102–110.

Booth, J.D. 1994. *Jasus edwardsii* larval recruitment off the east coast of New Zealand. Crustaceana 66:295–317.

Booth, J.D. 2006. *Jasus* species. Pages 340–358 in B.F. Phillips, editor. Lobsters: Biology, management, aquaculture and fisheries. Blackwell Publishing, UK.

Booth, J.D. 2008. Spiny lobsters: Fisheries and culture. Pages 78–89 *in* B. Phillips and J. Kittaka, editors. New Zealand's rock lobster fisheries, 2nd edition. Blackwell Science, UK.

Booth, J.D., and J.R. Ovenden. 2000. Distribution of *Jasus* spp. (Decapoda: Palinuridae) phyllosomas in southern waters: Implications for larval recruitment. Marine Ecology Progress Series 200: 241–255.

Booth, J.D., and W. Webber. 2001. All the pretty lobsters. Part 1. Seafood New Zealand 9:20–23.

Bradbury, I.R., and P.V.R. Snelgrove. 2001. Contrasting larval transport in demersal fish and benthic invertebrates: The roles of behaviour and advective processes in determining spatial pattern. Canadian Journal of Fisheries and Aquatic Sciences 58:811–823.

Breen, P.A., and T.H. Kendrick. 1997. A fisheries management success story: The Gisborne, New Zealand, fishery for red rock lobsters (*Jasus edwardsii*). Marine and Freshwater Research 48:1103–1110.

Breton, W.H. 1835. Scandinavian sketches, or, a tour in Norway. British Library, London, UK.

Buglass, S., H. Reyes, J. Ramirez-González, T.D. Eddy, P. Salinas-de-león, and J.M. Jarrin. 2018. Evaluating the effectiveness of coastal no-take zones of the Galapagos Marine Reserve for the red spiny lobster, *Panulirus penicillatus*. Marine Policy 88:204–212.

Calef, Z. 2016. Quantifying the impact of a network of small-scale MPAs on reproductive potential of European lobster (*Homarus gammarus*) in the Norwegian Skagerrak. Masters thesis, DTU-Aqua, Denmark.

Caputi, N., M. Kangas, A. Chandrapavan, A. Hart, M. Feng, M. Marin, and S. de Lestang. 2019. Factors affecting the recovery of invertebrate stocks from the 2011 Western Australian extreme marine heatwave. Frontiers in Marine Science 6:484.

Cau, A., A. Bellodi, R. Cannas, M. Fois, P. Guidetti, D. Moccia, C. Porcu, A. Pusceddu, and M.C. Follesa. 2019. European spiny lobster recovery from overfishing enhanced through active restocking in fully protected areas. Scientific Reports 9:13025.

Ceccaldi, H.J., and D. Latrouite. 2000. The French fisheries for the European spiny lobster *Palinurus elephas*. Pages 200–209 *in* B.F. Phillips and J. Kittaka, editors. Spiny lobsters: Fisheries and culture. Blackwell Science, UK.

Chiswell, S.M., and J.D. Booth. 2008. Sources and sinks of larval settlement in *Jasus edwardsii* around New Zealand: Where do larvae come from and where do they go? Marine Ecology Progress Series 354:201–217.

Chiswell, S.M., J. Wilkin, J.D. Booth, and B. Stanton. 2003. Trans-Tasman Sea larval transport: Is Australia a source for New Zealand rock lobsters? Marine Ecology Progress Series 247:173–182.

Cole, R.G., and D. Keuskamp. 1998. Indirect effects of protection from exploitation: Patterns from populations of *Evechinus chloroticus* (Echinoidea) in northeastern New Zealand. Marine Ecology Progress Series 173:215–226.

Cox, C., and J.H. Hunt. 2005. Change in size and abundance of Caribbean spiny lobsters *Panulirus argus* in a marine reserve in the Florida Keys National Marine Sanctuary, USA. Marine Ecology Progress Series 294:227–239.

Dannevig, A. 1936. Hummer og hummerkultur [in Norwegian]. Report on Norwegian Fisheries and Marine Investigations 4:1–60.

Davies, B.F.R., L. Holmes, A. Rees, M.J. Attrill, A.Y. Cartwright, and E.V. Sheehan. 2021. Ecosystem approach to fisheries management works—How switching from mobile to static fishing gear improves populations of fished and non-fished species inside a marine-protected area. Journal of Applied Ecology 58:2463–2478.

Davis, T.R., C. Champion, and M.A. Coleman. 2021. Climate refugia for kelp within an ocean warming hotspot revealed by stacked species distribution modelling. Marine Environmental Research. 166:105267.

Department of Conservation and Fisheries New Zealand. 2021. Sea change—Tai Timu Tai Pari Plan marine protected area (MPA) proposals: Agency analysis and advice on selection of MPAs towards development of the Hauraki Gulf Marine Park MPA network. Department of Conservation, New Zealand.

Díaz, D., M. Marí, P. Abelló, and M. Demestre. 2001. Settlement and juvenile habitat of the European spiny lobster *Palinurus elephas* (Crustacea: Decapoda: Palinuridae) in the Western Mediterranean Sea. Scientia Marina 65:347–356.

Díaz, D., S. Mallol, M. Parma, and R. Goñi. 2011. Decadal evolution of lobster reproductive output from a marine protected area. Marine Ecology Progress Series 433:149–157.

Díaz, D., S. Mallol, A. Parma, and R. Goñi. 2016. A 25-year marine reserve as proxy for the unfished condition of an exploited species. Biological Conservation 203:97–107.

Dillaman, R., R. Roer, T. Shafer, S. Modla. 2013. The crustacean integument: Structure and function. Pages 140–165 *in* L. Watling and M. Thiel, editors. The natural history of Crustacea, Volume 1: Functional morphology and diversity. Oxford University Press, New York, USA.

Eddy, T.D., T.J. Pitcher, A.B. MacDiarmid, T.T. Byfield, J.C. Tam, T.T Jones, J.J. Bell, and J.P. Gardner. 2014. Lobsters as keystone: Only in unfished ecosystems? Ecological Modelling 275:48–72.

Eisaguirre, J.H., J.M. Eisaguirre, K. Davis, P.M. Carlson, S.D. Gaines, and J.E. Caselle. 2020. Trophic redundancy and predator size class structure drive differences in kelp forest ecosystem dynamics. Ecology 101:e02993.

Ellis, C.D., D.J. Hodgson, C.L. Daniels, M. Collins, and A.G.F. Griffiths. 2017. Population genetic structure in European lobsters: Implications for connectivity, diversity, and hatchery stocking. Marine Ecology Progress Series 563:123–137.

Ellis, J.I., and D.C. Schneider. 1997. Evaluation of a gradient sampling design for environmental impact assessment. Environmental Monitoring and Assessment 48:157–172.

Enrichettia, R., S. Bavab, G. Bavestrelloa, F. Bettia, F. Lanteria, and M. Bo. 2019. Artisanal fishing impact on deep coralligenous animal forests: A Mediterranean case study of marine vulnerability. Ocean and Coastal Management 177:112–126.

Fernández-Chacón, A., D. Villegas-Ríos, E. Moland, M.L. Baskett, E.M. Olsen, and S.M. Carlson. 2020. Protected areas buffer against harvest selection and rebuild phenotypic complexity. Ecological Applications 30:e02108.

Fernández-Chacón, A., L. Buttay, E. Moland, H. Knutsen, and E.M. Olsen. 2021. Demographic responses to protection from harvesting in a long-lived marine species. Biological Conservation 257:1–10.

Florko, K.R.N., E.R. Davidson, K.J. Lees, L.J. Hammer, M.-F. Lavoie, R.J. Lennox, E. Simard, P. Archambault, M. Auger-Méthé, C.W. McKindsey, F.G. Whoriskey, and N.B. Furey. 2021. Tracking movements of decapod crustaceans: A review of a half-century of telemetry-based studies. Marine Ecology Progress Series 679:219–239.

Follesa, M.C., R. Cannas, A. Cau, D. Cuccu, A. Gastoni, A. Ortu, C. Pedoni, C. Porcu, and A. Cau. 2011. Spillover effects of a Mediterranean marine protected area on the European spiny lobster *Palinurus elephas* (Fabricius, 1787) resource. Aquatic Conservation: Marine and Freshwater Ecosystems 21:564–572.

Francis, R.C., M.A. Hixon, M.E. Clarke, S.A. Murawski, and S. Ralston. 2007. Ten commandments for ecosystem-based fisheries scientists. Fisheries 32:217–233.

Freeman, D.J., and A.B. MacDiarmid. 2009. Healthier lobsters in a marine reserve: Effects of fishing on disease incidence in the spiny lobster, *Jasus edwardsii*. Marine and Freshwater Research 60: 140–145.

Freeman, D.J., P.A. Breen, and A.B. MacDiarmid. 2012a. Use of a marine reserve to determine the direct and indirect effects of fishing on growth in a New Zealand fishery for the spiny lobster *Jasus edwardsii*. Canadian Journal of Fisheries and Aquatic Sciences 69:894–905.

Freeman, D.J., A.B. MacDiarmid, and R.B. Taylor. 2009. Habitat patches that cross marine reserve boundaries: Consequences for the lobster *Jasus edwardsii*. Marine Ecology Progress Series 388:159–167.

Freeman, D.J., A.B. MacDiarmid, R.B. Taylor, R.J. Davidson, R.V. Grace, T.R. Haggitt, S. Kelly, and N.T. Shears. 2012b. Trajectories of spiny lobster *Jasus edwardsii* recovery in New Zealand marine reserves: Is settlement a driver? Environmental Conservation 39:295–304.

Frid, A., M. McGreer, and A. Stevenson. 2016. Rapid recovery of Dungeness crab within spatial fishery closures declared under indigenous law in British Columbia. Global Ecology and Conservation 6:48–57.

Frisch, A.J., and J.P. Hobbs. 2012. Demography, fishery yield and potential management strategies of painted spiny lobster (*Panulirus versicolor*) at Northwest Island, Great Barrier Reef, Australia. Marine and Freshwater Research 63:387–396.

Galparasoro, I., A. Borja, J. Bald, P. Liria, and G. Chust. 2009. Predicting suitable habitat for the European lobster (*Homarus gammarus*), on the Basque continental shelf (Bay of Biscay), using ecological-niche factor analysis. Ecological Modelling 220:556–567.

Gibbs, M.T. 2008. The historical development of fisheries in New Zealand with respect to sustainable development principles. The Electronic Journal of Sustainable Development 1:23–33.

Gnanalingam, G., M.J. Butler IV, T.R. Matthews, E. Hutchinson, and R. Kilada. 2019. Directly ageing the Caribbean spiny lobster, *Panulirus argus* with validated band counts from gastric mill ossicles. ICES Journal of Marine Science 76:442–451.

Goñi, R. 2014. *Palinurus elephas*. The IUCN red list of threatened species. Version 2014.3. Accessed June 2, 2022. [https://www.iucnredlist.org/species/169975/1281221].

Goñi, R., and D. Latrouite. 2005. Biology, ecology and fisheries of *Palinurus* spp. species of European waters: *Palinurus elephas* (Fabricius, 1787) and *Palinurus mauritanicus* (Gruvel, 1911). Cahiers de Biologie Marine 46:127–142.

Goñi, R., O. Reñones, and A. Quetglas. 2001a. Dynamics of a protected Western Mediterranean population of the European spiny lobster *Palinurus elephas* (Fabricius, 1787) assessed by trap surveys. Marine and Freshwater Research 52:1577–1587.

Goñi, R., A. Quetglas, and O. Reñones. 2001b. Diet of the spiny lobster *Palinurus elephas* (Decapoda, Palinuridea) from the Columbretes Islands marine reserve (Northwestern Mediterranean). Journal of the Marine Biological Association of the United Kingdom 80:1–3.

Goñi, R., A. Quetglas, and O. Reñones. 2003. Size at maturity, fecundity and reproductive potential of a protected population of the spiny lobster *Palinurus elephas* (Fabricius, 1787) from the Western Mediterranean. Marine Biology 143:583–592.

Goñi, R., A. Quetglas, and O. Reñones. 2006. Spillover of lobster *Palinurus elephas* (Fabricius, 1787) from a Western Mediterranean marine reserve. Marine Ecology Progress Series 308:207–219.

Goñi, R., S. Adlerstein, D. Alvarez-Berastegui, A. Forcada, O. Reñones, G. Criquet, S. Polti, G. Cadiou, C. Valle, P. Lenfant, P. Bonhomme, A. Pérez-Ruzafa, J.L. Sánchez-Lizaso, J.A. García-Charton, G. Bernard, V. Stelzenmüller, and S. Planes 2008. Spillover from six Western Mediterranean marine protected areas: Evidence from artisanal fisheries. Marine Ecology Progress Series 366:159–174.

Goñi, R., R. Hilborn, D. Díaz, S. Mallol, and S. Adlerstein. 2010. Net contribution of spillover from a marine reserve to fishery catches. Marine Ecology Progress Series 400:233–243.

Guenther, C., D. López-Carr, and H.S. Lenihan. 2015. Differences in lobster fishing effort before and after MPA establishment. Applied Geography 59:78–87.

Guzman, H.M., and A. Tewfik. 2004. Population characteristics and co-occurrence of three exploited decapods (*Panulirus argus, P. guttatus* and *Mithrax spinosissimus*) in Bocas del Toro, Panama. Journal of Shellfish Research 23:575–580.

Hallbäck, H., and A. Warén. 1972. Food ecology of lobster, *Homarus vulgaris*, in Swedish waters. Some preliminary results. ICES shellfish and benthos communities, C.M. 1972/K 29:1–5.

Hanns, B.J, and N.T. Shears. 2023. Modelling spatial variation in an exploited species across marine reserve boundaries. Journal of Applied Ecology 60:1708–1719.

Harper, S., and D. Zeller. 2011. Fisheries catch reconstructions: Islands, Part II. Fisheries Centre Research Reports, Volume 19. The Fisheries Centre, University of British Columbia, Vancouver, Canada.

Haynes, P., P. Browne, L. Fullbrook, C.T. Graham, L. Hancox, M.P. Johnson, V. Lauria, and A.M. Power. 2016. Growth in *Nephrops norvegicus* from a tag-recapture experiment. Scientific Reports 6:35143.

Holthuis, L.B. 1991. FAO species catalogue: Marine lobsters of the world. An annotated and illustrated catalogue of species of interest to fisheries known to date. FAO Fisheries Synopsis No. 125, Volume 13 FIR/S125. Nationaal Natuurhistorisch Museum Leiden, The Netherlands.

Hoskin, M.G., R.A. Coleman, E. von Carlshausen, and C.M. Davis. 2011. Variable population responses to the establishment of a temperate no-take zone. Canadian Journal of Fisheries and Aquatic Sciences 68:185–200.

Howarth, L.M., P. Dubois, P. Gratton, M. Judge, B. Christie, J.J. Waggitt, J.P. Hawkins, C.M. Roberts, and B.D. Stewart. 2017. Trade-offs in marine protection: Multispecies interactions within a community-led temperate marine reserve. ICES Journal of Marine Science 74:263–276.

Huntsberger, C.J., R. Kilada, W.G. Ambrose Jr., and R.A. Wahle. 2020. Age-at-size relationships of the American lobster (*Homarus americanus*) from three contrasting thermal regimes using gastric mill band counts as a direct aging technique. Canadian Journal of Fisheries and Aquatic Sciences 77: 1733–1740.

Huserbråten, M.B.O., E. Moland, H. Knutsen, E.M. Olsen, C. André, and N.C. Stenseth. 2013. Conservation, spillover and gene flow within a network of Northern European marine protected areas. PLoS ONE 8:e73388.

IPBES. 2019. Summary for policymakers of the global assessment report on biodiversity and ecosystem services of the Intergovernmental Science-Policy Platform on Biodiversity and Ecosystem Services. S.M. Díaz, J. Settele, E.S. Brondízio, H.T. Ngo, M. Guèze, J. Agard, A. Arneth, P. Balvanera, K.A. Brauman, S.H.M. Butchart, K.M.A. Chan, L.A. Garibaldi, K. Ichii, J. Liu, S.M. Subramanian, G.F. Midgley, P. Miloslavich, Z. Molnár, D. Obura, A. Pfaff, S. Polasky, A. Purvis, J. Razzaque, B. Reyers, R. Roy Chowdhury, Y.J. Shin, I.J. Visseren-Hamakers, K.J. Willis, and C.N. Zayas, editors. IPBES secretariat, Bonn, Germany.

Jack, L., and S.R. Wing. 2010. Maintenance of old-growth size structure and fecundity of the red rock lobster *Jasus edwardsii* among marine protected areas in Fiordland, New Zealand. Marine Ecology Progress Series 404:161–172.

Jack, L., and S.R. Wing. 2013. A safety network against regional population collapse: Mature subpopulations in refuges distributed across the landscape. Ecosphere 4:1–16.

Jackson, J.B., M.X. Kirby, W.H. Berger, K.A. Bjorndal, L.W. Botsford, B.J. Bourque, R.H. Bradbury, R. Cooke, J. Erlandson, J.A. Estes, T.P. Hughes, S. Kidwell, C.B. Lange, H.S. Lenihan, J.M. Pandolfi, C.H. Peterson, R.S. Steneck, M.J. Tegner, and R.R. Warner. 2001. Historical overfishing and the recent collapse of coastal ecosystems. Science 293:629–637.

Jaco, E.M., and M.A. Steele. 2020. Pre-closure fishing pressure predicts effects of marine protected areas. Journal of Applied Ecology 57:229–240.

Jeffs, A.G., C. Gardner, and A. Cockcroft. 2013. *Jasus* and *Sagmariasus* species. Pages 259–288 *in* B.F. Phillips, editor. Lobsters: Biology, management, aquaculture and fisheries, 2nd edition. John Wiley and Sons Ltd., Oxford, UK.

Jones, E.V., D. Macintosh, S. Stead, and T. Gray. 2017. How effective are MPAs in conserving crab stocks? A comparison of fisheries and conservation objectives in three coastal MPAs in Thailand. Ocean and Coastal Management 149:186–197.

Kay, M.C., H.S. Lenihan, C.M. Guenther, J.R. Wilson, C.J. Miller and S.W. Shrout 2012a. Collaborative assessment of spiny lobster (*Panulirus interruptus*) population and fisheries responses to a marine reserve network. Ecological Applications 22:322–335.

Kay, M.C., H.S. Lenihan, M.J. Kotchen, and C.J. Miller. 2012b. Effects of marine reserves on California spiny lobster are robust and modified by fine-scale habitat features and distance from reserve borders. Marine Ecology Progress Series 451:137–150.

Kellner, J.B., I. Tetreault, S.D. Gaines, and R.M. Nisbet. 2007. Fishing the line near marine reserves in single and multispecies fisheries. Ecological Applications 17:1039–1054.

Kelly, S. 2001. Temporal variation in the movement of the spiny lobster *Jasus edwardsii*. Marine and Freshwater Research 52:323–331.

Kelly, S., and A.B. MacDiarmid. 2003. Movement patterns of mature spiny lobsters, *Jasus edwardsii*, from a marine reserve. New Zealand Journal of Marine and Freshwater Research 37:149–158.

Kelly, S., A. MacDiarmid, and R. Babcock. 1999. Characteristics of spiny lobster, *Jasus edwardsii*, aggregations in exposed reef and sandy areas. Marine and Freshwater Research 50:409–416.

Kelly, S., D. Scott, and A.B. MacDiarmid. 2002. The value of a spillover fishery for spiny lobsters around a marine reserve in northern New Zealand. Coastal Management 30:153–166.

Kelly, S., D. Scott, A.B. MacDiarmid, and R.C. Babcock. 2000. Spiny lobster, *Jasus edwardsii*, recovery in New Zealand marine reserves. Biological Conservation 92:359–369.

Keppel, E.A., R.A. Scrosati, and S.C. Courtenay. 2012. Ocean acidification decreases growth and development in American lobster (*Homarus americanus*) larvae. Journal of Northwest Atlantic Fishery Science 44:61–66.

Kleiven, A.R., S.H. Espeland, K. Ono, F. Zimmermann, S. Stiansen, and E.M. Olsen. 2022. Technological creep masks continued decline in a lobster (*Homarus gammarus*) fishery over a century. Scientific Reports 12:3318.

Kleiven, A.R., E.M. Olsen, and J.H. Vølstad. 2012. Total catch of a red-listed marine species is an order of magnitude higher than official data. PLoS ONE 7:e31216.

Knutsen, J.A., A.R. Kleiven, E.M. Olsen, H. Knutsen, S.H. Espeland, T.K. Sørdalen, S.H. Thorbjørnsen, J.A. Hutchings, A. Fernández-Chacón, M. Huserbråten, D. Villegas-Ríos, K.T. Halvorsen, P.J. Nillos Kleiven, T.K. Langeland, and E. Moland. 2022. Lobster reserves as a management tool in coastal waters: Two decades of experience in Norway. Marine Policy 136:104908.

Kolbusz, J., S. de Lestang, T. Langlois, and C. Pattiaratchi. 2021. Changes in *Panulirus cygnus* settlement along Western Australia using a long time series. Frontiers in Marine Science 8:628912.

Kough, A.S., C.B. Paris, and M.J. Butler IV. 2013. Larval connectivity and the international management of fisheries. PLoS ONE 8:e64970.

Kraufvelin, P., L. Bergström, F. Sundqvist, M. Ulmestrand, H. Wennhage, A. Wikström, and U. Bergström. 2023. Rapid re-establishment of top-down control at a no-take artificial reef. Ambio 52:556–570

Kruse, G.H., J. Zheng, and D.L. Stram. 2010. Recovery of the Bristol Bay stock of red king crabs under a rebuilding plan. ICES Journal of Marine Science 67:1866–1874.

Langlois, T.J., M.J. Anderson, and R.C. Babcock. 2005. Reef-associated predators influence adjacent soft-sediment communities. Ecology 86:1508–1519.

LaScala-Gruenewald, D.E., R.V. Grace, T.R. Haggitt, B.J. Hanns, S. Kelly, A. MacDiarmid, and N.T. Shears. 2021. Small marine reserves do not provide a safeguard against overfishing. Conservation Science and Practice 3:e362.

Leach, B.F., and A.J. Anderson. 1979. Prehistoric exploitation of crayfish in New Zealand. Birds of a feather. Osteological and archaeological papers from the South Pacific in honour of RJ Scarlett. New Zealand Archaeological Association Monograph 11:141–164.

Lenihan, H.S., J.P. Gallagher, J.R. Peters, A.C. Stier, J.K.K. Hofmeister, and D.C. Reed. 2021. Evidence that spillover from marine protected areas benefits the spiny lobster (*Panulirus interruptus*) fishery in southern California. Scientific Reports 11:2663.

Ley-Cooper, K., S. de Lestang, B.F. Phillips, and E. Lozano-Álvarez. 2014. An unfished area enhances a spiny lobster, *Panulirus argus*, fishery: Implications for management and conservation within a biosphere reserve in the Mexican Caribbean. Fisheries Management and Ecology 21: 264–274.

Ling, S., and C. Johnson. 2012. Marine reserves reduce risk of climate-driven phase shift by reinstating size- and habitat-specific trophic interactions. Ecological Applications 22: 1232–1245.

Ling, S., C. Johnson, S. Frusher, and K. Ridgway. 2009. Overfishing reduces resilience of kelp beds to climate-driven catastrophic phase shift. Proceedings of the National Academy of Sciences 106:22341–22345.

Linnane, A., McGarvey, R., Matthews, J.M., Feenstra, J.E., Jones, A. and N. Kimber. 2024. Adapting to climate change in a spiny lobster (*Jasus edwardsii*) fishery: A harvest strategy response. Fisheries Research 276:107054.

MacDiarmid, A.B. 1989a. Moulting and reproduction of the spiny lobster *Jasus edwardsii* (Decapoda: Palinuridae) in northern New Zealand. Marine Biology 103:303–310.

MacDiarmid, A.B. 1989b. Size at onset of maturity and size-dependent reproductive output of female and male spiny lobsters *Jasus edwardsii* (Hutton) (Decapoda, Palinuridae) in northern New Zealand. Journal of Experimental Marine Biology and Ecology 127:229–243.

MacDiarmid, A.B. 1991. Seasonal changes in depth distribution, sex ratio and size frequency of spiny lobster *Jasus edwardsii* on a coastal reef in northern New Zealand. Marine Ecology Progress Series 70:129–141.

MacDiarmid, A.B. 1994. Cohabitation in the spiny lobster *Jasus edwardsii* (Hutton, 1875). Crustaceana 66:341–355.

MacDiarmid, A.B., E. Abraham, C. Baker, E. Carroll, C. Chagué-Goff, P. Cleaver, M. Francis, J. Goff, P. Horn, and J. Jackson. 2016. Taking Stock—The changes to New Zealand marine ecosystems since first human settlement: Synthesis of major findings, and policy and management implications. New Zealand Aquatic Environment and Biodiversity Report 170.

MacDiarmid, A.B., and M.J. Butler IV. 1999. Sperm economy and limitation in spiny lobsters. Behavioural Ecology and Sociobiology 46:14–24.

MacDiarmid, A.B., D. Freeman, and S. Kelly. 2013. Rock lobster biology and ecology: Contributions to understanding through the Leigh marine laboratory 1962–2012. New Zealand Journal of Marine and Freshwater Research 47:313–333.

MacDiarmid, A.B., B. Hickey, and R. Maller. 1991. Daily movement patterns of the spiny lobster *Jasus edwardsii* (Hutton) on a shallow reef in northern New Zealand. Journal of Experimental Marine Biology and Ecology 147:185–205.

Mallol, S., D. Díaz, F. Sobrado, and R. Goñi. 2014. First v-notching experience of a spiny lobster: V-notch recovery and impacts on health and growth of *Palinurus elephas*. Journal of Crustacean Biology 34:25–30.

Mayfield, S., G.M. Branch, and A.C. Cockcroft. 2005. Role and efficacy of marine protected areas for the South African rock lobster, *Jasus lalandii*. Marine and Freshwater Research 56:913–924.

Mehrtens, F., M. Stolpmann, F. Buchholz, W. Hagen, and R. Saborowski. 2005. Locomotory activity and exploration behaviour of juvenile European lobsters (*Homarus gammarus*) in the laboratory. Marine and Freshwater Behaviour and Physiology 38:105–116.

Melville-Smith, R., and S. de Lestang. 2006. Spatial and temporal variation in the size at maturity of the western rock lobster *Panulirus cygnus* George. Marine Biology 150:183–195.

Miller, E., E. Spanier, R. Diamant, and R. Yahel. 2023. Nature reserves facilitate conservation of the Mediterranean slipper lobster *Scyllarides latus*. Marine Ecology Progress Series 714: 57–69.

Moksnes, P.-O., and P.R. Jonsson. 2020. Larval connectivity and marine protected area networks. Pages 408–436 *in* K. Anger, S. Harzsch, and M. Thiel, editors. The natural history of the Crustacea, Volume 7: Developmental biology and larval ecology. Oxford University Press, New York USA.

Moland, E., S.M. Carlson, D. Villegas-Ríos, J. Ree Wiig, and E. Moland Olsen. 2019. Harvest selection on multiple traits in the wild revealed by aquatic animal telemetry. Ecology and Evolution 9: 6480–6491.

Moland, E., A. Fernández-Chacón, T.K. Sørdalen, D. Villegas-Ríos, S.H. Thorbjørnsen, K.T. Halvorsen, M. Huserbråten, E.M. Olsen, P.J. Nillos Kleiven, A.R. Kleiven, H. Knutsen, S.H. Espeland, C. Freitas, and J.A. Knutsen. 2021. Restoration of abundance and dynamics of coastal fish and lobster within northern marine protected areas across two decades. Frontiers in Marine Science 8:674756.

Moland, E., E.M. Olsen, H. Knutsen, J. Knutsen, S. Enersen, C. André, and N. Stenseth. 2011a. Activity patterns of wild European lobster *Homarus gammarus* in coastal marine reserves: Implications for future reserve design. Marine Ecology Progress Series 429:197–207.

Moland, E., E.M. Olsen, K. Andvord, J.A. Knutsen, and N.C. Stenseth. 2011b. Home range of European lobster (*Homarus gammarus*) in a marine reserve: Implications for future reserve design. Fish and Fisheries 1210:1197–1210.

Moland, E., E.M. Olsen, H. Knutsen, P. Garrigou, S.H. Espeland, A.R. Kleiven, C. André, and J.A. Knutsen. 2013a. Lobster and cod benefit from small-scale northern marine protected areas: Inference from an empirical before-after control-impact study. Proceedings of the Royal Society B: Biological Sciences 280:20122679.

Moland, E., M. Ulmestrand, E.M. Olsen, and N.C. Stenseth. 2013b. Long-term decrease in sex-specific natural mortality of European lobster within a marine protected area. Marine Ecology Progress Series 491:153–164.

New Zealand Fisheries. 2020. Fisheries assessment plenary, November 2020: Stock assessments and stock status. Compiled by the Fisheries Science and Information Group, Fisheries New Zealand, Wellington, New Zealand.

Nillos Kleiven, P.J., S.H. Espeland, E.M. Olsen, R.A. Abesamis, E. Moland, and A.R. Kleiven. 2019. Fishing pressure impacts the abundance gradient of European lobsters across the borders of a newly established marine protected area. Proceedings of the Royal Society B: Biological Sciences 286:20182455.

Olsen, E.M., D. Johnson, P. Weaver, R. Goñi, M.C. Ribeiro, M. Rabaut, E. Macpherson, D. Pelletier, L. Fonseca, S. Katsanevakis, and T. Zaharia. 2013. Achieving ecologically coherent MPA networks in Europe: Science needs and priorities. Marine Board Position Paper 18. K.E. Larkin, and N. McDonough, editors. European Marine Board, Ostend, Belgium.

Oppenheim, N.G., R.A. Wahle, D.C. Brady, A.G. Goode, and A.J. Pershing. 2019. The cresting wave: Larval settlement and ocean temperatures predict change in the American lobster harvest. Ecological Applications 29:e02006.

Øresland, V., and M. Ulmestrand. 2013. European lobster subpopulations from limited adult movements and larval retention. ICES Journal of Marine Science 70:532–539.

Ovenden, J.R., D.J. Brasher, and R.W.G. White. 1992. Mitochondrial DNA analyses of the red rock lobster *Jasus edwardsii* supports an apparent absence of population subdivision throughout Australasia. Marine Biology 112:319–326.

Paulin, C.D. 2007. Perspectives of Mäori fishing history and techniques: Ngä ähua me ngä püräkau me ngä hangarau ika o te Māori. Tuhinga: Records of the Museum of New Zealand Te Papa Tongarewa 18:11–47.

Pavičić, M., I. Žužul, S. Matić-Skoko, A. Triantafyllidis, F. Grati, E.D.H. Durieux, I. Celić, and T. Šegvić-Bubić. 2020. Population genetic structure and connectivity of the European lobster *Homarus gammarus* in the Adriatic and Mediterranean Seas. Frontiers in Genetics 11:576023.

Peleg, O., C. Blain, and N.T. Shears. 2023. Long-term marine protection enhances kelp forest ecosystem stability. Ecological Applications 33:e2895.

Pérez-Ruzafa, A., J.A. García-Charton, and C. Marcos. 2017. Northeast Atlantic vs. Mediterranean marine protected areas as fisheries management tool. Frontiers in Marine Science 4:245.

Pettersen, A.R., E. Moland, and E.M. Olsen. 2009. Lobster reserves in coastal Skagerrak—An integrated analysis of the implementation process. Pages 178–188 *in* E. Dahl, E. Moksness, and J. Støttrup, editors. Integrated coastal zone management. Wiley-Blackwell Publishing, London, UK.

Pillans, S., R. Pillans, R. Johnstone, P. Kraft, M. Haywood, and H. Possingham. 2005. Effects of marine reserve protection on the mud crab *Scylla serrata* in a sex-biased fishery in subtropical Australia. Marine Ecology Progress Series 295:201–213.

Plinio il Vecchio. 77–78 AD. Trattato 1ª ed. originale: Naturalis historia. Editio prínceps, Venezia, Italy.

Pollock, D.E. 1986. Review of the fishery for and biology of the Cape rock lobster *Jasus lalandii* with notes on larval recruitment. Canadian Journal of Fisheries and Aquatic Sciences 43:2107–2117.

Pollock, D.E. 1995. Changes in maturation ages and sizes in crustacean and fish populations. South African Journal of Marine Science 15:99–103.

Roberts, C.M., B.C. O'Leary, D.J. McCauley, P.M. Cury, C.M. Duarte, J. Lubchenco, D. Pauly, A. Sáenz-Arroyo, U.R. Sumaila, R.W. Wilson, B. Worm, and J.C. Castilla. 2017. Marine reserves can mitigate and promote adaptation to climate change. Proceedings of the National Academy of Sciences 114:6167–6175.

Rowe, S. 2002. Population parameters of American lobster inside and outside no-take reserves in Bonavista Bay, Newfoundland. Fisheries Research 56:167–175.

Salmond, A. 2003. The trial of the cannibal dog: The remarkable story of Captain Cook's encounters in the south seas. Yale University Press, New Haven, Connecticut, USA.

Salomon, A.K., S.K. Gaichas, N.T. Shears, J.E. Smith, E.M. Madin, and S.D. Gaines. 2010. Key features and context-dependence of fishery-induced trophic cascades. Conservation Biology 24:382–394.

Salomon, A.K., N.T. Shears, T.J. Langlois, and R.C. Babcock. 2008. Cascading effects of fishing can alter carbon flow through a temperate coastal ecosystem. Ecological Applications 18:1874–1887.

Segura-García, I., L. Garavelli, M. Tringali, T. Matthews, L.M. Chérubin, J. Hunt, and S.J. Box. 2019. Reconstruction of larval origins based on genetic relatedness and biophysical modeling. Scientific Reports 9:7100.

Shears, N.T., and R.C. Babcock. 2002. Marine reserves demonstrate top-down control of community structure on temperate reefs. Oecologia 132:131–142.

Shears, N.T., and R.C. Babcock. 2003. Continuing trophic cascade effects after 25 years of no-take marine reserve protection. Marine Ecology Progress Series 246:1–16.

Shears, N.T., and R.C. Babcock. 2004. Community composition and structure of shallow subtidal reefs in northeastern New Zealand. Department of Conservation, New Zealand.

Shears, N.T., and R.C. Babcock. 2007. Quantitative description of mainland New Zealand's shallow subtidal reef communities. Science and Technical Publications. Department of Conservation, New Zealand.

Shears, N.T., R.C. Babcock, and A.K. Salomon. 2008. Context-dependent effects of fishing: Variation in trophic cascades across environmental gradients. Ecological Applications 18:1860–1873.

Shears, N.T., R.V. Grace, N.R. Usmar, V. Kerr, and R.C. Babcock. 2006. Long-term trends in lobster populations in a partially protected vs. no-take Marine Park. Biological Conservation 132: 222–231.

Sheehan, E.V., T.F. Stevens, S.C. Gall, S.L. Cousens, and M.J. Attrill. 2013. Recovery of a temperate reef assemblage in a marine protected area following the exclusion of towed demersal fishing. PLoS ONE 8:e83883.

Sheehy, M.R.J., R.C.A. Bannister, J.F. Wickins, and P.M.J. Shelton. 1999. New perspectives on the growth and longevity of the European lobster (*Homarus gammarus*). Canadian Journal of Fisheries and Aquatic Sciences 56:1904–1915.

Skerritt, D.J., P.A. Robertson, A.C. Mill, N.V.C. Polunin, and C. Fitzsimmons. 2015. Fine-scale movement, activity patterns and home-ranges of European lobster *Homarus gammarus*. Marine Ecology Progress Series 536:203–219.

Sköld, M., P. Börjesson, H. Wennhage, J. Hjelm, J. Lövgren, and K. Ringdahl. 2022. A no-take zone and partially protected areas are not enough to save the Kattegat cod, but enhance biomass and abundance of the local fish assemblage. ICES Journal of Marine Science 79:2231–2246.

Smith, I. 2013. Pre-European Maori exploitation of marine resources in two New Zealand case study areas: Species range and temporal change. Journal of the Royal Society of New Zealand 43:1–37.

Smith, I.P., A.C. Jensen, K.J. Collins, and E.L. Mattey. 2001. Movement of wild European lobsters *Homarus gammarus* in natural habitat. Marine Ecology Progress Series 222:177–186.

Smyth, Captain W.H., 1831. On the Columbretes, volcanic rocks near the coast of Valencia, in Spain. Journal of the Royal Geographical Society of London 1:58–62.

Spanier, E. 2024. Marine protected areas–a review of their potential effects on lobster population size & structure and fisheries management. Fisheries Research 275:107025.

Spanier, E., and K.L. Lavalli. 1998. Natural history of *Scyllarides latus* (Crustacea: Decapoda): A review of the contemporary biological knowledge of the Mediterranean slipper lobster. Journal of Natural History 32:1769–1786.

Spanier, E., K.L. Lavalli, J.S. Goldstein, J.C. Groeneveld, G.L. Jordaan, C.M. Jones, B.F. Phillips, M.L. Bianchini, R.D. Kibler, D. Díaz, S. Mallol, R. Goñi, G.I. van Der Meeren, A.-L. Agnalt, D.C. Behringer, W.F. Keegan, and A. Jeffs. 2015. A concise review of lobster utilization by worldwide human populations from prehistory to the modern era. ICES Journal of Marine Science 72:i7–i21.

Spyksma, A.J., R.B. Taylor, and N.T. Shears. 2017. Predation cues rather than resource availability promote cryptic behaviour in a habitat-forming sea urchin. Oecologia 183:821–829.

Starr, P.J. 2010. Rock lobster catch and effort data: Summaries and CPUE standardisations, 1979–80 to 2008–09. New Zealand Fisheries Assessment Report 17:131.

Stewart, B.D., L.M. Howarth, H. Wood, K. Whiteside, W. Carney, É. Crimmins, B.C. O'Leary, J.P. Hawkins, and C.M. Roberts. 2020. Marine conservation begins at home: How a local community and protection of a small bay sent waves of change around the UK and beyond. Frontiers in Marine Science 7:76.

Sørdalen, T.K., K.T. Halvorsen, H.B. Harrison, C.D. Ellis, L.A. Vøllestad, H. Knutsen, E. Moland, and E.M. Olsen. 2018. Harvesting changes mating behaviour in European lobster. Evolutionary Applications 11:963–977.

Sørdalen, T.K., K.T. Halvorsen, and E.M. Olsen. 2022. Protection from fishing improves body growth of an exploited species. Proceedings of the Royal Society B: Biological Sciences 289:20221718.

Sørdalen, T.K., K.T. Halvorsen, L.A. Vøllestad, E. Moland, and E.M. Olsen. 2020. Marine protected areas rescue a sexually selected trait in European lobster. Evolutionary Applications 13:2222–2233.

Tai, T.C., P. Calosi, H.J. Gurney-Smith, and W.W.L. Cheung. 2021. Modelling ocean acidification effects with life stage-specific responses alters spatiotemporal patterns of catch and revenues of American lobster, *Homarus americanus*. Scientific Reports 11:23330.

Templado, J., and M. Calvo 2002. Flora y fauna de la reserva marina de las Islas de Columbretes. Ministerio de Agricultura, Pesca y Alimentación, Madrid, España.

Thorbjørnsen, S.H., E. Moland, M.B. Olsen Huserbråten, J.A. Knutsen, H. Knutsen, and E.M. Olsen. 2018. Replicated marine protected areas (MPAs) support movement of larger, but not more, European lobsters to neighbouring fished areas. Marine Ecology Progress Series 595:123–133.

Triantafyllidis, A., A.P. Apostolidis, V. Katsares, E. Kelly, J. Mercer, M. Hughes, K.E. Jørstad, A. Tsolou, R. Hynes, and C. Triantaphyllidis. 2005. Mitochondrial DNA variation in the European lobster (*Homarus gammarus*) throughout the range. Marine Biology 146:223–235.

Tuffley, E-J., S. de Lestang. 2024. Density-dependence inside a marine protected area increases natural mortality and stunts the growth of a spiny lobster. Fisheries Research 280:107152.

Vigo, M., J. Navarro, G. Rotllant, N. Bahamon, M. Carretón, J. Quevedo, A. Rojas, and J.B. Company. 2023. Before–after control–impact (BACI) assessment of the effects of a deep-water no-take fishery reserve to recover Norway lobster (*Nephrops norvegicus*) overfished populations and coexisting megafauna. ICES Journal of Marine Science 80:2008–2023.

Webber, D., P. Starr, V. Haist, M. Rudd, and C. Edwards. 2018. The 2017 stock assessment and management procedure evaluation for rock lobsters (*Jasus edwardsii*) in CRA 2. New Zealand Fisheries Assessment Report 17:24.

Webster, S.G. 2015. Endocrinology of molting. Pages 1–35 *in* E.S. Chang. and M. Thiel, editors. The natural history of the Crustacea, Volume 4: Physiology. Oxford University Press, Oxford, New York, USA.

Wiig, J.R., E. Moland, T.O. Haugen, and E.M. Olsen. 2013. Spatially structured interactions between lobsters and lobster fishers in a coastal habitat: Fine-scale behaviour and survival estimated from acoustic telemetry. Canadian Journal of Fisheries and Aquatic Sciences 70:1–9.

Wing, S., M. Bowman, F. Smith, and S. Rutger. 2004. Analysis of biodiversity patterns and management decision making processes to support stewardship of marine resources and biodiversity in Fiordland—A case study. Ministry for the Environment, Wellington, New Zealand.

Wing, S., and L. Jack. 2014. Fiordland: The ecological basis for ecosystem management. New Zealand Journal of Marine and Freshwater Research 48:577–593.

16

CRUSTACEANS IN A LEGISLATIVE FRAMEWORK

D. Christopher Rogers

Abstract

This chapter provides a review of legislative measures affecting crustacean harvest, general conservation, specific taxa in need of conservation, and the limits of such legislation, with a focus on global practices. The prevalence of crustaceans in national and international lists of threatened or endangered species (e.g., CITES, IUCN, Red List) is evaluated. Examples of legislation and specific laws from all over the world are discussed in relation to crustaceans exploited in food fisheries, aquaculture, ornamental and hobbyist industries, invasive species, and taxa in need of conservation. Furthermore, bureaucracy, effects on research, and consequences to biodiversity of legislation regarding permits necessary for scientific investigations are also examined. The effectiveness and consequences of these legislative efforts regarding crustacean biodiversity are presented, including discussions on poaching, illegal and unregulated fisheries, invasive species introductions, and unenforced or selectively enforced legislation. Overall, most conservation legislation outside of fisheries is either inadequate or confounds conservation efforts. Conservation legislation that protects rare, threatened, and endangered species without circumscribing scientific assessments either through direct prohibition or excessive bureaucracy are necessary for the continued survival of crustaceans and other species, as well as maintaining biodiversity.

INTRODUCTION

Crustaceans are the most morphologically and ecologically diverse organisms on the planet and have been roaming its surface probably since the Precambrian (Vernberg and Vernberg 1983, Martin and Davis 2001). They are integral components of nearly every ecosystem (Coull and Bell

D. Christopher Rogers, *Crustaceans in a Legislative Framework*. In: *Ecology and Conservation*. Edited by: Lars Gutow and Martin Thiel, Oxford University Press. © Oxford University Press (2025). DOI: 10.1093/oso/9780197768242.003.0016

1983), often comprising the largest portion of biodiversity at a given site (Kawai and Rogers 2021, Martin and Davis 2001), with many taxa being keystone species or ecosystem engineers (e.g., Bauer 2004, Burukovsky 2009) and are as equally affected by anthropogenic impacts as are most organisms (Bauer 2004, Kawai and Rogers 2021, Plotnikov et al. 2021). Anthropogenic impacts that have reduced or eliminated crustacean populations include pollution, overfishing, habitat conversion, and, probably the most devastating, invasive species (Rigel 1959, Lewin et al. 2006, Acevedo-Alonso and Cumberlidge 2021, Daneliya and Wittmann 2021, de Mazancourt et al. 2021, Kawai and Rogers 2021). These problems translate directly from impaired ecosystems to economic problems, whether through environmental remediation needs, impaired or lost crustacean fisheries, or invasive crustaceans or crustacean diseases negatively affecting agriculture and aquaculture (Rigel 1959, Provenzano 1985, Bauer 2004, Acevedo-Alonso and Cumberlidge 2021, Kawai and Rogers 2021, Plotnikov et al. 2021). It is primarily for these reasons that legislation has been developed for conservation, whether specifically for crustaceans or not.

Although there are many pieces of legislation that affect crustaceans, very few focus, specifically on crustaceans. Ostensibly, most of these pieces of legislation are for crustacean conservation either of specific species (those in need of conservation) or of important fisheries. Other legislation that affects crustaceans deals with scientific collecting and research. Both conservation and research legislation occur in many local, national, and international situations.

Crustacean conservation legislation may take the form of (i) harvest limits and seasons for commercial species, (ii) protection for rare, threatened, and endangered species, (iii) attempts to control invasive species, or (iv) governing protected areas where crustaceans may live. Typically, conservation goals and objectives revolve around habitat and/or species protection, or harvest limits to protect wild stocks.

Fisheries legislation generally focuses on limiting commercial and recreational pressures on wild stocks. While most of this type of legislation is designed to protect the wild target populations from overharvest, and thus protect the various industries dependent on those populations (e.g., commercial fishing, restaurants, fishing equipment vendors, aquarium trade), they sometimes also provide direct conservation protection (e.g., gear limits, bycatch limits).

Legislation specifically for conservation is much more uncommon and is sometimes species targeted instead of ecosystem based. Legislation restricting the movement of invasive species is often very difficult to enforce, especially in a world where transportation of alien species across large distances (whether intentional or not) is relatively easy. Finally, various legislation tools developed to control and monitor scientific research have created far more problems than they have solved. The success of these various forms of legislation has varied tremendously and their effectiveness is sometimes a matter of debate. All are discussed below in more detail.

CRUSTACEAN HARVEST LEGISLATION

The vast majority of legislation that concerns crustaceans is focused on commercial harvest. Crustacean harvest of some type occurs in most countries, and legislation and regulations tend to govern both commercial and recreational harvest. Specific information on various crustacean fisheries can be found in Volume 9 of this series (Bauer 2020, Gardner et al. 2020, López 2020, Stevens and Miller 2020, Wahle et al. 2020). Harvest can occur either for food fisheries or for the aquarium trade. Legislation governing harvest generally provides one or more specific government agencies the ability to control and limit the seasons and equipment to be used, or the numbers, sizes, and/or sexes of crustaceans that can be harvested. Ostensibly, these harvest policies are to ensure the sustainability of the fishery, preventing its loss as an industry or food resource for local peoples.

For example, The Magnuson-Stevens Fishery Conservation and Management Act governs marine fisheries management in United States marine waters (NOAA 2007). The stated objectives of the Magnuson-Stevens Fishery Conservation and Management Act include preventing overfishing, rebuilding overfished stocks, increasing the economic and social benefits of fishing, and maintaining a safe and sustainable supply of seafood. Under this law, the US fisheries management process is supposed to be a transparent, public process, guided by science, management, innovation, and collaboration with industries involved.

The elimination of any fishery population can not only have dire consequences for the harvest target species, but also for the ecosystem that interacts with that species. Litzow et al. (2013) documented a collapse of Alaskan crustacean fisheries, which caused an ecosystem shift greatly increasing bottom fish and salmonid populations. The loss of a particular fishery can ultimately result in a severe economic impact, not just to a particular company or industry, but to the individual people employed in that industry. Furthermore, loss of a fishery also means a loss of license fees and taxes generated by the fishery. It is estimated that some 80 million metric tons of total fishery resources are harvested from the Earth's oceans annually, with an approximate value of US$800 billion, and the direct employment of roughly 200 million people globally (Sumaila and Munro 2008). Bauer (2004) summarized the Food and Agriculture Organization of the United Nations decapod fishery harvest data for 2000, giving the global decapod shrimp harvest catch as 3.1 million metric tons, with an estimated worth of more than US$900 million. And these numbers are growing. Boenish et al. (2021) demonstrated that crustacean (crab, shrimp, and lobster) harvests have increased more rapidly than any other seafood type. Boenish et al. (2021) found that this expansion of crustacean fisheries results from reduced predation and inherent productivity coupled with a high market value relative to other seafood products. However, the crustacean fisheries generate less food and nutritional value per unit volume, and overfishing is causing declines in crustacean populations. Various examples include Alaska (Orensanz et al. 1998) and Chesapeake Bay, USA (Sharov et al. 2003), Canada (Mullowney and Baker 2020), Japan (Shiota 1993), Brazil (Pioua et al. 2007), Russia (Dvoretsky and Dvoretsky 2017), and Argentina (Varisco et al. 2019).

The vast majority of commercially harvested crustaceans are marine decapods, comprising crabs (Anomura and Brachyura), shrimps, and lobsters (Calado 2020, Boenish et al. 2021). However, commercial fisheries also exist for many other crustacean groups. For example, the coconut crab *Birgus latro* (Linnaeus, 1767) is harvested for meat around the Indo-Pacific (Abadilla et al. 2020). Freshwater crayfish, prawns, and crabs are harvested in many parts of Asia, Australia, and the Americas (FAO 2021) for both human consumption and for the aquarium trade. Terrestrial (*Birgus latro, Coenobita* spp.) and freshwater (*Aegla* spp.) anomurans are harvested for the pet trade. Various barnacles, stomatopods, and amphipods are commercially harvested for human consumption (Bedecarratz et al. 2011, Sestelo and Roca-Pardiñas 2011), López 2020). In Chile and Canada, various amphipods are harvested for commercial and ornamental fish feed (López 2020). Various branchiopod and copepod planktonic crustaceans are used as live feed for commercial fisheries (Jepsen et al. 2020). Commercial fisheries for the brine shrimp *Artemia* (various species) occur nearly worldwide, for their use as an aquaculture feed, for increasing evaporation in salt ponds (through the consumption of phytoplankton), and for the aquarium hobbyist trade (e.g., Abatzopoulos et al. 2002), and various laws exist concerning their harvest (e.g., Administrative rule R657-52 in Utah, concerning the harvest of more than 200 pounds of Great Salt Lake brine shrimp eggs for personal use, or harvest of any amount for commercial use). In Mexico, spinicaudatan clam shrimp are harvested for aquarium fish food (Martínez-Pantoja et al. 2002) and tadpole shrimp have been harvested for human consumption (Creaser 1931), while the fairy shrimp *Streptocephalus sirindhornae* Sanoamuang, Murugan, Weekers, and Dumont, 2000 is cultivated for human consumption in Southeast Asia (Rogers et al.

2013). Although there are no laws in Thailand concerning the aquaculture of the fairy shrimp *Branchinella thailandensis* Sanoamuang, Saengphan, and Murugan, 2002 (Sanoamuang et al. 2002), the method is protected under patent law, the patent being held by the first author of the original description. In Thailand, I have eaten very popular and commercially available paste made from marine mysids as a food seasoning; there are no laws concerning the fishery, but several companies have their harvest methods protected by patent or their recipes protected by trademark.

Regulations limiting fisheries harvest are generally developed by government agencies and their biologists, but researchers, concerned environmentalists, nongovernmental organizations, and the fishing community may provide information or petition for regulations as well. Most commonly, governmental agencies concerned with natural resources (e.g., department of wildlife, department of fisheries, environmental departments, and game and fish departments) employ biologists either directly, or fund university or consultant researchers to monitor their commercial fishery populations. The recommendations of these biologists form the foundation of the regulations that control harvests provided by the legislation of that country.

Local efforts, including within commercial stakeholder communities, have driven legislation as well. For example, on the United States Atlantic Coast the commercial fishing community had been fighting legislation to expand mechanized fishing since the 1880s, and in 1912 the commercial fishing interests of Essex County, Massachusetts, pushed on the federal government for a bill to ban the use of otter trawls on the Georges Bank due to their devastating effects on nursery reefs (McKenzie 2013), although they ultimately failed.

In France, Portugal, and Spain, *Pollicipes pollicipes* (Gmelin, 1790) is an expensive delicacy called "percebes" (gooseneck barnacles). The fishery for this barnacle is heavily regulated, but with varied management and monitoring (Jacinto et al. 2011, Macho et al. 2013, López 2020, Geiger et al. 2022). Poaching is widespread and adapts to various strategies to circumvent it, although recent stakeholder workshops are developing new ideas to reduce or eliminate illegal percebes harvest (Geiger et al. 2022). New efforts at developing sustainable monitoring to maintain and support percebes fisheries have been proposed (Macho et al. 2013).

Recreational fishing (including for crustaceans) in industrialized nations has become one of the activities with the greatest impacts on coastal and inland fisheries (Arlinghaus et al. 2002), and directly impacts the associated ecosystems as well (e.g., Hoyle and Gibbons 2000, Lewin et al. 2006, Altieri et al. 2012, Anderson and Alford 2014). This sport is regulated as regards take limits, seasons, and gear, but the associated license fees, permits, and recreational fishing equipment industry support a large economy. The recreational fishing industry in the USA generates around $125 billion annually, employing more than 800,000 people (Cunningham 2019). Specific USA recreational harvest data are presented in Table 16.1.

Recreational fishing is also a persistent activity at the individual level: in Australia it is estimated that >92% of people engaging in recreational fishing have done so for more than 10 years of their lives (Sutton 2006, Baker 2017, Copeland et al. 2017).

Crustaceans are commercially farmed in many parts of the world, and the interests and markets are growing. Both marine and freshwater crustaceans are in aquaculture today, including freshwater crayfish and lobsters, and both freshwater and marine shrimps and crabs (Shelley and Lovatelli 2011, Calado 2020, Piper 2020, FAO 2021). Many nations have permitting legislation to control and monitor aquaculture activities, including its effects on the local ecology, at least in freshwater systems (e.g., Piper 2020). The benefit is that farming reduces harvest pressures on wild stocks. This is particularly good when native species are farmed. However, when nonnative species are farmed, escapees can lead to invasive species populations negatively affecting local ecosystems. For example, invasive crayfish species that have escaped or been intentionally introduced into new areas not only

Table 16.1 Recreational crustacean fishing annual average harvest data (pounds) from various US states (data collected from each state government's website).

State	Shrimp	Crayfish	Lobster	Blue Crab	Stone Crab	*Metacarcinus magister*
Alaska	15,000					~60,000
California			~210,000			~97,000,000
Florida	~2,000,000		100,000	190,000	105,000	
Maryland		~2,000		4,576,000		
Oregon						~1,100,000
Washington	~208,000					~1,500,000

affect ecosystems through their direct predation on other organisms but bring parasites and diseases with them (Rigel 1959, Weiperth et al. 2017, Gelder 2020, Jussila and Edsman 2020, Oficialdegui et al. 2020, Parpet and Gelder 2020, Kawai et al. 2021).

Unfortunately, the vast majority of laws prohibiting the importation of invasive crustacean species were only enacted after invasive crustaceans (typically crayfish) were already established (e.g., Oficialdegui et al. 2019; Fig. 16.1). Dresser and Swanson (2013) examined the regulations and spread of *Faxonius rusticus* (Girard, 1852) in the USA. They found that many state regulations did not explicitly prohibit the transport of crayfish species, were only applied to specific drainages, or prohibited only *F. rusticus*, and were not effective. However, proactive legislation

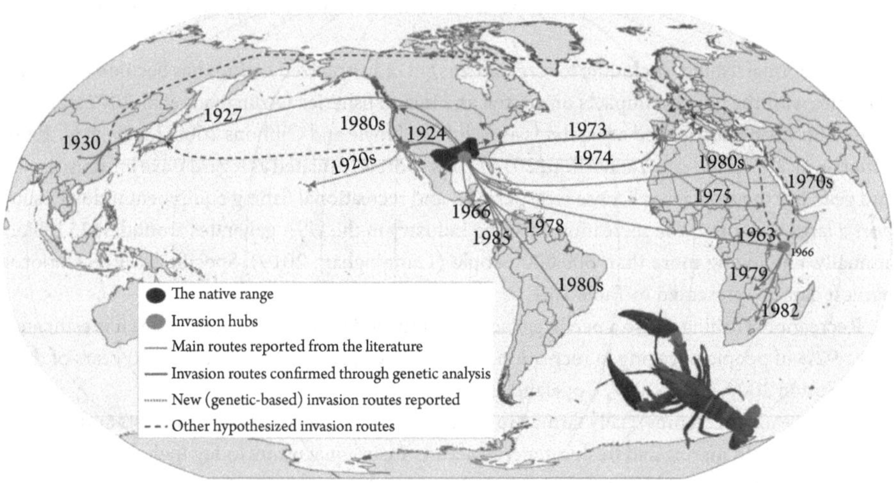

Fig. 16.1.
The global invasion routes of the red swamp crayfish, *Procambarus clarkii*, native of southern USA and northeastern Mexico. Relevant invasion hubs, which usually function as recipients and sources of new invasions, are shown as gray circles: Louisiana (in the native range), California, Kenya, and Spain. Reproduced from Oficialdegui et al. (2019), with permission from © John Wiley & Sons; see a color version of this figure in the centerfold.

prior to establishment of *F. rusticus* did reduce the likelihood of invasion when regulations prohibited the transport of all live crayfish species between water bodies. Legislation or regulations will not prevent dispersal of invasive species through shared drainages. But preemptive legislation may significantly reduce the anthropogenic spread of invasive species between drainages (Dresser and Swanson 2013). Furthermore, Dresser and Swanson's (2013) results suggested that the most effective pieces of legislation did not require individuals to identify species, but rather explicitly prohibited the transport of all live crayfish between water bodies. Ireland, for example, has banned all crayfish imports and is so far free of invasive crayfish species (Holdich 2003).

Commercial hobby aquarium fisheries also are regulated, not just in terms of harvest, but many nations also have import and export controls over various hobby species (e.g., Harriott 2001). A summary of the species I have found readily available for sale is presented in Table 16.2. The general availability of the internet and the various aquarium hobbyist blogs and chat rooms, coupled with advancements in speed of shipping, has allowed more and more species to be sold and shipped live around the world.

This whole industry was estimated to be worth more than US$1000 million globally in 2003 (Cato and Brown 2003) and is well documented to have had negative effects on wild populations, both in terms of harvest and the introduction of invasive species (e.g., Wood 1985, Young 1997, Holdich 2003, Calado 2020). However, various local and regional legislation coupled with industry controls have reduced the impacts in many areas, reducing the amount of destructive harvest techniques (such as cyanide or dynamite) (e.g., Tissot 1999, Harriott 2001), and efforts to use aquaculture to reduce pressure on wild stocks is increasing, although for marine species aquaculture has been less than successful (Tlusty 2002), mostly because of biological challenges such as long larval durations (Jeffs et al. 2020).

CRUSTACEAN CONSERVATION LEGISLATION

Specific conservation areas (such as parks or wilderness areas) rarely focus on crustaceans unless it is for the protection of specific species that are considered threatened or endangered under law (such as mitigation areas), which are treated in the next section. Most conservation areas protect crustacean species, but only as a component of the larger ecological system contained within the park or preserve.

At the international level there are marine protected areas (MPA) and Ramsar sites. MPAs are ocean habitat areas set aside for the protection of marine biodiversity, while Ramsar sites are for wetland habitat protection. Both are areas of "no take"; there is no harvesting or collecting allowed without special permits. These are international programs developed from the World Congress on National Parks in 1962 and the International Union for the Conservation of Nature (IUCN). Numerous nations have MPAs and/or Ramsar sites, with each nation managing their own. Although the level of protection, uses, and managements varies greatly from nation to nation (from full protection and monitoring to zero oversight at all) (Bacher 2010, Dureuil et al. 2018, Tibbett and Briggs 2019, Humphreys and Clark 2020), at least some sites have had positive effects, bolstering ecological services and promoting fisheries (Christie et al. 2010, Leenhardt et al. 2015, Moland et al. 2021; also see Chapter 15 in this volume).

At the national level, most countries have specific national parks or similar conservation areas, although protection and management vary tremendously, especially in nations where foreign researchers or commercial interests are heavily governed, while local activities are not (e.g., Ribeiro and Freitas 2014, Dureuil et al. 2018, Tibbett and Briggs 2019).

RARE, THREATENED, AND ENDANGERED CRUSTACEAN LEGISLATION

Some legislation is focused on crustacean species that are considered rare, threatened, or endangered. At the international level, there is one multilateral treaty: the Convention on International Trade in Endangered Species of Wild Fauna and Flora (CITES). Specifically, CITES was designed to protect wildlife populations from being driven to extinction through commercial trade. Like the MPAs, each signatory nation regulates the species listed under CITES differently. Under the CITES agreement, signatory nations agree to have a specific designation of management and scientific authorities, laws prohibiting the trade of species in violation of CITES, penalties for such trade, and laws providing for the confiscation of specimens. However, in practice, less than half the

Table 16.2 Freshwater and terrestrial crustaceans found in the internet pet trade under the scientific name used by vendor (many have been corrected).

Branchiopoda
Artemia franciscana Kellogg, 1906
Artemia tibetiana Abatzopoulos et al., 1998
Branchinecta lindahli Packard, 1883
Branchinecta mackini Dexter, 1956
Branchinecta gigas Lynch, 1937
Leptestheria compleximanus (Packard, 1887)
Ozestheria pilosa (Rogers et al., 2013)
Streptocephalus sirindhornae Sanoamuang et al., 2000
Streptocephalus sealli Ryder, 1879
Streptocephalus mackini Moore, 1966
Streptocephalus sp. (Taiwan)
Thamnocephalus platyurus Packard, 1877
Triops sp. (Australia)
Triops sp. (North America)

Isopoda
Armadillidium gestroi Tua, 1900
Armadillidium granulatum Brandt, 1833
Armadillidium klugii Brandt, 1833
Armadillidium maculatum Risso, 1816
Armadillidium peraccae Tua, 1900
Armadillidium vulgare (Latreille, 1804)
Cubaris murina Brandt, 1833
Oniscus asellus Linnaeus, 1758
Porcellio bolivari Dollfus, 1892
Porcellio expansus Dollfus, 1892
Porcellio flavomarginatus Lucas, 1853
Porcellio haasi Arcangeli, 1925
Porcellio hoffmannseggii Brandt, 1833
Porcellio laevis Latreille, 1804
Porcellio nicklesi Dollfus, 1892

Amphipoda
Arcitalitrus sylvaticus (Haswell, 1879)

Astacidea
Cambarellus texanus Albaugh and Black, 1972
Cambarellus patzcuarensis Villalobos, 1943
Cambarellus shufeldti Faxon, 1884
Cherax albidus Clark, 1936
Cherax alyciae Lukhaup et al., 2018
Cherax boesemani Lukhaup and Pekny, 2008
Cherax cainii Austin and Ryan, 2002
Cherax destructor (Clark, 1936)
Cherax holthuisi Lukhaup and Pekny, 2006
Cherax mosessalossa Lukhaup et al., 2018
Cherax peknyi Lukhaup and Herbert, 2008
Cherax pulcher Lukhaup, 2015
Cherax quadricarinatus (von Martens, 1868)
Cherax tenuimanus (Smith, 1912)
Faxonius limosus (Rafinesque, 1817)
Faxonius neglectus (Faxon, 1885)
Faxonius rusticus (Girard, 1852)
Faxonius virilis (Hagen, 1870)
Orconectes nais (Faxon, 1885)
Pacifastacus leniusculus (Dana, 1852)
Procambarus alleni (Faxon, 1884)
Procambarus acutus (Girard, 1852)
Procambarus braswelli Cooper, 1998
Procambarus clarkii (Girard, 1852)
Procambarus fallax (Hagen, 1870)
Procambarus hirsutus Hobbs, 1958
Procambarus pygmaeus Hobbs, 1942
Procambarus spiculifer (LeConte, 1856)
Procambarus virginalis Lyko, 2017

Caridea
Atya gabonensis Giebel, 1875
Atyoida pilipes (Newport, 1847)
Atyopsis spinipes (Newport, 1847)
Caridina babaulti Bouvier, 1918
Caridina cantonensis Yü, 1938
Caridina dennleri von Rintelen and Cai, 2009
Caridina formosae Hung et al., 1993
Caridina gracilirostris De Man, 1892
Caridina longirostris H. Milne-Edwards, 1837
Caridina multidentata Stimpson, 1860
Caridina propinqua De Man, 1908
Caridina serrata Stimpson, 1860
Caridina villadolidi Blanco, 1939
Caridina weberi De Man, 1892

(Continued)

Table 16.2 Continued

Macrobrachium assamensis (Tiwari, 1958)
Macrobrachium digueti (Bouvier, 1895)
Macrobrachium gracilirostre (Miers, 1875)
Macrobrachium placidulum (De Man, 1892)
Macrobrachium rosenbergii (De Man, 1879)
Macrobrachium spinipes (Schenkel, 1902)
Neocaridina davidi (Bouvier, 1904)
Neocaridina zhangjiajensis (Cai, 1996)
Palaemon kadiakensis (Rathbun, 1902)
Palaemon paludosus (Gibbes, 1850)
Palaemon texanus (Strenth, 1976)
Palaemon modestus (Heller, 1862)
Paracaridina sp.
Aegla sp.
Birgus latro (Linnaeus, 1767)
Coenobita clypealis (Fabricius, 1787)
Coenobita perlatus Milne-Edwards, 1837
Cardisoma armatum Herklots, 1851
Cardisoma carinifex (Herbst, 1794)
Gecarcinus quadratus (Saussure, 1853)
Gecarcoidea lalandi Milne-Edwards, 1837
Geosesarma bicolor Ng and Davie, 1995
Geosesarma dennerle Ng et al., 2015
Geosesarma hageni Ng et al., 2015
Geosesarma notophorum Ng and Tan, 1995
Geosesarma pontianak Ng, 2015
Geosesarma rouxi (Serène, 1968)
Geosesarma tiomanicum Ng, 1986
Geothelphusa dehanni White, 1847
Hainanpotamon daiae Yeo and Naruse, 2007
Indochinamon chinghungense (Dai et al., 1975)
Lepidothelphusa sp.
Limnopilos naiyenetri (Chuang and Ng, 1991)
Metasesarma aubryi (Milne-Edwards, 1869)
Metasesarma obesum (Dana, 1851)
Nanhaipotamon aculatum Dai, 1997
Neosarmatium rotundfrons Milne-Edwards, 1869
Neosarmatium smithi Milne-Edwards, 1853
Neotiwaripotamon whiteheadi (Parisi, 1916)
Parathelphusa ferruginea Chia and Ng, 2006
Parathelphusa pantherina (Schenkel, 1902)
Perisesarma bidens DeHaan, 1835
Ptychognathus barbatus Milne-Edwards, 1873
Somanniathelphusa sinensis (Milne-Edwards, 1853)
Tiwaripotamon vietnamicum Dang and Hô, 2002

signatory nations provide any such legal framework (Zimmerman 2003). CITES is the oldest piece of international conservation legislation. Regardless, no crustacean species are currently protected under CITES but this could change in the future.

The IUCN, as an official consulting nongovernmental organization to the United Nations, developed the Red List of Threatened Species (Red List) in 1964, which has been frequently updated (IUCN 2001). The Red List was the first international list of rare, threatened, and endangered species that had scientifically defensible criteria for the inclusion of taxa (IUCN 2001). While not having any legal status on its own, several nations use the Red List and/or the CITES lists to determine what species are protected under that country's law. One unique aspect of the Red List is that species may be assessed not only according to a threat level, but also be designated "Least Concern" or, if information is lacking, "Data Deficient." Some 2036 crustaceans are currently listed as assessed according to IUCN Red List criteria (https://www.iucnredlist.org/; see Table 16.3). An additional 1153 crustacean species have been determined to be "Data Deficient." Regardless, (excluding Hexapoda) with an underestimated 52,000 crustacean species globally (Martin and Davis 2001), only 3.92% of described crustacean species have had their conservation status assessed under Red List criteria. In fact, arthropods as a whole are have had few species assessed at all (Redak 2000).

Attempts to assess different crustacean groups under the IUCN Red List criteria are underway, although the focus to date has been on freshwater decapods and large branchiopods. Five decapod species were assessed by Fellowes et al. (2002). Cumberlidge et al. (2009) conducted an assessment of the freshwater crabs of the world, and De Grave et al. (2015) provided an assessment of the global freshwater Caridea. Magalhães et al. (2016) assessed the freshwater Decapoda of South America, and Acevedo-Alonso and Cumberlidge (2021) updated the assessments for the freshwater crabs of Colombia. Rogers includes Red List assessments for all the new branchiopod taxa he describes and reviews (e.g., Rogers et al. 2018, 2019). More catalogs and published species conservation assessments may provide more protection for more species through legislation over time.

Many nations or provinces have some form of endangered or threatened species conservation legislation. Typically, this legislation focuses on the protection of individual species that are determined to be at risk of extinction. Protection generally extends to prohibiting the take or harvest of such species and may also include protection for their habitat. The take or elimination of these species or their habitat may be permitted under most national and local laws if there is a public safety concern, such as the need to remove or kill a protected large predator, or (more likely in the case of crustaceans) the application of some sort of pesticide to control a public health problem, such as malaria or Zika virus-carrying mosquitoes or sleeping sickness-carrying tsetse flies.

Permitted activities that may result in the "take" (in the legal sense) of protected crustaceans or the destruction of their habitat also includes development or restoration projects, but the laws generally require mitigation of the impacted habitats and species. In some countries, the law requires either no net loss, and sometimes a net gain, of protected species populations, or their habitat before the permitted development or restoration may be completed. Typically, these projects require many years of monitoring for this type of mitigation, with specific success criteria that must be met, and penalties if they are not, although enforcement has not been consistent.

SOME SPECIFIC CASES OF PROTECTED CRUSTACEAN TAXA

All the specifically protected threatened or endangered crustacean species under federal or multinational legislation for species with conservation needs (i.e., rare, threatened, and endangered species) are branchiopods (fairy shrimp and tadpole shrimp) and malacostracans (amphipods, isopods, and decapods). Various nations and the European Union have different laws and acts

Table 16.3 IUCN Red List crustacean statistics as of 2021. One branchiopod, 22 maxillopods, two isopods, and 1128 decapods are listed as "Data Deficient."

Taxon	Least Concern	Near Threatened	Lower Risk/ Conservation Dependent	Vulnerable	Endangered	Critically Endangered	Extinct in the Wild	Extinct	Total
Branchiopoda	1	1	1	22	13	8			**46**
Maxillopoda			8	71		7		2	**88**
Ostracoda				9		2		2	**13**
Mictacea						1			**1**
Euphasiacea	1								**1**
Mysida						2			**2**
Anaspidacea				4					**4**
Isopoda	1			23	10	8	1		**43**
Amphipoda				57	6	9		1	**73**
Decapoda	1197	71		225	145	121		6	**1765**
TOTAL	1200	72	9	411	174	158	1	11	**2036**

protecting rare, threatened, and endangered species. However, the specific definitions and conservation needs categories are not necessarily directly comparable. These species are presented in Table 16.4.

Conservation of Branchiopods

Other crustacean species have been afforded protection under specific acts for those species individually. The first large branchiopod preserve was established in Austria in 1982 to protect *Chirocephalus shadini* Smirnov, 1928 (Rogers 2021). One anostracan and one notostracan species are protected under law in the United Kingdom, one anostracan in Australia, one notostracan in Spain, and one anostracan in Brazil under limited local legislative acts (Rogers 2021). In the Republic of Korea (South Korea), nonnative Nearctic populations of the notostracan *Triops* sp. were legally protected from 2004 until 2012 (Rogers 2021). In the USA, four fairy shrimp in the genus *Branchinecta, B. lynchi* Eng et al., 1990 (Fig. 16.2A), *B. conservatio* Eng et al., 1990, *B. longiantenna* Eng et al., 1990, and *B. sandiegonensis* Fugate 1993 (Fig. 16.2B), the fairy shrimp *Streptocephalus woottoni* Eng et al., 1990, and the tadpole shrimp *Lepidurus packardi* Simon, 1886 (Fig. 16.3B) are protected under the US Endangered Species Act.

The Socorro Isopod

Thermosphaeroma thermophilum (Richardson, 1897) was probably endemic to a marsh fed by three hot springs in New Mexico, USA. Two of the springs were capped for use as a water supply for the

Fig. 16.2.
Two anostracan species protected under the USA Endangered Species Act. (A) *Branchinecta lynchi;* (B) *Branchinecta sandiegonensis.* Photos courtesy of ©Joel Sartore/National Geographic Photo Ark; see a color version of this figure in the centerfold.

Table 16.4 Specific rare, threatened, or endangered crustacean taxa protected under federal or international laws and the protecting legislation.

Taxon	Protection by
Branchinecta conservatio *Branchinecta lynchi* *Branchinecta longiantenna* *Branchinecta sandiegonensis* *Streptocephalus woottoni* *Lepidurus packardi*	USA Endangered Species Act
Chirocephalus diaphanus	United Kingdom Wildlife and Countryside Act
Armadillidium ghardalamensis	European Union Habitats Directive
Antrolana lira *Lirceus usdagalun* *Thermosphaeroma thermophilus*	USA Endangered Species Act
Gammarus acherondytes *Gammarus hyalleloides* *Gammarus desperatus* *Gammarus pecos* *Spelaeorchestia koloana* *Stygobromus hayi* *Stygobromus pecki*	USA Endangered Species Act
Astacopsis gouldi	Australia Environment Protection and Biodiversity Conservation Act
Astacus astacus *Austropotamobius pallipes* *Austropotamobius torrentium*	European Union Bern Convention; European Union Habitats Directive
Cambaroides japonicus	Japan Act on Conservation of Endangered Species of Wild Fauna and Flora
Cambarus aculabrum *Cambarus callainus* *Cambarus veteranus* *Cambarus zophonastes*	USA Endangered Species Act
Cherax tenuimanus *Engaeus granulatus* *Engaeus martigener* *Engaeus orramakunna* *Engaeus spinicaudatus* *Engaeus yabbimunna* *Engaewa pseudoreducta* *Engaewa reducta* *Engaewa walpolea* *Euastacus bindal* *Euastacus bispinosus* *Euastacus dharawalus*	Australia Environment Protection and Biodiversity Conservation Act

(Continued)

Taxon	Protection by
Faxonius (Orconectes) shoupi *Pacifastacus fortis*	USA Endangered Species Act
Jasus lalandii	Republic of South Africa National Environmental Management: Biodiversity Act
Palaemonetes cummingi *Palaemonias alabamae* *Palaemonias ganteri* *Procaris hawaiana* *Syncaris pacifica* *Vetericaris chaceorum*	USA Endangered Species Act
Paratya boninensis	Japan Act on Conservation of Endangered Species of Wild Fauna and Flora
Birgus latro	Republic of China (Taiwan) Wildlife Conservation Act
Charybdis japonica	Russian Federation Red Data Book
Paraleptuca boninensis *Amamiku occulta* *Geothelphusa levicervix* *Geothelphusa miyakoensis* *Geothelphusa tenuimanus*	Japan Act on Conservation of Endangered Species of Wild Fauna and Flora

city of Socorro, with the last spring, Sedillo Spring, being diverted to a commercial bath house, resulting in the loss of the marsh (USFWS 1982). Today, this species is only known from the remains of Sedillo Spring, which empties into two concrete pools connected by a narrow ditch between them. The site is located on private property. The species was listed as endangered under the US Endangered Species Act in 1978 (USFWS 1982). Twice, the population has nearly gone extinct and both times captive-bred populations were needed to restore the wild population (NMDGF 2004, Lang et al. 2006).

Asian Freshwater Crabs

In Thailand, three semi-terrestrial potamid crab species *Thaipotamon chulabhorn* Naiyanetr, 1993 (Fig. 16.3A) (Pu Pang, Pu Toon Kramom, or Mealy Crab), *Phricotelphusa sirindhorn* Naiyanetr, 1989 (Pu Chao Fa, or Panda Crab), and *Thaiphusa sirikit* (Naiyanetr, 1992) are protected animals under Thai law (Wild Animal Conservation and Protection Act 1992, revised version in 2019) that were named in honor of members of the Thai royal family, Queen Sirikit and her daughters, Princess Sirindhorn and Princess Chulabhorn. Because of this specific distinction, these crabs are protected under Thai law; as these species are each only known from a single location, they meet certain standards of the IUCN Red List criteria as an endangered species, and thus this protection is probably warranted for conservation reasons as well. Singapore's freshwater crab *Johora singaporensis* (Ng, 1986) has a specific act protecting it. It was listed by the IUCN as one of the world's 100 most endangered species.

Fig. 16.3.
(A) *Thaipotamon chulabhorn*, named in honor of the beloved Thai Princess Chulabhorn and therefore protected under Thailand law; photo reproduced courtesy of the author. (B) *Lepidurus packardi*, protected under the USA Endangered Species Act. Photo reproduced courtesy of J. Olesen; see a color version of this figure in the centerfold.

Christmas Island Crabs

Australia's Christmas Island is probably the only place in the world with specific infrastructure for crustaceans. On Christmas Island barriers have been built to funnel migrating land crabs, *Gecarcoidea natalis* Pocock, 1888, through specifically built roadway underpasses and bridges just for the millions of crabs carrying their eggs to the sea (Orchard 2012) and their subsequent march back. Some roadways are closed to traffic during migration, which occurs between October and January depending upon rains (Orchard 2012). *Gecarcoidea natalis* is specifically protected under a legislative management plan, which makes provisions for the infrastructure that protects this species (Hutchings and Brown 2014).

DEFICIENCIES OF CONSERVATION LEGISLATION

Insufficient Scientific Evidence

Legislation is intrinsically hampered because law usually only defines what is wrong; it seldom defines what is right. As a result, legislation that affects crustaceans and crustacean research can seriously affect both the crustaceans and the researchers studying them. But legislation to protect species in conservation need is also limited by taxonomy. Adequate taxonomic knowledge is the basis for most endangered species legislation: law makers need a name and definition for each taxon to include on the lists (see Chapter 12 in this volume). Yet more and more species, especially cryptic species, are usually identified with molecular tools but never described according to modern standards (Delić et al. 2017).

In the same vein is the need for legislators to understand scientific methods, processes, and definitions, but it must be remembered that legislators are not scientists; it is up to scientists to properly inform them when needed, through proper scientific standards, methods, and peer reviewed literature. For example, the efforts of certain animal rights groups to claim that crustaceans are "sentient" and feel pain in similar ways as vertebrates is not based on proper scientific processes or definitions (Puri and Faulkes 2010, Browman et al. 2019, Diggles 2019, Diggles et al. 2023). Numerous laws have been enacted that state that animals are "sentient" (see Drinkwater et al. [2019] for a summary of such legislation), and as a result, there are often severe restrictions on research in relation to many animal (primarily vertebrate) species. Some of this follows the Treaty on the Functioning of European Union, of which Article 13 says: "In formulating and implementing the Union's agriculture, fisheries, transport, internal market, research and technological development and space policies, the Union and the Member States shall, since animals are sentient beings, pay full regard to the welfare requirements of animals [. . .]"

The United Kingdom's Animal Welfare (Sentience) Act 2022 formally recognizes decapod crustaceans as "sentient," and there are calls for more of the same type of legislation in other nations, covering more invertebrate taxa (e.g., Orzechowski 2015, Shroff 2021). However, detailed reviews (e.g., Puri and Faulkes 2010, Browman et al. 2019, Diggles 2019, Diggles et al. 2023) demonstrate that studies on the ability of crustaceans to perceive pain and/or stress arrive at erroneous and conflicting conclusions. Furthermore, the definitions of "pain" used in the literature seem to be moving targets; to quote Diggles (2019): "The scientific literature on the subject of welfare and pain in crustaceans is immature." It is obvious that further, properly conducted, and replicated research is needed to understand how crustaceans perceive stressors. Science can support or refute legislation, but science should never be used as a means to a political end: facts are what they are and must be addressed as such, regardless of opinion or ideology.

Lack of Efficient Administration

Legislation is further hampered by the fact that, ultimately, it is lawmakers and not biologists who write the laws. For example, several species of branchiopod crustaceans are listed as endangered or threatened under the US federal Endangered Species Act (see Table 16.4). As a result, no one can "take" those species; within the Act, "take" is defined in the law as "to harass, harm, pursue, hunt, shoot, wound, kill, trap, capture, or collect, or to attempt to engage in any such conduct." Large branchiopod crustaceans are well known to use aquatic birds (such as ducks) as their primary dispersal mechanism (for a review, see Rogers [2014] and papers cited therein). However, any attempt by the US federal government to regulate duck hunting in an effort to prevent take of the listed large branchiopods would cause huge pushback from the waterfowl hunting community, which is itself responsible for large amounts of wetland conservation in North America (e.g., Boudart 2022).

Collecting permits are commonly required in many countries or reserves, both for researchers and private or commercial interests. Permits make good sense; the permitting government or agency can track what species are found, their status, and what research is done in each region, commercial collecting can be limited, and collection data are vital for conservation planning. But excessive bureaucratic permitting requirements and delays can ultimately prevent research.

For example, Brazil is well known for having difficult collecting permit regulatory procedures (e.g., Antonelli and Rodriguez 2009, Alves et al. 2018). The Brazilian regulations for collecting by foreign researchers require joint participation and responsibility of a Brazilian institutional scientific collaborator with a high level of scientific competence and reputation related to the specific research proposed, and that all collections will be deposited in a Brazilian institution (CNPQ 2022). However, Brazil has so few taxonomists, while at the same time harboring some of the greatest biodiversity in the world (Antonelli and Rodriguez 2009), that finding any appropriate, let alone willing, local, native collaborators can be difficult. As a result, because the permitting process is so very time consuming and onerous, Brazil's biodiversity is poorly understood (Antonelli and Rodriguez 2009, Alves et al. 2018), and is being lost as all penalties for illegal land clearing by local landowners were voided and protected land areas were reduced by 29 million hectares (Ribeiro and Freitas 2014). Thus, it is far easier for researchers to work in other Neotropical nations, while research in Brazil and the conservation needs of its natural resources are left unstudied (Antonelli and Rodriguez 2009, Alves et al. 2018). A similar situation is developing in the Republic of South Africa (Alexander et al. 2021). This greatly restricts the advancement of biodiversity research and conservation.

The Convention on Biological Diversity is a multilateral treaty with numerous signatory nations (CBD 2022). The convention has three primary objectives: the conservation of biological diversity, the sustainable use of biodiversity, and the fair and equitable sharing of the benefits arising from the use of genetic resources. This last objective is the specific purview of the Nagoya Protocol on Access to Genetic Resources and the Fair and Equitable Sharing of Benefits Arising from their Utilization to the Convention on Biological Diversity, which was ratified in Nagoya, Japan, in 2014, and is a supplement to the United Nations Convention on Biological Diversity. The stated goal of the Nagoya Protocol is "sharing benefits arising from the utilization of genetic resources and of the traditional knowledge associated with genetic resources held by indigenous and local communities in a fair and equitable way, including by appropriate access to genetic resources and by appropriate transfer of relevant technologies," and it is specifically aimed at the commercial use of genetic resources and their relation to local Indigenous peoples (Tobin 2013, Morgera et al. 2015). However, in practice the Nagoya Protocol is not implemented consistently across various nations (Tobin 2013, Smith et al. 2017, Cueni 2020), with many countries using it to create excessive red tape and exact taxes and fees from institutions where biodiversity research is being conducted, and not commercial endeavors (Tobin 2013, Smith et al. 2017, Knauf et al. 2019, Alexander et al. 2021). In fact, the excessive negotiations, paperwork, and permitting required greatly hampers basic international collaboration on biodiversity research (e.g., Smith et al. 2017, Prathapan et al. 2018, Knauf et al. 2019). Since the ratification of the Nagoya Protocol, I have spoken with nine separate researchers, who tell me that they only collaborate with researchers in, or tailor their phylogenetic and conservation research studies to, nations where the Nagoya Protocol is limited or not implemented, because otherwise it is just not worth the effort. Thus, the Nagoya Protocol has the unintended consequence of obstructing research that ultimately would support and promote biodiversity and conservation (Prathapan et al. 2018).

Legislation Without Public Education

In the US state of California, among many other states and various nations, importation of crayfish is illegal without very specific permits (CNRA 2019). However, in several of these states and countries, I have found many nonnative crayfish species for sale without permits in local pet shops, or for sale by local citizens, via the internet. The internet provides easy communication between aquarium hobbyists and suppliers around the world, and modern shipping generally allows live specimens to be received within a few days of posting. These people are readily finding each other through hobbyist websites, chat rooms, and social media. Conversely, it took me more than three months to assemble what little data I present in Table 16.4; government websites seem to bury information on their laws and on which species are protected, and many links on government websites to protected species lists were nonfunctional. Thus, the hobbyists may possibly have little or no idea that they are breaking any laws by purchasing or swapping specimens with others, and agency programs are inadequate to enforce the laws prohibiting the activities (Galil et al. 2011).

Education of the general public and those participating in the ornamental pet trade concerning the laws and both the personal and ecological consequences of their actions, coupled with government agencies doing more to make regulations and laws more readily accessible, could help to alleviate this problem (Bakker et al. 2020). People do not want to damage their local ecosystem, either through invasive species or pathogen introductions (e.g., Mrugała et al. 2015). However, considering the increasing costs for different species and the equipment to keep them, with demand rising, the financial benefits of ignoring laws that are poorly enforced may outweigh the costs (Patoka et al. 2016). Especially when it is so easy in today's social media for the average hobbyist to electronically meet others globally and, using today's shipping services, quickly trade specimens, rather than pay a seller's price.

Deficient Monitoring of the Natural Environment

Most legislation designed to protect specific rare, threatened, or endangered species focuses on the species themselves. Typically, these laws require that before any land use changes or conversions (such as development) occur on a particular piece of land, surveys must be done to determine whether certain protected species are present. Generally, it is the presence of habitat that could potentially support one or more protected species that triggers the requirement for surveys. If the species are found, either development is stopped, or some sort of mitigation is required. However, if the species is not present, little or nothing is required in terms of mitigation.

The US Endangered Species Act was spearheaded by President Richard Nixon. In his 1972 Presidential Address, he pushed the US Congress to prepare an act of law that "would make the taking of endangered species a federal offense for the first time." In 1973, Nixon signed the bill into law. However, this law, as so many others like it, focuses on listed species instead of habitats. Endangered species are "canaries in a coal mine," because they are used as an indicator organism of a larger habitat or ecosystem that is in trouble. When these systems are stressed, these species are just the first to go. As shown above, the MPAs and Ramsar sites seem to be doing a far better job of protecting species because they are protecting the entire system that the rare, threatened, and endangered species, among others, rely upon. It is the large, contiguous protected habitat areas that show the greatest biodiversity. These areas need fundamental biological monitoring in order to assess the stability, long-term sustainability, functionality, and ecological health of these preserves. It is vital to know if indicator species or focus populations are thriving, recovering, collapsing, or worse being replaced by invasive or nonnative taxa. Only quantitative monitoring and scientific research efforts can provide the data necessary to inform preservation area management and establish success criteria, let alone tell us if those criteria are met or not. Basic monitoring, management, and scientific research is vital; if we do not know what we are protecting, we cannot protect it.

When potentially suitable habitat that is unoccupied by rare, threatened, and endangered species is not protected, then those species have fewer places to colonize and fewer opportunities to recover. But more than that, there are fewer new, unoccupied habitats for the evolution of new species to occur.

CONCLUSIONS

Overall, current conservation legislation is inadequate, underfunded, unenforceable, reactionary, and in too many cases impedes the very conservation that is its goal. Legislation affecting crustaceans and crustacean research needs to be informed by biologists who understand the organisms and the ecology in which they belong, and will do so without political agendas, instead focusing on the ecological needs of the ecosystem as a whole unit. Conservation legislation for protected areas and fisheries, when properly implemented, with adequate scientific monitoring and research, with quantitative monitoring, with quantifiable goals and progress measures, is demonstrably more successful than single-species conservation plans. As stated above: species in need of conservation are indicator organisms of a larger ecosystem that is in trouble and should be tracked. Furthermore, legislation affecting biodiversity research needs to be developed or informed by biologists and law makers so that the legislation does not ultimately prevent the advancement of science, biodiversity, and conservation. Legislation must be properly enforced with adequate support to ensure that the resources are effectively protected. The most successful conservation is proactive and focuses on large ecosystems, with unimpeded research and monitoring, so that we can know if the biological resources under protection are flourishing or are in urgent need of additional protections (such as from invasive species or poaching) and adaptive management. Finally, with crustaceans rapidly becoming the largest group of exploited aquatic animals on the planet, crustacean biologists need to find solutions for crustacean conservation problems where legislation can ultimately be avoided.

ACKNOWLEDGMENTS

I am grateful to Ben Diggles for discussions on the crustacean "sentience" legislation. Special thanks to Tom Groening, editor of "The Working Waterfront and Island Journal," Island Institute, in help tracking down a reference. Thanks to Sammy De Grave for information on legally protected caridean shrimp. I especially want to thank Martin Thiel, Lars Gutow, and Miguel Penna for their patience with me, their guidance, and good humor, as I wrote this very difficult chapter.

REFERENCES

Abadilla, M., W.R. McClain, T. Sato, L.M. Mejía-Ortíz, and M.A. Penna-Díaz. 2020. Inland crustacean fisheries. Pages 181–206 *in* G. Lovrich and M. Thiel, editors. The natural history of the Crustacea, Volume 9: Fisheries and aquaculture. Oxford University Press, New York, USA.

Abatzopoulos, T.J., J.A. Beardmore, J.S. Clegg, and P. Sorgeloos. 2002. *Artemia*: Basic and applied biology. Kluwer Academic Publishers, Dordrecht, The Netherlands.

Acevedo-Alonso, A., and N. Cumberlidge. 2021. Updated extinction risk assessment of the Colombian freshwater crabs (Brachyura: Pseudothelphusidae, Trichodactylidae) reveals and increased number of threatened species. Pages 405–424 *in* T. Kawai and D.C. Rogers, editors. Recent advances in freshwater crustacean biodiversity and conservation. CRC Press, Boca Raton, Florida, USA.

Alexander, G.J., K.A. Tolley, B. Maritz, A. McKechnie, P. Manger, R.L. Thomson, C. Schradin, A. Fuller, L. Meyer, R.S. Hetem, M. Cherry, W. Conradie, A.M. Bauer, D. Maphisa, J. O'Riain, D.M. Parker, M.C. Mlambo, G. Bronner, K. Madikiza, A. Engelbrecht, A.T.K. Lee, B.J. van Vuuren, T.G. Mandiwana-Neudani, D. Pietersen, J.A. Venter, M.J. Somers, R. Slotow, W.M. Strauss, M.S. Humphries, P.G. Ryan, and G.I.H. Kerley. 2021. Excessive red tape is strangling biodiversity research in South Africa. South African Journal of Science 117:10787.

Altieri, A.H., M.D. Bertness, T.C. Coverdale, N.C. Herrmann, and C. Angelini. 2012. A trophic cascade triggers collapse of a salt-marsh ecosystem with intensive recreational fishing. Ecology 93:1402–1410.

Alves, R.J.V., M. Weksler, J.A. Oliveira, P.A. Buckup, J.P. Pombal Jr., H.R.G. Santana, A.L. Peracchi, A.W.A. Kellner, A. Aleixo, A. Langguth, A.M.P. De Almeida, A.L. Albernaz, C.C. Ribas, C. Zilberberg, C.E.V. Grelle, C.F.D. Rocha, C.J.E. Lamas, C.F.B. Haddad, C.R. Bonvicino, C.P.A. Prado, D.O. De Lima, D.C. Rossa-Feres, F.R. Dos Santos, F.R.G. Salimena, F.A. Perini, F.A. Bockmann, F.L. Franco, G.M.L. Del Giudice, G.R. Colli, I.C.G. Vieira, J. Marinho-Filho, J.M.C.F. Werneck, J.A.D. Dos Santos, J.L. Do Nascimento, J.L. Nessimian, J.L.P. Cordeiro, K. Del Claro, L.O. Salles, L. Casatti, L.H.R. Py-Daniel, L.F. Silveira, L.F. Toledo, L.F. De Oliveira, L.R. Malabarba, M.D. Da Silva, M.S. Couri, M. Martins, M.D.S. Tavares, M.E.G. Sobral, M.V. Vieira, M.L.A. Oliveira, M. De Pinna, M.J.G. Hopkins, M. Solé, N.A. Menezes, P. Passos, P.S. D'andrea, P.C.E.A. Pinto, P.L. Viana, P.M. Toledo, R.E. Reis, R. Vilela, R.P. Bastos, R.G. Collevatti, R. Cerqueira, S. Castroviejo-Fisher, and U. Caramaschi. 2018. Brazilian legislation on genetic heritage harms biodiversity convention goals and threatens basic biology research and education. Anais da Academia Brasileira de Ciências 90:1279–1284.

Anderson, J.A., and A.B. Alford. 2014. Ghost fishing activity in derelict blue crab traps in Louisiana. Marine Pollution Bulletin 79:261–267.

Antonelli, A., and V. Rodriguez. 2009. Brazil should facilitate research permits. Society for Conservation Biology 23:1068–1069.

Arlinghaus, R., T. Mehner, and I.G. Cowx. 2002. Reconciling traditional inland fisheries management and sustainability in industrialized countries, with emphasis on Europe. Fish and Fisheries 3:261–316.

Bacher, D. 2010. WikiLeaks reveals U.S., British use marine reserves as tool of imperialism, December 23, 2010. San Francisco Bay Area Independent Media Center. Accessed December 3, 2023. [https://www.indybay.org/newsitems/2010/12/23/18667425.php].

Bauer, R.T. 2004. Remarkable shrimps: Adaptations and natural history of the carideans. University of Oklahoma Press, Norman, USA.

Bauer, R.T. 2020. Shrimp fisheries. Pages 91–116 *in* G. Lovrich and M. Thiel, editors. The natural history of the Crustacea, Volume 9: Fisheries and aquaculture. Oxford University Press, New York, USA.

Baker, E. 2017. Recreational fishers in Australia: A social snapshot. Fisheries Research and Development Corporation, Australia. Accessed December 3, 2023. [https://ozfish.org.au/wp-content/uploads/2017/05/Recreational-Fishers-in-Australia-a-social-snapshot-sml.pdf].

Bakker, F.T., A. Antonelli, J.A. Clarke, J.A. Cook, S.V. Edwards, P.G. Ericson, S. Faurby, N. Ferrand, M. Gelang, R.G. Gillespie, and M. Irestedt. 2020. The global museum: Natural history collections and the future of evolutionary science and public education. PeerJ, 8:e8225.

Bedecarratz, P.C., D.A. López, B.A. López, and O.A. Mora. 2011. Economic feasibility of aquaculture of the giant barnacle *Austromegabalanus psittacus* in southern Chile. Journal of Shellfish Research 30:147–157.

Boenish, R., J.P. Kritzer, K. Kleisner, R.S. Steneck, K.M. Werner, W. Zhu, F. Schram, D. Rader, W. Cheung, J. Ingles, Y. Tian, and J. Mimikakis. 2021. The global rise of crustacean fisheries. Frontiers in Ecology and The Environment 20:102–110.

Boudart, J. 2022. Wetlands and water: By capturing, storing, and filtering water, wetlands provide important benefits for wildlife and people. Ducks Unlimited, Memphis, Tennessee, USA. Accessed December 3, 2023. [https://www.ducks.org/conservation/waterfowl-research-science/wetlands-and-water?poe=conservation].

Browman, H.I., S.J. Cooke, I.G. Cowx, S.W.G. Derbyshire, A. Kasumyan, B. Key, J.D. Rose, A. Schwab, A.B. Skiftesvik, E.D. Stevens, C.A. Watson, and R. Arlinghaus. 2019. Welfare of aquatic animals: Where things are, where they are going, and what it means for research, aquaculture, recreational angling, and commercial fishing. ICES Journal of Marine Science 76:82–92.

Burukovsky, R.N. 2009. Feeding and trophic interactions of shrimps. FGOU VPO, Kaliningrad Technological State University Publishing Office, Kaliningrad, Russia.

CBD. 2022. Convention on biological diversity. Secretariat of the Convention on Biological Diversity, Montreal, Quebec, Canada. Accessed December 3, 2023. [https://www.cbd.int].

Calado, R. 2020. Marine ornamental decapods—Collection, culture, and conservation. Pages 313–340 *in* G. Lovrich and M. Thiel, editors. The natural history of the Crustacea, Volume 9: Fisheries and aquaculture. Oxford University Press, New York, USA.

Cato, J.C., and C.L. Brown. 2003. Marine ornamental species: Collection, culture, and conservation. Iowa State Press. Ames, Iowa, USA.

Christie, M.R., B.N. Tissot, M.A. Albins, J.P. Beets, Y. Jia, D.M. Ortiz, S.E. Thompson, and M.A. Hixon. 2010. Larval connectivity in an effective network of marine protected areas. PLoS ONE 5:e15715.

CNRA. 2019. Restricted species laws and regulations importation, transportation, and possession of wild animals—Manual 671. California Natural Resources Agency; Department of Fish and Wildlife. California, USA. Accessed December 3, 2023. [https://nrm.dfg.ca.gov/FileHandler.ashx?DocumentID=28427].

CNPQ. 2022. Conselho nacional de desenvolvimento científico e tecnológico. Ministério da Ciência, Tecnologia e Inovações, Brasilia, Brasil. Accessed December 3, 2023. [https://www.gov.br/mcti/pt-br/composicao/rede-mcti/conselho-nacional-de-desenvolvimento-cientifico-e-tecnologico].

Copeland, C., E. Baker, J.D. Koehn, S.G. Morris, and I.G. Cowx. 2017. Motivations of recreational fishers involved in fish habitat management. Fisheries Management and Ecology 24:82–92.

Coull, B.C., and S.S. Bell. 1983. Biotic assemblages: Populations and communities. Pages 283–320 *in* D.E. Bliss, editor. The biology of Crustacea, Volume 7: Behavior and ecology. Academic Press, Cambridge, Massachusetts, USA.

Creaser, E.P. 1931. North American phyllopods. Science 74:267–268.

Cueni, T. 2020. First Opinion: Novel coronavirus 2019-nCoV exposes a flaw in the Nagoya Protocol. Stat, 5 Feb. 2020. Stat—Reporting from the frontiers of health and medicine. Accessed December 3, 2023. [https://www.statnews.com/2020/02/05/novel-coronavirus-exposes-nagoya-protocol-flaw/].

Cumberlidge, N., P.K.L. Ng, D.C.J. Yeo, C. Magalhães, M.R. Campos, F. Alvarez, T. Naruse, S.R. Daniels, L.J. Esser, F.Y.K. Attipoe, F.-L. Clotilde-Ba, W. Darwall, A. McIvor, J.E.M. Baillie, B. Collen, and M. Ram. 2009. Freshwater crabs and the biodiversity crisis: Importance, threats, status, and conservation challenges. Biological Conservation 142:1665–1673.

Cunningham, R. 2019. Economics of recreational fishing. Saltwater Sportsman, Florida, USA, May 23, 2019. Accessed December 3, 2023. [https://www.saltwatersportsman.com/economics-recreational-fishing/].

Daneliya, M.E., and K.J. Wittmann. 2021. Conservation of continental Mysida and Stygiomysida. Pages 307–345 *in* T. Kawai and D.C. Rogers, editors. Recent advances in freshwater crustacean biodiversity and conservation. CRC Press, Boca Raton, Florida, USA.

De Grave, S., K.G. Smith, N.A. Adeler, D.J. Allen, F. Alvarez, A. Anker, Y. Cai, S.F. Carrizo, W. Klotz, F.L. Mantelatto, T.J. Page, J.-Y. Shy, J.L. Villalobos, and D. Wowor. 2015. Dead shrimp blues: A global assessment of extinction risk in freshwater shrimps (Crustacea: Decapoda: Caridea). PLoS ONE 10:e0120198.

Delić, T., P. Trontelj, M. Rendoš, and C. Fišer. 2017. The importance of naming cryptic species and the conservation of endemic subterranean amphipods. Scientific Reports 7:1–12.

de Mazancourt, V., W. Klotz, G. Marquet, B. Mos, D.C. Rogers, and P. Keith. 2021. New insights on biodiversity and conservation of amphidromous shrimps of the Indo-Pacific Islands (Decapoda: Atyidae: *Caridina*). Pages 381–404 *in* T. Kawai and D.C. Rogers, editors. Recent advances in freshwater crustacean biodiversity and conservation. CRC Press, Boca Raton, Florida, USA.

Diggles, B.K. 2019. Food for thought: Review of some scientific issues related to crustacean welfare. ICES Journal of Marine Science 76:66–81.

Diggles, B.K., R. Arlinghaus, H.I. Browman, S.J. Cooke, R.L. Cooper, I.G. Cowx, C.D. Derby, S.W. Derbyshire, P.J.B. Hart, B. Jones, A.O. Kasumyan, B. Key, J.G. Pepperell, D.C. Rogers, J.D. Rose, A. Schwab, A.B. Skiftesvik, D. Stevens, J.D. Shields, and C. Watson. 2023. Reasons to be skeptical about sentience and pain in fishes and aquatic invertebrates. Reviews in Fisheries Science and Aquaculture 32:127–150.

Dresser, C., and B. Swanson. 2013. Preemptive legislation inhibits the anthropogenic spread of an aquatic invasive species, the rusty crayfish (*Orconectes rusticus*). Biological Invasions, 15:1049–1056.

Drinkwater, E., E.J. Robinson, and A.G. Hart. 2019. Keeping invertebrate research ethical in a landscape of shifting public opinion. Methods in Ecology and Evolution 10:1265–1273.

Dureuil, M., K. Boerder, K.A. Burnett, R. Froese, and B. Worm. 2018. Elevated trawling inside protected areas undermines conservation outcomes in a global fishing hot spot. Science 362:1403–1407.

Dvoretsky, A.G., and V.G. Dvoretsky. 2017. Red king crab (*Paralithodes camtschaticus*) fisheries in Russian waters: Historical review and present status. Reviews in Fish Biology and Fisheries 28:331–353.

FAO. 2021. Cultured aquatic species information programme. Text prepared by Weimin, M. for Food and Agriculture Organization of The United Nations, Fisheries Division. Rome, Italy. Updated. Accessed December 3, 2023. [http://www.fao.org/fishery/docs/DOCUMENT/aquaculture/CulturedSpecies/file/en/en_fishlist.htm].

Fellowes, J.R., M.W. Lau, D. Dudgeon, G.T. Reels, G.W. Ades, G.J. Carey, B.P.L. Chan, R.C. Kendrick, K.S. Lee, M.R. Leven, K.D.P. Wilson, and Y.Y. Tung. 2002. Wild animals to watch: Terrestrial and freshwater fauna of conservation concern in Hong Kong. Memoirs of the Hong Kong Natural History Society 25:123–159.

Galil, B.S., P.F. Clark, and J.T. Carlton. 2011. In the wrong place—Alien marine crustaceans: Distribution, biology, and impacts. Invading Nature—Springer Series in Invasion Ecology 6. Journal of Crustacean Biology 32:337–339.

Gardner, C., R.A. Watson, A.D. Jayanti, J. Suadi, M. AlHusaini, and G.H. Kruse. 2020. Crustaceans as fisheries resources: General overview. Pages 1–20 *in* G. Lovrich and M. Thiel, editors. The natural history of the Crustacea, Volume 9: Fisheries and aquaculture. Oxford University Press, New York, USA.

Geiger, K., A. Rivera, A. Aguión, J. Álvarez, J. Arrontes, Y.J. Borrell, T. Cruz, J. Dubert, M.E. Feis, J.N. Fernandes, C. Fernandez, L. García-Florez, D. Jacinto, D. Jollivet, G. Macho, E. Mateo, D. Mateus, P. Moran, C. Muniz, A. Nicolle, R. Nolasco, M. Parrondo, H. Queiroga, J. Rico, A. Sousa, S. Roman, T. Silva, E. Thiebaut, E. Vazquez, and J. Luis Acuna. 2022. Coping with poachers in European stalked barnacle fisheries: Insights from a stakeholder workshop. Marine Policy 135:104826.

Gelder, S.R. 2020. An ongoing saga: Endemic branchiobdellidans (Annelida: Clitellata) on translocated commercial North American crayfish. Zoosymposia 17:141–158.

Harriott, V.J. 2001. The sustainability of Queensland's coral harvest fishery. CRC Reef Research Centre Technical Report No. 40. CRC Reef Research Centre, Townsville, Australia.

Holdich, D.M. 2003. Crayfish in Europe—An overview of taxonomy, legislation, distribution, and crayfish plague outbreaks. Pages 15–34 *in* D.M. Holdich and P.J. Sibley, editors. Management and conservation of crayfish. Proceedings of a conference held on 7th November 2002. Environment Agency, Bristol, England, UK.

Hoyle, M.E., and J.W. Gibbons. 2000. Use of a marked population of diamondback terrapins (*Malaclemys terrapin*) to determine impacts of recreational crab pots. Chelonian Conservation and Biology 3:735–737.

Humphreys, J., and R.W.E. Clark. 2020. Marine protected areas: Science, policy and management. Elsevier, London, UK.

Hutchings, E., and D. Brown. 2014. Christmas Island national park management plan, 2014–2024. Australian Government, Director of National Parks, F2014L00168. Canberra, Australia.

IUCN. 2001. IUCN red list categories and criteria: Version 3.1. IUCN Species Survival Commission. IUCN, Gland, Switzerland/Cambridge, UK.

Jacinto, D., T. Cruz, T. Silva, and J.J. Castro. 2011. Management of the stalked barnacle (*Pollicipes pollicipes*) fishery in the Berlengas Nature Reserve (Portugal): Evaluation of bag and size limit regulation measures. Scientia Marina 75:439–445.

Jeffs, A., C. Daniels, and K. Heasman. 2020. Aquaculture of marine lobsters. Fisheries and Aquaculture 9:285.

Jepsen, P.M., K. Syberg, G. Drillet, and B.W. Hansen. 2020. Planktonic crustacean culture—Live planktonic crustaceans as live feed for finfish and shrimps in aquaculture. Pages 341–366 *in* G. Lovrich and M. Thiel, editors. The natural history of the Crustacea, Volume 9: Fisheries and aquaculture. Oxford University Press, New York, USA.

Jussila, J., and L. Edsman. 2020. Relaxed attitude towards spreading of alien crayfish species affects protection of native crayfish species: Case studies and lessons learnt from a Fennoscandian viewpoint. Freshwater Crayfish 25:39–46.

Kawai, T., and D.C. Rogers. 2021. Recent advances in freshwater crustacean biodiversity and conservation. CRC Press, Boca Raton, Florida, USA.

Kawai, T., B. Grubb, and F. Grandjean. 2021. Conservation of the Japanese endangered crayfish *Cambaroides japonicus* (De Hann, 1841) (Decapoda: Cambaroididae). Pages 487–504 *in* T. Kawai and D.C. Rogers, editors. Recent advances in freshwater crustacean biodiversity and conservation. CRC Press, Boca Raton, Florida, USA.

Knauf, S., L. Abela, and L.K. Hallmaier-Wacker. 2019. The Nagoya protocol and research on emerging infectious diseases. Bulletin of the World Health Organization 97:379.

Lang, B.K., D.A. Kelt, and S.M. Shuster. 2006. The role of controlled propagation on an endangered species: Demographic effects of habitat heterogeneity among captive and native populations of the Socorro isopod (Crustacea: Flabellifera). Biodiversity and Conservation 15:3909–3935.

Leenhardt, P., N. Low, N. Pascal, F. Micheli, and J. Claudet. 2015. The role of marine protected areas in providing ecosystem services. Pages 211–239 *in* A. Belgrano, G. Woodward, and U. Jacob, editors. Aquatic functional biodiversity: An ecological and evolutionary perspective. Academic Press, London, UK.

Lewin, W.-C., R. Arlinghaus, and T. Mehner. 2006. Documented and potential biological impacts of recreational angling: Insights for conservation and management. Reviews in Fisheries Science 14:305–367.

Litzow, M.A., F.J. Mueter, and J.D. Urban. 2013. Rising catch variability preceded historical fisheries collapses in Alaska. Ecological Applications 23:1475–1487.

López, B.A. 2020. Marginal marine crustacean fisheries. Pages 159–180 *in* G. Lovrich and M. Thiel, editors. The natural history of the Crustacea, Volume 9: Fisheries and aquaculture. Oxford University Press, New York, USA.

Macho, G., I. Naya, J. Freire, S. Villasante, and J. Molares. 2013. The key role of the barefoot fisheries advisors in the co-managed TURF system of Galicia (NW Spain). AMBIO 42:1057–1069.

Magalhães, C., M.R. Campos, P.A. Collins, and F.L. Mantelatto. 2016. Diversity, distribution and conservation of freshwater crabs and shrimps in South America. Pages 303–322 *in* T. Kawai and N. Cumberlidge, editors. A global overview of the conservation of freshwater decapod crustaceans. Springer, Cham, Switzerland.

Martin, J.W., and G.E. Davis. 2001. An updated classification of the recent Crustacea. Science Series 39. Natural History Museum of Los Angeles County, Los Angeles, California, USA.

Martínez-Pantoja, M., J. Alcocer, and A.M. Maeda-Martínez. 2002. On the Spinicaudata (Branchiopoda) from Lake Cuitzeo, Michoacán, México: First report of a clam shrimp fishery. Hydrobiologia 486:207–213.

McKenzie, M. 2013. The Gardner Bill. The Island Journal 29:48–51. Island Institute, Rockland, Maine, USA. Accessed January 12, 2022. [https://www.islandjournal.com/business/the-gardner-bill/].

Moland, E., A. Fernández-Chacón, T.K. Sørdalen, D. Villegas-Ríos, S.H. Thorbjørnsen, K.T. Halvorsen, M. Huserbråten, E.M. Olsen, P.J. Nillos Kleiven, A.R. Kleiven, H. Knutsen, S.H. Espeland, C. Freitas, and J.A. Knutsen. 2021. Restoration of abundance and dynamics of coastal fish and lobster within northern marine protected areas across two decades. Frontiers in Marine Science 8:674756.

Morgera, E., E. Tsioumani, and M. Buck. 2015. Unraveling the Nagoya protocol: A commentary on the Nagoya protocol on access and benefit-sharing to the convention on biological diversity. Brill, The Netherlands.

Mrugała, A., E. Kozubíková-Balcarová, C. Chucholl, S. Cabanillas Resino, S. Viljamaa-Dirks, J. Vukić, and A. Petrusek. 2015. Trade of ornamental crayfish in Europe as a possible introduction pathway for important crustacean diseases: Crayfish plague and white spot syndrome. Biological Invasions 17:1313–1326.

Mullowney, D.R.J., and K.D. Baker. 2020. Gone to shell: Removal of a million tonnes of snow crab since cod moratorium in the Newfoundland and Labrador fishery. Fisheries Research 230:105680.

NMDGF (New Mexico Department of Game and Fish). 2004. Threatened and endangered species, 2004 biennial review, final draft recommendations. New Mexico Department of Game and Fish, Santa Fe, New Mexico, USA.

NOAA. 2007. Magnuson-Stevens Fishery Conservation and Management Act, Public Law 94–265, as amended by the Magnuson-Stevens Fishery Conservation and Management, Reauthorization Act (P.L. 109–479). Accessed December 3, 2023. [https://media.fisheries.noaa.gov/dam-migration/msa-amended-2007.pdf].

Oficialdegui, F.J., M. Clavero, M.I. Sánchez, A.J. Green, L. Boyero, T.C. Michot, K. Klose, T. Kawai, and C. Lejeusne. 2019. Unravelling the global invasion routes of a worldwide invader, the red swamp crayfish (*Procambarus clarkii*). Freshwater Biology 64:1382–1400.

Oficialdegui, F.J., M.I. Sánchez, and M. Clavero. 2020. One century away from home: How the red swamp crayfish took over the world. Reviews in Fish Biology and Fisheries 30:121–135.

Orensanz, J.M., J. Armstrong, D. Armstrong, and R. Hilborn. 1998. Crustacean resources are vulnerable to serial depletion—The multifaceted decline of crab and shrimp fisheries in the Greater Gulf of Alaska. Reviews in Fish Biology and Fisheries 8:117–176.

Orchard, M. 2012. Crabs of Christmas Island. Christmas Island Natural History Association, Australia.

Orzechowski, K. 2015. The importance of sentience in animal legislation. April 16. Accessed November 25, 2024. [https://faunalytics.org/the-importance-of-sentience-in-animal-legislation/].

Parpet, J.F., and S.R. Gelder. 2020. North American Branchiobdellida (Annelida: Clitellata) or crayfish worms in France: The most diverse distribution of these exotic ectosymbionts in Europe. Zoosymposia 17:121–140.

Patoka, J., M. Bláha, L. Kalous, and A. Kouba. 2016. Irresponsible vendors: Non-native, invasive and threatened animals offered for garden pond stocking. Aquatic Conservation Marine and Freshwater Ecosystems 2016:1–6.

Pioua, C., U. Berger, H. Hildenbrandt, V. Grimm, K. Diele, and C. D'Lima. 2007. Simulating cryptic movements of a mangrove crab: Recovery phenomena after small scale fishery. Ecological Modelling 205:110–122.

Piper, L. 2020. Potential for expansion of the freshwater crayfish industry in Australia. Publication No. 00/142, Project No. CSU-2A. Rural Industries Research and Development Corporation, Australia.

Plotnikov, I.S., N.V. Aladin, J. Mossin, and J. Høeg. 2021. Crustacean fauna of the Aral Sea and its relation to ichthyofauna during the modern regression crisis and efforts at restoration. Zoological Studies 60:25.

Prathapan, K.D., R. Pethiyagoda, K.S. Bawa, P.H. Raven, and P.D. Rajan. 2018. When the cure kills—CBD limits biodiversity research. Science 360:1405–1406.

Provenzano, A.J. 1985. Economic aspects: Fisheries and culture. *In* D.E. Bliss, editor. The Biology of Crustacea, Volume 10. Academic Press, Cambridge, Massachusetts, USA.

Puri, S., and Z. Faulkes. 2010. Do decapod crustaceans have nociceptors for extreme pH? PLoS ONE 5:e10244.

Redak, R.A. 2000. Arthropods and multispecies habitat conservation plans: Are we missing something? Environmental Management 26:S97–S107.

Ribeiro, D.B., and V.L. Freitas. 2014. Letter: Brazil's new laws, bug collectors. Science 345:1571.

Rigel, J.A. 1959. The systematics and distribution of crayfishes in California. California Fish and Game 45:29–50.

Rogers, D.C. 2014. Larger hatching fractions in avian dispersed anostracan eggs (Branchiopoda). Journal of Crustacean Biology 34:135–143.

Rogers, D.C. 2021. Conservation status of the large branchiopods (Branchiopoda: Anostraca, Notostraca, Laevicaudata, Spinicaudata, Cyclestherida). Pages 221–238 *in* T. Kawai and D.C. Rogers, editors. Recent advances in freshwater crustacean biodiversity and conservation. Advances in Crustacean Research 22. CRC Press, Boca Raton, Florida, USA.

Rogers, D.C., W. Thaimuangphol, N. Saengphan, and L. Sanoamuang. 2013. Current knowledge of the South East Asian large branchiopod Crustacea (Anostraca, Notostraca, Laevicaudata, Spinicaudata, Cyclestherida). Journal of Limnology 72:69–80.

Rogers, D.C., M. Archangelsky, and P. Pessacq. 2018. A new genus and species of thamnocephalid fairy shrimp (Branchiopoda: Anostraca) from Argentina (Chubut Province). Nauplius 26:e2018021.

Rogers, D.C., A. Dunn, and W.W. Price. 2019. A review of *Dendrocephalus* (*Dendrocephalinus*) (Crustacea: Anostraca) with the first records of male-male anostracan aggressive competition. European Journal of Taxonomy 509:1–14.

Sanoamuang, L.O., G. Murugan, P.H. Weekers, and H.J. Dumont. 2000. *Streptocephalus sirindhornae*, new species of freshwater fairy shrimp (Anostraca) from Thailand. Journal of Crustacean Biology 20:559–565.

Sanoamuang, L., N. Saengphan, and G. Murugan. 2002. First record of the family Thamnocephalidae (Crustacea: Anostraca) from Southeast Asia and description of a new species of *Branchinella*. Hydrobiologia 486:63–69.

Sestelo, M., and J. Roca-Pardiñas. 2011. A new approach to estimation of the length–weight relationship of *Pollicipes pollicipes* (Gmelin, 1789) on the Atlantic coast of Galicia (Northwest Spain): Some aspects of its biology and management. Journal of Shellfish Research 30:939–948.

Sharov, A.F., J.H. Vølstad, G.R. Davis, B.K. Davis, R.N. Lipcius, and M.M. Montane. 2003. Abundance and exploitation rate of the blue crab (*Callinectes sapidus*) in Chesapeake Bay. Bulletin of Marine Science 72:543–565.

Shelley, C., and A. Lovatelli. 2011. Mud crab aquaculture—A practical manual FAO fisheries and aquaculture technical paper. No. 567. Food and Agriculture Organization of The United Nations, Fisheries Division, Rome, Italy.

Shiota, K. 1993. Relationship between annual catch fluctuations and reproduction in the swimming crab in Hiuchi Nada, the Seto Inland Sea. Nippon Suisan Gakkaishi 59:1709–1715.

Shroff, V. 2021. Recognizing animals in Canada as sentient. Accessed November 25, 2024. [https://nationalmagazine.ca/en-ca/articles/law/opinion/2021/recognizing-animals-in-canada-as-sentient].

Smith, D., M. da Silva, J. Jackson, and C. Lyal. 2017. Explanation of the Nagoya protocol on access and benefit sharing and its implication for microbiology. Microbiology 163:289–296.

Sumaila, U.R., and G.R. Munro. 2008. Fisheries economics. Pages 491–498 *in* J.H. Steele, editor. Encyclopedia of ocean sciences, 2nd edition. Academic Press, Cambridge, Massachusetts, USA.

Stevens, B.G., and T.J. Miller. 2020. Crab fisheries. Pages 21–54 *in* G. Lovrich and M. Thiel, editors. The natural history of the Crustacea, Volume 9: Fisheries and aquaculture. Oxford University Press, New York, USA.

Sutton, S.G. 2006. An assessment of the social characteristics of Queensland's recreational fishers, CRC Reef Research Centre Technical Report No. 65. CRC Reef Research Centre, Townsville, Australia.

Tibbett, A., and B. Briggs. 2019. Fines to rogue fishermen fall and illegal fishing escapes prosecution. October 6, 2019. The Ferret. Accessed December 3, 2023. [https://theferret.scot/fines-fishermen-prosecution/].

Tissot, B.N. 1999. Adaptive management of aquarium fish collecting in Hawaii. SPC Live Reef Fish Information Bulletin 6:16–19.

Tlusty, M. 2002. The benefits and risks of aquacultural production for the aquarium trade. Aquaculture 205:203–219.

Tobin, B.M. 2013. Bridging the Nagoya compliance gap: The fundamental role of customary law in protection of indigenous peoples' resource and knowledge rights. Law, Environment and Development Journal 9:142–162.

USFWS (U.S. Fish and Wildlife Service). 1982. Socorro isopod (*Thermosphaeroma thermophilum*) recovery plan. U.S. Fish and Wildlife Service, Albuquerque, New Mexico, USA.

Varisco, M., J. Colombo, P. Di Salvatore, P. Balzi, N. Bovcon, G. Lovrich, and J. Vinues. 2019. Fisheries-related variations in the fecundity of the southern king crab in Patagonia. Fisheries Research 218:105–111.

Vernberg, F.J., and W. Vernberg. 1983. Environmental adaptations. *In* D.E. Bliss, editor. The Biology of Crustacea, Volume 8. Academic Press, Cambridge, Massachusetts, USA.

Wahle, R.A., A.J. Linnane, and A.M. Harrington. 2020. Lobster fisheries. Pages 55–90 *in* G. Lovrich and M. Thiel, editors. The natural history of the Crustacea, Volume 9: Fisheries and aquaculture. Oxford University Press, New York, USA.

Weiperth, A., B. Gál, P. Kuříková, M. Bláha, A. Kouba, and J. Patoka. 2017. *Cambarellus patzcuarensis* in Hungary: The first dwarf crayfish established outside of North America. Biologia 72:1529–1532.

Wood, E. 1985. Exploitation of coral reef fishes for the aquarium trade. Report to Marine Conservation Society, UK.

Young, L.G.L. 1997. Sustainability issues in the trade for wild and cultured aquarium species. Pages 145–151 *in* B. Paust and J.B. Peters, editors. Marketing and shipping live aquatic products, Seattle, Washington, October 13–15, 1996. Northeastern Regional Agricultural Engineering Service Cooperative Extension, Ithaca, New York, USA.

Zimmerman, M.E. 2003. The black market for wildlife: Illegal wildlife trade. Vanderbilt Journal of Transnational Law 36:1657–1689.

17

CRUSTACEANS IN SOCIETY: THE MEDIA, POPULAR CULTURE, AND PUBLIC AWARENESS

Christopher B. Boyko, and Shane T. Ahyong

Abstract

Crustaceans have "served" humans as an important protein source since prehistoric times, but they have also assumed a cultural importance that transcends mere seafood, as evidenced by their representation in popular culture. Here, we explore varied ways in which crustaceans have been represented in Eastern and Western popular cultures, how they are perceived, and highlight species that have become popular via social media or through print, film, and videogame subcultures. Although overlapping, cultural perceptions of crustaceans in the East and West broadly differ, in part as a result of familiarity, and perhaps also as a result of different underlying historical conceptions of nature. Even so, we contend that greater popular familiarity with crustaceans can only be beneficial to arthropods and humans alike. The ecological place of crustaceans, and sometimes even their survival, however, is now currently threatened not only by direct overharvesting and habitat destruction, but also the wider impacts of climate change. Therefore, conservation efforts using any and all methods to place crustaceans in the public eye should be pursued, including tapping into human emotional responses (via the "cute and cuddly" factor), gastronomic interest (the plight of overfished species and cost of seafood), and even a renewed emphasis on promoting the basic science of discovery and documentation (taxonomy, systematics) that underlies our understanding of the evolutionary history of this ancient, diverse, and fascinating group.

INTRODUCTION

The interest of humans in all things crustacean likely dates back to the earliest days of our species, particularly as food items as inferred from shell middens (e.g., Mitchell and Foster 2000,

Christopher B. Boyko, and Shane T. Ahyong, *Crustaceans in Society: The Media, Popular Culture, and Public Awareness.*
In: *Ecology and Conservation*. Edited by: Lars Gutow and Martin Thiel, Oxford University Press.
 DOI: 10.1093/oso/9780197768242.003.0017

Gutiérrez-Zugasti et al. 2016). Until recent times, however, there have been few direct records of any non-food related interests in crustaceans aside from some cave petroglyphs (e.g., Boyko 2003). In the past couple of centuries there has been an abundance of documented crustacean influence on human art and culture, albeit not at the level of influence and inspiration provided by vertebrates (e.g., Silberman 1989), insects (e.g., Dicke 2000), or mollusks (e.g., Mujiono et al. 2021). The geometric and aesthetic properties of mollusk shells have long caught the eye of mathematicians, artists, and collectors alike, and even spawned a major global shell trade (Dias et al. 2011). The same level of cultural influence cannot generally be claimed of crustaceans, but they have had a clear impact nevertheless, as we will discuss in this chapter.

Why should we be interested in how humans consider crustaceans in a cultural context? An important reason is that the quantity and diversity of crustacean representations in popular culture reflect not only the general degree of public awareness of species but may also reinforce how people in different regions continue to view crustaceans, especially in the area of conservation. Effective conservation of species, in general, is often linked to public empathy for specific taxa or to some cultural value that highlights their importance (McNeely and Watchel 1988). From time to time, crustaceans become highlighted in the public eye through a combination of curious new scientific discoveries, active social media, engaging images, and sometimes an additional "je ne sais quoi." One such example is the deep-sea yeti crab (*Kiwa hirsuta* Macpherson and Jones, 2005) discovered in 2005 by researchers in the submersible *Alvin* (see http://www.deepseanews.com/2011/12/yeti-crab-roundup/) and named after the legendary abominable snowman of the Himalayas because of their "furry" (actually setose) pereopods. After an initial report on this "crab" (in reality an anomuran squat lobster) by the BBC (http://news.bbc.co.uk/2/hi/science/nature/4785482.stm), the news media generated a large number of similar stories, likely due to the charismatic nature of the crab's "furry legs," as well a large number of cute plush toys inspired by real specimens. Another example of crustaceans that have become elevated in the public eye are the giant isopods (*Bathynomus* spp.) and so-called tongue-eating isopods (*Cymothoa* spp.). While giant isopods grabbed the attention of Japanese consumers, who responded with affection by purchasing a wide range of *Bathynomus*-inspired toys and figurines (see below under "Replicas—Toys and Figurines"; Fig. 17.1), the interest in *Cymothoa* spp., mostly in the English-speaking world, has generally been confined to news reports that emphasize the general "ick" factor of their parasitic lifestyle (see https://www.snopes.com/fact-check/parasite-fish-tongue-mouth/). Public interest in some other species, such as mantis shrimps (e.g., https://theoatmeal.com/comics/mantis_shrimp; https://www.youtube.com/watch?v=F5FEj9U-CJM) has more of a scientific bent and focuses on their complex vision and raptorial tendencies. These bursts of public interest sometimes spawn substantial merchandising (e.g., for *Bathynomus*) that can far outlast the initial public exposure via news reports.

Heightened public awareness of crustaceans coupled with their place in historical and popular culture can only improve the efficacy of public policy implemented to protect crustacean species under anthropogenic threat, although in certain cases, this very popularity may negatively impact crustaceans via overharvesting for the aquarium trade and gastronomics. The fact that many crustaceans are considered culinary delicacies that are part of the gastronomical culture and cultural identity in many countries is, therefore, a double-edged sword.

Some species that form a basis for regional and cultural identities are highly regulated and not currently considered substantially impacted by overfishing. Examples of these include the Chesapeake Bay blue crabs (*Callinectes sapidus* Rathbun, 1896), a species that serves as the Official State Crustacean of Maryland, USA, and which appears on state license plates as well as countless souvenirs. Blue crabs are both a source of income and the basis of a way of life for people on the eastern shore of Maryland, as exemplified by the "National Hard Crab Derby" that is held each Labor Day, complete with crab races, cooking and meat picking contests, and a beauty pageant where a "Miss Crustacean" is crowned (Horton 1996).

Fig. 17.1.

(A–F) Selected *Bathynomus* models and figurines from Japan: (A) *B. giganteus* A. Milne Edwards, 1879, life size (Kitan Club). (B) *B. giganteus*, with fully articulated body parts (Bandai). (C) *B. doederleini* Ortmann, 1894 (Kaiyodo). (D–E) *B. giganteus* (Kaiyodo). (F) *B. giganteus* (T-Arts). (G–I) T-shirt designs: (G–H) Mantis shrimp designs (*Odontodactylus scyllarus* (Linnaeus, 1758)) from The Oatmeal. (I) Tourist souvenir promoting the famous dish, Singapore chilli crab (*Scylla* spp.). (J) *Wasser* by Giuseppe Arcimboldo (1566), oil on canvas (Kunsthistorisches Museum, Vienna) featuring (K) a penaeid prawn, (L) *Squilla mantis* Linnaeus, 1758, (L–M) *Cancer pagurus* Linnaeus, 1758 and *Astacus* sp. See a color version of this figure in the centerfold.

Another regionally important species is the American lobster (*Homarus americanus* H. Milne Edwards, 1837). The image of the lobster is indelibly linked with the state of Maine, USA, and 80% of the commercial catch for this species originates there (Abel 2020). American lobsters were once considered a low-class food only fit for servants (Corson 2004), but times and tastes have changed and the species is the focus of a 485 million dollar fishery market as of 2019 (Abel 2020).

There are also numerous locally important seafood festivals, such as the "Festa do Percebe," celebrating the stalked barnacle *Pollicipes pollicipes* (Gmelin, 1791) in Spain. Analogous to the history of the American lobster, these cirripedes were once peasant food but now command high prices and are a gastronomic delicacy. More than six thousand people come each year to the festival in Corme alone (https://visitacostadamorte.com/en/post/seafood-from-the-coast-of-the-morte). In North America, as in most parts of the world, barnacles do not attain sizes suitable for eating and are considered pests as marine foulers (Wahl 1997), although attempts have been made to improve their public image (https://www.youtube.com/watch?v=UZQE0Z2aZHE).

Numerous crayfish festivals, featuring "crawfish boils" take place in the United States each year (https://www.welovecrawfish.com/list-of-crawfish-festivals-outside-la-tx.htm), where literal boatloads of *Procambarus clarkii* (Girard, 1852) are boiled and served with potatoes and corn; as with the stalked barnacles noted above, it takes a lot of these organisms to make a full meal as both the barnacles and crayfish contain relatively small amounts of meat. In the fall of each year, the City of Otaru, Japan, celebrates the *Otaru Shako Matsuri* (おたる しゃこ まつり), or the Otaru Mantis Shrimp Festival, where large numbers of *Oratosquilla oratoria* De Haan, 1844 are prepared and typically served in a soup.

In contrast to the examples given above, however, other commercial species (king crabs; Szuwalski et al. 2020, blue crabs; Lipcius and Stockhausen 2002) and many trawl fisheries worldwide have seen declining catches, which may reflect a combination of climate change/habitat degradation and overfishing of subsequently reduced populations. A television show such as *The Deadliest Catch* (Discovery channel, from 2005 onwards), which focuses on crab fisheries in the Bering Sea, is an example of a cultural product that serves to bring more awareness to a primarily Western audience not accustomed to seeing crustaceans on the screen.

Crustaceans and humans may have coexisted for the whole of human history, but only recently have we channeled our interest in our chitinous companions into so many different categories of cultural expression. In importance to humans, the culinary role of crustaceans is perhaps the most obvious and utilitarian aspect as a protein source. Crustaceans as seafood, however, is not the focus of this survey. Instead, we explore the varied ways crustaceans have been represented in Eastern and Western popular cultures, how they are perceived, and highlight those species that have become popular via social media or through print, film, and videogame subcultures.

THE FINE ARTS

Most art involving images of single species of crabs falls into two categories: blue or rock crabs (*Callinectes sapidus* and *Cancer pagurus* Linnaeus, 1758, or reasonable facsimiles thereof; Fig. 17.2A) and hermit crabs (usually vaguely similar to *Coenobita* spp.). This focus on a relatively few species out of the many others that are equally or more charismatic or colorful reflects the trend of focusing on the things of most importance to humans, in this case food and pets. The same is true of other crustaceans represented in art, such as lobsters and crayfish. Occasionally, crustaceans in fine art are the main subject of a work, as in Albrecht Dürer's (1471–1528) "Krebs" (depicting *Eriphia spinifrons*, Herbst 1785) and "Ein Hummer" (depicting *Homarus gammarus*, Linnaeus 1758), but crustaceans more often form part of larger artistic displays of culinary extravagance, particularly in European Renaissance paintings (Fig. 17.2A, B). An unusual

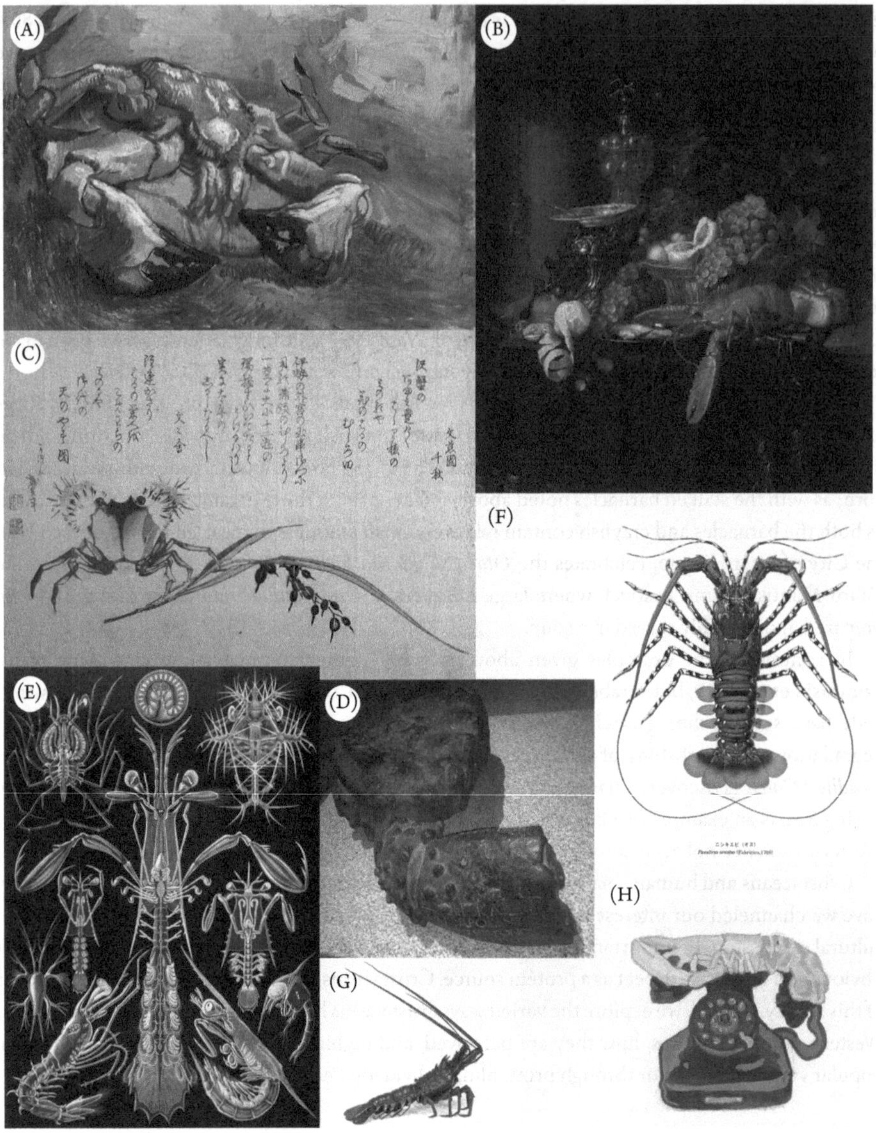

Fig. 17.2.
Crustaceans in fine art. (A) *Cancer pagurus* Linnaeus, 1758 in *A Crab on its Back* by Vincent Van Gogh (1887), oil on canvas (Van Gogh Museum, Amsterdam). (B) *Homarus gammarus* (Linnaeus, 1758) and *C. pagurus* in *Still Life with Golden Goblet* by Pieter de Ring (ca. 1640–1660), oil on canvas (Rijksmuseum, Leiden). (C) *Eriocheir* sp. in *Crab and Rice Plant* by 葛飾 北斎 (Katsushika Hokusai) (19th century). (D) Statue of *Eriphia verrucosa* (Forskål, 1775) in the collection of the Metropolitan Museum of Art, New York; formerly part of a set surrounding the base of the obelisk, Cleopatra's Needle. Photograph by Christopher Boyko. (E) Thoracostraca (Haeckel 1904: pl. 76). (F) *Panulirus ornatus* (Fabricius, 1798) (Sugiura and Asakura 2012). (G) Spiny lobster (Palinuridae) metal kusshin jizai (fully articulated) okimono model, unknown artist (Meiji period) (photo: © Bonhams, used with permission). (H) Gouache on paper interpretation of Salvador Dalí's *Lobster Telephone* (*Homarus gammarus* (Linnaeus, 1758)) by Maria Spector (2009). See a color version of this figure in the centerfold.

departure from the usual culinary subjects, however, is in Giuseppe Arcimboldo's (1527–1593) "Wasser" (Fig. 17.1J) from his "Four Elements" series, which includes a penaeid prawn (Fig. 17.1K), *Cancer pagurus* (Fig. 17.1M), and as far as we are aware, the only stomatopod (*Squilla mantis* Linnaeus, 1758) depicted in Renaissance art (Fig. 17.1L). Even so, there is more diversity in crustacean representation in art in Asia, where the culinary interest spans a greater number of species (e.g., *Scylla* spp., *Eriocheir* spp., *Macrobrachium* spp.) and those species also are represented in art originating from China, Japan, and neighboring countries (Fig. 17.2C).

Some of the oldest and largest crustacean sculptures are the crab statues at the base of the obelisk called Cleopatra's Needle that was erected in Heliopolis (Cairo) circa 1443 BC.; these four crabs (originals of only two of which still exist in the Metropolitan Museum in New York; Fig. 17.2D) appear to represent *Eriphia verrucosa* (Forskål, 1775), a species whose range includes the Mediterranean waters near Egypt. The four crabs currently surrounding the obelisk in Central Park, New York, are modern reproductions.

Many crustacean illustrations appearing in scientific books and journals could also be classified as art but were intended for a specialist audience rather than for popular consumption. Illustrations made for scientific publications, such as those by Ernst Haeckel (Fig. 17.2E), tend to be anatomically correct compared to crustacean fine art, but some artists such as Chisato Sugiura (Sugiura and Asakura 2012; Fig. 17.2F), while creating art for its own sake, produce beautiful images every bit as detailed as the best seen in scientific journals and books.

While fewer in number than paintings or drawings, there are some very interesting sculptures with crustacean themes, ranging from hyperrealistic Japanese okimono sculptures (Fig. 17.2G) to surrealistic lobster telephones (Fig. 17.2H).

CURRENCY—STAMPS AND COINS

Postage stamps were first issued in May 1840, and crustaceans debuted in 1870 as stylized freshwater crayfish on several Russian issues. From 1870 to 1950, fewer than two dozen crustacean-themed stamps were issued worldwide, but since then, the number has grown to more than 1400 from more than 200 countries (Holthuis 1967, Omori and Holthuis 2000, 2005, Hong and Wei 2010, Omori 2014, 2021; Fig. 17.3). Crustaceans from almost all major taxonomic groups have been featured, including branchiopods, hexanauplians (mostly copepods and barnacles), and all manner of malacostracans, ranging from stomatopods and euphausiaceans to peracarids and decapods. Not surprisingly, decapods, particularly crabs and lobsters, dominate. Crustacean depictions are sometimes stylized and emblematic or elements of a wider theme, but in most cases, they are the primary subject and taxonomically recognizable. Fishery-related species commonly feature, as well as colorful or charismatic deep-sea forms. In an unusual twist, a colorful species of Australian amphipod that first featured on a postage stamp in 1984 was subsequently and fittingly formally named *Amaryllis philatelica* (Lowry and Stoddart 2002; Fig. 17.3).

The frequency of crustacean stamp depictions is not evenly distributed among issuing countries. Very few crustaceans feature on stamps from Europe and North America; in the past 150 years, only seven were issued by the USA, three each by Canada and Germany, two each by the United Kingdom and France, six from Sweden, and a few others from other parts of Europe (Omori 2014). Conversely, equatorial and maritime, primarily non-Western countries, especially in the tropical Pacific and Caribbean, account for the vast majority of crustacean stamp issues. In these areas, crustaceans are substantially more familiar to local people, not only as food but because they are frequently encountered in daily life, whether in freshwater or along the seashore. Stamps from these places frequently include colorful coral reef, freshwater, and terrestrial species in addition to those regularly fished. It is probably no coincidence that crustaceans figure more frequently on postage

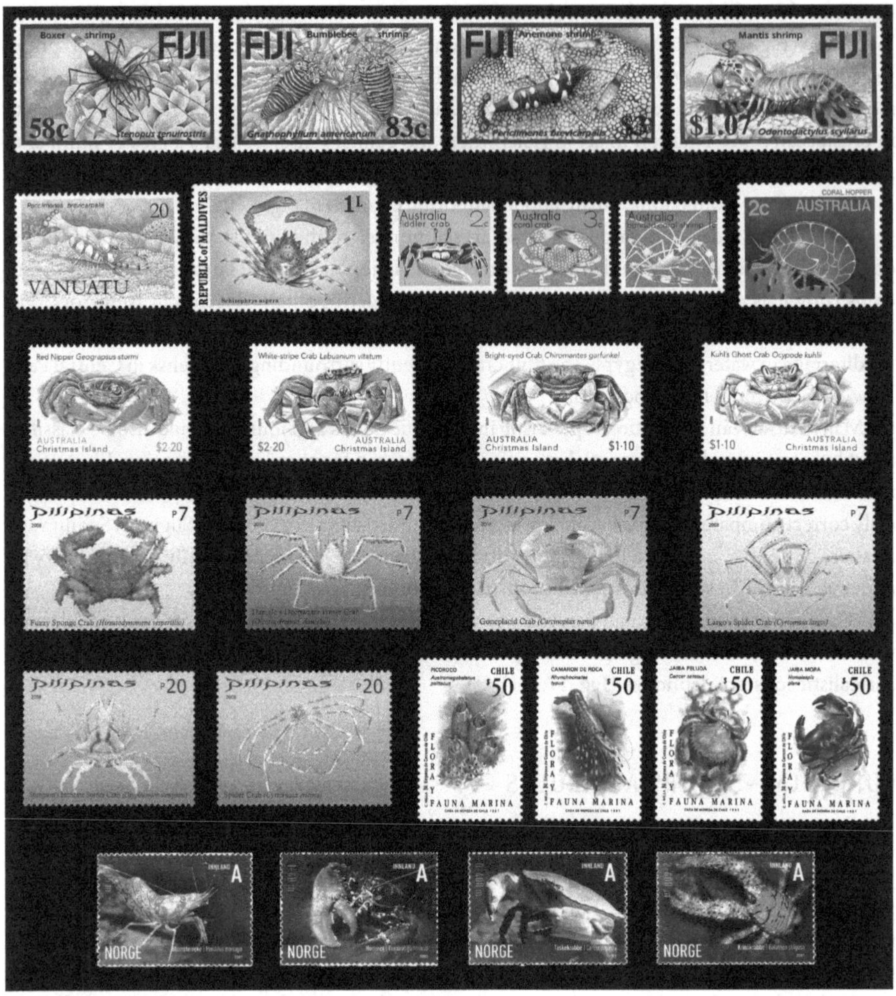

Fig. 17.3.
Crustaceans on stamps (identifications using current taxonomy). Line 1 (top; left to right): coral shrimp, *Stenopus tenuirostris* de Man, 1888; bumblebee shrimp, *Gnathophyllum americanum* Guérin-Méneville, 1855; anemone shrimp, *Ancylocaris brevicarpalis* Schenkel, 1902 (female); mantis shrimp, *Odontodactylus scyllarus* (Linnaeus, 1758). Line 2: anemone shrimp, *Ancylocaris brevicarpalis* Schenkel, 1902 (male); horned spider crab, *Schizophrys aspera* (H. Milne Edwards, 1831); fiddler crab, *Gelasimus vomeris* (McNeill, 1920); reticulated coral crab, *Trapezia septata* Dana, 1852; banded coral shrimp, *Stenopus hispidus* (Olivier, 1811); amphipod, *Amaryllis philatelica* Lowry and Stoddart 2002. Line 3: red nipper crab, *Geograpsus stormi* de Man, 1895; white-strip crab, *Circulium vitatum* (Ng and Davie, 2011); bright-eyed crab, *Chiromantes garfunkel* Davie and Ng, 2013; Kuhl's ghost crab, *Ocypode kuhlii* de Haan, 1835. Line 4: hairy crab, *Hirsutodynomene vespertilio* McLay and Ng, 2005; Cameron's spider crab, *Pleisticanthoides cameroni* Ng and Richer de Forges, 2012 (misidentified as *Dicranodromia danielae* Ng and McLay, 2005); small deep-water crab, *Carcinoplax nana* Guinot, 1989; Largo's spider crab, *Cyrtomaia largoi* Richer de Forges and Ng, 2007. Line 5: Stimpson's spider crab, *Oxypleurodon stimpsoni* Miers, 1886; *Cyrtomaia ericina* Guinot and Richer de Forges, 1982; giant barnacle, *Austromegabalanus psittacus* (Molina, 1782); hinged-beak shrimp, *Rhynchocinetes typus* H. Milne Edwards, 1837; hairy crab, *Romaleon setosum* (Molina, 1782); blackberry crab, *Homalaspis plana* (H. Milne Edwards, 1834). Line 6: pink shrimp, *Pandalus montagui* Leach, 1814; European lobster, *Homarus gammarus* (Linnaeus, 1758); brown crab, *Cancer pagurus* Linnaeus, 1758; blue-striped squat lobster, *Galathea strigosa* (Linnaeus, 1761). © Posten Bring Group Norway, Design: Enzo Finger. See a color version of this figure in the centerfold.

stamps from these regions. Whatever the cause, the frequency of crustacean stamp depictions correlates with local familiarity and regional crustacean diversity.

Crustaceans appear on coins used in the ancient Greek-speaking world, dating to at least the 4th century BC, with depictions typically resembling the common freshwater crab in the region, *Potamon fluviatile* (Herbst, 1758), a species that has been depicted even in modern times (Fig. 17.4A). Despite their long history, during the last century, as far as we can determine, crustaceans on monetary currency are rare, certainly in comparison to postage stamps. Crustaceans on coins are typically large edible species, such as the mangrove crab *Scylla serrata* (Forskål, 1775) (Seychelles; see Fig. 17.4C), the brown crab *Cancer pagurus* (Guernsey; see Fig. 17.4D), kuruma prawn *Marsupenaeus japonicus* (Spence Bate, 1888) (Mozambique; see Fig. 17.4F), Nordic shrimp *Pandalus borealis* (Krøyer, 1838) (Iceland; see Fig. 17.4G), clawed lobster *Homarus* sp. (San Marino; see Fig. 17.4J), spiny lobster *Panulirus* sp. (Turks and Caicos Islands and Cayman Islands; see Fig. 17.4H, I), and coconut crab *Birgus latro* (Linnaeus, 1767) (Vanuatu; see Fig. 17.4E), but smaller, regionally common species coins have also been minted, such as the European green crab *Carcinus maenas* (Linnaeus, 1758) (Iceland; see Fig. 17.4B). As with postage stamps, most crustacean coin issues are from equatorial and maritime countries, especially in the tropical Indo-Pacific and Caribbean.

LITERATURE

Crustaceans are traditionally underrepresented in adult literature, probably owing to their primarily aquatic nature, typically small sizes, and limited interactions with humans except as food. When they appear in literature, they tend to be as supporting characters/objects or with little to no personality but who impart fear or apprehension into the story's human characters, such as the "lobstrosities" in Stephen King's *The Drawing of the Three* (King 1987) or the court of crabs that judge humanity in Jean-Paul Sartre's *The Condemned of Altona* (Sartre 1961). There are exceptions, usually in folk tales, such as Kipling's *The Crab that Played with the Sea* and the traditional Japanese story *The Crab and the Monkey* (Fig. 17.5A). In the realm of novels, Guy N. Smith (1939–2020) was a one-man crab book industry, writing a dozen sex- and gore-laden books featuring ravenous killer crabs from the deep (Fig. 17.5B). These chitinous killers are literary descendants of a host of killer crabs that ran riot on the covers and interiors of US science fiction and "men's" magazines in the 1950s (Fig. 17.5C, D), an era when Americans apparently feared nature just as much as the Russians.

Most books with crustacean content are those written for small children interested in wildlife, who are perhaps less predisposed to consider crustaceans as "other" and tend to feature familiar species; Eric Carle's *A House for Hermit Crab* (Fig. 17.5E) is perhaps the most popular example but there are others, many of which are more lavishly illustrated, such as *Pagoo* (Fig. 17.5F).

In the realm of comic books, there are almost no crustacean-themed heroes or villains, in contrast to those from many other arthropod groups (e.g., Spider-Man, Black Widow, The Fly, the Blue Beetle); the Marvel Comics' D-list villain Cancer, a member of the Zodiac, is the only one of note. Assorted crab/human hybrid creatures appear in numerous comics, usually fighting against the heroes of sword and sorcery tales (Fig. 17.5G). However, there are a few non-superhero crustacean-centric comic stories of note, again with the theme of crustaceans as "other," such as Titano (Fig. 17.5H), apparently some kind of quasi-grapsoid crab (albeit with only eight legs), the humanoid crab monsters that crawl by night (Fig. 17.5I), and the gruesome ending to an otherwise historical drama (Fig. 17.5J).

Fig. 17.4.
Crustaceans on coins. (A) Freshwater crab, *Potamon fluviatile* (Herbst, 1758) (Malta). (B) European green crab, *Carcinus maenas* (Linnaeus, 1758) (Iceland). (C) Mangrove crab, *Scylla serrata* (Forskål, 1775) (Seychelles). (D) Brown crab, *Cancer pagurus* (Linnaeus, 1758) (Guernsey). (E) Coconut crab, *Birgus latro* (Linnaeus, 1767) (Vanuatu). (F) Kuruma prawn, *Marsupenaeus japonicus* (Bate, 1888) (Mozambique). (G) Nordic shrimp, *Pandalus borealis* Krøyer, 1838 (Iceland). (H) Spiny lobster, *Panulirus* sp. (Turks and Caicos Islands). (I) Spiny lobster, *Panulirus* sp. (Cayman Islands). (J) Clawed lobster, *Homarus* sp. (San Marino). (Photos: ©Daniel Clements, used with permission, www.coinzoo.net). See a color version of this figure in the centerfold.

Recently, the series *The March of the Crabs* (Fig. 17.5k) has brought crustaceans into the realm of the graphic novel and features a whole society of crabs with distinctly human personality traits.

MUSIC

Although a diverse assortment of crustaceans, including crabs, crayfish, lobsters, shrimp, krill, barnacles, and isopods, are mentioned in popular music (Coelho 2011), few songs are really about these crustaceans. In fact, most references to crustaceans are not germane to the rest of the song; examples of these types of lyrics include "Italian mobster shoots a lobster" (*The Magnificent Seven*

Fig. 17.5.
Crustaceans in literature. (A) Mitten crab (*Eriocheir* sp.) featured in a depiction of the Japanese fable *The Crab and the Monkey* by 渓斎 英泉 (Keisai Eisen) (Edo period), woodblock print (Princeton University Art Museum). (B) Slavering brachyurans, loosely based on species of Ocypodidae and Gecarcinidae on the covers to *Night of the Crabs* (Smith 1989) and *Crabs on the Rampage* (Smith 1988). (C) Menacing giant lobster on the cover to *Fantastic* (July 1957 issue) (Valigursky 1957). (D) Vicious *Cancer* sp. and *Birgus latro* (Linnaeus, 1767) ripped their flesh on the covers to *All Man* (Nov 1959) and *Man's Conquest* (Nov 1956) issues (Collins and Hagenauer 2008). (E) A paguroid on the cover to *A House for Hermit Crab* (Carle 1987). (F) Cover to *Pagoo* (Holling 1957). (G) Crab-men on the cover of *Conan the Barbarian* #99 (Buscema and Chan 1979). (H) The titular monster (cf. Grapsoidea) from "I was trapped by Titano, the monster that time forgot!" (Leiber and Kirby 1960). (I) Two panels from the title pages of "They crawl by night!" (Keyes and Wolverton 1953). (J) The denouement of the story "Rough Riders!" featuring the attack of the *Cardisoma guanhumi* Latreille, 1828; from the original art (Kurtzman and Estrada 1953). (K) Panel from *The March of the Crabs* (de Pins 2015). See a color version of this figure in the centerfold.

by The Clash, released 1981) or "plastic fantastic lobster telephone" (*Aphrodisiac Jacket* by The Cult, released 1987). Additionally, there are bands whose names are derived from crustacean groups (e.g., Isopoda, Stomatopoda) but whose songs, lyrics, and album art do not reference their namesakes. There are also albums whose covers dramatically feature crustaceans in either the art (e.g., Fig. 17.6A) or title and art (Fig. 17.6B, D–I) but which otherwise lack any crustacean content in the music itself. The American progressive rock band Kansas made Arcimboldo's crustacean laden 1566 work, *Wasser* (Fig. 17.1J–M), the cover image of their 1975 album, *Masque*. One exception to this is the folk rock band Noggins, who put six tracks each named *Crab Tune* on their album *Crab Tunes* (released 1971; Fig. 17.6C). Another exception is Decapod Claw, who in the year 2000 released their eponymous album of electronic music with only an image of a hyperiid amphipod on the cover but whose subsequent album, *Necrocrab Re-Sequencing Error* (2003), contained instrumental tracks titled "Amphipod Lover" and "Car[a]pace" (the album art and song title being rare cases of amphipods in popular culture).

There are, however, a few crustacean references in classical music, such as the *canon cancrizans* (or "crab canon"), which is an arrangement of two lines of music that are like a palindrome where one is the reverse of the other. The name presumably derives from the side to side reading of the lines, not unlike the sideways walking of a crab. A more taxonomically specific piece of classical music is *Embryons desséchés* ("Desiccated embryos") by Erik Satie (composed 1913) which features two of its parts entitled "d'Edriophthalma" and "de Podophthalma," both named for taxonomic groupings of sessile-eyed and stalk-eyed crustaceans no longer in use today.

The one pop song that fully embraces its inner crustacean is, of course, *Rock Lobster* by the B-52s (released 1978; Fig. 17.6J). This song also gives its name to a dance, although it is not the only crustacean to inspire a named type of dance; other examples include the "Crawfish Pie Line Dance" and "The Crab Dance." One other piece, the electronic instrumental *Crab Rave* by Noisestorm (released 2018; Fig. 17.6K), would be in the same category as the aforementioned albums by Decapod Claw (i.e., crustacean-named electronic instrumental songs) but for the addictive video released to promote the song (https://youtu.be/cE0wfjsybIQ). The vivid imagery of hundreds of animated red land crabs (clearly derived from the famous Christmas Island *Gecarcoidea natalis* Pocock, 1889) dancing and clicking along to the beat of this earworm song is quite memorable.

Perhaps the oddest example of crustacean music is that of music actually played on a crustacean. A video from Nippon TV News (https://youtu.be/0vW9VJRq6Gg) shows didgeridoo player Kazuki Kitagawa (Fig. 17.6L) using the body and limbs of a Japanese giant spider crab (*Macrocheira kaempferi* Temminck, 1836) as a wind instrument.

FILM AND TELEVISION

The crustacean presence in film and television is one where they are usually portrayed as sinister, threatening and, many times, as exceptionally large crabs or lobsters. This portrayal is most likely due to their non-vertebrate morphology and often conspicuous size of these decapods. The most abundant source of crustaceans on the screen is in films and TV shows from Japan, where crustacean-themed beasts run rampant in the kaiju genre. Examples of such oversized threats include Ebirah, the giant lobster from ゴジラ.エビラ.モスラ 南海の大決闘 (a.k.a. *Ebirah, Horror of the Deep*, 1966; Fig. 17.7A; known to Western audiences as *Godzilla versus the Sea Monster*) and numerous crab-like kaiju (Zanika, Crabgan, Black Dome, among others) that ran amok in the various *Ultraman* series (Fig. 17.7B–D).

Fig. 17.6.
Crustaceans in music. (A) Half-man, half-lobster (*Homarus americanus* H. Milne Edwards, 1837) on the cover of the self-titled album by Nantucket (released 1978). (B) Truth in advertising on the cover of *A Giant Crab Comes Forth* by Giant Crab (released 1968). The title track is the same name as the album itself. (C) Cover of *Crab Tunes* by Noggins (released 1971); fittingly, the album contains six songs each named *Crab Tune* (#1–6). (D) Spiny lobsters in a crate on the cover of *A White Sport Coat and a Pink Crustacean* by Jimmy Buffet (released 1973). (E) Artist's rendition of a male "*Uca*" sp. (Ocypodidae) winning the race against another, indeterminate, crab on the cover of *Crab Race* by The Morwells (released 1978). (F) Giant anatomically inaccurate crab monster on the cover of *Crab Society North* by Stormtroopers of Death (released 2011). (G) A portunid crab covered in honey on the cover of aptly named album *Crab in Honey* by Fever the Ghost (released 2014). (H) Cover for the self-titled album by *Obey the Lobsters* (released 2017), although the image is a crayfish, not a lobster. (I) Album cover for *Attack of the Giant Purple Lobsters!* by Washington Dead Cats (released 2019), although the "purple lobsters" pictured are actually images of the lobster-like Ultraman villain Alien Baltan. (J) 1986 12" single rerelease of *Rock Lobster* by the B-52's; the original 1978 single release had no crustacean-themed cover art. (K) One of the pseudo-*Gecarcoidea natalis* Pocock, 1889 featured in the music video for Noisestorm's *Crab Rave* (released 2018). (L) Kazuki Kitagawa plays the Japanese giant spider crab (*Macrocheira kaempferi* Temminck, 1836) like a didgeridoo. See a color version of this figure in the centerfold.

In Western cinema, we have the giant *Cancer pagurus* in *Mysterious Island* (1961; Fig. 17.7E) as one of the few cinematic crustaceans that can be unequivocally identified with a real species; unlike the Japanese creatures portrayed by men in rubber suits, this Ray Harryhausen-animated crab was constructed with actual crab parts fitted over a wire armature (Fig. 17.7F). At the other end of the realism spectrum, Onibaba in the American film *Pacific Rim* (2013; Fig. 17.7G) appears to be a Western analog to the crab-like kaiju of Japanese films and television, although some aspects of his anatomy are decidedly spider-like. Likewise fanciful are the high-domed, bulbous-eyed decapods in Roger Corman's *Attack of the Crab Monsters* (1957; Fig. 17.7H) that eat humans and absorb their intellects, leading to that deathless line of dialogue: "Once they were men; now they are land crabs." Slightly more realistic crustaceans, emphasis on the "slightly," provide some of the horror in the Japanese/American film *The House Where Evil Dwells* (1982), although the crabs in question appear to be oversized samurai crabs (*Heikeopsis japonica* von Siebold, 1824); colloquially known as heikegani) with a bit of spider crab thrown in for good measure; these crabs, uncharacteristically for dorippids, are terrestrial and mutter in the voices of Japanese samurai ghosts. They are accompanied in their evil deeds by a large number of small, live terrestrial crabs that are never seen clearly enough to be able to make out the species. In keeping with cinematic creature-feature portrayals in the horror genre, the latent menace of crustaceans is realized by their becoming unnaturally giant-sized, usually through some freak of nature or industrial accident. *Queen Crab* (2015), features a giant genetically modified crab that terrorizes a town, and *Aquarium of the Dead* (2021) draws crustaceans into the zombie genre, featuring giant spider crabs along with other public aquarium creatures that have become zombies and wreak the usual havoc. The horror comedy *Crabs* (2021) features hordes of giant mutated horseshoe crabs (not true crabs, of course) that terrorize a sleepy town following a nuclear spill. The generally negative cinematic portrayals of crustaceans carry through in Neill Blomkamp's 2009 political allegory *District 9*, which features an alien race of humanoid-prawns, stranded on Earth and subjugated by fearful humans.

In a rare case of cinematic species specificity, the isopod *Cymothoa exigua* (Schioedte and Meinert, 1884) is name-checked in the horror film *The Bay* (2012) as having mutated from fish parasitism to infesting humans with predictably graphic results. Within the horror genre, the terrifying xenomorph queen in James Cameron's *Aliens* (1986 and sequels) has been claimed to be inspired by the hyperiid amphipod *Phronima* sp. (e.g., O'Dwyer 2014). Despite an undeniable resemblance between amphipod and xenomorph, however, we could not find any direct evidence to confirm such a connection. In contrast to more conventional depictions of crustaceans as threatening, the black comedy *The Lobster* (2015) promises much with its title and tells the story of a man who wishes to become a lobster if he fails to find a romantic partner in 45 days but, alas, no actual lobsters appear in the film. Similarly, the 2022 murder mystery, *Where the Crawdads Sing*, based on the novel of the same name by Delia Owens, does not feature crayfishes, although we have a fleeting glimpse of a brachyuran crab and hermit crab. On the other hand, the crustacean content in the 2022 Norwegian post-apocalyptic film, *Black Crab*, is limited to its title and very brief dialogue alluding to a military strategy.

Even the animated crustaceans, at least those with personalities, can be threatening. Such is the case of Sheldon J. Plankton (Fig. 17.7I), first seen in a 1999 episode of the cartoon series *SpongeBob* (Nickelodeon, 1999–present). Plankton, a copepod, is an anomaly among cartoon and live-action crustaceans in not being a decapod (or even a malacostracan). Another menacing character is the giant selfish shellfish Tamatoa in Disney's *Moana* (2016; Fig. 17.7J). This animated crab was purportedly based on *Birgus latro* but an actual coconut crab would, of course, drown if kept continually submerged as seen in the film. Assorted undersea decapods, as well as perhaps a stomatopod, caused trouble for *Marine Boy* (1965), an early TV anime with distribution outside of Japan. Episode 1 of the 1976 Australian television adventure drama *The Lost Islands* featured actual mud crabs (*Scylla serrata*), which appeared en masse during a rising tide, threatening to kill and consume the human heroes trapped in a cave.

Fig. 17.7.
Crustaceans in film and television. (A) Ebirah, the giant lobster (Decapoda: cf. Nephropidae) who battled Godzilla in ゴジラ.エビラ.モスラ 南海の大決闘 (*Ebirah, Horror of the Deep*). (B) Zanika, a giant alien resembling a gall crab (Decapoda: Cryptochiridae) (from *Return of Ultraman* [1971]). (C) Crabgan, an alien resembling a giant crayfish (Decapoda: Astacoidea), shown here combined with Anemos, an alien resembling a large polychaete with which it forms a symbiotic relationship (from *Ultraman Gaia* [1998]). (D) Black Dome, a giant crab alien resembling a flying chocolate cake (from *Ultraman Leo* [1974]). (E) Giant *Cancer pagurus* animated by Ray Harryhausen from *The Mysterious Island*. (F) *Cancer pagurus* model (armature inside crab body) from *The Mysterious Island* as it looks today (Photo: ©Leonard Hawksmoor, used with permission). (G) Artist's conception of Onibaba from *Pacific Rim*. (H) Crab monster (*gen. et sp. indet.*) from Roger Corman's *Attack of the Crab Monsters*. (I) Sheldon J. Plankton (Copepoda, left) and Eugene H. Krabs (possible *Ocypode* sp. based on the eyestalks, right) from *SpongeBob SquarePants*. (J) Tamatoa (a fully aquatic coconut crab, *Birgus latro* (Linnaeus, 1767)) from Disney's *Moana*. (K) Sebastian, the hermit crab (Paguroidea) from Disney's *The Little Mermaid*. See a color version of this figure in the centerfold.

A rare example of a beneficent crustacean is seen in Disney's *The Little Mermaid* (1989) where Sebastian (Fig. 17.7K) is one of the titular character's helpmates. Although consistently referred to as a crab, he has only four pairs of legs visible and is actually an anomuran, perhaps a porcelain crab (although animated humans are often portrayed with only four fingers on each hand, suggesting that perhaps he really is a brachyuran; in this film, the humans have five fingers). In the 2023 live-action remake of the film, Sebastian is an ocypodid crab and, as such, he could not possibly live completely submerged in the water at all times. In a similar vein, the comical Tuck and Roll in Disney's *A Bug's Life* (1998) are clearly terrestrial isopods, apparently of the family Armadillidiidae because they can conglobulate, but they have only five pairs of legs instead of seven. Likewise, although having only a small role, the cleaner shrimp Jacques in Disney's *Finding Nemo* (2003) is a helpful, non-threatening character; he also lacks a pair of legs compared to a real shrimp. Eugene H. Krabs (Fig. 17.7I), owner of The Krusty Krab in the *SpongeBob* series, is sometimes benevolent, but his mood shifts episodically, and Dr. John A. Zoidberg from *Futurama* (20th Century Fox Television 1999–2004; Comedy Central 2010–2013; Hulu 2023–present) attempts benevolence but his lack of understanding of human physiology usually ends in poor medical treatment for his patients. The only friendly crustacean from a live-action television series that comes to mind is the giant humanoid Lobstercules from *The Tick* (3rd series, season 2, 2019) but even here, she initially appeared as a bank-robbing villain, committing crimes under threat to her children (who are portrayed as normal-sized, albeit singing lobsters rather than as a true larval stage).

VIDEO GAMES

Most of the crustaceans found in video games are, like in so many other media, brachyuran crabs. The vast majority fall into two categories: crab-like giant monsters (usually with other arthropodal elements added) that are semi-realistic, or rather cute anime-style crabs, often of very stylized forms. Examples in the former category include Crawmerax (Fig. 17.8A) from *Borderlands: The Secret Armory of General Knoxx* (Gearbox Software 2010), The Invincible Son of Crawmerax the Invincible (Fig. 17.8B) from Borderlands 2 (Gearbox Software 2012), and Giant Enemy Crab (Fig. 17.8C) from *Genji: Days of the Blade* (Game Republic 2006). In the latter category, we have Crabmeat from *Sonic the Hedgehog* (Sega 1991) and Krabby from *Pokémon Mystery Dungeon: Explorers of Time* (Nintendo 2008); the Pokémon character also appears in the Pokémon trading card game (Fig. 17.8D) as well as in the associated manga and anime derivative works. An excellent summary of this and other crustacean-themed Pokémon, including those derived from crab, shrimp, barnacle, and isopod species, is given by Rosa et al. (2020).

In contrast to stylized and "cartoonish" representations, some video games depict crustaceans in a much more realistic and anatomically correct manner. The mudcrabs (Fig. 17.8E) in *The Elder Scrolls* series (Bethesda Softworks, 2002–2017) are aptly named, as they appear to literally be mud crabs (Panopeidae). While in the *Fallout* game series (Bethesda Softworks, Xbox Game Studios 1997–2018), a group of aquatic nuclear-mutated creatures called mirelurks appear to be some kind of cross between a decapod, a cirolanid isopod, and a horseshoe crab, the most common type is named by the scientific faction in the videogame lore as a *Scylla serrata horrendous*. *Neo Aquarium—The King of Crustaceans* (Calappa Games, 2015) (Fig. 17.8F) features a wide variety of crustaceans, such as spiny lobsters, king crabs, frog crabs, hermit crabs, mantis shrimp, and barnacles(!), all accurately depicted and with the correct scientific names. The player controls the activities of the crustaceans and makes them forage, molt, and, of course, fight each other or human characters. As a bonus, and we are not making this up, players can have a gigantic *Anomalocaris* (not a crustacean, of course) attack a submarine. In a similar vein, a wide diversity of brachyurans and anomurans can be made to fight each other in *King of Crabs* (Robot Squid, 2019)

Fig. 17.8.
Crustaceans in video games and toys. (A) Crawmerax from *Borderlands: The Secret Armory of General Knoxx.* (B) The Invincible Son of Crawmerax the Invincible from Borderlands 2. (C) Giant Enemy Crab from *Genji: Days of the Blade.* (D) Krabby from the *Pokémon* card game. (E) Mudcrab (Panopeidae) from *The Elder Scrolls* series. (F) Denizens of *Neo Aquarium—The King of Crustaceans*: *Chionoecetes opilio* (Fabricius, 1788), Paguroidea, *Ranina ranina* (Linnaeus, 1758), and *Panulirus* sp. (G) Characters from *King of Crabs*: *Pseudocarcinus gigas* (Lamarck, 1818), Cancridae sp., and a fiddler crab (Ocypodidae). (H) Not only crabs fight in *Fight Crab*: A *P. gigas* hefting a chainsaw versus an *Homarus americanus* H. Milne Edwards, 1837 brandishing a pair of pistols. (I) *Mantis*, card game with educational booklet on mantis shrimp. See a color version of this figure in the centerfold.

(Fig. 17.8G) and the recently released *Crab Champions* (Noisestorm, 2023), the main difference in these games is that the crustaceans carry weapons. The tendency to weaponize crustaceans is radically expanded upon in *Fight Crab* (Fig. 17.8H) (Calappa Games 2019), a game where crabs fight (what else?). Apparently, claws are not enough as the game features a wide assortment of mayhem caused by grapsoids with halberds, xanthoids with chain saws, goneplacids with light sabers, and, yes, calappids with swords, among others. Most of the characters are brachyurans, but there are some anomurans, lobsters, and a stomatopod or two thrown in for variety.

REPLICAS—TOYS AND FIGURINES

Animal toys, replicas, and collectible figurines are perennially popular, whether representing farm horses or prehistoric dinosaurs. Crustacean toys, however, are uncommon, at least in Western countries, and those to be found are usually of the few familiar edible types: portunid or cancrid crabs, clawed or spiny lobsters, and sometimes prawns. Occasionally, crustacean-themed educational games can be found, as in *Crazy Crustaceans*, a combination puzzle card set featuring common commercial species and crustacean "fun facts" (Fig. 17.9A), and *Mantis* (Fig. 17.8I), self-described as a "colorfully cutthroat card game of rainbows and revenge" based on the mantis shrimp, and including a 16 page educational booklet. Of course, most crustacean-themed toys are much simpler (Fig. 17.9B). In Australia in the late 1990s, Cadbury, a confectionary company, marketed their chocolate "Yowies" with several series of collectible animal figurines, including a number of highly stylized crustaceans (but bearing species names together with educational notes). These Cadbury "Yowie" crustaceans included a mantis shrimp (Fig. 17.9C), a freshwater crayfish (Fig. 17.9D), spiny lobster (Fig. 17.9E), hermit crab (Fig. 17.9F), and several brachyurans (Fig. 17.9G–I). Even so, the Western crustacean figurine fauna can be considered as relatively depauperate. Diversity of crustacean figurines is highest in East Asia, particularly Japan, exceeding that of Western countries by an order of magnitude. Here, figurines of almost all crustacean groups can be found (Fig. 17.10), including branchiopods (e.g., *Daphnia* spp., *Triops* spp., *Branchinella* spp.), copepods, stomatopods, isopods, and most commonly, decapods. Decapod infraorders are widely represented, including clawed and spiny lobsters, freshwater crayfish, polychelidans, anomurans (mostly hermit crabs), and numerous families of Brachyura, both commercial and noncommercial, freshwater and marine, from the shore and the deep sea. Not surprisingly, given their Asian origins, species include the Japanese giant spider crab (*Macrocheira kaempferi*), Japanese mantis shrimp (*Oratosquilla oratoria*), and Japanese endemic freshwater crayfish (*Cambaroides* spp.), but non-Japanese examples exist as well, including the Australian giant crab (*Pseudocarcinus gigas*, Lamarck 1818) (Fig. 17.10D), and the fossil polychelidan *Eryon* sp. These figurines, mostly by major Japanese toy manufacturers (e.g., Kaiyodo, Yujin, Kitan, Epoch, and Bandai), are often exquisitely crafted, carefully painted, usually identifiable to species, and often sized (3–5 cm) for sale through toy vending machines (Fig. 17.10A–D). Figurines sold in this way are particularly common in Japan and East Asia, spanning almost every subject imaginable, including anime characters, food, construction machinery, kitchen appliances, toilet bowls, whitegoods, and all types of animals and plants. The crustacean figurines are usually released as part of larger, rolling themed series, such as deep-sea fauna, freshwater endemics, commercial species, endangered species, prehistoric animals, plankton, and coral reef animals. No doubt, the popularity of detailed miniature figurines in part reflects a cultural propensity in Japan that is less evident in the West. However, that so many animal groups, including familiar through arcane crustaceans, are regularly represented in a mass-produced medium indicates a much wider public appreciation of Crustacea in the East than the West. Why this is so is not immediately obvious, and explanations probably require much wider considerations of social and cultural history, well beyond the scope of our survey. Undoubtedly,

Fig. 17.9.
Crustaceans as toys (A, B) and Cadbury "Yowie" figurines (C–I). (A) *Crazy Crustaceans* combination puzzle card set with educational facts. (B) Rubber toy mangrove crab, *Scylla serrata* (Forskål, 1775), apparently "cooked" (C) Mantis shrimp. (D) Freshwater crayfish "Yabby." (E) Spiny lobster. (F) Hairy hermit crab. (G) Sponge crab. (H) Blue swimmer crab. (I) Soldier crab. See a color version of this figure in the centerfold.

the explanation is multifaceted with both deeper and more proximate causes, but we suspect it relates to more than mere familiarity. Instead, historical differences in underlying conceptions on the place of humans in the world may be at the root of this disparity. Broadly speaking, the metanarrative of the post-enlightenment West emphasizes discontinuity between humans and nature (man vs. wild) more so than traditional Eastern perspectives, which have generally upheld greater continuity (humans as part of nature) (Glacken 1992, Nakamura 1992). The import of such traditional differences should not be overstated, but they perhaps make intrinsic appreciation of crustaceans (and other biota) an easier default in the East than in the West.

Although myriad crustaceans are modeled in Japan, occasionally some species disproportionally captures the public imagination, spawning what might be termed a "craze." One such craze involves the giant deep-sea isopods (Fig. 17.1A–E), *Bathynomus* spp., of which the Caribbean *B. giganteus* A. Milne-Edwards, 1879, reaching at least 450 mm in length (Lowry & Dempsey 2006) is perhaps the best known. How and why the craze started is not clear, and may be owed in part to the great size and "mysterious" deep-sea habitat of *Bathynomus* spp. Either way, the strangeness of the giant isopod was clearly not seen as threatening, but rather embraced, resulting in numerous *Bathynomus* representations spanning key chains, plush toys, pillows, tea infusers (Fig. 17.11F), iPhone covers (Fig. 17.11G), and detailed figurines in multiple sizes, including a highly realistic life-sized static model of *B. giganteus* by Kitan in 2014 (Fig. 17.1A), as well as an articulated version by Bandai in 2019 that can be enrolled in defensive posture (Fig. 17.1B). Subsequently, terrestrial isopods have also become popular in these media.

ORNAMENTS, TRINKETS, MONUMENTS, AND OTHER PARAPHERNALIA

Crustaceans are frequently seen in common, even banal and mass-produced "everyday" consumer items, such as household ornaments, refrigerator magnets (Fig. 17.11B), trinkets (Fig. 17.11E), clothing and homewares (Figs. 17.1G–I, 17.11D), kitchenware (Fig. 17.11I), and garden ornaments, often cast in iron, copper, or brass. Stylized crustacean forms are also seen in folk art and tourist souvenirs, jewelry, and artisanal sculptures. As with fine art, crustaceans represented in more pedestrian circumstances are those of immediate importance to humans, and again, crabs are the most common, being both edible and representing the zodiac sign Cancer (Fig. 17.11H). Not surprisingly, tourist souvenirs often highlight culinary favorites such as Chesapeake Bay blue crab (*Callinectes sapidus*) (Fig. 17.11I) and Singapore chilli crab (*Scylla* spp.) (Fig. 17.11I). When taxonomically recognizable, depicted crustaceans are usually the larger, familiar commercial species, such as the swimming crabs, *Scylla* spp. (Fig. 17.11A) and *Portunus* spp., or in the case of lobsters, *Panulirus* spp. and *Homarus* spp. Nevertheless, unusual crustaceans occasionally turn up. Some of the most unusual items we have encountered, all from Japan, include a mantis shrimp cigarette lighter (Fig. 17.11C), probably modeled on *Oratosquilla oratoria*, and a tea infuser (Fig. 17.11F) and iPhone case (Fig. 17.11G), both depicting the giant isopod *Bathynomus* sp. The tea infuser is even thoughtfully molded with a pair of enlarged antennae that hang over the lip of the teacup.

In public spaces, representations of crustaceans are mainly on business-related shopfronts of seafood restaurants, pet shops, or public aquaria. Occasionally, however, crustaceans figure as massive freestanding monuments to locally important fishing industries, and over time become well-known tourist attractions. For example, the outdoor crab sculptures at the aptly named Thai town of Krabi (although meaning "sword" in Thai, alluding to the legend that an ancient sword was unearthed from that location), depict *Scylla* spp. (Fig. 17.12A), which are major fishery species in the region. Similarly, the "Big Prawn" at Ballina, Australia, recognizes the important local penaeid fishery and simultaneously advertises a local seafood business (Fig. 17.12B). The Canadian town of Shediac claims possession of the world's largest lobster, an 11 m long reinforced concrete statue of *Homarus americanus* by Canadian artist Winston Bronnum, commissioned by the local Rotary Club in tribute to the town's lobster fishery (Fig. 17.12C).

PETS AND AQUARIA

Freshwater crayfish and terrestrial hermit crabs (*Coenobita* spp.) are among the most popular of crustacean pets. Branchiopods, primarily brine shrimp (*Artemia salina* Linnaeus, 1758) marketed under the trade name "Sea Monkeys," and to a lesser extent, the shield shrimp (*Triops* spp.),

Fig. 17.10.
Crustacean replicas from various Japanese manufacturers. (A) Assorted. (B) Spanner crab, *Ranina ranina* (Linnaeus, 1758) (Kaiyodo). (C) Ornate spiny lobster, *Panulirus ornatus* (Fabricius, 1798) (Kitan). (D) Australian giant crab, *Pseudocarcinus gigas* (Lamarck, 1818) (Kaiyodo). See a color version of this figure in the centerfold.

Fig. 17.11.
Crustaceans as ornaments and paraphernalia. (A) Brass crab home ornament (*Scylla* sp.). (B) Refrigerator magnets. (C) Mantis shrimp cigarette lighter. (D–E) Various souvenirs and homewares. (F) *Bathynomus* tea infuser. (G) *Bathynomus* iPhone case. (H) Cigarette ash tray. (I) Kitchenware. See a color version of this figure in the centerfold.

have long been sold in toy and pet shops, but mostly as short-term science experiment curiosities. Notably, the "Sea Monkey" caricatures in advertisements bore little resemblance to actual brine shrimp (Fig. 17.12D, E) and no doubt disappointed more than one young child who ordered them! In the last decade or so, terrestrial isopods (Oniscoidea) have also become available, mostly in North America (McMonigle 2013). The growing popularity of marine aquaria since the 1960s, followed by the increasing sophistication of marine "reef aquaria" and freshwater "planted aquaria" since the late 1980s has made aquatic crustaceans more popular than ever as long-term aquarium subjects. Appreciation of crustaceans in captivity has very much shifted from a casual to a scientific interest. Volumes have been published on keeping live crustaceans (along with other aquatic invertebrates), ranging from basic introductory guides (e.g., Nash 1976, Michael 2008) to detailed treatises on captive husbandry using the latest technology and often referencing current scientific research (e.g., Debelius 1983, Debelius and Baensch 1994, Werner 1998, Fosså and Nilsen 2000, Karge and Klotz 2007, Lukhaup and Pekny 2015, McMonigle 2018, Calado 2020). Some of the most popular crustaceans in reef aquaria are hermit crabs, commensal porcelain crabs (*Neopetrolisthes* spp.), dwarf lobsters (*Enoplometopus* spp.), and colorful decapod shrimps, especially *Lysmata* spp., *Rhynchocinetes* spp., and *Periclimenes* spp. (Calado 2020). Mantis shrimps, too, are in demand, especially the spectacular peacock mantis shrimp (*Odontodactylus scyllarus* Linnaeus, 1758), which is typically kept alone owing to its large size and predatory behaviors, although large individuals have been known to break the glass of their aquaria with their raptorial appendages (Caldwell and Dingle 1976).

Among freshwater aquarists, especially in Europe and Asia, crabs and crayfish have long been the mainstay crustaceans, but the last two decades have seen an explosion of interest around the world in small freshwater atyid shrimps (*Caridina* and *Neocaridina* spp.) that live in large groups in densely planted aquaria; popular selectively-bred strains include Crystal Red, Red Cherry, and Sakura Red shrimp.

The body of knowledge accumulated and shared by hobbyists through books, magazine articles, aquaristic clubs and societies, and now accelerated by the internet, is substantial. Leading hobbyists are committed to acquiring the most up-to-date scientific knowledge, including accurate taxonomic identifications, so much so that some regularly interact and collaborate with the scientific community (e.g., Karge et al. 2013, Klotz et al. 2021). Suffice it to say, the "hard core" aquarists, many of whom keep crustaceans, do represent a niche, but a sufficiently large one able to support a $25 billion-per-year global industry, albeit one focused largely in the USA, Europe, and Japan (Padilla and Williams 2004).

The aquarium hobby has certainly raised popular awareness of crustaceans, but sourcing specimens can raise conservation issues. Many traded freshwater species are captive bred but most marine species are wild caught (Calado 2008, Calado et al. 2017). In some areas, harvesting for the aquarium trade is well regulated and sustainable, but overharvesting is commonplace in many parts of the tropical Indo-Pacific where poverty and minimal environmental regulations intersect. The majority of marine aquarium species are exported from the Philippines and Indonesia, with lesser numbers from the Solomon Islands, Australia, Fiji, Palau, Sri Lanka, and the Maldives (Wabnitz et al. 2003, Calado et al. 2017).

Additional dimensions attending the global wildlife trade are the unintended effects of translocations. Numerous crustaceans are now invasive worldwide, some having been translocated for fishery purposes, such as the North American red swamp crayfish (*Procambarus clarkii* Girard, 1852) and the Australian freshwater red claw crayfish (*Cherax quadricarinatus* von Martens, 1868), of which the latter is also invasive in Singapore having arrived through the aquarium trade (Ahyong and Yeo 2007, Loureiro et al. 2015, Haubrock et al. 2021). More recently, the "Marmorkrebs," an enigmatic parthenogenetic North American freshwater crayfish, *Procambarus virginalis* Lyko, 2017, originally identified in German aquaria in the 1990s, has, via anthropogenic releases, become

Fig. 17.12.
Crustaceans as public monuments (A–C); "Sea monkeys" (D–E). (A) Crab sculptures (*Scylla* sp.) at Krabi, Thailand. (B) The "Big Prawn" (Penaeidae), Ballina, Australia. (C) Lobster (*Homarus americanus* H. Milne Edwards, 1837), Shediac, Canada. (D) Advertisement for "Sea Monkeys" from 1970s *Archie and Me* comic book; art by Joe Orlando (1927–1998). (E) Live brine shrimp, *Artemia salina* (Linnaeus, 1758). (Photos: (A) Stefan Fussan, Wikimedia Commons, CC-BY-SA 3.0; (B) public domain; (C) Mike Young, Wikimedia Commons, CC-BY-SA 3.0; (E) Hans Hillewaert, Wikimedia Commons, CC-BY-SA-4.0). See a color version of this figure in the centerfold.

invasive in parts of Europe, Japan, and Madagascar, potentially threatening native crayfish populations (Scholtz et al. 2003, Martin et al. 2010, Vogt et al. 2018).

DISCUSSION

The degree of human interest in crustaceans appears to relate to the history and nature of human interactions with them, and this is geographically correlated. Regions with a greater diversity of crustaceans bring a wider array and a longer history of human interactions, including as food, and not surprisingly, crustaceans are more widely embraced and enjoy greater presence in the popular culture in those regions (Table 17.1). For example, in East and Southeast Asia, where there is a wide variety of crustacean species, especially decapods, there are far more crustacean-related consumer products than seen in North America or Europe, where there are many fewer such items and those that are available mostly focus on a few commercial species and where a monoculture of food

Table 17.1 Crustaceans represented in selected popular cultural categories, their context, and geographic region of depiction. Decapods are the crustaceans most familiar to the general public, and include crabs, prawns, lobsters, crayfish, and shrimps.

Category	Taxa	Context of depiction	Regions
Stamps	All taxa, mainly decapods	As main subject, often colorful species	Worldwide (few in North America and Europe)
Coins	Decapods	Edible species	Mostly equatorial countries
Fine art	Mainly decapods	Edible species as part of still-life works	Europe; East Asia
Domestic trinkets	Decapods	Edible species, often stylized	Worldwide (most diverse in East Asia)
Modern public sculptures	Decapods	Edible species	Worldwide
Children's toys	Decapods, usually crabs and lobsters	Edible species, often stylized	Worldwide
Figurines	All taxa, mainly decapods	Diverse range of species, often with realistic depictions	Worldwide (most diverse, most intricate in East Asia)
Television	Decapods, usually crabs	Monster genre; anthropomorphic decapods	Europe; North America; East Asia
Cinema	Decapods, usually crabs	Usually monster genre, rarely "friendly"	Europe; North America; East Asia
Literature	Decapods, usually crabs	Monster genre	Worldwide
Music	Decapods, isopods, Stomatopods	Band namesake; cover art	North America; Europe
Video games	Decapods, stomatopods	Anthropomorphic decapods, often as player avatars	North America; East Asia
Aquarium subjects	Decapods, isopods, stomatopods	Live display, mainly coral reef species	Worldwide

(e.g., blue crabs, clawed lobsters) is unsurprisingly reflected in a monoculture of crustacean-themed products.

In a related observation, it is probably no coincidence that many representations of crabs in North American art and products are highly stylized and often have too few limbs, reflecting a more limited experience of the artists or craftsmen with actual crabs (or intrinsic regard for their subjects). Perhaps the more common utilitarian Western view of crustaceans as simply "another thing we eat" also limits the attention to details, or even interest per se. Other than as seafood, depictions in popular media suggest that general Western perceptions of crustaceans tend to be as objects of suspicion, if not fear, much like insects and spiders, or at best as unusual curiosities. In contrast, the more intimate association between humans and crustaceans, particularly crabs, in Eastern cultures has resulted in more anthropomorphic depictions of crustaceans than in the West, with crustaceans sometimes imbued with human qualities rather than humans being separate and distinct and having less connection with nature. This, for instance, may partly explain why crustaceans, whether as depictions on stamps, toys, or domestic objects, or as media "celebrities," can be more easily assimilated into popular culture in the East than in the West. At a general level, in the East, crustaceans are perhaps seen more as friend than foe. As a result, crustaceans, while sometimes invoked in the monster genre of science fiction, are, in popular media, usually depicted more neutrally, or even more favorably than in the West. Additionally, Asian non-humanized representations of crustaceans may reflect a greater perception of the continuity of animals and humans as part of nature, even if they are no less desirable as culinary subjects.

Today, many species of crustaceans face significant threats from habitat degradation or loss, over-exploitation, and the amplifying effects of climate change. These effects not only impact crustaceans directly, but the trophic and ecological web that includes humans, and as such, have a bearing on human well-being (Behringer and Duermit-Moreau 2021). More crustaceans are now being wild harvested than ever before, with some six million tonnes or more taken annually since 2016 (Boenish et al. 2022). Public awareness of crustaceans as something beyond being mere menu items must be a starting point for effective conservation efforts and appreciation of their intrinsic value. Paradoxically, the ornamental trade in crustaceans may assist this endeavor in many respects but can also have a negative effect, by generating interest both in the species themselves and a concern for their future, yet simultaneously potentially threatening their survival in the wild. A robust oversight of the ornamental trade, as well as crustacean fisheries, is needed to ensure that appreciation does not morph into over-exploitation.

One way to increase the interest in conservation of crustaceans is to make some of the commercially or culturally important species the emphasis of public outreach campaigns for broader appreciation of local biodiversity issues (Novacek 2008). "It's the economy, stupid" is an oft quoted message regarding political campaign outcomes in the USA, and reaching out to the public via exemplar organisms they already have a vested interest in, both economically and gastronomically, is one way to reach their minds via their stomachs. It would also be helpful to tie specific crustacean conservation issues into broader environmental concerns; for example educating on the importance of lobsters as keystone species that can help to shape their whole ecosystem (Bond 1994). A synthetic approach that ties selected crustaceans to other organisms that are perhaps easier for humans to "connect" to (e.g., fish, dolphins, whales), be they as predators of crustaceans or beneficiaries of the effects of crustaceans on the local environment, could also motivate more people to support crustacean conservation.

There is a diverse literature on the interface between human and nonhuman animals. Some are scholarly and pedantic (e.g., Mandel [1995], on mollusks) and some more tongue-in-cheek (e.g., Cavallari [2015], also on mollusks) but all point to the myriad representations of nonhuman animals in human art and culture. The only reasonable conclusion is that as nonhuman animals become more important to us (as food, pets, and symbols), they become better represented in our cultural output. Conversely, it is logical to conclude that those nonhuman animals that are more often represented in culture are also made to become more important to us in something of a positive feedback loop. Therefore, while it might seem rather unimportant how often a relatively "minor" taxon, such as Onychophora (Monge-Nájera and Morera-Brenes 2015), is included in cultural output, such a recognition by itself can generate interest in a taxon that was previously "invisible" and therefore less likely to engender conservation interest.

CONCLUSIONS AND OUTLOOK

Out of sight is out of mind, so raising awareness of crustaceans—their diversity, complexity, beauty, ecology, and intrinsic values—as creatures of value rather than suspicion must be a priority. Some of this can come via education, especially from academia, through popularizing new discoveries and highlighting little-known facts. Nevertheless, we submit that anything, including the ornamental trade and the myriad categories including art, music, or hobbies discussed above, that increases public appreciation of crustaceans can only serve to improve the prospects for effective conservation of crustacean species across the globe in the face of climate change, overfishing, and habitat degradation. As things stand, crustaceans do not hold a metaphorical candle to mollusks when it comes to cultural representation; as a crude comparison, at the time of this writing, the Wikipedia entry for "Molluscs in culture" has 74 references while a search for "Crustaceans in culture" redirects to "Arthropods in culture," which has only 48 references and where crustaceans are mentioned by name only eight times. More tangibly, some 1400 postage stamps figure crustaceans compared to at least 5080 mollusks (https://www.conchology.be/?t=43). As there are approximately twice as many species of extant mollusks than crustaceans (www.molluscabase.org; www.marinespecies.org), it seems fitting that crustaceans are represented in popular culture substantially less often than mollusks, but the current disparity is rather disproportionate. More public exposure of crustaceans, and indeed a greater diversity of crustaceans, is likely to engender more interest in this group and to increase the likelihood that crustaceans receive a level of attention proportional to their numbers and diversity. Regional differences in cultural outlook should also be considered when calibrating conservation strategies, since differing emphasis on intrinsic versus pragmatic aspects of crustaceans may gain more purchase in some regions than others. Nevertheless, any and all methods of increasing their exposure to a wider audience should be pursued, including tapping into people's emotional response (i.e., the "cute and cuddly" factor as exemplified by giant isopod plush toys), gastronomic implications (the plight of overfished species and increasing cost of seafood), and even a renewed emphasis on promoting the basic science (taxonomy, systematics) that underlies our understanding of the evolutionary history of this ancient, diverse, and fascinating group.

As humans evolve and continue to develop new ways to express their cultural interests, crustaceans will, barring extirpation, always be included as part of that expression.

ACKNOWLEDGMENTS

Thanks go to Daniel Clements (https://www.coinzoo.net) for kindly permitting use of his images of the crustacean coins.

REFERENCES

Abel, D. 2020. Fact checking Trump's claims about the Maine lobster industry. Boston Globe, June 25, 2020. Accessed August 8, 2022 [https://www.bostonglobe.com/2020/06/25/nation/fact-checking-trumps-claims-about-maine-lobster-industry/].

Ahyong, S.T., and D.C.J. Yeo. 2007. Feral populations of the Australian red-claw crayfish (*Cherax quadricarinatus* von Martens) in water supply catchments of Singapore. Biological Invasions 9:943–946.

Behringer, D.C., and E. Duermit-Moreau. 2021. Crustaceans, one health and the changing ocean. Journal of Invertebrate Pathology 186:107500.

Boenish, R., J.P. Kritzer, K. Kleisner, R.S. Steneck, K.M. Werner, W. Zhu, F. Schram, D. Rader, W. Cheung, J. Ingles, Y. Tian, and J. Mimikakis. 2022. The global rise of crustacean fisheries. Frontiers in Ecology and the Environment 20:102–110.

Bond, W.J. 1994. Keystone species. Pages 237–253 *in* E.-D. Schulze and H.A. Mooney, editors. Ecological studies 99: Biodiversity and ecosystem function. Springer-Verlag, Berlin, Germany.

Boyko, C.B. 2003. The endemic marine invertebrates of Easter Island: How many species and for how long? Pages 155–175 *in* J. Loret and J.T. Tanacredi, editors. Easter Island. Scientific exploration into the world's environmental problems in microcosm. Kluwer Academic/Plenum Publishers, New York, USA.

Buscema, J., and E. Chan. 1979. [Cover art]. *Conan the Barbarian* #99. Marvel Comics, New York, USA.

Calado, R. 2008. Marine ornamental shrimp: Biology, aquaculture and conservation. Wiley-Blackwell, Oxford, UK.

Calado, R. 2020. Marine ornamental decapods—Collection, culture, and conservation. Pages 313–340 *in* G. Lovrich and M. Thiel, editors. The natural history of the Crustacea, Volume 9: Fisheries and aquaculture. Oxford University Press, New York, USA.

Calado, R., I. Olivotto, M.P. Oliver, and G.J. Holt. 2017. Marine ornamental species aquaculture. Wiley-Blackwell, West Sussex, UK.

Caldwell, R.L., and H. Dingle. 1976. Stomatopods. Scientific American 234:80–89.

Carle, E. 1987. A house for hermit crab. Picture Book Studio, Saxonville, USA.

Cavallari, D.C. 2015. Shells and bytes: Mollusks in the 16-bit era. Journal of Geek Studies 2:28–43.

Coelho, J.R. 2011. Noninsect arthropods in popular music. Insects 2:253–263.

Collins, M.A., and G. Hagenauer. 2008. Men's adventure magazines in postwar America. Taschen, Hong Kong.

Corson, T. 2004. The secret life of lobsters. Harper Collins, New York, USA.

Debelius, H. 1983. Armoured knights of the sea. Kernan Verlag, Essen, Germany.

Debelius, H., and H.A. Baensch. 1994. Baensch marine atlas. Mergus Verlag GmbH, Melle, Germany.

de Pins, A. 2015. The march of the crabs, Volume 1: The Crabby Condition. Archaia Entertainment, Los Angeles, USA.

Dias, T.L.P., N.A. Neto Leo, and R.R.N. Alves. 2011. Molluscs in the marine curio and souvenir trade in NE Brazil: Species composition and implications for their conservation and management. Biodiversity Conservation 20:2393–2405.

Dicke, M. 2000. Insects in western art. American Entomologist 46:228–237.

Fosså, S.A., and A.J. Nilsen. 2000. The modern coral reef aquarium, Volume 3. Birgit Schmettkamp Verlag, Bornheim, Germany.

Glacken, C.J. 1992. Reflections on the history of western attitudes to nature. GeoJournal 26:103–111.

Gutiérrez-Zugasti, I., E. Tong, A. García-Escárzaga, D. Cuenca-Solana, G.N. Bailey, and M.R. González-Morales. 2016. Collection and consumption of echinoderms and crustaceans at the mesolithic shell midden site of El Mazo (northern Iberia): Opportunistic behaviour or social strategy? Quaternary International 407:118–130.

Haubrock, P.J., F.J. Oficialdegui, Y. Zeng, J. Patoka, D.C.J. Yeo, and A. Kouba. 2021. The redclaw crayfish: A prominent aquaculture species with invasive potential in tropical and subtropical biodiversity hotspots. Reviews in Aquaculture 13:1–43.

Holling, H.C. 1957. Pagoo. Houghton Mifflin Company, Boston, USA.

Holthuis, L.B. 1967. Schaaldieren (Crustacea) afgebeeld op postzegels. Zoologische Bijdragen 8:1–21. (in Dutch)

Hong, M., and Wei, S. 2010. Stamps of the world's crab appreciation [in Chinese]. Gao Tan Cultural Publishing Co., Ltd., New Taipei City, China.

Horton, T. 1996. An island out of time. W.W. Norton & Company, New York, USA.

Karge, A., and W. Klotz. 2007. Süßwassergarnelen aus aller Welt [in German]. Dähne Verlag GmbH, Ettlingen, Germany.

Karge, A., T.J. Page, and W. Klotz. 2013. A new species of freshwater shrimp of the genus *Micratya* (Decapoda: Atyidae: Caridea) from Puerto Rico. Zootaxa 3608:357–368.

Keyes, D., and B. Wolverton. 1953. They crawl by night! Journey into unknown worlds 15. Marvel Comics, New York, USA.

King, S. 1987. The dark tower II: The drawing of the three. Donald M. Grant, West Kingston, USA.
Klotz, W., T. von Rintelen, D. Wowor, C. Lukhaup, and K. von Rintelen. 2021. Lake Poso's shrimp fauna revisited: The description of five new species of the genus *Caridina* (Crustacea, Decapoda, Atyidae) more than doubles the number of endemic lacustrine species. ZooKeys 1009:81–122.
Kurtzman, H., and R. Estrada. 1953. Rough riders! Frontline combat 11. EC, New York, USA.
Leiber, L., and J. Kirby. 1960. I was trapped by Titano, the monster that time forgot. Tales to Astonish 10. Marvel Comics, New York, USA. [the authorship of the script is probably Leiber but this is not certain; see https://www.comics.org/issue/15735/#125685].
Lipcius, R.N., and W.T. Stockhausen. 2002. Concurrent decline of the spawning stock, recruitment, larval abundance, and size of the blue crab *Callinectes sapidus* in Chesapeake Bay. Marine Ecology Progress Series 226:45–61.
Loureiro, T.G., P.M.S.G. Anastácio, P.B. Araujo, C. Souty-Grosset, and M.P. Almerão. 2015. Red swamp crayfish: Biology, ecology and invasion—An overview. Nauplius 23:1–19.
Lowry, J.K., and Dempsey, K. 2006. The giant deep-sea scavenger genus *Bathynomus* (Crustacea, Isopoda, Cirolanidae) in the Indo-West Pacific. Mémoires du Muséum national d'Histoire naturelle, Paris 193: 163-192.
Lowry J.K., and H.E. Stoddart. 2002. The Amaryllididae of Australia (Crustacea: Amphipoda: Lysianassoidea). Records of the Australian Museum 54:129–214.
Lukhaup, C., and R. Pekny. 2015. Invertebrates: Shrimp, crayfish, crabs & snails in freshwater aquaria. Dähne Verlag GmbH, Ettlingen, Germany.
Mandel, C. 1995. Miró's mystical mollusks. Mediterranean Studies 5:113–133.
Martin, P., N.J. Dorn, T. Kawai, C. van der Heiden, and G. Scholtz. 2010. The enigmatic Marmorkrebs (marbled crayfish) is the parthenogenetic form of *Procambarus fallax* (Hagen, 1870). Contributions to Zoology 79:107–118.
McMonigle, O. 2013. Pillbugs and other isopods: Cultivating vivarium cleanup crews and feeders for dart frogs, arachnids, and insects. Coachwhip Publications, Greenville, USA.
McMonigle, O. 2018. Breeding the vampire and other crabs (Brachyura and Anomura in captivity): Husbandry, reproduction, biology, and diversity. Coachwhip Publications, Greenville, USA.
McNeely, J.A., and P.S. Watchel. 1988. Soul of the tiger: Searching for nature's answers in exotic Southeast Asia. Doubleday, New York, USA.
Michael, S.W. 2008. The 101 best marine invertebrates: How to choose and keep hardy, brilliant, fascinating species that will thrive in your home aquarium. T.F.H. Publications, Neptune City, New Jersey, USA.
Mitchell, D.R., and M.S. Foster. 2000. Hohokam shell middens along the Sea of Cortez, Puerto Peñasco, Sonora, Mexico. Journal of Field Archaeology 27:27–41.
Monge-Nájera, J., and B. Morera-Brenes. 2015. Velvet worms (Onychophora) in folklore and art: Geographic pattern, types of cultural reference and public perception. British Journal of Education, Society and Behavioural Science 10:1–9.
Mujiono, N., A. Alfiah, R. Prihandini, and P.H. Santoso. 2021. Identifikasi gambar hewan moluska dalam media cetak dua dimensi (identification of molluscan animal image in two-dimensional print media). Jurnal Moluska Indonesia 5:25–33.
Nakamura, H. 1992. The idea of nature in the east in comparison with the west. GeoJournal 26:113–128.
Nash, P.J. 1976. Land hermit crabs. T.F.H. Publications, Neptune City, New Jersey, USA.
Novacek, M.J. 2008. Engaging the public in biodiversity issues. PNAS 105:11571–11578.
O'Dwyer, K. 2014. Meet *Phronima*, the barrel-riding parasite that inspired the movie Alien. The Conversation. Accessed August 8, 2022. [https://theconversation.com/meet-phronima-the-barrel-riding-parasite-that-inspired-the-movie-alien-22555].
Omori, M. 2014. Crustaceans on postage stamps from 1871 through 2002: The complete checklist. Journal of the Tokyo University of Marine Science and Technology 10:40–86.
Omori, M. 2021. Shrimp and crab museum—World crustacean stamps. Tsukiji Shokan.
Omori, M., and L.B. Holthuis. 2000. Crustaceans on postage stamps from 1870 to 1997. Report of Tokyo University of Fisheries 35:1–87.

Omori, M., and L.B. Holthuis. 2005. Crustaceans on postage stamps from 1870 to and including 2002: Revised article for our paper in 2000 and addendum. Journal of the Tokyo University of Marine Science and Technology 1:1–39.

Padilla, D.K., and S.L. Williams. 2004. Beyond ballast water: Aquarium and ornamental trades as sources of invasive species in aquatic ecosystems. Frontiers in Ecology and the Environment 2:131–138.

Rosa, R.M., D.C. Cavallari, and A.L. Vera-Silva. 2020. Pokécrustacea: The crustacean-inspired Pokémon. Journal of Geek Studies 7:97–113.

Sartre, J.P. 1961. The condemned of Altona. Alfred A. Knopf, New York, USA.

Scholtz, G., A. Braband, L. Tolley, A. Reimann, B. Mittmann, C. Lukhaup, F. Steuerwald, and G. Vogt. 2003. Parthenogenesis in an outsider crayfish. Nature 421:806.

Silberman, R. 1989. Why a duck? Birds in art and bird art. American Art 3:64–79.

Smith, G.N. 1988. Crabs on the rampage. Dell, New York, USA.

Smith, G.N. 1989. Night of the crabs. Dell, New York, USA.

Sugiura, C., and A. Asakura. 2012. Natural history illustrations of Chisato Sugiura. Seizanō Shoten, Tokyo, Japan.

Szuwalski, C., W. Cheng, R. Foy, A.J. Hermann, A. Hollowed, K. Holsman, J. Lee, W. Stockhausen, and J. Zheng. 2020. Climate change and the future productivity and distribution of crab in the Bering Sea. ICES Journal of Marine Science 78:502–515.

Valigursky, E. 1957. [Cover art.] Fantastic 6(6). Ziff-Davis Publishing Co., New York, USA.

Vogt, G., C. Lukhaup, B.W. Williams, M. Pfeiffer, N.J. Dorn, R. Schulz, and A. Schrimpf. 2018. Morphological characterization and genotyping of the marbled crayfish and new evidence on its origin. Zootaxa 4524:329–350.

Wabnitz, C., M.L. Taylor, E. Green, and T. Razak. 2003. From ocean to aquarium: The global trade in marine ornamental species. UNEP-WCMC, Cambridge, UK.

Wahl, M. 1997. Living attached: Aufwuchs, fouling, epibiosis. Pages 31–83 *in* R. Nagabhushanam and M.-F. Thompson, editors. Fouling organisms of the Indian Ocean: Biology and control technology. A.A. Balkema, Rotterdam, The Netherlands.

Werner, U. 1998. Shrimps, crayfishes, and crabs in the freshwater aquarium. Aqualog Special. Verlag A.C.S. GmbH.

INDEX